VOLUME

2

# ADVANCED MECHANISM DESIGN:

## Analysis and Synthesis

**GEORGE N. SANDOR**

Professor of Mechanical Engineering
University of Minnesota
Minneapolis, MN

**ARTHUR G. ERDMAN**

Research Professor
of Mechanical Engineering
University of Florida
Gainesville, FL

**PRENTICE-HALL, INC.,** *Englewood Cliffs, New Jersey 07632*

*Library of Congress Cataloging in Publication Data*
(Revised for vol. 2)

ERDMAN, ARTHUR G. (date)
Mechanism design.

Vol. 2 has title: Advanced mechanism design: analysis and synthesis.
Authors' names in reverse order in v. 2.
Includes bibliographies and indexes.
1. Machinery—Design. 2. Machinery, Dynamics of. 3. Machinery, Kinematics of. I. Sandor, George N. II. Title. III. Title: Advanced mechanism design.
TJ230.E67 1984 621.8'15 83-3148
ISBN 0-13-572396-5 (v. 1)
ISBN 0-13-011437-5 (v. 2)

Editorial/production supervision
and interior design: **Karen Skrable**
Manufacturing buyer: **Anthony Caruso**
Cover design: **Photo Plus Art, Celine Brandes**



Volume 1 published under the title *Mechanism Design: Analysis and Synthesis, Vol. 1.*



Printed in the United States of America

10 9 8 7 6 5 4 3 2 1

ISBN 0-13-011437-5 01

PRENTICE-HALL INTERNATIONAL, INC., *London*
PRENTICE-HALL OF AUSTRALIA PTY. LIMITED, *Sydney*
EDITORA PRENTICE-HALL DO BRASIL, LTDA., *Rio de Janeiro*
PRENTICE-HALL CANADA INC., *Toronto*
PRENTICE-HALL OF INDIA PRIVATE LIMITED, *New Delhi*
PRENTICE-HALL OF JAPAN, INC., *Tokyo*
PRENTICE-HALL OF SOUTHEAST ASIA PTE. LTD., *Singapore*
WHITEHALL BOOKS LIMITED, *Wellington, New Zealand*

*George Sandor*
*dedicates this work*
*to his wife*
**Magdi**

*Art Erdman*
*dedicates this work*
*to his daughters*
**Kristy** *and* **Kari**

# Contents

## 4 CURVATURE THEORY 301



## 5 DYNAMICS OF MECHANISMS: ADVANCED CONCEPTS 366

# Preface

This two-volume work, consisting of Volume 1, *Mechanism Design: Analysis and Synthesis,* and Volume 2, *Advanced Mechanism Design: Analysis and Synthesis,* was developed over a 15-year period chiefly from the teaching, research, and consulting practice of the authors, with contributions from their working associates and with adaptations of published papers. The authors represent a combination of over 30 years of teaching experience in mechanism design and collectively have rendered consulting services to over 35 companies in design and analysis of mechanical systems.

The work represents the culmination of research toward a general method of kinematic, dynamic, and kineto-elastodynamic analysis and synthesis, starting with the dissertation of Dr. Sandor under the direction of Dr. Freudenstein at Columbia University, and continuing through a succession of well over 100 publications. The authors' purpose was to present texts that are timely, computer-oriented, and teachable, with numerous worked-out examples and end-of-chapter problems.

The topics covered in these two textbooks were selected with the objectives of providing the student, on one hand, with sufficient theoretical background to understand contemporary mechanism design techniques and, on the other hand, of developing skills for applying these theories in practice. Further objectives were for the books to serve as a reference for the practicing designer and as a source-work for the researcher. To this end, the treatment features a computer-aided approach to mechanism design (CAD). Useful and informative graphically based techniques are combined with computer-assisted methods, including applications of interactive graphics, which provide the student and the practitioner with powerful mechanism design tools. In this manner, the authors attempted to make all contemporary kinematic analysis and synthesis readily available for the student as well as for the busy practicing designer, without the need for going through the large number of pertinent papers and articles and digesting their contents.

Many actual design examples and case studies from industry are included in the books. These illustrate the usefulness of the complex-number method, as well as other techniques of linkage analysis and synthesis. In addition, there are numerous end-of-chapter problems throughout both volumes: over 250 multipart problems in each volume, representing a mix of SI and English units.

The authors assumed only a basic knowledge of mathematics and mechanics on the part of the student. Thus Volume 1 in its entirety can serve as a first-level text for a comprehensive one- or two-semester undergraduate course (sequence) in Kinematic Analysis and Synthesis of Mechanisms. For example, a one-semester, self-contained course of the subject can be fashioned by omitting Chapters 6 and 7 (cams and gears). Volume 2 contains material for a one- or two-semester graduate course. Selected chapters can be used for specialized one-quarter or one-semester courses. For example, Chapter 5, Dynamics of Mechanisms—Advanced Concepts, with the use of parts from Chapters 2 and 3, provides material for a course that covers kinetostatics, time response, vibration, balancing, and kineto-elastodynamics of linkage mechanisms, including rigid-rotor balancing.

The foregoing are a few examples of how the books can be used. However, due to the self-contained character of most of the chapters, the instructor may use other chapters or their combinations for specialized purposes. Copious reference lists at the end of each book serve as helpful sources for further study and research by interested readers. Each volume is a separate entity, usable without reference to the other, because Chapters 1 and 8 of Volume 1 are repeated as Chapters 1 and 2 of Volume 2.

The contents of each volume may be briefly described as follows. In Volume 1, the first chapter, Introduction to Kinematics and Mechanisms, is a general overview of the fundamentals of mechanism design. Chapter 2, Mechanism Design Philosophy, covers design methodology and serves as a guide for selecting the particular chapter(s) of these books to deal with specific tasks and problems arising in the design of mechanisms or in their actual operation. Chapter 3, Displacement and Velocity Analysis, discusses both graphical and analytical methods for finding absolute and relative velocities, joint forces, and mechanical advantage; it contains all the necessary information for the development of a complex-number-based computer program for the analysis of four-bar linkages adaptable to various types of computers. Chapter 4, Acceleration Analysis, deals with graphical and analytical methods for determining acceleration differences, relative accelerations and Coriolis accelerations; it explains velocity equivalence of planar mechanisms, illustrating the concept with examples. Chapter 5 introduces dynamic and kinetostatic analyses with various methods and emphasizes free-body diagrams of mechanism links. Chapter 6 presents design methods for both simple cam-and-follower systems as well as for cam-modulated linkages. Chapter 7 acquaints the student with involute gears and gear trains, including the velocity ratio, as well as force and power-flow analysis of planetary gear trains. The closing chapter of Volume 1, Chapter 8, is an introduction to dimensional synthesis of planar mechanisms using both graphical and closed-form linear analytical methods based on a "standard-form" complex-number approach. It treats the synthesis of single and multiloop mechanisms as function, path, and motion generators, with first- and higher-order approximations.

Volume 2 starts with the same introductory chapter, Chapter 1, as the first volume. Chapter 2 of Volume 2 is the same as Chapter 8 of Volume 1. Chapter 3 extends planar analytical kinematic synthesis to greater than three-condition precision, accomplished by way of closed-form nonlinear methods, including Burmester theory, and describes a computer package, "LINCAGES," to take care of the computational burden. Cycloidal-crank and geared linkages are also included. Chapter 4 presents a new computer-oriented, complex-number approach to planar path-curvature theory with new explicit forms of the Euler-Savary Equation (ESE) and describes all varieties of Bobillier's Construction (BC), demonstrating the equivalence of the ESE and BC methods. Chapter 5 is a comprehensive treatment of the dynamics of mechanisms. It covers matrix methods, the Lagrangian approach, free and damped vibrations, vibration isolation, rigid-rotor balancing, and linkage balancing for shaking-forces and shaking-moments, all with reference to computer programs. Also covered is an introduction to kineto-elastodynamics (KED), the study of high-speed mechanisms in which the customary rigid-link assumption must be relaxed to account for stresses and strains in elastic links due to inertial forces. Rigid-body kinematics and dynamics are combined with elastic finite-element techniques to help solve this complex problem. The final chapter of Volume 2, Chapter 6, covers displacement, velocity, and acceleration analysis of three-dimensional spatial mechanisms, including robot manipulators, using matrix methods. It contains an easily teachable, visualizable treatment of Euler-angle rotations. The chapter, and with it the book, closes with an introduction to some of the tools and their applications of spatial kinematic synthesis, illustrated by examples.

In view of the ABET accreditation requirements for increasing the design content of the mechanical engineering curriculum, these books provide an excellent vehicle for studying mechanisms from the design perspective. These books also fit in with the emphasis in engineering curricula placed on CAD/CAM and computer-aided engineering (CAE). Many computer programs are either included in the texts as flow charts with example input-output listings or are available through the authors.

The complex-number approach in this book is used as the basis for interactive computer programs that utilize graphical output and CRT display terminals. The designer, without the need for studying the underlying theory, can interface with the computer on a graphics screen and explore literally thousands of possible alternatives in search of an optimal solution to a design problem. Thus, while the burden of computation is delegated to the computer, the designer remains in the "loop" at each stage where decisions based on human judgment need to be made.

The authors wish to express their appreciation to the many colleagues and students, too many to name individually, who have made valuable contributions during the development of this work by way of critiques, suggestions, working out and/or checking of examples, and providing first drafts for some of the sections. Among the latter are Dr. Ashok Midha (prepared KED section), Dianne Rekow (Balancing section), Dr. Robert Williams (Spatial Mechanisms), and Dr. Donald R. Riley, who taught from the preliminary versions of the texts and offered numerous suggestions for improvements. Others making significant contributions are John Gustafson, Lee Hunt, Tom Carlson, Ray Giese, Bill Dahlof, Tom Chase, Sern Hong Wang, Dr. Sanjay G. Dhande, Dr. Patrick Starr, Dr. William Carson, Dr. Charles F. Reinholtz, Dr. Manuel Hernandez, Martin Di Girolamo, Xirong Zhuang, Shang-pei Yang, and others.

Acknowledgment is also due to the Mechanical Systems Program, Civil and Mechanical Engineering Division, National Science Foundation, for sponsoring Research Grant No. MEA-8025812 at the University of Florida, under which parts of the curvature chapter were conceived and which led to the publication of several journal articles. Sources of illustrations and case studies are acknowledged in the text and in captions. Other sponsors are acknowledged in many of the authors' journal papers (listed among the references), from which material was adapted for this work.

The authors and their collaborators continue to develop new material toward possible inclusion in future editions. To this end, they will appreciate comments and suggestions from the readers and users of these texts.

George N. Sandor
Arthur G. Erdman

# 1 Introduction to Kinematics and Mechanisms

## 1.1 INTRODUCTION

Engineering is based on the fundamental sciences of *mathematics, physics,* and *chemistry.* In most cases, engineering involves the analysis of the conversion of energy from some source to one or more outputs, using one or more of the basic principles of these sciences. *Solid mechanics* is one of the branches of physics which, among others, contains three major subbranches: *kinematics,* which deals with the study of relative motion; *statics,* which is the study of forces and moments, apart from motion; and *kinetics,* which deals with the action of forces on bodies. The combination of kinematics and kinetics is referred to as *dynamics.* This text describes the appropriate mathematics, kinematics, and dynamics required to accomplish mechanism design.

A *mechanism* is a mechanical device that has the purpose of transferring motion and/or force from a source to an output. A *linkage* consists of links (or bars) (see Table 1.1), generally considered rigid, which are connected by joints (see Table 1.2), such as pins (or revolutes) or prismatic joints, to form open or closed chains (or loops). Such *kinematic chains,* with at least one link fixed, become (1) *mechanisms* if at least two other links retain mobility, or (2) *structures* if no mobility remains. In other words, a mechanism permits relative motion between its "rigid" links; a structure does not. Since linkages make simple mechanisms and can be designed to perform complex tasks, such as nonlinear motion and force transmission, they will receive much attention in this book. Some of the linkage design techniques presented here are the result of a resurgence in the theory of mechanisms based on the availability of the computer. Many of the design methods were discovered before the 1960s, but long, cumbersome calculations discouraged any further development at that time.

## 1.2 MOTION

A large majority of mechanisms exhibit motion such that all the links move in parallel planes. This text emphasizes this type of motion, which is called *two-dimensional, plane,* or *planar* motion. Planar rigid-body motion consists of *rotation* about axes perpendicular to the plane of motion and *translation*—where all points in the body move along identical straight or curvilinear paths and all lines embedded in the body remain parallel to their original orientation. *Spatial* mechanisms, introduced in Chap. 6, allow movement in three dimensions. Combinations of rotation about three nonparallel axes and translation in three directions are possible depending on the constraints imposed by the joints (spherical, helical, cylindrical, etc.; see Table 6.1).

## 1.3 THE FOUR-BAR LINKAGE

Mechanisms are used in a great variety of machines and devices. The simplest closed-loop linkage is the four-bar, which has three moving links (plus one fixed link)* and four pin joints (see Fig. 1.1). The link that is connected to the power source or prime mover is called the *input* link ($A_0A$). The *output* link connects the moving

**Figure 1.1**

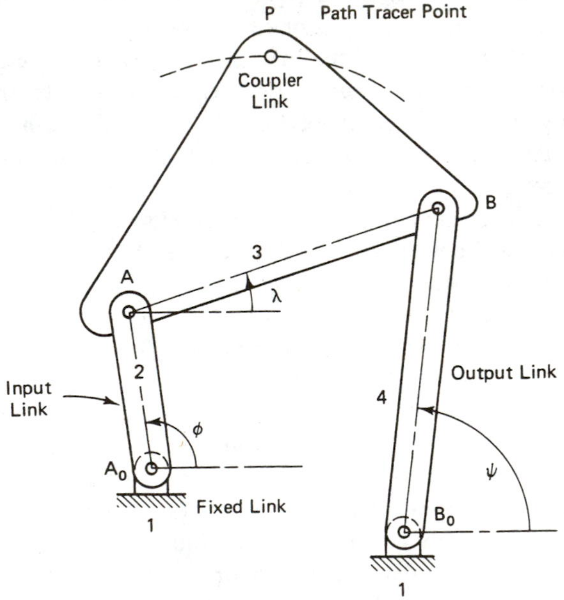


* A *linkage* with one link fixed is a *mechanism.*

pivot $B$ to ground pivot $B_0$. The *coupler* or *floating* link connects the two moving pivots, $A$ and $B$, thereby "coupling" the input to the output link.

Figure 1.2 shows three applications where the four-bar has been used to accomplish different tasks. The *level luffing crane* of Fig. 1.2a is a special type of four-bar that generates approximate *straight-line motion* of the *path tracer* point (point $P$). Cranes of this type can be rated at 50 tons capacity and typically have an approximate straight-line travel of the coupler tracer point about 9 m long.

Figure 1.2b is a drive linkage for a lawn sprinkler, which is adjustable to obtain different ranges of oscillation of the sprinkler head. This *adjustable linkage* can be

**Figure 1.2**

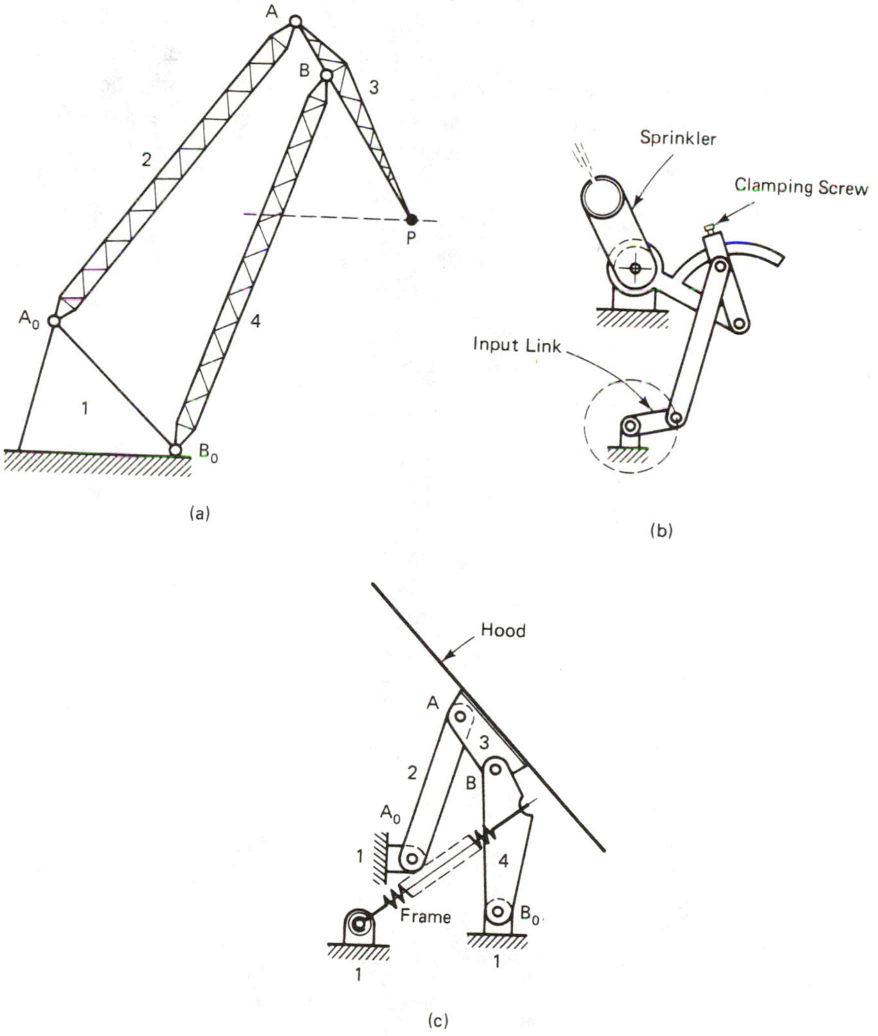

varied in its function by changing the length and angle of the output link by way of the clamping screw. Figure 1.2c shows a four-bar automobile hood linkage design. The linkage controls the relative motion between the hood and the car frame. [120]

The three applications shown in Fig. 1.2 are quite different and in fact represent three different *tasks* by which all mechanisms may be classified by application: path generation, function generation, and motion generation (or rigid-body guidance). In *path generation* (Fig. 1.2a), we are concerned with the path of a tracer point. A *function generator* (Fig. 1.2b) is a linkage in which the relative motion (or forces) between links (generally) connected to ground is of interest. In *motion generation* (Fig. 1.2c), the entire motion of the coupler link is of concern. These tasks are discussed in greater depth in Chaps. 2 and 3.

The four-bar has some special configurations created by making one or more links infinite in length. The slider-crank (or crank and slider) mechanism of Fig. 1.3 is a four-bar chain with a slider replacing an infinitely long output link. The internal combustion engine is built around this mechanism—the crank is link 2, the connecting rod is the coupler (link 3), and the piston is the slider (link 4).

**Figure 1.3**

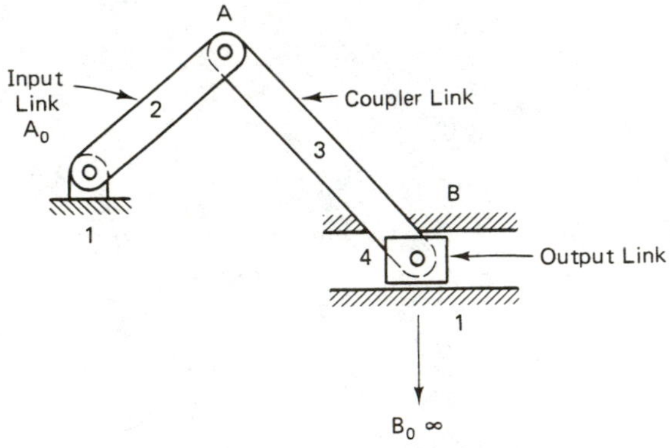


(a)

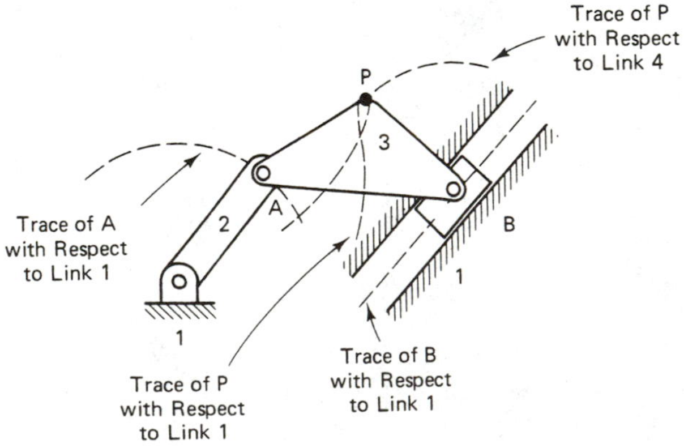


(b)

Other forms of four-link mechanisms exist in which a slider is guided on a moving link rather than on the fixed link. These are called *inversions* of the slider-crank, produced when another link (the crank, coupler, or slider) is the fixed link. Section 3.1 (Vol. 1) shows some applications of the inversions of the slider-crank.

## 1.4 THE SCIENCE OF RELATIVE MOTION

All motion observed in nature is relative motion; that is, the motion of the observed body relative to the observer. For example, the seated passenger on a bus is moving relative to the waiting observer at the bus stop, but is at rest relative to another seated passenger. Conversely, the passenger moving along the isle of the bus is in motion relative to the seated passenger as well as relative to the waiting observer at the bus stop.

The study of motion, kinematics, has been referred to as the science of relative motion. Design and analysis of machinery and mechanisms relies on the designer's ability to visualize relative motion of machinery components. One major objective of this chapter is to familiarize the reader with motion generated by a variety of linkage mechanisms and thus prepare for topics in both analysis and synthesis based on this fundamental understanding. Figure 1.3b shows a slider-crank linkage with a triangular coupler link *ABP*. Each point of the coupler link traces a different path, called *coupler curves,* with respect to ground (link 1). Point *A* traces out a circular arc centered at $A_0$, point *B* travels in a straight line, and point *P* traces out a more complex curve. All these coupler curves are part of the *absolute motion** of link 3. Suppose that the path of point *P* with respect to link 4 instead of link 1 is desired. This relative motion may be found by envisioning oneself sitting on link 4 and observing the motion of link 3, in particular point *P* of link 3. In other words, we invert the mechanism, fixing link 4 (the slider) instead of link 1, and move the rest of the mechanism (including the former fixed link) with respect to link 4. Here the relative path of point *P* with respect to link 4 is a circular arc centered at *B.* Thus absolute motion is a special case of relative motion.

## 1.5 KINEMATIC DIAGRAMS

Although the four-bar and slider-crank are very useful linkages and are found in thousands of applications, we will see later that these linkages have limited performance levels. Linkages with more members are often used in more demanding circumstances.

Figure 1.4 shows a typical application of a multiloop mechanism in which a mechanical linkage is required. A casement window must open 90° outward from the sill and be at sufficient distance from one side to satisfy the egress codes and from the other side to provide access to the outside of the window pane for cleaning.

* In mechanism analysis it is convenient to define one of the links as the fixed frame of reference. All motion with respect to this link is then termed absolute motion.

**Figure 1.4** (*Courtesy of Truth Inc.*)

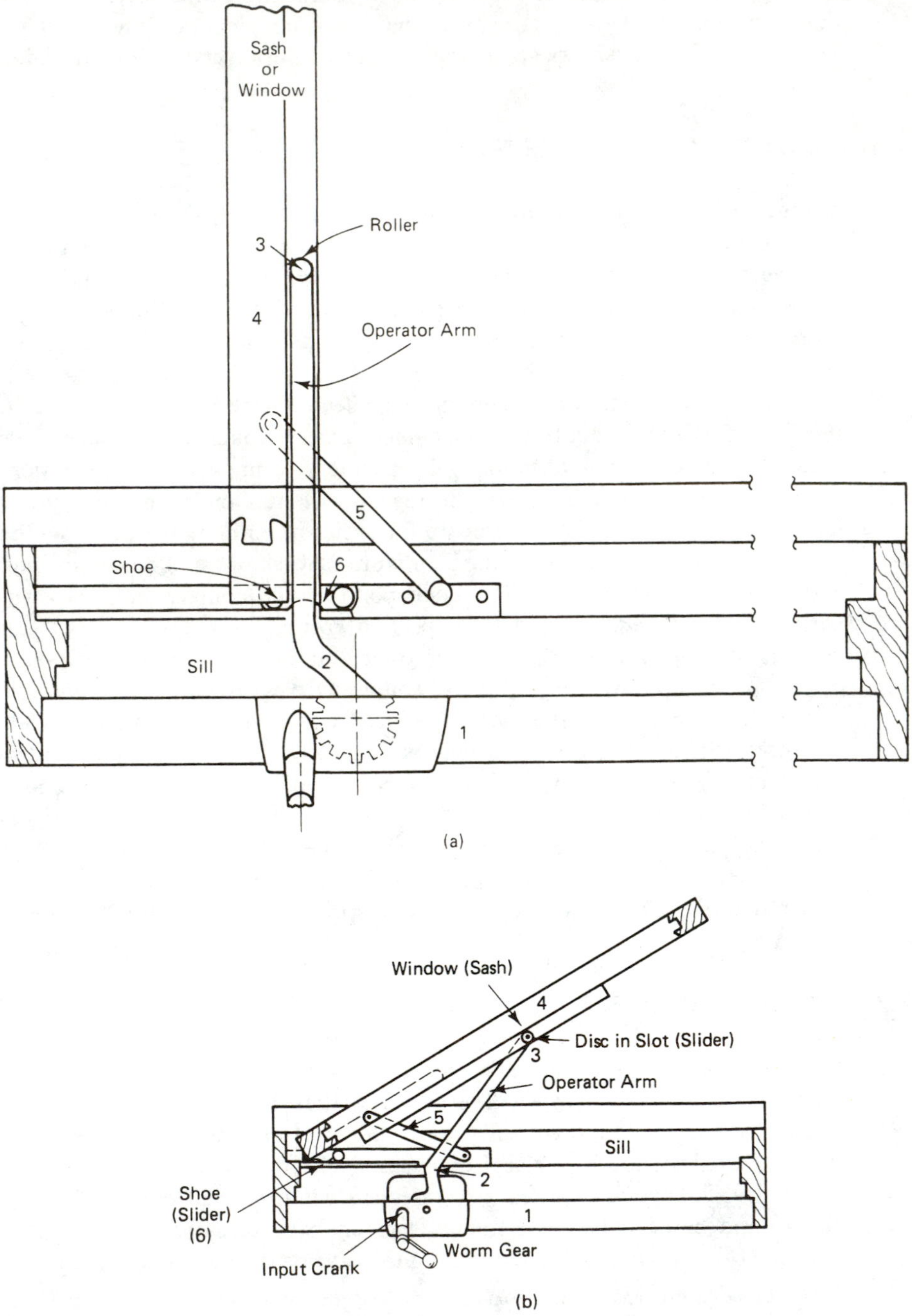

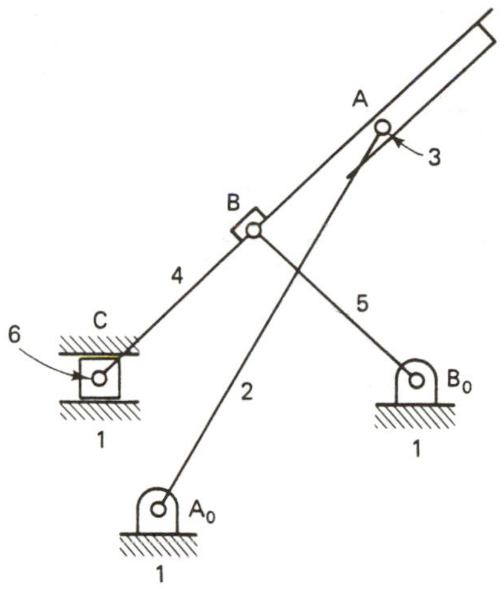


**Figure 1.5** Equivalent *kinematic diagram* (sketch) of Fig. 1.4.

Also, the force required to drive the linkage must be reasonable for hand operation. Figure 1.4a and b show one of the popular casement window operator mechanisms in the 90° and 30° positions, respectively.

It is often difficult to visualize the movement of a multiloop linkage such as that shown in Fig. 1.4, especially when other components appear in the same diagram. The first step in the motion analysis of more complicated mechanisms is to sketch the equivalent *kinematic* or *skeleton diagram*. This requires a "stripped-down" stick diagram, such as that shown in Fig. 1.5. The skeleton diagram serves a purpose similar to that of the electrical schematic or circuit diagram in that it displays only the essential skeleton of the mechanism, which, however, embodies the key dimensions that affect its motion. The kinematic diagram takes one of two forms: a sketch (porportional but not exactly to scale), and the scaled kinematic diagram (usually used for further analysis: position, displacement, velocity, acceleration, force and torque transmission, etc.). For convenient reference, the links are numbered (starting with ground link as number 1), while the joints are lettered. The input and output links are also labeled. Table 1.1 shows typical skeleton diagrams of planar links. One purpose of the skeleton diagram is to provide a kinematic schematic of the relative motions in the mechanisms. For example, a pin joint depicts relative rotation, a slider depicts relative straight-line translation, and so on.

Figure 1.5 shows the kinematic diagram (sketch) for the casement window linkage. Notice that there are six links, five pin joints, one slider joint, and one roller in this sketch. Note also that one loop of the mechanism contains a slider-crank linkage (1,5,4,6). Connected to the slider crank is a bar and a roller (2,3), which provides the input for opening and closing the window. The kinematic diagram simplifies the mechanism for visual inspection and, if drawn to scale, provides the means for further analysis.

Another application where a multiloop mechanism has been suggested is a proposed variable-stroke engine [204] (Fig. 1.6). This linkage varies the piston stroke in response to power requirements. The operation of the stroke linkage is shown in Fig. 1.7.

For each piston, the lower end of a control link is adjusted along an arc prescribed by the control yoke shown. The top of the control link is connected to the main link

**TABLE 1.1** PLANAR LINK TYPES

| Link types | Typical form | Skeleton diagram(s) |
|---|---|---|
| Binary |  |  |
| Ternary | 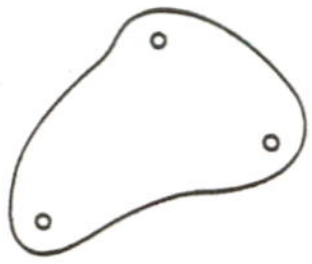 | 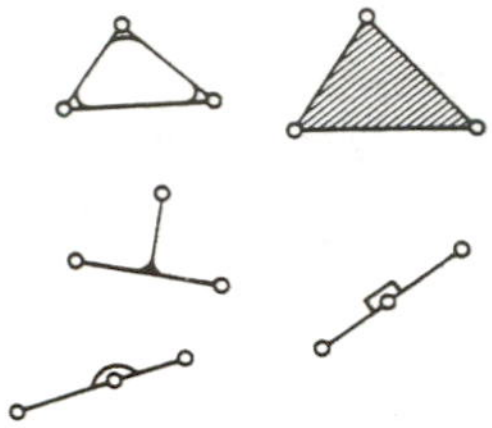 |
| Quaternary | 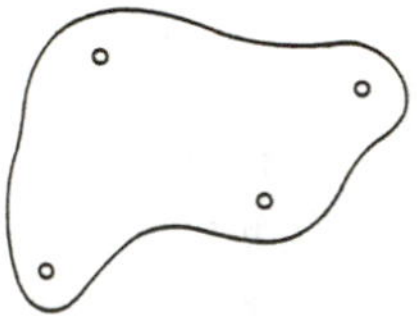 | 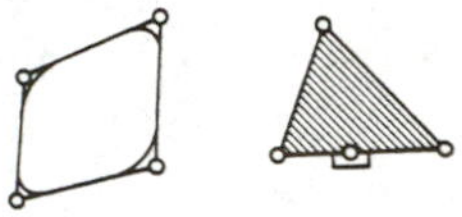 |

which, in turn, connects to a component that plays the role of a conventional connecting rod. In essence, the result is an engine with variable crank-throw.

When control-yoke divergence from vertical is slight [Fig. 1.7a] the main link is restricted in its movement, and the resulting piston stroke is small. As the control nut moves inward on its screw, the angle between the control yoke and "the axis of the cylinder" is increased. This causes the main link to move in a broader arc, bringing about a longer stroke. The angle between the control yoke and the cylinder axis varies between 0 and 70°; the resulting stroke varies from 1 in. to 4.25 in. "The linkage is designed so that the compression ratio stays approximately the same, regardless of piston stroke."

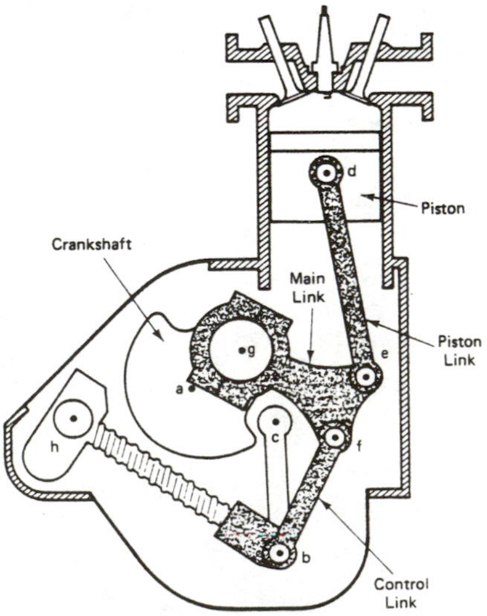


**Figure 1.6** Section view of variable displacement engine showing crankshaft, main link, piston link, and stroke control link. Stroke is varied by moving the lower end of the control link.

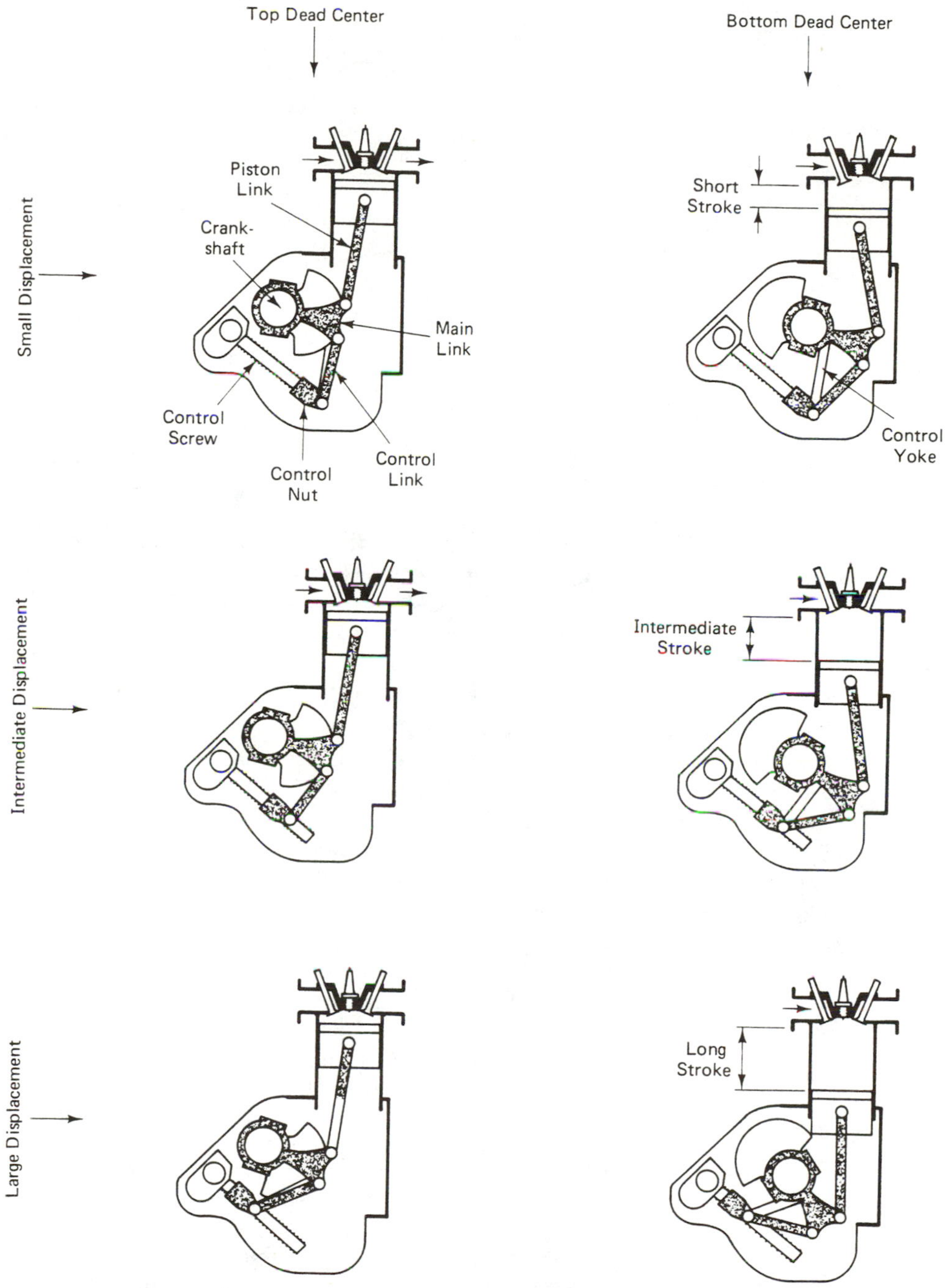


**Figure 1.7** Variable-displacement linkage; stroke is varied by moving lower end of control link.

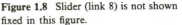

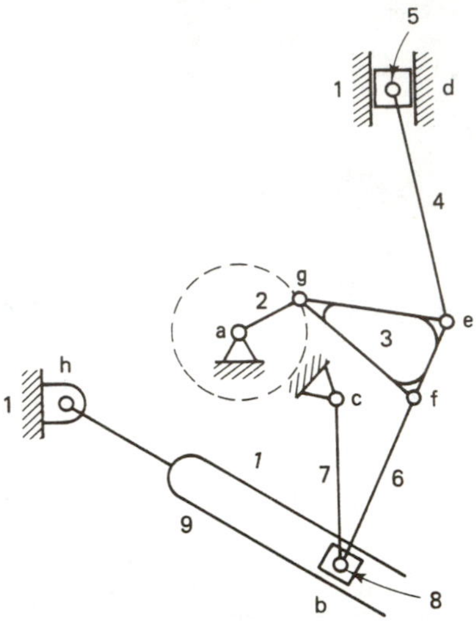


**Figure 1.8** Slider (link 8) is not shown fixed in this figure.

The equivalent kinematic diagram of this adjustable mechanism is shown in Fig. 1.8. Notice that there are nine links, nine pins, and two sliders in this sketch.

## 1.6 SIX-BAR CHAINS

If a four-bar linkage does not provide the type of performance required by a particular application, one of two single-degree-of-freedom six-bar linkage types (with seven revolute joints) is usually considered next: the *Watt chain* or the *Stephenson chain* (see Sec. 1.7 and Figs. 1.9 to 1.13). These classifications depend on the placement of the ternary* links (members with three revolute joints; see Table 1.1). In the Watt chain, the ternary links are adjacent; in the Stephenson chain, the ternary links are separated by binary links (links with only two revolute joints). Several applications where six-bar chains have been employed will help us become familiar with these linkages.

**Example 1.1 [80]**

In the manufacture of cassette tape cartridges, it is sometimes necessary to thread a leader tape contained in the assembled cassette onto a device which winds blank magnetic tape into the cassette. A mechanical linkage is sought to thread the leader tape.

Figure 1.14 shows the position of the cassette, the leader tape, and the device through which the tape must be threaded at the time it is desired that the guiding linkage begin operation. The dashed line is the final configuration of the threaded leader tape.

The tape unwinds from both sides of the cassette and so forms a loop as it is pulled out. The numbers 1 to 5 indicate the successive positions the tape must pass.

* Notice in Figs. 1.9 to 1.13 that some of the triangular-shaped links are truly ternary, while others are shown as triangular to indicate possible path tracer points on floating links.

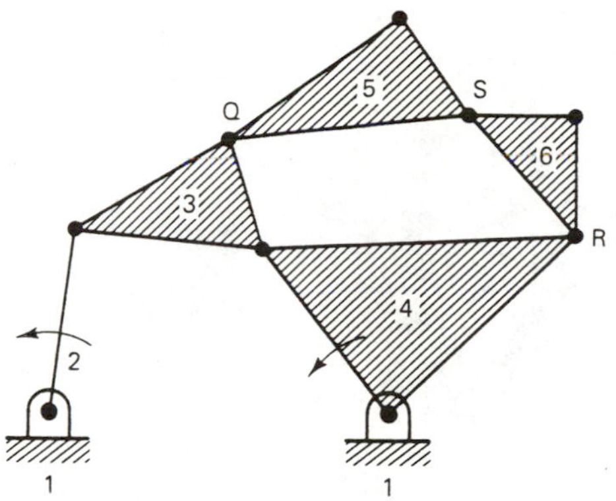


**Figure 1.9** Watt I six-bar linkage.

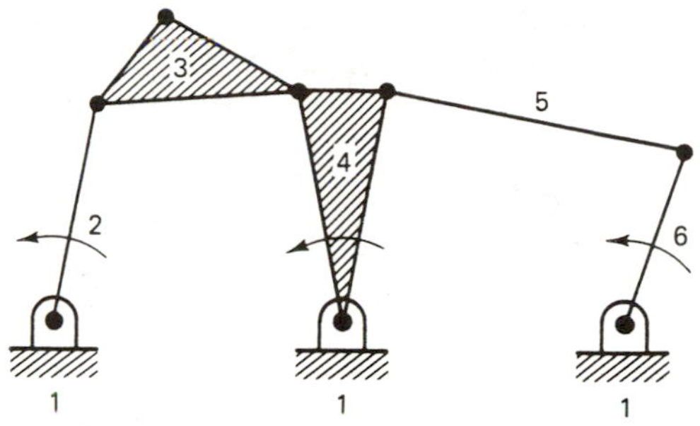


**Figure 1.10** Watt II six-bar linkage.

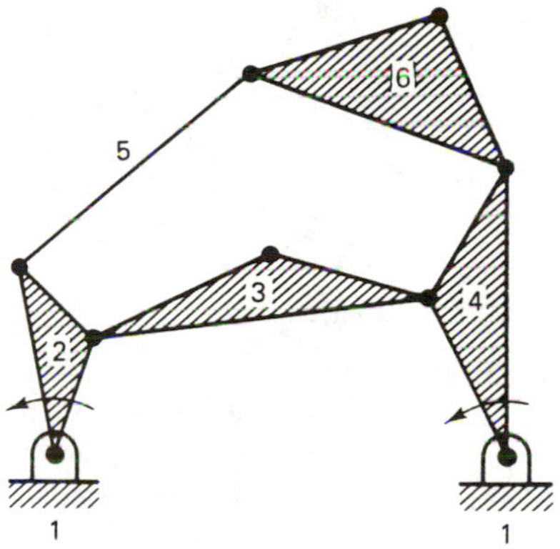


**Figure 1.11** Stephenson I six-bar linkage.

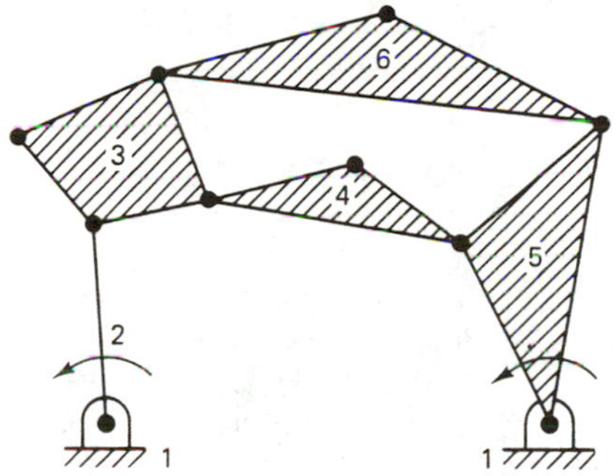


**Figure 1.12** Stephenson II six-bar linkage.

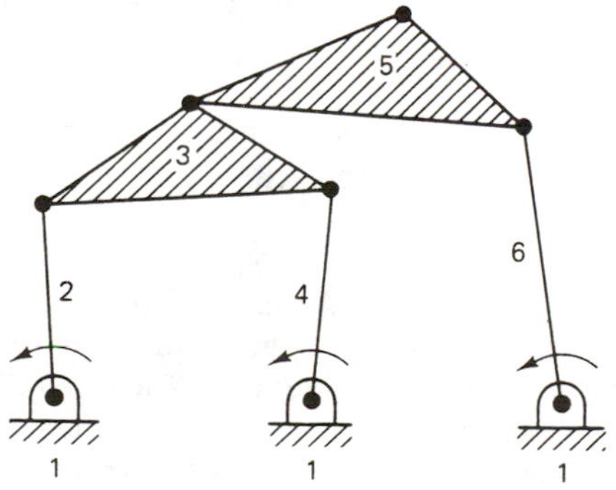


**Figure 1.13** Stephenson III six-bar linkage.

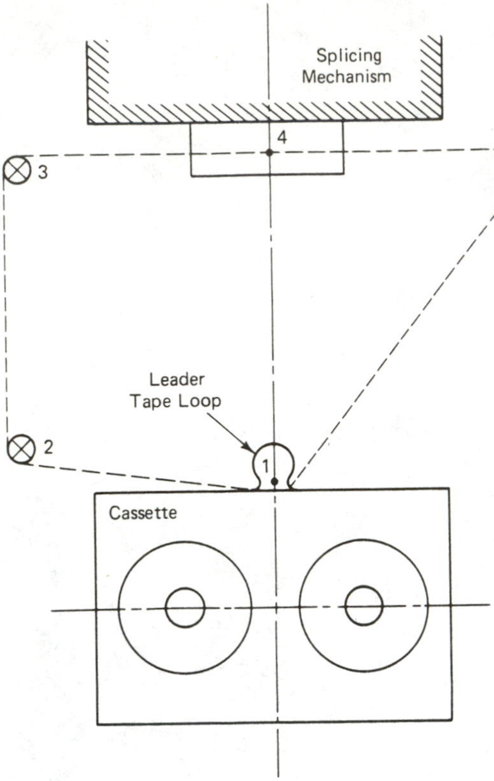


Figure 1.14

The crossed circles at positions 2, 3, and 5 (Fig. 1.14) are posts that hold the tape loop. These posts are initially below the tape deck and each comes up to hold the tape once the tape loop has passed over its position. The tape loop must be guided clear of the posts at positions 2, 3, and 5 in the proper direction of travel.

The following were considered necessary requirements for the linkage:

1. 360° input crank rotation.
2. Input rotation must be timed to the positions of the path point (tape loop threading guide) to allow posts 2, 3, and 5 to be brought up at the correct time.
3. Angular orientation of the coupler link containing the path point must be specified at each prescribed position.

The Stephenson III chain was chosen for this example. The computer-aided techniques of Chap. 3 were used to produce the final design shown in Fig. 1.15.

**Example 1.2 [208,209]**

Mechanisms are extremely useful in the design of biomechanical devices. For example, in the design of an external prosthesis for a through-knee amputee, it is desirable to duplicate the movement of the relative center of rotation (see Chap. 3 of Vol. 1) between the thigh and the leg bones (femur and tibia) so as to maintain stability in walking. Figures 1.16 and 1.17 show a Stephenson I six-bar motion generator designed for this

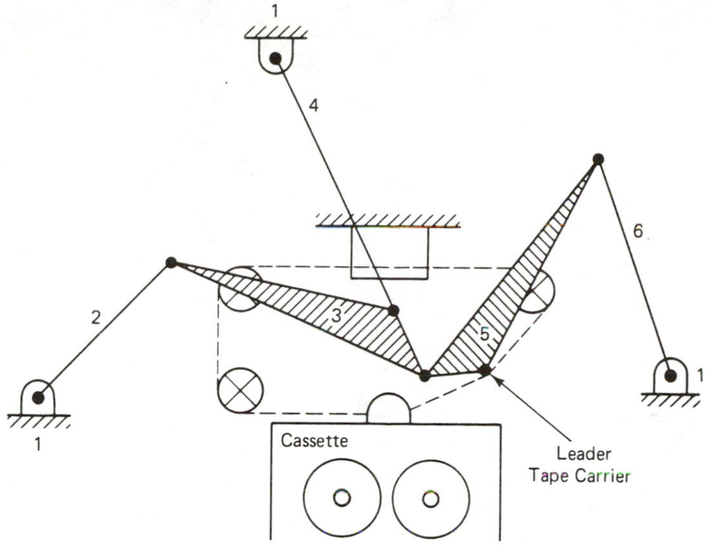


**Figure 1.15**

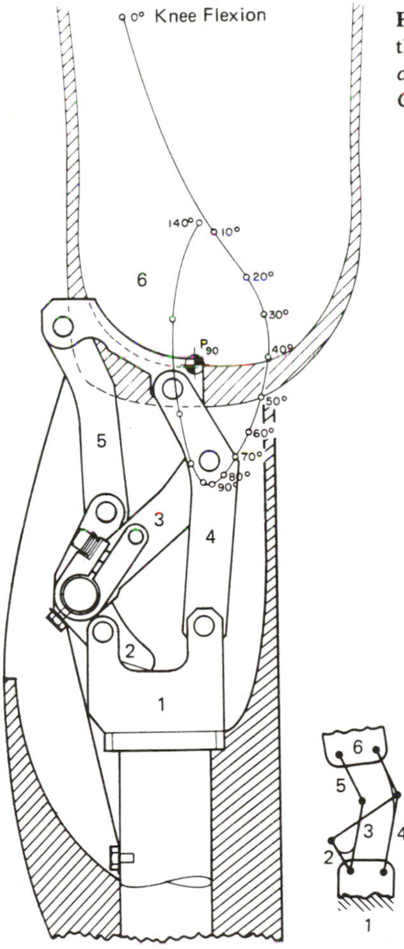


(a) (b)

**Figure 1.16** Six-bar linkage prosthetic knee mechanism. (*Biomechanics Laboratory, University of California, Berkeley.*)

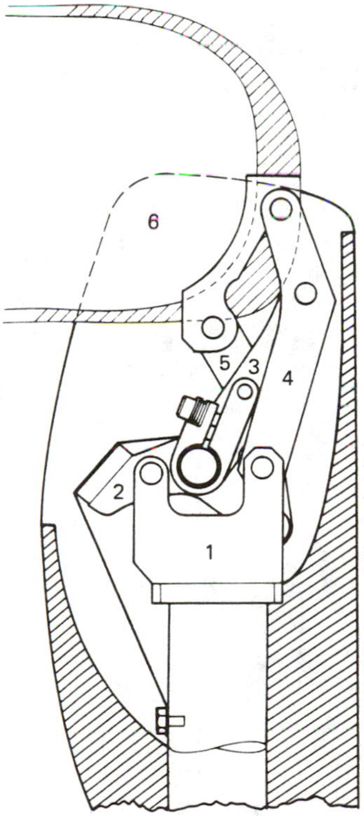


**Figure 1.17** Six-bar linkage prosthetic knee mechanism. (*Biomechanics, Laboratory, University of California, Berkeley.*)

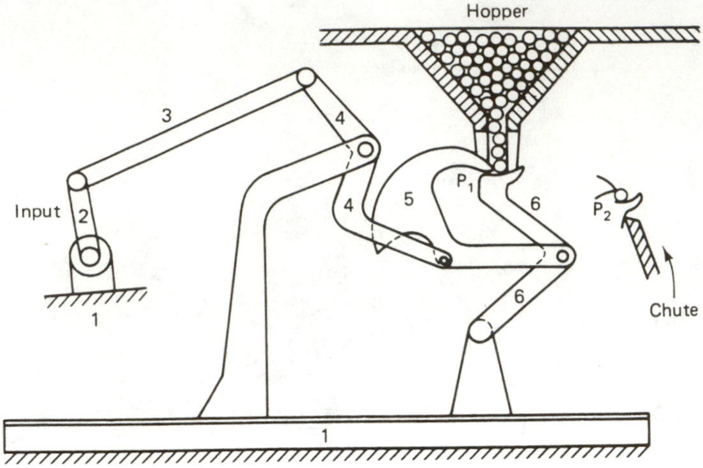


Figure 1.18

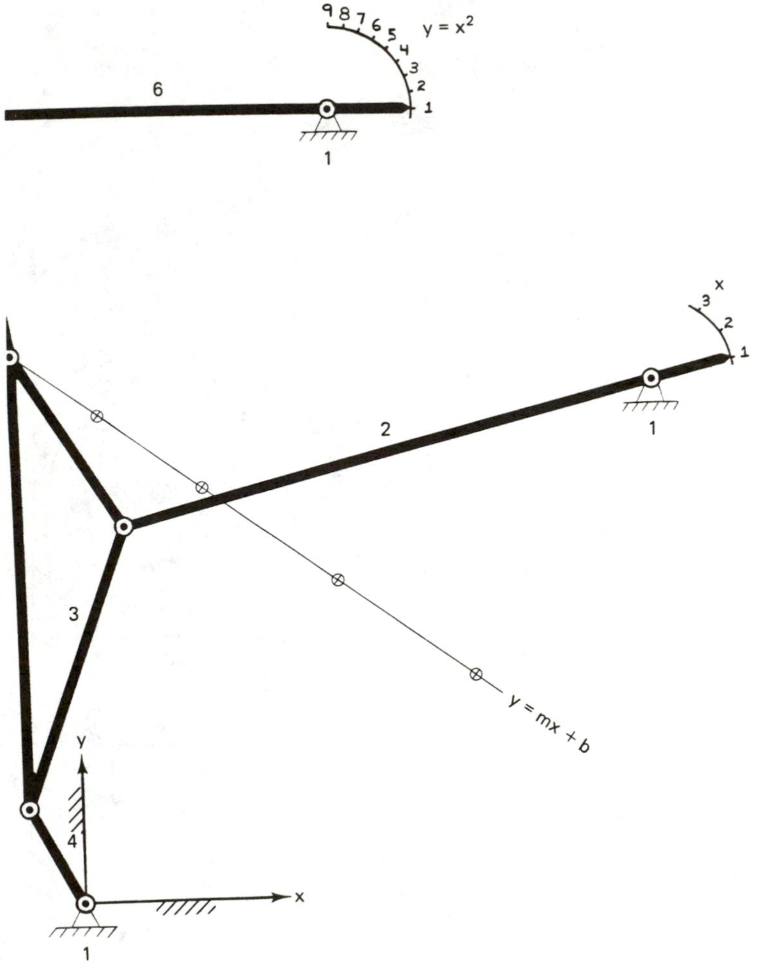


Figure 1.19

purpose. The zero-degree flexion (fully extended) position is shown in Fig. 1.16a together with the trajectory of the instant center of rotation of link 1, the artificial leg, with respect to the femur (link 6). The 90° flexion (bent knee) position is shown in Fig. 1.17, and the kinematic diagram (sketch) of this linkage is shown in Fig. 1.16b.

**Example 1.3 [119]**

A feeding mechanism (see Fig. 1.18) is required to transfer cylindrical parts from a hopper to a chute for further machining. A Watt II mechanism was chosen for this task. The timing link (6) provides rotation to the cupped platform (whose rotation is a prescribed function of input-link rotation) which transfers the cylinder from the hopper to the chute, while the prescribed path of the output coupler (point $P$) positions the cylinder on the platform and then pushes the cylinder into the chute.

**Example 1.4 [213]**

Another example of a dual-task requirement for a linkage is simultaneous path and function generation (Fig. 1.19). The specified function is $y = x^2$ for $1 \leq x \leq 3$, while the required path is an approximate straight line. Figure 1.19 shows the first prescribed position of a Stephenson III linkage synthesized using the techniques described in Chap. 3. The other four prescribed positions are shown in Fig. 1.20.

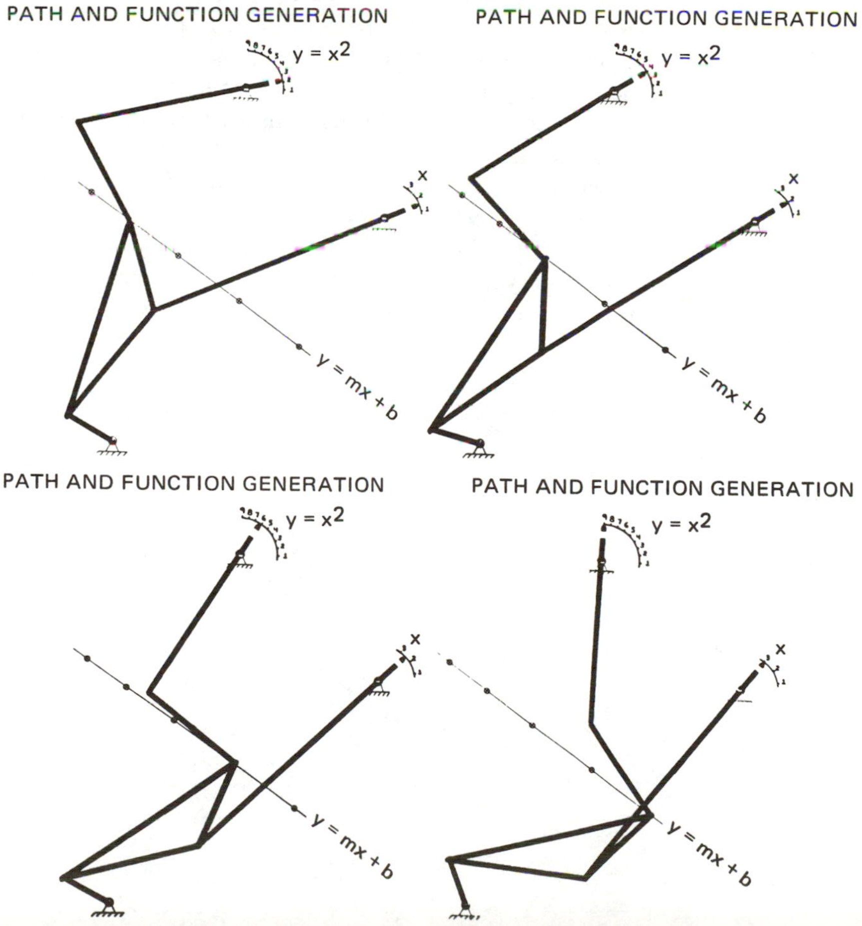


**Figure 1.20**

The previous four examples consist of links with pin (or revolute) connections. If one or more of the links in one of the six-bars in Figures 1.9 to 1.13 is changed to a slider, different six-link mechanisms are obtained. Numerous possible six-link mechanisms exist with combinations of links, pins, and sliders. (See the appendix to Chap. 2 for a sample case study.)

## 1.7 DEGREES OF FREEDOM

The next step in the kinematic analysis of mechanisms, following the drawing of the schematic, is to determine the number of *degrees of freedom* of the mechanism. By degrees of freedom we mean the number of independent inputs required to determine the position of all links of the mechanism with respect to ground. There are hundreds of thousands of different linkage types that one could invent. Envision a bag containing a large variety of linkage components from Tables 1.1 and 1.2; binary, ternary, quaternary, and so on, links; pin joints, slider joints; cams and cam followers; gears, chains, sprockets, belts, pulleys, and so on. (Spherical and helical as well as other connections that allow three-dimensional relative motion are not included here, as only planar motion in parallel planes is discussed in this portion of the book. Three-dimensional motion is covered in Chap. 6.) Furthermore, imagine the possibility of forming all sorts of linkage types by putting these components together. For example, several binary links might be connected by pin joints. Are there any rules that help govern how these mechanisms are formed? For instance, is the linkage in Fig. 1.21 usable as a function generator, where we wish to specify the angular relationship between $\phi$, the independent variable, and $\psi$, the dependent variable?

The obvious problem with the linkage of Fig. 1.21 is that, if a motor is attached to the shaft of the input link, the output link may not respond directly—there appear to be too many intervening links. Clearly, there is a need for some rule of mobility by which linkages are put together. We can start to develop such a rule by examining a single link.

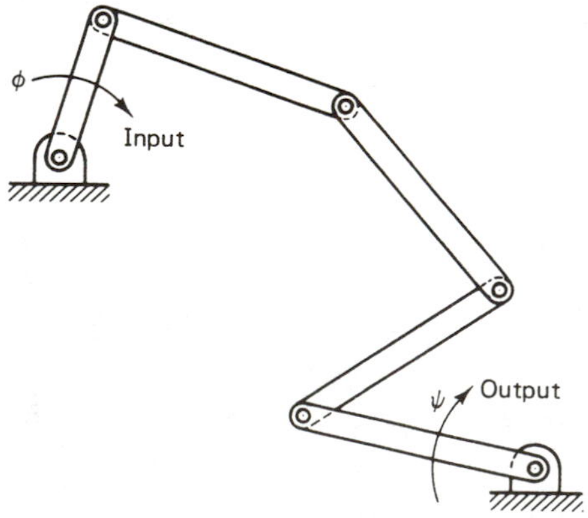


**Figure 1.21**

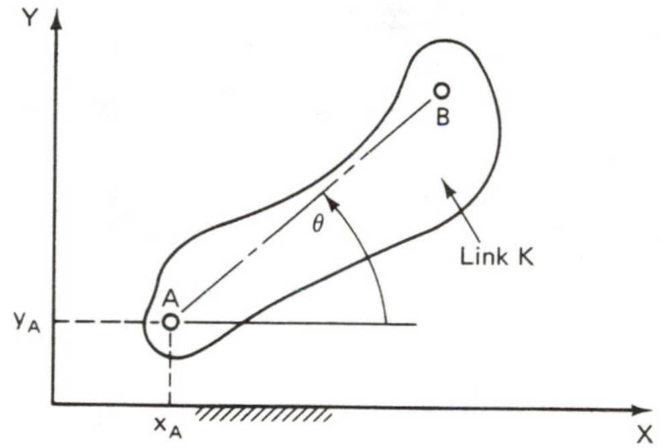


**Figure 1.22**

Suppose that the exact position of rigid link $K$ is required in coordinate system $XY$ as depicted in Fig. 1.22. How many independent variables will completely specify the position of this link? The location of point $A$ can be reached, say, from the origin by first moving along the $X$ axis by $x_A$ and then $y_A$ in the direction of the $Y$ axis. Thus, these two coordinates, representing two translations, locate point $A$. More information is required, however, to define completely the position of link $K$. If the angle of the line of points $A$ and $B$ with respect to the $X$ axis is known, the position of link $K$ is specified in the plane $XY$. Thus there are three independent variables: $x_A$, $y_A$, and $\theta$ (two translations and one rotation, or three independent coordinates) associated with the position of a link in the plane. In other words, an unconstrained rigid link in the plane has *three degrees of freedom.*

If there is an assembly of $n$ links, they possess a total of $3n$ degrees of freedom before they are joined to form a linkage system. Connections between links result in the loss of degrees of freedom of the total system of links. A pin (or revolute) joint is called a *lower-pair* connector—defined in some previous literature as one that has surface contact between its elements such as the pin and the bushing. How many degrees of freedom does a pin joint subtract from the previously unconstrained link which it connects? If point $A$ on the link in Fig. 1.22 is a pin joint between link $K$ and ground, then two independent variables, $x_A$ and $y_A$ are fixed, leaving $\theta$ as the one remaining degree of freedom of link $K$.

In an assembly of links such as that in Fig. 1.21, each pin connection will remove two degrees of freedom of relative motion between successive links. This observation suggests an equation that will determine the degrees of freedom of an $n$-link chain connected by $f_1$ pin joints, with ground (the fixed link) considered as one of the links:

$$\text{degrees of freedom} = F = 3(n-1) - 2f_1 \tag{1.1}$$

Equation (1.1) is known as *Gruebler's equation.* The number of mobile links is $(n - 1)$. The pin joint permits *one* degree of relative freedom between two links—thus the notation $f_1$. This equation is one of the most popular mobility equations used in practice. For other versions, see Ref. 105.

Most mechanism tasks require a single input to be transferred to a single output. Therefore, single-degree-of-freedom mechanisms, those that have *constrained motion,* are the forms used most frequently. For example, it is easy to see intuitively that the four-bar of Fig. 1.1 is a single-degree-of-freedom linkage. An intuitive degree-of-freedom analysis may proceed as follows. Once the independent variable $\phi$ is specified, the position of point $A$ is known with respect to $A_0$ and $B_0$; since the lengths of the coupler base $AB$ and the output link $B_0B$ are known, $B_0AB$ is a triangle with no further mobility (zero degrees of freedom) and the position of the rest of the linkage is determined.*

Using Gruebler's equation to determine the number of degrees of freedom of the linkage in Fig. 1.1, we have

$$n = 4, \qquad f_1 = 4$$

$$F = 3(4-1) - 2(4) = +1$$

The "+1" indicates a single degree of freedom for the linkage. As a further demonstration of the use of Gruebler's equation, refer to the Watt I six-bar in Fig. 1.9:

$$n = 6, \qquad f_1 = 7$$

$$F = 3(6-1) - 2(7) = +1$$

Intuitively, one can be satisfied that this linkage has a single degree of freedom as predicted by the equation. Once assembled, links 1 to 4 form a four-bar linkage which has already been demonstrated to have a single degree of freedom. Observe that links 3, 4, 5, and 6 form a second four-bar linkage with positions of two of the links (3 and 4) already determined. Since the positions of points $Q$ and $R$ are determined, $QSR$ forms a "rigid" triangle and the position of the entire mechanism is specified.

Determine the degrees of freedom of the trench hoe of Fig. 1.23. This linkage system has an element that has not been included in the degree-of-freedom discussion up to this point—the slider (hydraulic cylinder in this case). Let us therefore determine how many degrees of freedom of relative motion a sliding connection subtracts between adjacent links: in other words, how many relative constraints a slider imposes. In Fig. 1.3, the slider (link 4) is constrained with respect to ground (link 1) against moving in the vertical direction as well as being constrained from rotating in the plane. Thus the slider joint allows movement only along the slide and subtracts two degrees of freedom of relative motion: one rotation and one translation. Equation (1.1) may now be expanded in scope so that $f_1$ *equals the sum of the number of pin joints plus the number of slider joints*—since they both allow only one degree of relative motion.

* Actually, there are two possible *branches* for the rest of the four-bar ($ABB_0$), mirror images about the diagonal $B_0A$, in which the links $AB$ and $BB_0$ could be assembled. However, the linkage cannot move from one branch to the other without disassembly. Thus the number of degrees of freedom of a linkage is independent of the fact that the mechanism may have several different branches. Therefore, one may formulate the intuitive definition of degrees of freedom thus: When, after specifying $n$ coordinates ($x$, $y$, and/or $\theta$) of link positions, the possible positions of the remaining links are finite, the number of degrees of freedom is $n$. The concept of branching is addressed in Chap. 3 of Vol. 1 (see especially Figs. 3.18 and 3.19).

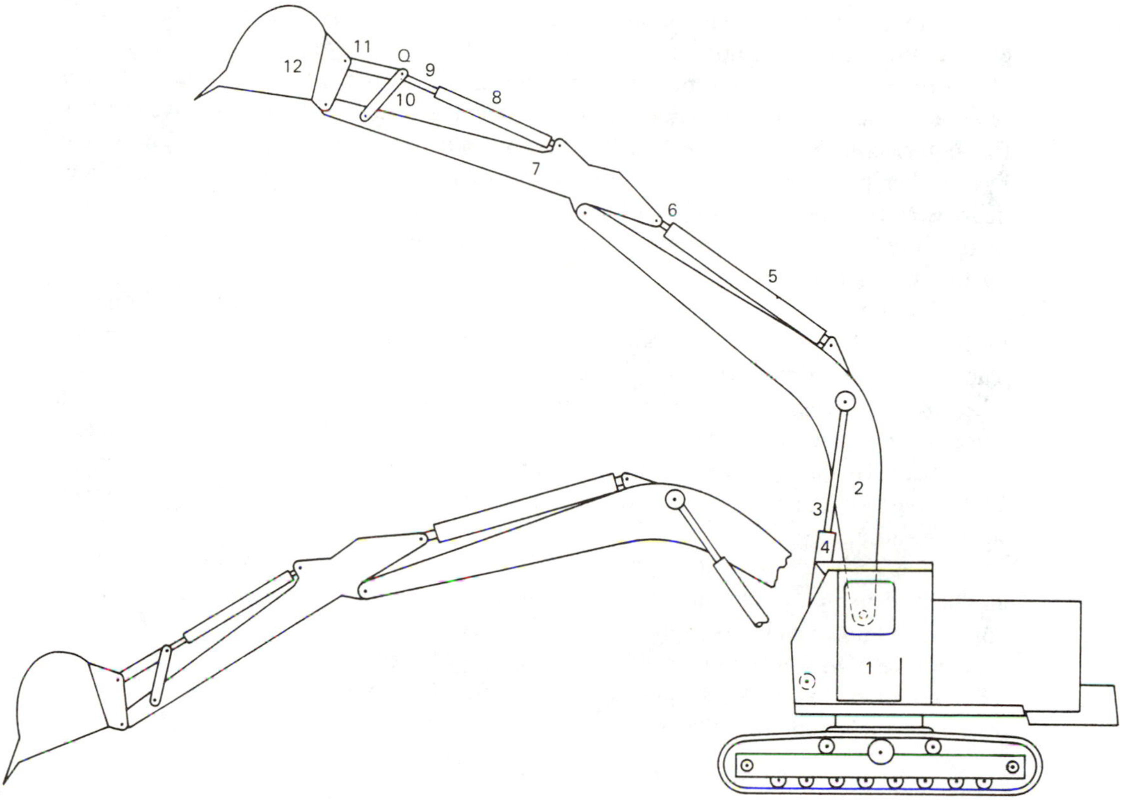


Figure 1.23

The trench hoe has 12 links (consider the cab as the ground link), 12 pin joints, and the three slider joints (the piston–cylinder combinations). If you counted only 11 pin connections, look more carefully at point $Q$ in the figure. Three links are connected by the same pin constraint. There are two pin joints at $Q$, one connecting links 9 and 10, the other connecting links 10 and 11. In general the number of pin joints at a common connection is

$$f_1 = m - 1 \tag{1.2}$$

where $m$ is the number of links joined by a single revolute joint.

The number of degrees of freedom of the trench hoe is therefore

$$F = 3(12 - 1) - 2(15) = +3$$

Thus the trench hoe linkage requires three input coordinates to determine the position of all its links relative to the cab. These are supplied by the three hydraulic cylinders that are attached along the boom.

Are there other types of joints besides pins and sliders that can be used to connect mechanism members in plane motion? If so, do they all subtract two degrees of freedom? Five other types of planar joints are shown in Table 1.2. Whereas pin and slider joints (lower pairs) allow only one degree of freedom of relative motion, the *higher-pair* joints (joints that have been defined in the literature as joints that have either point or line contact only) may permit a higher number (two or three) degrees of freedom of relative motion. Each has a lower-pair equivalent, consisting of as many lower pairs as the number of degrees of freedom of relative motion allowed by the higher-pair joint.

*Rolling contact with no sliding* allows only one degree of freedom of relative motion, due to the absence of sliding, which leaves only the relative rotation $\theta$ (see Table 1.2). The pure rolling joint can therefore be included as an $f_1$-type joint. The lower-pair equivalent for instantaneous velocity equivalence* is simply a pin joint at the relative instant center (see Chap. 3 of Vol. 1), which is the contact point between the two links for rolling contact with no sliding. Thus this essentially higher-pair joint allows only one degree of freedom because of the additional constraint against sliding.

The *roll-slide contact* constrains only one degree of freedom (relative motion in the $y$ direction in Table 1.2). First, let us consider the lower-pair combination for instantaneous velocity equivalence, which is a slider and pin joint combination. This allows two degrees of freedom ($n = 3$, $f_1 = 2$) of relative motion. The degrees of freedom of the roll-slide joint can be verified by a modified *Gruebler's equation extended to include roll-slide joints:*

$$F = 3(n-1) - 2f_1 - 1f_2 \tag{1.3}$$

where $f_2$ is the number of roll-slide contact joints (those that permit *two* degrees of relative motion across the joint).

Using Eq. (1.3) for the higher-pair model itself:

$$F = 3(2-1) - 1 = +2$$

For the gear set shown in Table 1.2, in which link 1 is fixed, the gear bearings are pin joints and the *gear tooth contact* is roll-slide. Therefore, $f_1 = 2$ and $f_2 = 1$, so that

$$F = 3(3-1) - 2(2) - 1 = +1$$

The lower-pair linkage for instantaneous velocity equivalence is a four-bar with fixed pivots located at the center of the gears, and moving pivots at centers of curvature of tooth profiles. The coupler goes through the pitch point $P$ along the line of action of the gear mesh, perpendicular to the common tangent of the contacting

* Average velocity is a measure of displacement in the interval in which it occurs, $\Delta s/\Delta t$. The limiting value of this as $\Delta t \to 0$, namely $ds/dt$, is the instantaneous velocity at a point in time. "Instantaneous velocity equivalence" means that, if a higher-pair joint in a mechanism is replaced by its lower-pair equivalent, the instantaneous velocity of the relative motion allowed between the two original links of the higher pair will remain the same but the relative acceleration will, in general, be different.

**TABLE 1.2** PLANAR KINEMATIC PAIRS—LINK JOINTS

| Joint name | Diagram | | Lower pair with equivalent instantaneous velocity | |
|---|---|---|---|---|
| Pin (revolute) |  | n = 2<br>$f_1$ = 1<br>F = +1<br>F = degrees of freedom | — | |
| Slider (prismatic) | 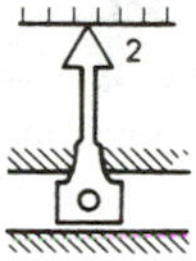 | n = 2<br>$f_1$ = 1<br>F = +1 | 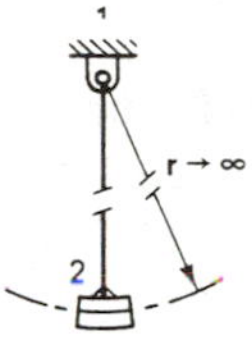 | n = 2<br>$f_1$ = 1<br>F = +1 |
| Rolling contact (no sliding) | 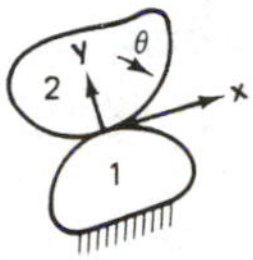 | n = 2<br>F = +1 | 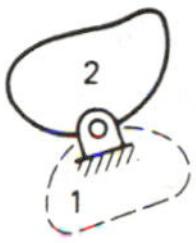 | n = 2<br>$f_1$ = 1<br>F = +1 |
| Roll-slide contact | 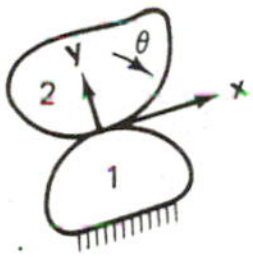 | n = 2<br>$f_1$ = 0<br>$f_2$ = 1<br>F = +2 | 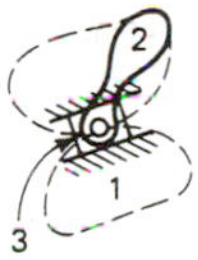 or<br>n = 3<br>$f_1$ = 2<br>$f_2$ = 0<br>F = +2 |  |
| Gear contact (includes roll-slide contact between gear-teeth and rolling contact between pitch circles) | 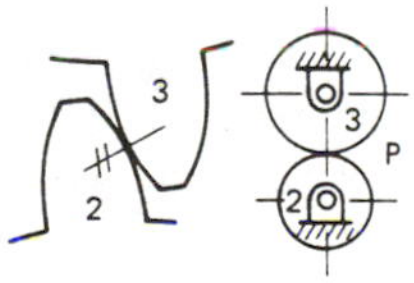 | n = 3<br>$f_1$ = 2<br>$f_2$ = 1<br>F = 1 | 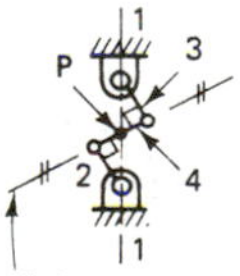<br>Parallel to common normal of contacting tooth flanks; P, pitch point | n = 4<br>$f_1$ = 4<br>$f_2$ = 0<br>F = +1 |
| Spring | 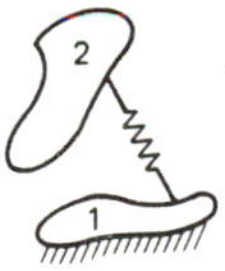 | n = 2<br>$f_1$ = 0<br>$f_2$ = 0<br>F = +3 | 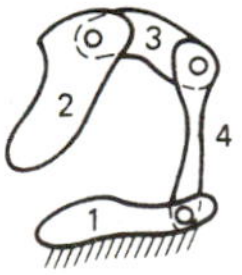 | n = 4<br>$f_1$ = 3<br>$f_2$ = 0<br>F = +3 |
| Belt and pulley (no sliding) or chain and sprocket | F = +1<br>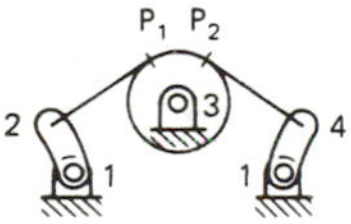<br>$P_1$, $P_2$: points of tangency of approaching and receding belt (chain) leads | | 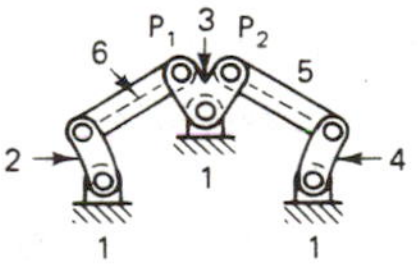 | n = 6<br>$f_1$ = 7<br>$f_2$ = 0<br>F = +1 |

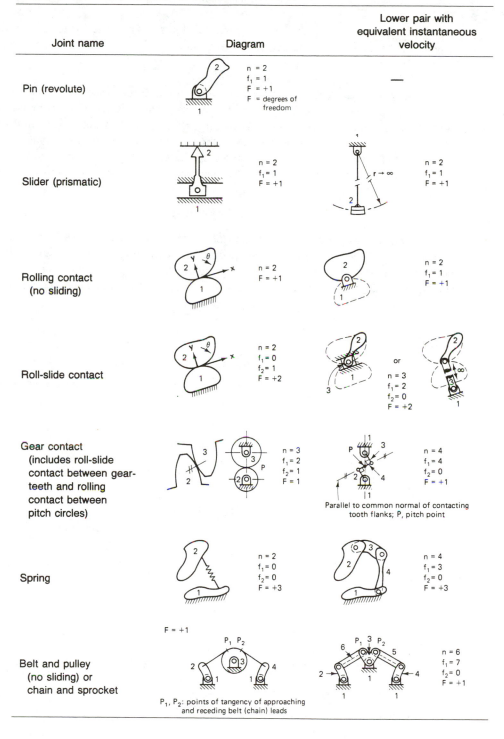

tooth flank surfaces and to the two grounded links. Thus the lower-pair model of a gear set predicts the same degrees of freedom:

$$F = 3(4 - 1) - 2(4) = +1$$

A *spring connection* (Table 1.2) produces a mutual force between the two links it connects, but it does not kinematically constrain the relative motion between the two links (assuming that the spring is within its range of extension and compression). Two binary links and three pin joints form the instantaneous-velocity-equivalent lower-pair model to the spring connection, allowing the same degrees of freedom of relative motion between the links connected by the spring. Thus for the equivalent lower-pair linkage, the number of degrees of freedom is

$$F = 3(4 - 1) - 2(3) = +3$$

The *belt and pulley* or *chain and sprockets* (see Table 1.2), where the belt or chain is maintained tight, are also possible planar connections. A ternary link with three pin joints is the instantaneous-velocity-equivalent lower-pair connection to the belt and pulley (no sliding allowed). Using Eq. (1.3) for the equivalent six-bar linkage,

$$F = 3(6 - 1) - 2(7) = +1$$

**Example 1.5**

Determine the degrees of freedom of the mechanism shown in Fig. 1.24.

**Solution** There are seven links, seven lower pairs, one roll-slide contact, and one spring connection. From Eq. (1.3),

$$F = 3(7 - 1) - 2(7) - 1(1) = +3$$

Let us check this by way of the velocity-equivalent lower-pair connections shown in Fig. 1.25. The spring has been replaced by two binary links, and the *fork joint* (or *pin-in slot joint*—rolling contact with sliding) has been replaced by a pin and slider. Therefore,

$$F = 3(10 - 1) - 2(12) = +3$$

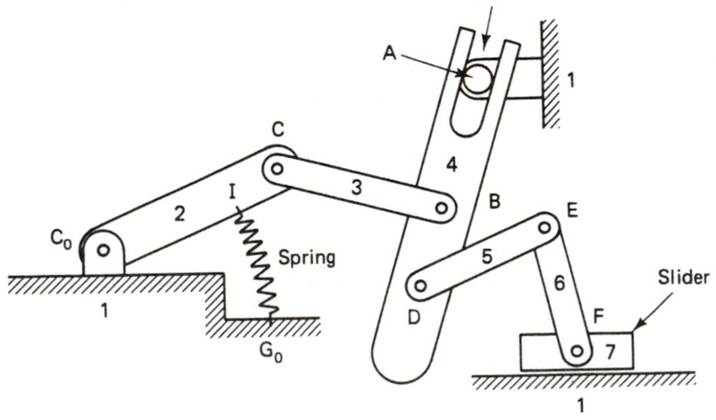


Figure 1.24

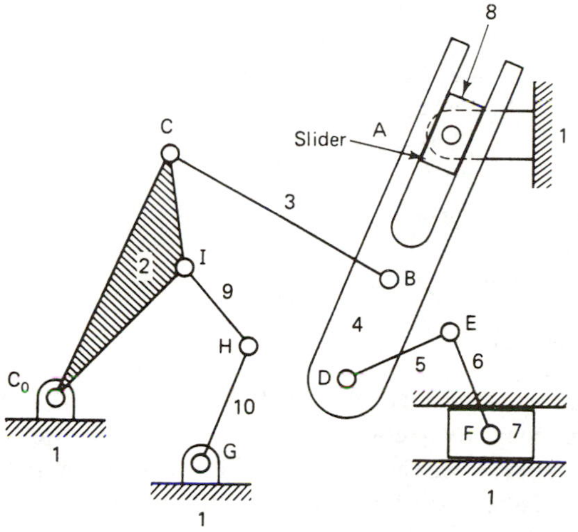


**Figure 1.25**

This answer can be verified by intuition. If both the pin and slider joint at point *A* are fixed (taking away two degrees of freedom), link 4 is fixed in the plane. The slider-crank (*DEF*) is still free to move, however. Since the slider-crank has a single degree of freedom, the entire mechanism has a total of three degrees of freedom.

Before leaving the subject of degrees of freedom, it should be pointed out that there are linkages whose computed degrees of freedom may be zero (indicating a structure) or negative (indicating an indeterminate structure). They can, nevertheless, move due to special linkage proportions. For example, for the five-bar of Fig. 1.26, $F = 3(5 - 1) - 2(6) = 0$, but because of the parallelogram configuration, the linkage can still move. This is called an *overconstrained* linkage, in which the third grounded link provides a *redundant* constraint. If there are manufacturing errors in link lengths or pivot locations, this linkage will jam. Figure 1.27 shows another overconstrained example. Here we see two "nip rolls" in pure rolling contact. Gruebler's equation yields $F = 3(3 - 1) - 2(3) - 1(0) = 0$. This simple mechanism does move, owing simply to the fact that the sum of the radii of the points of contact of the rolls equals the distance between the ground pivots for all positions of the rolls.

**Figure 1.26**

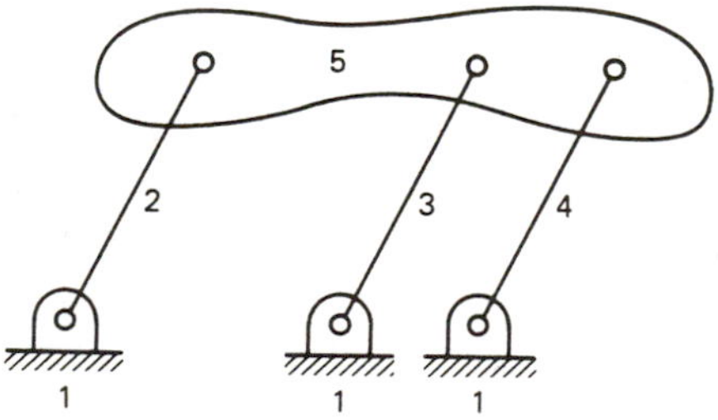


**Figure 1.27**

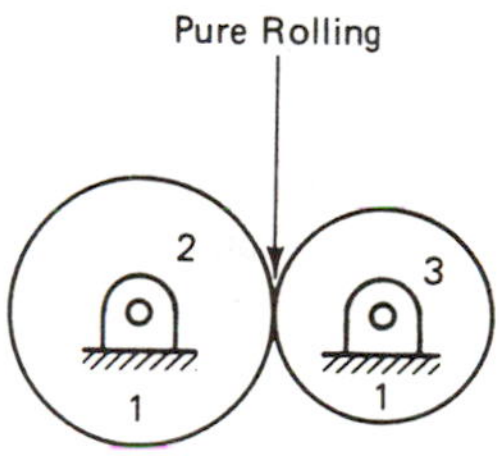

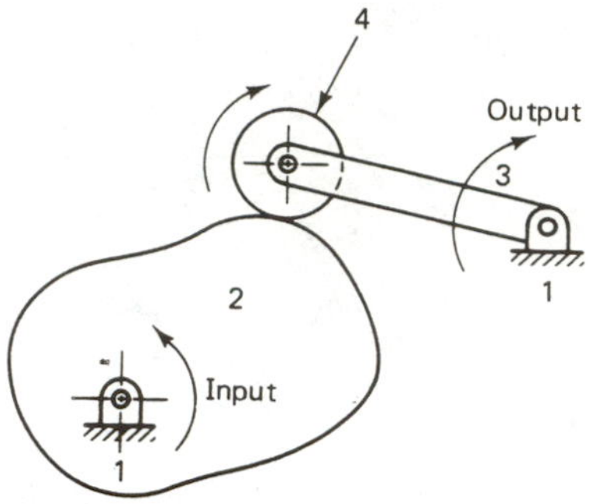


**Figure 1.28**

There are also instances when Gruebler's formula yields a seemingly excessive number of freedoms. This may involve a "passive or redundant degree of freedom" (see also Sec. 6.9) and does not alter the constraint between the output versus input motions of a mechanism. Take, for example, the cam-and-follower mechanisms of Fig. 1.28.

Here the rotation of the follower roller 4 does not affect the output oscillation of the follower arm 3. Even if roller 4 were "welded" to arm 3, the motion of arm 3 would remain unchanged. Checking this condition by Gruebler's equation, regarding the cam and roller contact as a "roll-slide,"

$$F = 3(4 - 1) - 2(3) - 1(1) = +2$$

Now, if we "weld" roller 4 to arm 3,

$$F = 3(3 - 1) - 2(2) - 1(1) = +1$$

Also, if slipping (the redundant freedom between cam and roller) is prevented by friction—in other words, their contact becomes rolling only:

$$F = 3(4 - 1) - 2(4) = +1$$

The reader can verify that the casement window mechanism in Figs. 1.14 and 1.15 contains a similar passive degree of freedom.

## 1.8 ANALYSIS VERSUS SYNTHESIS

The process of drawing kinematic diagrams and determining degrees of freedom of more complex mechanisms are the first steps in both the kinematic analysis and synthesis process. In *kinematic analysis,* a particular given mechanism is investigated based on the mechanism geometry plus possibly other known characteristics (such as input angular velocity, angular acceleration, etc.). *Kinematic synthesis,* on the other hand, is the process of designing a mechanism to accomplish a desired task. Here, both choosing the type as well as the dimensions of the new mechanism can be part of kinematic synthesis (see Chaps. 2 and 3).

The fundamentals described in this chapter are most important in the initial stages of either analysis or synthesis. The ability to visualize relative motion, to reason why a mechanism is designed the way it is, and the ability to improve on a particular design are marks of a successful kinematician. Although some of this ability comes in the form of innate creativity, much of it is a learned skill that improves with practice. Chapter 2 (Vol. 1) helps put mechanism design into perspective. The structure or methodology of design is described, including the place of kinematic analysis and synthesis.

## PROBLEMS

**1.1.** As described in the chapter, all mechanisms fall into the categories of motion generation (rigid-body guidance), path generation, or function generation (including input–output force specification). Find and sketch an example of each task type (different from those presented in this book). Identify the type of linkage (four-bar, slider-crank, etc.), its task, and why this type of linkage was used for this task.

**1.2.** A linkage used for a drum foot pedal is shown in Fig. P1.1. Identify the linkage type. Why is this linkage used for this task? Can you design another simple mechanism for this task?

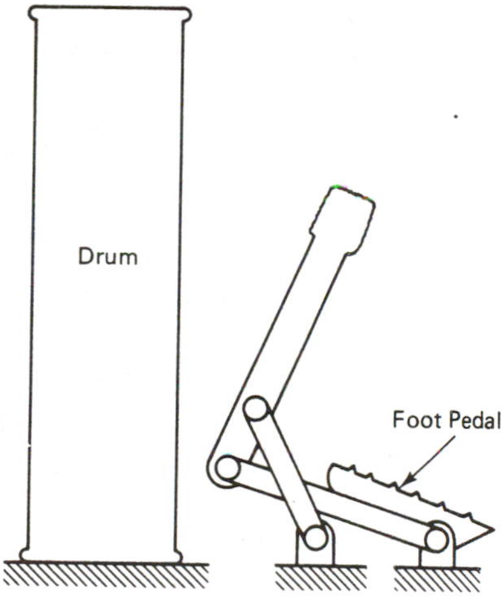


**Figure P1.1**

**1.3.** Figure P1.2 shows a pair of locking toggle pliers. Identify the type of linkage (four-bar, slider-crank, etc.), its task, and why this type of linkage was used for this task. Notice that there is an adjusting screw on the mechanism. What is its function? Why is it located where it is?

**1.4.** A desolventizer (see Fig. P1.3) receives a fluid, pulpy food material (e.g., "spent" soybean flakes) and passes it on to the successive trays by gravity. These trays are heated by

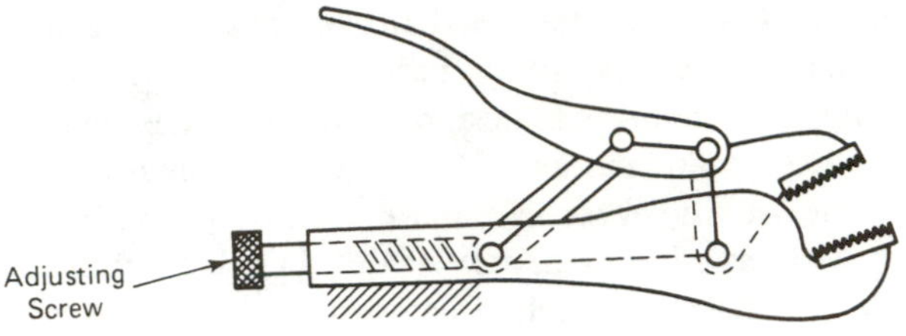


**Figure P1.2**

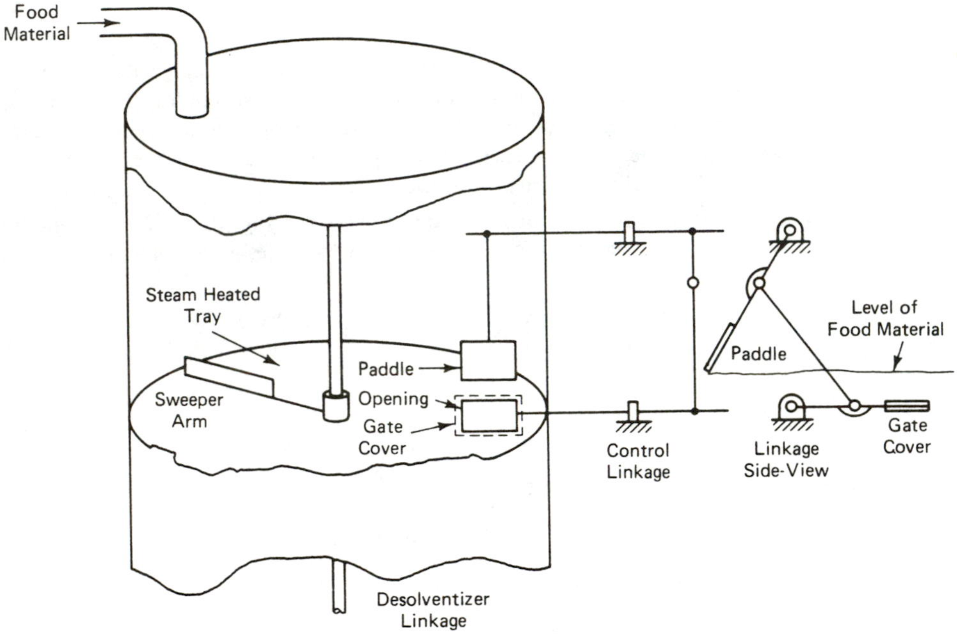


**Figure P1.3**

passing steam through them. The flakes are completely dried up by the time they come out of the last tray. The material is forced through an opening on the bottom of each plate to the next tray. The task of helping the material through the opening is accomplished by a "sweeping arm" attached to a central rotating shaft that runs vertically through the desolventizer. A control is needed for the gate opening to correspond with the rise in the level of the food material; that is, the gate opening should increase as the level of the material rises. A linkage* is used to perform this task. In order to sense any increase in the level of the material, a paddle is rigidly connected to the input link of the linkage while the gate on the bottom of the plate is attached to the output link.

* Suggested by P. Auw, S. Royle, and F. Kwong [81].

What is the task of this linkage (motion, path, or function generation)? Why was this linkage chosen for this task?

**1.5.** Figure P1.4 shows a proposed speed-control device† that could be mounted on an automobile engine and would serve a twofold purpose:

1. To function as a constant-speed governor for cold mornings so that the engine will race until the choke is reset. This speed control would enable the engine speed to be regulated, thus maintaining a preset idle speed. The idle speed would be selected on the dash-mounted speed-control level.
2. To function as an automatic cruise control for freeway driving. The desired cruising speed could be selected by moving the indicator lever on the dashboard to the desired speed.

   **(a)** Sketch the kinematic diagram (unscaled) of the portion of this linkage that moves in planar motion.

   **(b)** Sketch the lower-pair equivalent linkage.

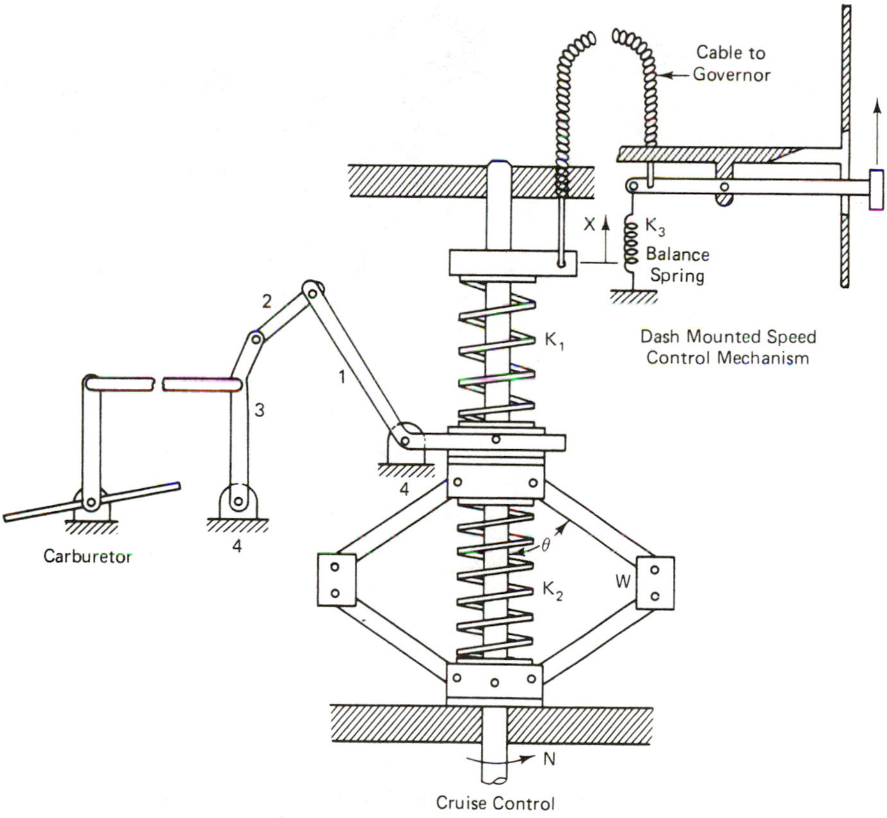


**Figure P1.4**

**1.6.** Those who are participating in the two-wheel revolution are aware that a derailer mechanism helps to change speeds on a 10-speed bicycle. A 10-speed, as the name implies,

† Suggested by G. Anderson, R. Beer, and W. Gullifer [81].

has 10 gear ratios that may be altered while the bike is in operation. The rear wheel has a five-sprocket cluster and the crank has two sprockets. The gear ratio is altered by applying a side thrust on the drive chain, the thrust causing the chain to "derail" onto the adjacent sprocket. The operation of the rear derailer is adequate. However, the front derailer is less efficient due to the larger step necessary in transferring from one sprocket to the other.

This calls for a design that would enhance the life of the chain–sprocket system by reducing the side thrust on it and by enabling more teeth to be in contact with the chain during the initial stages of the transfer. These objectives were accomplished* by lifting the chain off one sprocket, moving it along the path as shown in Fig. P1.5, and then setting it down onto the adjacent sprocket.

**(a)** Draw the scaled kinematic diagram of this mechanism.
**(b)** Is this a motion-, path-, or function-generator linkage?

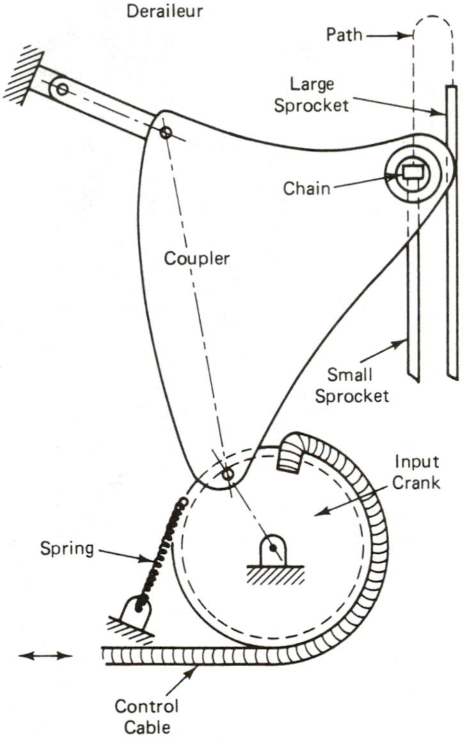


**Figure P1.5**

**1.7.** As steam enters into a steam trap, it is condensed and allowed to flow out of the trap in liquid form. The linkage in Fig. P1.6 has been suggested† to be a feedback control valve for the steam trap. The float senses the level of the condensate while the linkage adjusts the exit valve.

* By G. Fichtinger and R. Westby [92,95].

† By M. L. Pierce, student, University of Minnesota.

(a) Draw the unscaled kinematic diagram for this linkage.
(b) Is this a function, path, or motion generator?
(c) Can you design another simple linkage for this task?

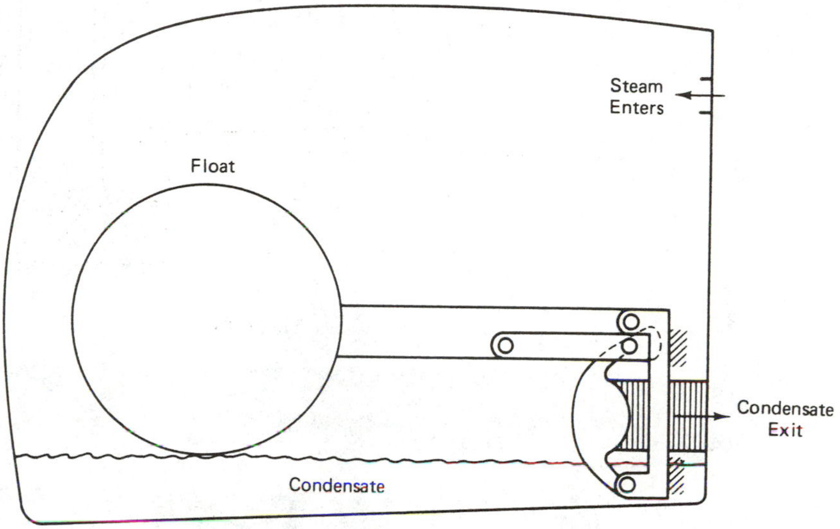


**Figure P1.6**

**1.8.** A typical automotive suspension system is shown in Fig. P1.7. A cross-sectional schematic is shown in Fig. P1.8.

(a) What type of linkage is this (motion, path, or function generator)?
(b) Why is a linkage used in this application?
(c) If the dimensions of the linkage were changed, what would be the effect on the vehicle?

**Figure P1.7**

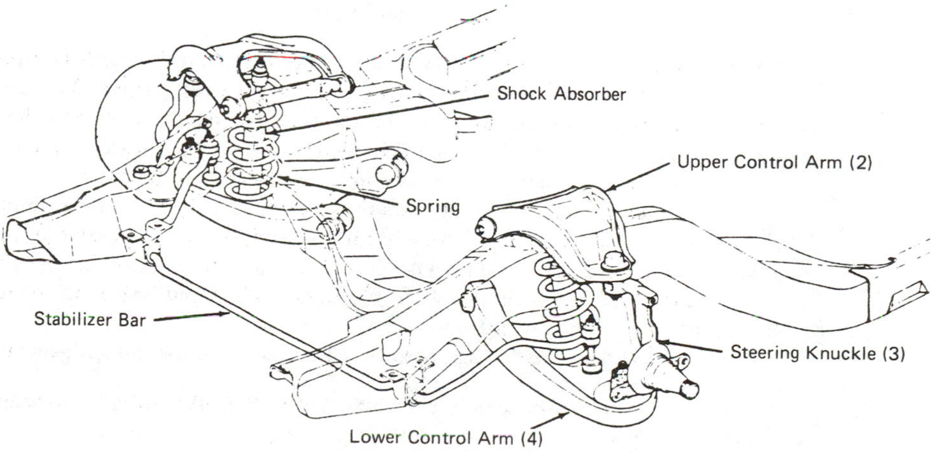

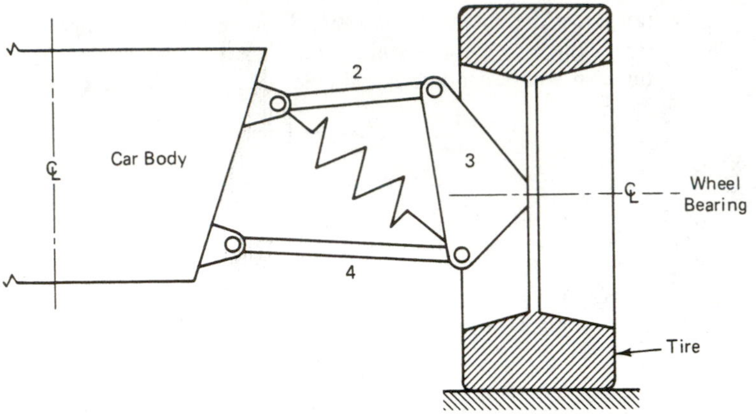


Figure P1.8

Figure P1.9

**1.9.** Frequently in the control of fluid flow, a valve is needed that will regulate flow proportional to its mechanical input. Unfortunately, very few valves possess this characteristic. Gate valves, needle valves, ball valves, and butterfly valves, to name a few, all have nonlinear flow versus mechanical input characteristics. A valve with linear characteristics would do much to simplify the proportional control of fluid flow.

The linkage* in Fig. P1.9 appears to be one means of providing a simple, durable, and inexpensive device to transform a linear mechanical control signal into the nonlinear valve positions which will produce a flow proportional to the control signal. The butterfly valve is connected to the short link on the right. The input link is on the left.

**(a)** What type of linkage is this?

**(b)** Which task does this linkage perform (function, path, or motion generation)?

* Designed by B. Loeber, B. Scherer, J. Runyon, and M. Zafarullah using the technique described in Sec. 2.16 (see Ref. 81).

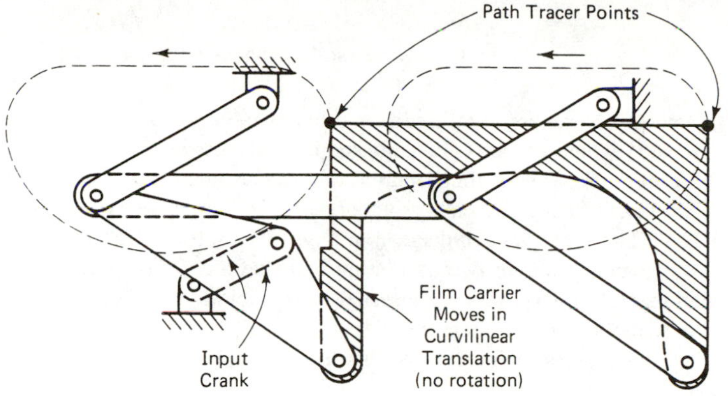


Eight-Bar Transport Mechanism

**Figure P1.10**

**1.10.** In converting x-ray film from the raw material to a finished product, a multiloop mechanism was designed to transport the film from the sheeting operation, to the stenciling operation, to a conveyor belt.

The linkage shown in Fig. P1.10 must pick up the film from beneath the stenciling and sheeting devices with a vertical or nearly vertical motion to prevent sliding between the film and mechanism. The mechanism follows a horizontal path (with no appreciable rotation) slightly above the stenciling and sheeting devices while transporting the film from pickup to delivery.

Although the double-parallelogram-based linkage in Fig. P1.10 accomplished the task adequately, a simpler linkage* (shown in Fig. P1.11) was synthesized using the techniques of Chap. 3.

**(a)** Draw the unscaled kinematic diagrams of both linkages.

**(b)** What type of six-bar is shown in Fig. P1.11?

* Designed by D. Bruzek, J. Love, and J. Riggs [81].

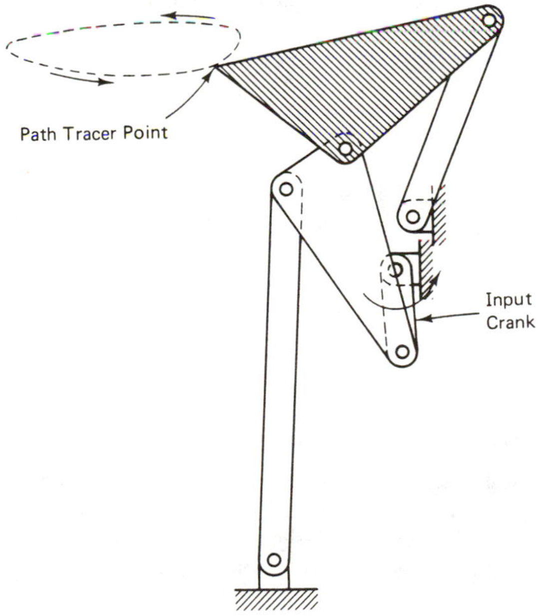


**Figure P1.11** Six-bar transport mechanism.

**1.11.** In order to emboss characters onto credit or other data cards, a multiloop linkage has been designed [23] which exhibits high mechanical advantage (force out/force in; see Chap. 3 of Vol. 1). Separately timed punch and die surfaces are required so that the card is not displaced during the embossing process. (The desired motions are shown in Fig. P1.12.) The embossing linkage (see Figs. P1.13 to P1.15) makes use of an interposer arrangement wherein two oscillating bail shafts drive respective punches and dies, provided that the interposers are inserted in the keyways on top of the shafts.

**(a)** Draw the scaled kinematic diagram of this linkage.

**(b)** Determine the degrees of freedom of this linkage by both intuition and Gruebler's equation.

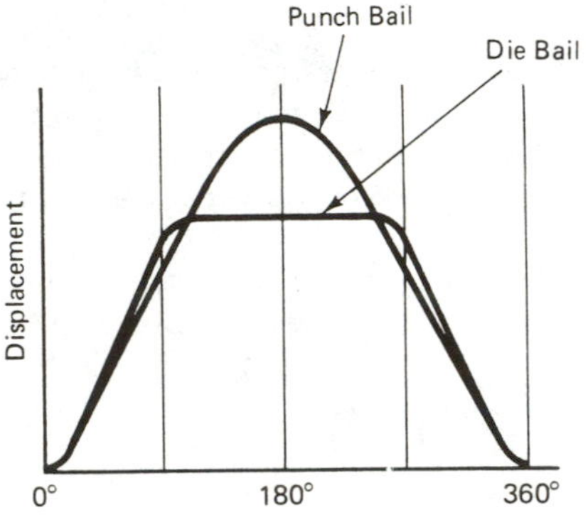


**Figure P1.12**

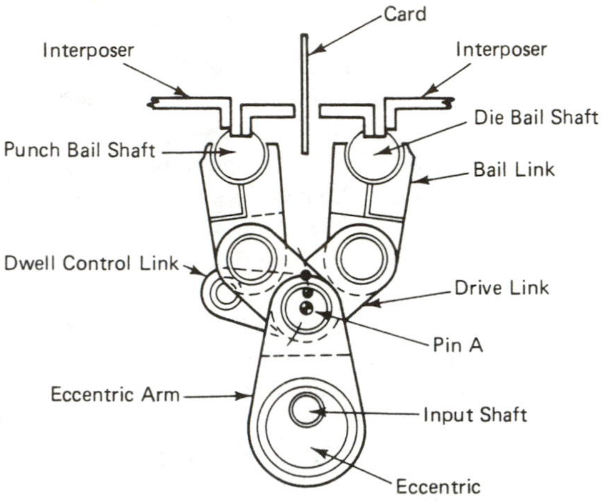


**Figure P1.13** 90° position. (*Courtesy of Data Card Corporation.*)

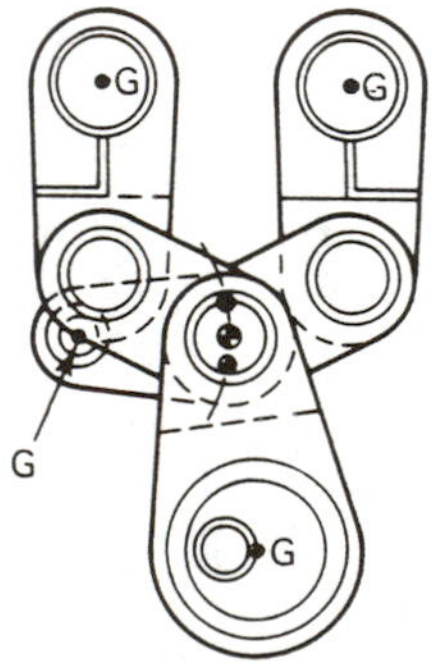


**Figure P1.14** 180° position (*G*, grounded pivot).

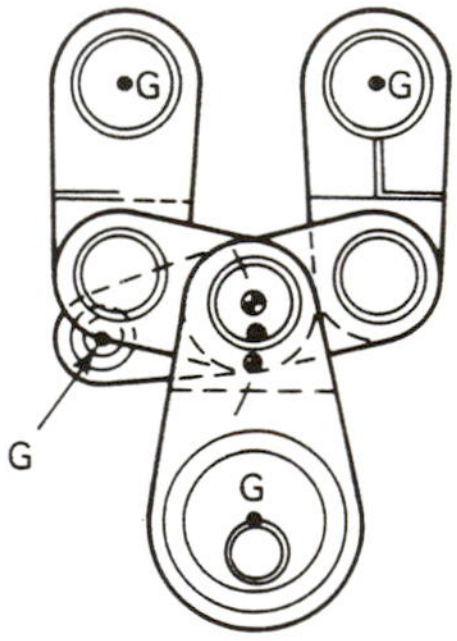


**Figure P1.15** 270° position (*G*, grounded pivot).

**1.12.** An idea came to mind to design and build a mechanism inside a box that, once turned on, would send a finger out of the box, turn itself off, and return back into the box [23,27]. Two different types of linkages were designed for this task. The linkage shown in Figs. P1.16 to P1.18 was created by D. Harvey while the mechanism in Figs. P1.19

to P1.21 was invented by T. Bjorklund. (Note that the external switch and the internal limit switch are in parallel, so that the latter keeps the motor running until the finger has been withdrawn into the box.)

**(a)** Draw the kinematic diagrams of these linkages.

**(b)** Show (by intuition and Gruebler's equation) that both these mechanisms have a single degree of freedom. (Disregard the lid in Fig. P1.16.)

**(c)** In Fig. P1.16, what type of six-bar linkage is this? What is its task?

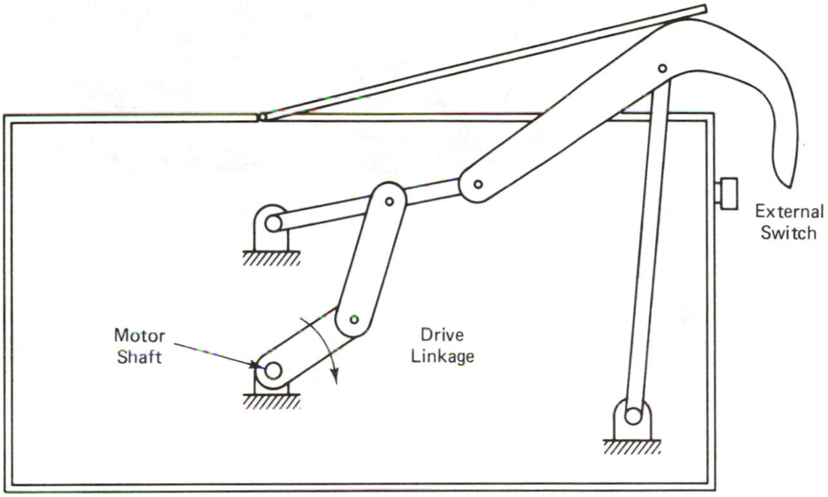


**Figure P1.16**

**Figure P1.17**

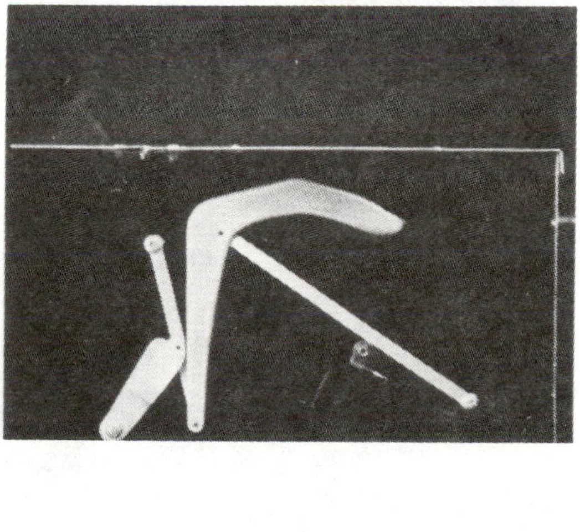

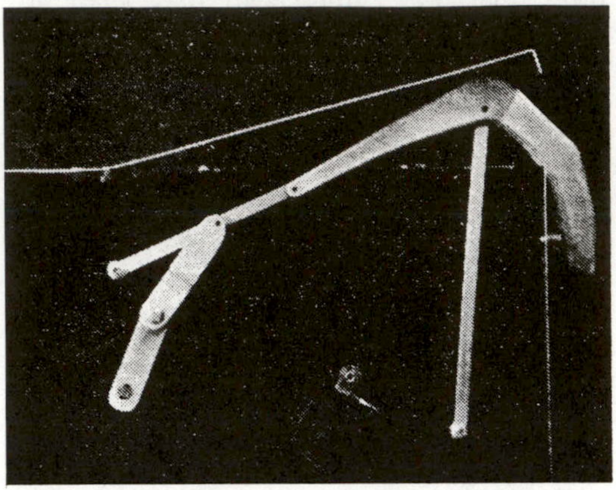

Figure P1.18

Figure P1.19

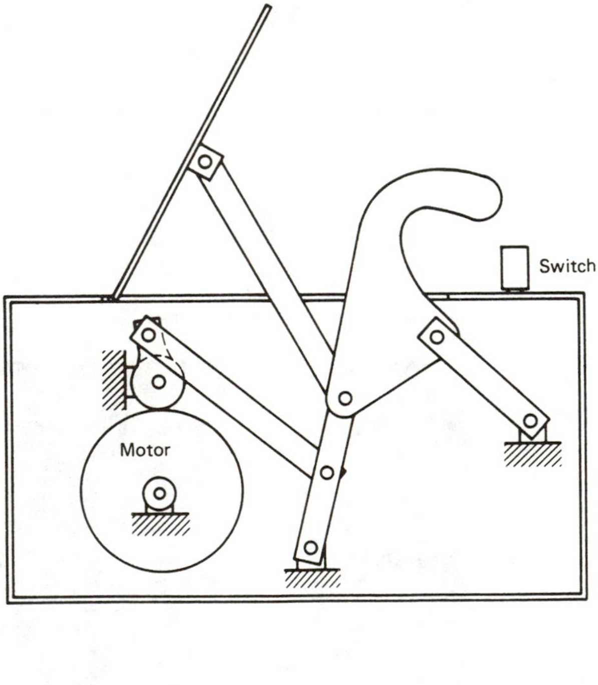

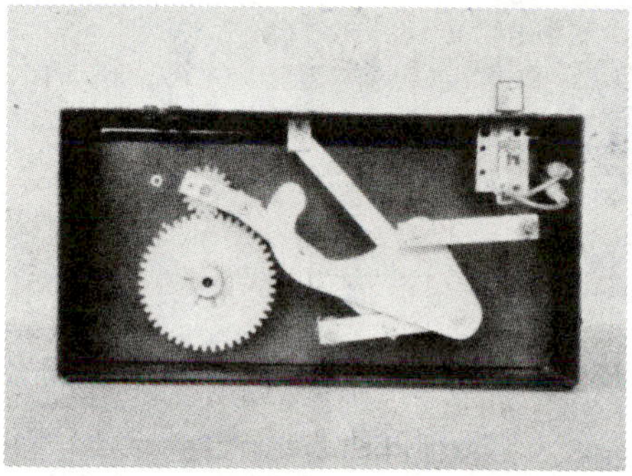

Figure P1.20

Figure P1.21

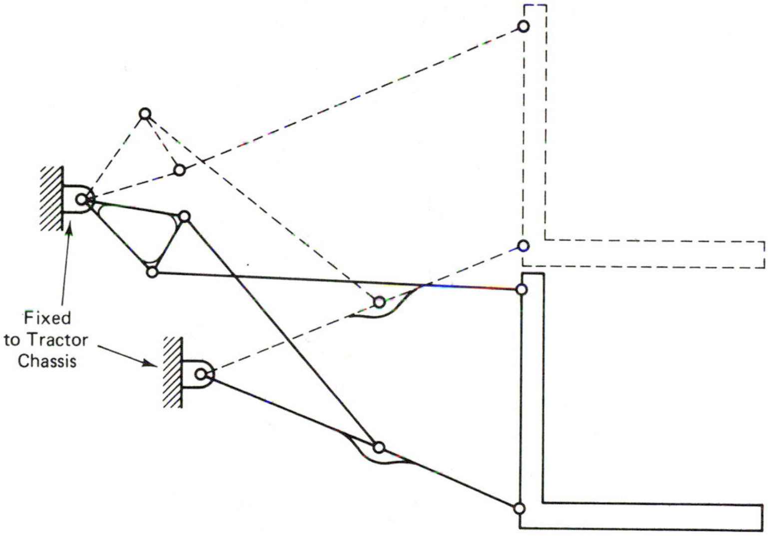


Figure P1.22

**1.13.** A six-bar lift mechanism for a tractor is shown in two positions (solid and dashed lines) in Fig. P1.22. What type of six-bar is this? What task does it satisfy?

**1.14.** An agitator linkage for a washing machine is shown in Fig. P1.23 (ground pivots are identified by G's).

(a) What type of six-bar is this?

(b) What task does this linkage fulfill (motion, path, or function generation)?

(c) Why use a six-bar linkage in this application?

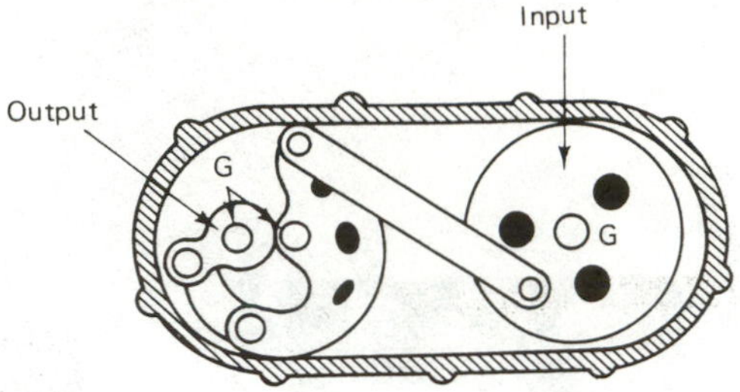


**Figure P1.23** *G* signifies a ground pivot.

**1.15.** An automobile hood linkage is shown in Fig. P1.24. Notice the difference from the linkage in Fig. 1.2c.

**(a)** Neglecting the spring, what type of six-bar is this linkage?

**(b)** Draw the instantaneous-velocity-equivalent lower-pair diagram of this linkage (including the spring).

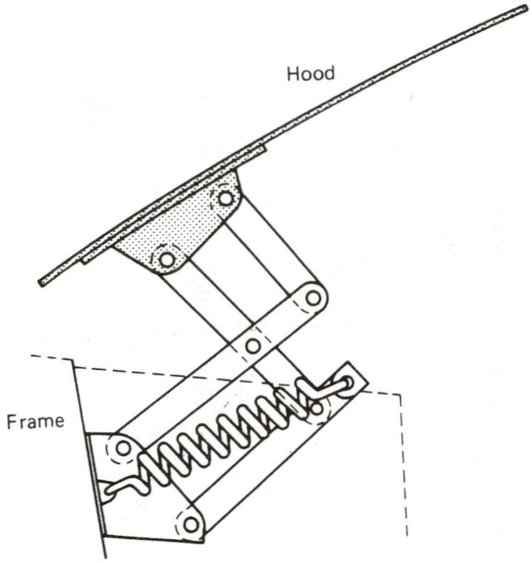


**Figure P1.24**

**1.16.** Figure P1.25 shows a cutaway view of a Zero-Max variable-speed drive [73,166]. This drive yields stepless variable speed by changing the arc through which four one-way clutches rotate the output shaft when they move back and forth successively. Figure P1.26 shows one of these linkages, which is referred to as a "single lamination." The drive has sets of equally spaced out-of-phase linkages which use three common fixed shafts, $A_0$, $C_0$, and $D_0$. The rotation of the input $A_0A$ causes the output link $DD_0$ to oscillate, thus rotating the output shaft $D_0$ in one direction (due to the one-way clutch assembly). The position of pivot $B_0$ is adjusted by rotating the speed control arm about $C_0$ to change the output speed of the drive. As $B_0$ approaches the centerline $L_5$ the output speed decreases since the center of curvature of the trajectory of $B$ will approach point $D$, causing link 6 to become stationary.

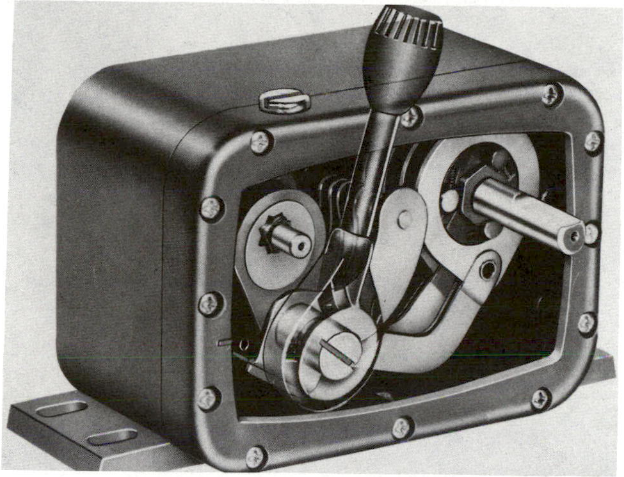

**Figure P1.25** (*Courtesy of Zero-Max Inc.*)

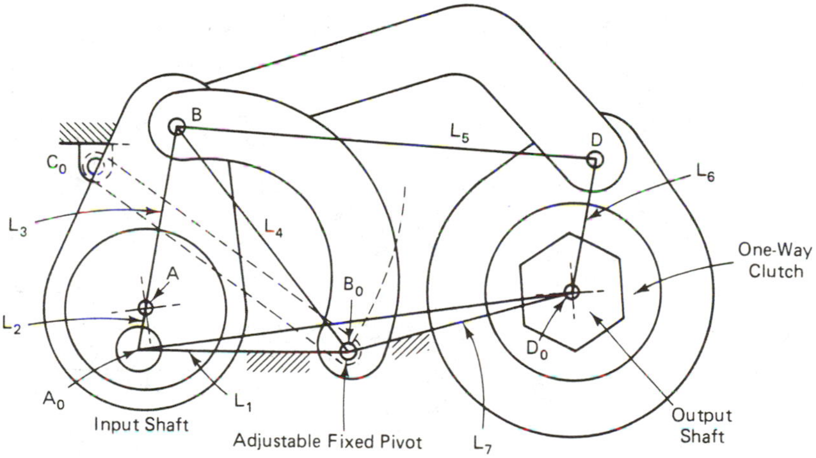


**Figure P1.26**

**(a)** What type of six-bar is this (with $B_0$ considered fixed)? What is its task?

**(b)** If link $C_0B_0$ is considered mobile, how many degrees of freedom does the linkage have? (Use Gruebler's equation.)

**1.17.** Based on the concept that mechanisms can be things of beauty besides having functional value, a mechanism clock was conceived.* Mechanisms would manipulate small cubes with numbers on them in such a way as to indicate the time. It was determined that three sets of cubes would be used to read minutes and hours (two for minutes). The

* By Jim Turner [293].

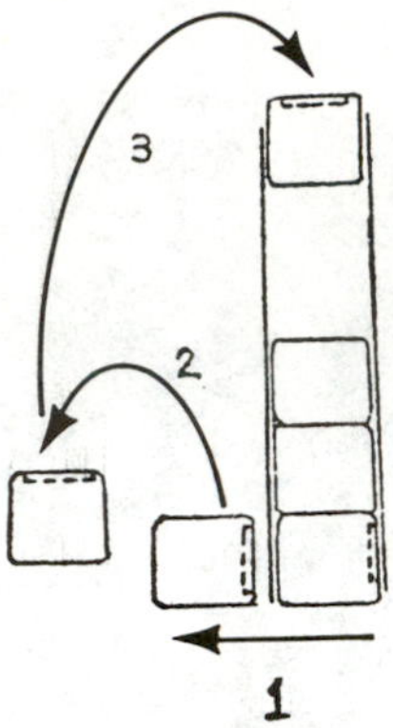


Figure P1.27

Figure P1.28

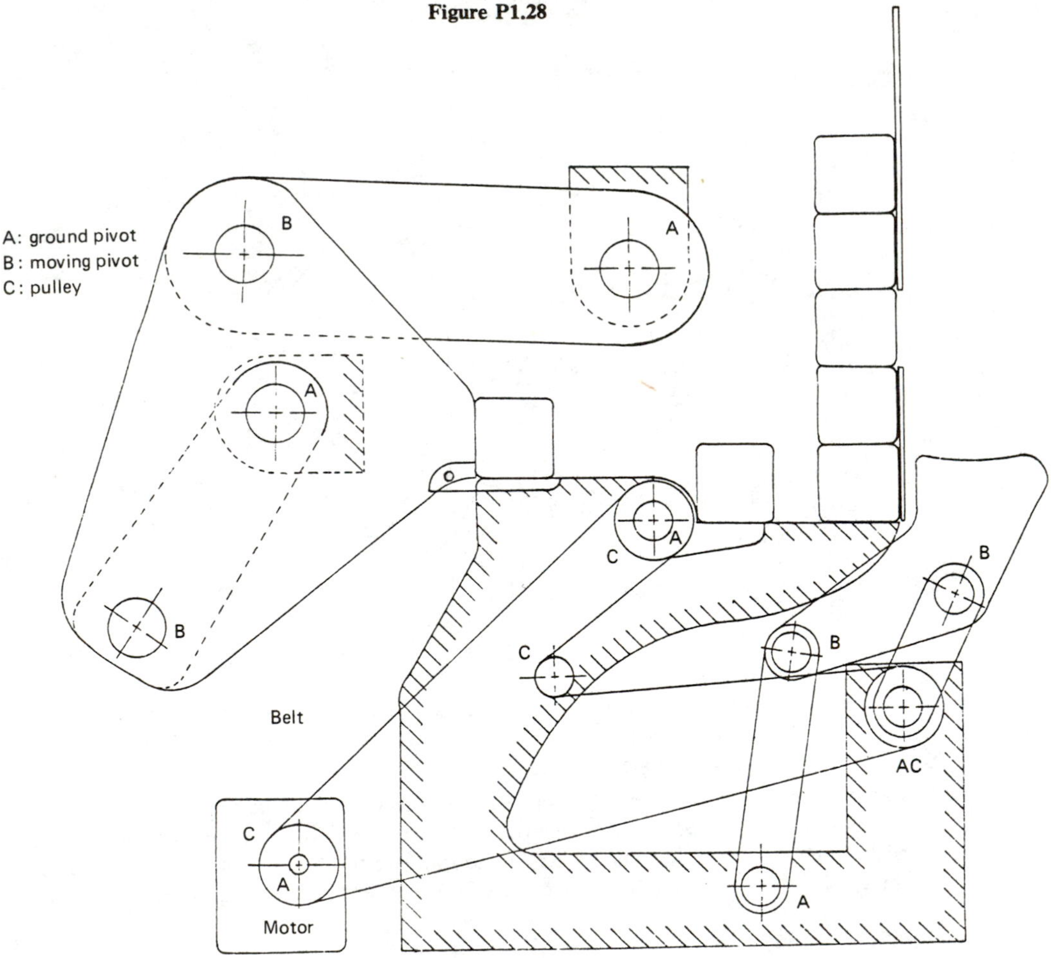

cubes would be turned over 90° to use four sides of each for numbers—thus five cubes for the 0–9 set and three cubes each for the 0–5 and 1–12 set. Rather than work against nature, it was decided to remove the bottom cube and allow gravity to settle other cubes into their place, while the cube was placed on top of the stack. The devices had to be reasonable in size, be reliable, and be manufacturable. The motion was separated into three steps, as shown in Fig. P1.27. The problem: (1) how to remove the bottom cube from the stack, (2) how to rotate the cube, and (3) how to transport the cube to the top of the stack. Figures P1.28 and P1.29 show the final design for the mechanism clock.

**(a)** Draw the unscaled kinematic diagram for (1) each of three steps separately; (2) the entire mechanism.

**(b)** Determine which task (motion, path, or function generation) is accomplished in each step.

**(c)** Determine the degrees of freedom of the entire linkage.

**(d)** Design your own mechanism clock and show it in a conceptual diagram.

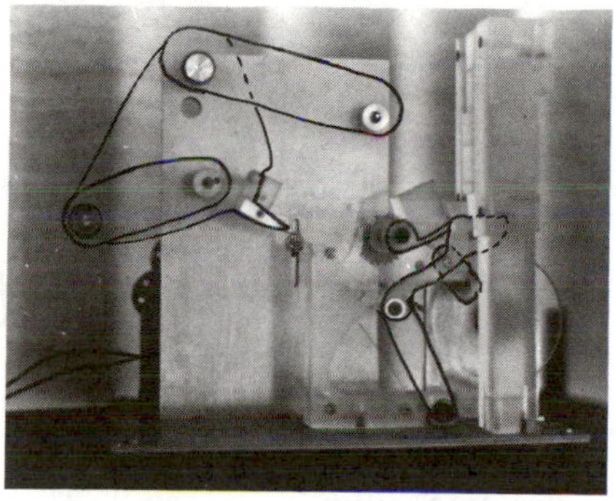

**Figure P1.29**

**1.18.** Multiloop mechanisms have numerous applications in assembly line operations. For example, in a soap-wrapping process, where a piece of thin cardboard must be fed between rollers which initiate the wrapping operation, a seven-link mechanism* is employed such as that shown schematically in Fig. P1.30.

The motion of the suction cups is prescribed in order to pick up one card from a gravitation feeder (the suction cups mounted on the coupler approach and depart from the card in the vertical direction) and insert the card between the rollers (the card is fed in a horizontal direction). The input timing is prescribed in such a fashion that the cups pick up the card during a dwell period (a pause in the motion) and also in a way that the card is fed into the rollers at approximately the same speed as the tangential velocity of the rollers.

**(a)** Sketch the instantaneous-velocity-equivalent lower-pair diagram of this mechanism.

**(b)** Determine the degrees of freedom of this linkage as shown in Fig. P1.30 and verify your answer by determining the degrees of freedom of the lower-pair equivalent linkage.

* This application was brought to the authors' attention by D. Tesar of the University of Florida.

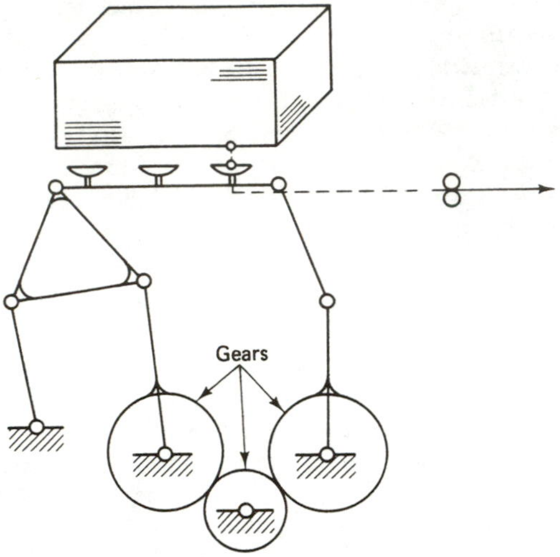


**Figure P1.30**

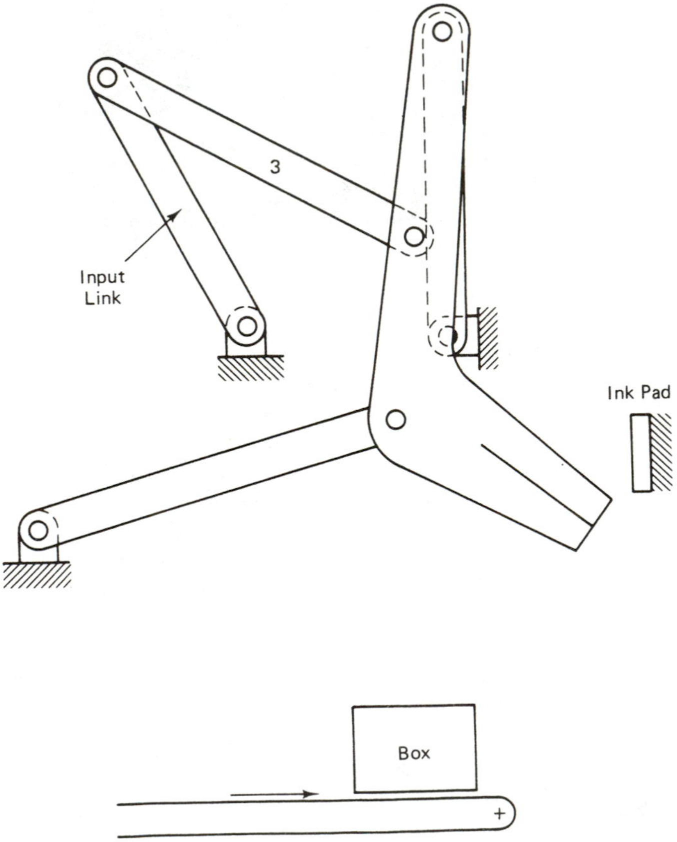


**Figure P1.31**

Figure P1.32

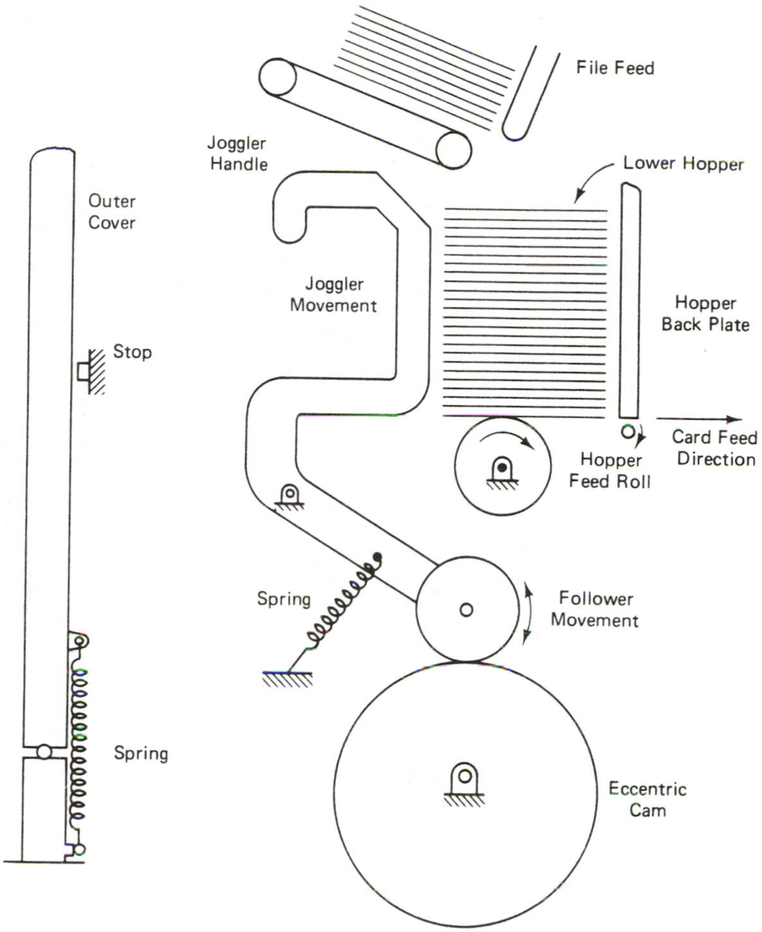


**1.19.** The linkage in Fig. P1.31 has been suggested* for stamping packages automatically at the end of an assembly line. The ink pad is located at the initial linkage position while the packages will travel along an assembly line and stop at the "final position" in order to be stamped. It is desirable that the linkage have straight-line motion into the box so that the stamp imprint will not be smudged. A solenoid will drive the input link through its range of motion.

**(a)** What type of six-bar linkage is this?

**(b)** Draw this linkage in at least four other positions in order to (1) determine the range of rotation of the input link; (2) check if the linkage indeed does hit the ink pad in a straight-line motion approaching and receding from the box.

**(c)** Is this type of linkage a good choice for this task? Why?

**(d)** If the input link and link 3 are changed in length and orientation and the input pivot location moved, what are the consequences on the performance of the entire linkage?

**1.20.** Figure P1.32 shows a schematic diagram of a computer card feeder mechanism in its initial configuration. Computer cards are placed in the "file feed" by the machine opera-

* By J. Sylind (synthesized by the techniques presented in Chap. 3).

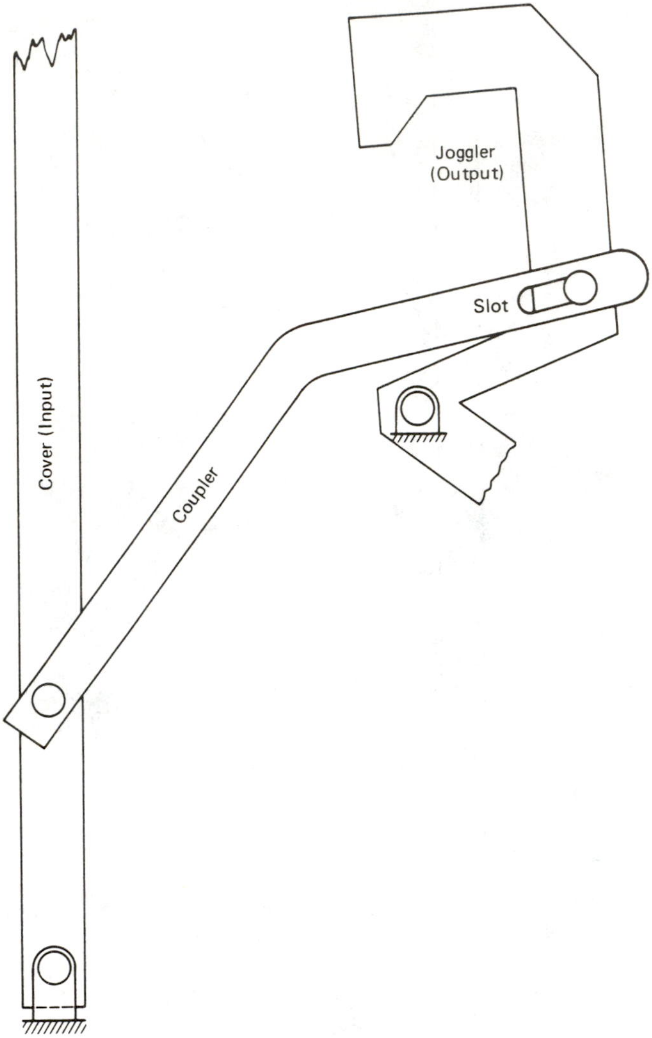


**Figure P1.33**

tor. The file feed then intermittently feeds cards into the lower hopper. The cards must be joggled to align them against the hopper back plate so that they will feed out of the hopper properly when they reach the hopper feedroll. A cam causes the joggler movement.

If there should be a jam or misfeed in the hopper, the operator must pull on the joggler handle to pivot it open so that cards can be removed from the hopper. To get at the joggler handle, the operator must first open an outer cover. (The purpose of this outer cover is to reduce noise levels from the machine.)

The linkage in Fig. P1.33 has been suggested* to avoid the inconvenience for the operator to have to open the outer cover as well as to open the joggler.

* By R. E. Baker of IBM, Rochester, MN.

(a) Draw the kinematic diagram of the linkage in Fig. P1.32.
(b) Add the change suggested in Fig. P1.33 to Fig. P1.32 and draw the new kinematic diagram.
(c) Determine the degrees of freedom of both mechanisms [parts (a) and (b)].
(d) Determine by graphical construction the total rotation of the joggler if the cover is rotated 90° counterclockwise.

**1.21.** Double-boom cranes and excavation devices are commonly used in the building construction industry. Their popularity is due primarily to their versatility, mobility, and high load-lifting capacity. This type of equipment is typically actuated by means of hydraulic cylinders. Figure P1.34 shows a typical knuckle boom crane [260].
(a) Draw the unscaled kinematic diagram for this mechanism.
(b) Determine the degrees of freedom for this linkage.

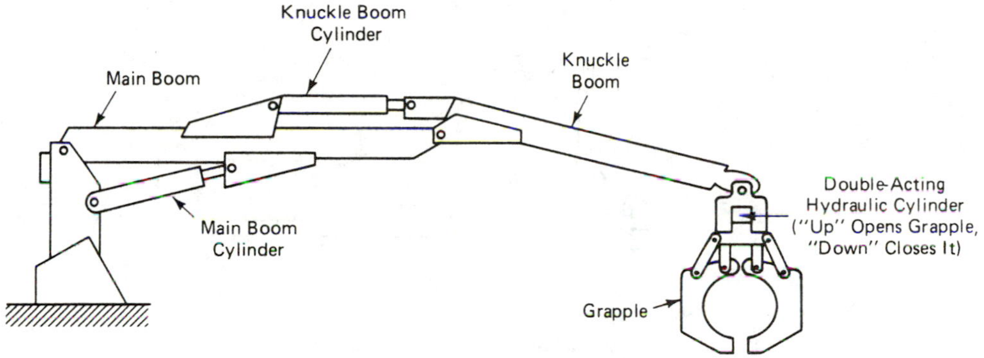


**Figure P1.34**

**1.22.** (a) Draw the unscaled kinematic diagram of the linkage in Fig. P1.35.
(b) Determine the degrees of freedom of both the original linkage and a lower-pair equivalent kinematic diagram.

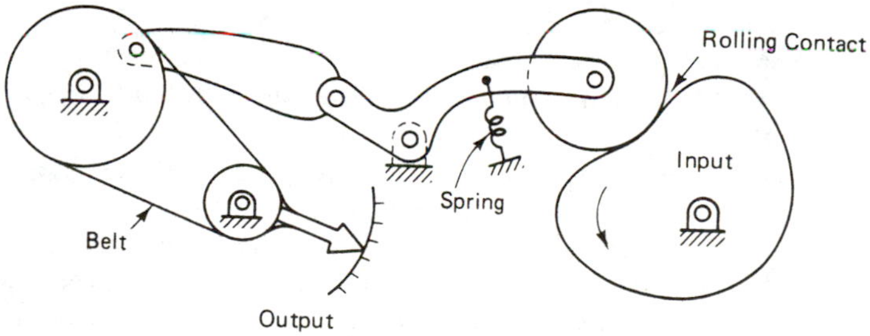


**Figure P1.35**

**1.23.** (a) Draw the unscaled lower-pair equivalent kinematic diagram of the linkage in Fig. P1.36.
(b) Determine the degrees of freedom of both the original linkage and the lower-pair equivalent.

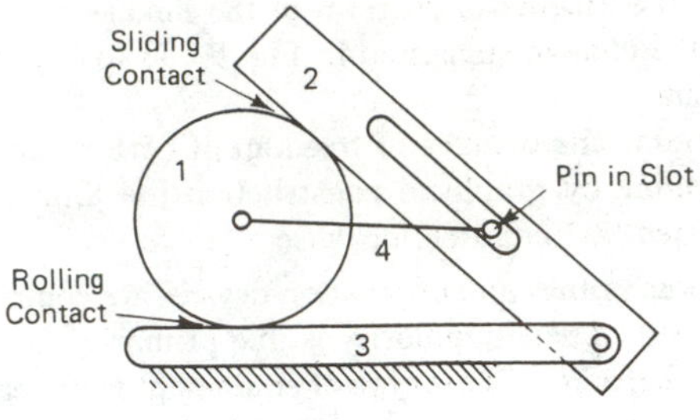


**Figure P1.36**

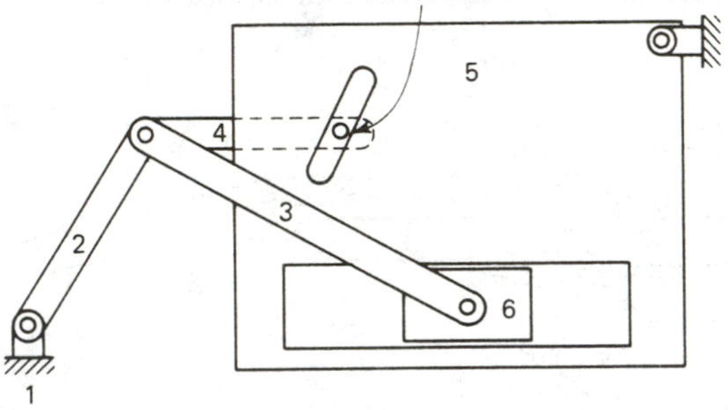


**Figure P1.37**

**1.24.** **(a)** Draw the unscaled kinematic diagram of the linkage in Fig. P1.37.
**(b)** Determine the degrees of freedom of this linkage.

**1.25.** An end loader is to be designed for attachment onto crawler-type tractors.* The linkage must have two degrees of freedom: one allowing the system to lift the bucket and the other to allow the bucket to be tipped while the first is held fixed. Assume that all joints will be those allowing only one relative degree of freedom.
**(a)** What is the minimum number of binary links that your design must contain?
**(b)** What is the minimum number of links that would produce a linkage with $F = 2$? [*Hint:* Consider part (a).]
**(c)** Assuming that a linkage with the fewest number of links in part (b) will not work, what is the fewest number of links in the next more complicated linkage that will work?

**1.26.** Could one build a 10-link linkage in Prob. 1.25 with the desired degrees of freedom? If the chosen linkage has nine links ($n = 9$):
**(a)** How many pairs (joints) will have to be purchased or designed?
**(b)** What is the maximum number of elements that can occur on one link?
**(c)** How many links can have the maximum number of elements?

**1.27.** The linkage in Fig. P1.38 was designed as a function generator ($y = -x + 2$, for $1 \leq x \leq 4$) using the methods described in Chap. 8. Verify that this geared mechanism has a single degree of freedom.

**1.28.** For the linkages in Figs. P1.38 to P1.48, determine the number of degrees of freedom of the mechanisms by intuition and Gruebler's equation.

* This question is used by courtesy of Wm. Carson, University of Missouri, Columbia.

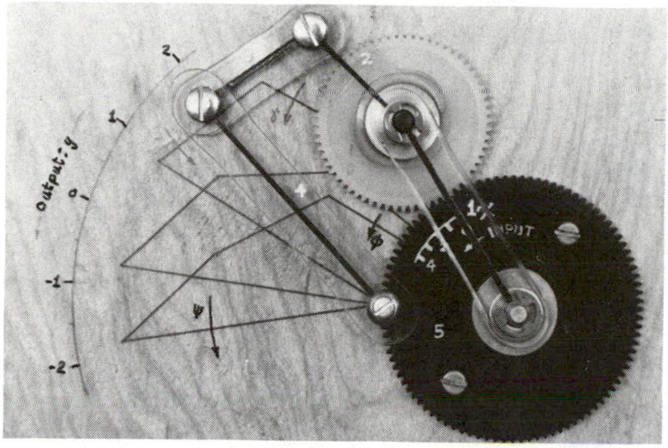


Figure P1.38

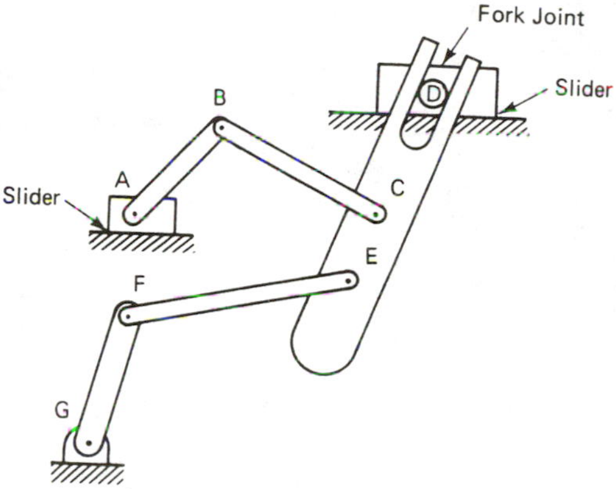


Figure P1.39

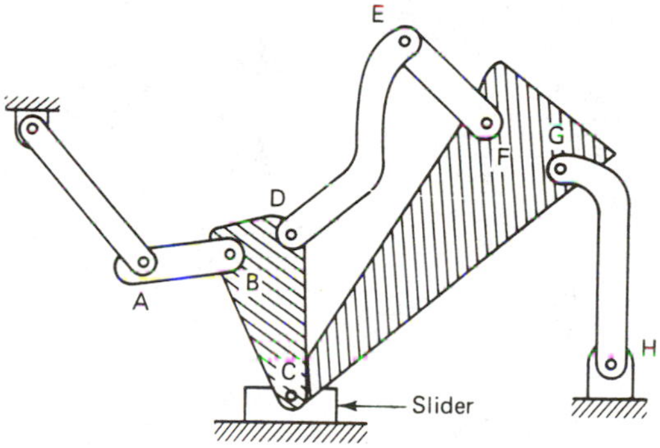


Figure P1.40

Figure P1.41

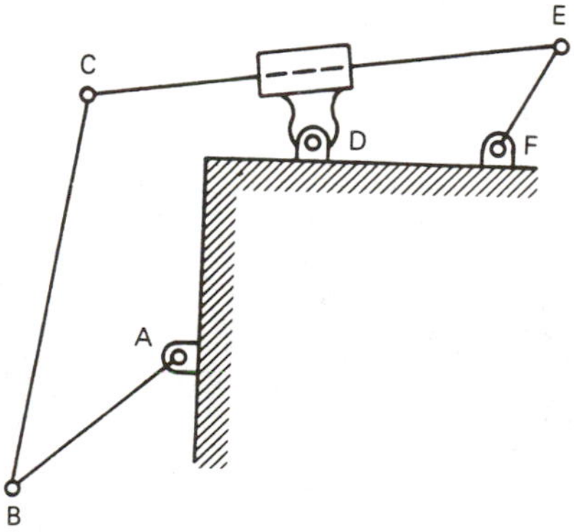


Figure P1.42

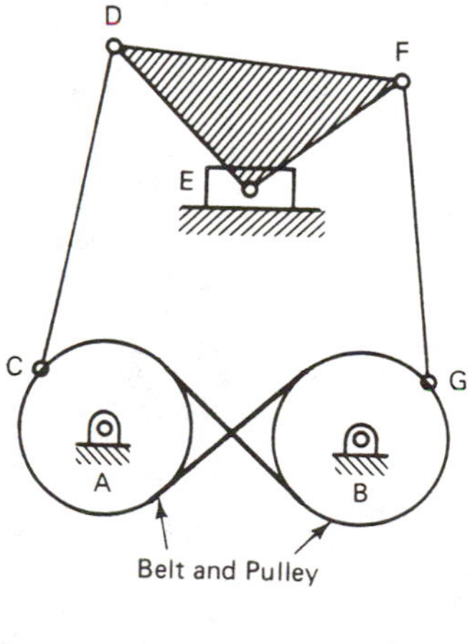

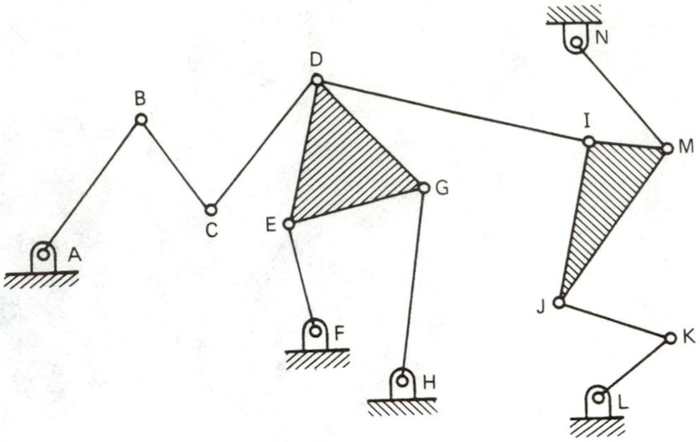


Figure P1.43

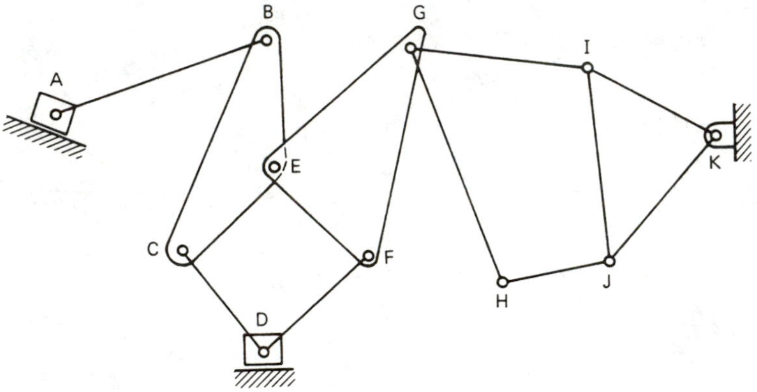


Figure P1.44

Figure P1.45

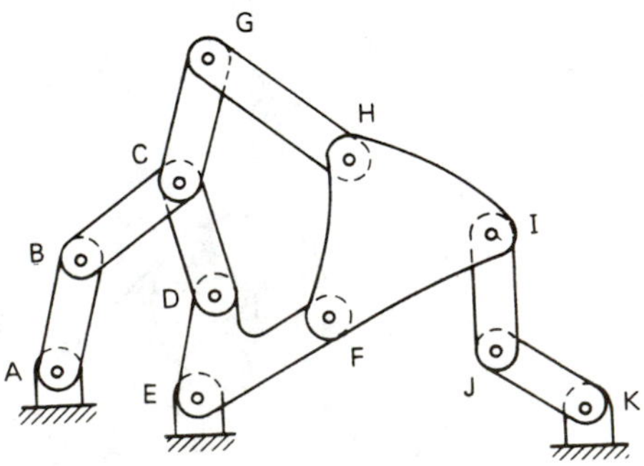


Figure P1.46

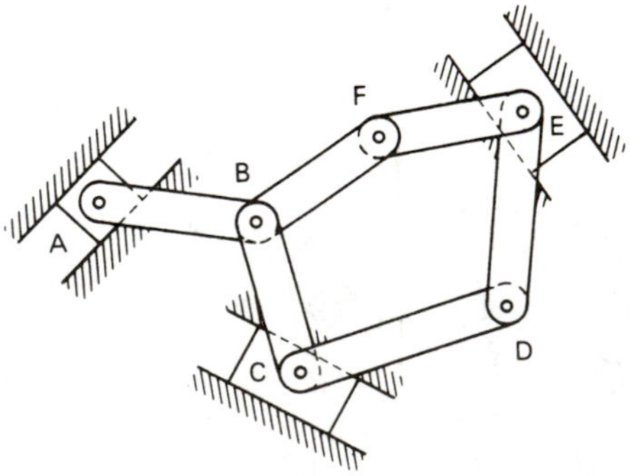

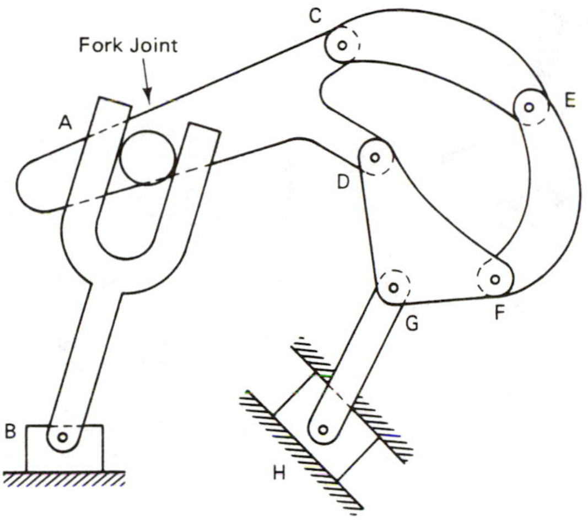


**Figure P1.47**

**Figure P1.48**

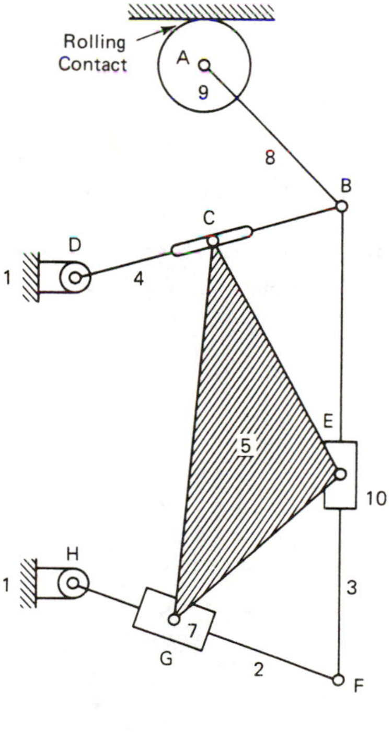

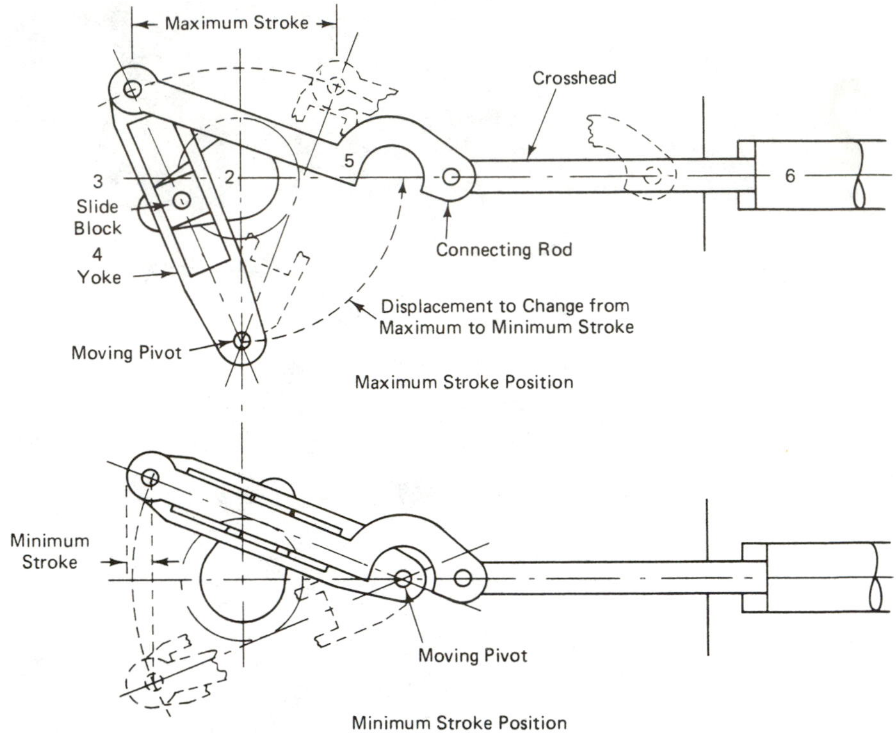


**Figure P1.49**

**1.29.** A metering pump [120] was designed such that a moving pivot controls the stroke of the slider (see Fig. P1.49). The pivot is adjustable to any position through a 90° arc about the center of the crank. When the crank-pivot line is perpendicular to the crosshead motion, the stroke is maximum. When it is in line with the crosshead motion, the stroke is minimum. Draw the unscaled kinematic diagram of this linkage. Determine the degrees of freedom of the linkage.

**1.30.** What type of six-bar is shown in Fig. P1.42?

# 2 Introduction to Kinematic Synthesis—Graphical and Linear Analytical Methods

## 2.1 INTRODUCTION

Ampère defined kinematics as "the study of the motion of mechanisms and methods of creating them." The first part of this definition deals with kinematic *analysis*. Given a certain mechanism, the motion characteristics of its components will be determined by kinematic analysis (as described in Chap. 3 of Vol. 1). The statement of the task of analysis contains all principal dimensions of the mechanism, interconnections of its links, and the specification of the input motion or method of actuation. The objective is to find the displacements, velocities, accelerations, shock or jerk (second acceleration), and perhaps higher accelerations of the various members, as well as the paths described and motions performed by certain elements. In short, *in kinematic analysis we determine the performance of a given mechanism.* The second part of Ampère's definition may be paraphrased in two ways:

1. The study of methods of creating a given motion by means of mechanisms
2. The study of methods of creating mechanisms having a given motion

In either version, the *motion* is given and the mechanism is to be found. This is the essence of *kinematic synthesis.* Thus kinematic synthesis deals with the *systematic design of mechanisms for a given performance.*

The areas of synthesis may be grouped into two categories.

1. *Type synthesis.* Given the required performance, what type of mechanism will be suitable? (Gear trains? Linkages? Cam mechanisms?) Also: How many links should the mechanism have? How many degrees of freedom are required?

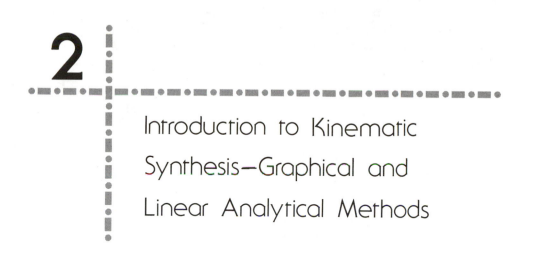

What configuration is desirable? And so on. Deliberations involving the number of links and degrees of freedom are often referred to as the province of a subcategory of type synthesis called *number synthesis,* pioneered by Gruebler (see Chap. 1). One of the techniques of type synthesis which utilizes the "associated linkage" concept is described in Sec. 2.3.

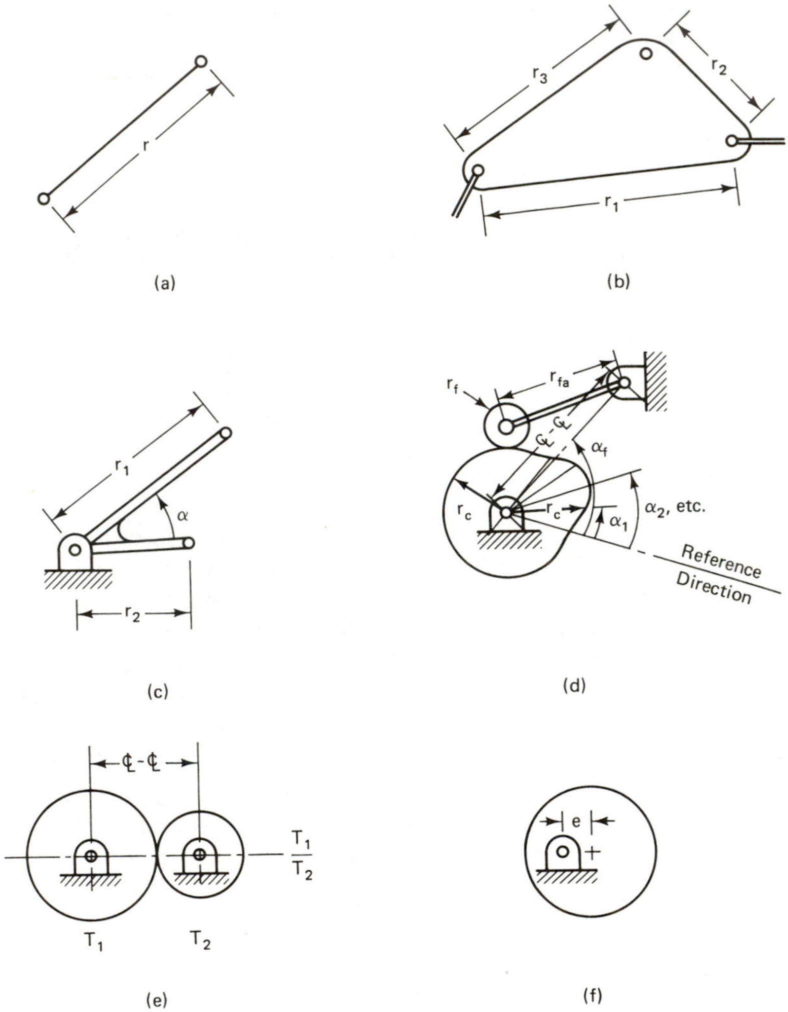


**Figure 2.1** Significant dimensions; (a) binary link: has one length only; (b) ternary link: 3 lengths, 2 lengths and one angle, or 1 length and two angles; (c) bell crank: same as for ternary link; (d) cam and roller follower: centerline distance, follower arm length $r_{fa}$, follower radius $r_f$, and an infinite number of radial distances to the cam surface, $r_c$, at angles $\alpha_1$, $\alpha_2$, etc., specified from a reference direction; (e) gear pair: centerline distance and gear tooth ratio; (f) eccentric: eccentricity only (this is a binary link).

2. *Dimensional synthesis.* The second major category of kinematic synthesis is best defined by way of its objective:

> *Dimensional synthesis seeks to determine the significant dimensions and* the *starting position* of a mechanism of *preconceived type* for a *specified task* and *prescribed performance.*

Principal or *significant dimensions* mean link lengths or pivot-to-pivot distances on binary, ternary, and so on, links, angle between bell-crank levers, cam-contour dimensions and cam-follower diameters, eccentricities, gear ratios, and so forth (Fig. 2.1). Configuration or *starting position* is usually specified by way of an angular position of an input link (such as a driving crank) with respect to the fixed link or frame of reference, or the linear distance of a slider block from a point on its guiding link (Fig. 2.2).

A *mechanism of preconceived type* may be a slider-crank, a four-bar linkage, a cam with flat follower, or a more complex linkage of a certain configuration defined

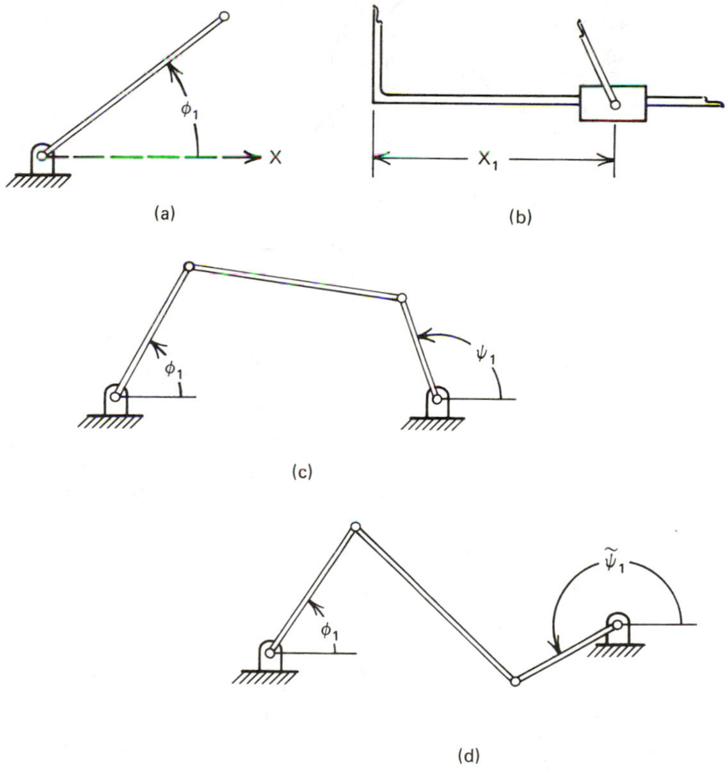


**Figure 2.2** Configuration or starting position; (a) starting position of a crank; (b) starting position of a slider; (c) starting position of a four-bar linkage requires two crank angles, because one crank angle leaves two possibilities for the other crank, as shown in Fig. 2.2(d).

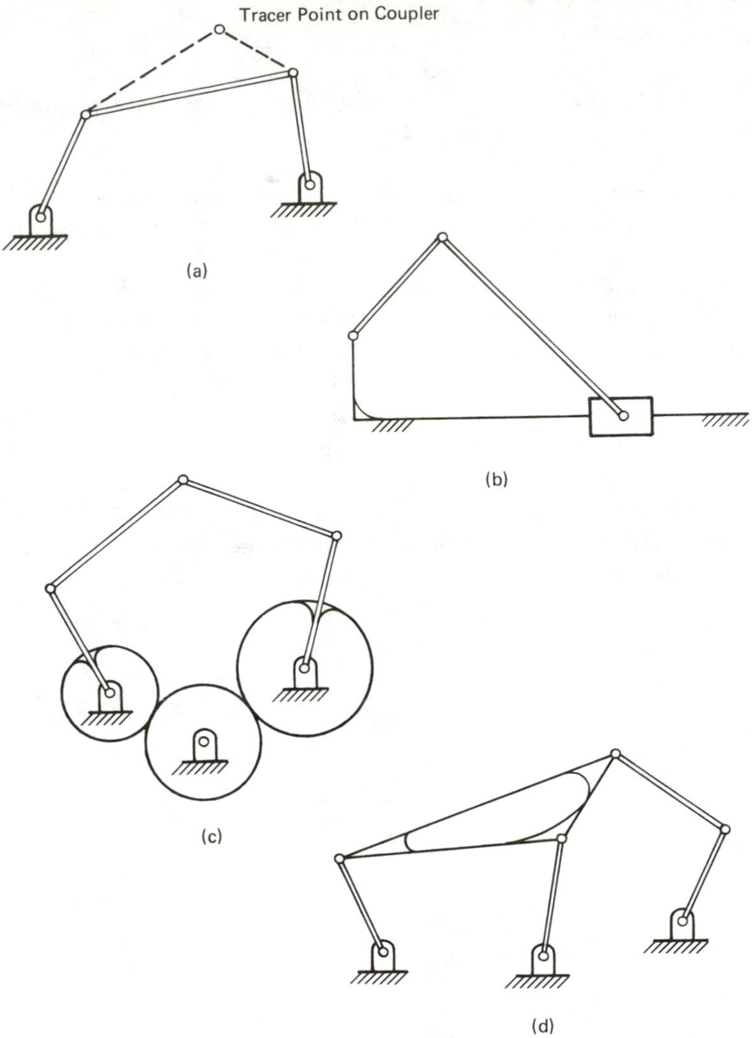


**Figure 2.3** Some mechanisms of preconceived type; (a) four-bar linkage; (b) slider-crank; (c) geared five-bar linkage; (d) Stephenson III six-link mechanism.

topologically but not dimensionally (geared five-bar, Stevenson or Watt six-bar linkage, etc.), as depicted in Fig. 2.3.

## 2.2 TASKS OF KINEMATIC SYNTHESIS

Recall from Chap. 1 that there are three customary *tasks* for kinematic synthesis: *function, path,* and *motion* generation.

In *function generation* rotation or sliding motion of input and output links must be correlated. Figure 2.4 is a graph of an arbitrary function $y = f(x)$. The kinematic synthesis task may be to design a linkage to correlate input and output

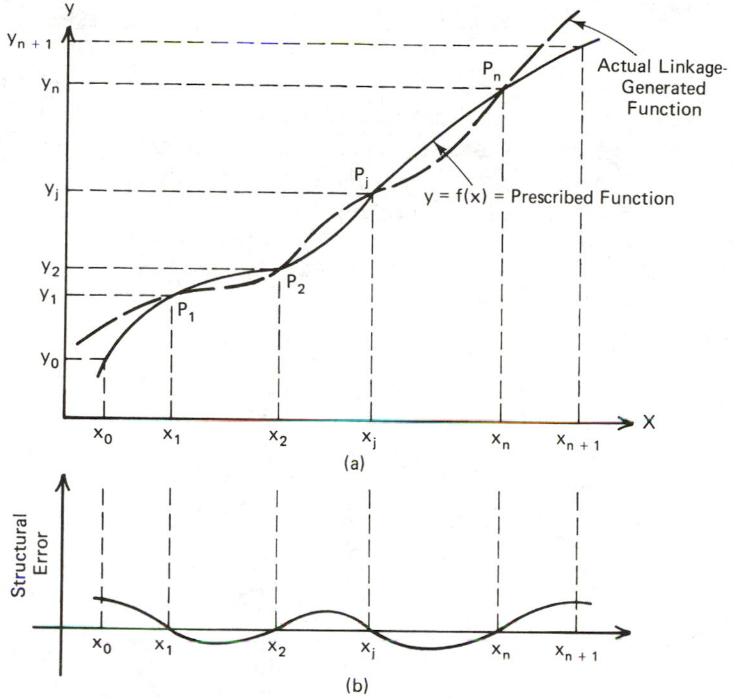


**Figure 2.4** Function-generation synthesis; (a) ideal function and generated function; (b) structural error.

such that as the input moves by $x$, the output moves by $y = f(x)$ for the range $x_0 \leq x \leq x_{n+1}$. Values of the independent parameter, $x_1, x_2, \ldots, x_n$ correspond to prescribed *precision points* $P_1, P_2, \ldots, P_n$ on the function $y = f(x)$ in a range of $x$ between $x_0$ and $x_{n+1}$. In the case of rotary input and output, the angles of rotation $\phi$ and $\psi$ (Fig. 2.5a) are the linear analogs of $x$ and $y$, respectively. When the input is rotated to a value of the independent parameter $x$, the mechanism in the "black box" causes the output link to turn to the corresponding value of the dependent variable $y = f(x)$. This may be regarded as a simple case of a mechanical analog computer.

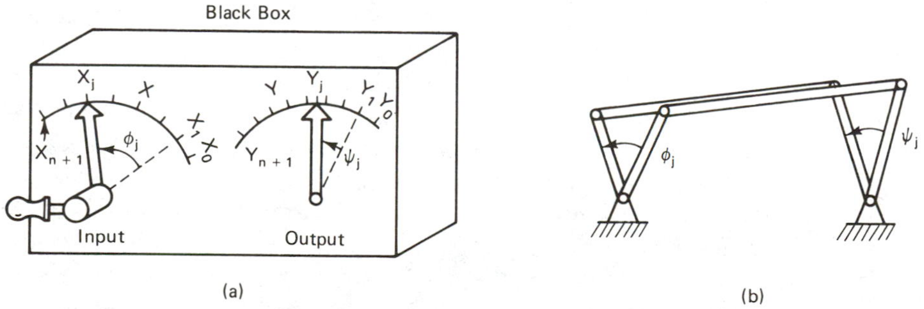


**Figure 2.5** Function-generator mechanism; (a) exterior view; (b) schematic of the mechanism inside.

The subscript $j$ indicates the $j$th prescribed position of the mechanism; the subscript 1 refers to the *first* or *starting* prescribed position of the mechanism, and $\Delta\phi$, $\Delta x$, $\Delta\psi$, and $\Delta y$ are the desired *ranges* of the respective variables, $\phi$, $x$, $\psi$, and $y$ (e.g., $\Delta x \equiv | x_{n+1} - x_0 |$, $\Delta\phi \equiv | \phi_{n+1} - \phi_0 |$, etc.). Since there is a linear relationship between the angular and linear changes,

$$\frac{\phi_j - \phi_1}{x_j - x_1} = \frac{\Delta\phi}{\Delta x} \tag{2.1}$$

where $\phi_1$ is the datum for $\phi_j$, and therefore $\phi_1 = 0$. It follows that

$$\begin{aligned} \phi_j &= \frac{\Delta\phi}{\Delta x}(x_j - x_1) \\ \psi_j &= \frac{\Delta\psi}{\Delta y}(y_j - y_1) \end{aligned} \tag{2.2}$$

These relationships may also be written as

$$\phi_j = R_\phi(x_j - x_1) \tag{2.3}$$

$$\psi_j = R_\psi(y_j - y_1) \tag{2.4}$$

where $R_\phi$ and $R_\psi$ are the *scale factors* in degrees per unit variable defined by

$$R_\phi = \frac{\Delta\phi}{\Delta x} \tag{2.5}$$

$$R_\psi = \frac{\Delta\psi}{\Delta y} \tag{2.6}$$

The four-bar linkage is not capable of error-free generation of an arbitrary function and can match the function at only a limited number of precision points (see Fig. 2.4a). It is, however, widely used in industry in applications where high precision at many points is not required because the four-bar is simple to construct and maintain. The number of precision points that are used in the dimensional synthesis of the four-bar linkage varies in general between two and five.* It is often desirable to space the precision points over the range of the function in such a way as to minimize the *structural error* of the linkage. Structural error is defined as the difference between the generated function (what the linkage actually produces) and the prescribed function for a certain value of the input variable (see Fig. 2.4).

Notice that the first precision point ($j = 1$) is not at the beginning of the range (see Fig. 2.4). The reason for this is to reduce the extreme values of the structural error. It is also evident from Eq. 2.1) that angles of rotation are measured from the first position (e.g., $\phi_1 = 0$).

* Function generation synthesis up to 7 and path generation synthesis up to 9 precision points are possible, but they generally require numerical rather than the preferable closed-form methods of synthesis.

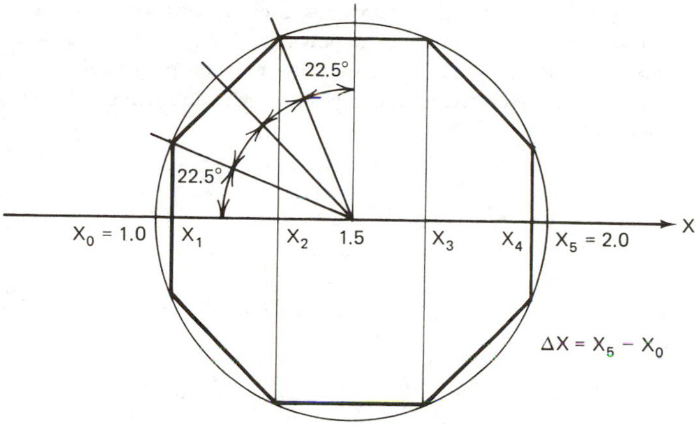


**Figure 2.6** Chebyshev spacing of four precision points.

Chebyshev determined that the best linkage approximation to a function occurs when the absolute value of the maximum structural error between precision points and at both ends of the range are equalized. *Chebyshev spacing* [125] of precision points is employed to minimize the structural error. This technique, based on *Chebyshev* polynomials [43,125], is often used as a "first guess," although it is applicable only in special cases (such as symmetric functions). After the synthesis is completed, the resultant structural error of the mechanism can be determined, followed by assessment and alteration of the placement of precision points to improve the mechanism accuracy. Two techniques for locating precision points for minimized structural error are the *Freudenstein respacing formula* [104] and the *Rose–Sandor direct optimal technique* [222]. Both are based on the fact that reducing the space between adjacent precision points reduces the extreme error between them, and vice versa.

A simple construction is available for determining Chebyshev spacing as an initial guess (see Fig. 2.6). Precision points may be located graphically; a circle is drawn whose diameter is proportional to the range of the independent parameter ($\Delta x$). A regular equilateral polygon having $2n$ sides (where $n$ = the number of prescribed precision points) is then inscribed in the circle such that two sides of the polygon are vertical. Lines drawn perpendicular to the horizontal diameter through each corner of the polygon intersect the diameter at points spaced at distances proportional to Chebyshev spacing of precision points.* This procedure is now explained by way of examples.

**Example 2.1**

Determine the Chebyshev spacing for a four-bar linkage generating the function $y = 2x^2 - 1$, in the range of $1 \leq x \leq 2$, where four precision points are to be prescribed ($n = 4$).

* Hand calculator program available from the first author.

**Solution** The first step is to draw a circle with diameter $\Delta x = x_{n+1} - x_0 = 2.0 - 1.0 = 1.0$. Next, construct a polygon of $2n = 8$ sides, with two sides vertical, as shown in Fig. 2.6. The corners of the polygon projected vertically onto the horizontal axis are the prescribed precision points. Measurements from the geometric construction yield

$$x_0 = 1.00, \qquad x_3 = 1.69$$
$$x_1 = 1.04, \qquad x_4 = 1.96$$
$$x_2 = 1.30, \qquad x_5 = 2.00$$

The foregoing construction for Chebyshev spacing is tantamount to the following formulas:

$$\Delta x_j = x_j - x_0 = \tfrac{1}{2}\Delta x \left[1 - \cos\left(\frac{\pi(2j-1)}{2n}\right)\right], \qquad j = 1, 2, \ldots, n$$

and

$$x_j = x_0 + \Delta x_j, \qquad j = 1, 2, \ldots, n$$

where $\Delta x_j$ = distance from the beginning of the $x$ range to the $j$th precision point

$\Delta x = x_{n+1,} - x_0$ = range in $x$

$j$ = precision point number, $j = 1, 2, \ldots, n$

$n$ = total number of precision points

Thus, in this example,

$$\Delta x_1 = \tfrac{1}{2}(1)\left[1 - \cos\left(\frac{\pi}{8}\right)\right] = 0.038$$

$$x_1 = 1.04$$

and

$$\Delta x_2 = 0.309, \qquad \Delta x_3 = 0.691, \qquad \Delta x_4 = 0.962$$

so that

$$x_2 = 1.31$$
$$x_3 = 1.69$$
$$x_4 = 1.96$$

**Example 2.2**

Given the Chebyshev precision points derived in Ex. 2.1 and the ranges in the input and output link rotations $\Delta\phi = 60°$, $\Delta\psi = 90°$, find $\phi_2$, $\phi_3$, $\phi_4$, $\psi_2$, $\psi_3$, and $\psi_4$.

**Solution** $y_j$ is found by substituting the values of $x_j$ into the function $y = 2x^2 - 1$:

$$y_0 = 1.00, \qquad y_3 = 4.71$$
$$y_1 = 1.16, \qquad y_4 = 6.68$$
$$y_2 = 2.43, \qquad y_5 = 7.00$$

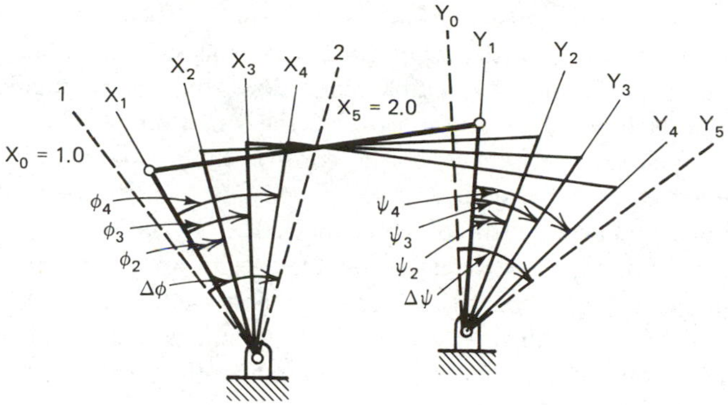


**Figure 2.7** Not-to-scale schematic of a function-generator four-bar mechanism with four precision positions of the input and output links $x_i$ and $y_i$, i = 1, 2, 3, 4, within the range $\Delta x = x_5 - x_0$ and $\Delta y = y_5 - y_0$. Input rotations $\phi$ and output rotations $\psi$ are the analogs of independent and dependent variables $x$ and $y$, respectively.

Using Eqs. (2.1) and (2.2), where $\Delta x = 1$, $\Delta y = 6$, $\Delta\phi = 60°$, and $\Delta\psi = 90°$, we have

$$\phi_2 = 16.2°, \qquad \psi_2 = 19.1°$$
$$\phi_3 = 39.0°, \qquad \psi_3 = 54.3°$$
$$\phi_4 = 55.2°, \qquad \psi_4 = 82.8°$$

Figure 2.7 shows a not-to-scale schematic of the input and output links of a four-bar function generator mechanism in the four precision positions, illustrating the relationship between $x_j$ and $\phi_j$ as well as $y_j$ and $\psi_j$. The dimensional synthesis techniques described later in this chapter and Chap. 3 will show us how to use such precision-point data for the synthesis of four-bar linkages and other mechanisms for function generation.

A variety of different mechanisms could be contained within the "black box" of Fig. 2.5a. In this case, Fig. 2.5b shows a four-bar linkage function generator. A typical example of a function generator is shown schematically in Fig. 2.8. A four-bar linkage connects a cam follower driven by the cam to a type bar of a typewriter

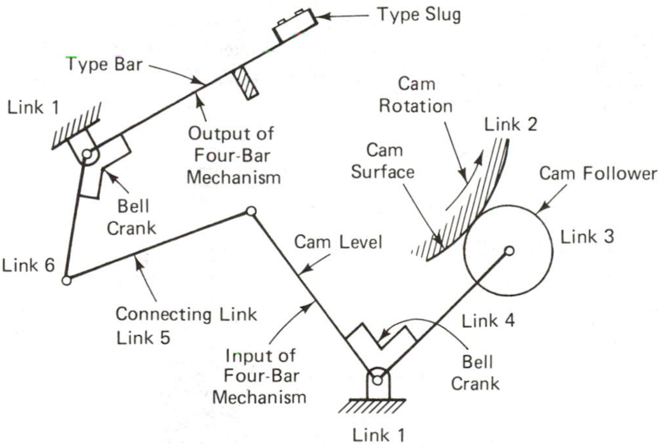


**Figure 2.8** Four-bar mechanism used as the impact printing mechanism in an electric typewriter.

printing mechanism. Here the type must be moved, first by smaller then by larger angles per increment of input rotation, in order to throw the type against the platen roller with an impact. Another application of function generation would be an engine where the mixing ratios of fuel to oxidant might vary as the function $y = y(x)$. Here $\phi$ might control the fuel valve while $\psi$ would control the oxidant valve. Flow characteristics of the valves and the required ratio at various fuel rates would dictate the nature of the functional relationship to be generated.

Yet another example is a linkage to correlate steering positions of the front wheels of an all-terrain vehicle with the relative speed at which each individually driven wheel should rotate to avoid scuffing. Here the input crank is connected to the steering arm, while the output adjusts a potentiometer controlling the relative speed of the two front-drive wheels.

Mechanical function generators may also be of the type shown in Fig. 2.9, in which a *linear* displacement may be the analog of one variable and the crank rotation may be the analog of another, a functionally related variable. As illustrated in Fig. 2.10, a function generator may have more degrees of freedom than one; an output variable may be a function of two or more inputs. For example, such a linkage might be used to simulate the addition, subtraction, multiplication, or any other algebraic or transcendental functional correlation of several variables. Figure 2.11 shows a six-link function generator mechanism in which two four-link mechanisms are joined in a series. The objective in this linkage is to provide a measure of flow rate through the weir where the input is the vertical translation $x$ of the water level.

In *path generation* a point on a "floating link" (not directly connected to the fixed link) is to trace a path defined with respect to the fixed frame of reference.

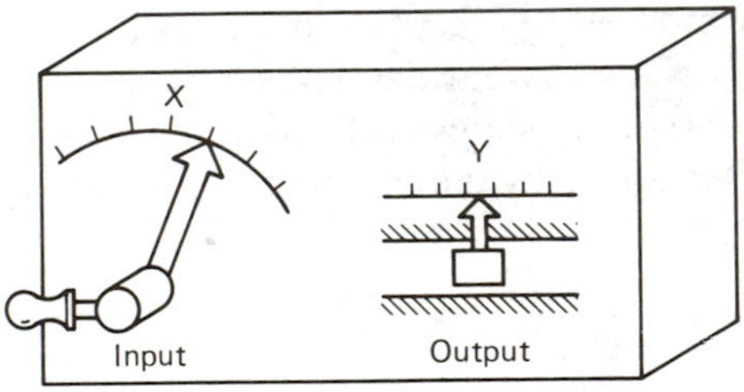


**Figure 2.9** Function generator with rotary input and translational output, analogs of the independent and dependent variables of the function $y = f(x)$.

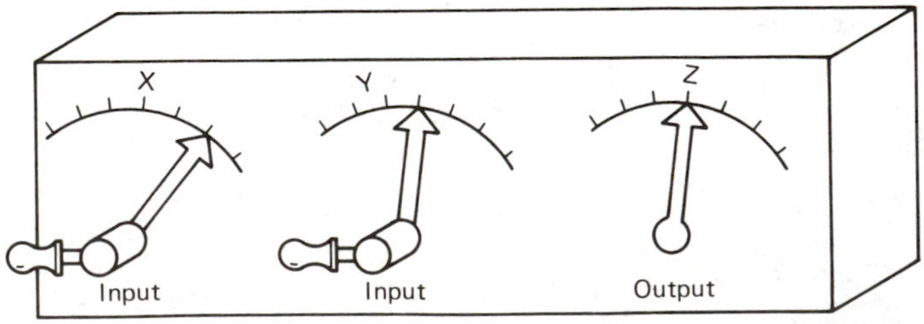


**Figure 2.10** Two-degree-of-freedom function generator for generating the function $z = f(x,y)$.

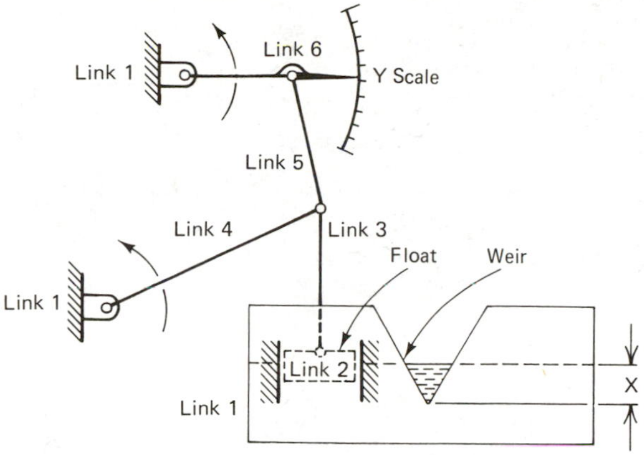


**Figure 2.11** Flow-rate-indicator mechanism, $y = K_1 x^{K_2}$, where $K_1$ and $K_2$ are constants.

If the path points are to be correlated with either time or input-link positions, the task is called *path generation with prescribed timing.* An example of path generation is a four-bar linkage designed to pitch a baseball or tennis ball. In this case the trajectory of point $P$ would be sure as to pick up a ball at a prescribed location and to deliver the ball along a prescribed path with prescribed timing for reaching a suitable throw-velocity and direction.

In Fig. 2.12, a linkage whose floating link will contain point $P$ is desired such that point $P$ will trace $y = f(x)$ as the input crank turns. Typical examples are where $y = f(x)$ is the path desired for a thread guiding eye on a sewing machine (Fig. 2.13) or the path to advance the film in a camera (Fig. 2.14). Various straight-line mechanisms, such as Watt's and Robert's linkages, are examples of a special kind of path generator (see Fig. 2.15) in which geometric relationships assure the generation of straight-line segments within the cycle of the linkages motion.

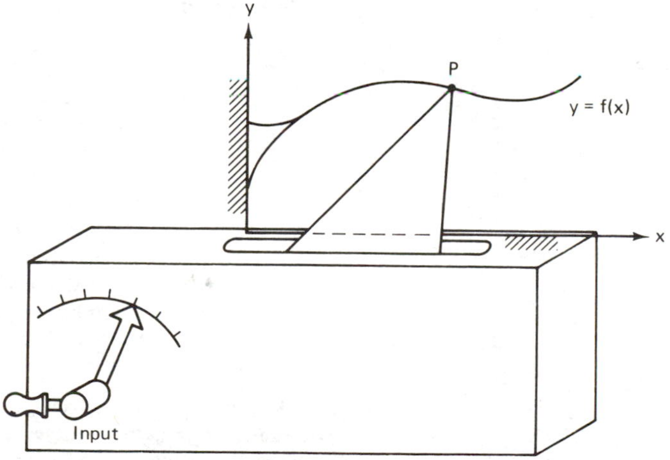


**Figure 2.12** A path generator linkage.

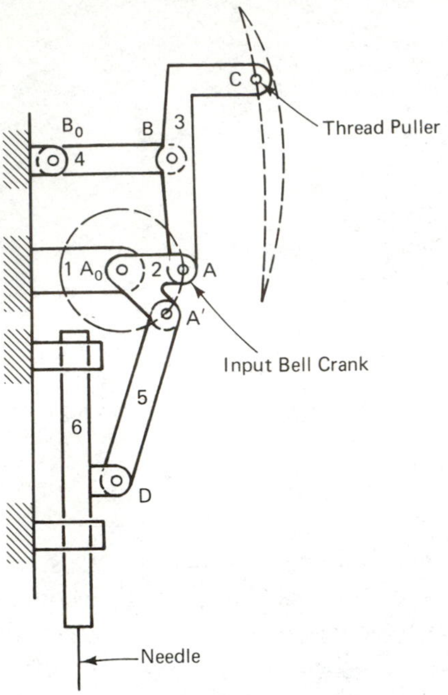


**Figure 2.13** In a sewing machine, one input (bell crank 2) drives a path generator (four-bar mechanism 1,2,3,4) and a function generator slidercrank (1,2,5,6). The first generates the path of thread-guide C and the second generates the straight-line motion of the needle, whose position is a function of crank rotation.

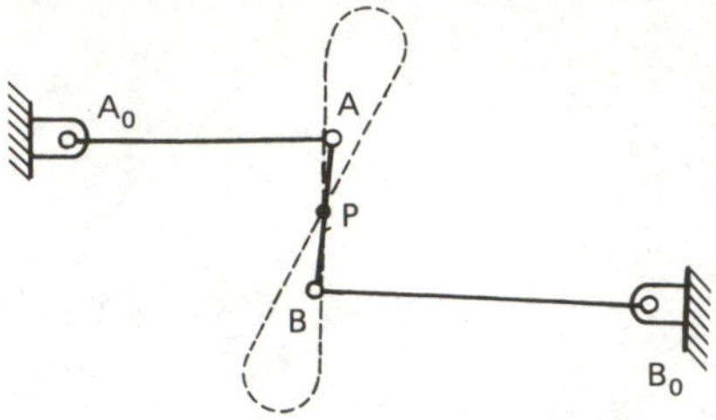


(a)

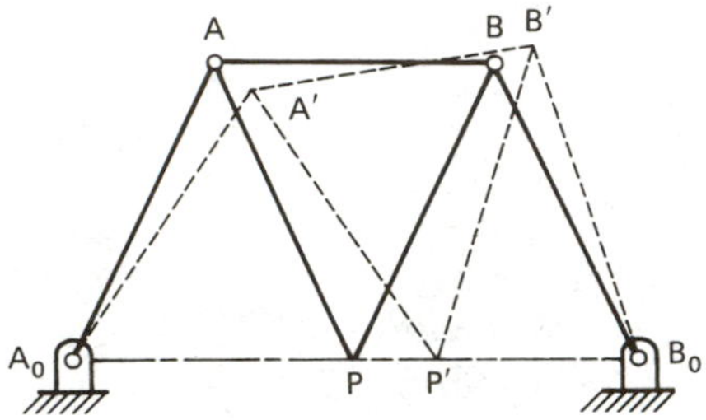


(b)

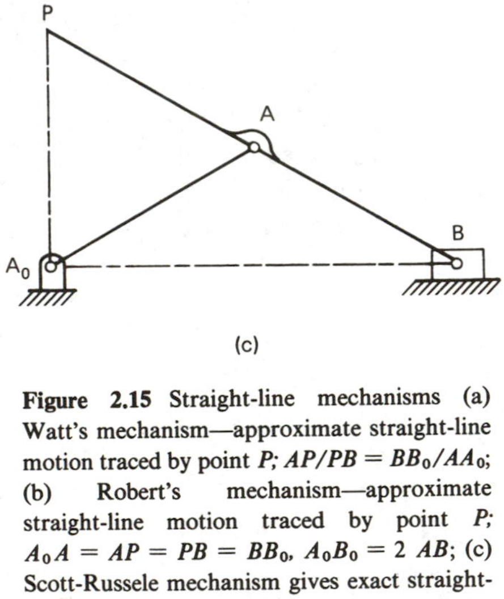


(c)

**Figure 2.15** Straight-line mechanisms (a) Watt's mechanism—approximate straight-line motion traced by point $P$; $AP/PB = BB_0/AA_0$; (b) Robert's mechanism—approximate straight-line motion traced by point $P$; $A_0A = AP = PB = BB_0$, $A_0B_0 = 2\ AB$; (c) Scott-Russele mechanism gives exact straight-line motion traced by point $P$. Note the equivalence to Cardan motion (see chapt. 3); $A_0A = AB = AP$.

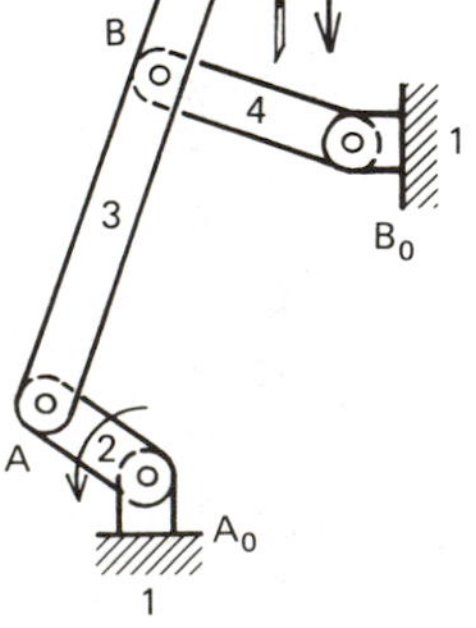

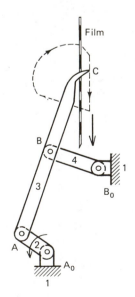

**Figure 2.14** Film-advance mechanism of a movie camera or projector generates the path of point C as a function of the angle of rotation of crank 2.

*Motion generation* or *rigid-body guidance* requires that an entire body be guided through a prescribed motion sequence. The body to be guided usually is a part of a floating link. In Fig. 2.16 not only is the path of point *P* prescribed, but also the rotations $\alpha_j$ of vector **Z** embedded in the moving body. The corresponding input rotations may or may not be prescribed. For instance, vector **Z** might represent a carrier link in automatic machinery where a point located on the carrier link (the tipe of **Z**) has a prescribed path while the carrier has a prescribed angular orientation (see Fig. 2.17). Prescribing the movement of the bucket for a bucket loader is another example of motion generation. The path of the tip of the bucket is critical since

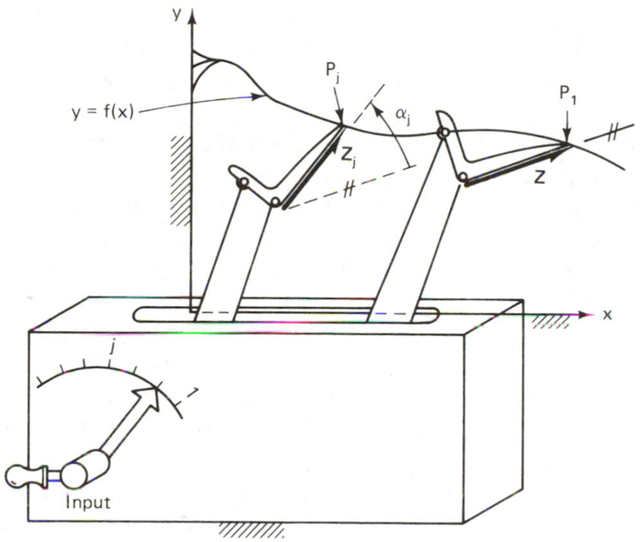


**Figure 2.16** Motion-generator mechanism.

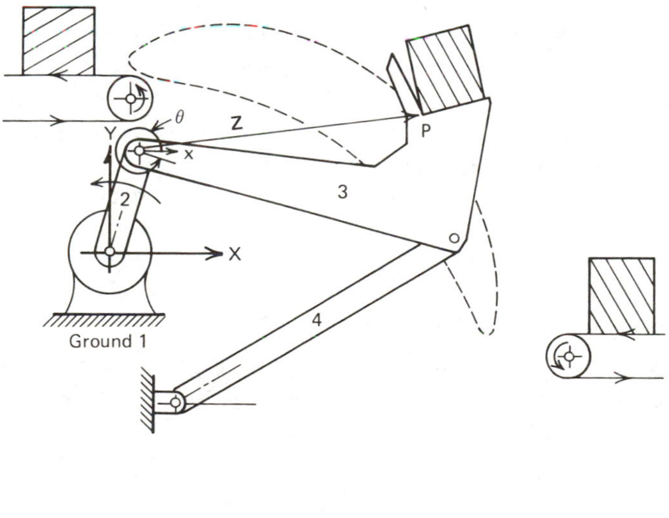


**Figure 2.17** Carrier mechanism in an assembly machine.

the tip must perform a scooping trajectory followed by a lifting and a dumping trajectory. The rotations of the bucket are equally important to ensure that the load is dumped from the correct position.

Since a linkage has only a finite number of significant dimensions, the designer may only prescribe a finite number of *precision conditions;* that is, we may only *prescribe* the *performance* of a linkage at a finite number of *precision points.* There are three methods of specifying the *prescribed performance* of a mechanism: *first-order* or *point approximation, higher-order approximation,* and combined *point-order approximation.* *

In *first-order approximation* for function and path generation, discrete points on the prescribed (or ideal) function or path are specified. Recall that Fig. 2.4a showed precision points $P_1$ to $P_n$ of the ideal function. The synthesized mechanism will generate a function that will coincide with the ideal function at the precision points but will generally deviate from the ideal function between these points (Fig. 2.4b).

Structural error for path generation may be defined as the vector from the ideal to the generated path perpendicular to the ideal path or it may be defined as the vector between corresponding points on an ideal and a generated path taken at the same value of the independent variable. The latter definition is used when there is prescribed timing. In motion generation there will be both a path and an angular structural error curve to analyze.

In some cases a mechanism is desired to generate not only a position but also the velocity, acceleration, shock, and so on, at one or more positions (see Fig. 2.18). For example, the blade of a cutter that must slice a web of paper into sheets while the web is in motion would not only be required to match the correct position at the instant of the cut, but also several derivatives at that position in order to cut straight across and to preserve the sharpness of the blade. For *higher-order approximation,* the first derivative, $dy/dx_2$, prescribes the slope of function (or path) at that point; the second derivative, $d^2y/dx^2$, implies prescribing the radius of curvature; the third derivative, $d^3y/dx^3$, prescribes the rate of change of curvature; and so on (see Sec. 2.24).

The combination of both point and order approximations is called *point-order approximation* or approximation by *multiply separated precision points* [284]. For example, one might desire to prescribe a position and a velocity at one precision point, only a position at a second precision point, and a position and velocity at a third point. Figure 2.19 shows such an application where a mechanism is desired to pick up an item from conveyor belt 1 traveling at velocity $\mathbf{V}_1$ and deposit it on a conveyor belt 2 traveling at $\mathbf{V}_2$, having traversed the intervening space in such a way as to avoid some machinery components. Typical application of this occurs in bookbinding, where "signatures" (32- or 64-page sections) of a book from conveyor 1 are to be stacked on conveyor 2 to form the complete book.

* Approximate (rather than precise) generation of greater numbers of prescribed conditions are possible by the use of least squares or non-linear programming methods. These, however, are numerical procedures rather than closed-form solutions.

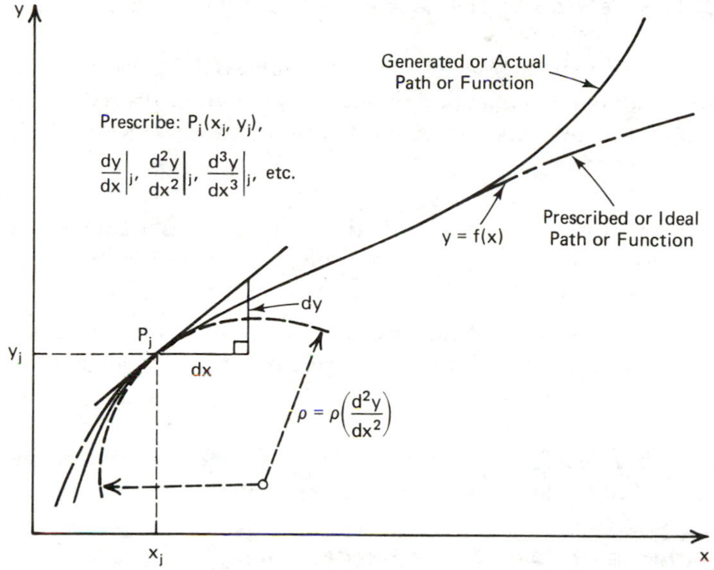


**Figure 2.18** Higher-order approximation of function or path.

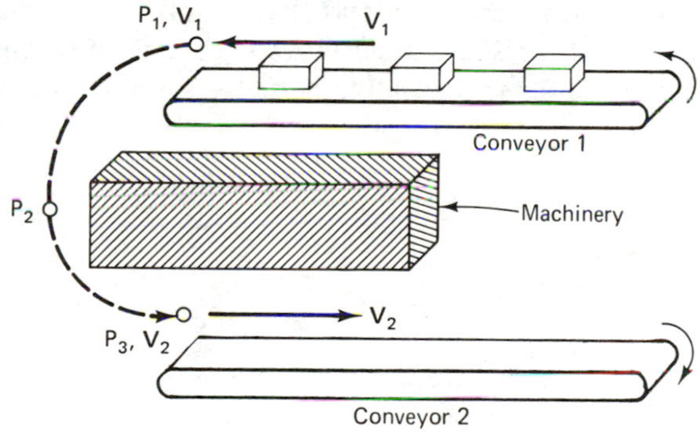


**Figure 2.19** Point-order approximation for path generation. There are five prescribed conditions: three path points and velocities at two of these, tantamount to two infinitesimally close prescribed position at $P_1$ and $P_3$.

Kinematic synthesis has been defined here as a combination of type and dimensional synthesis. Most of the rest of this chapter and Chap. 3 are devoted to dimensional synthesis (an introduction to type synthesis and number synthesis appeared in Chap. 1). Before moving on to dimensional synthesis, however, one of the methods to creatively discover suitable types of linkages for a prescribed task will be introduced. The method is based on "structural models" or "associated linkages." A case study of type synthesis using another method can be found in the appendix to this chapter.

## 2.3 NUMBER SYNTHESIS: THE ASSOCIATED LINKAGE CONCEPT

Several different theories of number synthesis have been suggested to assist in the *creative* design of mechanical devices. One of these procedures, the *associated linkage concept,* was developed by R. C. Johnson and K. Towligh [143] and consists of the following procedure:

1. The determination of rules that must be satisfied for the selection of a suitable "associated linkage." These rules are derived by observing the specific design application.
2. The application of suitable associated linkages to the synthesis of different types of devices. (See Table 1.2 for equivalent lower-pair joints for velocity matching of higher-pair connections.)

This technique of applying number synthesis to the creative design of practical devices will be illustrated by several examples.

### Synthesis of Some Slider Mechanisms

Suppose that it is desired to derive types of mechanisms for driving a slider with rectilinear translation along a fixed path in a machine. Assume that the drive shaft will be fixed against translation and that it must rotate with unidirectional rotation. Also, assume that the slider must move with a reciprocating motion.

A basic rule for this example is that a suitable associated linkage must have a

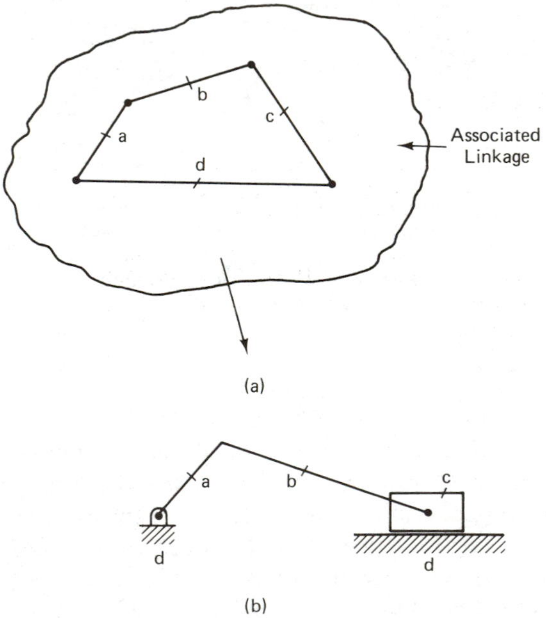


**Figure 2.20** Slider-crank mechanism and its associated linkage; (a) four-bar chain; (b) slider-crank mechanism.

single degree of freedom ($F = +1$) when one link is fixed. Let us start with the least complicated associated linkage chain (which is the four-bar) since simplicity is an obvious design objective (Fig. 2.20a). The four-bar associated linkage has four revolute joints. If one of the revolutes (joint *c-d*) is replaced by a slider, the slider-crank mechanism is derived as shown in Fig. 2.20b.

Increasing the degree of complexity, a Stephenson six-bar chain (in which ternary links are not directly connected) is considered next as a suitable associated linkage (Fig. 2.21a). By varying the location of the slider one creates the slider mechanisms of Fig. 2.21b–f, different from the slider-crank of Fig. 2.20. Finally, in Fig. 2.22,

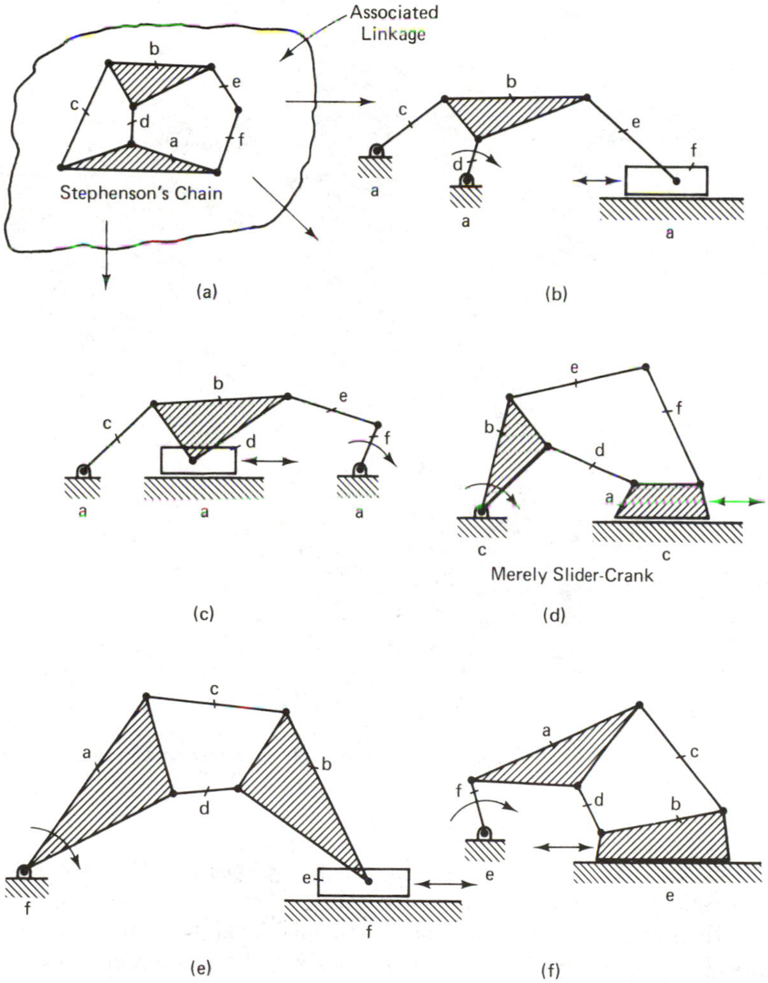


**Figure 2.21** Slider mechanisms derived from Stephenson's six-bar chain as the associated linkage. Note that (d) shows merely a slider crank with redundant (superfluous) links, the passive dyad consisting of links e and f.

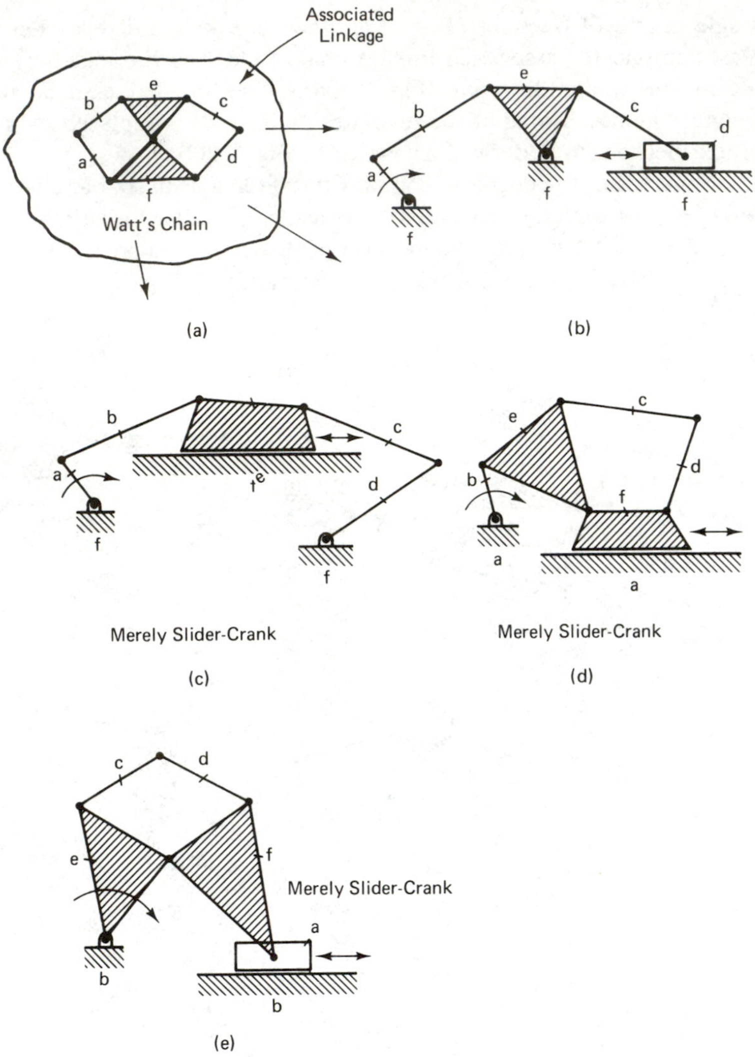


**Figure 2.22** Slider mechanisms derived from Watt's six-bar chain as the associated linkage.

from the Watt six-bar chain (in which the ternary links are direct connected) we derive only one new mechanism (Fig. 2.22b), which is of the same degree of complexity as those in Fig. 2.22c, d, and e are merely slider-cranks, with an added passive dyad. Thus five different six-link mechanisms, each having only a single slider joint, can be derived for this problem.

This general procedure could be extended to other suitable linkages of greater complexity, including those containing *higher pairs.* Thus cams and sliding pivots

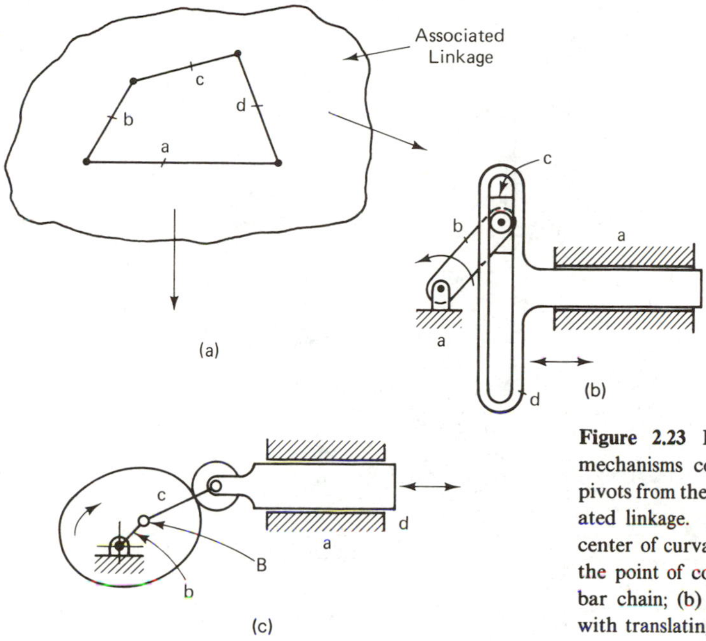


**Figure 2.23** Derivation of some slider mechanisms containing cams and sliding pivots from the four-bar chain as the associated linkage. Notice that point $B$ is the center of curvature of the cam contour at the point of contact of the cam; (a) four-bar chain; (b) Scotch yoke; (c) disk cam with translating follower.

may be incorporated in the derivations of different types of mechanisms, such as those illustrated in Fig. 2.23, derived from the four-bar chain as the associated linkage.

### Synthesis of Some Gear-Cam Mechanisms

A typical meshing gear set is shown in Fig. 2.24 with two typical teeth in contact. At the instant of observation the meshing gear set is equivalent to a hypothetical quadric chain (see Table 1.2). Hence, as shown in Fig. 2.24, a meshing gear set has a four-bar chain as an associated linkage. The basic rules for a suitable associated linkage involved in the synthesis of a mechanism containing a meshing gear set are as follows:

1. The number of degrees of freedom with one link fixed must be $F = +1$.
2. The linkage must contain at least one four-sided closed loop. This is true since the meshing gear set corresponds to a four-sided closed loop containing two centers of rotation, $R_{p/f}$ and $R_{g/f}$, and two base points, $B_p$ and $B_g$, which are the instantaneous centers between gear $p$ and the fictitious coupler $C$ and between gear $g$ and $C$, respectively. In the gear set, coupler $C$ is replaced by the higher-pair contact between the tooth profiles. Hence $B_p$ and $B_g$ coincide with the centers of curvature of the respective involute tooth profiles at their point of contact. In traversing this four-sided closed loop, the two centers of rotation must be encountered in succession, such as $RRBB$ rather than $RBRB$.

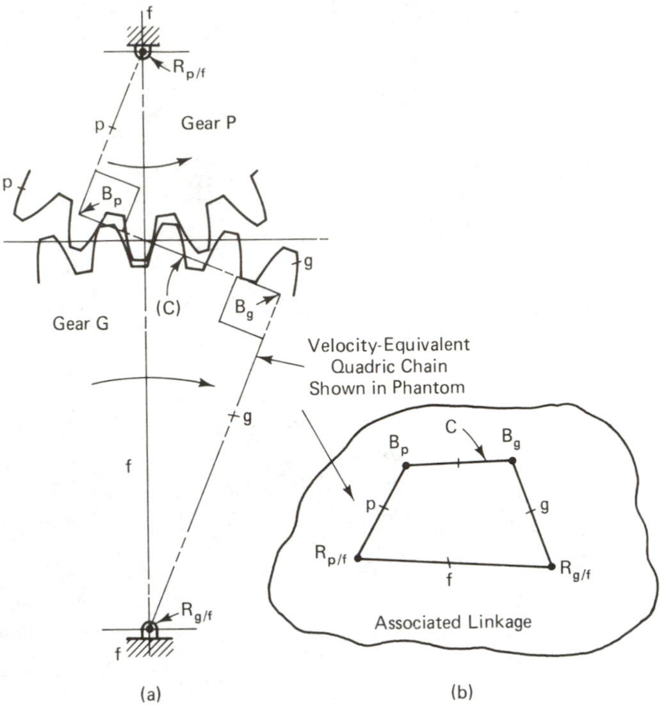


**Figure 2.24** Meshing gear set with its associated linkage. $B_g$ and $B_p$ are the centers of curvature of the involutes at the contact point of gear $G$ and gear P, respectively; (a) gear pair; (b) associated linkage.

3. The four-sided closed loop must contain at least one binary link. This is true because in the four-sided closed loop the link connecting the two base points must be a binary link. This is evident since the base points on the meshing gears are instantaneous and they are joined by a hypothetical connecting rod in the equivalent quadric chain.

Suppose that it is required to design a gear mechanism for driving a slider with arbitrary motion along fixed ways in a machine. Assume that the driving shaft must have unidirectional rotation and that the slider must have a reciprocating motion. One possible design would be the mechanism shown in Fig. 2.25, where the driving cam provides arbitrary motion and a gear and rack drive the slider. In Fig. 2.26 the equivalent linkage for this mechanism is shown together with its associated linkage. Incidentally, a gear and rack is a special gear type with one base point and one center of rotation at infinity.

Simplicity in design is a practical goal worth striving for. Suppose that we wish to explore different simpler mechanism types for the basic problem described in the preceding paragraph (assuming that a cam, follower, gear, and rack are to

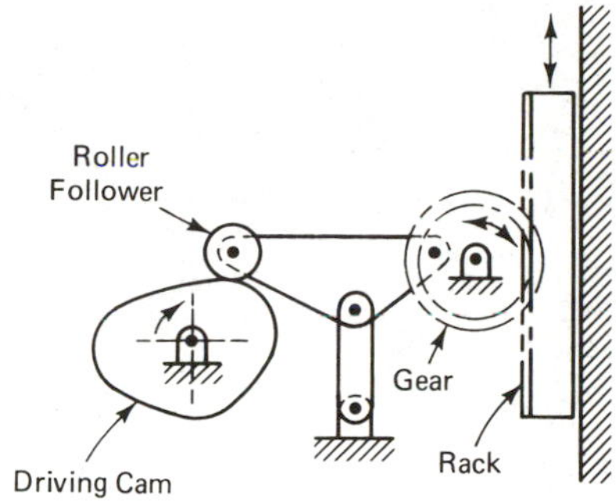


**Figure 2.25** Slider mechanism with cam and gear.

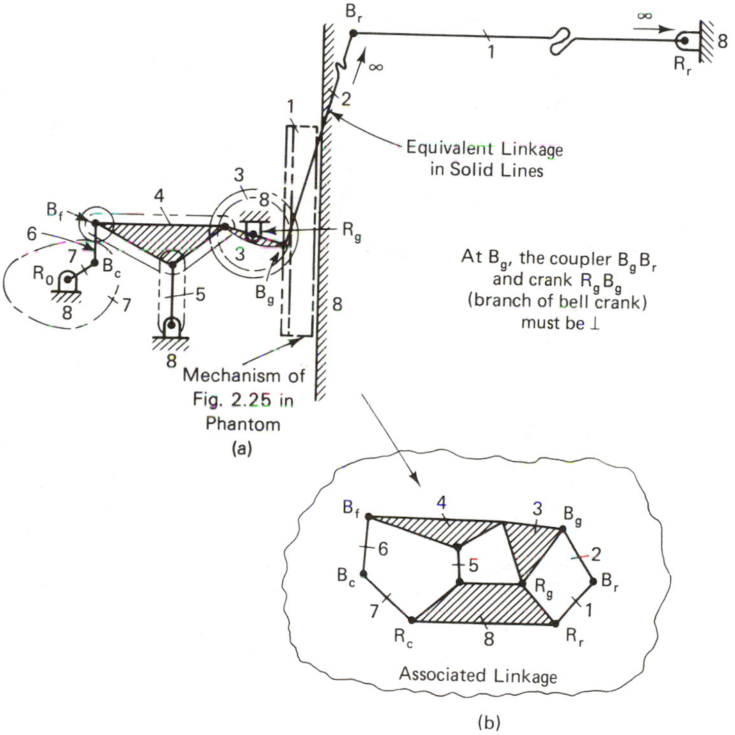


**Figure 2.26** Slider mechanism of Fig. 2.25 with equivalent linkage (a) and associated linkage (b) from which it was derived.

be employed for driving the slider). The simplest suitable associated linkage for this application would be either a Watt chain or a Stephenson chain. From these chains three different mechanism types are derived (Figs. 2.27 and 2.28), where Fig. 2.28c would require a flexible shaft for driving the cam.

**Figure 2.27** Cam-gear-slider mechanism derived from Watt's chain.

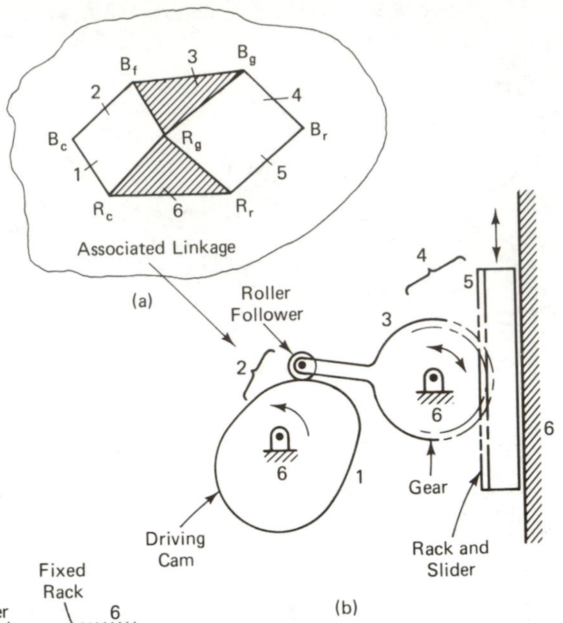


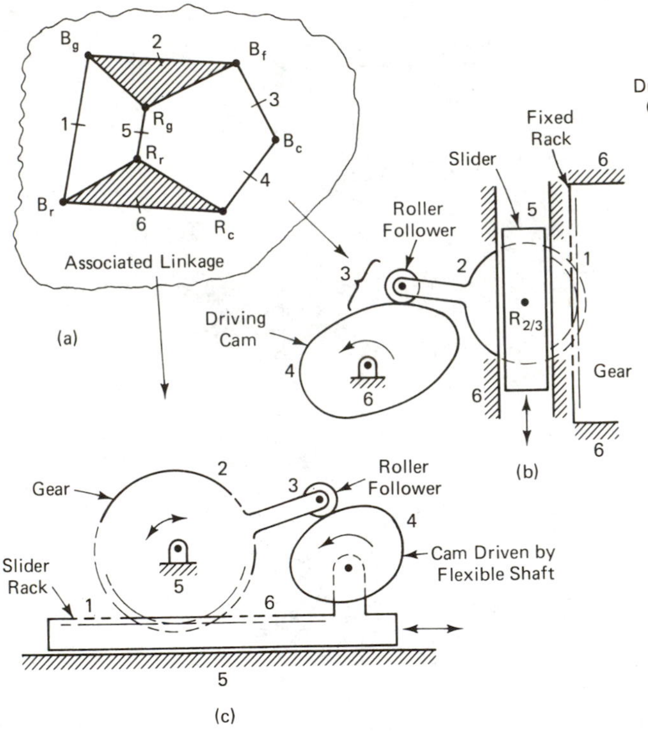


**Figure 2.28** Cam-gear-slider mechanisms derived from Stephenson's chain.

## Synthesis of Some Internal-Force-Exerting Devices

Kurt Hain [119] has applied number synthesis to the design of differential brakes and differential clamping mechanisms by recognizing the analogy with preloaded structures. This analogy shows that, for the synthesis of internal-force-exerting devices in general, a suitable associated linkage must have $F = -1$ for the number of degrees of freedom with one link fixed. Also, *forces* exerted by the device on the work piece correspond to *binary links* in the associated linkage, recognizing that a binary link is a two-force member. Let us apply this technique to the synthesis of two practical devices. First, different types of compound lever snips are explored, followed by several types of yoke riveters.

**Synthesis of compound-lever snips.** Simply constructed compound-lever snips are to be designed for cutting through tough materials with a relatively small amount of effort. The actuating force is designated by $P$ and the resisting force by $F_r$. We will assume that the compound-lever snips should be hand-operated and mobile. Hence there will be no ground link in the construction. However, a high amplification of force is required in the device. Therefore, in the associated linkage, binary links $P$ and $F_r$ must not be connected by a single link; otherwise, a simple lever type of construction will result in relatively low force amplification.

In summary, for application to the synthesis of compound-lever snips, the rules or requirements for a suitable associated linkage are as follows:

1. $F = -1$.
2. There must be at least two binary links because of $P$ and $F_r$.
3. Two binary links $P$ and $F_r$ must not connect the same link, because in that case the snips will be simple instead of compound.

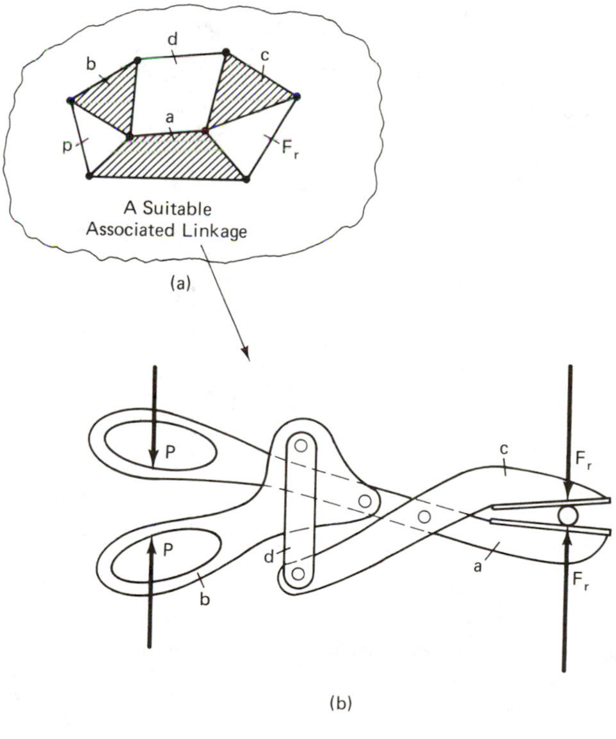


**Figure 2.29** Synthesis of compound lever snips from a suitable associated linkage.

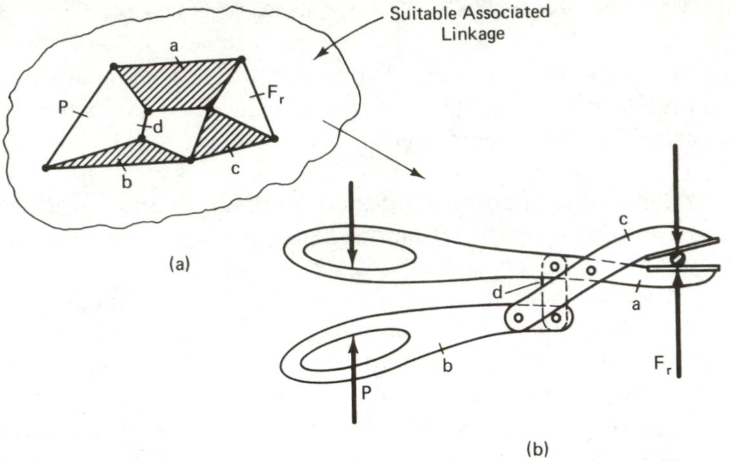


**Figure 2.30** Synthesis of compound-lever snips from a suitable associated linkage.

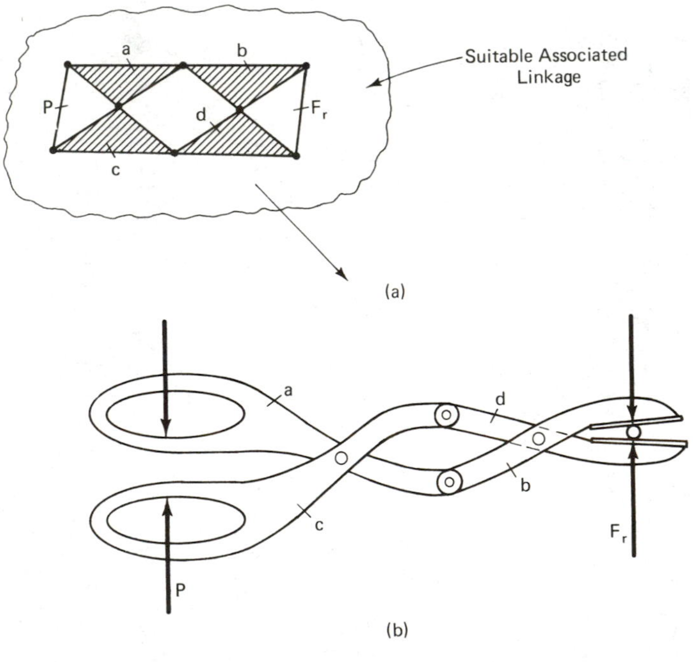


**Figure 2.31** Different design derived from another suitable associated linkage for compound-lever snips.

The associated linkages in Figs. 2.29, 2.30, and 2.31 satisfy the requirements. Each suitable associated linkage yields a different mechanism for compound-lever snips.

**Synthesis of yoke riveters.** The configuration for an existing yoke-riveter design [143] is shown in Fig. 2.32. Let us apply number synthesis in the creation of other types of yoke-riveter designs.

The following characteristics are assumed to be requirements for a suitable yoke riveter in our particular application:

1. Simple features of construction.
2. Self-contained, portable unit.
3. High force amplification between power piston and rivet die.
4. One part of the two-part rivet die and the relatively large pneumatic power cylinder are fixed to the frame link.
5. Another part of the rivet die and the power piston are to slide relative to the frame link.

From Fig. 2.32 of the existing yoke-riveter design the associated plane linkage with single pin joints is derived as shown in Fig. 2.33. Applying Gruebler's equation (Chap. 1) to the linkage in Fig. 2.33, we obtain $F = -1$, which is expected, since this value of $F$ is characteristic of the associated linkage for any internal-force-exerting device: $F = 3(n - 1) - 2f_1$. Note that $n = 10$, including the binary links representing

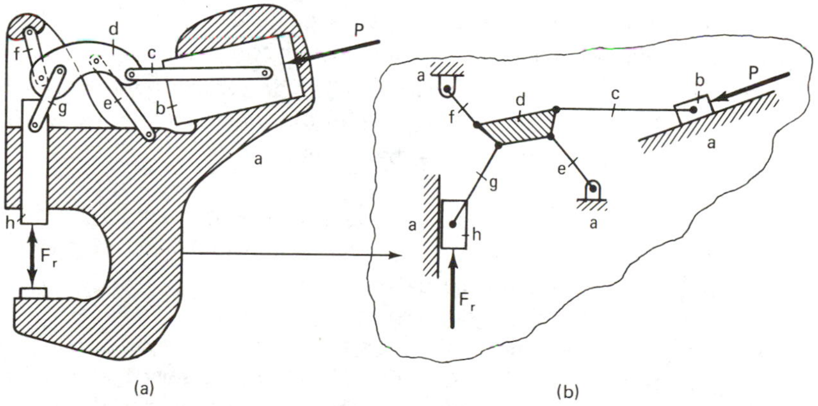


**Figure 2.32** Existing yoke riveter (a) and the equivalent toggle linkage in inset (b).

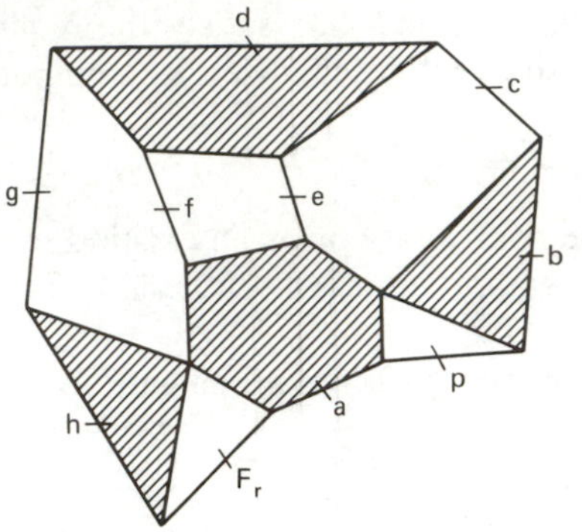


**Figure 2.33** Associated linkage for the existing yoke riveter of Fig. 2.32.

$P$ and $F_r$, connecting $a$ with $b$ and $a$ with $h$, respectively. Note also that the number of pin joints, $f_1$ is 14. Therefore,

$$F = 3(10 - 1) - 2(14) = -1$$

In the synthesis of new configurations of yoke riveters, it will be necessary to reverse the procedure just illustrated in going from Fig. 2.32 to Fig. 2.33. Thus first it will be necessary to select a suitable associated linkage for a new yoke-riveter design. From a careful study of Figs. 2.32 and 2.33, and from a consideration of the desired features of a suitable yoke riveter listed previously, the following rules or requirements for a suitable associated linkage are obtained.

1. $F = -1$.
2. Must have at least two binary links (for $P$ and $F_r$).
3. The binary links corresponding to $P$ and $F_r$ must be connected to the same link at one end, which is the frame link, and to different *ternary* links at their other end. This assures simple construction of the linkage with high force amplification between the rivet die set and the power piston.
4. The frame link must be at least a quaternary link for $P$, $F_r$, and two lower-pair sliding joints for the rivet die and power piston.
5. The different *ternary* links mentioned in requirement 3 must be connected to the frame link, since the power piston and rivet die are to have a lower-pair sliding connection with the frame link.

Since simplicity of construction is a feature of practical importance, the simpler associated linkage in the inset of Fig. 2.34a is a suitable choice. From this associated linkage the simple toggle-type riveter is derived.

The associated linkage method for type synthesis is one of the useful techniques used for synthesizing mechanism *types*. Similar methods of analysis are sometimes employed in patent cases in determining whether a device is of the same or different type than others. Another type synthesis method is described in the appendix to this chapter by way of a case study.

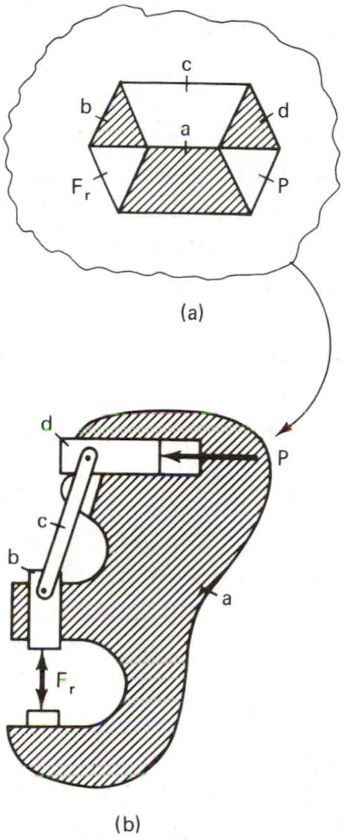


**Figure 2.34** Simple toggle-type riveter; (a) associated linkage; (b) the mechanism derived from (a).

Observe that nothing yet has been said regarding actual dimensions of these type-synthesized mechanisms. The specific dimensions will control the relative motions and the force transmission characteristics of the examples given above.

## 2.4 TOOLS OF DIMENSIONAL SYNTHESIS

The two basic tools of dimensional synthesis are geometric construction and analytical (mathematical) calculation.

*Geometric* or graphical methods of synthesis provide the designer with a fairly quick, straightforward method of design. Graphical techniques do have limitations of accuracy (due to drawing error, which can be very critical) and complexity of solution because, to achieve suitable results, the geometric construction may have to be repeated many times.

*Analytical* methods of synthesis are suitable for computer simulation and have the advantages of accuracy and repeatability. Once a mechanism is modeled mathematically and coded on a computer, mechanism parameters are easily manipulated to create new solutions without further programming. Although this text emphasizes analytical synthesis, it is important to have experience in graphical techniques for use in the initial phases of kinematic synthesis. The next section presents a review of useful geometric approaches before moving on to analytical synthesis.

## 2.5 GRAPHICAL SYNTHESIS—MOTION GENERATION: TWO PRESCRIBED POSITIONS [234]

Suppose that we wish to guide a link in a mechanism in such a way that it will assume several arbitrarily prescribed distinct (finitely separated) positions. For two positions of motion generation, this can be accomplished by a simple rotation (Fig. 2.35) about a suitable center of rotation. This *pole* (see Sec. 4.2), $P_{12}$, is found graphically by way of the *midnormals* $a_{12}$ and $b_{12}$, of two *corresponding* positions each of points $A$ and $B$, namely $A_1$, $A_2$ and $B_1$, $B_2$.

If pole $P_{12}$ happens to fall off the frame of the machine, we may use a four-bar linkage to guide link $AB$ from position 1 to position 2 (Fig. 2.36). Two fixed pivots, one each anywhere along the two midnormals, will accomplish this task. The construction is as follows.

Draw the perpendicular bisector (or midnormal) to $A_1A_2$, the first and second positions of the "circle point"—so named because a circular arc can be drawn through its corresponding positions. Any point along this midnormal, say $A_0$, is a possible fixed pivot or "center point," conjugate to circle point $A$. A link between a center and circle point will guide $A$ from $A_1$ and $A_2$. This construction is now repeated for another circle point, $B$, to yield $B_0$.

Figure 2.36 shows one of the possible four-bar linkages that will act as a motion generator for two positions. Notice that the construction of each circle point/center point pair involved *three free choices:* For two prescribed positions, a circle point $A$ may be chosen anywhere in the plane or its extension, located by two independent coordinates along the $x$ and $y$ axes of a Cartesian system fixed in the moving body,

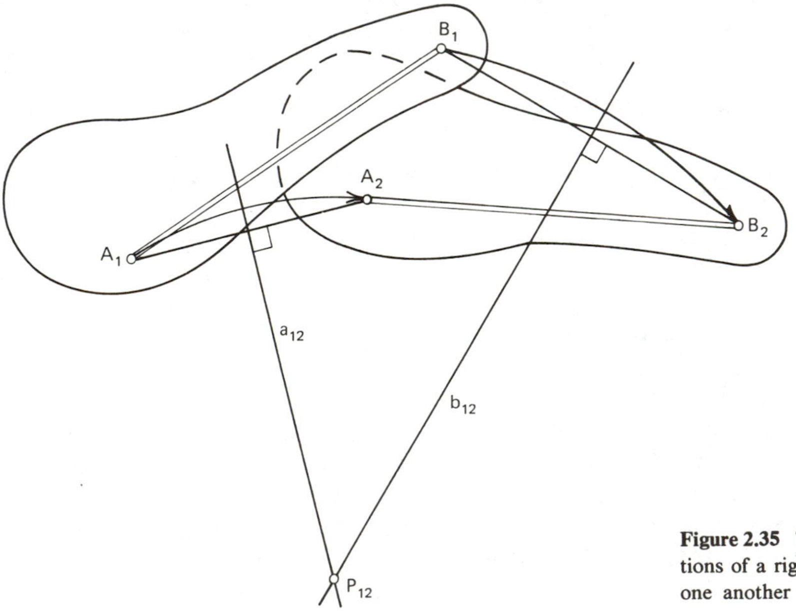


**Figure 2.35** Two prescribed coplanar positions of a rigid body can be reached from one another by rotation about pole $P_{12}$.

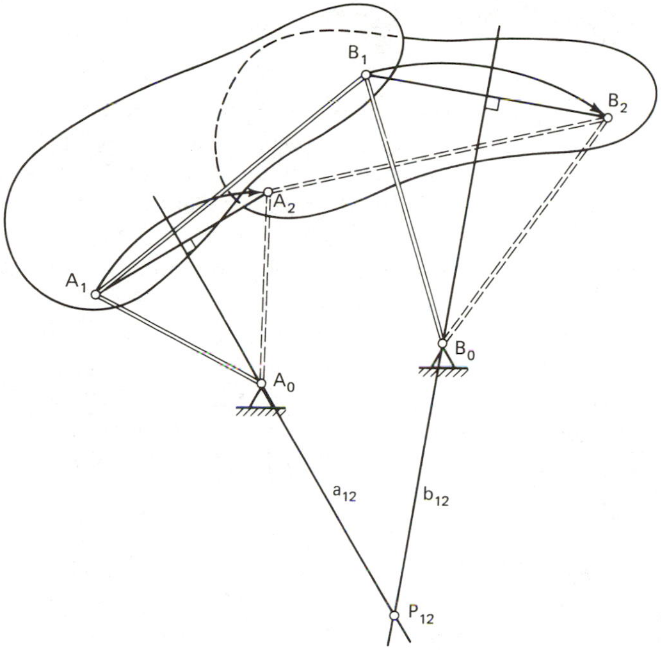


**Figure 2.36** Two-position graphical synthesis of a four-bar motion generator mechanism. Fixed pivots $A_0$ and $B_0$ can be located anywhere along the midnormals between $A_1A_2$ and $B_1B_2$, respectively.

and the conjugate center point may be selected anywhere along the midnormal of its corresponding positions. Thus there are ideally *three* infinites of solutions (for each pair of center and circle points) to build a four-bar linkage. For instance, if the entire midnormal $a_{12}$ represent undesirable locations for fixed pivots, we can rigidly attach point $C$ to $A$ and $B$ by means of a triangle in the plane of the moving (or "floating") link and use $C$ as a crank pin. Figure 2.37 shows the construction yielding an alternative linkage replacing the $A_1A_0$ link of Fig. 2.36 with $C_1C_0$.

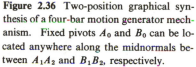

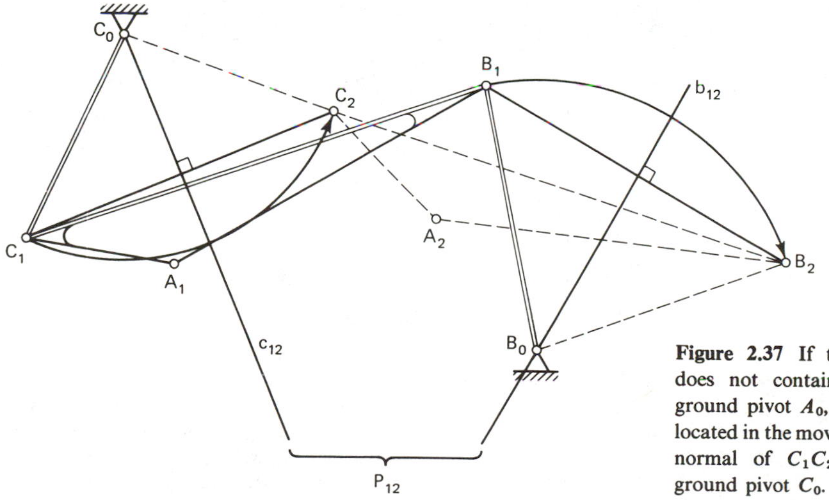


**Figure 2.37** If the midnormal of $A_1A_2$ does not contain suitable locations for ground pivot $A_0$, another point $C$ can be located in the moving body. Then the midnormal of $C_1C_2$ may yield a suitable ground pivot $C_0$.

## 2.6 GRAPHICAL SYNTHESIS—MOTION GENERATION: THREE PRESCRIBED POSITIONS

Let us now consider three arbitrary positions of a plane, $A_1B_1$, $A_2B_2$, and $A_3B_3$ (Fig. 2.38). There will be three poles associated with these positions, $P_{12}$, $P_{23}$, $P_{31}$ (note that $P_{ij} \equiv P_{ji}$). Here the poles can no longer be used as fixed pivots even if they are accessible, because each would lead $AB$ through only two of the three prescribed positions.

Two circle points $A$ and $B$ are chosen and their three corresponding positions are located. The midnormal construction of the preceding section is repeated twice for point $A$ ($a_{12}$ and $a_{23}$). Since the center point for each pair of two positions may lie anywhere along their midnormal, the intersection of the two midnormals locates the common center point $A_0$ for all three positions. Figure 2.38 shows the resulting unique four-bar mechanism synthesized for the choices of circle points $A$ and $B$. Notice that there are, however, *two infinities* of possibilities for each circle point, and thus for each center point/circle point pair.

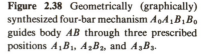

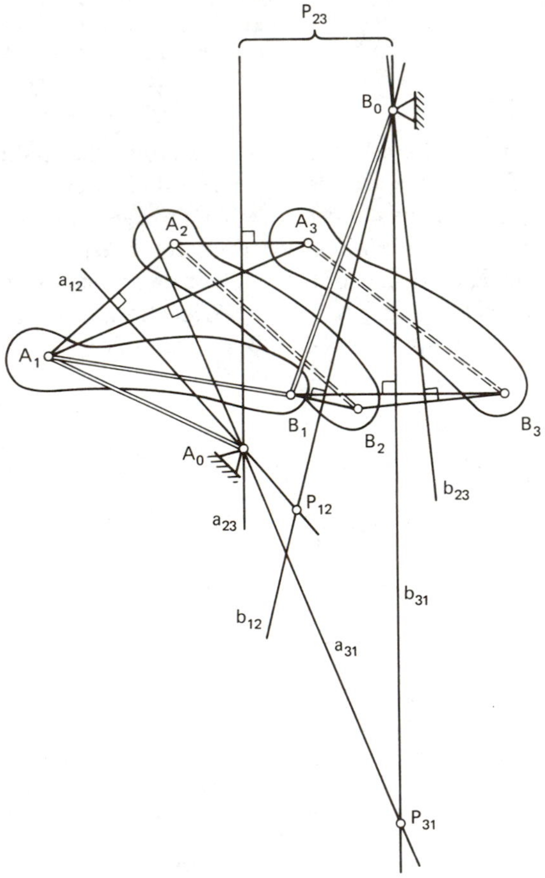


**Figure 2.38** Geometrically (graphically) synthesized four-bar mechanism $A_0A_1B_1B_0$ guides body $AB$ through three prescribed positions $A_1B_1$, $A_2B_2$, and $A_3B_3$.

The following sections illustrate how four-bar path and function generators can be constructed for three positions. The very same technique of intersection of the perpendicular bisectors is used, but only after a kinematic inversion is performed. In path generation the coupler is fixed, in function generation the input crank is fixed, and in path generation with prescribed timing, first the input link is fixed, then the coupler is fixed. The following sections clarify these procedures.

## 2.7 GRAPHICAL SYNTHESIS FOR PATH GENERATION: THREE PRESCRIBED POSITIONS [169]

A very similar construction is involved for graphical synthesis of a four-bar path generator for three positions. Let us design a four-bar mechanism so that a path point $P$ on the coupler link will pass through three selected positions, $P_1$, $P_2$, and $P_3$ (Fig. 2.39).

In designing for three prescribed positions, the positions of $A_0$ and $B_0$ (length and inclination of the fixed link) are free choices. Also, the length of the input crank and the distance between $A$ and $P$ are arbitrary. (As the number of design positions is increased, restrictions are imposed on some of these free choices.) The construction is as follows (Fig. 2.39):

1. Select locations for $A_0$ and $B_0$, establishing the fixed link with respect to prescribed path points $P_1$, $P_2$, and $P_3$.
2. Choose a length for the crank and draw in the path of $A$ (a circle). Pick a point for $A_1$ (position of $A$ for position $P_1$).
3. With $AP$ established, locate $A_2$ and $A_3$. $A$, $P$, and $B$ are all points on the coupler and thus remain the same distance apart in all positions.
4. The position of $B$ is found by means of a kinematic inversion (see Sec. 3.1 of Vol. 1). This is accomplished by fixing the coupler in position 1. The rest of the mechanism, including the frame, must move so that the same relative motion exists between all links. The relative positions of $B_0$ with respect to

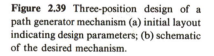

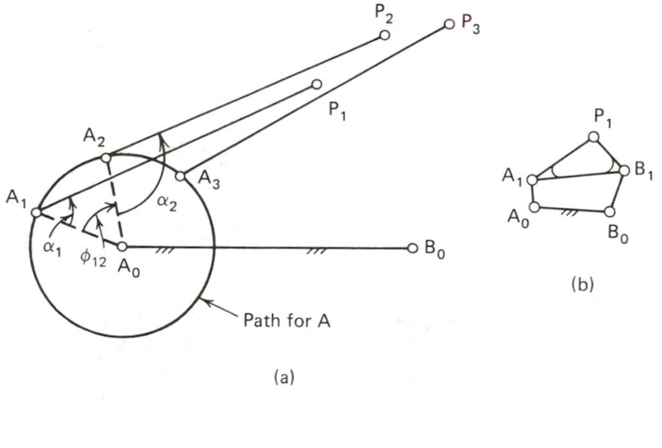


**Figure 2.39** Three-position design of a path generator mechanism (a) initial layout indicating design parameters; (b) schematic of the desired mechanism.

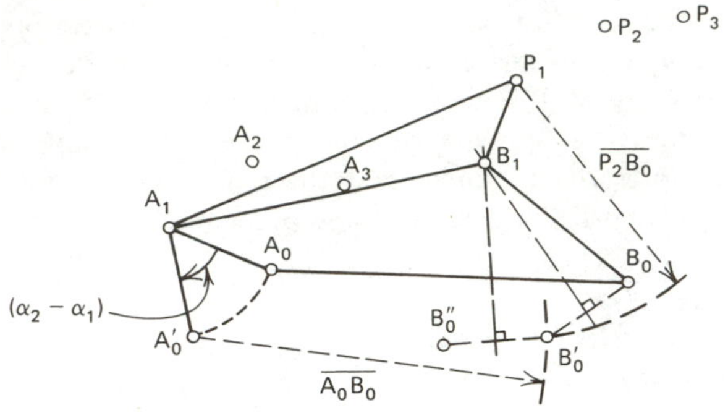


**Figure 2.40** Three-position path-generator design. Inversion to locate $B_1$.

position 1 of the coupler are obtained by the construction shown in Fig. 2.40 as follows (see Figs. 2.39 and 2.40). Rotate $A_0$ about $A_1$ by $(\alpha_2 - \alpha_1)$ (where $\alpha_2 = \measuredangle A_0A_2P_2$ and $\alpha_1 = \measuredangle A_0A_1P_1$ of Fig. 2.39) to $A_0'$. Draw an arc about $A_0'$ with radius $\overline{A_0B_0}$. Draw an arc about $P_1$ with radius $\overline{P_2B_0}$ measured in Fig. 2.39. The intersection of these two arcs locates $B_0'$. The construction of $B_0''$ follows the same procedure with $A_0''$ (rotated about $A_1$ from $A_0$ by $(\alpha_3 - \alpha_1)$) as the center of arc with radius $\overline{A_0B_0}$, and with $\overline{P_3B_0}$ as the radius of a second arc from center $P_1$.

5. Erect perpendicular bisectors to lines $B_0B_0'$ and $B_0'B_0''$. The point of intersection locates $B_1$ as the center of the circle that will pass through the three relative positions of $B_0$: $B_0$, $B_0'$, and $B_0''$.
6. Draw the mechanism in all three positions to check the design (Fig. 2.41). If the design is not satisfactory, these steps can be repeated with different choices for $A_0$, $B_0$, and $A_1$.

Notice that there are ideally six infinities of four-bar linkages that will accomplish this path generation task, since location of $A_0$ ($x$, $y$ coordinates) and the vectors $\overrightarrow{A_0B_0}$ and $\overrightarrow{A_0A_1}$ were arbitrarily chosen in the fixed plane of reference. This is tantamount to *three infinities* of solutions for each side of the linkage for path generation compared with two infinities of solutions for motion generation. If path generation with prescribed timing [i.e., prescribed rotations of the input link ($\phi_{12}$ and $\phi_{13}$) correlated with the path points] is the objective, there are two infinities of solutions for

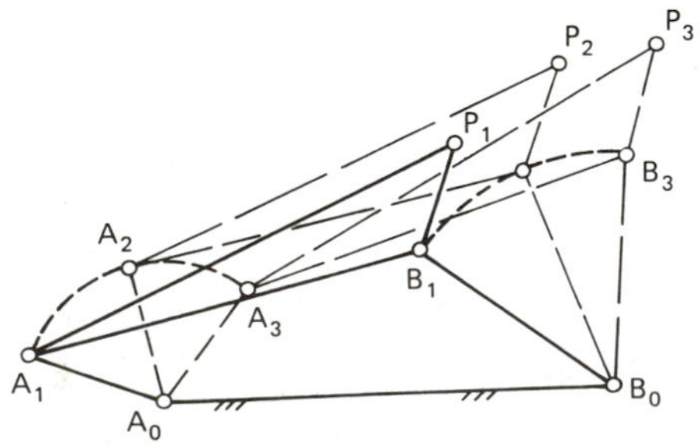


**Figure 2.41** Three-position path-generator design. Checking the completed mechanism.

each side, or a total of four infinities for the four-bar linkages, as shown in the following section.

An important point should be made here that has relevance to all the graphical techniques. In step 5, the interesection of the perpendicular bisectors located $B_1$. Slight error in locating $B_0$, $B_0'$, or $B_0''$ will result in a magnified error in the location of $B_1$. In fact, as lines $B_0B_0'$ and $B_0'B_0''$ become close to being parallel, the error magnification is very large. The designer must be aware of these inherent drawbacks of graphical construction.

## 2.8 PATH GENERATION WITH PRESCRIBED TIMING: THREE PRESCRIBED POSITIONS

The preceding construction must be modified in order to prescribe input crank rotations which are to correspond with the prescribed path positions. The same example will be used as in Fig. 2.39, except that input crank rotations are prescribed: 58° clockwise (cw) corresponding to the movement of point $P$ from $P_1$ to $P_2$ and 108° cw movement from $P_1$ to $P_3$ (see Figure 2.42). The construction, shown in Fig. 2.43, is as follows:

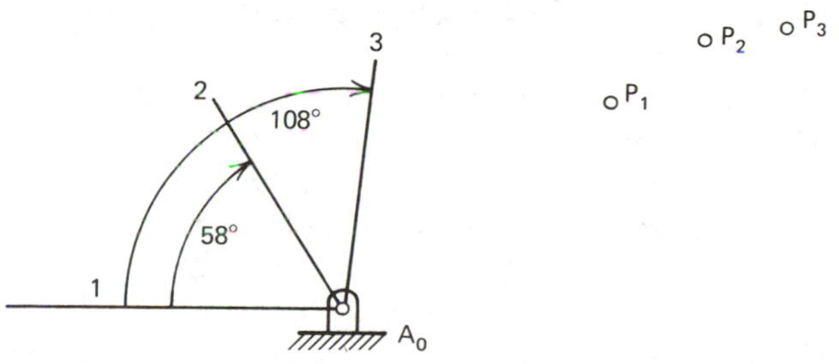


**Figure 2.42** Prescribed path points and crank rotations for path generation with prescribed timing with three finitely separated precision points.

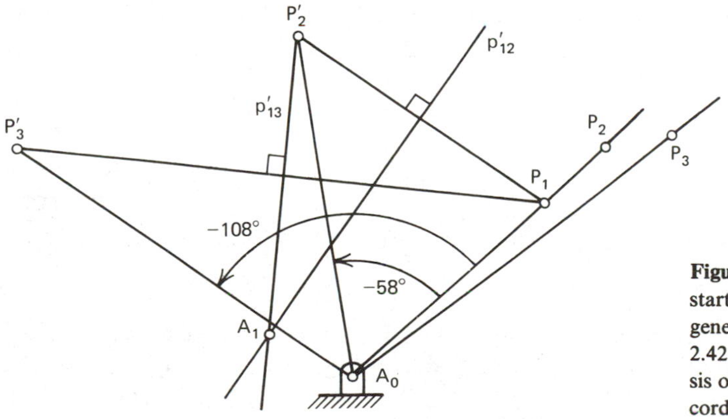


**Figure 2.43** Graphical construction of the starting position of crank $A_0A_1$ for the path generator with the prescribed data of Fig. 2.42. Completion of the geometric synthesis of the four-bar mechanism proceeds according to Figs. 2.39 and 2.40.

1. Pick the fixed pivot of the input link ($A_0$) with respect to the prescribed path precision points $P_1P_2P_3$ (two infinities of choices, one for $x$ and one for $y$ of $A_0$).
2. Draw lines $\overline{P_2A_0}$ and $\overline{P_3A_0}$.
3. Inverting the motion (by fixing the yet unknown input link $A_0A$), rotate $\overline{P_2A_2}$ 58° counterclockwise (ccw) about $A_0$ and $\overline{P_3A_0}$ 108° ccw around $A_0$ locating $P'_2$ and $P'_3$.
4. Draw lines $\overline{P'_2P_1}$ and $\overline{P'_3P_1}$.
5. The intersection of the perpendicular bisectors $p'_{12}$ and $p'_{13}$ locates $A_1$, the first position of $A$.
6. The rest of the construction is found as illustrated in the preceding section. Thus path generation with prescribed timing involves two free choices for the left side of the four-bar (the $x$ and $y$ location of $A_0$ with respect to $P_1$) and therefore ideally yields *two infinities* of solutions.

## 2.9 GRAPHICAL SYNTHESIS FOR PATH GENERATION (WITHOUT PRESCRIBED TIMING): FOUR POSITIONS

A design procedure similar to that of Fig. 2.40 may be employed for path generation (without prescribed timing) for four precision points using the *point-position reduction method* [119,169,268].

The point-position reduction method is based on the fact that a circle can be drawn through three points. Three different relative positions for a point on a link are determined, then a circle is drawn through the points. The center and radius of the circle determine the position and lengths of the remaining links of the mechanism. (Up to six precision points [169] can be satisfied in this method. However, the design parameters are chosen so that some corresponding positions of a design point, usually a pin joint, coincide and thereby the total number of distinct positions is reduced to three.) This is demonstrated in the following. Designs 1 and 2, in which the number of distinct positions is reduced from four to three. This is accomplished by locating either point $B_0$ or $B$ at one of the poles of the coupler. Designs will be presented first with $B_0$ and then with $B$ at the pole.

### Design 1

**The task.** Design a four-bar mechanism such that the coupler point $P$ will pass through four arbitrarily selected positions in the order $P_1$, $P_2$, $P_3$, $P_4$, (Fig. 2.44). Locate the fixed pivot $B_0$ at one of the poles of the coupler motion. *The procedure* is as follows.

1. Choose two positions to make coincident in the inversion. Positions 1 and 4 were picked so that $B_0$ is positioned at pole $P_{14}$. The pole is located on the perpendicular bisector of the line $P_1P_4$ (any convenient point on this line will

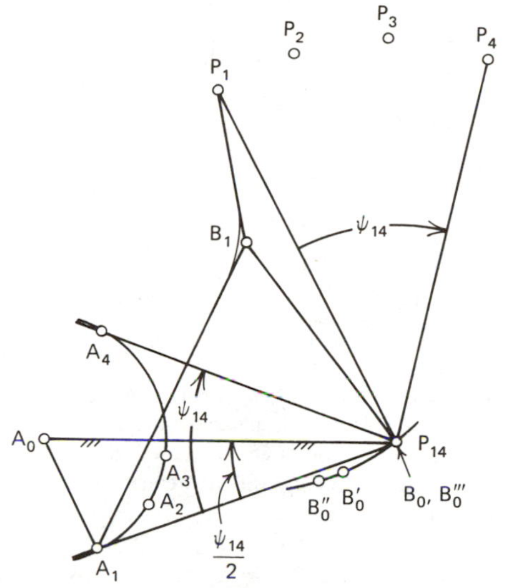


**Figure 2.44** Four-position design. Layout showing parameters and design procedure. Pivot at pole.

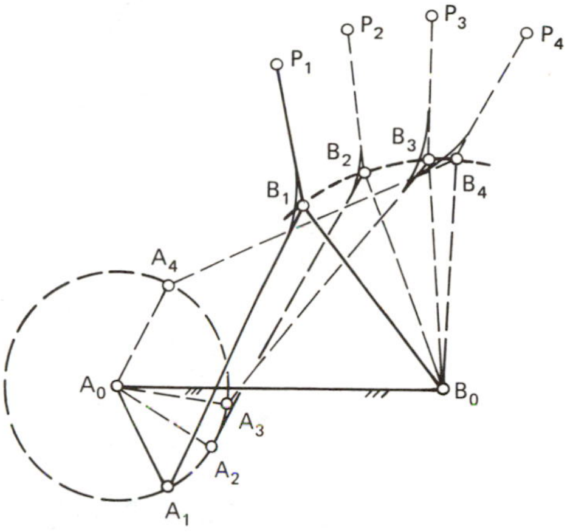


**Figure 2.45** Four-position design. Check of completed mechanisms.

do). This determines the angle $\psi_{14}$, rotation of the follower link from position 1 to 4.

2. Since $B_0$ is at the pole $P_{14}$, the coupler can be rotated about $B_0$ from position 1 to position 4. This means that $A$ and $B$, both points on the coupler, must also rotate the same angle $\psi_{14}$.
3. Select some direction for $A_0B_0$ and draw two lines through $B_0$ at angles $\pm \psi_{14}/2$ from $B_0A_0$ (Fig. 2.44). $A_1$ and $A_4$ must lie on these lines equidistant from $B_0$.
4. Choose positions for $A_1$ and $A_0$. This establishes $A_0$ and the lengths of the fixed and input links and the distance $AP$.
5. Locate $A_2$ and $A_3$ on the arc about $A_0$ with radius $A_0A_1 = A_0A_4$, such that $P_2A_2 = P_3A_3 = P_1A_1$.
6. $B_0$ and $B_0'''$ are located at $P_{14}$. Fix the coupler (a kinematic inversion) and locate the relative position of $B_0$ for positions 2 and 3 ($B_0'$, $B_0''$) by constructing $\Delta A_1P_1B_0' = \Delta A_2P_2B_0$ and $\Delta A_1P_1B_0'' = \Delta A_3P_3B_0$. The center of the circle that passes through $B_0$, $B_0'$, and $B_0''$ is $B_1$. This establishes the lengths of coupler and output links and completes the design.
7. Figure 2.45 shows the mechanism in all four positions as a check on the design.

## Design 2

**The task.** Design a four-bar mechanism such that the coupler point $P$ will pass through the prescribed positions $P_1$, $P_2$, $P_3$, and $P_4$ in that order (Fig. 2.46). Locate the coupler point $B$ at one coupler pole. *The procedure* is as follows:

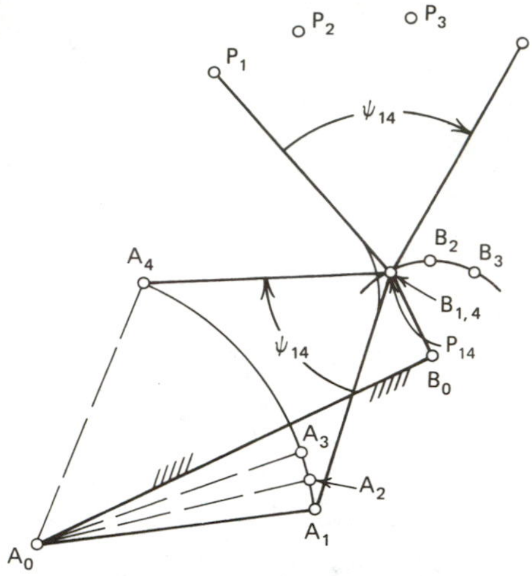


**Figure 2.46** Four-position geometric synthesis of four-link path generator mechanism. First position of coupler point, $B_1$, at pole $P_{14}$ (point-position reduction method).

1. Locate the pole $P_{14}$ on the perpendicular bisector of the line $P_1P_4$ arbitrarily. Let $B_1$ and $B_4$ be collocated with $P_{14}$. Angle $P_1P_{14}P_4 = \psi_{14}$.
2. Since the coupler triangle $ABP$ is rigid, the angle $A_1B_1P_1$ must equal angle $A_4B_4P_4$. With $B_1$ and $P_1$ located, a line can be drawn from $B_1$ in an arbitrary direction to establish a locus for $A_1$. The distance $B_1A_1$ is arbitrary.
3. Locate $A_4$ so that angle $A_1B_1A_4 = \psi_{14}$ in both magnitude and sense and $A_4B_1 = A_1B_1$.
4. Select the pivot $A_0$ for the input link on the bisector of angle $A_1B_1A_4$. Thus $A_0A_1 = A_0A_4$. Draw the circular arc path of $A$ from $A_1$ to $A_4$.
5. Locate $A_2$ so that $A_2P_2 = A_1P_1$ and $A_3$ so that $A_3P_3 = A_1P_1$.
6. $\Delta A_1B_1P_1 = \Delta A_2B_2P_2 = \Delta A_3B_3P_3 = \Delta A_4B_4P_4$. Use this information to locate $B_2$ and $B_3$.
7. Since $B_1$ and $B_4$ are collocated, a circle that passes through $B_1$, $B_4$, $B_2$, and $B_3$ can be drawn. The center of this circle is the fixed pivot $B_0$. The radius is the length of the output link $B_0B$. This establishes the mechanism.

These two designs show how the pole is used in reducing the number of four point positions to three. The graphical procedure is somewhat simpler when the coupler point $B$ is at the pole than when the pivot $B_0$ is at the pole. The design situation may dictate which to use.

Notice that each of these designs involved choosing four parameters (e.g., in design 1 we pick arbitrarily the position of $B_0$ along the perpendicular bisector of $P_1P_4$, the $x$ and $y$ coordinate of $A_0$, and the radius $A_0A$). Thus there are *two infinities* of solutions per side for path generation for four prescribed positions. If path generation with prescribed timing (i.e., rotations of $A_0A_1$, $A_0A_2$, $A_0A_3$, and $A_0A_4$) were the objective, there would be *one infinity* of solutions per side. Lindholm

[168,169] has also presented the graphical procedures for five and six prescribed path positions using point-position reduction procedures.

## 2.10 FUNCTION GENERATOR: THREE PRECISION POINTS

The graphical procedure for three-precision-point function generation is very similar to that of motion and path generation for the same number of precision points. Again, kinematic inversion and the intersection of midnormals is used. An illustrative example [61] will be employed to demonstrate the method.

A mechanism will be synthesized to generate the function $y = \sin(x)$ for $0° \leq x \leq 90°$. The input range is chosen arbitrarily to be $\Delta\phi = 120°$ and the output range is similarly chosen to be $\Delta\psi = 60°$. For this case the scale factors $R_\phi$ and $R_\psi$ are found to be

$$R_\phi = \frac{\Delta\phi}{\Delta x} = \frac{120°}{90°} = \frac{4}{3} \tag{2.7}$$

$$R_\psi = \frac{\Delta\psi}{\Delta y} = \frac{60°}{1} = 60° \tag{2.8}$$

The next task is to pick three precision points, $x_1$, $x_2$, and $x_3$. Chebyshev spacing (discussed previously in this chapter) is used for these precision points. Referring to Fig. 2.47, we find that

$$x_0 = 0°, \qquad x_3 = 84°$$
$$x_1 = 6°, \qquad x_4 = 90°$$
$$x_2 = 45°,$$

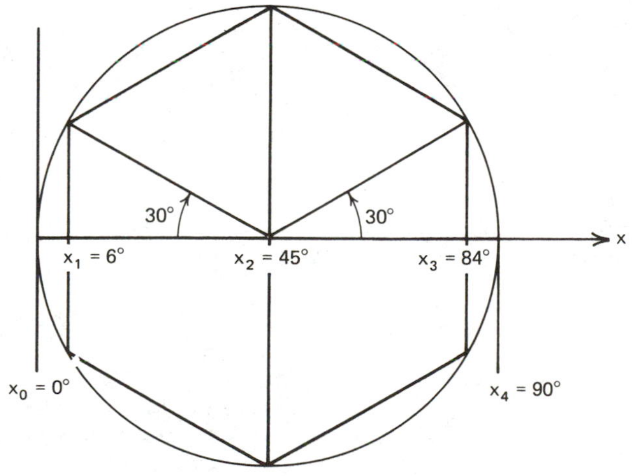


**Figure 2.47** Graphical determination of three precision points with Chebyshev spacing.

Recall from Eqs. (2.3) and (2.4) that

$$\phi_j = R_\phi (x_j - x_1)$$

$$\psi_j = R_\psi (y_j - y_1)$$

so that

$$\phi_2 = \tfrac{4}{3}(45 - 6) = 52°, \qquad \psi_2 = 60(0.7071 - 0.1045) = 36.15°$$

$$\phi_3 = \tfrac{4}{3}(84 - 6) = 104°, \qquad \psi_3 = 60(0.9945 - 0.1045) = 53.40°$$

See Fig. 2.48 for a geometric interpretation of the function-generation synthesis task. The graphical construction procedure is as follows (refer to Figs. 2.48 and 2.49).

1. Pick the position of the ground pivots ($A_O$, $B_O$) and the output link ($B_OB$). Here the ground pivots are along the $x$ axis, the length of the fixed link $A_OB_O$ = 1 unit, the length of the output link $B_OB$ = 0.75 unit, and $\psi_0 = 60°$. Notice that the initial position of the output link is therefore $\psi_0 + 60°(0.1045 - 0) = 66.27°$ (see Fig. 2.48).
2. Using inversion, fix the input link (although the position of the input link is unknown). The mechanism will now be moved through the specified precision points preserving the same relative motion between links. Therefore, in the second precision position the fixed link rotates by $-\phi_2 = -52°$ about $A_0$, locating $B'_0$, while the output rotates by $\psi_2 = 36.15°$ about $B'_0$ locating $B'$. The third precision-point position may be generated by rotations (from the first position)

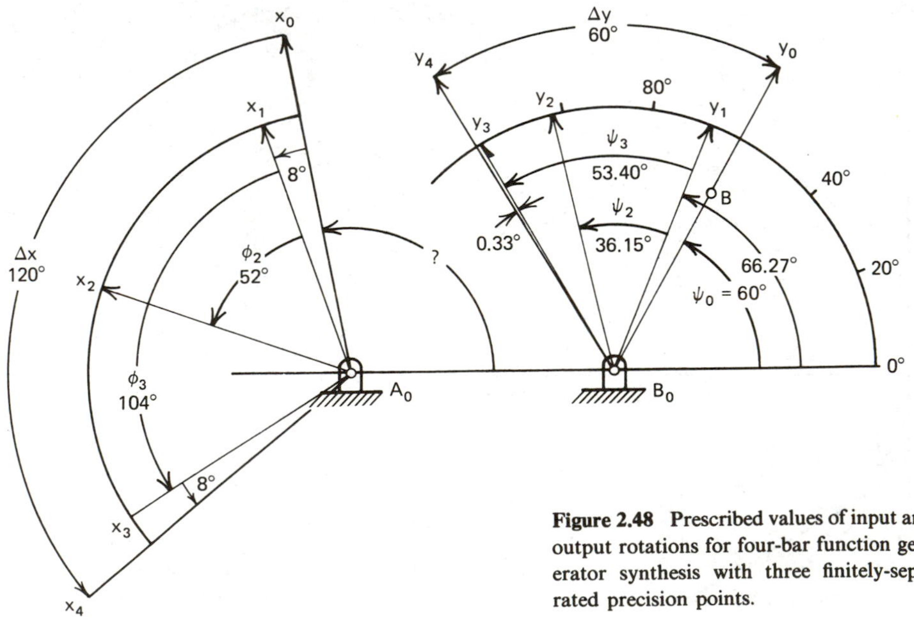


**Figure 2.48** Prescribed values of input and output rotations for four-bar function generator synthesis with three finitely-separated precision points.

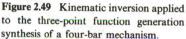

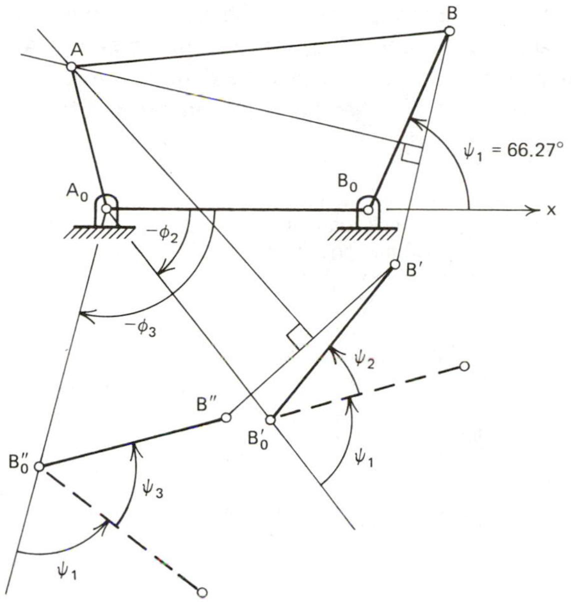


**Figure 2.49** Kinematic inversion applied to the three-point function generation synthesis of a four-bar mechanism.

of −104° for the fixed link about $A_0$, locating $B_0''$, and 53.40° for the output link about $B_0''$, locating $B''$.

3. Lines $B_0B$, $B_0'B'$, and $B_0''B''$ represent the actual precision positions of the output link relative to the input link. The center of the circular arc $B - B' - B''$ will locate $A$, found by intersection of the perpendicular bisectors of $BB'$ and $B''B'$ (see Fig. 2.49).

*Two infinities* of solutions are available for each side of the four-bar for function generation for three prescribed finitely separated positions, since both the positions of $A_0$ and $B$ relative to $B_0$ (four parameters) and thus the ground and output links, were picked arbitrarily in the construction.

Before moving on to analytical methods, another popular function-generation technique, the overlay method will be described.

## 2.11 THE OVERLAY METHOD

Another graphical method often used for kinematic synthesis (primarily for function generation) is the overlay technique. The technique consists basically of constructing a part of the solution to a problem on transparent paper and another part of the solution on a separate sheet. The transparency ("overlay") is placed over the separate sheet and a search is made by moving the transparency until precision points are matched between the transparency and the separate sheet.

The technique can be used for the synthesis of mechanisms involving two to five positions although the solution procedure is more difficult as the number of required precision points increases. The method will be demonstrated by way of a five-precision-point design [61,168]. A four-bar function generator is to be designed for the following precision points:

| Precision point number | Crank rotation from starting position (deg) | |
|---|---|---|
| | Input (cw) | Output (cw) |
| 1 | 0 | 0 |
| 2 | $\phi_2 = 15°$ | $\psi_2 = 20°$ |
| 3 | $\phi_3 = 30$ | $\psi_3 = 35$ |
| 4 | $\phi_4 = 45$ | $\psi_4 = 50$ |
| 5 | $\phi_5 = 60$ | $\psi_5 = 60$ |

### Method

1. On tracing paper lay out the input crank positions and select lengths for the input and coupler links (see Fig. 2.50). Draw a family of circular arcs with centers at successive crank pin position with a radius equal to the arbitrarily chosen coupler length.
2. On a second piece of paper (Fig. 2.51) lay out the output crank positions and add several arcs, indicating possible lengths of link 4.
3. Place the first layout on the second and move until the family of arcs of Fig. 2.50 falls on the respective positions of the output crank as shown in Fig. 2.52. This establishes the lengths of the ground link and the output link.

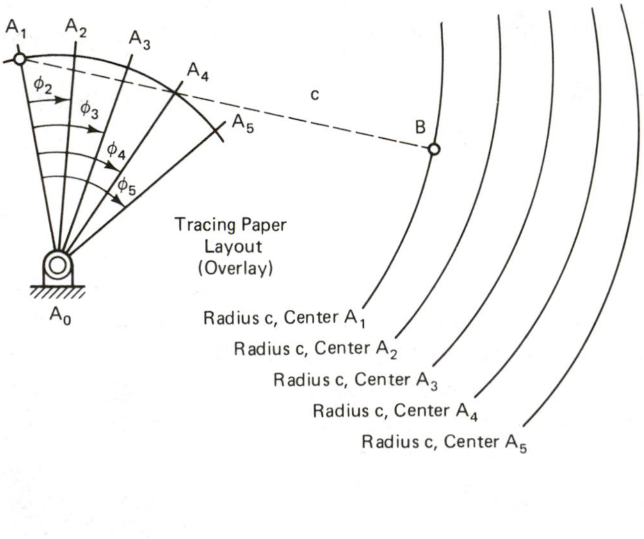


**Figure 2.50** Overlay technique. A five-position design. Input crank and connecting rod side.

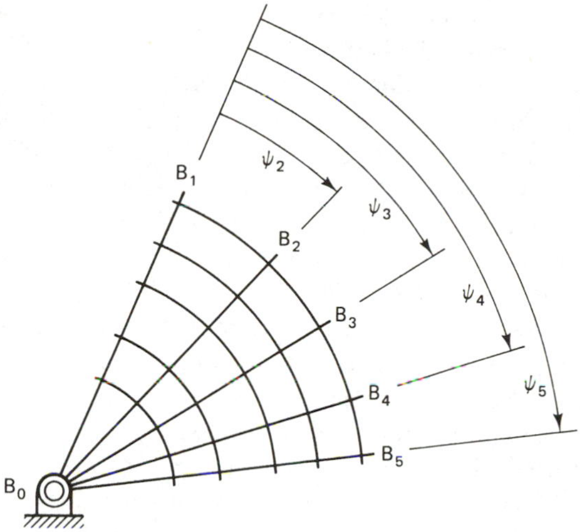


**Figure 2.51** Five-position design. Layout of output crank possibilities.

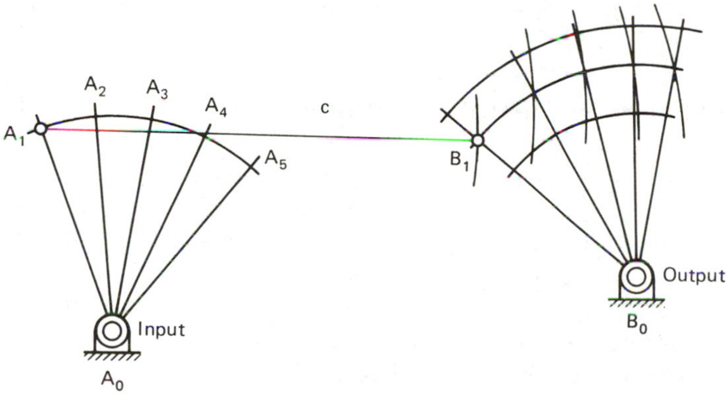


**Figure 2.52** Five-position overlay technique design. Fitting of the overlay and the resulting mechanism.

It may be necessary to try different lengths for the coupler link in order to achieve a match between the overlay of Fig. 2.50 and the layout of Fig. 2.51. With practice, this method should be accurate within 1°.

Notice that the scale factors ($\Delta\phi$, $\Delta\psi$) and the coupler length are free choices here. There is no guaranteed solution, however. This is not a closed-form solution; it is a trial-and-error technique.

## 2.12 ANALYTICAL SYNTHESIS TECHNIQUES

Figures 2.36 to 2.38 show that the geometric construction of four-bar motion generators for two and three prescribed positions is a fairly simple task. Suppose, however, that we wish to find an "optimal" four-bar motion generator for a specific application—

perhaps a case that has constraints on ground and moving pivot locations, transmission angle, link-length ratio, and/or mechanical advantage. The construction of Fig. 2.38, although simple, may be too time consuming to repeat until a suitable solution is obtained. A graphical search through two infinities of solutions is inconceivable. What other alternatives are available? By choosing the position of the circle point $A_1$ in Fig. 2.38, we have arbitrarily picked two free choices—those free choices in turn specify the corresponding center point $A_0$. These two free choices for the three-precision-point motion-generation synthesis of one side of the four-bar linkage can be picked with different strategies in mind toward various design objectives.

In order to obtain a handle on the design variables and free choices, an analytical model of the linkage must be developed. Several mathematical techniques for modeling linkages have been utilized for planar synthesis objectives. These include algebraic methods, matrix methods, and complex numbers. For planar linkages, the complex-number technique is the simplest, yet the most versatile method. In this text we therefore concentrate on the latter method. Before exploring the question of free choices versus synthesis options, the complex-number technique will be reviewed,* especially as it relates to modeling linkages for synthesis.

## 2.13 COMPLEX-NUMBER MODELING IN KINEMATIC SYNTHESIS

Any planar mechanism can be represented by a general chain, consisting of one or more loops of successive bar-slider members (Fig. 2.53). For example, the offset slider-crank mechanism of Fig. 2.54a may be derived from the general chain (Fig. 2.54b) by fixing the sliders to their respective bars between members 1 and 4, 4 and 3, and 3 and 2 as well as fixing bars 1 and 4 to ground.

Complex numbers readily lend themselves as an ideal tool for modeling linkage members as parts of planar chains. For each bar-slider member of Fig. 2.53, the position of the pivot on the slider with respect to the pivot of the bar can be defined by the relative position vector $\mathbf{Z}_k$ (Fig. 2.55a) expressible as a complex number. The first or starting position of the $k$th bar can be written as

$$\mathbf{Z}_k = Z_k e^{i\theta_1} = Z_k(\cos\theta_1 + i\sin\theta_1) \tag{2.9}$$

where $i \equiv \sqrt{-1}$

$k = k$th bar of the chain

$Z_k = |\mathbf{Z}_k| =$ length between the pivot of the bar and the pivot on the slider in the first position

$\theta_1 = \arg \mathbf{Z}_k =$ angle measured to vector $\mathbf{Z}_k$ from the real axis of a fixedly oriented rectangular coordinate system translating with the pivot of the bar (counterclockwise rotations are positive)

* A more complete review of complex numbers is given in the appendix to Chap. 3 of Vol. 1.

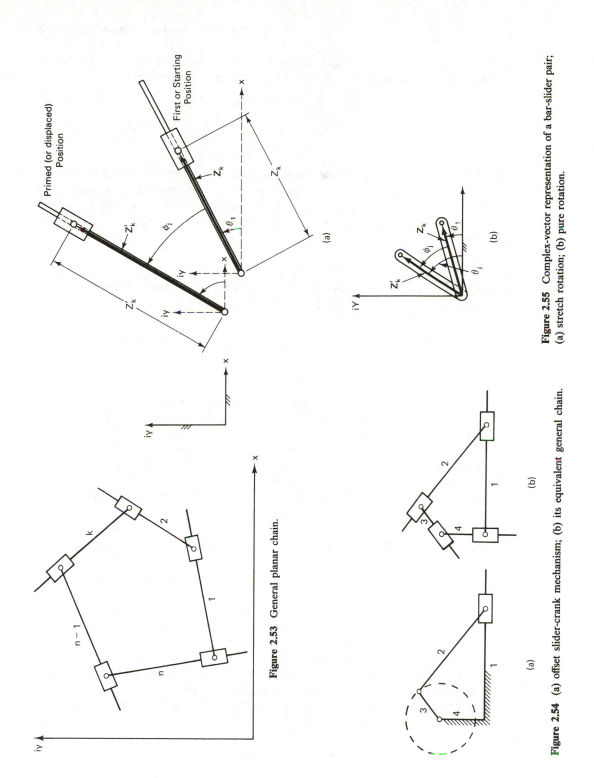

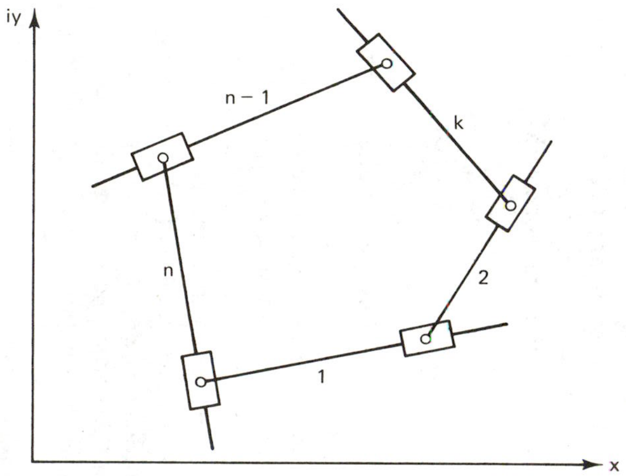


Figure 2.53 General planar chain.

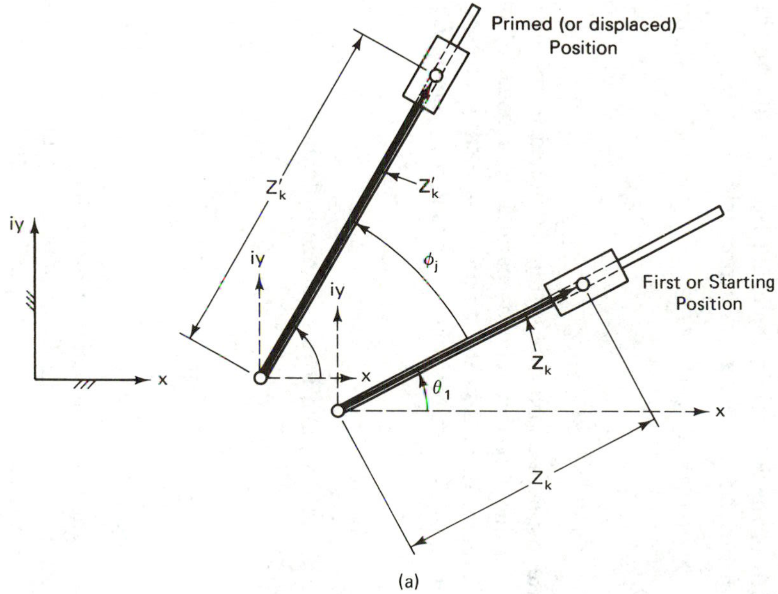


Figure 2.55 Complex-vector representation of a bar-slider pair; (a) stretch rotation; (b) pure rotation.

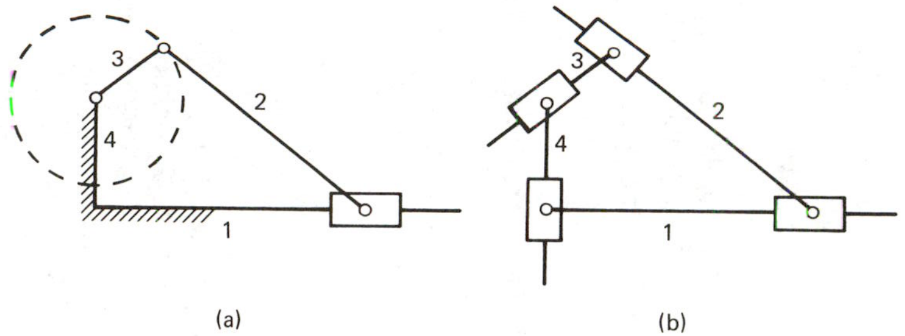


Figure 2.54 (a) offset slider-crank mechanism; (b) its equivalent general chain.

If there is no change in the length of the $k$th bar in the chain from the first to the primed ($j$th) position as shown in Fig. 2.55b, then $\mathbf{Z}'_k$ is expressible as

$$\mathbf{Z}'_k = Z_k e^{i(\theta_1+\phi_j)} = Z_k e^{i\theta_1} e^{i\phi_j} \tag{2.10}$$

where

$$\phi_j = \theta_j - \theta_1 \tag{2.11}$$

Notice that as a link moves in the plane, a coordinate system is pinned to the base of the link (Fig. 2.55a). This coordinate system remains parallel to a fixed set of coordinates so that $\theta_j$ and $\theta_1$ are arguments of $\mathbf{Z}$ in the $j$th and first positions respectively, while angle $\phi_j$ is the rotation of $\mathbf{Z}$ from position 1 to position $j$. Using Eq. (2.9) yields

$$\mathbf{Z}'_k = \mathbf{Z}_k e^{i\phi_j} \tag{2.12}$$

If there is a change in length of the $k$th bar, and if this change is defined by

$$\rho_j \equiv \frac{Z'_k}{Z_k} \tag{2.13}$$

then

$$\mathbf{Z}'_k = \mathbf{Z}_k \rho_j e^{i\phi_j} \tag{2.14}$$

$e^{i\phi_j}$ in Eqs. (2.12) and (2.14) is termed the *rotational operator* [233,237] and will rotate a vector from its initial position by the angle $\phi_j$ without changing the length of the vector. The factor $\rho_j$ is the *stretch ratio,** while $\rho_j e^{i\phi_j}$ is called the *stretch rotation operator* [233,237]. We may now model any bar-slider member in a planar mechanism by a vector and express its motion with respect to any reference in terms of an initial position, a stretch, and a rotation. How can we collect the links of the mechanism into one model and develop some equations to work with?

## 2.14 THE DYAD OR STANDARD FORM

The great majority of planar linkages may be thought of as combinations of vector pairs called *dyads.* For example, the four-bar linkage in Fig. 2.56 can be perceived as two dyads: the left side of the linkage represented as a vector pair (**W** and **Z**) shown in solid lines, and the right side represented by the dashed dyad (**W*** and **Z***). The vectors that represent the coupler $\overrightarrow{AB}$ and the ground link $\overrightarrow{A_0B_0}$ are easily determined by vector addition when these dyads are synthesized [see Eqs. (2.25) and (2.26)]. The path point of the coupler link moves along a path from position $P_1$ to $P_j$ defined in an arbitrary complex coordinate system by $\mathbf{R}_1$ and $\mathbf{R}_j$.

All vector rotations are measured from the starting position, positive counterclockwise (Figs. 2.56 and 2.57). Angle $\beta_2$ is the rotation of vector **W** from the

* See, for example, the slider crank of Fig. 2.81 and Eq. (2.83).

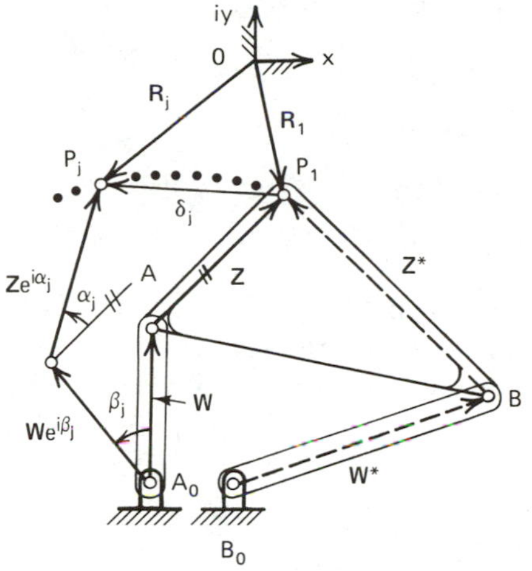


**Figure 2.56** Notation associated with a dyad shown as it would model the left half of a four-bar linkage. The dyad ($W$ and $Z$) is drawn in its first and $j$th positions.

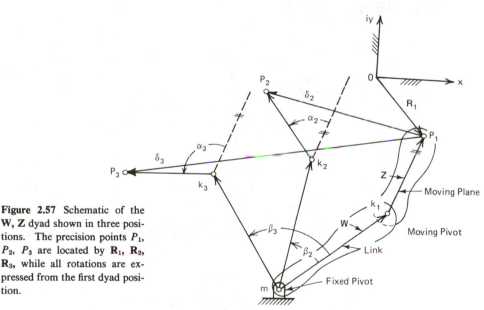

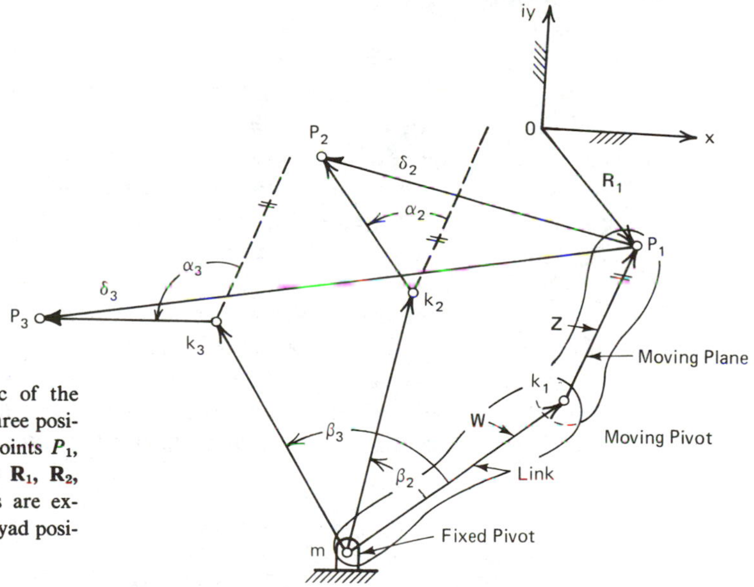


**Figure 2.57** Schematic of the **W, Z** dyad shown in three positions. The precision points $P_1$, $P_2$, $P_3$ are located by $\mathbf{R}_1$, $\mathbf{R}_2$, $\mathbf{R}_3$, while all rotations are expressed from the first dyad position.

first to the second position, while $\beta_3$ is the rotation from the first to the third position. Similarly, angles $\alpha_j$ are rotations of the vector **Z** from its first to its $j$th position (see Fig. 2.57).

Suppose that we specify two positions for an unknown dyad by prescribing the values of $\mathbf{R}_1$, $\mathbf{R}_j$, $\alpha_j$, and $\beta_j$ (Fig. 2.56). To find the unknown starting position vectors of the dyad, **W** and **Z**, a loop-closure equation may be derived by summing the vectors clockwise around the loop containing $\mathbf{W}e^{i\beta_j}$, $\mathbf{Z}e^{i\alpha_j}$, $\mathbf{R}_j$, $\mathbf{R}_1$, **Z**, and **W**:

$$\mathbf{W}e^{i\beta_j} + \mathbf{Z}e^{i\alpha_j} - \mathbf{R}_j + \mathbf{R}_1 - \mathbf{Z} - \mathbf{W} = 0 \tag{2.15}$$

or

$$\mathbf{W}(e^{i\beta_j} - 1) + \mathbf{Z}(e^{i\alpha_j} - 1) = \boldsymbol{\delta}_j \tag{2.16}$$

where the displacement vector along the prescribed trajectory from $P_1$ to $P_j$ is

$$\boldsymbol{\delta}_j \equiv \mathbf{R}_j - \mathbf{R}_1 \tag{2.17}$$

Equation (2.16) is the *standard-form* equation. This equation is simply the vector sum around the loop containing the first and $j$th positions of the dyad forming the left side of the four-bar linkage. As we will see, Eq. (2.16) is called the standard form if $\boldsymbol{\delta}_j$ and either $\alpha_j$ or $\beta_j$, are prescribed or known. This requirement is consistent with the definitions of the usual tasks of kinematic synthesis: motion generation, path generation with prescribed timing, and function generation.

## 2.15 NUMBER OF PRESCRIBED POSITIONS VERSUS NUMBER OF FREE CHOICES

For how many positions can we synthesize a four-bar linkage for motion, path, or function generation? A finite number of parameters (the two components of each vector) completely describe this linkage in its starting position. Therefore, there are only a finite number of prescribable parameters which can be imposed in a synthesis effort. The four-bar motion generator will be used to determine how many positions may actually be prescribed. In Fig. 2.56, the path displacement vectors $\boldsymbol{\delta}_j$ and coupler rotations $\boldsymbol{\alpha}_j$ will be prescribed in a *motion-generation task.*

Table 2.1 illustrates how to determine the maximum number of prescribable positions for the synthesis of a four-bar motion generator. Although Table 2.1 is based on the left side of the linkage of Fig. 2.56 [Eq. (2.16)], the right side of the linkage will yield the same results [see Eq. (2.24)]. The table shows that for two positions there are two independent scalar equations contained in the vector equation Eq. (2.16): the summation of $x$ components and the summation of the $y$ components

**TABLE 2.1** MAXIMUM NUMBER OF SOLUTIONS FOR THE UNKNOWN DYAD **W, Z** WHEN $\boldsymbol{\delta}_j$ AND $\alpha_j$ ARE PRESCRIBED IN THE EQUATION:

$$\mathbf{W}(e^{i\beta_j} - 1) + \mathbf{Z}(e^{i\alpha_j} - 1) = \boldsymbol{\delta}_j \tag{2.16}$$

| Number of positions ($n$): $j = 2, 3, \ldots, n$ | Number of scalar equations | Number of scalar unknowns | Number of free choices (scalars) | Number of solutions |
|---|---|---|---|---|
| 2 | 2 | 5(**W, Z**, $\beta_2$) | 3 | $O(\infty^3)$ |
| 3 | 4 | 6(above + $\beta_3$) | 2 | $O(\infty^2)$ |
| 4 | 6 | 7(above + $\beta_4$) | 1 | $O(\infty^1)$ |
| 5 | 8 | 8(above + $\beta_5$) | 0 | Finite |

of the vectors. These are called the *real* and *imaginary* parts of the equation, each a scalar equation in itself. This system of two scalar equations contains five scalar unknowns: two coordinates each of the vectors **W** and **Z** ($W_x$, $W_y$, $Z_x$, and $Z_y$) and the input rotation $\beta_2$. If three of the five unknowns are chosen arbitrarily, the equations can be solved for the remaining two unknowns. Since in general there is an infinite number of choices for each of the three free choices, the number of possible solutions for the two-position synthesis problem is on the order of *infinity cubed,* symbolized by $O(\infty^3)$.

In the case of three prescribed positions of the moving plane, specified by three precision points $P_1$, $P_2$, and $P_3$ and two angles of rotation, $\alpha_2$ and $\alpha_3$, there are two more real equations but only one more scalar unknown ($\beta_3$). Thus two free choices can be made and $O(\infty)^2$ solutions are available. Each additional prescribed position in Table 2.1 adds two scalar equations and one scalar unknown. Thus, for four positions, there is one free choice and a single infinity of solutions. For five prescribed positions there are no free choices available, and at best a finite number of solutions will exist (see Chap. 3). Five prescribed positions is therefore the maximum number of precision points possible for the standard-form solutions for the motion-generation dyad of Fig. 2.56.

In review, Table 2.1 correlates the number of prescribed positions, the number of free choices, and the number of closed-form solutions expected for the standard form. However, Table 2.1 does not say anything directly about the difficulty in solving the sets of standard-form equations in closed form. An important question is this: Can a linear equation-solver technique be applied for two, three, four, and five prescribed positions?

The answer lies in the form of the respective sets of equations of synthesis: Are they linear or nonlinear in the unknown reals? A nonlinearity test will be applied to Eq. (2.16) for each row of Table 2.1.

**Two positions.** There are three free choices to be made in Eq. (2.16). For example, if $\boldsymbol{\delta}_2$ and $\alpha_2$ were prescribed, **Z** and $\beta_2$ could be chosen arbitrarily, yielding a simple linear solution for the remaining unknown, **W**:

$$\mathbf{W} = \frac{\boldsymbol{\delta}_2 - \mathbf{Z}(e^{i\alpha_2} - 1)}{e^{i\beta_2} - 1} \tag{2.18}$$

This case of motion generation for two positions for a dyad is analogous to the graphical technique described in Sec. 2.5. *In both cases there are three infinities of solutions.*

**Three positions.** Here, according to Table 2.1, two free choices must be made, so one may expect *two infinities of solutions,* as in the graphical method (Sec. 2.6). The system of equations for three positions is

$$\begin{aligned} \mathbf{W}(e^{i\beta_2} - 1) + \mathbf{Z}(e^{i\alpha_2} - 1) &= \boldsymbol{\delta}_2 \\ \mathbf{W}(e^{i\beta_3} - 1) + \mathbf{Z}(e^{i\alpha_3} - 1) &= \boldsymbol{\delta}_3 \end{aligned} \tag{2.19}$$

If $\boldsymbol{\delta}_2$, $\boldsymbol{\delta}_3$, $\alpha_2$, and $\alpha_3$ are prescribed, $\beta_2$ and $\beta_3$ can be picked arbitrarily. Thus system (2.19) is a set of complex equations, *linear* in the complex unknowns **W** and **Z** (the vectors representing the dyad in its first position) with known coefficients. The system can be solved by Cramer's rule:

$$\mathbf{W} = \frac{\begin{vmatrix} \boldsymbol{\delta}_2 & e^{i\alpha_2} - 1 \\ \boldsymbol{\delta}_3 & e^{i\alpha_3} - 1 \end{vmatrix}}{\begin{vmatrix} e^{i\beta_2} - 1 & e^{i\alpha_2} - 1 \\ e^{i\beta_3} - 1 & e^{i\alpha_3} - 1 \end{vmatrix}} \tag{2.20}$$

$$\mathbf{Z} = \frac{\begin{vmatrix} e^{i\beta_2} - 1 & \boldsymbol{\delta}_2 \\ e^{i\beta_3} - 1 & \boldsymbol{\delta}_3 \end{vmatrix}}{\begin{vmatrix} e^{i\beta_2} - 1 & e^{i\alpha_2} - 1 \\ e^{i\beta_3} - 1 & e^{i\alpha_3} - 1 \end{vmatrix}} \tag{2.21}$$

Equations (2.20) and (2.21) are readily programmed on a hand calculator.*

Thus the three-position motion-synthesis case yields a linear solution if $\beta_2$ and $\beta_3$ are free choices. The two free choices in system (2.16) may be made with different strategies (as will be explored in Sec. 2.20 and 2.21), but will always involve two infinities of solutions.

**Four positions.** The system of equations for four prescribed positions of the moving plane is as follows:

$$\begin{aligned} \mathbf{W}(e^{i\beta_2} - 1) + \mathbf{Z}(e^{i\alpha_2} - 1) &= \boldsymbol{\delta}_2 \\ \mathbf{W}(e^{i\beta_3} - 1) + \mathbf{Z}(e^{i\alpha_3} - 1) &= \boldsymbol{\delta}_3 \\ \mathbf{W}(e^{i\beta_4} - 1) + \mathbf{Z}(e^{i\alpha_4} - 1) &= \boldsymbol{\delta}_4 \end{aligned} \tag{2.22}$$

Table 2.1 allows only one free choice from among the seven real unknowns: coordinates of **W**, **Z** and angles $\beta_2$, $\beta_3$, and $\beta_4$. Recall that $\boldsymbol{\delta}_j$ and $\alpha_j$, $j = 2, 3, 4$, are prescribed. Thus only one of the rotations or one coordinate of a link vector can be picked arbitrarily. System (2.22) contains three unknown angles $\beta_j$ in transcendental expressions. Even if we pick one $\beta_j$ as a free choice, system (2.22) requires a nonlinear equation-solving technique. Thus three precision points comprise the maximum number which may be prescribed and yet obtain a linear solution. For the four-position problem, Chap. 3 presents a closed-form nonlinear solution for equations (2.22), yielding up to an infinity of solutions.

**Five positions.** The system of equations for five positions [one additional equation added to system (2.22) with $j = 5$] is also nonlinear in the unknown $\beta_j$, and there are no free choices available. This case is also solved in closed form in Chap. 3.

* A copy of a hand calculator program can be obtained from the first author.

## 2.16 THREE PRESCRIBED POSITIONS FOR MOTION, PATH, AND FUNCTION GENERATION

This chapter concentrates on kinematic synthesis objectives that yield linear solutions—those easily solved graphically, on a hand calculator, or by a simple computer program. In the preceding section we discovered that, for motion generation of a dyad, three positions were the limit for a linear solution.

### Three-Position Motion-Generation Example

The following example* will help demonstrate the correlation between the graphical and complex number methods for three-precision-point motion synthesis. Figure 2.58a shows three positions of a coupler link for which a four-bar guiding linkage is desired. Using points $A$ and $B$ as proposed circle points, Fig. 2.58b shows the ground pivots $A_0$ and $B_0$ found by the intersection of the perpendicular bisectors (refer to Fig. 2.38). The resulting arguments of the input and output links, $\mathbf{W}_A$ and $\mathbf{W}_B$, at the precision positions are

$$\theta_1 = 330°, \qquad \sigma_1 = 235°$$

$$\theta_2 = 9°, \qquad \sigma_2 = 156°$$

$$\theta_3 = 64°, \qquad \sigma_3 = 135°$$

Let us now try to match the graphically generated solution with the standard-form method. Solving for the left-hand side first using point $A$ as the path tracer point, we compute from Fig. 2.58:

$$\boldsymbol{\delta}_2 = \mathbf{R}_2 - \mathbf{R}_1 = (1.75 + 0.30i) - (1.55 - 0.90i)$$

$$= 0.20 + 1.20i$$

$$\boldsymbol{\delta}_2 = \mathbf{R}_3 - \mathbf{R}_1 = -0.75 + 2.5i$$

$$\alpha_2 = 138° - 293° = 205° \qquad \beta_2 = 9° - 330° = 39°$$

$$\alpha_3 = 348° - 293° = 55° \qquad \beta_3 = 64° - 330° = 94°$$

Since the path tracer point is located at the circle point $A$, $\mathbf{W}_A$ should result in the vector between $A_0$ and $A$. Vector $\mathbf{Z}_A$ should be zero length. Using Eq. (2.20) yields

$$\mathbf{W}_A = \frac{\boldsymbol{\delta}_2(e^{i\alpha_3} - 1) - \boldsymbol{\delta}_3(e^{i\alpha_2} - 1)}{(e^{i\beta_2} - 1)(e^{i\alpha_3} - 1) - (e^{i\beta_3} - 1)(e^{i\alpha_2} - 1)}$$

$$= \frac{\boldsymbol{\delta}_2 e^{i\alpha_3} - \boldsymbol{\delta}_3 e^{i\alpha_2} + \boldsymbol{\delta}_3 - \boldsymbol{\delta}_2}{e^{i(\beta_2+\alpha_3)} - e^{i(\beta_3+\alpha_2)} - e^{i\beta_2} - e^{i\alpha_3} + e^{i\beta_3} + e^{i\alpha_2}}$$

* Contributed by Ray Giese.

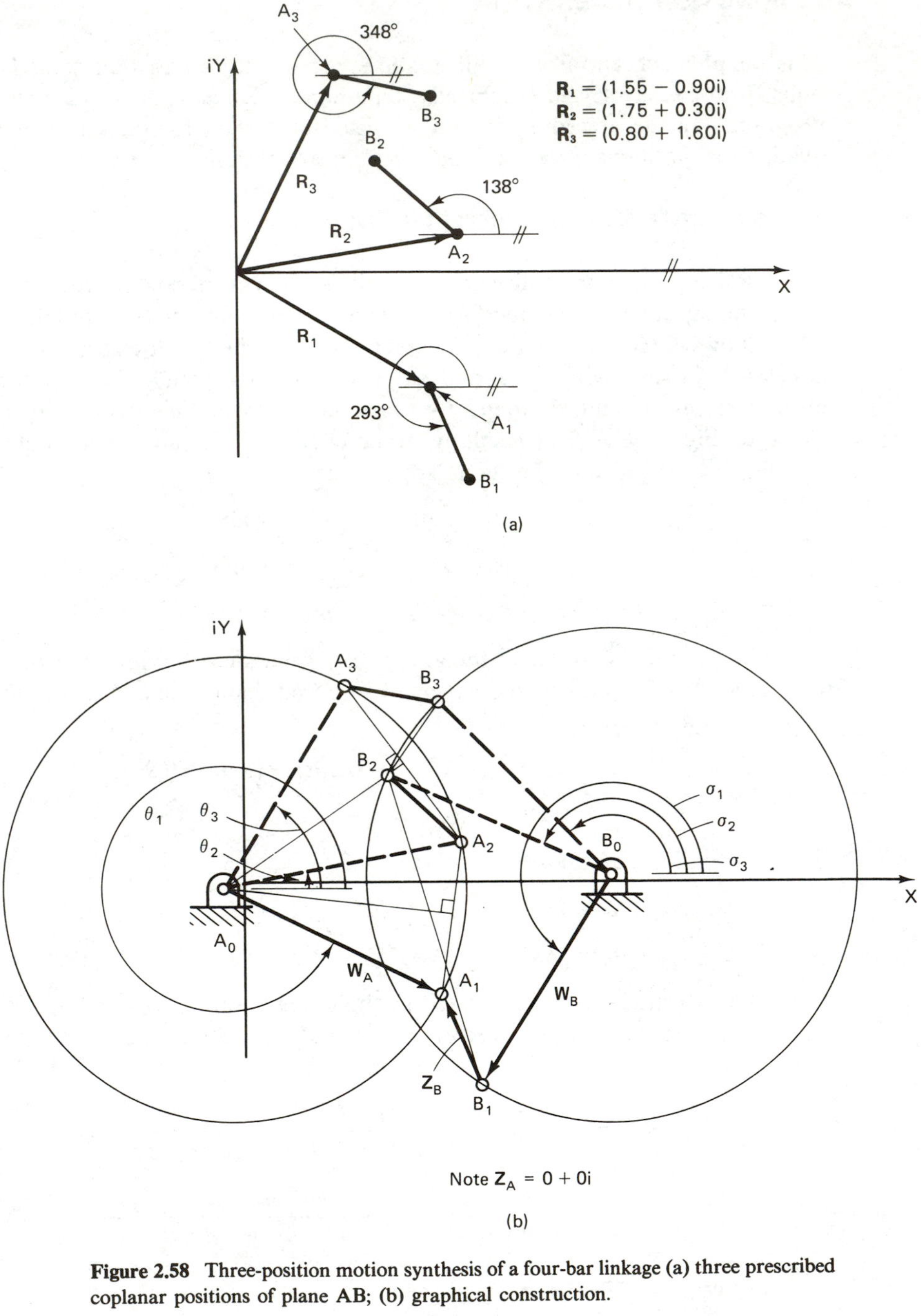


**Figure 2.58** Three-position motion synthesis of a four-bar linkage (a) three prescribed coplanar positions of plane AB; (b) graphical construction.

Using Euler's equation ($e^{i\theta} = \cos\theta + i \sin\theta$) and substituting the values for the variables; we obtain

$$\mathbf{W}_A = \frac{-3.556 + 4.104i}{-2.882 + 1.004i}$$

$$= 1.779 \measuredangle 330.1°$$

From Fig. 2.58b,

$$\mathbf{W}_A = 1.55 - 0.90i$$

$$= 1.792 \measuredangle 329.9°$$

which is less than a 1% difference, well within graphical error. Vector $\mathbf{Z}_A$ is now checked using Eq. (2.21).

$$\mathbf{Z}_A = \frac{\boldsymbol{\delta}_3(e^{i\beta_2} - 1) - \boldsymbol{\delta}_2(e^{i\beta_3} - 1)}{e^{(i\beta_2+\alpha_3)} - e^{i(\beta_3+\alpha_2)} - e^{i\beta_2} - e^{i\alpha_3} + e^{i\beta_3} + e^{i\alpha_2}}$$

$$= \frac{\boldsymbol{\delta}_3 e^{i\beta_2} - \boldsymbol{\delta}_2 e^{i\beta_3} - \boldsymbol{\delta}_3 + \boldsymbol{\delta}_2}{e^{i(\beta_2+\alpha_3)} - e^{i(\beta_3 + \alpha_2)} - e^{i\beta_2} - e^{i\alpha_3} + e^{i\beta_3} + e^{i\alpha_2}}$$

$$= \frac{0.008 - 0.006i}{3.052e^{i160.8°}}$$

$$= 0.0033 \measuredangle 162°$$

$$\simeq 0 + 0i$$

The right-hand side dyad ($\mathbf{W}_B = B_0B$ and $\mathbf{Z}_B = B_1A_1$) can be synthesized using the same $\boldsymbol{\delta}_2$, $\boldsymbol{\delta}_3$, $\alpha_2$, $\alpha_3$, but with

$$\beta_2 = 156° - 235° = 281°$$

$$\beta_3 = 135° - 235° = 260°$$

The same procedure is followed for the right side, resulting in

$$\mathbf{W}_B = \frac{5.430e^{i130.9°}}{2.703e^{i255.6°}}$$

$$= 2.01 \measuredangle 235.3°$$

From Fig. 2.58b the graphical construction produced

$$\mathbf{W}_B = 2.00 \measuredangle 235.5°$$

and the coupler link

$$\mathbf{Z}_B = \frac{2.135e^{i8.6°}}{2.703e^{i255.6°}}$$

$$= 0.79 \measuredangle 113°$$

again within graphical error.

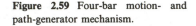

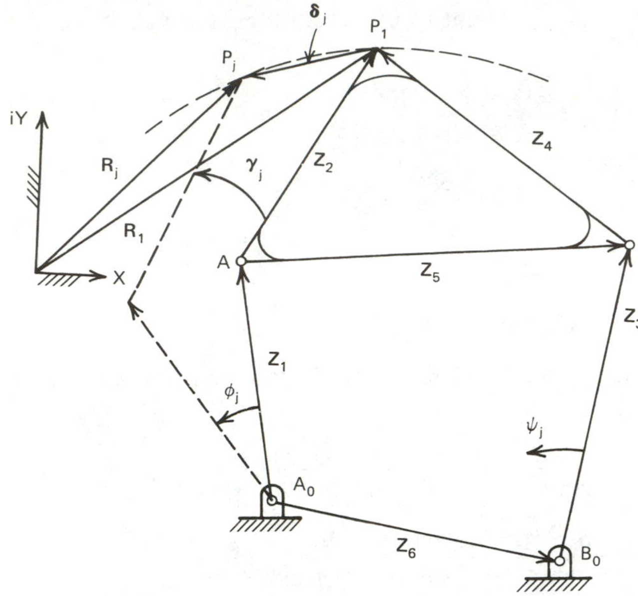


**Figure 2.59** Four-bar motion- and path-generator mechanism.

Notice that the balancing of the number of equations and the number of unknowns in Table 2.1 was based on motion synthesis. Equation (2.16) has been termed the *standard form* with the understanding that both $\boldsymbol{\delta}_j$ and $\alpha_j$ or $\beta_j$ were prescribed. The numbers in Table 2.1 will be the same if $\beta_j$ were prescribed instead of $\alpha_j$, which, as we know from Sec. 2.2, is the case for path generation with prescribed timing.

Let us look at the four-bar of Fig. 2.59, as well as a six-bar linkage, and attempt to express the synthesis of these linkages in the standard form for motion, path, and function generation. If this can be accomplished, only one computer program will be needed to synthesize these linkages for either of these tasks. (This generality of the standard form also extends to the nonlinear solutions of Chap. 3.)

## Synthesis of a Four-Bar Motion Generator for Three Precision Points

The four-bar linkage of Fig. 2.59 is to be synthesized for motion generation. As suggested in Sec. 2.14, there are two independent dyads in the four-bar linkage, which will be called the *left-hand side* and the *right-hand side*. Each dyad connects a ground pivot (a center point) to the path point on the coupler by way of joint $A$ or $B$ of the coupler (the circle point). The equations describing the displacements of the left-hand side have already been derived, but in the notation of Fig. 2.59 the standard form is

$$\mathbf{Z}_1(e^{i\phi_j} - 1) + \mathbf{Z}_2(e^{i\gamma_j} - 1) = \boldsymbol{\delta}_j, \qquad j = 2, 3 \tag{2.23}$$

where $\boldsymbol{\delta}_j$ and $\gamma_j$ are prescribed.

The displacement equations for the right-hand side of the linkage may be written as

$$\mathbf{Z}_3(e^{i\psi_j}-1)+\mathbf{Z}_4(e^{i\gamma_j}-1)=\boldsymbol{\delta}_j, \qquad j=2,\ 3 \tag{2.24}$$

where $\boldsymbol{\delta}_j$ and $\gamma_j$ are prescribed.

If we assume $\phi_j$ and $\psi_j$ arbitrarily, Eqs. (2.23) and (2.24) can be solved by Cramer's rule for $\mathbf{Z}_1$, $\mathbf{Z}_2$, $\mathbf{Z}_3$, and $\mathbf{Z}_4$ [see the form of solution in Eqs. (2.20) and (2.21). The other two linkage vectors are simply

$$\mathbf{Z}_5=\mathbf{Z}_2-\mathbf{Z}_4 \tag{2.25}$$

and

$$\mathbf{Z}_6=\mathbf{Z}_1+\mathbf{Z}_5-\mathbf{Z}_3 \tag{2.26}$$

### Synthesis of a Four-Bar Path Generator with Prescribed Timing

Suppose that the four-bar linkage of Fig. 2.59 is to be synthesized for path generation with prescribed timing. The very same equations as derived for motion generation [(2.23) to (2.26)] will apply in this case, but the prescribed angles will be different. Instead of $\gamma_j$ in Eq. (2.23), $\phi_j$ will be prescribed and $\gamma_j$, $j=2, 3$, are free choices. Thus Eq. (2.23) will still be in the standard form. As for Eq. (2.24), in order to connect the right-hand side with the left side, vectors $\mathbf{Z}_4$ must rotate by the same rotations ($\gamma_j$) as $\mathbf{Z}_2$. Thus the same $\gamma_j$, $j=2, 3$, that were picked as free choices for Eq. (2.23), are prescribed in Eq. (2.24). Therefore, the four-bar path generator with prescribed timing has the same solution procedure as the four-bar motion generator.

### Synthesis of a Four-Bar Function Generator

The standard form for a four-bar function generator can be derived from Fig. 2.59 as follows. Recall that in function generation we wish to correlate the prescribed rotations of the input link ($\phi_j$) and the output link ($\psi_j$). Therefore, the upper portion of the coupler link ($\mathbf{Z}_2$ and $\mathbf{Z}_4$) is of no concern for this task. Figure 2.60 shows the basic four-bar of Fig. 2.59 in the first and $j$th position. The vector loop containing $\mathbf{Z}_1$, $\mathbf{Z}_5$, and $\mathbf{Z}_3$ is

$$\mathbf{Z}_1(e^{i\phi_j}-1)+\mathbf{Z}_5(e^{i\gamma_j}-1)-\mathbf{Z}_3(e^{i\psi_j}-1)=0 \tag{2.27}$$

Since this vector equation is not in the standard form, Table 2.2 is formulated to help correlate the number of free choices and the number of prescribed positions. The same development as that done in connection with Table 2.1 is repeated here. Notice that the maximum number of prescribed positions is seven when a triad (three

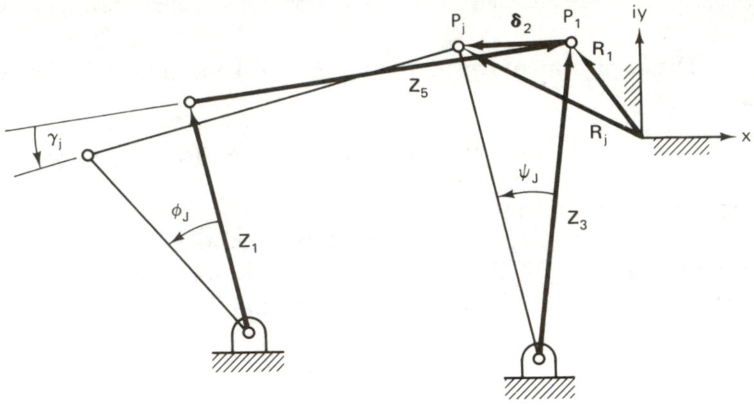


**Figure 2.60** Four-bar function-generator mechanism.

links) is used and when two of the three rotations are prescribed, as they must be for function generation.

Picking $\mathbf{Z}_3$ as an arbitrary choice ($\mathbf{Z}_1$ could be picked instead) will convert Eq. (2.27) into the standard form:

$$\mathbf{Z}_1(e^{i\phi_j}-1)+\mathbf{Z}_5(e^{i\gamma_j}-1)=\boldsymbol{\delta}_j=\mathbf{Z}_3(e^{i\psi_j}-1) \qquad (2.28)$$

The justification for picking $\mathbf{Z}_3$ is two fold: first, by comparing Tables 2.1 and 2.2, the latter becomes equivalent to the first if two of the original seven real unknowns of Table 2.2 are picked arbitrarily; second, by choosing $\mathbf{Z}_3$, we are actually specifying the scale and orientation of the function generator. In fact, once a four-bar is synthesized for function generation, the entire linkage may be scaled up or down and oriented in any direction without changing the functional relationship between input and output link rotations [$\psi_j = f(\phi_j)$]. Therefore, new function-generation solutions do not result from allowing $\mathbf{Z}_3$ to be an unknown. (Path and motion generators *do change* their prescribed path with a change in scale; therefore, only in function generation synthesis of four-bar linkages do we pick one of the link vectors arbitrarily.)

Equation (2.28) is now in the standard form. In fact, function generation can

**TABLE 2.2** NUMBER OF AVAILABLE SOLUTIONS IN THE SYNTHESIS OF FOUR-BAR FUNCTION GENERATORS. (FIG. 2.60) ACCORDING TO THE EQUATIONS

$$\mathbf{Z}_1(e^{i\phi_j}-1)+\mathbf{Z}_5(e^{i\gamma_j}-1)-\mathbf{Z}_3(e^{i\psi_j}-1)=0,\ j=2,3,\ldots,7 \qquad (2.27)$$

| Number of positions ($n$): $j=2, 3, \ldots, n$ | Number of scalar equations | Number of scalar unknowns | Number of free choices (scalars) | Number of solutions |
|---|---|---|---|---|
| 2 | 2 | 7($\mathbf{Z}_1, \mathbf{Z}_5, \mathbf{Z}_3, \gamma_2$) | 5 | $O(\infty)^5$ |
| 3 | 4 | 8(above + $\gamma_3$) | 4 | $O(\infty)^4$ |
| 4 | 6 | 9(above + $\gamma_4$) | 3 | $O(\infty)^3$ |
| 4 | 8 | 10(above + $\gamma_5$) | 2 | $O(\infty)^2$ |
| 6 | 10 | 11(above + $\gamma_6$) | 1 | $O(\infty)^1$ |
| 7 | 12 | 12(above + $\gamma_7$) | 0 | Finite |

be thought of as a special case of path generation with prescribed timing, the path of $\mathbf{Z}_3$ being along a circular arc. Note also that only one dyad needs to be synthesized [Eq. (2.28) for $j = 2, 3$] for function generation.

Sections 2.22 and 2.23 show other techniques for generating design equations for function generation: Freudenstein's equation and the loop-closure-equation technique. The number of free choices here does coincide with the latter method. These other techniques do not necessarily yield the standard form, however, although they can also be so formulated (see Chap. 3).

## 2.17 THREE-PRECISION-POINT SYNTHESIS PROGRAM FOR FOUR-BAR LINKAGES

A program can be written to synthesize a four-bar motion, path, or function generator mechanism for three finitely separated precision points utilizing the notation of Figure 2.59 (Fig. 2.61 is a flowchart for this program). The system of equations [Eq. (2.23),

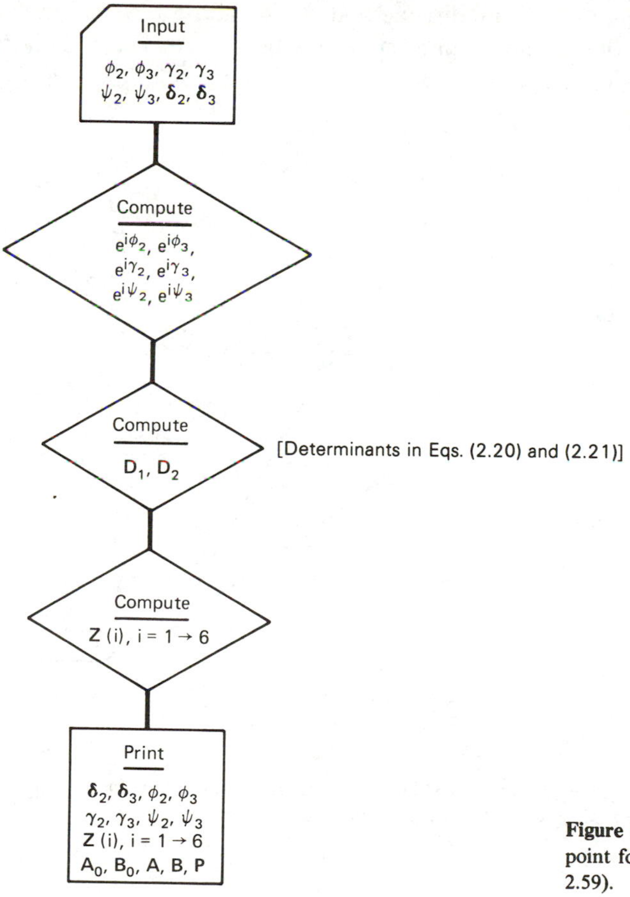


**Figure 2.61** Flowchart of three-precision-point four-bar synthesis program (see Fig. 2.59).

$j = 2, 3$] for the left side of the four-bar and the equations for the right side [Eq. (2.24), $j = 2, 3$] are solved by Cramer's rule as suggested in Eqs. (2.20) and (2.21). The input data required are the rotations of the input, output, and coupler links: PHI2, PHI3, GAM2, GAM3, PSI2, PSI3, ($\phi_2$, $\phi_3$, $\gamma_2$, $\gamma_3$, $\psi_2$, $\psi_3$) and the path displacements: XDEL2, YDEL2, XDEL3, YDEL3 (the $x$ and $y$ coordinates of $\boldsymbol{\delta}_2$ and $\boldsymbol{\delta}_3$). As can be seen by the examples below, the output of the program can include a repeat of the input data, link vectors in the starting position of the synthesized linkage in both Cartesian and polar form, as well as the coordinates of the coupler bar points: $A$, $B$, and $P$ with respect to $A_0$. Figures 2.62, 2.63, and 2.64 show linkages that have been synthesized for motion, path, and function generation, respectively.

Notice that arbitrary choices must be made for all three examples according to Table 2.1. Thus for motion generation, $\phi_2$, $\phi_3$, $\psi_2$, and $\psi_3$ are free choices. For path generation with prescribed timing $\gamma_2$, $\gamma_3$, $\psi_2$, and $\psi_3$ are free choices. The procedure of Sec. 2.16 is not used here in the function-generation case. Rather than expanding the program of Fig. 2.61, the function generator is synthesized by prescribing $\phi_2$, $\phi_3$, $\psi_2$, $\psi_3$. With this method $\boldsymbol{\delta}_2$, $\boldsymbol{\delta}_3$ and $\gamma_2$, $\gamma_3$ are free choices. The only portion of the output of interest would be $\mathbf{Z}_1$, $\mathbf{Z}_5$, $\mathbf{Z}_3$, and $\mathbf{Z}_6$ (see Fig. 2.59 and the example of Fig. 2.64).

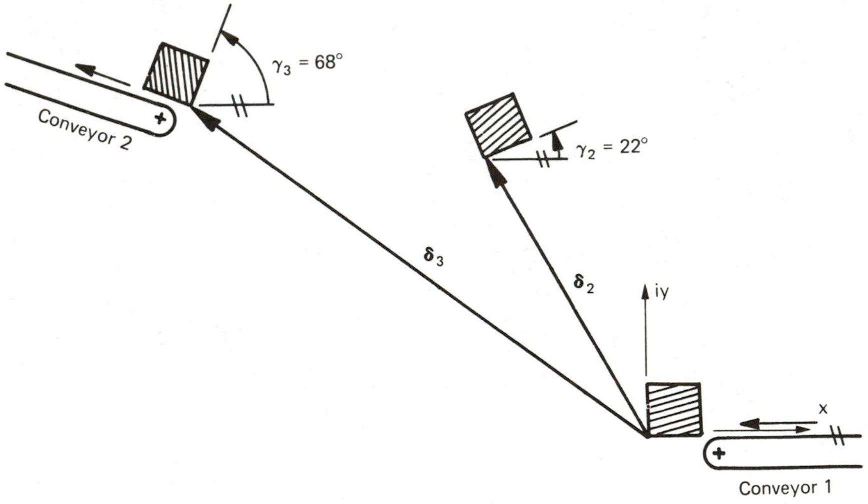


(a)

**Figure 2.62** (a) three prescribed positions for four-bar motion synthesis; (b) synthesized conveyor linkage of Ex. 8.3 using the program of Fig. 2.61.

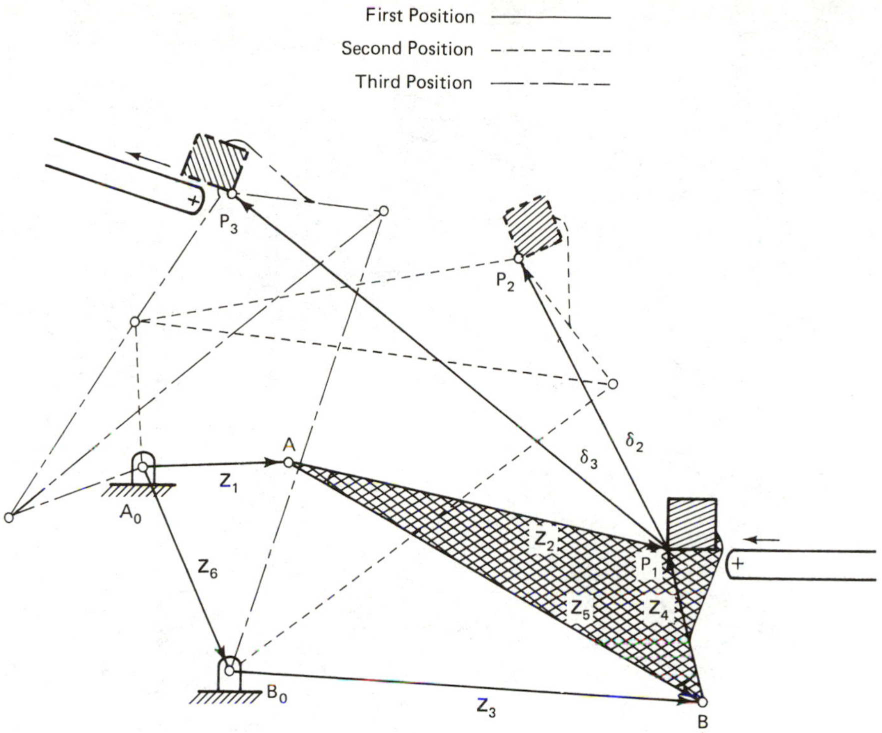


(b)

**Figure 2.62 (continued)**

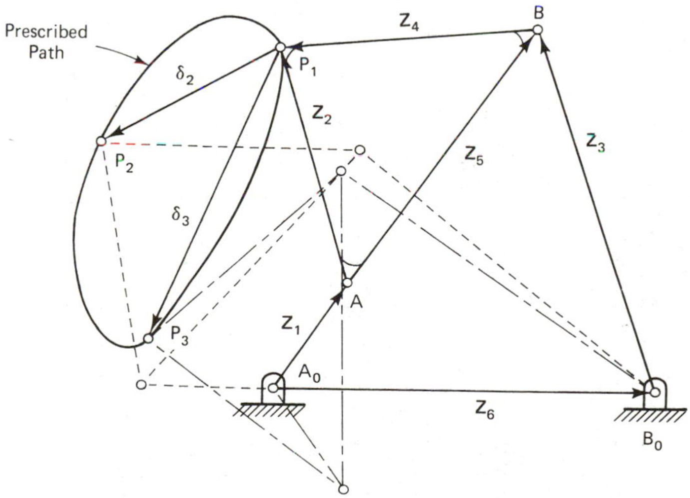


**Figure 2.63** Four-bar path generator with three prescribed path points.

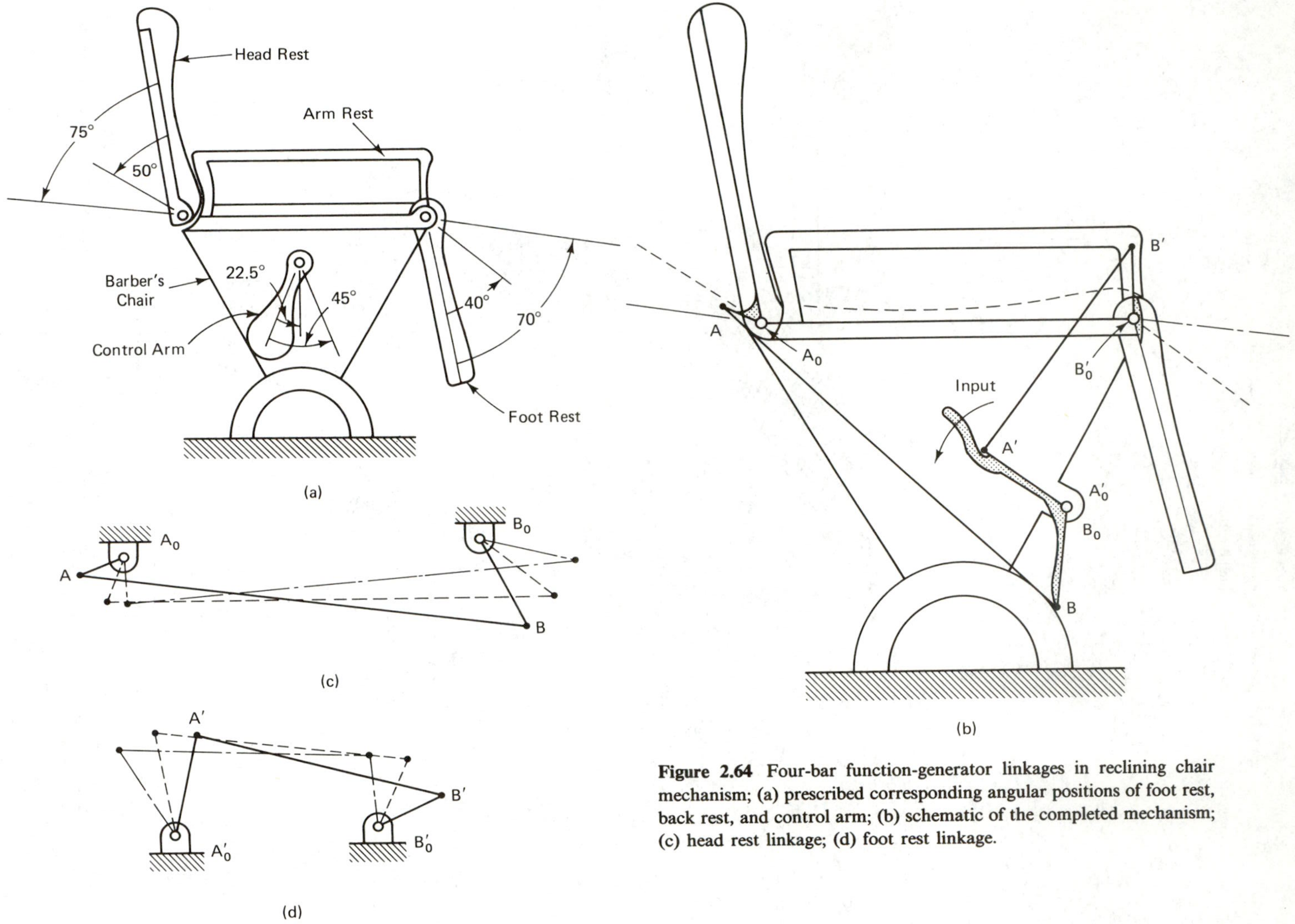


**Figure 2.64** Four-bar function-generator linkages in reclining chair mechanism; (a) prescribed corresponding angular positions of foot rest, back rest, and control arm; (b) schematic of the completed mechanism; (c) head rest linkage; (d) foot rest linkage.

**Example 2.3: Motion Generation**

Completion of an assembly line requires the synthesis of a motion generator linkage to transfer boxes from one conveyor belt to another as depicted in Fig. 2.62a. A pickup and release position plus an intermediate location are specified. For simplicity a four-bar linkage (Fig. 2.59) is the type of linkage chosen for the task. From Fig. 2.62a prescribed quantities for the motion generation are

$$\boldsymbol{\delta}_2 = -6 + 11i, \qquad \gamma_2 = 22^\circ$$

$$\boldsymbol{\delta}_3 = -17 + 13i, \qquad \gamma_3 = 68^\circ$$

The free choices are arbitrarily set* as

$$\phi_2 = 90^\circ, \qquad \psi_2 = 40^\circ$$

$$\phi_3 = 198^\circ, \qquad \psi_3 = 73^\circ$$

Table 2.3 shows a copy of the computer printout for this example, while Fig. 2.62b shows the solution drawn in three positions. Section 3.9 of Chap. 3 (of Vol. 1) analyzes this linkage throughout its cycle of motion (see Ex. 3.12 of Vol. 1).

**TABLE 2.3** COMPUTER PRINTOUT OF THREE-POSITION MOTION-GENERATOR SYNTHESIS OF FOUR-BAR LINKAGE (SEE EX. 2.3).

INPUT DATA

| | X COMPONENT | Y COMPONENT |
|---|---|---|
| DELTA 2 = | −6.0000 | 11.0000 |
| DELTA 3 = | −17.0000 | 13.0000 |

| | | | | | |
|---|---|---|---|---|---|
| PHI 2 = | 90.000 | GAMMA 2 = | 22.000 | PSI 2 = | 40.000 |
| PHI 3 = | 198.000 | GAMMA 3 = | 68.000 | PSI 3 = | 73.000 |

COMPUTED VECTORS

| | X COMPONENT | Y COMPONENT | LENGTH | DIRECTION(DEG) |
|---|---|---|---|---|
| Z(1) = | 5.7550 | .4809 | 5.7751 | 4.777 |
| Z(2) = | 14.6106 | −3.4698 | 15.0169 | −13.359 |
| Z(3) = | 18.3746 | −.6611 | 18.3864 | −2.061 |
| Z(4) = | −1.4207 | 5.9518 | 6.1190 | 103.426 |
| Z(5) = | 16.0313 | −9.4215 | 18.5948 | −30.443 |
| Z(6) = | 3.4118 | −8.2796 | 8.9550 | −67.605 |

LINKAGE PIVOT AND COUPLER LOCATIONS

| | X COMPONENT | Y COMPONENT |
|---|---|---|
| $A_0$ = | 0 | 0 |
| $B_0$ = | 3.4118 | −8.2796 |
| A = | 5.7550 | .4809 |
| B = | 21.7863 | −8.9407 |
| P = | 20.3656 | −2.9889 |

* By specifying input rotations about twice as large as output rotations, these choices are meant to bring about a crank-rocker type of four-bar solution. The ability to bring this about is a valuable attribute of the three-position standard-form method.

**Example 2.4: Path Generation with Prescribed Timing** (Fig. 2.63)

A stirring operation requires the generation of an elliptical path. A four-bar linkage is picked for this task. Since a crank rocker is required, the input rotations are also to be prescribed. Specified quantities are

$$\delta_2 = -1.4 - 0.76i, \qquad \phi_2 = 126°$$

$$\delta_3 = -1.0 - 2.3i, \qquad \phi_3 = 252°$$

Arbitrarily chosen variables are

$$\gamma_2 = -6°, \qquad \psi_2 = 33°$$

$$\gamma_3 = 37°, \qquad \psi_3 = 37°$$

Table 2.4 is a copy of the computer-generated output for this example. Figure 2.63 illustrates the computer-generated linkage solution in its three prescribed positions. Section 3.9 of Chap. 3 (Vol. 1) analyzes this linkage in Ex. 3.13.

**TABLE 2.4** COMPUTER PRINTOUT OF SYNTHESIS OF FOUR-BAR LINKAGE FOR PATH GENERATION WITH PRESCRIBED TIMING (SEE EX. 2.4).

INPUT DATA

| | X COMPONENT | Y COMPONENT |
|---|---|---|
| DELTA 2 = | −1.4000 | −.7600 |
| DELTA 3 = | −1.0000 | −2.3000 |

| | | | | | |
|---|---|---|---|---|---|
| PHI 2 = | 126.000 | GAMMA 2 = | −6.000 | PSI 2 = | 33.000 |
| PHI 3 = | 252.000 | GAMMA 3 = | 37.000 | PSI 3 = | 37.000 |

COMPUTED VECTORS

| | X COMPONENT | Y COMPONENT | LENGTH | DIRECTION(DEG) |
|---|---|---|---|---|
| Z(1) = | .5919 | .8081 | 1.0017 | 53.777 |
| Z(2) = | −.5182 | 1.8246 | 1.8967 | 105.856 |
| Z(3) = | −.9412 | 2.8331 | 2.9854 | 108.376 |
| Z(4) = | −1.9958 | −.1888 | 2.0047 | −174.596 |
| Z(5) = | 1.4776 | 2.0134 | 2.4974 | 53.725 |
| Z(6) = | 3.0107 | −.0117 | 3.0107 | −.223 |

LINKAGE PIVOT AND COUPLER LOCATIONS

| | X COMPONENT | Y COMPONENT |
|---|---|---|
| $A_0$ = | 0 | 0 |
| $B_0$ = | 3.0107 | −.0117 |
| A = | .5919 | .8081 |
| B = | 2.0695 | 2.8214 |
| P = | .0737 | 2.6326 |

**Example 2.5: Function Generation**

Figure 2.64a shows a barber's chair in which a single control arm is to actuate both the foot rest and the head rest. Notice the nonlinear relationship between the angles of rotation of the three members in the three specified positions. The type of linkage chosen for this task is a Watt II six-bar, which is simply two four-bars in series (usually connected through a bell crank). Specified quantities for the *first four-bar* function generator (between the head rest and the control arm) are

$$\phi_2 = 50°, \qquad \psi_2 = 22.5°$$
$$\phi_3 = 75°, \qquad \psi_4 = 45°$$

Arbitrarily chosen are

$$\delta_2 = -0.07 + 0.4i, \qquad \gamma_2 = 7°$$
$$\delta_3 = -0.3 + 0.7i, \qquad \gamma_3 = 12°$$

The *second four-bar* function generator (between the control arm and the foot rest) has specified variables of

$$\phi_2 = 22.5°, \qquad \psi_2 = 40°$$
$$\phi_3 = 45°, \qquad \psi_3 = 70°$$

Arbitrarily chosen are

$$\delta_2 = -0.07 + 0.4i, \qquad \gamma_2 = 8°$$
$$\delta_3 = -0.3 + 0.7i, \qquad \gamma_3 = 13°$$

Table 2.5 is a printout for both sides of the six-bar linkage. Figure 2.64b illustrates how both these solutions are put together by appropriate rescaling and reorientation as one of the numerous possible Watt's II six-bar solutions to this problem, while Fig. 2.64c and d show the two four-bar halves in their three design positions.

**TABLE 2.5** COMPUTER OUTPUT OF FOUR-BAR SYNTHESIS OF RECLINER MECHANISM (SEE FIG. 2.64 AND EX. 2.5).

HEAD REST LINKAGE

INPUT DATA

| | X COMPONENT | Y COMPONENT |
|---|---|---|
| DELTA 2 = | −.0700 | .4000 |
| DELTA 3 = | −.3000 | .7000 |

| | | | | | |
|---|---|---|---|---|---|
| PHI 2 = | 50.000 | GAMMA 2 = | 7.000 | PSI 2 = | 22.500 |
| PHI 3 = | 75.000 | GAMMA 3 = | 12.000 | PSI 3 = | 45.000 |

COMPUTED VECTORS

| | X COMPONENT | Y COMPONENT | LENGTH | DIRECTION(DEG) |
|---|---|---|---|---|
| Z(1) = | .0404 | −.4640 | .4657 | −85.022 |
| Z(2) = | 1.8676 | 3.2580 | 3.7554 | 60.178 |
| Z(3) = | 1.0009 | .2777 | 1.0388 | 15.506 |
| Z(4) = | .2552 | −.9384 | .9725 | −74.788 |
| Z(5) = | 1.6124 | 4.1965 | 4.4956 | 68.982 |
| Z(6) = | .6518 | 3.4548 | 3.5158 | 79.315 |

LINKAGE PIVOT AND COUPLER LOCATIONS

| | X COMPONENT | Y COMPONENT |
|---|---|---|
| $A_0$ = | 0 | 0 |
| $B_0$ = | .6518 | 3.4548 |
| A = | .0404 | −.4640 |
| B = | 1.6528 | 3.7325 |
| P = | 1.9080 | 2.7941 |

FOOT REST LINKAGE

INPUT DATA

| | X COMPONENT | Y COMPONENT |
|---|---|---|
| DELTA 2 = | −.0700 | .4000 |
| DELTA 3 = | −.3000 | .7000 |

| | | | | | |
|---|---|---|---|---|---|
| PHI 2 = | 22.500 | GAMMA 2 = | 8.000 | PSI 2 = | 40.000 |
| PHI 3 = | 45.000 | GAMMA 3 = | 13.000 | PSI 3 = | 70.000 |

COMPUTED VECTORS

| | X COMPONENT | Y COMPONENT | LENGTH | DIRECTION(DEG) |
|---|---|---|---|---|
| Z(1) = | .9642 | .2270 | .9906 | 13.247 |
| Z(2) = | .3001 | −.6696 | .7338 | −65.859 |
| Z(3) = | .5189 | −.4332 | .6759 | −39.857 |
| Z(4) = | −.1359 | 1.6410 | 1.6466 | 94.733 |
| Z(5) = | .4360 | −2.3105 | 2.3513 | −79.315 |
| Z(6) = | .8813 | −1.6503 | 1.8709 | −61.897 |

LINKAGE PIVOT AND COUPLER LOCATIONS

| | X COMPONENT | Y COMPONENT |
|---|---|---|
| $A_0$ = | 0 | 0 |
| $B_0$ = | .8813 | −1.6503 |
| A = | .9642 | .2270 |
| B = | 1.4002 | −2.0835 |
| P = | 1.2643 | −.4426 |

## 2.18 THREE-PRECISION-POINT SYNTHESIS: ANALYTICAL VERSUS GRAPHICAL

Thus far, both graphical and analytical approaches have been presented for three finitely separated positions of motion-, path with timing- and function-generation synthesis of a four-bar linkage. Both techniques are straightforward. Which is better? The answer: both are equally important. Graphical techniques are extremely useful in the initial stages of synthesis. If a graphical construction does not yield an "optimal" solution in a reasonable amount of time or if the error sensitivity is high (e.g., the need to locate the intersection of lines that form an acute angle), then the analytical standard-form method is very attractive. In such cases the preliminary graphical solution will yield reasonable values for arbitrarily assumed (free choice) quantities, which will help obtain workable computer solutions. The Cramer's rule solution described above is easily programmed for digital computation (the flowchart of a three-precision-point program is shown in Fig. 2.61) and numerous accurate solutions can be obtained in a fraction of the time required for a graphical construction. (Section 2.20 shows an alternative computer graphics technique for three precision points, which is a combined graphical and analytical method.)

A notable correlation between the graphical and analytical methods should be emphasized at this point. In both techniques, for the three-position synthesis of each dyad, there are two infinities of solutions for motion–path generation with prescribed timing and function generation. As pointed out above, a function generator four-bar linkage actually appears to require two additional scalars as free choices: the two components of the starting position vector of one of the links. However, picking that link specifies only the scale and orientation of the linkage. No new function generators are obtained by varying this link, because the functional relationship of the input and output rotations is not affected by this choice.

A very useful reference for linkage design is an atlas of four-bar coupler curves by *Hrones and Nelson's* [131]. Approximately 7300 coupler curves of crank-rocker four-bar linkages are displayed (e.g., see Fig. 2.65). The small circles represent coupler points whose coupler curves are plotted. These can be used as "tracer points" in a path or motion generator linkage. Each dash on the coupler curves represents 10° of input crank rotation to provide an indication of coupler point velocity. The crank always has length 1 while the lengths of the coupler *A,* follower *B,* and the fixed link *C* vary from page to page, yielding a variety of families of coupler curves. This atlas is extremely useful in the initial stages of a path or in some cases motion-generation synthesis effort. A designer may be able to find several coupler curve forms that nearly accomplish the task at hand and then use these linkages to come up with proper "free choices" (see Sec. 2.15) to help find a more nearly optimal linkage in a shorter time. Also, it may turn out that there are no crank-rocker four-bars that satisfy the design requirements and the *type* synthesis step may have to be reconsidered.

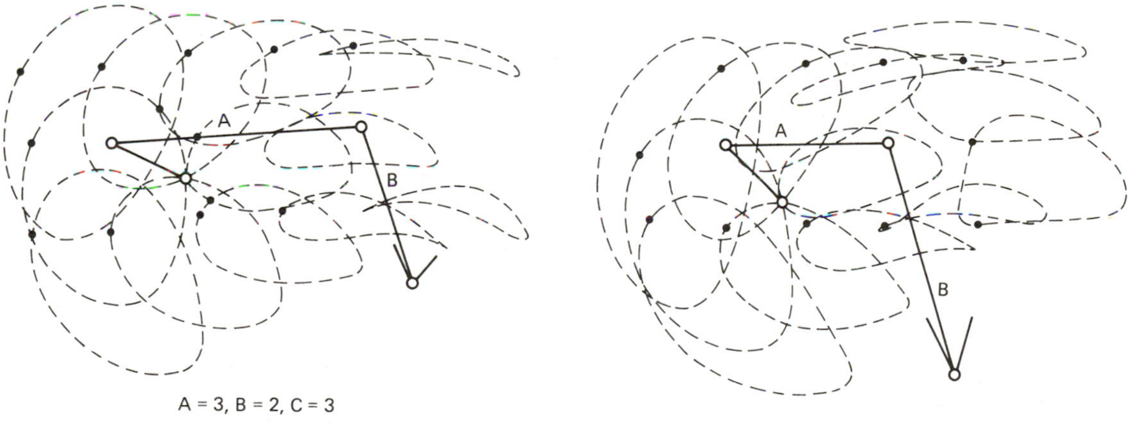


**Figure 2.65** Sample pages from the atlas of four-bar coupler curves by Hrones and Nelson [131]. In [131], lengths of dashes of the curves indicate 10° increments of crank rotations.

## 2.19 EXTENSION OF THREE-PRECISION-POINT SYNTHESIS TO MULTILOOP MECHANISMS

In addition to the four-bar linkage, multiloop planar mechanisms can also be synthesized by recognizing key dyads that yield equations of the *same form as Eq.* (2.16.) In addition to the Watt six-bar of Ex. 2.5, the Stephenson III linkage of Fig. 2.66a will be used to demonstrate the extension of the dyad approach beyond four-bar linkages.

Inspection of Fig. 2.66a will yield three independent loops; one is a dyad loop and two are triad loops (see Fig. 2.66b):

Loop 1:

$$\mathbf{Z}_1(e^{i\phi_j}-1)+\mathbf{Z}_2(e^{i\gamma_j}-1)=\boldsymbol{\delta}_j \tag{2.29}$$

Loop 2:

$$\mathbf{Z}_5(e^{i\psi_j}-1)+\mathbf{Z}_4(e^{i\beta_j}-1)-\mathbf{Z}_3(e^{i\gamma_j}-1)=\boldsymbol{\delta}_j \tag{2.30}$$

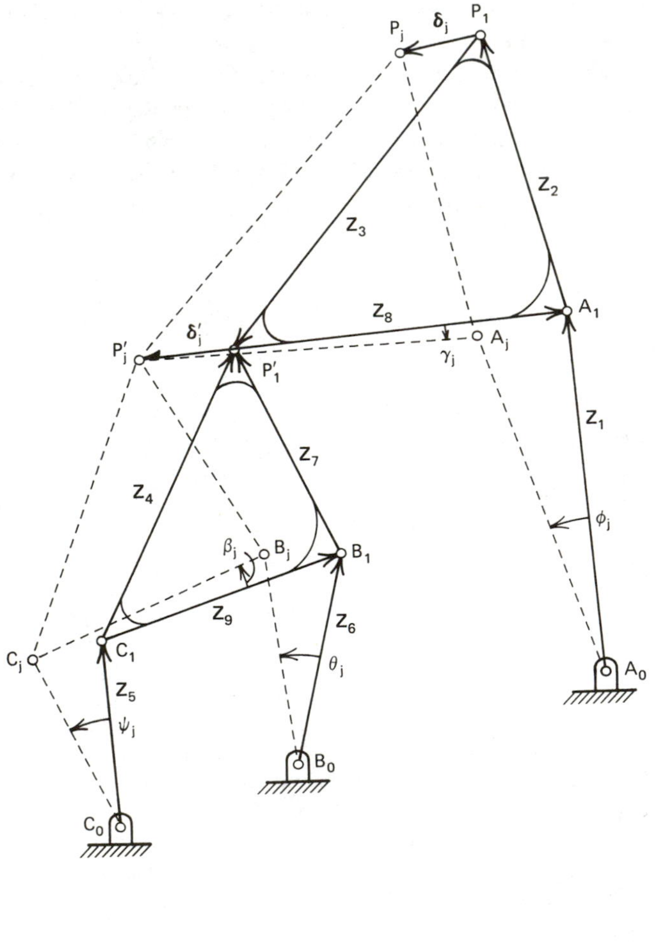


**Figure 2.66** Dyadic synthesis of a Stephenson III six-bar mechanism for motion generation, path generation with prescribed timing and function generation. $P_1P_j$ is path cord $\boldsymbol{\delta}_j$. For motion generation, $\boldsymbol{\delta}_j$ is prescribed. For path generation with prescribed timing, $\phi_j$ is also prescribed. For additional function generation, not only $\phi_j$ but also $\psi_j$ or $\theta_j$ is prescribed. Loop closure equations are written for $A_0A_jP_jP_1A_1A_0$, $B_0B_jP_j'P_jP_1P_1'B_1B_0$, and $C_0C_jP_j'P_jP_1P_1'C_1C_0$.

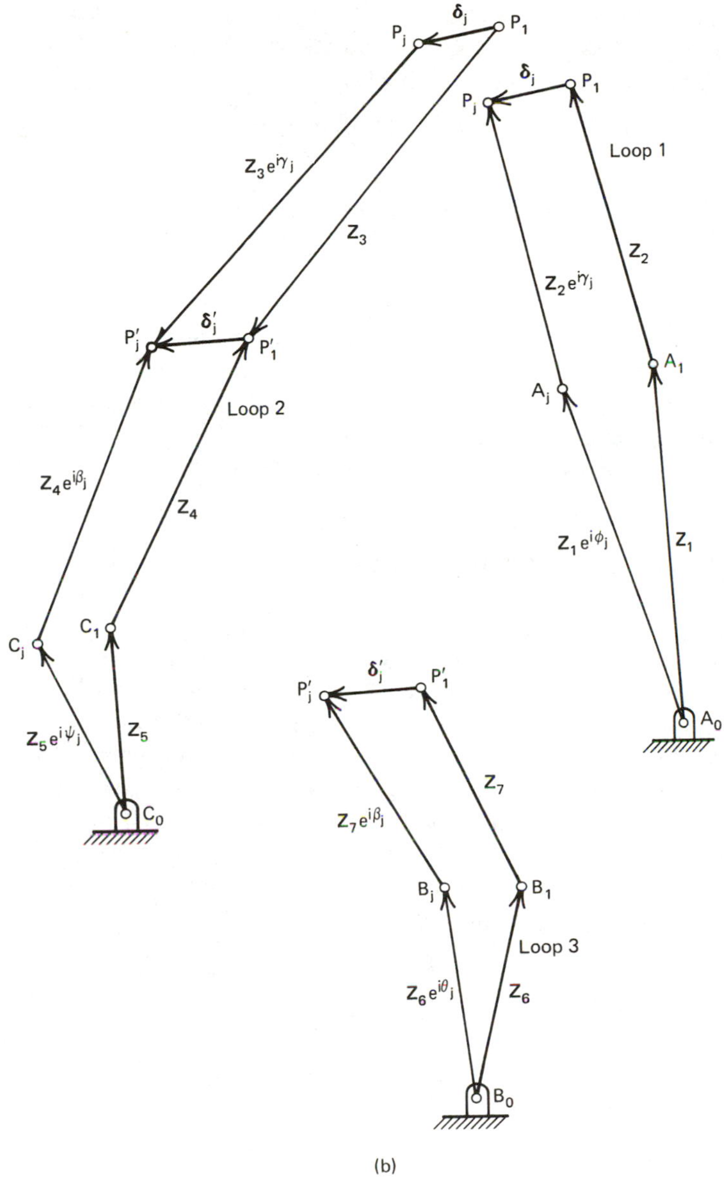


**Figure 2.66 (continued)**

Loop 3:

$$\mathbf{Z}_6(e^{i\theta_j}-1)+\mathbf{Z}_7(e^{i\beta_j}-1)-\mathbf{Z}_3(e^{i\gamma_j}-1)=\boldsymbol{\delta}_j \tag{2.31}$$

Loop 1 is in the standard form: $\boldsymbol{\delta}_j$ and either $\phi_j$ or $\gamma_j$ are prescribed. Neither loop 2 or 3 is in the standard form as they are written. But they are in the same form as Eq. (2.27), which was analyzed in Table 2.2, except that instead of zero on the right-hand side, there is a known path displacement vector $\boldsymbol{\delta}_j$. With two free choices available, vector $\mathbf{Z}_3$ can be chosen arbitrarily. If $\gamma_j$ is prescribed or is known from loop 1, Eqs. (2.30) and (2.31) can be rewritten with the aid of Eq. (2.32) and solved as standard-form equations. If we let

$$\boldsymbol{\delta}_j' = \boldsymbol{\delta}_j + \mathbf{Z}_3(e^{i\gamma_j} - 1) \tag{2.32}$$

then, for loop 2,

$$\mathbf{Z}_5(e^{i\psi_j} - 1) + \mathbf{Z}_4(e^{i\beta_j} - 1) = \boldsymbol{\delta}_j' \tag{2.33}$$

and for loop 3,

$$\mathbf{Z}_6(e^{i\theta_j} - 1) + \mathbf{Z}_7(e^{i\beta_j} - 1) = \boldsymbol{\delta}_j' \tag{2.34}$$

where $\boldsymbol{\delta}_j'$ is the displacement of $P'$. The free choice of $\mathbf{Z}_3$ offers the designer the possibility of picking the shape of the coupler link or the generation of different solutions by varying $\mathbf{Z}_3$. Other multiloop linkages may be synthesized by repeated use of the same standard-form solution method by employing a similar procedure as described in this section (see Chap. 3 for Stephenson III and other multiloop examples) [79, 80].

## 2.20 CIRCLE-POINT AND CENTER-POINT CIRCLES

This section describes an alternative approach to choosing the two free choices indicated in Table 2.1. The angular unknowns will be considered as candidates for parameters on which the locations of the fixed and moving pivots of the solution dyads will depend. Loerch [170] discovered that, if an arbitrary value is chosen for one unprescribed angular parameter while the other angular parameter is allowed to assume all possible values, the resulting loci of corresponding fixed pivots $m$ and moving pivots $k_1$ are found to be pairs of circles. For example, in Fig. 2.57 if $\boldsymbol{\delta}_2$, $\boldsymbol{\delta}_3$, $\alpha_2$, $\beta_2$, and $\beta_3$ are chosen to have fixed values, then points $m$ and $k_1$ trace circular loci as $\alpha_3$ ranges between 0 and $2\pi$. These will be referred to as $M$ and $K_1$ circles, respectively. A complex-number formulation will be used to generate these circles analytically.

### Dyad Equations

The dyad vectors are defined in Fig. 2.67. The loop-closure equations for the dyad in three finitely separated positions are:

First position:

$$\mathbf{R} + \mathbf{W} + \mathbf{Z} = 0 \tag{2.35}$$

Second position:

$$\mathbf{R} + \mathbf{W}e^{i\beta_2} + \mathbf{Z}e^{i\alpha_2} = \boldsymbol{\delta}_2 \tag{2.36}$$

Third position:

$$\mathbf{R} + \mathbf{W}e^{i\beta_3} + \mathbf{Z}e^{i\alpha_3} = \boldsymbol{\delta}_3 \tag{2.37}$$

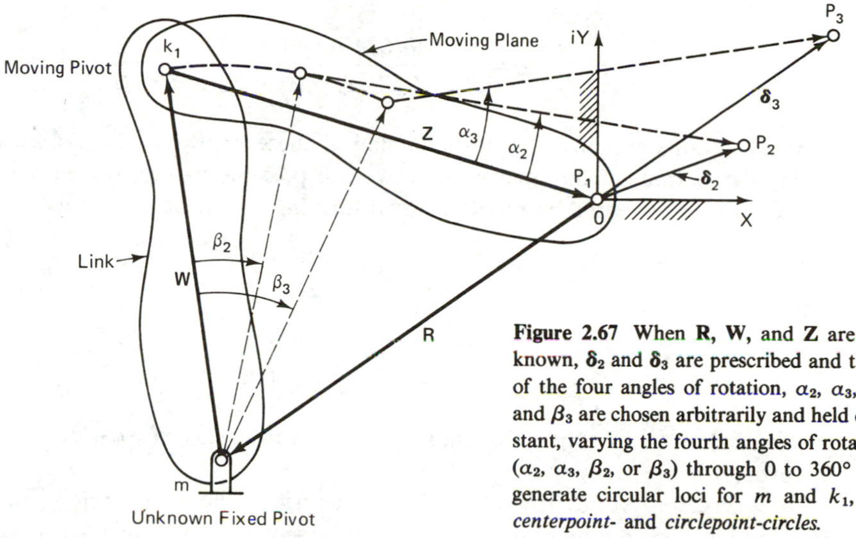


**Figure 2.67** When **R**, **W**, and **Z** are unknown, $\boldsymbol{\delta}_2$ and $\boldsymbol{\delta}_3$ are prescribed and three of the four angles of rotation, $\alpha_2$, $\alpha_3$, $\beta_2$, and $\beta_3$ are chosen arbitrarily and held constant, varying the fourth angles of rotation ($\alpha_2$, $\alpha_3$, $\beta_2$, or $\beta_3$) through 0 to 360° will generate circular loci for $m$ and $k_1$, the *centerpoint-* and *circlepoint-circles.*

The unknown location of the moving pivot $k_1$ is defined by the vector $-\mathbf{Z}$ with respect to $P_1$, the origin of the fixed coordinate system (as shown in Fig. 2.67), which coincides with the given initial position $P_1$ of the tracer point of the moving plane. The yet unknown fixed pivot $m$ is located by vector **R**. Synthesis problems can be formulated by specifying $\boldsymbol{\delta}_2$ and $\boldsymbol{\delta}_3$, plus the appropriate angular parameters. Vectors **R** and $-\mathbf{Z}$ may be obtained from Eqs. (2.35) to (2.37) (using Cramer's rule). With $e^{i\alpha_j} = \boldsymbol{\alpha}_j$ and $e^{i\beta_j} = \boldsymbol{\beta}_j$, these equations yield

$$\mathbf{R} = \frac{\begin{vmatrix} 0 & 1 & 1 \\ \boldsymbol{\delta}_2 & \boldsymbol{\beta}_2 & \boldsymbol{\alpha}_2 \\ \boldsymbol{\delta}_3 & \boldsymbol{\beta}_3 & \boldsymbol{\alpha}_3 \end{vmatrix}}{\begin{vmatrix} 1 & 1 & 1 \\ 1 & \boldsymbol{\beta}_2 & \boldsymbol{\alpha}_2 \\ 1 & \boldsymbol{\beta}_3 & \boldsymbol{\alpha}_3 \end{vmatrix}} \tag{2.38}$$

or

$$\mathbf{R} = \frac{\boldsymbol{\delta}_2(\boldsymbol{\beta}_3 - \boldsymbol{\alpha}_3) - \boldsymbol{\delta}_3(\boldsymbol{\beta}_2 - \boldsymbol{\alpha}_2)}{\boldsymbol{\alpha}_2 - \boldsymbol{\alpha}_3 + \boldsymbol{\beta}_3 - \boldsymbol{\beta}_2 + \boldsymbol{\beta}_2\boldsymbol{\alpha}_3 - \boldsymbol{\alpha}_2\boldsymbol{\beta}_3} \tag{2.39}$$

and

$$-\mathbf{Z} = \frac{\begin{vmatrix} 1 & 1 & 0 \\ 1 & \boldsymbol{\beta}_2 & \boldsymbol{\delta}_2 \\ 1 & \boldsymbol{\beta}_3 & \boldsymbol{\delta}_3 \end{vmatrix}}{\begin{vmatrix} 1 & 1 & -1 \\ 1 & \boldsymbol{\beta}_2 & -\boldsymbol{\alpha}_2 \\ 1 & \boldsymbol{\beta}_3 & -\boldsymbol{\alpha}_3 \end{vmatrix}} \tag{2.40}$$

or

$$-\mathbf{Z} = \frac{-\boldsymbol{\delta}_2(\boldsymbol{\beta}_3 - 1) + \boldsymbol{\delta}_3(\boldsymbol{\beta}_2 - 1)}{\boldsymbol{\beta}_2 - \boldsymbol{\beta}_3 + \boldsymbol{\alpha}_3 - \boldsymbol{\alpha}_2 - \boldsymbol{\beta}_2\boldsymbol{\alpha}_3 + \boldsymbol{\alpha}_2\boldsymbol{\beta}_3} \tag{2.41}$$

If all parameters on the right-hand side of these expressions are fixed except for an angular parameter $\theta$, which ranges over all possible values, the equations for **R** and **−Z** can be expressed as functions of $\theta$ forming "bilinear mappings" [170]:

$$\mathbf{R}(\theta) = \frac{\mathbf{a}\boldsymbol{\theta} + \mathbf{b}}{\mathbf{c}\boldsymbol{\theta} + \mathbf{d}} \tag{2.42}$$

$$-\mathbf{Z}(\theta) = \frac{\mathbf{e}\boldsymbol{\theta} + \mathbf{f}}{\mathbf{g}\boldsymbol{\theta} + \mathbf{h}} \tag{2.43}$$

where $\boldsymbol{\theta} = e^{i\theta}$ [$\theta$ stands for the angle to be varied ($\beta_2$, $\alpha_2$, $\alpha_3$, or $\beta_3$)], and where **a** through **h** are known.

When $\theta$ varies from 0 to $2\pi$, $\boldsymbol{\theta}$ describes the unit circle. Equations (2.42) and (2.43) are tantamount to the following sequence of transformations:

$$\mathbf{p}(\theta) = \mathbf{a}\boldsymbol{\theta}, \quad \text{a stretch rotation,} \tag{2.44}$$

$$\mathbf{q}(\theta) = \mathbf{a}\boldsymbol{\theta} + \mathbf{b}, \quad \text{a change of origin,} \tag{2.45}$$

$$\mathbf{r}(\theta) = \mathbf{c}\boldsymbol{\theta}, \quad \text{another stretch rotation,} \tag{2.46}$$

$$\mathbf{s}(\theta) = \mathbf{c}\boldsymbol{\theta} + \mathbf{d}, \quad \text{another change of origin,} \tag{2.47}$$

$$\mathbf{t}(\theta) = \frac{\mathbf{q}(\theta)}{\mathbf{s}(\theta)} \quad \text{a "bilinear mapping"} \tag{2.48}$$

Since both $\mathbf{q}(\theta)$ and $\mathbf{s}(\theta)$ are circles, it can be shown that $\mathbf{t}(\theta)$ is also a circle [170]. Thus it is seen that the loci of $\mathbf{R}(\theta)$ and $-\mathbf{Z}(\theta)$ are circles, which in the limit can become straight lines. The complex constants **a** through **h** are found by appropriately rearranging Eqs. (2.40) and (2.41) in the form of Eqs. (2.42) and (2.43). The centers of the circular loci $C_M$ and $C_K$ can be found directly from the constants **a** through **h,** or more simply by evaluating **R** and **−Z** at three $\theta$ values, yielding three points each, which define the circles. Either way the solutions are within the realm of programmable hand calculators. Computer programs have been written to display the $M$ and $K_1$ circles on a computer graphics terminal in order to examine their properties. This leads to the manual graphical constructions presented below.

When the circles have been drawn, the associated fixed and moving pivots on a pair of circles remain to be coordinated. This is done using the pole relationship presented in Fig. 2.68. For example, rays emanating from $P_{12}$ defining an angle $\frac{1}{2}\alpha_2$ will intersect the $M$ and $K_1$ circles at fixed and moving pivot pairs $m$ and $k_1$. An angle meter (adjustable protractor) rotated about the coordinating pole serves as a convenient tool for constructing such pivot pairs.

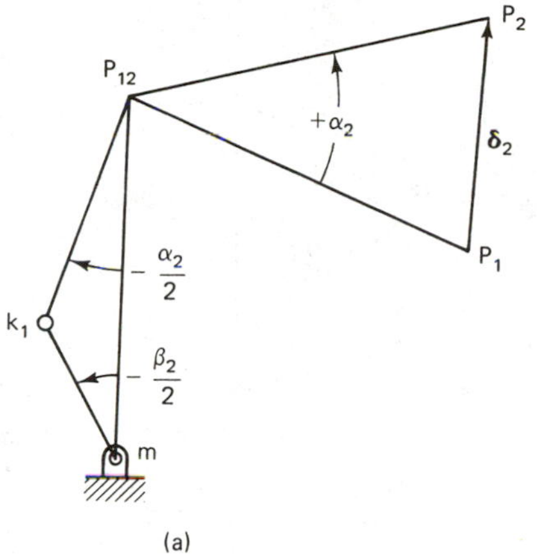


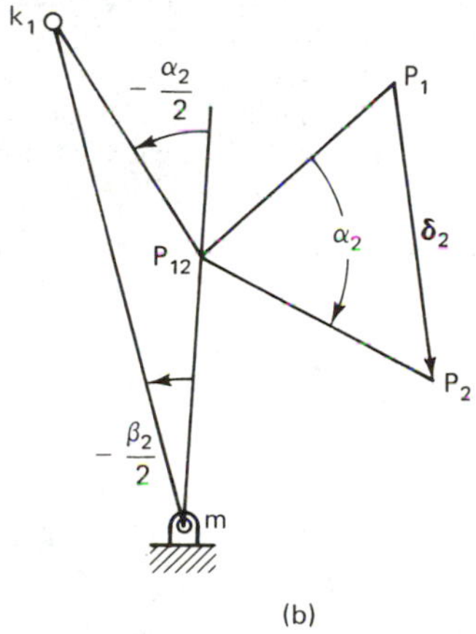


**Figure 2.68** Pole, center-, and circle-point configuration for $\alpha_2$ and $\beta_2$ having (a) opposite signs, and (b) equal signs ($k_1$ on $K_1$ circle, $m$ on $M$ circle).

## Graphical Constructions: Motion and Path Generation with Prescribed Timing

Examples of motion and path generation with prescribed timing are presented here to illustrate how ground- and moving-pivot circles may be generated on a graphics terminal.*

Sample computer plots have been generated for each problem type (Figs. 2.69 and 2.70). Points $m$ on the $M$ circles (with solid arcs) are labeled with the values of the angle held fixed while the circle was generated. A short line segment is directed from a sample fixed pivot $m$ on each $M$ circle toward the moving pivot

* More examples may be found in Ref. 170.

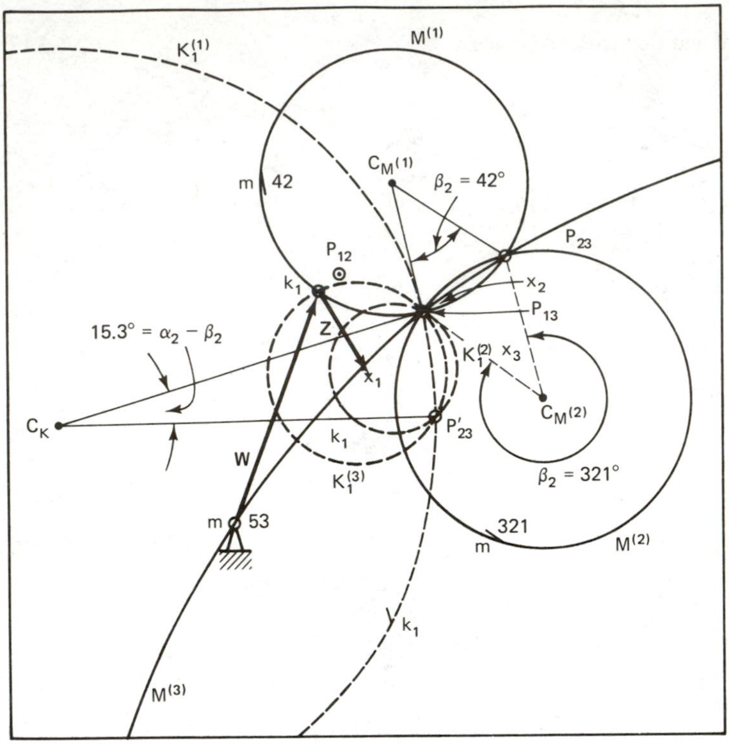


**Figure 2.69** *M* and $K_1$ circles for motion generation (see Ex. 2.6). Prescribed are $\boldsymbol{\delta}_2$, $\boldsymbol{\delta}_3$, $\alpha_2$, and $\alpha_3$, and **R**, **M** and $\mathbf{K}_1$ are the unknowns (see Fig. 2.67). Each pair of conjugate *M* and $K_1$ circles (say, $M^{(1)}$ and $K_1^{(1)}$) is generated by arbitrarily choosing a value for $\beta_2$, holding this constant, and varying $\beta_3$ from 0 to $2\pi$. For example, for $M^{(1)}$ and $K_1^{(1)}$, $\beta_2 = 42°$. A sample solution **W**, **Z** dyad is shown.

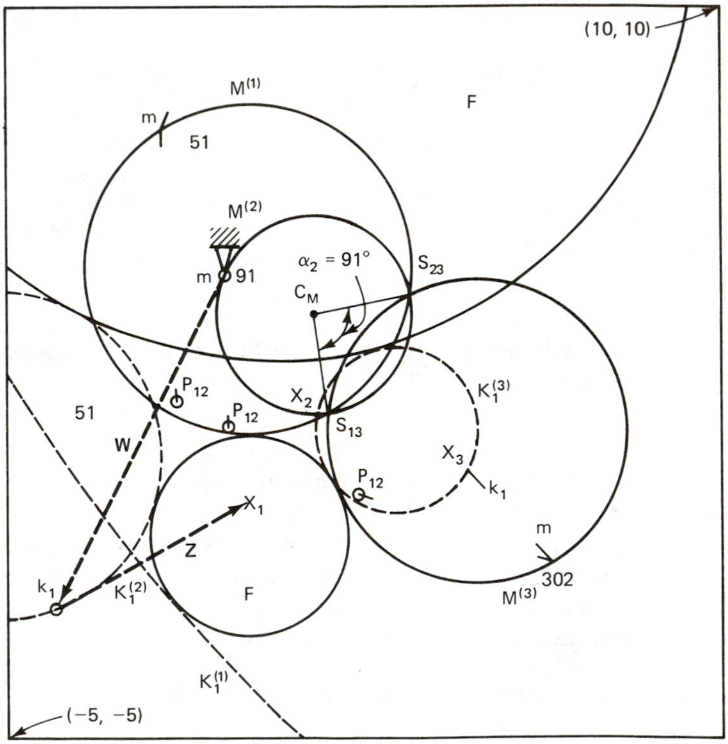


**Figure 2.70** M and $K_1$ circles for path generation with prescribed timing (see Ex. 2.7). Referring to Fig. 2.67, **R**, **W** and **Z** are unknown, $\boldsymbol{\delta}_2$, $\boldsymbol{\delta}_3$, $\beta_2$, and $\beta_3$ are prescribed. To generate a conjugate pair of *M* and $K_1$ circles, choose arbitrarily a value for $\alpha_2$ and let $\alpha_3$ vary from 0 to $2\pi$. A sample solution dyad is shown: **W**, **Z**.

$k_1$ on the conjugate $K_1$ circle (with dotted arcs), generated concurrently. The poles used in finding conjugate $m$–$k_1$ pairs are represented by small rings.

In path generation with prescribed timing (Fig. 2.70), the pole $P_{12}$ to be used for finding conjugate pivots on each $MK_1$ circle pair will be different for each such pair, since $\boldsymbol{\alpha}_2$ and $\boldsymbol{\alpha}_3$ are not the same for different pairs. A shorter line segment is then directed from the sample fixed pivot of each $M$ circle toward the associated pole $P_{12}$.

**Example 2.6: Motion Generation**

Figure 2.69 illustrates a motion-generation example. The intersections of the $X$ of $X_1$, $X_2$, and $X_3$ marks are the prescribed positions of the tracer point $P$, with $\boldsymbol{\delta}_2 = 1 + i$, $\boldsymbol{\delta}_3 = 2 + 0.5i$, $\alpha_2 = 1$ rad, and $\alpha_3 = 2$ rad. The $M$ and $K_1$ circles are generated for three values of the varied angle $\beta_2$ (42°, 321°, and 353°), with $\beta_3$ ranging from 0 to $2\pi$. These three circle-point circles and center-point circles are sufficient to display the properties of the diagram. It is interesting to note that the $M$ and $K_1$ circles intersect at points which can be shown to be the poles $P_{13}$ and $P_{23}$ for the $M$ circles and $P_{13}$, $P'_{23}$, for the $K_1$ circles, where $P'_{23}$ is the *image pole* of $P_{23}$ obtained by reflecting $P_{23}$ about the line $P_{12}P_{23}$. Similarly notable are the angles subtended at the circle centers by the line connecting the intersection (see Fig. 2.69). The steps for obtaining a set of circular $M$ and $K_1$ loci for a motion generation problem are therefore:

1. Find the circle-intersection poles: $P_{13}$ and $P_{23}$ for the $M$ circles: $P_{13}$ and $P'_{23}$ for the $K_1$ circles.
2. Bisect the lines between the intersection pole pairs to find the lines of centers for the $M$ and $K_1$ circles.
3. For each value of the angle $\beta_2$, lay off the circle centers so that $\measuredangle P_{13}C_MP_{23} = \beta_2$ and $\measuredangle P_{13}C_KP'_{23} = \alpha_2 - \beta_2$.
4. Draw the circle pairs through the intersection poles with centers $C_M$ and $C_K$.

It can be shown (Fig. 2.71a) that the complex-number expressions for the pole are

$$\mathbf{P}_{12} = \frac{\boldsymbol{\delta}_2}{1 - \boldsymbol{\alpha}_2} \tag{2.49}$$

$$\mathbf{P}_{13} = \frac{\boldsymbol{\delta}_3}{1 - \boldsymbol{\alpha}_3} \tag{2.50}$$

$$\mathbf{P}_{23} = \frac{\boldsymbol{\delta}_3\boldsymbol{\alpha}_2 - \boldsymbol{\delta}_2\boldsymbol{\alpha}_3}{\boldsymbol{\alpha}_2 - \boldsymbol{\alpha}_3} \qquad \text{(see Fig. 2.71b)} \tag{2.51}$$

where $\mathbf{P}_{ij}$ is the vector from the origin of $\mathbf{R}_i$ (Fig. 2.67) to the pole $P_{ij}$ and where $\boldsymbol{\alpha}_j = e^{i\alpha_j}$.

**Example 2.7: Path Generation with Prescribed Timing**

$M$ and $K_1$ circles for an example of path generation with prescribed timing are shown in Fig. 2.70, with $\boldsymbol{\delta}_2 = 1 + 2i$, $\boldsymbol{\delta}_3 = 4 + i$, $\beta_2 = 1$ rad, and $\beta_3 = 2$ rad. The $M$ and $K_1$ circles are generated for $\alpha_2 = 51°$, 91°, and 302°, with $\alpha_3$ ranging from 0 to 360°. The $M$ circles all have common intersections at the pseudopoles* $S_{13}$ and $S_{23}$, and

* The pseudopoles are defined as poles that would be obtained if the moving plane rotations were those of the grounded link, namely $\beta$ rather than $\alpha$ (see Fig. 2.67).

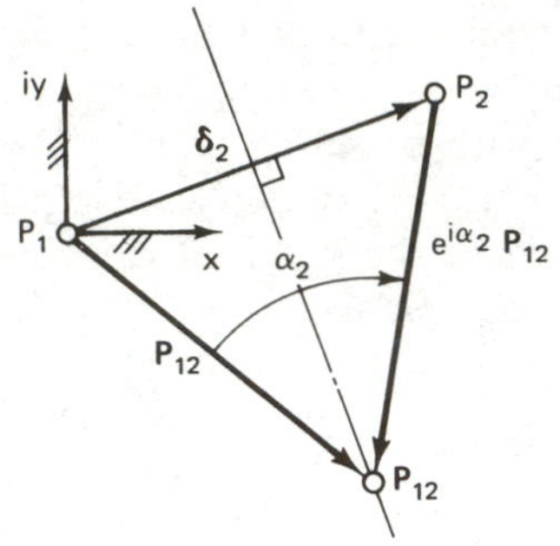


(a)

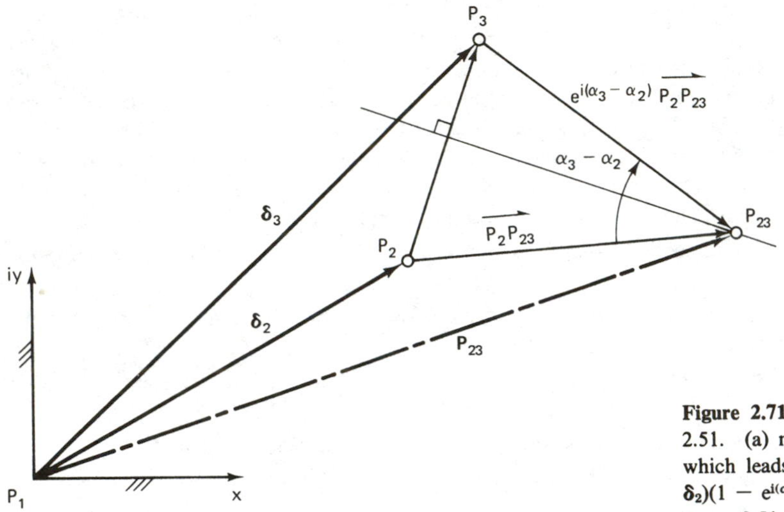


(b)

**Figure 2.71** Derivation of eqs. 2.49 and 2.51. (a) note that $\mathbf{P_{12}}\ (1 - e^{i\alpha_2}) = \boldsymbol{\delta}_2$, which leads to eq. 2.49; (b) here $(\mathbf{P_{23}} - \boldsymbol{\delta}_2)(1 - e^{i(\alpha_3 - \alpha_2)}) = \boldsymbol{\delta}_3 - \boldsymbol{\delta}_2$, which leads to eq. 2.51.

exhibit properties that allow $M$-circle construction with the steps used in motion generation, except that for each conjugate pair of $M$ and $K_1$ circles the varied angle $\alpha_2$ is chosen arbitrarily and then held constant, rather than $\beta_2$.

The $K_1$ circles have no intersections, but a useful property exists that permits easy construction: the $M$- and $K_1$-circle centers are coordinated about the poles, as are the $m$ and $k_1$ pivots (see Fig. 2.70). Referring to Fig. 2.72, $C_K$ and $k_1$ (the center of the $K_1$ circle and the moving pivot conjugate to either fixed pivot on the $M$ circle diameter through $P_{12}$) are found accordingly, allowing the $K_1$ circle to be drawn about $C_K$ with radius $|C_K K_1|$. Thus the $m$ and $K_1$ circles of path generation with prescribed timing can be found as follows:

1. Construct the pseudopoles $S_{13}$ and $S_{23}$.
2. Choose arbitrarily a value of $\alpha_2$ and find $C_M$ on the perpendicular bisector of $S_{13}S_{23}$ such that $\measuredangle\ S_{13}C_M S_{23} = \alpha_2$.
3. For each $M$ circle (see Fig. 2.72):
   a. Construct the pole $P_{12}$.

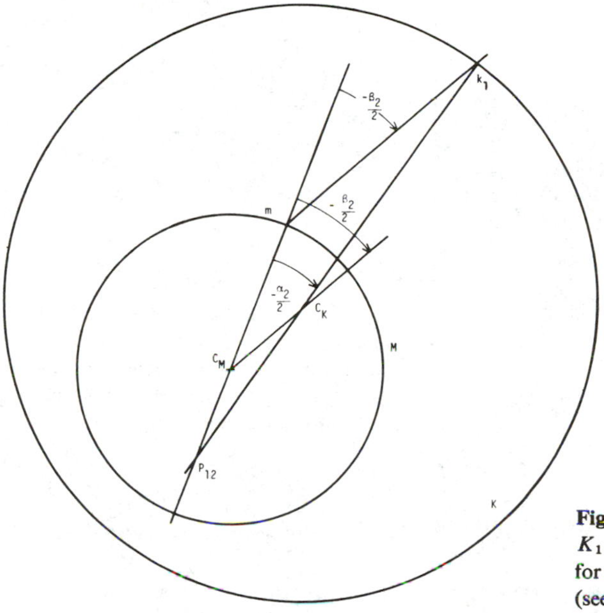


**Figure 2.72** Graphical construction of a $K_1$ circle conjugate to a known $M$ circle for path generation with prescribed timing (see Ex. 2.7, steps 1, 2, and 3).

b. Draw the diameter of the $M$ circle through $P_{12}$ and extend it.

c. From this line, lay off the angle $-\alpha_2/2$ with $P_{12}$ as the apex and the angle $-\beta_2/2$ with $C_M$ as the apex. The intersection of these two lines is $C_K$. Choose one of the intersections of the $C_M P_{12}$ line with the $M$ circle to be the fixed pivot $m$. With this point as the apex, lay off the angle $-\beta_2/2$ from the $P_{12}C_M m$ line. The intersection of this line with the $-\alpha_2/2$ line is $k_1$, and the $K_1$ circle can now be drawn.

## Dyad Moving-Pivot Existence: Three Precision Points

When computer plots of the $M$ and $K_1$ circles were made, it was found that moving pivots cannot exist within certain regions of the plane in path generation with prescribed timing: Two circles exist within which no moving pivots will be found. One of the circles surrounds the first path-point position $X_1$. It is possible to define analytically a distance bound from this point, within which no moving pivots can exist [170]. All $K_1$ circles are tangent to both nonexistence circles and are arranged so that two $k_1$ pivot solutions occur for each point outside the nonexistence circles. Figures 2.70 and 2.73 show the nonexistence circles (labeled as $F$, for "forbidden regions"). Of these, Fig. 2.73 is drawn for the example:

$$\delta_2 = 2 + 2i; \qquad \beta_2 = 0.5 \text{ rad}$$

$$\delta_3 = 4 + i; \qquad \beta_3 = 1 \text{ rad}$$

The circles shown in Fig. 2.73 are the $K_1$ circles which are tangent to the two nonexistence $F$ circles. The smaller $F$ circle surrounds the initial precision point, $X_1$.

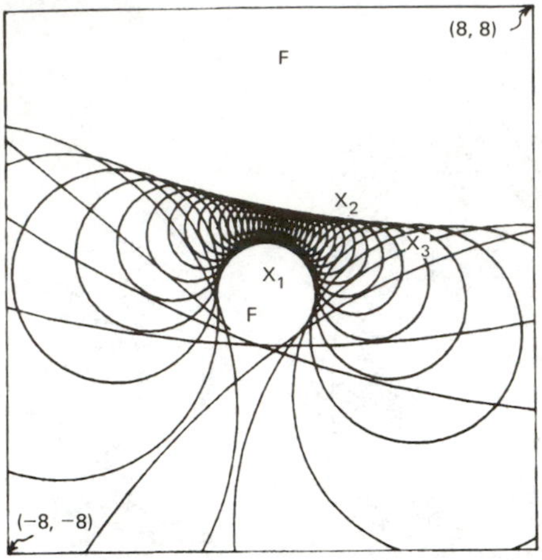


**Figure 2.73** Circular non-existence regions F for path generation with prescribed timing, within which $K_1$ circles cannot exist. All existing $K_1$ circles are tangent to these "forbidden" F regions.

## 2.21 GROUND-PIVOT SPECIFICATION [171]

There is yet another useful strategy for choosing the two free choices in the system of equations for three finitely separated precision positions of a dyad (Table 2.1). Recall that in each case in Sec. 2.16 two rotation angles were chosen arbitrarily, yielding a simple set of linear equations. In the preceding section, it was observed that varying one of the free-choice angles as a parameter produces circular loci of center and circle points. With two free choices available, however, one of the link vectors (**W** or **Z**) can be assumed arbitrarily instead. In fact, by writing the dyad equations in a different form, a ground- or moving-pivot location may be specified directly.

Figure 2.74 shows a dyad in three finitely separated positions. The synthesis equations can be written as*

$$\mathbf{W} + \mathbf{Z} = \mathbf{R}_1 \tag{2.52}$$

$$\mathbf{W}e^{i\beta_2} + \mathbf{Z}e^{i\alpha_2} = \mathbf{R}_2 \tag{2.53}$$

$$\mathbf{W}e^{i\beta_3} + \mathbf{Z}e^{i\alpha_3} = \mathbf{R}_3 \tag{2.54}$$

Suppose that we wish to synthesize the dyad in Fig. 2.74 for motion generation. According to Sec. 2.4, this requires that $\boldsymbol{\delta}_2 = (\mathbf{R}_2 - \mathbf{R}_1)$, $\boldsymbol{\delta}_3 = (\mathbf{R}_3 - \mathbf{R}_1)$, $\alpha_2$, and $\alpha_3$ be specified. Subtracting Eq. (2.52) from (2.53) and (2.52) from (2.54) will in fact yield the standard form (2.16). Table 2.1 requires two additional free choices for this system of equations. Let us specify $\mathbf{R}_1$ (which locates the ground pivot). Thus the coefficients of **Z** and $\mathbf{R}_j$ are known.

If we view Eqs. (2.52) to (2.54) temporarily as three complex equations linear and nonhomogeneous in the two complex unknowns **W** and **Z**, this set has a solution

* Note that in Fig. 2.74 $\mathbf{R}_1 = -\mathbf{R}$, as defined in Fig. 2.67.

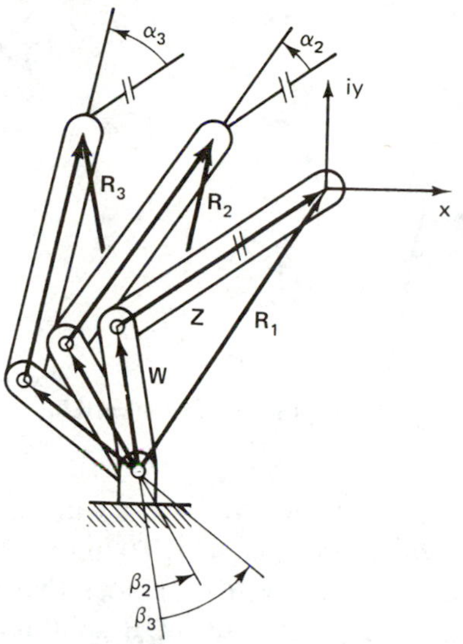


**Figure 2.74** Three discrete positions of the unknown dyad **W,Z.** For synthesis for motion generation with specified ground pivot, $\mathbf{R}_j$, $j = 1,2,3$, $\alpha_2$ and $\alpha_3$ are prescribed and $\beta_j$, $j = 2,3$, are to be found before the system of Eqs. 2.52 to 2.54 can be solved for **W** and **Z.**

for **W** and **Z** only if the determinant of the augmented matrix of the coefficient is identically zero:

$$\begin{vmatrix} 1 & 1 & \mathbf{R}_1 \\ e^{i\beta_2} & e^{i\alpha_2} & \mathbf{R}_2 \\ e^{i\beta_3} & e^{i\alpha_3} & \mathbf{R}_3 \end{vmatrix} = 0 \tag{2.55}$$

Equation (2.55) represents a complex equation with two unknowns, $\beta_2$ and $\beta_3$. Since the unknowns are in the first column, the determinant is expanded about this column:

$$(\mathbf{R}_3 e^{i\alpha_2} - \mathbf{R}_2 e^{i\alpha_3}) + e^{i\beta_2}(-\mathbf{R}_3 + \mathbf{R}_1 e^{i\alpha_3}) + e^{i\beta_3}(\mathbf{R}_2 - \mathbf{R}_1 e^{i\alpha_2}) = 0 \tag{2.56}$$

or

$$\mathbf{D}_1 + \mathbf{D}_2 e^{i\beta_2} + \mathbf{D}_3 e^{i\beta_3} = 0 \tag{2.57}$$

which is transcendental in the unknowns $\beta_2$ and $\beta_3$, and where

$$\begin{aligned} \mathbf{D}_1 &= \mathbf{R}_3 e^{i\alpha_2} - \mathbf{R}_2 e^{i\alpha_3} \\ \mathbf{D}_2 &= \mathbf{R}_1 e^{i\alpha_3} - \mathbf{R}_3 \\ \mathbf{D}_3 &= \mathbf{R}_2 - \mathbf{R}_1 e^{i\alpha_2} \end{aligned} \tag{2.58}$$

are known from prescribed data.

A simple graphical construction aids in solving Eq. (2.57) for $\beta_2$ and $\beta_3$. Figure 2.75 shows a geometric solution where the knowns $\mathbf{D}_1$, $\mathbf{D}_2$, and $\mathbf{D}_3$ are represented as vectors. Notice that $\mathbf{D}_3$ and $\mathbf{D}_2$ are pinned to $\mathbf{D}_1$ but vector $\mathbf{D}_1$ is fixed. Note in Eq. (2.57) that vectors $\mathbf{D}_2$ and $\mathbf{D}_3$ are multiplied by $e^{i\beta_2}$ and $e^{i\beta_3}$, respectively. These quantities are regarded as *rotation operators.*

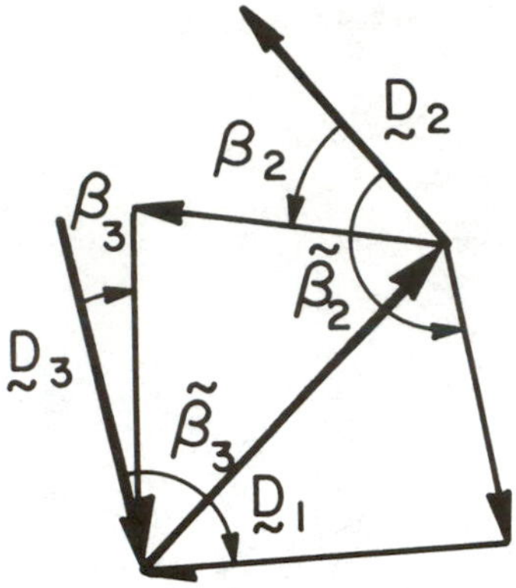


**Figure 2.75** Graphical solution of Eq. 2.56 for $\beta_2$ and $\beta_3$.

In Fig. 2.75, if the vectors form a closed loop, Eq. (2.57) will be satisfied. Thus $\mathbf{D}_2$ and $\mathbf{D}_3$ are rotated about their pin connection with $\mathbf{D}_1$ until they meet. The rotations required to close the loop are then $\beta_2$ and $\beta_3$. Notice that there are two solutions for the triangle: $\beta_2$, $\beta_3$ and $\tilde{\beta}_2$, $\tilde{\beta}_3$. One set of $\beta$ solutions will be trivial, however. The trivial solution is $\beta_2 = \alpha_2$ and $\beta_3 = \alpha_3$. This can be verified by plugging the trivial roots back into Eq. (2.57).*

Once the nontrivial set of $\beta$'s is found (either graphically or analytically based on Fig. 2.75), plugging them into Eqs. (2.53) and (2.54) will make the set (2.52) to (2.54) compatible. Thus, **W** and **Z** can be found as the simultaneous solution of any two of the set, and the synthesis of the motion generator dyad with prescribed ground-pivot location is completed.

**Example 2.8**

An engineering student who had recently purchased a tape unit for his sports car was concerned with possible theft of his investment. Therefore, the student envisioned synthesis of a four-bar linkage to hide the tape player behind the glove compartment when not is use. Figure 2.76 shows a cross section of the area of interest, including the glove compartment and heating duct as well as the three prescribed positions for the tape unit. Since there is a small acceptable area for possible ground pivots, the method of Sec. 2.21 is used as a synthesis tool.

**Solution** Two ground-pivot locations are chosen: $A_0$ and $B_0$, as shown in Fig. 2.76. The three position vectors for the first dyad of the four-bar are

$$\mathbf{R}_1 = 2.14 - 3.68i$$

$$\mathbf{R}_2 = 4.46 - 0.63i$$

$$\mathbf{R}_3 = 4.10 + 3.22i$$

while the rotations of the coupler (the tape player) are

$$\alpha_2 = 50.7°$$

$$\alpha_3 = 91.9°$$

* Hand-calculator program listing for this solution is available from the first author.

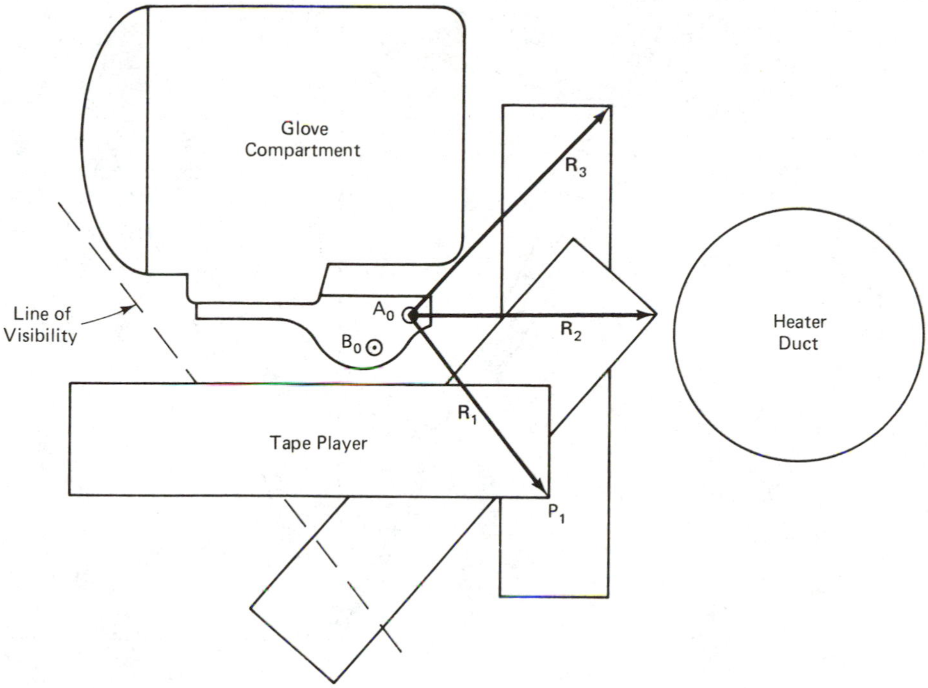


**Figure 2.76** Four-bar motion-generation synthesis with prescribed ground pivots. See Example 2.8.

Using Eqs. (2.58) yields

$$\mathbf{D}_1 = -0.377 + 0.734i$$

$$\mathbf{D}_2 = -0.493 - 0.959i$$

$$\mathbf{D}_3 = 0.257 + 0.0448i$$

and with these, Eq. (2.57) gives

$$\beta_2 = 58.09°$$

$$\beta_3 = 122.70° \qquad \text{(see Fig. 2.77a)}$$

with which the results of simultaneous solution of any two of Eqs. (2.52), (2.53), and (2.54) will be

$$\mathbf{W} = -1.42 - 1.45i$$

$$\mathbf{Z} = 3.56 - 2.23i$$

Figure 2.77b shows the synthesized linkage in its starting, final, and two intermediate positions. The other side of the linkage is designed the same way.*

* Hand-calculator program for this method is available from the second author.

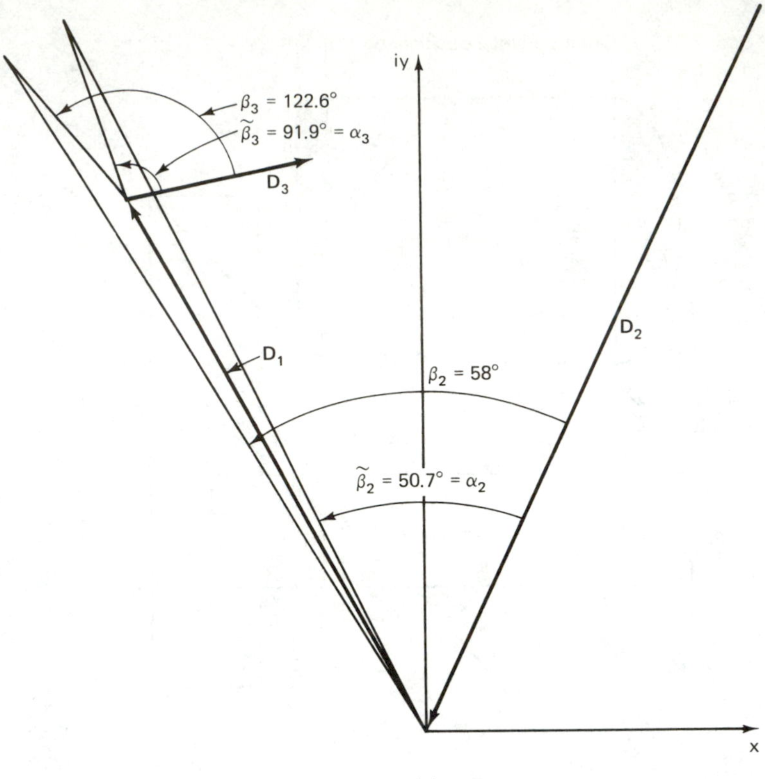


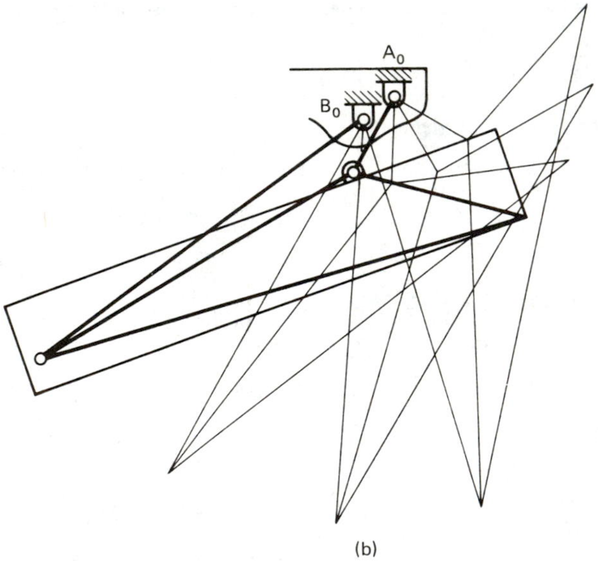


**Figure 2.77** Four-bar motion generator synthesized to the requirements shown in Fig. 2.76. Note the specified ground-pivot locations; (a) graphical check of calculated values of $\beta_2$ and $\beta_3$ for the first dyad pivoted at $A_0$ [Eq. (2.56) and Fig. 2.75]; (b) the synthesized four-bar mechanism in four positions.

## 2.22 FREUDENSTEIN'S EQUATION FOR THREE-POINT FUNCTION GENERATION

Another well-known analytical synthesis method is based on *Freudenstein's Equation technique* [102, 257]. This algebraic method utilizes Freudenstein's displacement equations for three-precision-point function generation. This technique has been extended to four and five precision points and, by regarding the scale factors $R_\phi$ and $R_\psi$ of input and output rotations as unknowns, also to six and seven precision points. It has also been extended to other linkages, but these cases are not presented here.

The equation can be derived from the loop-closure equation written for Fig. 2.78 (notice that in this section all angles are the arguments of the link vectors in all positions):

$$\mathbf{Z}_1 + \mathbf{Z}_2 + \mathbf{Z}_3 - \mathbf{Z}_4 = 0 \tag{2.59}$$

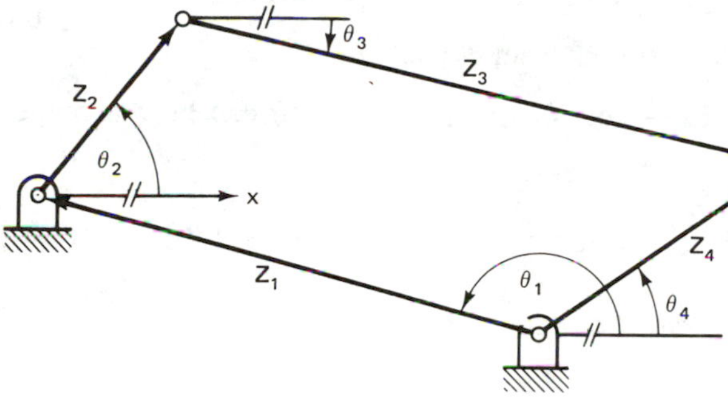


**Figure 2.78** Freudenstein's equation is based on the closure of the four-bar loop [Eq. (2.59), etc.].

If this complex equation is separated into real and imaginary components, two algebraic equations are produced:

$$Z_1 \cos\theta_1 + Z_2 \cos\theta_2 + Z_3 \cos\theta_3 - Z_4 \cos\theta_4 = 0 \tag{2.60}$$

$$Z_1 \sin\theta_1 + Z_2 \sin\theta_2 + Z_3 \sin\theta_3 - Z_4 \sin\theta_4 = 0 \tag{2.61}$$

Assuming that the ground link is along the $x$ axis (as in Fig. 2.79), $\theta_1 = 180°$ and

$$-Z_1 + Z_2 \cos\theta_2 + Z_3 \cos\theta_3 - Z_4 \cos\theta_4 = 0 \tag{2.62}$$

$$Z_2 \sin\theta_2 + Z_3 \sin\theta_3 - Z_4 \sin\theta_4 = 0 \tag{2.63}$$

Since we wish to synthesize a function generator, $\theta_3$ is not of interest and will be eliminated by transferring the $Z_3$ terms to the right-hand side of Eqs. (2.62) and (2.63):

$$-Z_1 + Z_2 \cos\theta_2 - Z_4 \cos\theta_4 = -Z_3 \cos\theta_3 \tag{2.64}$$

$$Z_2 \sin\theta_2 - Z_4 \sin\theta_4 = -Z_3 \sin\theta_3 \tag{2.65}$$

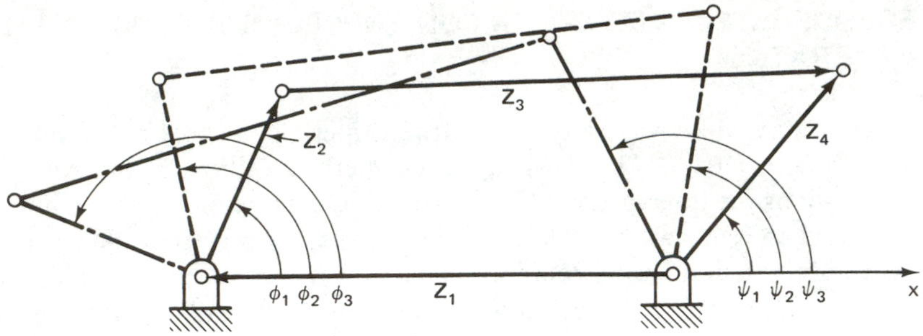


**Figure 2.79** Notation for the four-bar mechanism for writing Freudenstein's equation [Eq. (2.70)].

Next, Eqs. (2.64) and (2.65) are squared and added together to eliminate $\theta_3$. The resulting equation is

$$\begin{aligned} Z_3^2 = Z_1^2 + Z_2^2 + Z_4^2 - 2Z_1Z_2 \cos \theta_2 + 2Z_1Z_4 \cos \theta_4 \\ - 2Z_2Z_4(\cos \theta_2 \cos \theta_4 + \sin \theta_2 \sin \theta_4) \end{aligned} \tag{2.66}$$

Since $\cos \theta_2 \cos \theta_4 + \sin \theta_2 \sin \theta_4 = \cos (\theta_2 - \theta_4)$, Eq. (2.66) can be rearranged as

$$\frac{Z_3^2 - Z_1^2 - Z_2^2 - Z_4^2}{2Z_2Z_4} + \frac{Z_1}{Z_4} \cos \theta_2 - \frac{Z_1}{Z_2} \cos \theta_4 = -\cos (\theta_2 - \theta_4) \tag{2.67}$$

In a more compact form, *Freudenstein's Equation* reads

$$K_1 \cos \theta_2 + K_2 \cos \theta_4 + K_3 = -\cos (\theta_2 - \theta_4) \tag{2.68}$$

where

$$\begin{aligned} K_1 &= \frac{Z_1}{Z_4} \\ K_2 &= \frac{-Z_1}{Z_2} \\ K_3 &= \frac{Z_3^2 - Z_1^2 - Z_2^2 - Z_4^2}{2Z_2Z_4} \end{aligned} \tag{2.69}$$

Notice that the $K$'s are three independent algebraic expressions containing the three unknown lengths of the links. Freudenstein's equation is a displacement equation for the four-bar linkage which holds true for each position of the linkage. Thus, for three prescribed positions, the equation can be written for each position. The notation will be changed at this point to avoid double subscripts: the three angles for the three prescribed positions of $Z_2$ with respect to the fixed $x$ axis will be $\phi_1$, $\phi_2$, and $\phi_3$, while those of $Z_4$ will be $\psi_1$, $\psi_2$, and $\psi_3$, as in Fig. 2.79. Thus Freudenstein's equation for three prescribed positions is

$$\begin{aligned} K_1 \cos \phi_1 + K_2 \cos \psi_1 + K_3 &= -\cos (\phi_1 - \psi_1) \\ K_1 \cos \phi_2 + K_2 \cos \psi_2 + K_3 &= -\cos (\phi_2 - \psi_2) \\ K_1 \cos \phi_3 + K_2 \cos \psi_3 + K_3 &= -\cos (\phi_3 - \psi_3) \end{aligned} \tag{2.70}$$

Cramer's rule may be used to solve system (2.70). To find the $Z$'s, one length, say $Z_1$, is arbitrarily picked to scale the function generator (as was done in the previous function-generator techniques).

Dealing with third-order determinants may be avoided by first subtracting the second and third equations from the first, eliminating $K_3$:

$$K_1(\cos \phi_1 - \cos \phi_2) + K_2(\cos \psi_1 - \cos \psi_2) = -\cos (\phi_1 - \psi_1) + \cos (\phi_2 - \psi_2) \tag{2.71}$$

$$K_1(\cos \phi_1 - \cos \phi_3) + K_2(\cos \psi_1 - \cos \psi_3) = -\cos (\phi_1 - \psi_1) + \cos (\phi_3 - \psi_3) \tag{2.72}$$

and solving the resulting system of two equations for $K_1$ and $K_2$:

$$K_1 = \frac{\omega_3\omega_5 - \omega_2\omega_6}{\omega_1\omega_5 - \omega_2\omega_4}, \qquad K_2 = \frac{\omega_1\omega_6 - \omega_3\omega_4}{\omega_1\omega_5 - \omega_2\omega_4} \tag{2.73}$$

in which

$$\begin{aligned} \omega_1 &= \cos \phi_1 - \cos \phi_2, & \omega_4 &= \cos \phi_1 - \cos \phi_3 \\ \omega_2 &= \cos \psi_1 - \cos \psi_2, & \omega_5 &= \cos \psi_1 - \cos \psi_3 \\ \omega_3 &= -\cos (\phi_1 - \psi_1) + \cos (\phi_2 - \psi_2), & \omega_6 &= -\cos (\phi_1 - \psi_1) \\ & & &\quad + \cos (\phi_3 - \psi_3) \end{aligned} \tag{2.74}$$

Substituting values of $K_1$ and $K_2$ into either of Eqs. (2.70) yields

$$K_3 = -\cos (\phi_i - \psi_i) - K_1 \cos \phi_i - K_2 \cos \psi_i, \qquad i = 1, 2, \text{ or } 3 \tag{2.75}$$

The link lengths may be expressed in terms of the known $K$'s by using Eqs. (2.69) (having chosen the length of $Z_1$):

$$\begin{aligned} Z_4 &= \frac{Z_1}{K_1} \\ Z_2 &= \frac{Z_1}{K_2} \\ Z_3 &= \sqrt{2K_3Z_2Z_4 + Z_1^2 + Z_2^2 + Z_4^2} \end{aligned} \tag{2.76}$$

By using Freudenstein's equation, as in the other synthesis methods, one may obtain *two infinities* of solutions for the same set of precision points. All that is required is to shift the precision points so that the starting position input and output angles, $\phi_1$ and $\psi_1$ vary between 0 and 360°. Each new $\phi_1$ or $\psi_1$ will yield a new solution.

**Example 2.9**

Let us synthesize the same function generator as in Sec. 2.10 using the Freudenstein equation. Recall that the function to be synthesized was $y = \sin(x)$ for $0° \le x \le 90°$. The range in $\phi$ is 120° and the range in $\psi$ is 60° (see Fig. 2.48). Chebyshev spacing yielded the following precision points:

$$\phi_2 - \phi_1 = 52°, \qquad \psi_2 - \psi_1 = 36.15°$$

$$\phi_3 - \phi_1 = 104°, \qquad \psi_3 - \psi_1 = 53.40°$$

In order to obtain the same four-bar solution as Fig. 2.49, we set $\phi_1 = 105°$ and $\psi_1 = 66.27°$. Thus the absolute precision points (required for this method) are

$$\phi_1 = 105°, \qquad \psi_1 = 66.27°$$

$$\phi_2 = 157°, \qquad \psi_2 = 102.42°$$

$$\phi_3 = 209°, \qquad \psi_3 = 119.67°$$

Thus

$$\omega_1 = \cos(105.0) - \cos(157.0) = 0.662$$

$$\omega_2 = \cos(66.27) - \cos(102.42) = 0.618$$

$$\omega_3 = -\cos(105.0 - 66.27) + \cos(157.0 - 102.42) = -0.201$$

$$\omega_4 = \cos(105.0) - \cos(209.0) = 0.616$$

$$\omega_5 = \cos(66.27) - \cos(119.67) = 0.897$$

$$\omega_6 = \cos(105.0 - 66.27) + \cos(209.0 - 119.67) = -0.768$$

Solving for $K_1$, $K_2$, and $K_3$ from Eqs. (2.73) and (2.75) yields

$$K_1 = \frac{-0.180 + 0.475}{0.594 - 0.381} = 1.385$$

$$K_2 = \frac{-0.504 + 0.124}{0.594 - 0.381} = -1.803$$

$$K_3 = 0.304$$

If $Z_1 = 52.5$ mm, then from Eqs. (2.76)

$$Z_2 = 29.0 \text{ mm}$$

$$Z_4 = 38.0 \text{ mm}$$

$$Z_3 = 75.6 \text{ mm}$$

which agrees with the graphical solution of Fig. 2.49.

## 2.23 LOOP-CLOSURE-EQUATION TECHNIQUE

An alternative method for synthesizing function generators is the loop-closure-equation technique. In the case of function generation, the bar-slider members of the general chain of Fig. 2.53 form one or more closed polygons. Therefore, using the notation

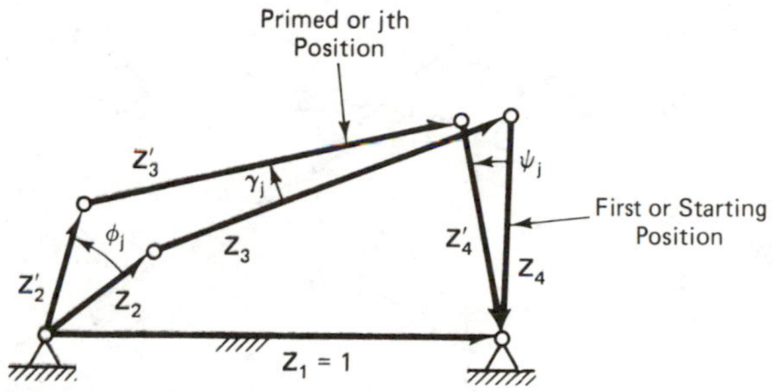


**Figure 2.80** Notation for the four-bar mechanism for writing its loop-closure equation [Eq. (2.81)].

of Fig. 2.80, with the link vectors $\mathbf{Z}_k$ identifying the mechanism loop, one can write the *equation of closure* (or *loop closure equation*) for the $j$th position of the four-bar chain as follows:

$$\sum_{k=1}^{4} \mathbf{Z}_k = 0 \tag{2.77}$$

Specifically, the equation of closure for the four-bar linkage (Fig. 2.80) in its first position will be

$$\mathbf{Z}_2 + \mathbf{Z}_3 + \mathbf{Z}_4 - \mathbf{Z}_1 = 0 \tag{2.78}$$

In case of function generation, only angular relationships of the four-bar linkage are of interest; doubling the size of the mechanism, or even a stretch-rotation, does not alter the rotations between the links of that mechanism. Therefore, we may set $\mathbf{Z}_1 = 1$. Then

$$\mathbf{Z}_2 + \mathbf{Z}_3 + \mathbf{Z}_4 - 1.0 = 0 \tag{2.79}$$

The $j$th position of the four-bar linkage of Fig. 2.80 can be expressed as

$$\mathbf{Z}_2' + \mathbf{Z}_3' + \mathbf{Z}_4' - 1.0 = 0 \tag{2.80}$$

or, using Eq. (2.12),

$$\mathbf{Z}_2 e^{i\phi_j} + \mathbf{Z}_3 e^{i\gamma_j} + \mathbf{Z}_4 e^{i\psi_j} - 1.0 = 0 \tag{2.81}$$

Equation (2.81) is an example of a *displacement equation* which is nonhomogeneous and linear in the complex unknowns $\mathbf{Z}_k$, $k = 2, 3, 4$, and has complex coefficients $e^{i\phi_j}$, $e^{i\psi_j}$ and $e^{i\gamma_j}$.

In some cases the stretch-rotation operator is useful in writing a displacement equation. The displacement equations for the first and $j$th position for the offset slider crank of Fig. 2.81 are

$$\mathbf{Z}_4 + \mathbf{Z}_3 + \mathbf{Z}_2 - \mathbf{Z}_1 = 0 \tag{2.82}$$

and

$$\mathbf{Z}_4 + \mathbf{Z}_3 e^{i\phi_j} + \mathbf{Z}_2 e^{i\gamma_j} - \rho_j \mathbf{Z}_1 = 0 \tag{2.83}$$

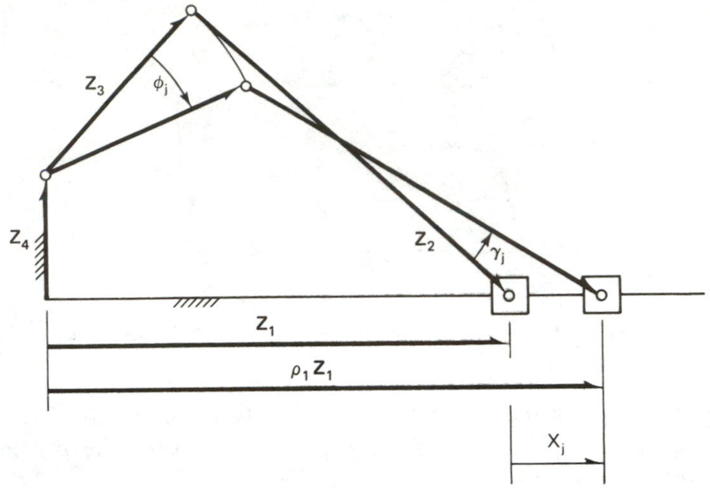


**Figure 2.81** Notation for the offset slider-crank for writing its loop-closure equation [Eq. (2.83)].

where

$$\rho_j \equiv \frac{Z_1 + x_j}{Z_1}$$

is a stretch ratio.

## Applications of the Loop-Closure Method for Function Generation (Three Precision Points)

Suppose that we wish to synthesize a four-bar linkage for function generation where three precision points are to be prescribed (three-point approximation). From Eq. (2.81), the three equations that represent the loop equations of Fig. 2.80 for an initial position and two displacements are

$$\begin{aligned} \mathbf{Z}_2 + \mathbf{Z}_3 + \mathbf{Z}_4 &= 1.0 \\ \mathbf{Z}_2 e^{i\phi_2} + \mathbf{Z}_3 e^{i\gamma_2} + \mathbf{Z}_4 e^{i\psi_2} &= 1.0 \\ \mathbf{Z}_2 e^{i\phi_3} + \mathbf{Z}_3 e^{i\gamma_3} + \mathbf{Z}_4 e^{i\psi_3} &= 1.0 \end{aligned} \tag{2.84}$$

Since we are synthesizing this linkage for function generation where the rotations of $\mathbf{Z}_2$ and $\mathbf{Z}_4$ are to be prescribed according to a functional relationship, $\phi_2$, $\phi_3$, $\psi_2$, and $\psi_3$ are prescribed. The unknowns in the system of equations (2.84) are the vectors $\mathbf{Z}_2$, $\mathbf{Z}_3$, and $\mathbf{Z}_4$ (which represent the four-bar linkage in its first position) and the two rotations $\gamma_2$ and $\gamma_3$ of the coupler link. There are six independent equations (real and imaginary parts of each complex equation) and eight unknown reals. If we choose some arbitrary values for $\gamma_2$ and $\gamma_3$, the system of Eqs. (2.84) is nonhomogeneous and linear in the unknown link vectors $\mathbf{Z}_2$, $\mathbf{Z}_3$, and $\mathbf{Z}_4$. The complex coefficients of the link vectors are now known and the number of unknowns equals the number of equations. Cramer's rule may be used to solve these equations:

$$\mathbf{Z}_2 = \frac{\begin{vmatrix} 1 & 1 & 1 \\ 1 & e^{i\gamma_2} & e^{i\psi_2} \\ 1 & e^{i\gamma_3} & e^{i\psi_3} \end{vmatrix}}{\mathbf{D}} \tag{2.85}$$

$$\mathbf{Z}_3 = \frac{\begin{vmatrix} 1 & 1 & 1 \\ e^{i\phi_2} & 1 & e^{i\psi_2} \\ e^{i\phi_3} & 1 & e^{i\psi_3} \end{vmatrix}}{\mathbf{D}} \tag{2.86}$$

$$\mathbf{Z}_4 = \frac{\begin{vmatrix} 1 & 1 & 1 \\ e^{i\phi_2} & e^{i\gamma_2} & 1 \\ e^{i\phi_3} & e^{i\gamma_3} & 1 \end{vmatrix}}{\mathbf{D}} \tag{2.87}$$

where

$$\mathbf{D} = \begin{vmatrix} 1 & 1 & 1 \\ e^{i\phi_2} & e^{i\gamma_2} & e^{i\psi_2} \\ e^{i\phi_3} & e^{i\gamma_3} & e^{i\psi_3} \end{vmatrix} \tag{2.88}$$

Notice that in Sec. 2.16, the standard form resulted in $2 \times 2$ determinants rather than the $3 \times 3$ determinants here. The same four-bar solution will result from either method, except that the scale and orientation of the synthesized linkages will differ. The standard form may be derived easily from the loop-closure technique by subtracting Eq. (2.79) from Eq. (2.81) and choosing one of the link vectors arbitrarily.

## 2.24 ORDER SYNTHESIS: FOUR-BAR FUNCTION GENERATION

In many instances a kinematic synthesis objective involves specifying not only finitely separated positions but velocities, accelerations, or higher derivatives as well. This is called *order synthesis,* and it may be accomplished using the complex-number loop-closure method by taking derivatives of the position equations.

Figure 2.78 shows a four-bar function generator in which we wish to specify the relative angular velocity, angular acceleration, and so on, of the output link with respect to those of the input link. Recall from the preceding section that the loop-closure (or position) equation is written for the starting position as follows:

$$\mathbf{Z}_2 + \mathbf{Z}_3 - \mathbf{Z}_4 = -\mathbf{Z}_1 \tag{2.89}$$

Since it is more convenient to take derivatives of vectors in the polar form, Eq. (2.89) is expressed as

$$\textit{Position equation: } \mathbf{Z}_2 e^{i\theta_2} + \mathbf{Z}_3 e^{i\theta_3} - \mathbf{Z}_4 e^{i\theta_4} = \mathbf{Z}_1 e^{i\theta_1} \tag{2.90}$$

The velocity equation is formed by taking the derivative of Eq. (2.90) with respect to time ($\omega_i = d\theta_i/dt$):

$$\textit{Velocity equation: } Z_2\omega_2 i e^{i\theta_2} + Z_3\omega_3 i e^{i\theta_3} - Z_4\omega_4 i e^{i\theta_4} = 0 \tag{2.91}$$

or

$$\mathbf{Z}_2\omega_2 + \mathbf{Z}_3\omega_3 - \mathbf{Z}_4\omega_4 = 0 \tag{2.92}$$

where the fixed link $\mathbf{Z}_1$ has dropped out because $\omega_1 = 0$.

Notice that Eq. (2.91) is equivalent to the velocity polygon technique of Chap. 3 of Vol. 1, in which all position vectors are rotated by 90°. The acceleration equation requires the second derivative of Eq. (2.90) obtained by taking the time derivative of Eq. (2.91) ($\alpha_i \equiv d\omega_i/dt$):

$$\begin{aligned}\textit{Acceleration equation: } & Z_2(i\alpha_2 - \omega_2^2)e^{i\theta_2} + Z_3(i\alpha_3 - \omega_3^2)e^{i\theta_3} \\ & - Z_4(i\alpha_4 - \omega_4^2)e^{i\theta_4} = 0\end{aligned} \tag{2.93}$$

or

$$\mathbf{Z}_2(i\alpha_2 - \omega_2^2) + \mathbf{Z}_3(i\alpha_3 - \omega_3^3) - \mathbf{Z}_4(i\alpha_4 - \omega_4^2) = 0 \tag{2.94}$$

Further derivatives may be obtained following the same procedure. The question arises: For how many derivatives can we synthesize the four-bar function generator? Following the logic laid out in Sec. 2.15, only a finite number of derivatives may be specified. Table 2.6 shows some not so surprising results for the four-bar function generator. With $\mathbf{Z}_1$ specified as $(-1.0 + 0.0i)$ (which establishes the linkage scale and orientation) as well as the derivatives of $\theta_2$ and $\theta_4$, the maximum number of positions plus derivatives is five.* However, if the scale factors $R_{\theta_2}$ and $R_{\theta_4}$ are also regarded as unknowns and the system of equations is written in terms of the independent and dependent variables $x$ and $y$† [where $y = f(x)$ is the function to

* Note that each added derivative adds an "infinitesimally separated" position.

† This is accomplished by solving the defining equations of $R_{\theta_2}$ and $R_{\theta_4}$ for the input and output angles, respectively, as follows:

$$R_{\theta_2} = \frac{\Delta\theta_2}{\Delta x} = \frac{\theta_2 - (\theta_2)_0}{x - x_0}$$

$$R_{\theta_2}(x - x_0) = \theta_2 - (\theta_2)_0, \qquad \theta_2 = (\theta_2)_0 + R_{\theta_2}(x - x_0)$$

$$\dot{\theta}_2 = R_{\theta_2}\dot{x}, \qquad \ddot{\theta}_2 = R_{\theta_2}\ddot{x}, \qquad \ldots, \qquad \overset{(j)}{\theta_2} = R_{\theta_2} \overset{(j)}{x}$$

Similarly,

$$\theta_4 = (\theta_4)_0 + R_{\theta_4}(y - y_0)$$

$$\dot{\theta}_4 = R_{\theta_4}\dot{y}, \qquad \ddot{\theta}_4 = R_{\theta_4}\ddot{y}, \qquad \ldots, \qquad \overset{(j)}{\theta_4} = R_{\theta_4} \overset{(j)}{y}$$

Substituting these in the position equation and its derivatives will yield a set of equations in which $R_{\theta_2}$ and $R_{\theta_2}$ are present and can be regarded as unknowns.

**TABLE 2.6** NUMBER OF POSSIBLE SOLUTIONS USING EQ. 2.90, NAMELY $\mathbf{Z}_2 e^{i\theta_2} + \mathbf{Z}_3 e^{i\theta_3} - \mathbf{Z}_4 e^{i\theta_4} = -\mathbf{Z}_1$, AND ITS TIME-DERIVATIVES IN HIGHER ORDER APPROXIMATE SYNTHESIS OF A FOUR-BAR MECHANISM. NOTE THAT FOR FUNCTION GENERATION $\theta_2$, $\theta_4$, $\omega_2$, $\omega_4$, $\alpha_2$, $\alpha_4$ AND SO ON ARE PRESCRIBED AND THAT $\mathbf{Z}_1 = -1$

| Number of positions plus derivatives | Number of scalar equations | Number of scalar unknowns | Number of free choices (scalars) | Number of solutions |
|---|---|---|---|---|
| 1 | 2 | 6($\mathbf{Z}_2$, $\mathbf{Z}_3$, $\mathbf{Z}_4$) | 4 | $0(\infty)^4$ |
| 2 | 4 | 7(above) + $\omega_3$ | 3 | $0(\infty)^3$ |
| 3 | 6 | 8(above) + $\alpha_3$ | 2 | $0(\infty)^2$ |
| 4 | 8 | 9(above) + $\dot{\alpha}_3$ | 1 | $0(\infty)^1$ |
| 5 | 10 | 10(above) + $\ddot{\alpha}_3$ | 0 | Finite |

be generated], the number of infinitesimally separated prescribed positions can be increased to 7. In other words, seventh-order approximation of function generation is possible with the four-bar linkage. This is the same result as for the synthesis of the four-bar function generators with finitely separated precision points. In fact, one would find that any mixture of point- and order-precision synthesis equations (i.e., any case of "multiply separated" precision points) will yield a similar table as Table 2.1. Also, the maximum number of positions, finitely or infinitesimally separated, that yields a set of linear equations is the same in each case. The maximum number of infinitesimally separated prescribed positions for order synthesis of path and motion generators will be found as in Table 2.1 for finitely separated positions.

## Position, Velocity, and Acceleration Specification for the Four-Bar Function Generator

As was the case with the finite theory of Sec. 2.16, three multiply separated positions is the maximum number of positions available that still yields a set of linear equations in the starting position vectors of the movable links $\mathbf{Z}_j$, $j = 2, 3, 4$. The position, velocity, and acceleration equations for the four-bar function generator (Fig. 2.80) from Eqs. (2.89), (2.92), and (2.94) are

$$\begin{aligned} &\mathbf{Z}_2 + \mathbf{Z}_3 - \mathbf{Z}_4 = 1.0 \\ &\mathbf{Z}_2\omega_2 + \mathbf{Z}_3\omega_3 - \mathbf{Z}_4\omega_4 = 0.0 \\ &\mathbf{Z}_2(i\alpha_2 - \omega_2^2) + \mathbf{Z}_3(i\alpha_3 - \omega_3^2) - \mathbf{Z}_4(i\alpha_4 - \omega_4^2) = 0.0 \end{aligned} \tag{2.95}$$

where $\mathbf{Z}_1$ was specified as $(-1.0 + 0.0i)$. Using Cramer's rule,

$$\begin{aligned} \mathbf{Z}_2 &= \frac{-\omega_3(i\alpha_4 - \omega_4^2) + \omega_4(i\alpha_3 - \omega_3^2)}{\mathbf{D}} \\ \mathbf{Z}_3 &= \frac{\omega_2(i\alpha_4 - \omega_4^2) - \omega_4(i\alpha_2 - \omega_2^2)}{\mathbf{D}} \\ \mathbf{Z}_4 &= \frac{\omega_2(i\alpha_3 - \omega_3^2) - \omega_3(i\alpha_2 - \omega_2^2)}{\mathbf{D}} \end{aligned} \tag{2.96}$$

where

$$\mathbf{D} = \omega_2[-\omega_4^2 + \omega_3^2 + i(-\alpha_3 + \alpha_4)] + \omega_3[-\omega_2^2 + \omega_4^2 + i(-\alpha_4 + \alpha_2)] \\ -\omega_4[\omega_3^2 - \omega_2^2 + i(\alpha_2 - \alpha_3)] \tag{2.97}$$

**Example 2.10**

Synthesize a four-bar linkage for function generation where the angular velocity and acceleration of the moving links are prescribed according to

$$\omega_2 = 2 \text{ rad/sec}, \qquad \alpha_2 = 0 \text{ rad/sec}^2$$

$$\omega_3 = 3.5 \text{ rad/sec}, \qquad \alpha_3 = 2 \text{ rad/sec}^2$$

$$\omega_4 = 5 \text{ rad/sec}, \qquad \alpha_4 = 4 \text{ rad/sec}^2$$

**Solution** Using Eqs. (2.96) and (2.97), we have

$$\mathbf{D} = +6.75 + 0.0i \qquad \mathbf{Z}_3 = -4.44 + 1.19i,$$

$$\mathbf{Z}_2 = 3.89 - 0.59i, \qquad \mathbf{Z}_4 = -1.56 + 0.59i$$

## APPENDIX: CASE STUDY—TYPE SYNTHESIS OF CASEMENT WINDOW MECHANISMS [89]

A powerful alternative method of type synthesis to the associated linkage approach presented in this chapter is applied here in an industrial application.

### Structure Phase of Type Synthesis

Freudenstein and Maki [105] suggest separation of *structure* and *function* in the conceptual phase of mechanism synthesis. They point out that the degree of freedom of a mechanism imposes constraint on the structure of the mechanism. Rather than using Gruebler's equation (see Chap. 1) for degrees of freedom of mechanisms, they suggest the following forms:

$$F = \lambda(l - j - 1) + \sum_{i=1}^{j} f_i \tag{2.98}$$

and

$$L_{\text{IND}} = j - l + 1 \tag{2.99}$$

where $F$ = degree of freedom of mechanism

$l$ = number of links of mechanism (including the fixed link; all links are considered as rigid bodies having at least two joints)

$j$ = number of joints of mechanism; each joint is assumed as binary (i.e., connecting two links); if a joint connects more than two links, the number of joints $j = N - 1$, where $N$ = the number of links at the common joint

$f_i$ = degree of freedom of $i$th joint; this is the freedom of the relative motion between the connected links

$\lambda$ = degree of freedom of the space within which the mechanism operates; for plane motion and motion on a surface $\lambda = 3$ and for spatial motions $\lambda = 6$

$L_{\text{IND}}$ = number of independent circuits or closed loops in the mechanism

Combining (2.98) and (2.99), we obtain

$$\Sigma f_i = F + \lambda L_{\text{IND}} \tag{2.100}$$

Since we are dealing with planar motion and a single degree of freedom,

$$F = 1, \qquad \lambda = 3$$

For the case of one closed loop,

$$L_{\text{IND}} = 1$$

From Eq. (2.100),

$$\Sigma f_i = 4$$

For example, if we investigate the four-link chain from Eq. (2.99),

$$j = L_{\text{IND}} + l - 1 = 4 \tag{2.101}$$

Thus the number of joints is four (as would be expected). Equation (2.98) shows that a five-bar chain with joints (with pin and slider joints) yields a mechanism with two degrees of freedom:

$$F = 3(5 - 5 - 1) + 5 = +2$$

If four joints allow a single degree of freedom between connecting links and one joint allows two degrees of freedom of relative motion (e.g., gear connection), then a five-bar chain has a single degree of freedom. In the case of a six-bar with two loops ($L_{\text{IND}} = 2$), from Eq. (2.101),

$$j = 2 + 6 - 1 = 7$$

and from Eq. (2.98) (assuming pin and slider joints),

$$F = 3(6 - 7 - 1) + 7 = 1$$

## Design Objective

Casement-type windows (see Fig. 1.4) are generally defined as vertically pivoted, outward-swinging, ventillating windows. Screens, where used, are placed to the inside of the room. Casement windows were initially operated without screens and were merely pushed open and closed. Development of linkage operators for this window brought various improvements, such as:

1. A method to lock the window at various open and closed positions
2. Screens through which an opener could function
3. Concealed hinges
4. Weather stripping
5. Geared operators to control window position

These developments were achieved by about 1906. Since the year 1906, at least 44 U.S. patents have been issued (as discovered in a prior art search) which make further improvements to casement window mechanisms. These improvements are necessary because over 5,000,000 operating casement-type windows are sold annually in the United States and Canada.

Consumer demands have brought about better insulation by (1) double- and triple-glazed sashes and (2) multiple weather stripping. These changes result in significantly increased loading on the casement window operator and its associated linkages, causing objectionable operating characteristics. Existing casement window mechanisms function satisfactorily on windows up to about 50 lb weight but require excessive operating torque on larger windows.

Some popular operators have good pull-in (closing) characteristics but they lose mechanical advantage as the window approaches the 90° open position. The transmission angle (also a critical factor in mechanism design) is also poor near the 90° position. Other operators found in use today have good (low) torque requirement at open positions but have poor pull-in characteristics. A low, uniform torque from fully open to the fully closed position is desirable. Low torque gives user satisfaction and long operator life. The prior art search disclosed many window mechanism concepts, but no satisfactory scheme was available to evaluate and compare the various designs.

Consequently, a plan was developed to analyze and categorize past mechanism concepts and new designs which would hopefully lead to an improved casement window mechanism.

### Design Constraints

The casement window operator linkage design has many challenging constraints. The most important considerations are:

1. The sash (window) must open 90° from the sill.
2. The end of the sash must slide at least 10.16 cm (4 in.) in order to allow washing on both sides of the window from the inside.
3. An open sash must leave 50.8 cm (20 in.) for egress codes (exit in case of emergency). Some local codes require even greater opening.
4. The operator linkage must support the weight of the window with minimal sag of the sash.

5. The operator linkage must have a single actuator arm (the mechanism should have one degree of freedom).
6. A new operator linkage must have improved transmission angle and better mechanical advantage than the present mechanism.
7. When the sash is in the closed position, all portions of the mechanism must be below the sill cover, not extend beyond the plane of the sash toward the outside, and have minimum extension into the room.
8. During deployment, all parts of the operator linkage must be between the sill and the sash (so as not to interfere with the weather stripping) and cannot extend farther into the sill.
9. The casement operator must be as simple as possible due to economic considerations (e.g., pin and slider joints are preferred over gear and cam connections for both initial cost and maintenance considerations).
10. A maximum number of parts must be interchangeable for both left- and right-handed operators. (A sash may be hung at either the left or right side.)

Note that the need for selflocking was not mentioned. This caused problems later, as will be seen.

### Analysis of the Current Operator

Before looking for new casement window linkages, the performance of the current operator will be investigated.

**Mechanical advantage analysis.** The instant center technique (Chap. 3 of Vol. 1) is used to perform a static force analysis (or mechanical advantage analysis) which is useful in determining possible improvements in the current operator. Figure 1.4b shows the current casement linkage with the operator arm in an intermediate position. The kinematic diagram (unscaled) of the mechanism in Fig. 1.4b is shown in Fig. 1.5. Notice that this mechanism is a six-bar chain. The pertinent instant centers have been located on Fig. 2.82 (operator arm at 60° to the horizontal of the drawing) and also in Fig. 2.83 (fully open position).

Consider the operator arm (link 2) as the input and the window (link 4) as the output. If one assumes that the energy losses in a linkage as it moves are small, then (as described in Chap. 3 of Vol. 1) power in should be equal to power out:

$$P_{in} = P_{out}$$

or

$$\omega_{in} T_{in} = \omega_{out} T_{out}$$

where $\omega$ = angular velocity (rad/sec)

$T$ = torque (in·lb)

In this case

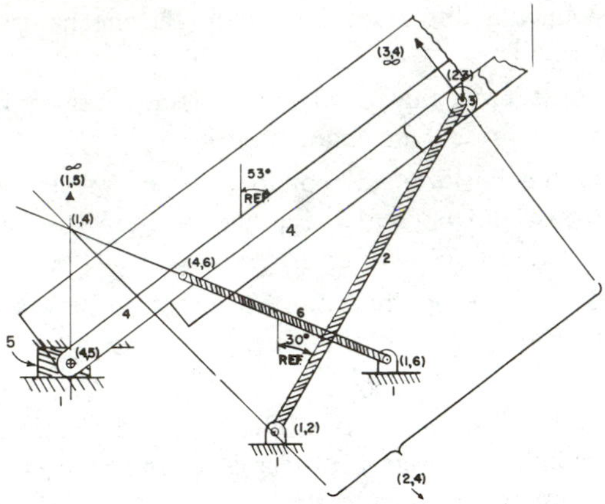


**Figure 2.82** Current casement window at 60° operator arm position showing locations of instant centers.

$$|\omega_2 T_2| = |\omega_4 T_4|$$

or

$$|\omega_2| F_2 R_2 = |\omega_4| F_4 R_4$$

where $F$ = a force (lbf)

$R$ = a radius from the instant center of the link (with respect to the ground) perpendicular to the line of action of the force (in.)

Using instant centers, the angular velocity ratio may be expressed in terms of linkage geometry (see Chap. 3 of Vol. 1):

$$\left|\frac{\omega_2}{\omega_4}\right| = \frac{(2{,}4 - 1{,}4)}{(2{,}4 - 1{,}2)}$$

where 2,4; 1,4; and 1,2 are the location of the instant centers between links 2 and 4,1 and 4, and 1 and 2, respectively. The right-hand side of the foregoing equation, Eq. (2.101), is negative if center 2,4 lies between 1,4 and 1,2. From Fig. 2.82,

$$\frac{\omega_2}{\omega_4} \simeq +1.1$$

The mechanical advantage (M.A.) is defined as the (force out)/(force in) or in this case,

$$\text{M.A.} = \frac{F_4}{F_2}$$

or

$$\frac{F_4}{F_2} = \frac{\omega_2}{\omega_4}\frac{R_2}{R_4} \simeq 1.1\frac{R_2}{R_4}$$

Given a unit radius of the input force on the operator handle, a 28∶1 gear reduction in the worm gear, a 30% efficiency in the worm gear, and a length of 3.5 in. between the shoe (4,5) which locates the resistance load on the window) and the instant center (1,4), the mechanical advantage of the system is (at 60° of the operator arm)

$$\text{M.A.} = \frac{F_{\text{out}}}{F_{\text{in}}} = \frac{(1{,}4-2{,}4)}{(1{,}2-2{,}4)}(\text{efficiency}) \times (\text{gear ratio}) \times \frac{\text{input radius}}{(1{,}4-4{,}5)}$$

$$= (1.1)(0.30)(28)\frac{1}{3.5} \tag{2.102}$$

$$= 2.64$$

The mechanical advantage analysis was also performed at the fully open position (see Fig. 1.4a). Here the angular velocity ratio is reduced to

$$\frac{\omega_2}{\omega_4} = \frac{3.5}{10} = 0.35$$

In the open position, (1,4 − 4,5) is about the same length as before and the mechanical advantage from Eq. (2.102) is

$$\text{M.A.} = (0.35)(0.30)(28)\frac{1}{3.5}$$

$$= 0.84$$

**Methods of increasing the mechanical advantage of the current operator.** According to Eq. (2.102), there are several adjustments that could lead to an increased mechanical advantage of the current operator linkage:

1. Increase the efficiency of the operator.
2. Increase the gear ratio.
3. Reduce the distance (4,6 − 4,5), which equals (1,4 − 4,5) in the open position.
4. Increase the ratio (1,4 − 2,4)/(1,2 − 2,4). Notice that these instant centers change their relationship with any change in linkage geometry.
5. Also, the operator will be more effective (i.e., $F_{\text{out}}$ will cause more motion) if the frictional load at the shoe is decreased by reducing the coefficient of friction between the shoe (link 5) and the sill (link 1).

Adjustments 1, 2, and 5 involve higher cost, and 2 will also require more turns of the operator handle to open the sash.

**Transmission-angle analysis.** Another measure of the mobility of the operator linkage is the transmission angle. For optimal mobility, the net force action on the output link (link 5 in this analysis) of a mechanism should be in the same direction as the output link velocity at the point of action of the resisting force. For our purposes, the transmission angle was defined in Chap. 3 of Vol. 1 as the smallest angle between the direction of the perpendicular to the shoe velocity ($\mathbf{V}_B$) and the direction of the net static force $\mathbf{F}_{54}$ (since $\mathbf{F}_{45} = \mathbf{F}_{54}$) acting on the sash (link 4). A free-body diagram of link 5 is drawn (see Fig. 2.83). Forces $\mathbf{F}_{34}$, $\mathbf{F}_{64}$, and $\mathbf{F}_{54}$ must intersect at a single point. Although not shown, the frictional forces at the shoe could be included (as well as the smaller frictional effects at *A* and *C*). The figure shows that the transmission angle in the open position is about 27°, which is marginal. As the linkage closes the sash, the transmission angle improves.

Figure 2.83 gives an indication of how the transmission angle can be improved in the present design. These improvements are:

1. Move the pivot of the crank (center 1,6) out from the sill and/or over to the right. (Center 1,6 cannot be moved beyond the edge of the sill, however.) Moving pivot $A_o$ to the right will have minor effect on the transmission angle and will introduce more cost due to the increased length of link 6.
2. Lower the connection between the crank (link 6) and the window (link 4). This will decrease the distance (4,6 — 4,5), which will also improve the mechanical advantage.

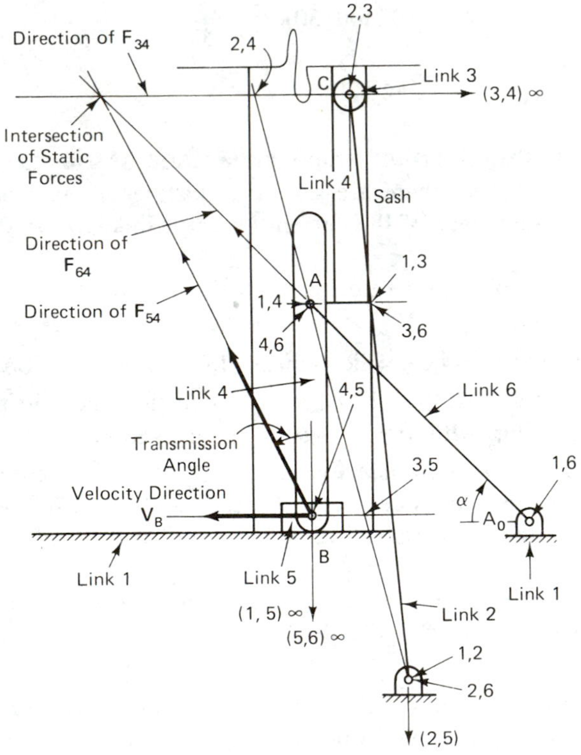


**Figure 2.83** Current casement window mechanism at fully open position showing locations of instant centers.

3. Decrease the length of the operator arm (link 2). This, however, is directly counter to what is advantageous for better mechanical advantage.

Thus the best change that would improve both the mechanical advantage and the transmission angle at the open position would be to follow suggestion 2. Unfortunately, a decrease in length (4,6 − 4,5) directly decreases washability (the distance the base of the window slides out from its closed position) since, when the window is fully open

$$\text{washability} = (4{,}6 - 1{,}6)(1 - \cos\alpha) \tag{2.103}$$

where $\alpha = \measuredangle AA_0B$.

The conclusion of both the mechanical advantage and the transmission-angle analyses is that even with some changes as suggested above, either the cost will increase or other desirable performance characteristics must be sacrificed. Thus this type of linkage has limited performance possibilities and other types of casement operators are to be investigated.

## Type Synthesis

Kinematic synthesis of mechanisms can be separated into two steps: *type* and *dimensional* synthesis. The first helps determine the best linkage types, while the latter produces the significant dimensions of the mechanism that will best perform the desired task.

The simplest linkage chain, the four-bar, is the logical initial choice for a mechanism to suggest as a casement window operator. The sash could be connected to the coupler link of the four-bar linkage, as shown in Fig. 2.84a, for example. The drawbacks of this design are (1) the interference of the links with weather stripping

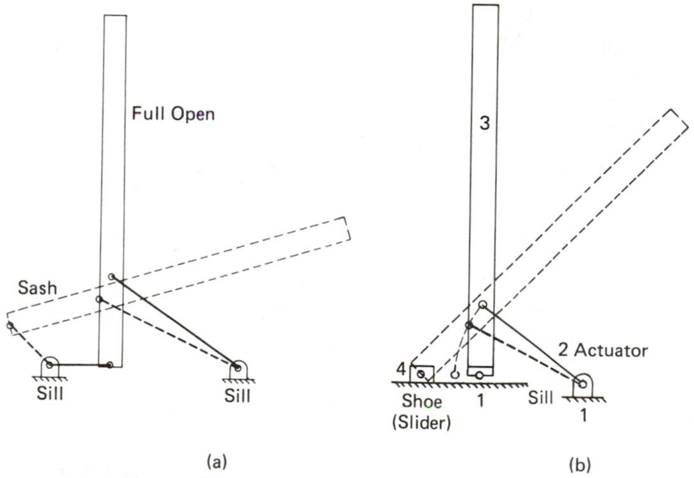


**Figure 2.84** (a) four-bar operator; (b) slider-crank operator.

in the open position, and (2) the inability of the sash to open to the 90° position without traveling through a toggle position if either of the links pinned to ground was designated as the input link. Here, a torque would be required on the sash to return it to the closed position (not possible without adding more linkage members). If the 90° opening requirement were relaxed (as in an awning window application), the four-bar design would be more acceptable. In both cases, however, the four-bar's ability to carry the weight of the sash is questionable since the entire sash moves away from the sill during its motion. In fact, the larger the requirement for washability, the longer the input link and the higher the probability for sash sag.

The slider-crank linkage is the next logical choice for a casement operator linkage since the slider could move along the sill and thereby support the weight of the window (see Fig. 2.84b). Unfortunately, a torque on the sash will still be required to close it from a 90° position. Link 2 could not be used effectively as an input link since the transmission angle is 0° in the open position. Link 4 is also undesirable as an input link since the transmission angle is 0° in the closed position.

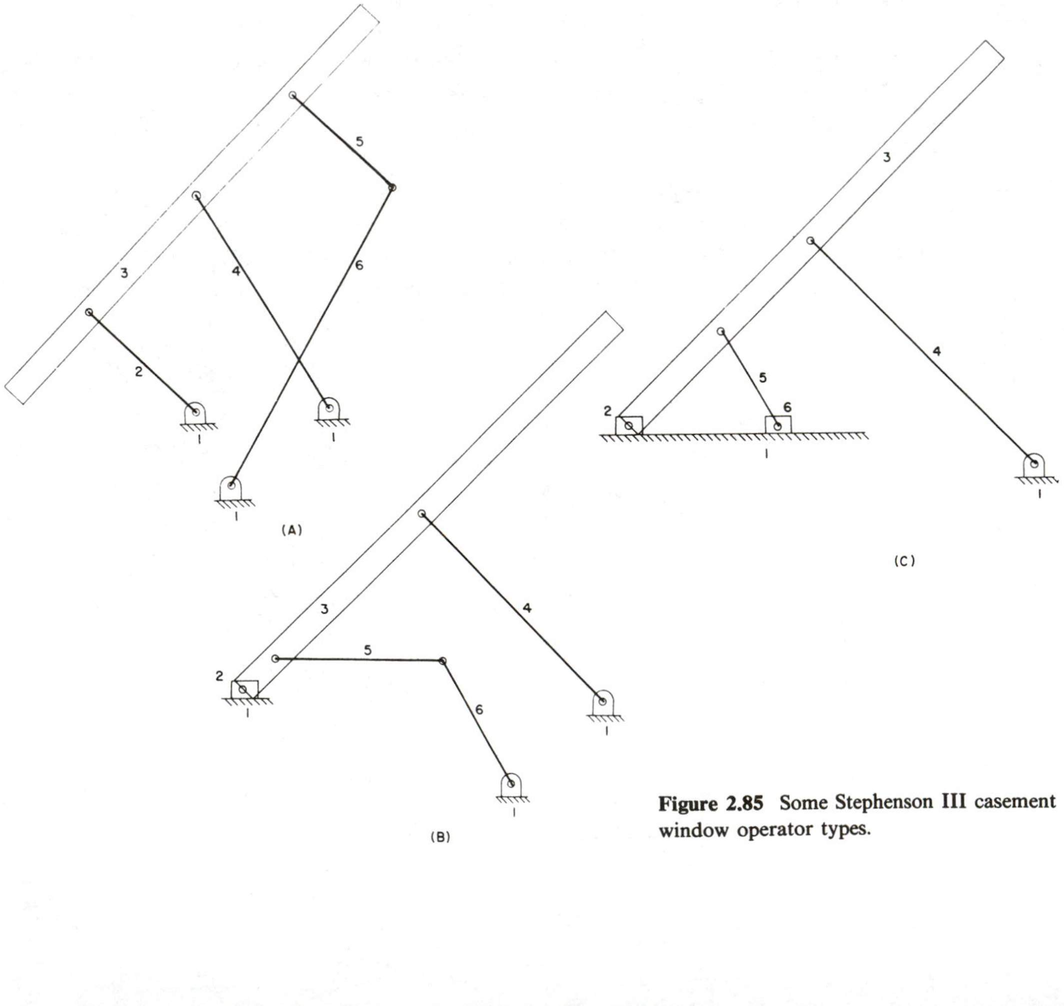


**Figure 2.85** Some Stephenson III casement window operator types.

Since the four-link chain looks to have limited acceptability to the casement window application, mechanisms with more links must be pursued. Since the degree of freedom of a casement mechanism must be 1, a type synthesis effort can be simplified. For example, unless gear- or cam-type connections between links are used (those that allow two degrees of freedom between connecting members), the five-bar and seven-bar chains will not be appropriate. This is because these chains will result in two degrees of freedom if only pin or slider joints are used. Thus the six-bar chain is the next logical chain to investigate. As introduced in Chap. 1 (Figs. 1.9 to 1.13), there are only five types of six-bar linkages with a single degree of freedom—variations of the Stephenson and the Watt chains. The Stephenson six-bar has nonadjacent ternary links; in the Watt chain, the ternary links are pinned together. Since the window would be connected to a floating link, there are only 11 possible six-bar combinations with only revolute joints. If, however, sliders are allowed to replace one or more links, many more possible combinations result. For example, Fig. 2.85 shows some of the possible Stephenson III linkages that can be used for a casement operator. One can now appreciate the need for a systematic type synthesis.

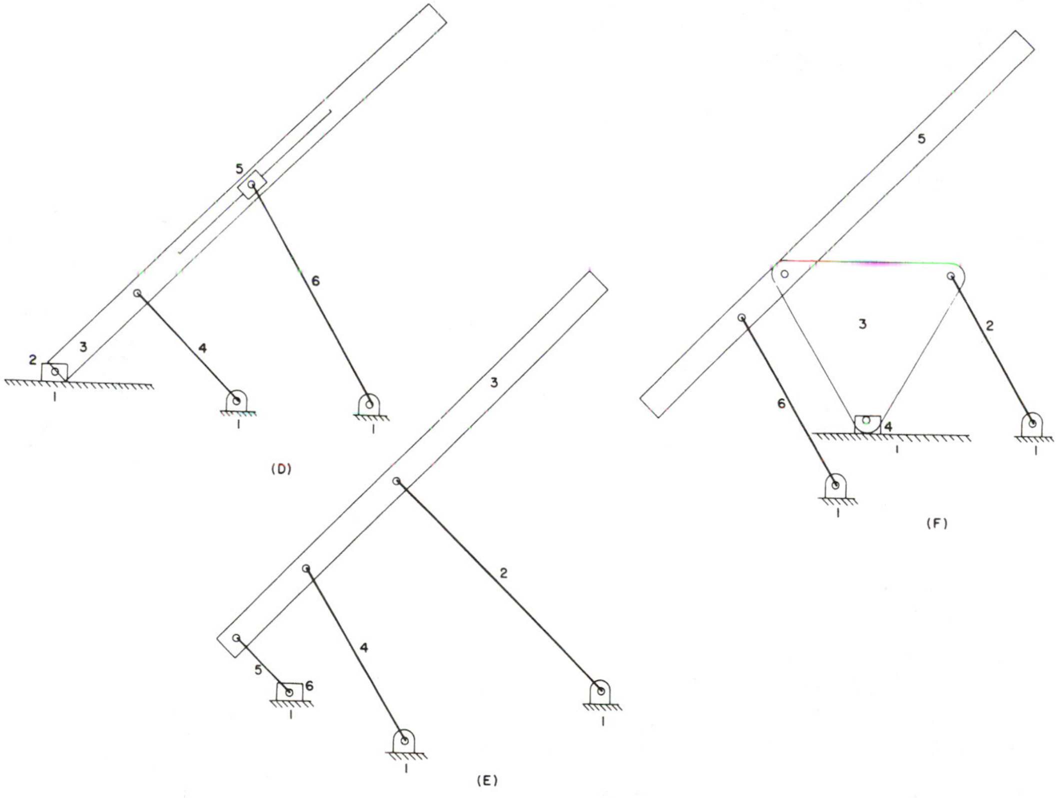


Figure 2.85 (continued)

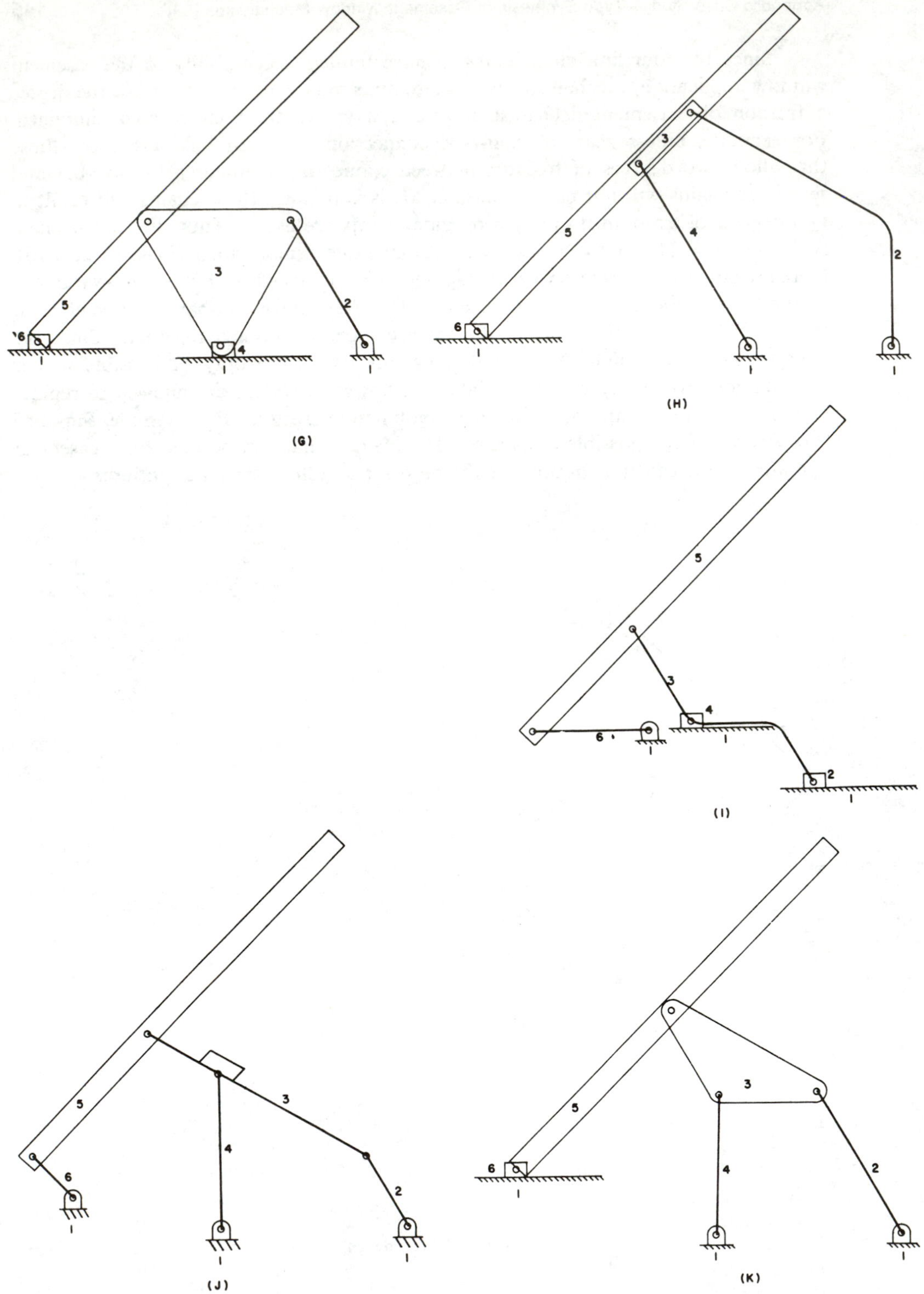


**Figure 2.85 (continued)**

## Other Window Operator Designs

Several popular casement and awning operators are six-bar chains. The "current" operator (Fig. 1.4) is a Stephenson III type (Fig. 2.85, linkage *D*). A metal awning linkage must move the window out from the frame before rotating the sash. Most do not allow 90° of rotation. The most popular awning linkage, the Anderberg (U.S. patent 2,784,459), shown in Figs. 2.86 and 2.87, is a Stephenson I type of

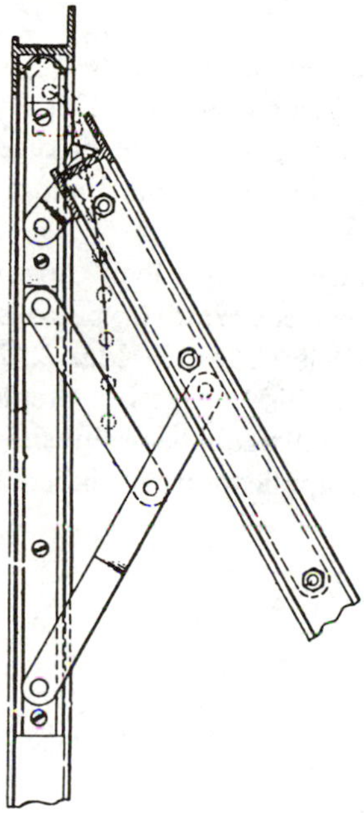

**Figure 2.86** Awning linkage. (*A. W. Anderberg, 1957.*)

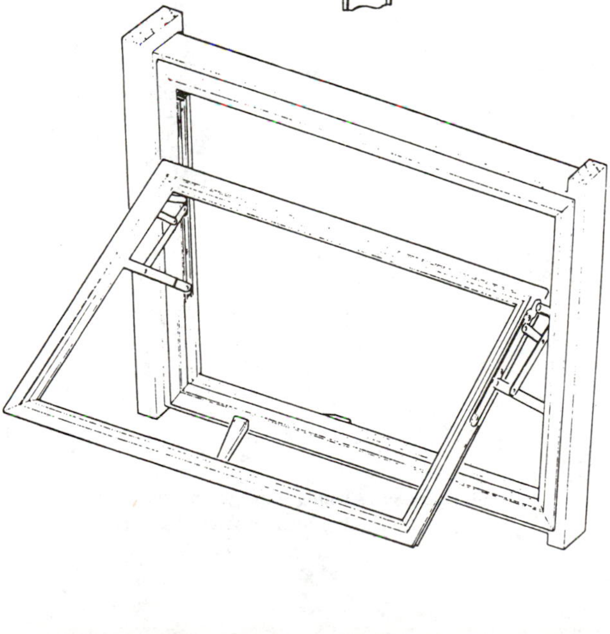

**Figure 2.87** Awning linkage shown in open position.

six-bar (see Fig. 1.11). There are several interesting variations of the basic Anderberg design. Figure 2.88 shows one (U.S. patent 3,345,777) where two pin joints on the slider are combined presumably for patent purposes (although kinematically, both of these linkages are the same structural type). Figure 2.89 shows another proposed variation, where another link is added (dashed) for the purpose of reducing sag of the sash. From Eq. (2.98) we see that this design has zero degrees of freedom. The linkage will, however, have mobility (although extremely dependent on manufacturing accuracy) due to the special geometry of the new link and the clearances in the linkage joints.

Figure 2.90 shows a geared version (U.S. patent 3,838,537) which has the objective of helping "pull in" the awning window. This "overconstrained" (4) linkage also maintains mobility due to geometry. The linkage is designed to have nearly constant angular velocity rotation between links 3 and 5 so that the gears connecting these two links do not bind.

Figure 2.91 shows the "gearless torque lock" awning window mechanism (U.S. patent 2,761,674). This mechanism is a Watt II type six-bar. The Pella-type casement operator (U.S. patent 3,438,151) is shown in Fig. 2.92. Until the engagement of the pin and slot (discussed later), this linkage is a Stephenson III type six-bar.

Equations (2.98) to (2.100) will again be useful shortly, but first some functional aspects of the casement window application will be enumerated.

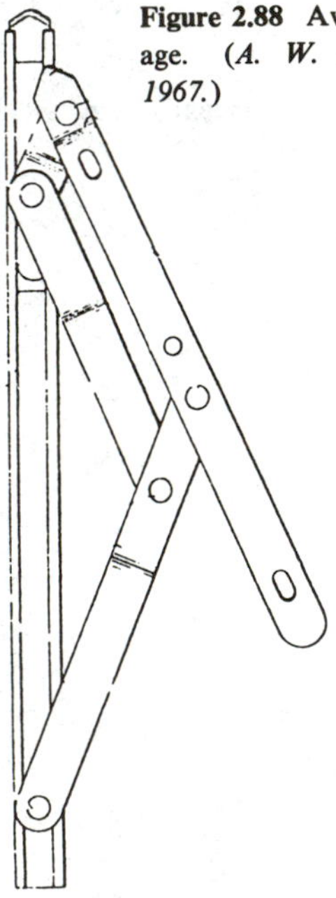

**Figure 2.88** Awning linkage. *(A. W. Anderberg, 1967.)*

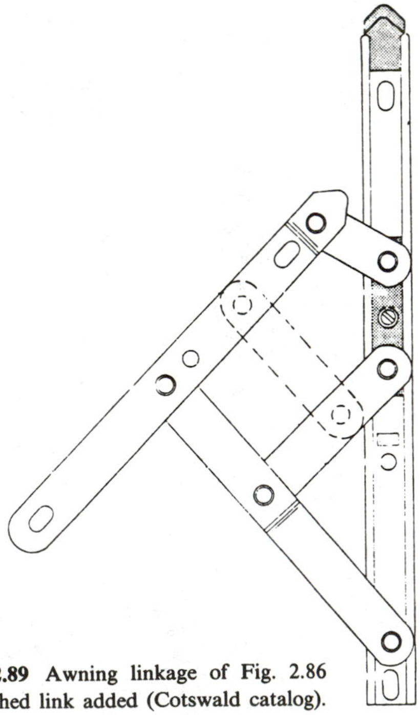

**Figure 2.89** Awning linkage of Fig. 2.86 with dashed link added (Cotswald catalog).

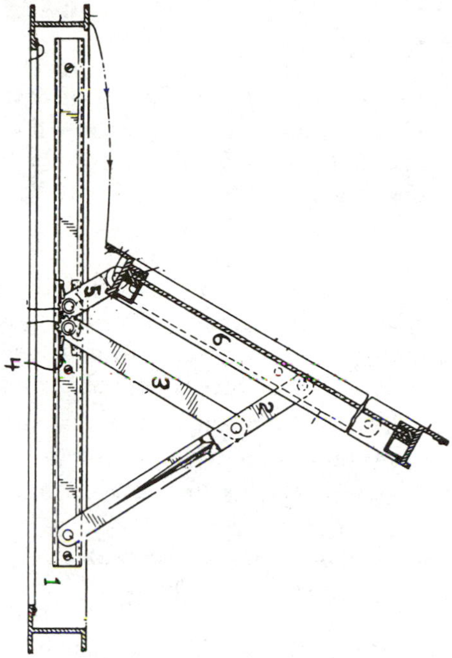


**Figure 2.90** Awning linkage. (*H. L. Starenau and W. C. Bates, 1973*)

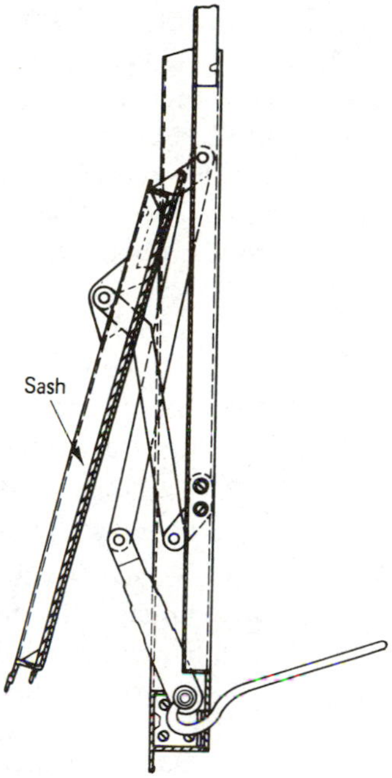


**Figure 2.91** Awning linkage. (*N. C. Walberg et al, U.S. patent 2,761,674, 1956.*)

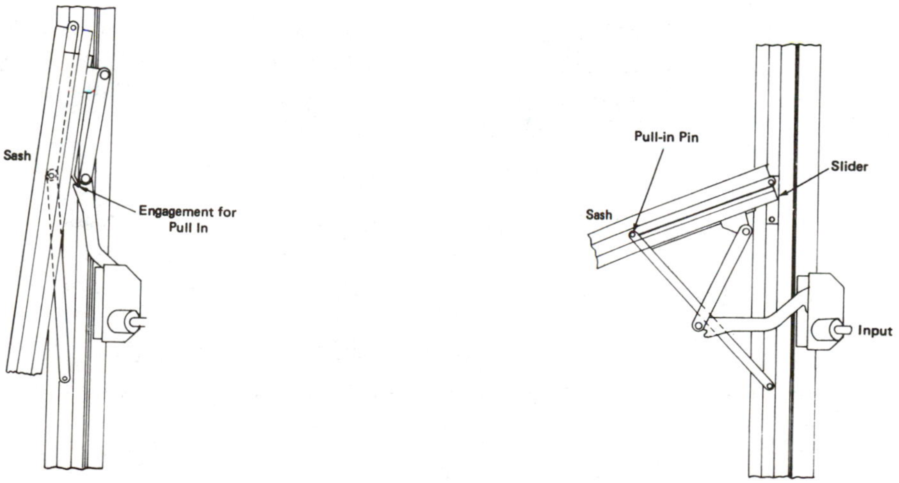


**Figure 2.92** Pella linkage. (*R. Rivers and M. Minter, U.S. patent 3,438,151, 1969.*)

### Functional Phase of Type Synthesis

The previous discussion clearly demonstrates that a four-link chain will not accomplish all the objectives of a new casement window operator. To help narrow the field of possible mechanisms with, say, five to eight links, we must investigate the function of an optimal mechanism. Based on observation of the four-bar chains and the current operator:

1. A multiloop linkage is preferred such that one loop guides the window through 90° while another loop acts as the driver or the operator. The concept followed here is that the guiding loop will probably run into transmission angle and/or mechanical advantage problems if one of its members is used as the input.
2. The sash should rest on a slider (which slides on the sill) so that the weight of the sash is supported in a simple manner as the window opens.
3. A slider-crank linkage (with the sash connected to its coupler) is the best guiding linkage, due to simplicity. Since the top of the window also must be guided, all one would need would be a duplicate slider-crank on top (without the driving loop) to keep the top of the sash coordinated with the bottom.

Based on these three simplifying decisions, the possible solutions have significantly decreased. For example, the five-bar chain (with one two-degree-of-freedom joint) is disregarded (since only a single loop is possible) in favor of a six-link chain. Also, several of the six-bar chain possibilities are set aside since they do not have a slider-crank loop with the window attached to the coupler link of that loop (e.g., mechanisms $A$, $C$, $E \longrightarrow K$ of Fig. 2.85 would be discarded in the Stephenson III six-bar options shown).

Two other important observations can now be made:

1. Any mechanism loop(s) that will be added to the slider-crank chain must have total joint freedoms of three (for single loop addition), six (for two loops), or nine (for three loops), as can be derived from Eq. (2.100). For example, with $F = 1$, $\lambda = 3$, and $L_{IND} = 2$,

$$\Sigma f_i = 1 + 6 = 7$$

   but since the slider-crank has four joint freedoms, the additional mechanism loop must have three total joint freedoms.
2. There are three ways of effectively driving the window through the 90° of motion:
   a. Push at or near the outside end of the sash (for high mechanical advantage).
   b. Pull at or near the shoe since the resistance to sash movements is at the shoe.

c. Combination of methods a and b, causing a torque on the sash.

In the following sections, driving mechanisms of all three types are discussed.

## New Casement Linkage Operators

**Push-type operators.** The current casement mechanism (Fig. 1.4 and linkage *D* of Fig. 2.85) has a push-type operator. The slider-crank is driven by a dyad (two links), a link pinned to a slider, in this case with $\Sigma f_i = 3$. The advantage of the current mechanism is good "pull-in" characteristics. Mechanisms *B* and *C* of Fig. 2.85 are two other possible single-loop dyad driver loops ($\Sigma f_i = 3$). Neither of these offers great improvement. Linkage *B* has much better characteristics at the open position but poor "pull in." Linkage *C* offers poor mechanical advantage in all positions.

Several two loop drivers with $\Sigma f_i = 6$ which had the possibility of extending the actuator arm, were investigated. Although the mechanical advantage of some of these extending actuator arm linkages was promising due to their complexity (eight links), other problems (higher cost and multiple transmission angles) make them unacceptable.

Based on the desirability to have a less complex extending arm actuator, mechanisms with higher-pair contact (gears) were investigated. The concept was that a cycloidal curve could possibly match the movement of the end of the window. This led to writing the equation of point *P* at the end of the window (Fig. 2.93):

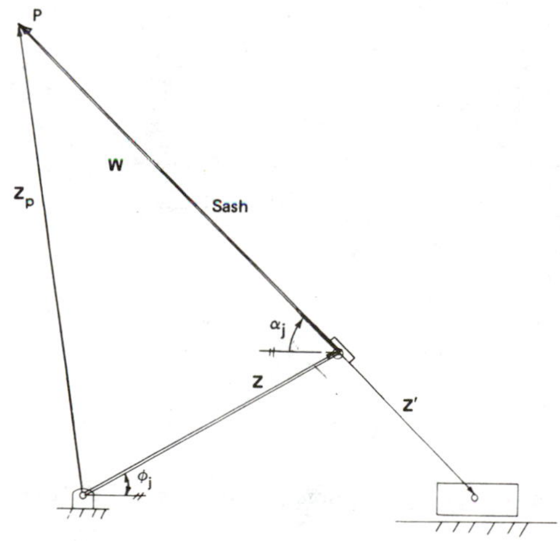


**Figure 2.93** Slider-crank guiding linkage notation.

$$\mathbf{Z}_p = \mathbf{Z}(e^{i\phi_j} + \rho e^{i\alpha'_j}), \qquad \text{where } \alpha'_j = 180° - \alpha_j \tag{2.104}$$

$$\alpha_j = \sin^{-1}\frac{Z \sin\phi_j}{Z'} \geq \frac{\pi}{2} \tag{2.105}$$

and

$$\rho = \frac{|\mathbf{W}|}{|\mathbf{Z}|} \tag{2.106}$$

An investigation of possible cycloidal driving mechanisms with $\Sigma f_i = 6$ or 9 (two or three loops) yielded a new hypocycloidal mechanism [82] in Figs. 2.94 and 2.95 (U.S. patent 4266371). The equation for the position vector of point $P'$ at the end of this operator is

$$\mathbf{Z}'_p = \mathbf{Z}'(e^{i\theta_j} + \rho e^{i\Psi_j}) \tag{2.107}$$

where

$$\rho = \frac{|\mathbf{W}'|}{|\mathbf{Z}'|} \tag{2.108}$$

and since the radius of the outer planet is one-half of that of the sun,

$$|\Psi_j| = |\theta_j| \tag{2.109}$$

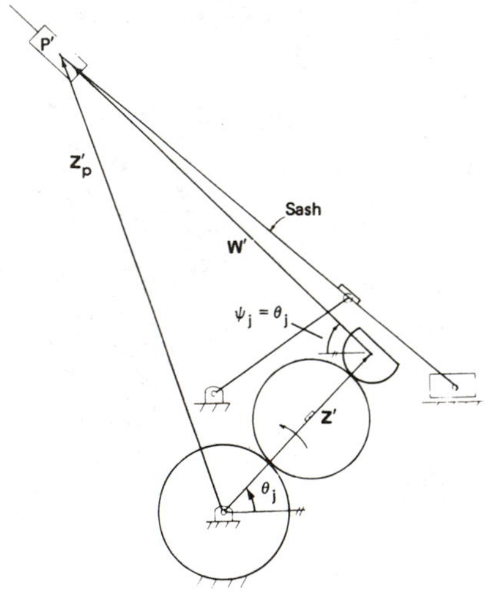


**Figure 2.94** Hypocycloidal crank driving linkage notation. (*A. Erdman and J. Peterson, U.S. patent 4,266,371*)

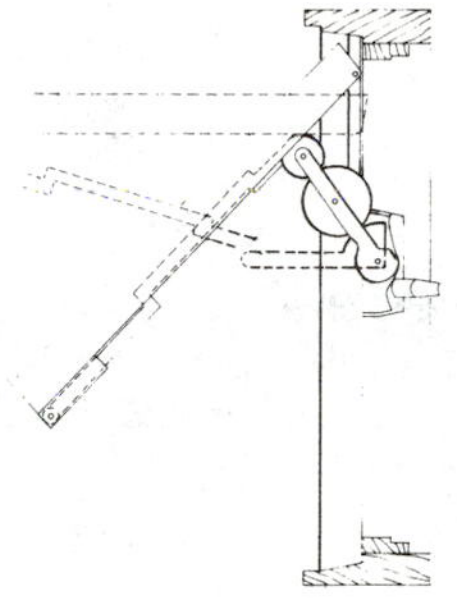

**Figure 2.95** Hypocycloidal mechanism (slider-crank not shown).

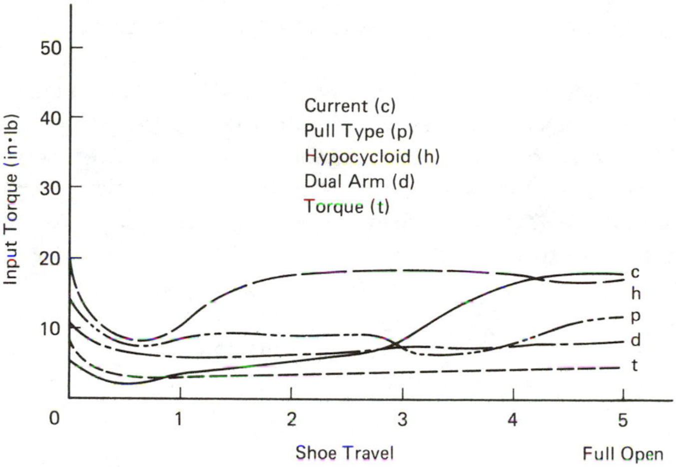


**Figure 2.96** Required input torque for current operator and other designs.

Equation (2.107) was set equal to Eq. (2.104) for a number of positions and the best solution built for testing purposes. The maximum difference between $|\mathbf{Z}'_p|$ and $|\mathbf{Z}_p|$ is only $\frac{1}{2}$ in. for a 24-in. window yielding maximum mechanical advantage through the driving linkage. Figure 2.96 shows the level of input torque (experimentally determined) of this operator compared with the current operator. The potential improvement* is promising, especially near the fully open position. A force analysis performed on the driving linkage yields high forces between gear teeth and at the bearings. These high force demands as well as the complexity of the hypocycloidal design detract from the promising results shown in Fig. 2.96.

* The hypocyclic model was binding during the test due to misalignment of gears.

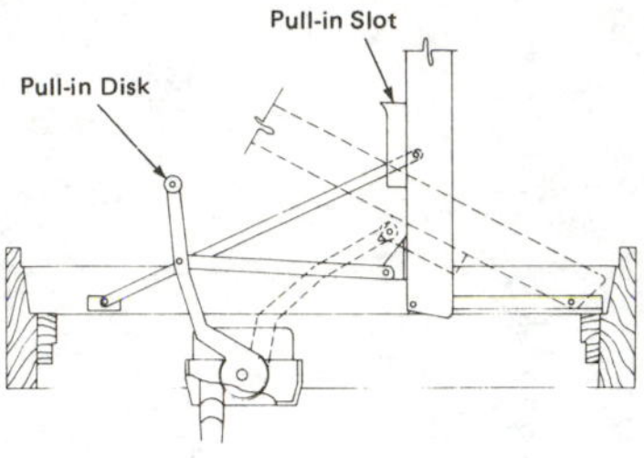


**Figure 2.97** Pull-type operator. A disk enters a slot in the dashed position to aid pull-in. (*J. Peterson and E. Nelson, U.S. patent 4,253,276.*)

**Pull-type operators.** One of the most efficient methods of opening a window is driving near the point of frictional load—the shoe. The simplest way to pull the shoe is a dyad (see Fig. 2.85, linkage *B*). Unfortunately, the "pull-in" characteristics of this design are poor.

To achieve the desired window control near the closed position, a mechanism [202] was proposed that operates in two modes. First, an overconstrained system ($F = 0$) was synthesized where a point on the crank follows a track on the sash for better pull-in and window control through the first 30° of motion (see the dashed position of Fig. 2.97). Beyond 30°, the mechanism has a single degree of freedom.

The dimensional synthesis of this new mechanism* (U.S. patent 4253276) was based on kinematics and kinetics. The relative sizes of the crank and connecting link were adjusted to allow good transmission angles throughout the operation and a toggle at closing. The cam slot was programmed to match the kinematic constraints of the rest of the mechanism. Experimental results for required input torque for the pull-type operator are shown in Fig. 2.96.

**Push–pull-type operators.** The concept that inspired the evolution of the push–pull-type operator was to simultaneously push the window and pull near the slider. The problem with this objective is that the slider-crank linkage constrains the relationship between the slider velocity and the sash angular velocity. It is extremely difficult to match these kinematic relationships with a single-degree-of-freedom driving mechanism.

Thus a drive linkage had to be synthesized that would either *match* these constraints (i.e., $F = 0$) or incorporate strategic linkage geometry to capitalize on the hinge timing ($F = 1$). The former suggestion could be accomplished by using noncircular gears to achieve the desired shoe position and window angular position coordination. Unfortunately, the cost of the gears would probably be prohibitive and the mechanism would be too sensitive to inaccuracies, improper mounting, and wear. To convert this concept into one that employs gears of constant radius and have a mechanism with $F = 1$, from Eqs. (2.98) to (2.100), $\ell = 9$, $L_{\text{IND}} = 4$, $j = 12$, and

* Somewhat similar to the Pella concept of Fig. 2.92.

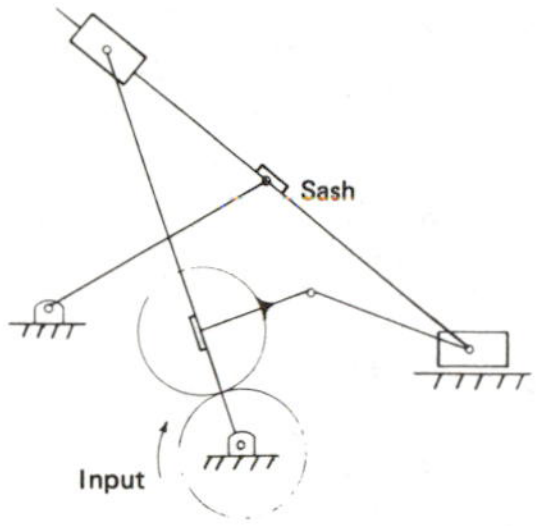


**Figure 2.98** Dual-arm operator. (*Van Klompenburg, J. Peterson, E. Nelson, U.S. patent 4,241,541.*)

$\Sigma f_i = 13$. A new casement operator linkage (shown in Fig. 2.98) called the "dual arm" operator [298] satisfies these constraints. The optimal dimensions of this mechanism were derived based on kinematic and kinetic constraints. For example, an attempt was made to balance the forces applied on the window by both arms of the operator. Figure 2.96 shows the required input torque of this operator measured from an optimized prototype.

The push-pull concept is intuitively most attractive. Another attempt was made to design such an operator. Instead of having separate arms extending to the window, a concept was pursued to provide a torque directly on a window by fixing a gear segment to the sash. With a gear as part of the kinematic chain, the minimum number of links is $\ell = 5$ with $\Sigma f_i = 7$. The resulting concept is shown in Fig. 2.99. Notice that the sum of the diameters of the gears is approximately equal to the required opening. As shown, this design does not satisfy the requirement for supporting the weight of the window. One can add the slider-crank guiding linkage (not shown) by carefully designing the geared five-bar portion so that the inner edge of the sash follows a straight line. The dimensional synthesis was accomplished by noticing that links 3, 4, 5, and 1 make up a four-bar chain (similar to Fig. 2.84). The objective then was to synthesize the four-bar such that the path of a point at the base of the window traveled a nearly straight line path. This can be approximated closely enough that the clearance in the slider will allow smooth operation of the entire mechanism. This new torque operator (patent pending) is shown in Fig. 2.99. Notice that $F = 1$ from Eq. (2.98) since $\ell = 7$ and $\Sigma f_i = 10$.

Figure 2.96 shows the experimentally obtained input torque requirements for the torque operator. These values are the lowest of the mechanisms tested. Unfortunately, the torque is too low. A nonperceived design constraint appeared after this mechanism was tested: a casement window mechanism should be self-locking. That is, when the input is set at a particular position, any reasonable wind load should not be able to close or further open the sash. This design overcame the mechanical advantage and transmission-angle objectives so well that it is not self-locking (without addition of auxiliary components). The other drawback of this design (although not as critical) is the size of the planet gear and thus the size of the required sill cover.

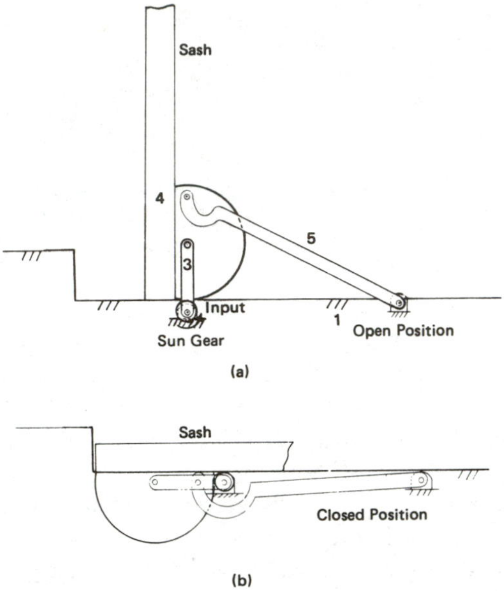


**Figure 2.99** Torque operator shown in two positions.

## Conclusions

A systematic type synthesis approach was applied to a casement window application which had many clearly defined design constraints. Separation of structure and function of mechanisms allowed a productive interchange of guiding rules that significantly narrowed the search for optimal solutions. The type synthesis yielded many casement window mechanism concepts, and the most feasible designs were investigated further in a dimensional synthesis step. Those that appeared to meet the constraints were designed and prototyped for evaluation. At least four new operator concepts were prototyped and exhibited favorable operating characteristics over operators currently on the market. At this point, marketing, manufacturing, and management input was utilized to select the best new mechanism. The "dual arm" concept was selected as the best alternative because:

1. It offered the best overall performance. (It pushed the window as well as pulled the window, to achieve a much lower operating torque from fully open to fully closed position.)
2. It could be manufactured at a reasonable cost.
3. It would require minimal changes for window manufacturers.
4. It would be compatible with the existing four-link slider-crank hinge.
5. Parts would be compatible with current manufacturing capabilities.

## PROBLEMS*

**2.1.** **(a)** Determine the Chebyshev spacing for a function $y = 2x^2 - x$ for the range $0 \leq x \leq 2$ where four precision points are required.

**(b)** Based on these precision points, find $\phi_2, \phi_3, \phi_4$ and $\psi_2, \psi_3, \psi_4$ if $\Delta\phi = 45°$ and $\Delta\psi = 90°$.

**2.2.** It is desired to generate $y = e^x$, $0 \leq x \leq 4$, and specify three precision points. Using Chebyshev spacing, find:

**(a)** $x_1, x_2, x_3$

**(b)** $\phi_j, \psi_j$ $(j = 2, 3)$ if $\Delta\phi = 80°$ and $\Delta\psi = 110°$

**2.3.** Find the Chebyshev spacing for three prescribed positions for the function $y = x^2 + 3x + 5$, $0 \leq x \leq 2$. Also find $\phi_j, \psi_j$ $(j = 2, 3)$ if $\Delta\phi = \Delta\psi = 45°$.

**2.4.** **(a)** Determine the three-point Chebyshev spacing for the function $y = x^2$, $0 \leq x \leq 10$, and $\Delta\phi = \Delta\psi = 60°$.

**(b)** Find $\phi_j, \psi_j$ for $j = 2, 3$.

**(c)** If a four-bar linkage is to be designed to generate this function, determine the starting position of the linkage (if $\gamma_2 = 1°$ and $\gamma_3 = 30°$) by the complex-number method.

**(d)** Draw the linkage in its three precision positions and determine if this is an acceptable linkage.

**2.5.** **(a)** Find the three-precision-point Chebyshev spacing for the function $y = x^{3/2}$, $0 \leq x \leq 100$ where $\Delta\phi = \Delta\psi = 60°$.

**(b)** Find $\phi_j, \psi_j$ $(j = 2, 3)$.

**(c)** If a four-bar linkage is required for this task, solve for the resulting linkage using complex numbers if $\gamma_2 = 0.01°$ and $\gamma_3 = 12°$.

**(d)** Draw the resulting four-bar in its three precision positions and determine if it is an acceptable linkage.

**2.6.** Design another pair of compound-lever snips from the suitable associated linkage of Fig. 2.30, which satisfies the objective set forth in the chapter and which is different from the design shown in Fig. 2.30.

**2.7.** Determine the associated linkage of the yoke-riveter configuration of Fig. P2.1.

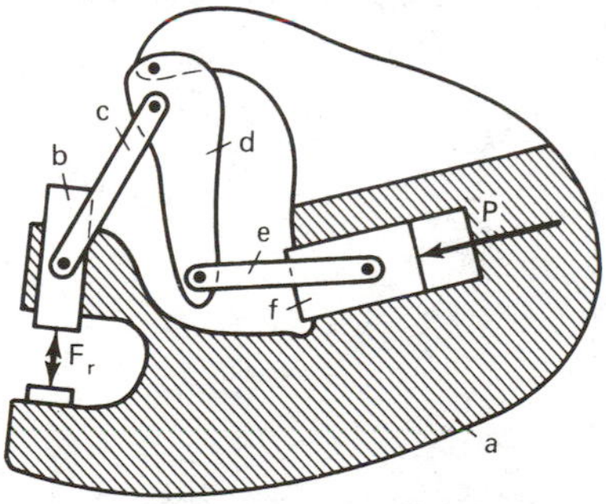


**Figure P2.1** Yoke riveter.

* Many of the examples in this and the next chapter were generated from projects submitted by former students at the University of Minnesota. The creative ideas of these students are acknowledged.

**2.8.** Determine the associated linkage of the yoke-riveter configuration of Fig. P2.2a.

**2.9.** Determine the associated linkage of the yoke-riveter configuration of Fig. P2.2b.

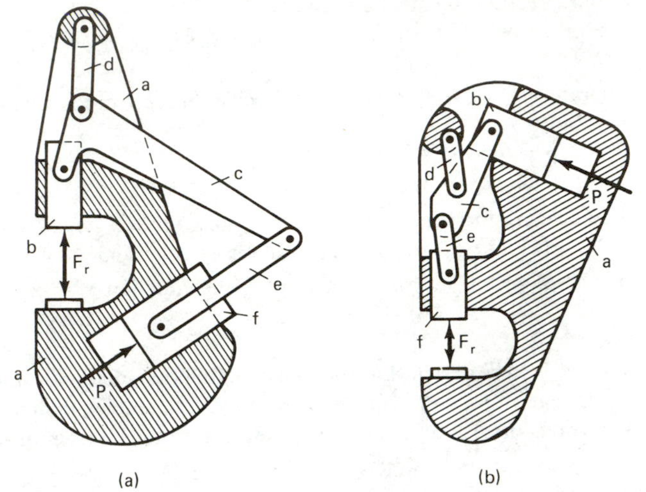


**Figure P2.2** Two different types of yoke riveters.

**2.10.** Create other designs for the compound-lever snips that are different from those in the text or in Prob. 2.6.

**2.11.** Create other designs for the yoke-riveter that are different from those in the text.

**2.12.** Figure P2.3 shows one side of a container with a removable top in two required positions. Find the locations of acceptable ground pivots ($A_0$, $B_0$) of a four-bar linkage that will guide the top through these two positions without interfering with the side of the container.

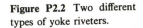

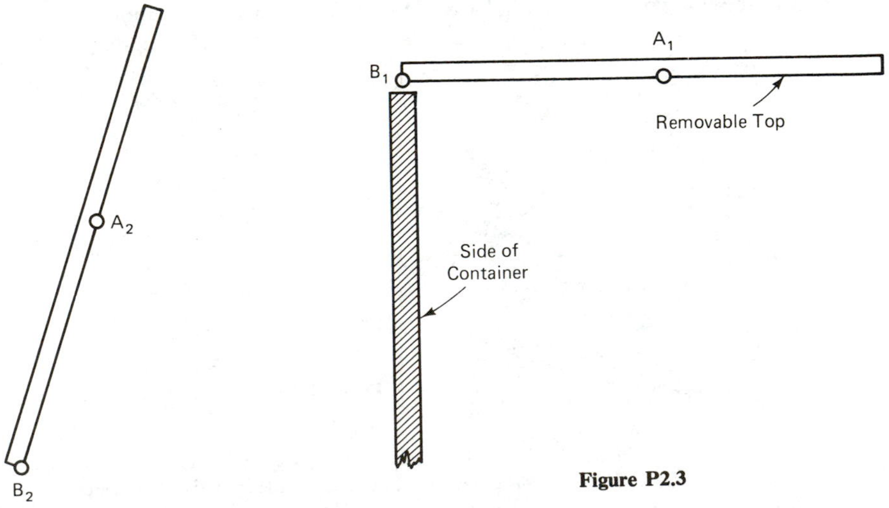


**Figure P2.3**

**2.13.** A student wishes to design a four-bar linkage that will store a bicycle above his bed. Two positions of the storage rack are shown in Fig. P2.4. Find acceptable ground and moving pivots for this design objective.

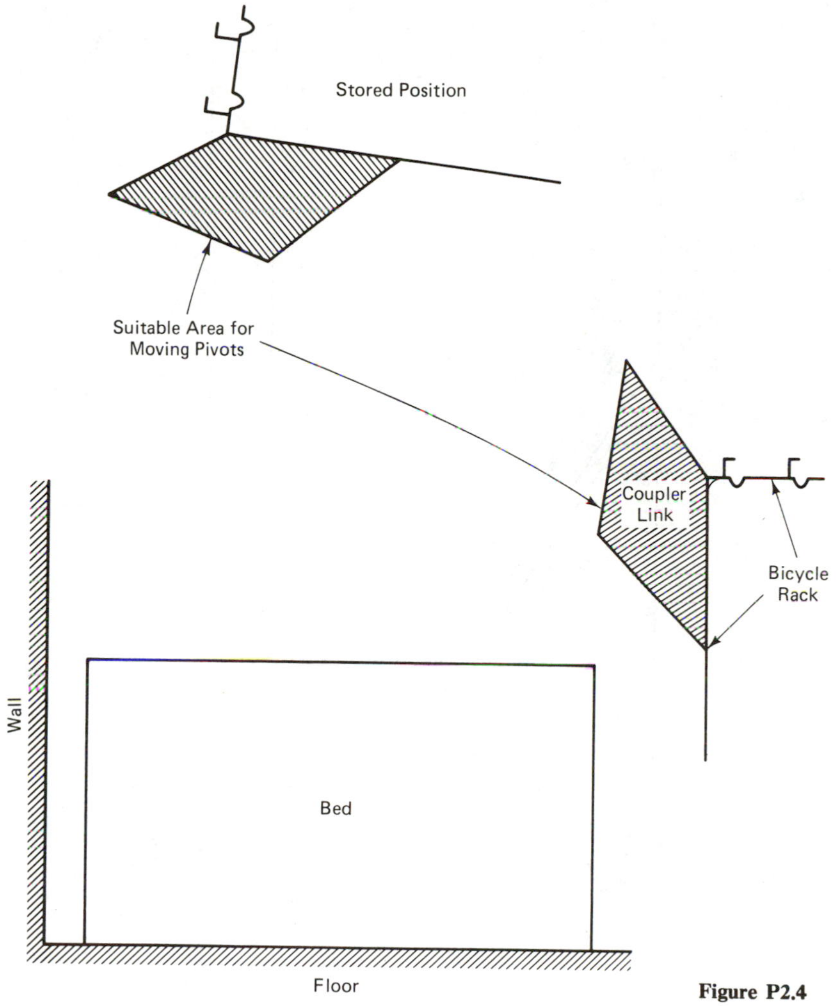


**Figure P2.4**

**2.14.** During maintenance a dust cup must be removed from a filter for dumping collected particles. Rather than bolting the dust cup to the filter, the dust cup is hinged at $A_0$ and acts as the output link of a four-bar function generator. Figure P2.5 shows the dust cup and the proposed coupler link in two positions. Determine an acceptable location of the fixed and moving pivot ($B_0$, $B$) such that $B$ is located along the coupler $AP$, the ground pivot $B_0$ is within the filter dimensions, and the rotation of $B_0B$ from position 1 to position 2 is 28° ccw.

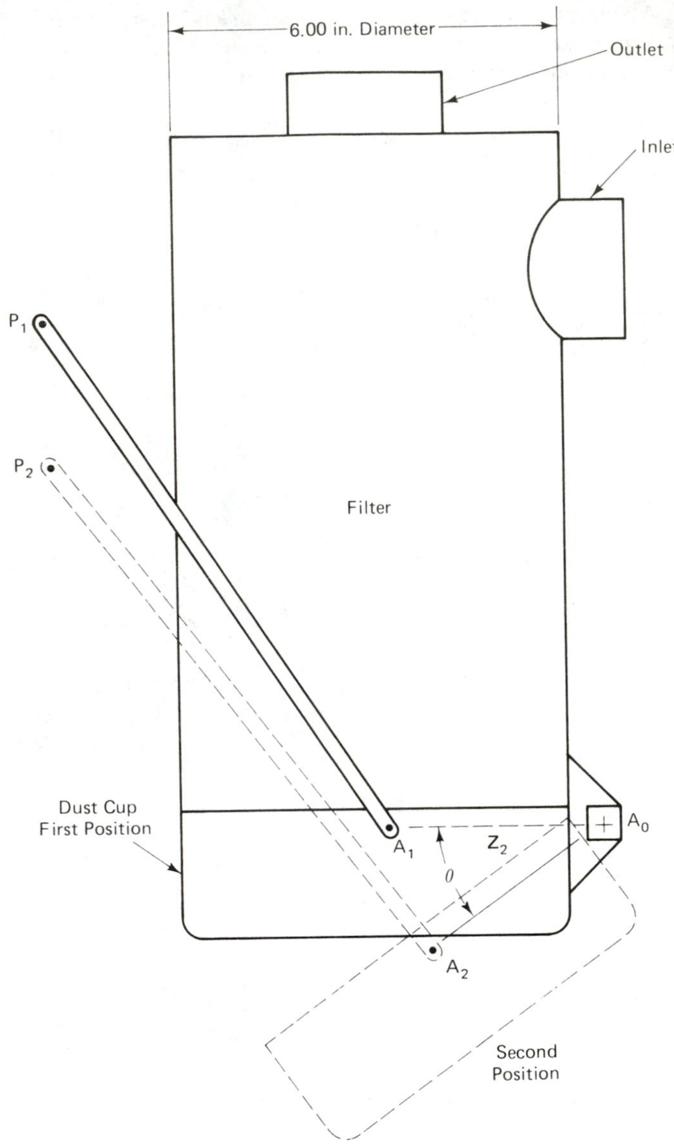


**Figure P2.5**

**2.15.** Part of the design of an assembly line requires removing a box from one conveyor belt, rotating it by 90°, and placing it on another conveyor belt. Find an acceptable four-bar linkage to guide the box carrier through the two positions shown in Fig. P2.6.

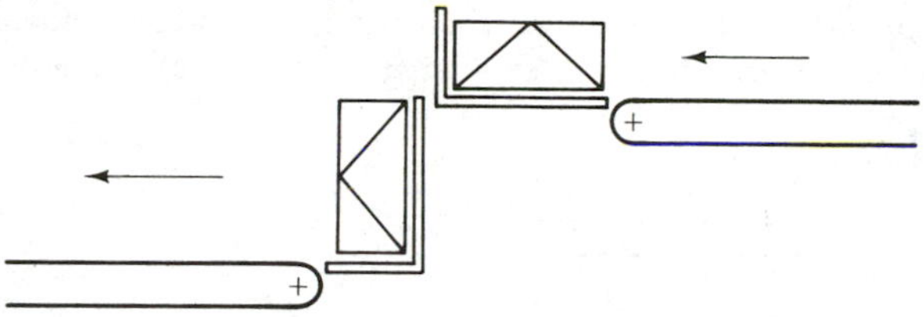

Figure P2.6

**2.16.** As part of an automation process, a four-bar linkage must be designed to remove boxes from one conveyor belt and deposit them on an upper conveyor belt as shown in Fig. P2.7 (three prescribed positions). Both ground and moving pivots must be located between the upper and lower conveyor belts.

**(a)** Design an acceptable four-bar by the graphical method.

**(b)** Design an acceptable four-bar by the complex-number method.

**(c)** Design an acceptable four-bar by the ground-pivot specification method.

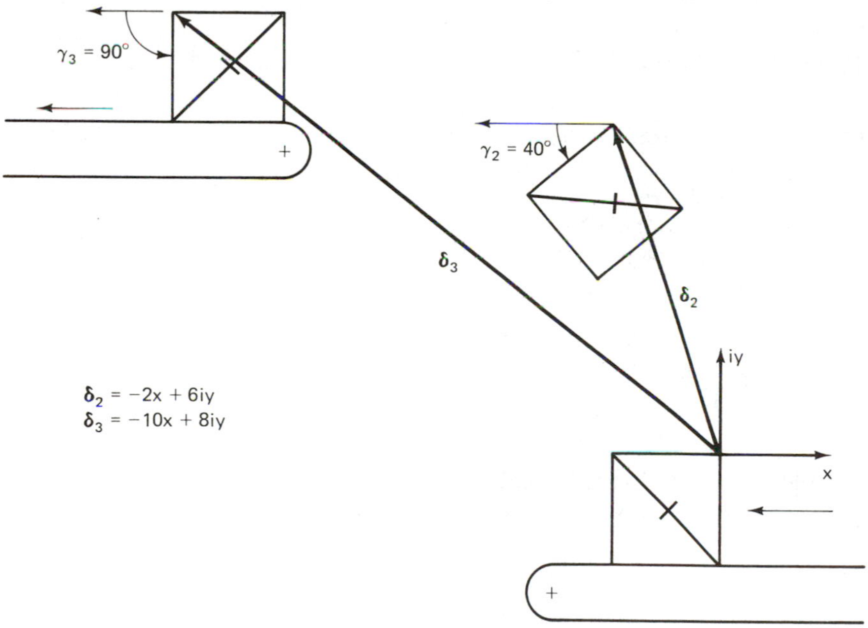


Figure P2.7

**2.17.** It is desired to synthesize a linkage to guide the movable shelf through the three positions shown in Fig. P2.8. The first position is level with the top of the cabinet for writing purposes, and the third position is a stored position for the shelf. Ground pivots should fall within the cabinet while the linkage size should be minimized so as to take up the least amount of cabinet space. Find acceptable locations of ground and moving pivots by (a) the graphical method; (b) the complex-number method; (c) the ground-pivot specification method.

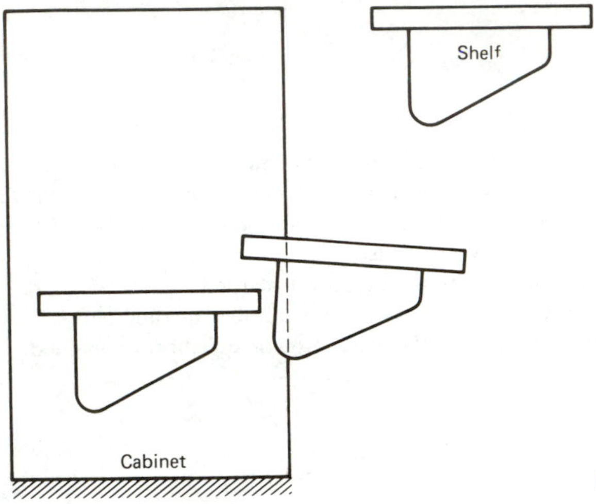


**Figure P2.8**

**2.18.** Design a compact linkage to be added to the farm vehicle of Fig. P2.9 so that the operator may maintain a vertical position as the tractor traverses the sloping terrain shown in Fig. P2.9.
**(a)** Use the graphical technique.
**(b)** Use the complex-number method.
**(c)** Use the ground-pivot specification method.

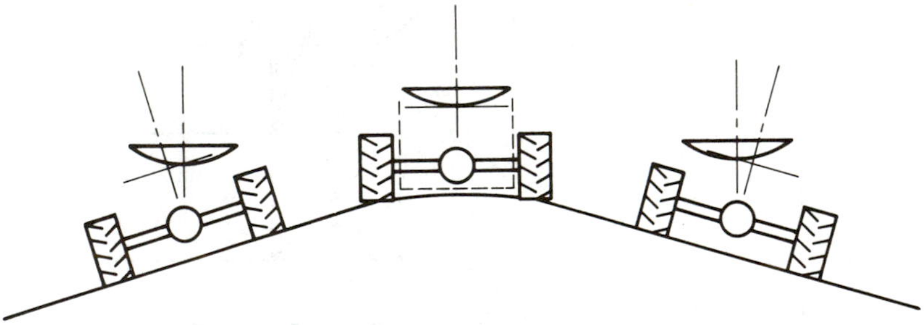

**Figure P2.9**

**2.19.** An avid foosball player wishes to design a ball-return linkage to be attached to the foosball table. Figure P2.10 shows three required positions for the coupler so that the ball may be guided from the slot on the side of the table to the holder on the top edge. Ground pivots should fall within the table and the linkage should be compact.
**(a)** Use the graphical method.
**(b)** Use the complex-number method.
**(c)** Use the ground-pivot specification method.

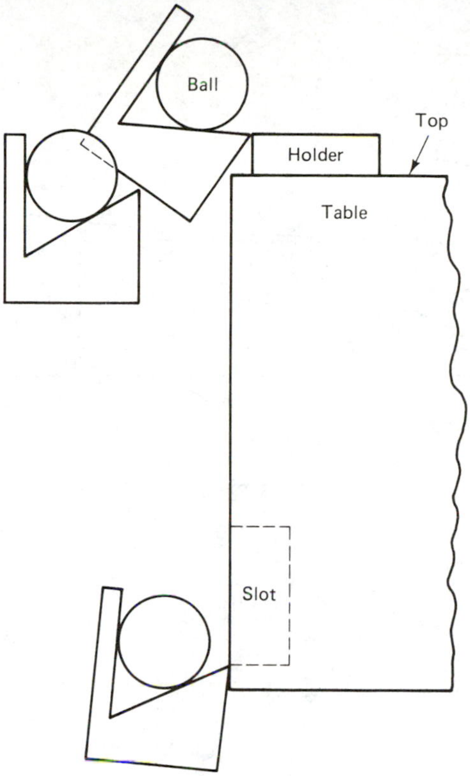


**Figure P2.10**

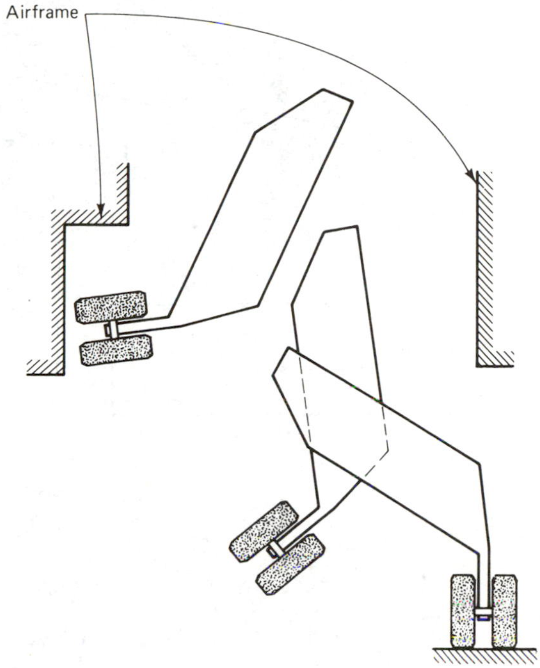


**Figure P2.11**

**2.20.** Figure P2.11 shows three prescribed positions for the wing landing gear of a remote-controlled model aircraft. Design a four-bar motion generator for this task such that the moving pivots are within the wheel supporting member and the ground pivots are within the airframe.

**(a)** Use the graphical method.
**(b)** Use the complex-number method.
**(c)** Use the ground-pivot specification method.

**2.21.** Rather than having to back a boat trailer into the water to unload a boat, a four-bar mechanism is sought to transfer the boat from the trailer to the water (see Fig. P2.12). Moving pivots should be connected to the cradle, and the fixed pivots should be close to the trailer platform. Design a four-bar motion generator for three positions.

**(a)** Use the graphical method.
**(b)** Use the complex-number method.
**(c)** Use the ground-pivot specification method.

**2.22.** A four-bar path generator (with prescribed timing) is required as part of an arm-actuated propulsion system for the wheelchair in Fig. P2.13. The three prescribed path points shown have been determined to be the most efficient arm motion by a set of patients. This movement of the coupler path point ($C_1$, $C_2$, $C_3$) provides the input, while the output is a rotation of the large wheel with a ground pivot at $A_0$. (A clutch located at $A_0$ will slip when the patient returns from $C_3$ to $C_1$ along the same path.) The other ground pivot $B_0$ is specified as well as the rotations of the wheel-driving link $A_0A$ ($\phi_2 = 38°$ cw, $\phi_3 = 80°$ cw). By the graphical method, find the initial position of an acceptable four-bar linkage for this task.

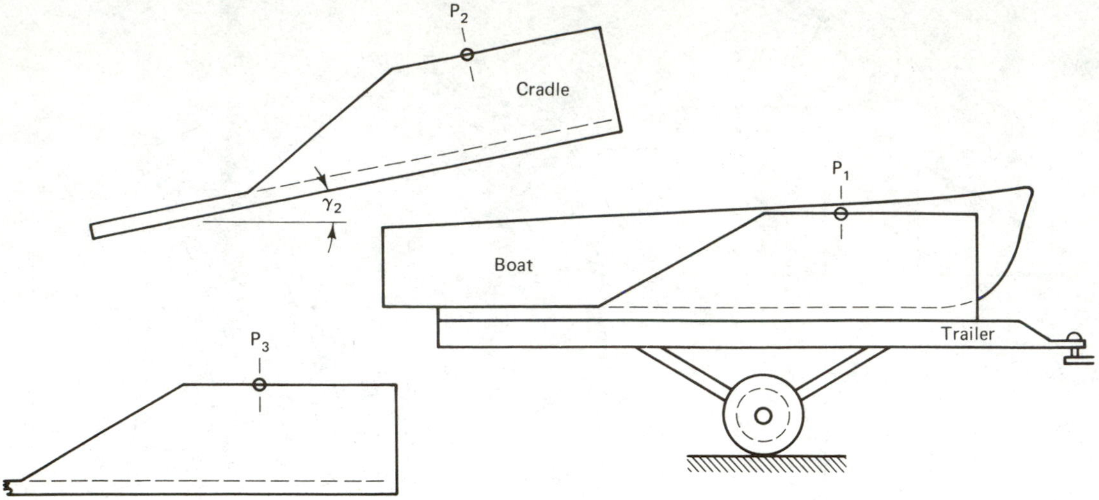


**Figure P2.12**

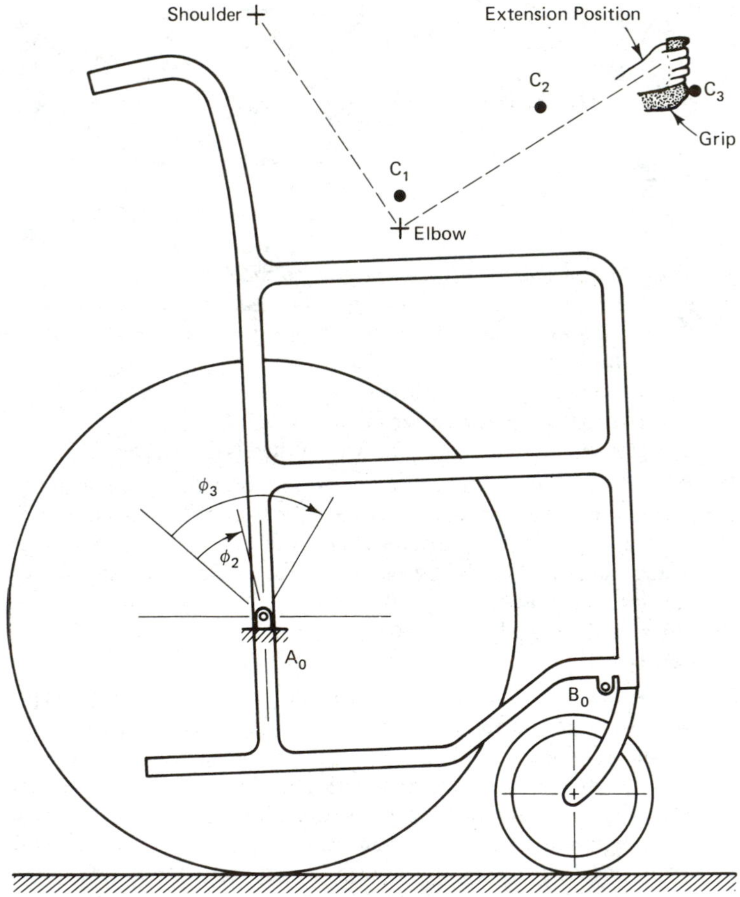


**Figure P2.13**

**2.23.** A four-bar linkage must be designed to accomplish one task in an automatic sewing machine (see Fig. P2.14). As input link ($A_0A$) rotates through $\phi_2 = 25°$ ccw, $\phi_3 = 135°$ ccw, the coupler point ($C$) must travel $C_1$, $C_2$, and $C_3$ to catch the thread loop.

**(a)** If the positions of $A_0$, $B_0$, and $A$ are prescribed (see the figure), find the location of $B$ by the graphical method and draw the linkage in its three design positions.

**(b)** Use the complex-number method to synthesize a new path generator (the same $C_1$, $C_2$, $C_3$, $\phi_2$, and $\phi_3$ are prescribed) with better transmission angles.

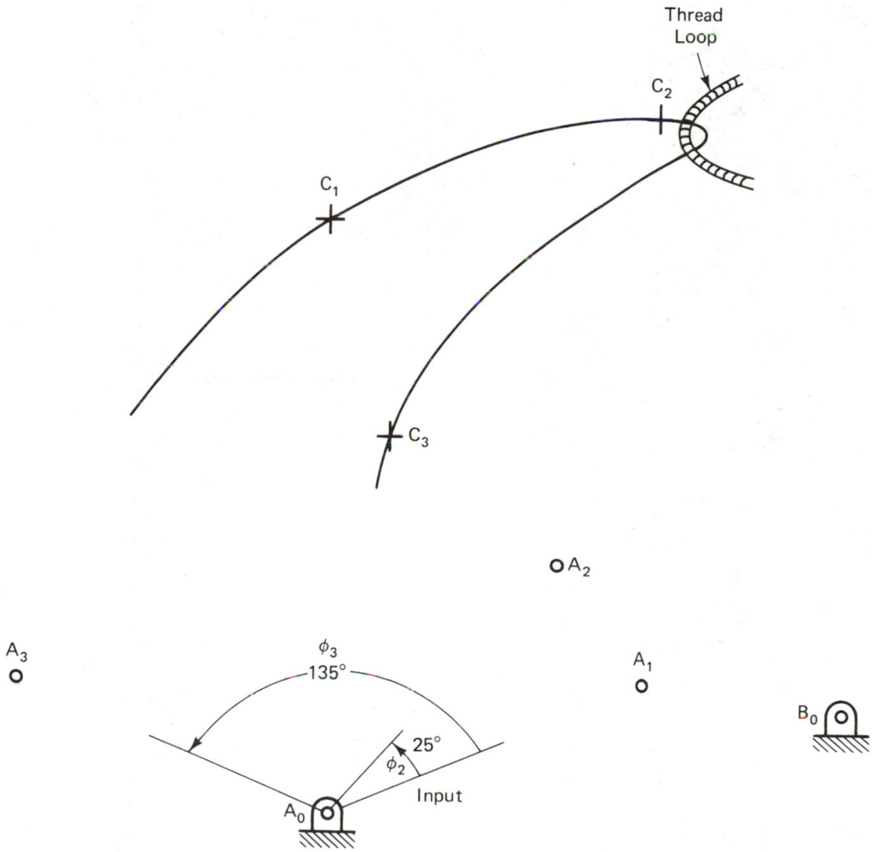


| | |
|---|---|
| $C_{1x}$ = 2.0 m | $C_{1y}$ = 11.4 m |
| $C_{2x}$ = 7.27 m | $C_{2y}$ = 13.03 m |
| $C_{3x}$ = 2.95 m | $C_{3y}$ = 8.1 m |
| $B_{0x}$ = 8.08 m | $B_{0y}$ = 1.55 m |
| $A_{0x}$ = 0.0 m | $A_{0y}$ = 0.0 m |

**Figure P2.14**

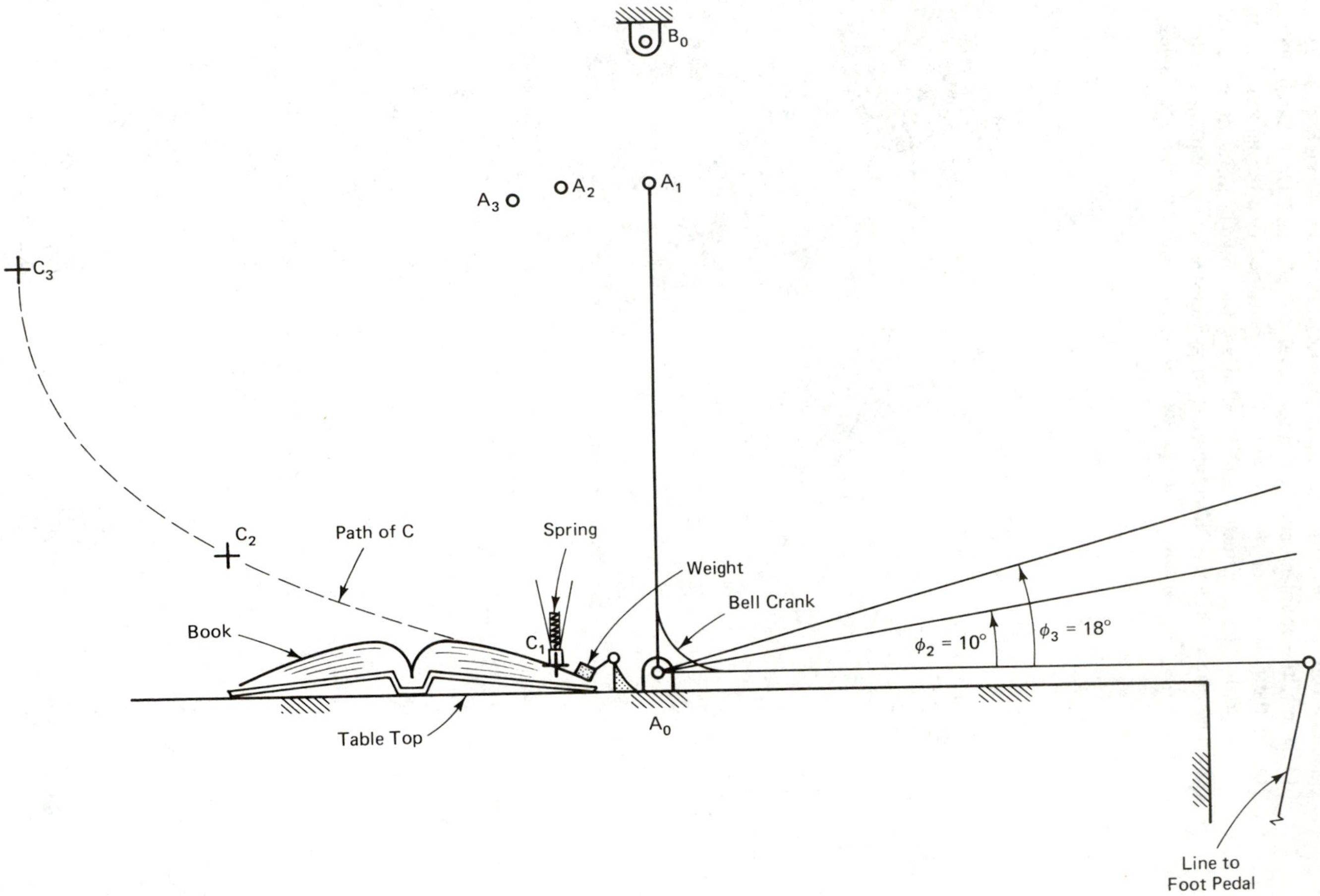
B0
A1
A2
A3
C3
C2
Path of C
Spring
Weight
Bell Crank
Book
C1
φ2 = 10°
φ3 = 18°
Table Top
A0
Line to
Foot Pedal

**Figure P2.15**

**2.24.** A handicapped person is unable to turn the pages of a book but is able to depress a foot pedal. A four-bar linkage path generator (with prescribed timing) is the main component of a mechanism that will turn a page when actuated by the foot pedal. As the coupler point $C$ of the four-bar moves from $C_1$ to $C_3$, one page flips over (see Fig. P2.15) and then $C$ returns along the same path for the next cycle.

**(a)** With $C_1$, $C_2$, $C_3$, $A_0$, $B_0$, $A_1$, $A_2$, and $A_3$ prescribed, find the rest of the linkage by the graphical method.

**(b)** With only $\phi_2$, $\phi_3$ and $C_1$, $C_2$, $C_3$ prescribed, find acceptable alternative four-bars for this task by the complex-number method.

**2.25.** A crank-rocker path generating four-bar is required to advance film in a camera as shown in Fig. P2.16.

**(a)** Using the graphical method, find the four-bar linkage if $A_0$, $A$, $B_0$, $C_1$, $C_2$, $C_3$, $\phi_2$, and $\phi_3$ are given.

**(b)** Using the complex-number method, find other acceptable four-bar linkages given $C_1$, $C_2$, $C_3$, $\phi_2$, and $\phi_3$.

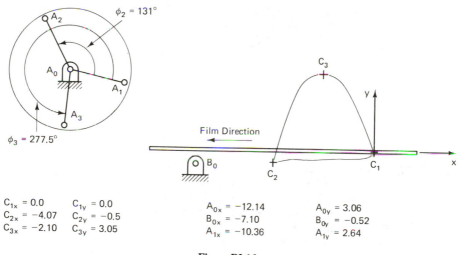

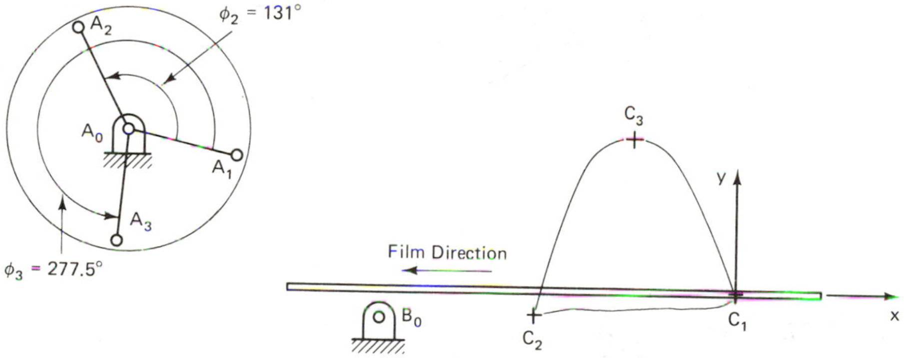


| | | | |
|---|---|---|---|
| $C_{1x} = 0.0$ | $C_{1y} = 0.0$ | $A_{0x} = -12.14$ | $A_{0y} = 3.06$ |
| $C_{2x} = -4.07$ | $C_{2y} = -0.5$ | $B_{0x} = -7.10$ | $B_{0y} = -0.52$ |
| $C_{3x} = -2.10$ | $C_{3y} = 3.05$ | $A_{1x} = -10.36$ | $A_{1y} = 2.64$ |

**Figure P2.16**

**2.26.** Figure P2.17a shows a butterfly valve in a tube that has a liquid flowing through it. A four-bar function generator is to be designed so that movement of the input link ($A_0A$) in equal increments will produce equal incremental changes of the flow through the butterfly valve (the output). Figure P2.17b shows the angles required for this objective ($\phi_2$, $\phi_3$, $\psi_2$, and $\psi_3$) as well as the location of $B_0$, $B$, and $A_0$.

**(a)** Using the graphical method of Fig. 2.49, find the location of point $A$.

**(b)** Use the overlay technique to find point $A$.

**(c)** Use the loop-closure method to find point $A$.

**(d)** Use Freudenstein's equation to solve for this linkage ($\phi_1 = 128.5°$).

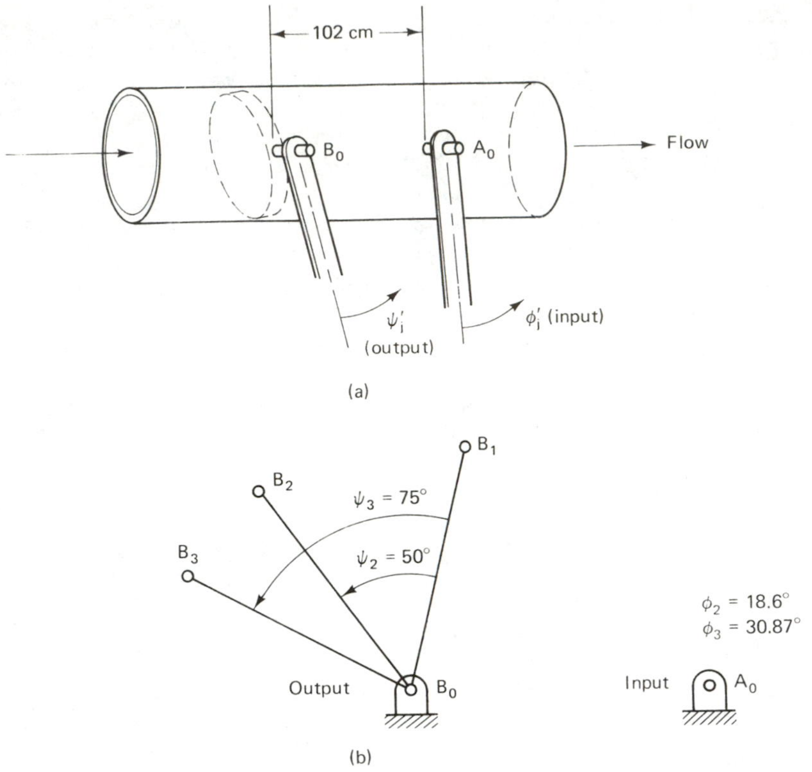


**Figure P2.17**

**2.27.** A four-bar function generator is to be designed to guide an attic stairway from its stored position down to its deployed position where half of the ladder can slide down to meet the floor below. Figure P2.18 shows the proposed location of $A_0$, $B_0$, $B$ and angles $\phi_2$, $\phi_3$, $\psi_2$, and $\psi_3$ as well as the space constraints (roof joists). Find an acceptable four-bar linkage by:

**(a)** The graphical method of Fig. 2.49.

**(b)** The overlay method.

**(c)** The loop-closure method (only $A_0$, $B_0$, and the angles are prescribed in this case).

**(d)** Freudenstein's equation (with $\phi_1 = 119°$).

**2.28.** Figure P2.19 shows a conceptual drawing of a leg-driven recreational vehicle. Series of four-bar linkages transmit the leg movements of two occupants to the rear wheels. Offsetting the function generators in phase should ensure smooth operation. If the input link ($A_0A$) rotates by $\phi_2 = 15°$ cw, $\phi_3 = 30°$ cw while the output ($B_0B$) rotates by $\psi_2 = 90°$ ccw, $\psi_3 = 180°$ ccw, find an acceptable four-bar function generator by:

**(a)** The graphical method of Fig. 2.49.

**(b)** The overlay method.

**(c)** The loop-closure method.

**(d)** Freudenstein's equation.

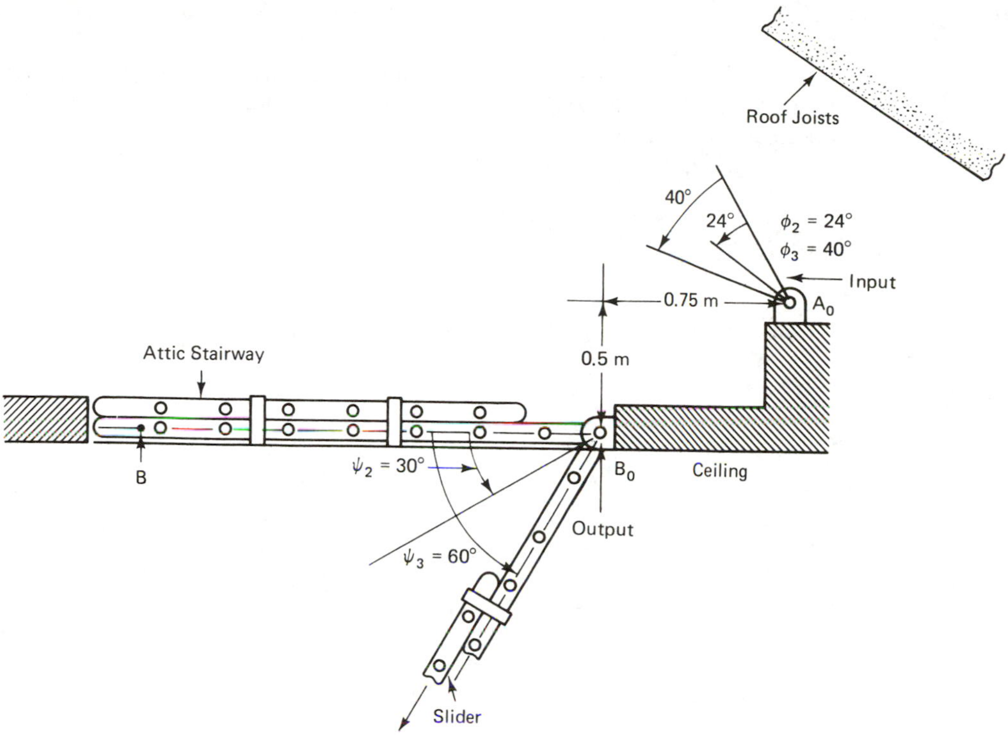


**Figure P2.18**

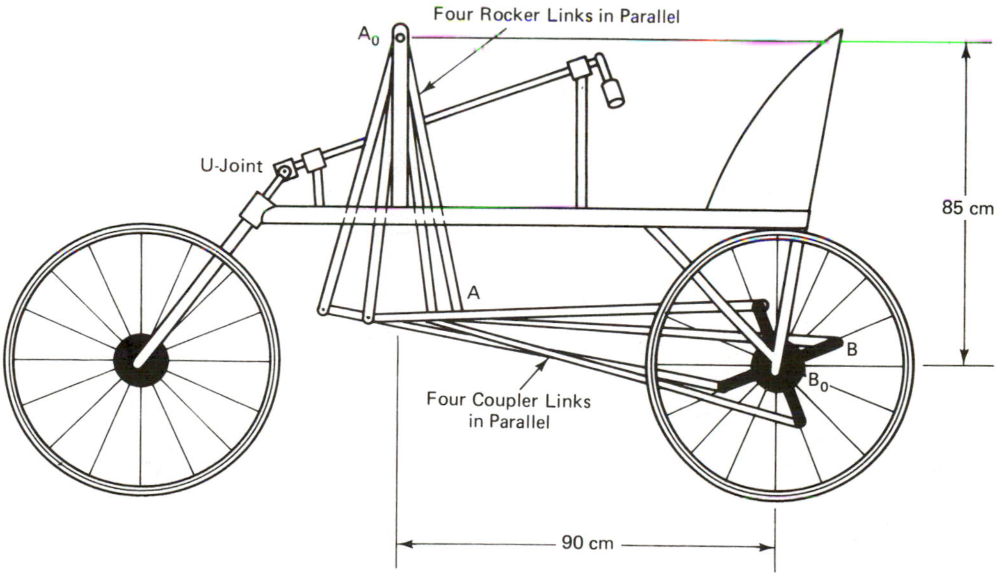


**Figure P2.19**

**2.29.** A four-bar linkage is to be designed to operate an artificial hand in the gripping operation. Figure P2.20 shows the angles that have been derived based on mechanical advantage principles. Design an acceptable four-bar linkage for this task by:

**(a)** The graphical method of Fig. 2.49.

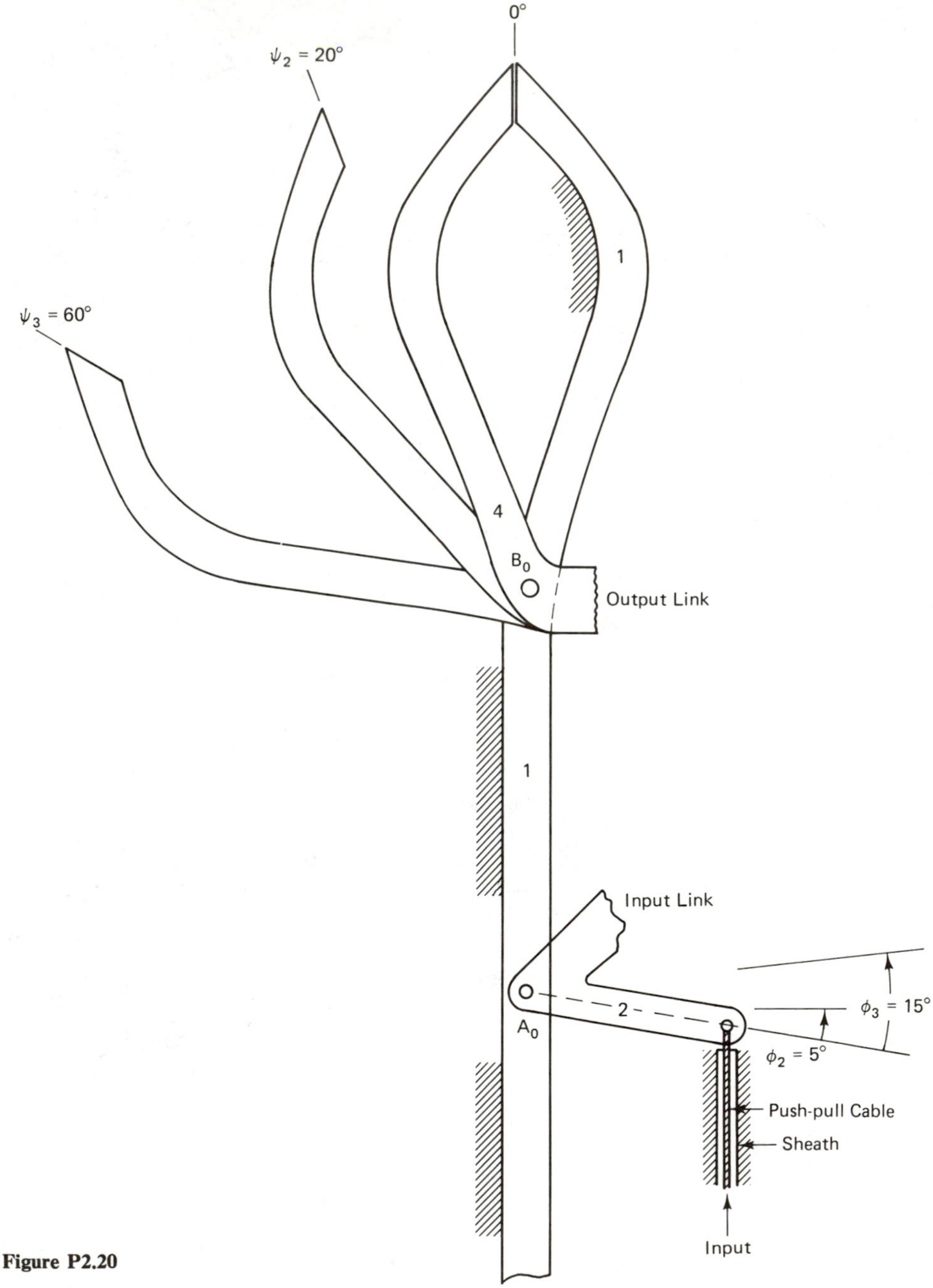


Figure P2.20

(b) The overlay method.
(c) The loop-closure method.
(d) Freudenstein's equation.

**2.30.** The problem of a binding accelerator cable led to a proposed direct linkage between the accelerator pedal and the carburetor using a six-bar linkage—two four-bars in series. Figure P2.21 shows the required link rotations as well as the location of the three ground pivots. Synthesize the six-bar function generator by:
(a) The graphical method of Fig. 2.49.
(b) The overlay method.
(c) The loop-closure method.
(d) Freudenstein's equation.

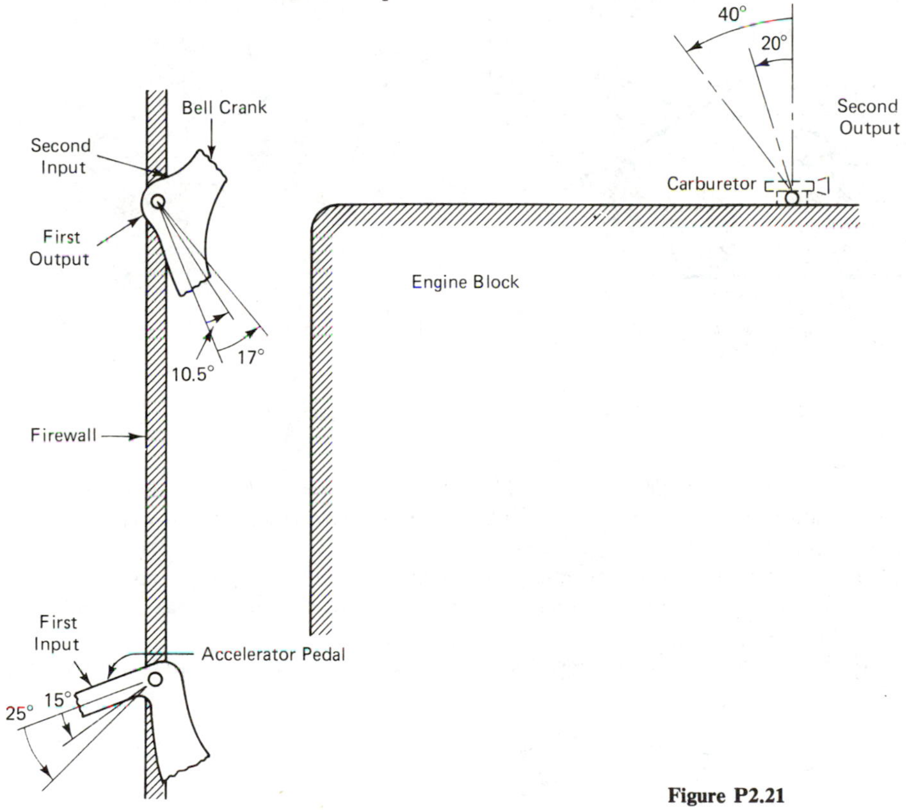


**Figure P2.21**

**2.31.** A proposed tachometer (Fig. P2.22), which uses a rotating governor principle as an engine speed indicator, requires a function-generator linkage to convert the movement of the rack into a linear movement around a tachometer dial. The geared five-bar linkage is chosen to be synthesized for this task. Known quantities are $\psi_j$, $\phi_j$, $T_2/T_1$, and $\mathbf{Z}_5$.
(a) Write the loop-closure equation for this linkage in its first and $j$th position.
(b) Determine the maximum number of positions that this linkage can be synthesized for.
(c) What is the maximum number of positions for which a linear solution is obtainable?

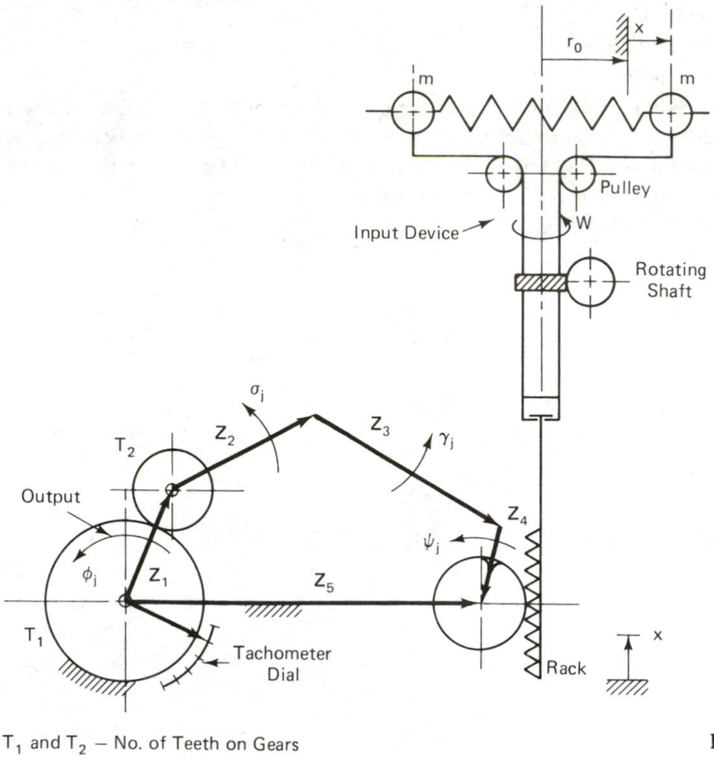


**Figure P2.22**

**2.32.** A geared six-bar function generator (Fig. P2.23) is synthesized for the maximum number of positions allowable by linear solution techniques. Vector $\mathbf{Z}_6$ as well as $T_2/T_1$, $T_4/T_5$, and $\phi_j = f(\psi_j)$ are known quantities.

**(a)** Write the loop-closure equation for this linkage in its first and $j$th positions.

**(b)** Determine the maximum number of positions that this linkage can be synthesized for.

**(c)** What is the maximum number of positions for which a linear solution is obtainable?

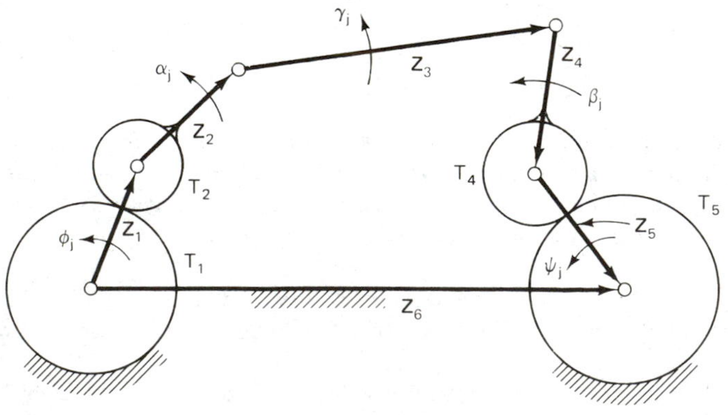


**Figure P2.23**

**2.33.** We wish to synthesize the linkage in Fig. P2.24 ($\mathbf{Z}_1$ and $\mathbf{Z}_2$) to guide a disk ($A$) in a slot defined by $\mathbf{R}_j$ ($j = 1, 2, \ldots, n$) such that the input angle ($\phi_j$) and positions $\mathbf{R}_j$ are given.

**(a)** Write the standard-form equations for this linkage in its first and $j$th positions.

**(b)** Determine the maximum number of positions that this linkage can be synthesized for.

**(c)** What is the maximum number of positions for which a linear solution is obtainable?

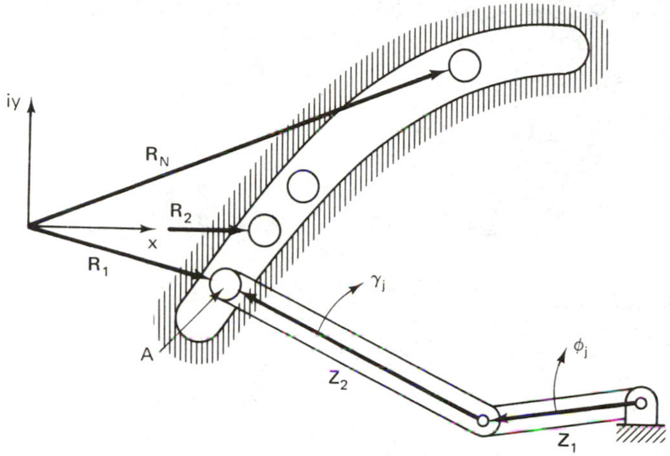


**Figure P2.24**

**2.34.** The geared linkage shown in Fig. P2.25 is to be used as a function generator where $\phi_j$ is the independent (input) variable (a rotation of arm $\mathbf{Z}_1$) and $S_j$ is the dependent (output) variable (a linear displacement of the slider). $\phi_j$ and $S_j$ are prescribed as well as $T_2/T_1$ and $\mathbf{Z}_5$.

**(a)** Write the loop-closure equation for this linkage in its first and $j$th positions.

**(b)** Determine the maximum number of positions that this linkage can be synthesized for.

**(c)** What is the maximum number of positions for which a linear solution is obtainable?

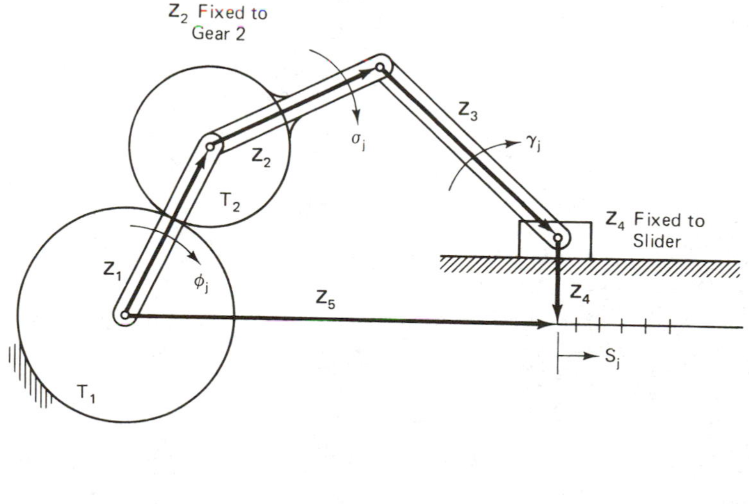


**Figure P2.25**

**2.35.** We wish to synthesize a function generator of Fig. P2.26 such that the rotations $\phi_j$ of the input crank and the displacement $S_j$ of the output slider are prescribed (i.e., $\phi_j$ and $S_j$ are known for $j = 1, 2, \ldots, n$). *Notice* that $\mathbf{Z}_4$ is fixed to the slider at an unknown constant angle $\alpha$. (Also, the *initial* position of the slider $\mathbf{Z}_1$ is given as $1.0 + 0.0i$.)

**(a)** Write the loop-closure equations (by two methods) for this linkage.

**(b)** Determine the maximum number of positions that this linkage can be synthesized for (both methods).

**(c)** What is the maximum number of positions (both methods) for which a linear solution is obtainable?

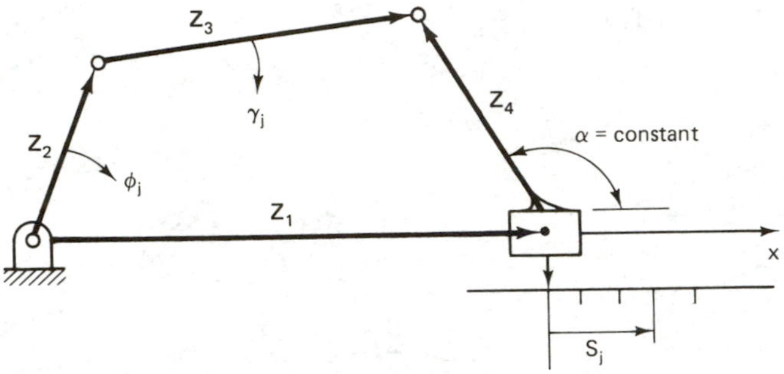


**Figure P2.26**

**2.36.** The six-bar linkage of Fig. P2.27 is to be synthesized for both path generation of point *P* and function generation $[\theta_j = f(\phi_j)]$. Write the standard-form equations for this so that the entire linkage may be synthesized by the standard form.

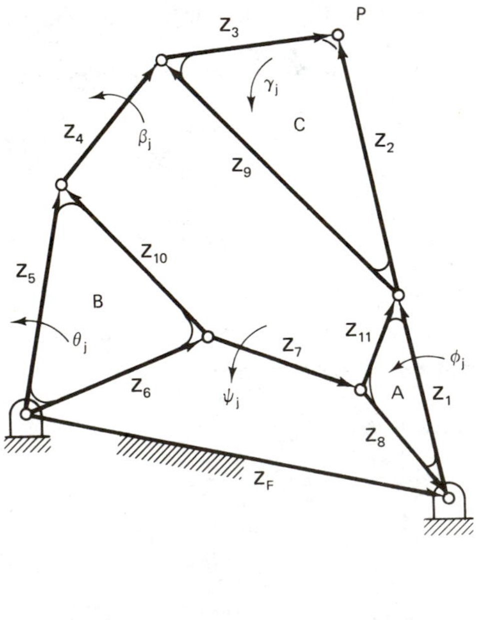


**Figure P2.27**

**2.37.** To comply with safety standards, the seven-bar motion generator linkage of Fig. P2.28 is being designed with two degrees of freedom so that the operator must use both hands to actuate the device. One hand-crank input is on link 3 (which rotates by specified angles $\beta_j$) and the other hand input is on link 7 (which rotates by specified angles $\psi_j$). These two simultaneous rotations will cause point $P$ to move along its path (specified by $\boldsymbol{\delta}_j$) while link 2 rotates by specified angles $\gamma_j$. Thus $\boldsymbol{\delta}_j$, $\beta_j$, $\psi_j$, and $\gamma_j$ are prescribed.

Write the equations that describe this synthesis task in the standard form. Be sure to cover all independent loops. (One or more vectors may have to be chosen arbitrarily in order to utilize the standard form for all loops.)

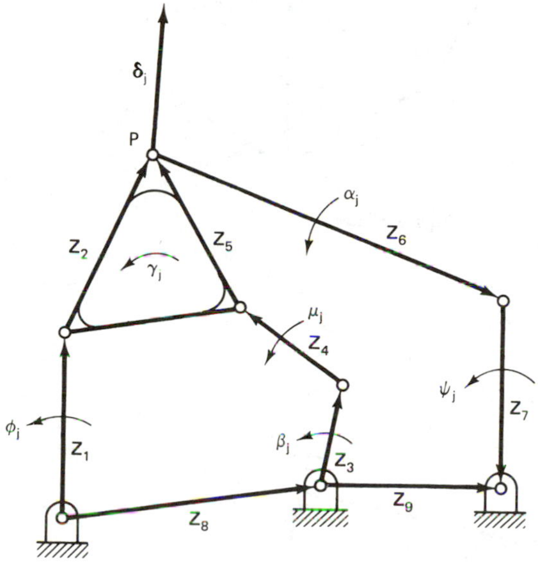


**Figure P2.28**

**2.38.** The seven-bar linkage of Fig. P2.29 is to be synthesized such that $\boldsymbol{\delta}_j$, $\alpha_{2j}$, $\alpha_{4j}$, and $\alpha_{7j}$ are prescribed.

**(a)** Write all the standard-form equations for this mechanism (making appropriate assumptions to assure the standard form).

**(b)** Can we prescribe $\boldsymbol{\delta}_j$, $\alpha_{3j}$, $\alpha_{4j}$, and $\alpha_{7j}$ instead? Why or why not?

**(c)** Can we prescribe $\boldsymbol{\delta}_j$, $\alpha_{2j}$, $\alpha_{3j}$, and $\alpha_{9j}$ instead? Why or why not?

**(d)** Can we prescribe $\boldsymbol{\delta}_j$, $\alpha_{3j}$, $\alpha_{4j}$, and $\alpha_{9j}$ instead? Why or why not?

**2.39.** Write the displacement, velocity, and acceleration equations describing the linkage in Fig. P2.30.

**2.40.** Synthesize a four-bar (see notation of Fig. 2.78) linkage such that the following are specified:

$$\mathbf{Z}_1 = -1.0 + 0.0i$$

$$\omega_2 = 1 \text{ rad/sec}, \qquad \omega_3 = -2 \text{ rad/sec}, \qquad \omega_4 = 3 \text{ rad/sec}$$

$$\alpha_2 = 3 \text{rad/sec}^2, \qquad \alpha_3 = 1 \text{ rad/sec}^2, \qquad \alpha_4 = 2 \text{ rad/sec}^2$$

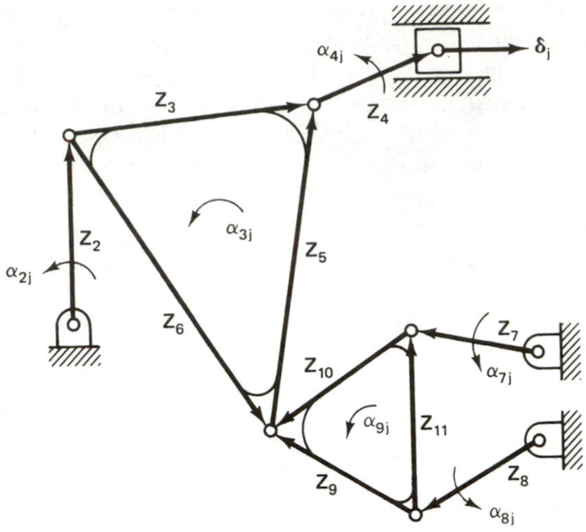


**Figure P2.29**

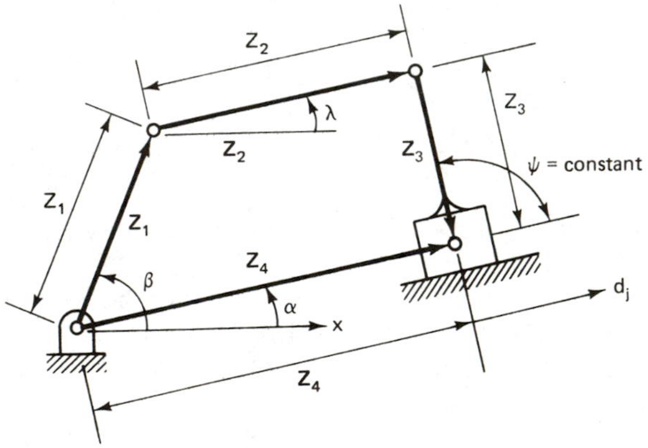


**Figure P2.30**

**2.41.** Design a four-bar linkage that will satisfy the following precision conditions:

$$\omega_2 = 8 \text{ rad/sec}, \qquad \alpha_2 = 0$$

$$\omega_3 = 1 \text{ rad/sec}, \qquad \alpha_3 = 20 \text{ rad/sec}^2$$

$$\omega_4 = 6 \text{ rad/sec}, \qquad \alpha_4 = 0$$

# 3

# Kinematic Synthesis of Linkages: Advanced Topics

## 3.1 INTRODUCTION

In Chap. 2, a variety of approaches to the synthesis of linkages were introduced. Motion, path, and function generation for three prescribed positions was emphasized because that number of precision points corresponds to the maximum number of prescribed positions for which a four-bar linkage may be synthesized by linear methods using the standard dyad form (see Table 2.1). This chapter expands on the complex-number technique and the standard dyad form introduced in Chap. 2 to investigate the four- and five-precision-point cases. Other methods of kinematic synthesis for these cases will not be pursued here since most planar linkages may be designed with the standard-form algorithm. It will be shown that in the four-precision-point case, the infinity of solutions (Table 2.1) may be surveyed at a glance by viewing computer graphics routines designed to take advantage of the capabilities of interactive displays.

## 3.2 FOUR PRESCRIBED POSITIONS: MOTION GENERATION

Figures 2.35 to 2.38 presented a geometric interpretation of synthesizing dyads for the two- and three-prescribed-position coplanar motion-generation case. Figure 3.1a and b show a moving plane $\pi$ in two and three positions. The notation in this chapter corresponds to the kinematics literature: ground pivots are designated by $m$ (for the German "Mittelpunkt" meaning "center point"), while moving pivots of ground-pivoted binary links are signified as $k$ (for "Kreispunkt" meaning "circle point").

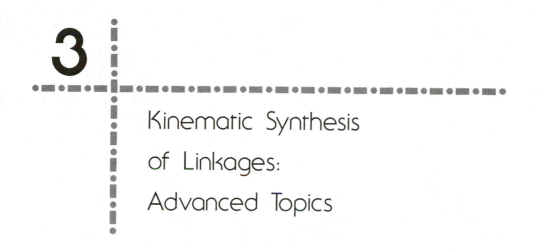

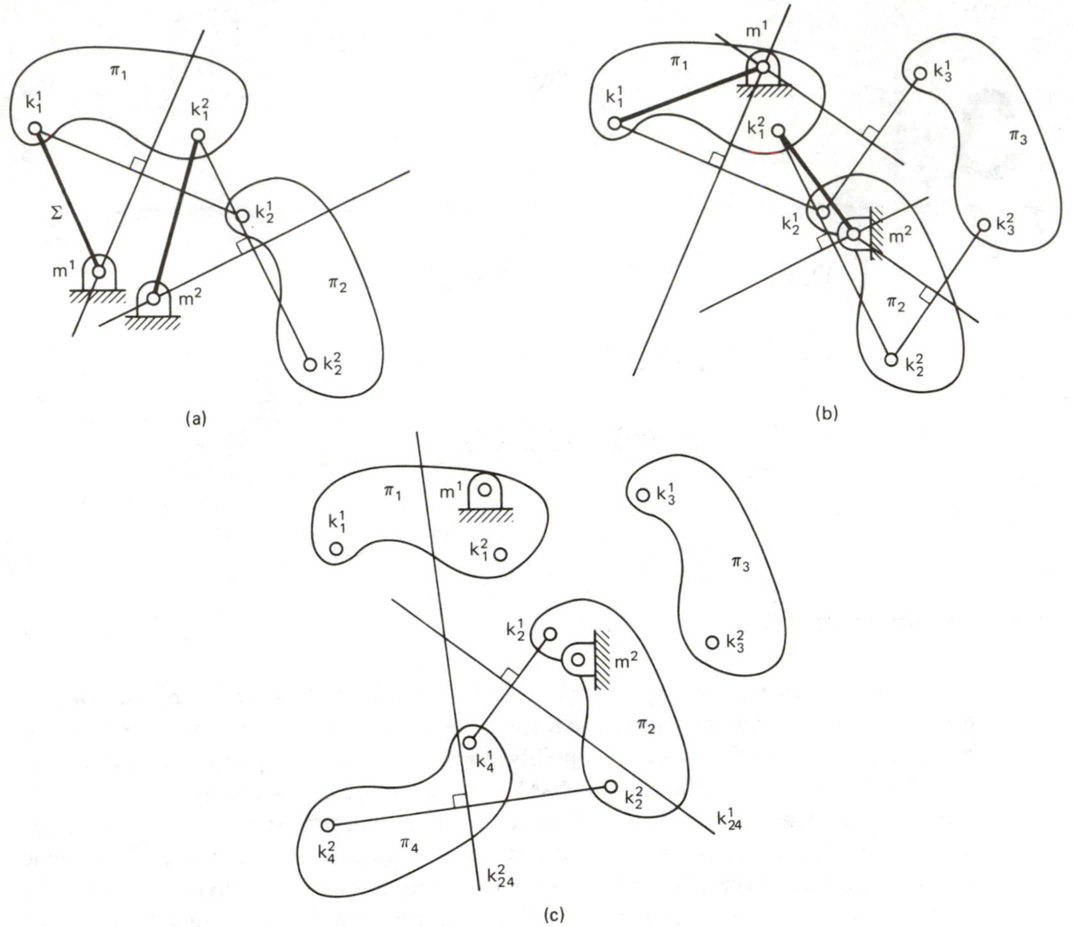


**Figure 3.1** (a) two coplanar prescribed positions of moving plane $\pi$; (b) three prescribed positions; (c) four prescribed positions: perpendicular bisectors ($k_{24}^1$ and $k_{24}^2$) constructed for the second and four positions do not pass through ground pivots $m^1$ and $m^2$ obtained from first three positions.

Recall that for two prescribed positions, there are three infinities of $k$ and $m$ point pairs, since $k_1$ (the moving pivot in its first position) may be located anywhere in the first position of plane $\pi$, and $m$ anywhere along the perpendicular bisector of $k_1$ and $k_2$, the first and second corresponding positions of $k$. For example, in Fig. 3.1a, two of these $k$–$m$ point pairs are shown: $k_1^1m^1$ and $k_1^2m^2$. For three positions (Fig. 3.1b), the location of $k_1^1$ represents two infinities of choices, but the intersection of perpendicular bisectors of $k_1^1k_2^1$ and $k_2^1k_3^1$ yield only one center point, $m^1$.

Figure 3.1c shows the three previously prescribed positions of a plane ($\pi_1$, $\pi_2$, and $\pi_3$) plus an additional position, $\pi_4$. Also shown are the two ground pivots

($m^1$ and $m^2$) corresponding to the moving pivots ($k^1$ and $k^2$) in the first three prescribed positions. Perpendicular bisectors $k_{24}^1$ and $k_{24}^2$ of $k_2^1$, $k_4^1$ and $k_2^2$, $k_4^2$ are also shown. Notice that these perpendicular bisectors do not pass through $m^1$ and $m^2$. This means that neither $k^1$ nor $k^2$ are acceptable moving pivots for these *four* positions. But Table 2.1 indicates an infinite number of solutions in general—is this a paradox? No. The question should read: *Are there any points k in body $\pi$ whose corresponding positions lie on a circle of the fixed plane for the four arbitrarily prescribed positions of $\pi$?* This same question was asked (and answered in the affirmative) by Burmester (1876). The following Burmester theory development by means of complex numbers will parallel what he discovered by geometric means.

The standard-form dyad expression (see Figs. 2.56 and 2.57) will be derived here again. Figure 3.2 shows a moving plane $\pi$ in two prescribed positions, $\pi_1$ and $\pi_j$. Positions of the plane are defined by locations of the embedded tracer point $P$ and of a directed line segment $Pa$, also embedded in the moving plane. Positions $P_1$ and $P_j$ may be located with respect to an arbitrary fixed coordinate system by $\mathbf{R}_1$ and $\mathbf{R}_j$ respectively. A path displacement vector $\boldsymbol{\delta}_j = \mathbf{R}_j - \mathbf{R}_1$ locates $P_j$ with respect to $P_1$. The rotation of the plane from position 1 to $j$ is equal to the rotation of the directed line segment $Pa$, signified as $\alpha_j$.

Let point $k_1$ (embedded in the moving plane) be the unknown location of a possible circle point and let point $m$ be the corresponding unknown center point (embedded in the fixed plane). Since both $k$ and $P$ are embedded in the moving plane, an unknown vector $\mathbf{Z}$ embedded in $\pi_1$, may be drawn from $k_1$ to $P_1$. Also, we may locate the circle point $k_1$ with respect to the center point $m$ by another unknown vector $\mathbf{W}$. Thus, as plane $\pi$ moves from $\pi_1$ to $\pi_j$, vector $\mathbf{W}$ rotates by the unknown angle $\beta_j$ about $m$, while $P_1a_1$ rotates to $P_ja_j$ by the angle $\alpha_j$.

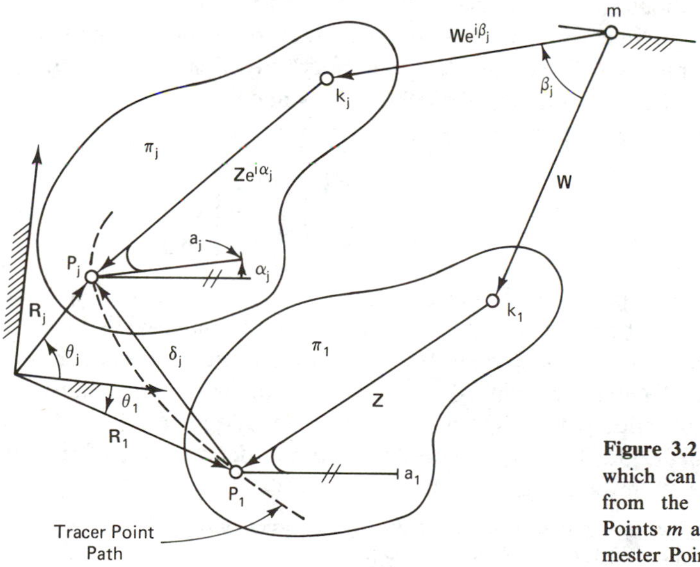


**Figure 3.2** The unknown dyad **W**,**Z**, which can guide the moving plane $\pi$ from the first to the $j$th position. Points $m$ and $k_1$ are an unknown Burmester Point Pair.

Notice that the vectors defined above form a closed loop, including the first and $j$th positions:

$$\mathbf{W}e^{i\beta_j} + \mathbf{Z}e^{i\alpha_j} - \boldsymbol{\delta}_j - \mathbf{Z} - \mathbf{W} = 0 \tag{3.1}$$

Combining terms, we obtain

$$\mathbf{W}(e^{i\beta_j} - 1) + \mathbf{Z}(e^{i\alpha_j} - 1) = \boldsymbol{\delta}_j \tag{3.2}$$

Note that this equation is the "standard form" [see Eq. (2.16)], since $\boldsymbol{\delta}_j$ and $\alpha_j$ are known from the prescribed positions of $\pi$. For four positions, there will be three equations like Eq. (3.2), ($j = 2, 3, 4$):

$$\begin{aligned} \mathbf{W}(e^{i\beta_2} - 1) + \mathbf{Z}(e^{i\alpha_2} - 1) &= \boldsymbol{\delta}_2 \\ \mathbf{W}(e^{i\beta_3} - 1) + \mathbf{Z}(e^{i\alpha_3} - 1) &= \boldsymbol{\delta}_3 \\ \mathbf{W}(e^{i\beta_4} - 1) + \mathbf{Z}(e^{i\alpha_4} - 1) &= \boldsymbol{\delta}_4 \end{aligned} \tag{3.3}$$

Recall from Table 2.1 that, for four prescribed positions, one free choice must be made in order to balance the number of equations and the number of unknowns. If one of the rotations of link **W** is chosen, say $\beta_2$, then the system must be solved for six unknown reals: **Z** and **W** and the angles $\beta_3$ and $\beta_4$. Equations (3.3) are nonlinear (transcendental) in $\beta_3$ and $\beta_4$.

## 3.3 SOLUTION PROCEDURE FOR FOUR PRESCRIBED POSITIONS

Let us for a minute consider Eqs. (3.3) to be a set of three complex equations linear and non-homogeneous in the two complex unknowns **Z** and **W**. In order for this set of three equations to have simultaneous solutions for **Z** and **W**, one of the complex equations must be linearly dependent on the other two; that is, the coefficients of the equations must satisfy certain "compatibility" relations. Satisfaction of these relations will lead to the solution of the equations above.

Equation (3.3) may be expressed in matrix form as

$$\begin{bmatrix} e^{i\beta_2} - 1 & e^{i\alpha_2} - 1 \\ e^{i\beta_3} - 1 & e^{i\alpha_3} - 1 \\ e^{i\beta_4} - 1 & e^{i\alpha_4} - 1 \end{bmatrix} \begin{bmatrix} \mathbf{W} \\ \mathbf{Z} \end{bmatrix} = \begin{bmatrix} \boldsymbol{\delta}_2 \\ \boldsymbol{\delta}_3 \\ \boldsymbol{\delta}_4 \end{bmatrix} \tag{3.4}$$

The second column of the coefficient matrix on the left side of the equation as well as the right side of the equation contain prescribed input data, while the first column contains unknown rotations $\beta_3$ and $\beta_4$. This system can have a solution only if the rank* of the "augmented matrix" of the coefficients is 2. The augmented matrix $M$ is formed by adding the right-hand column of system (3.4) to the coefficient matrix of the left side. Thus it is necessary that the determinant of the augmented matrix of this system be equal to zero:

* A matrix has rank $r$ if at least one of its $(r \times r)$-order square minors is nonzero, while all $[(r + 1) \times (r + 1)]$ and higher-order minors are zero.

$$\text{Det } M = \text{Det}\begin{bmatrix} e^{i\beta_2}-1 & e^{i\alpha_2}-1 & \boldsymbol{\delta}_2 \\ e^{i\beta_3}-1 & e^{i\alpha_3}-1 & \boldsymbol{\delta}_3 \\ e^{i\beta_4}-1 & e^{i\alpha_4}-1 & \boldsymbol{\delta}_4 \end{bmatrix} = 0 \tag{3.5}$$

Equation (3.5) is a complex equation (containing two independent scalar equations) and thus may be solved for the two scalar unknowns, $\beta_3$ and $\beta_4$. Observing that the unknowns appear in the first column of matrix $M$, the determinant is expanded about this column:

$$\boldsymbol{\Delta}_2 e^{i\beta_2} + \boldsymbol{\Delta}_3 e^{i\beta_3} + \boldsymbol{\Delta}_4 e^{i\beta_4} + \boldsymbol{\Delta}_1 = 0 \tag{3.6}$$

where

$$\boldsymbol{\Delta}_1 = -\boldsymbol{\Delta}_2 - \boldsymbol{\Delta}_3 - \boldsymbol{\Delta}_4 \tag{3.7}$$

and the $\boldsymbol{\Delta}_j$, $j = 2, 3, 4$, are the cofactors of the elements in the first column:

$$\begin{aligned} \boldsymbol{\Delta}_2 &= \begin{vmatrix} e^{i\alpha_3}-1 & \boldsymbol{\delta}_3 \\ e^{i\alpha_4}-1 & \boldsymbol{\delta}_4 \end{vmatrix} \\ \boldsymbol{\Delta}_3 &= -\begin{vmatrix} e^{i\alpha_2}-1 & \boldsymbol{\delta}_2 \\ e^{i\alpha_4}-1 & \boldsymbol{\delta}_4 \end{vmatrix} \\ \boldsymbol{\Delta}_4 &= \begin{vmatrix} e^{i\alpha_2}-1 & \boldsymbol{\delta}_2 \\ e^{i\alpha_3}-1 & \boldsymbol{\delta}_3 \end{vmatrix} \end{aligned} \tag{3.8}$$

The $\boldsymbol{\Delta}$'s are known, since they contain only known input data. Equation (3.6) is termed a compatibility equation, because sets of $\beta_2$, $\beta_3$, and $\beta_4$, which satisfy this equation, will render system (3.3) "compatible." This means that the system will yield simultaneous solutions for **W** and **Z.**

In the compatibility equation, the unknowns are located in the exponents of exponentials. This transcendental equation can be simplified through a graphical solution procedure, adaptable for computer programming, as shown in Table 3.1 (see also Fig. 3.3).

Equation (3.6) can be further simplified in notation as follows:*

$$\boldsymbol{\Delta}_3 e^{i\beta_3} + \boldsymbol{\Delta}_4 e^{i\beta_4} = -\boldsymbol{\Delta}, \qquad \text{where } -\boldsymbol{\Delta} = -\boldsymbol{\Delta}_1 - \boldsymbol{\Delta}_2 e^{i\beta_2} \tag{3.9}$$

Then, for an arbitrary choice of $\beta_2$, $-\boldsymbol{\Delta}$ as well as $\boldsymbol{\Delta}_3$ and $\boldsymbol{\Delta}_4$ are known and can be drawn to scale as illustrated in Fig. 3.3. Notice that in Eq. (3.9) $\boldsymbol{\Delta}_3$ is multiplied by $e^{i\beta_3}$, which is a rotation operator. This also holds for $\boldsymbol{\Delta}_4$ and $e^{i\beta_4}$. Equation (3.9) tells us that when $\boldsymbol{\Delta}_3$ is rotated by $\beta_3$ and $\boldsymbol{\Delta}_4$ is rotated $\beta_4$, these two vectors form a closed loop with $\boldsymbol{\Delta}$.

Equation (3.6) may be regarded as the "equation of closure" of a four-bar linkage, the so-called "compatibility linkage," with "fixed link" $\boldsymbol{\Delta}_1$, "movable links" $\boldsymbol{\Delta}_j$, $j = 2, 3, 4$, and "link rotations" $\beta_j$, measured from the "starting position" of

* Note that this equation is the same form as derived for the "ground-pivot specification" technique for three precision points [Eq. (2.57)].

**TABLE 3.1** ANALYTICAL SOLUTION OF COMPATIBILITY EQUATION EQ. (3.6) FOR $\beta_j$, $j = 3, 4$, BASED ON GEOMETRIC CONSTRUCTION.

Computation of $\beta_3$, $\beta_4$, $\tilde{\beta}_3$, and $\tilde{\beta}_4$ for a given value of $\beta_2$ (see Fig. 3.3)

$$\mathbf{\Delta} = \mathbf{\Delta}_1 + \mathbf{\Delta}_2 e^{i\beta_2}$$

$$\cos\theta_3 = \frac{\Delta_4^2 - \Delta_3^2 - \Delta^2}{2\Delta_3\Delta}, \qquad \text{where } \Delta_j = |\mathbf{\Delta}_j| \text{ and } \Delta = |\mathbf{\Delta}|$$

$$\sin\theta_3 = |(1 - \cos^2\theta_3)^{1/2}| \geq 0$$

Let $\cos\theta_3 = x$, $\sin\theta_3 = y > 0$ With these, use the ATAN2 function (FORTRAN IV or WATIV) to find $0 \leq \theta_3 \leq \pi$.

$$\beta_3 = \arg\mathbf{\Delta} + \theta_3 - \arg\mathbf{\Delta}_3$$

$$\tilde{\theta}_3 = 2\pi - \theta_3$$

$$\tilde{\beta}_3 = \arg\mathbf{\Delta} + \tilde{\theta}_3 - \arg\mathbf{\Delta}_3$$

$$\cos\theta_4 = \frac{\Delta_3^2 - \Delta_4^2 - \Delta^2}{2\Delta_4\Delta}$$

$$\sin\theta_4 = |(1 - \cos^2\theta_4)^{1/2}| \geq 0$$

Let $\cos\theta_4 = x$, $\sin\theta_4 = y > 0$ Use the ATAN2 function to find $0 \leq \theta_4 \leq \pi$; $\tilde{\theta}_4 = -\theta_4$.

$$\beta_4 = \arg\mathbf{\Delta} - \theta_4 - \arg\mathbf{\Delta}_4$$

$$\tilde{\beta}_4 = \arg\mathbf{\Delta} + \theta_4 - \arg\mathbf{\Delta}_4 + \pi$$

the compatibility linkage, defined by Eq. (3.7) as the closure equation at the start. This concept is illustrated in Fig. 3.3. Here the starting position is shown in full lines. Regarding $\mathbf{\Delta}_2$ as the "driving crank," the arbitrarily assumed $\beta_2$ amounts to

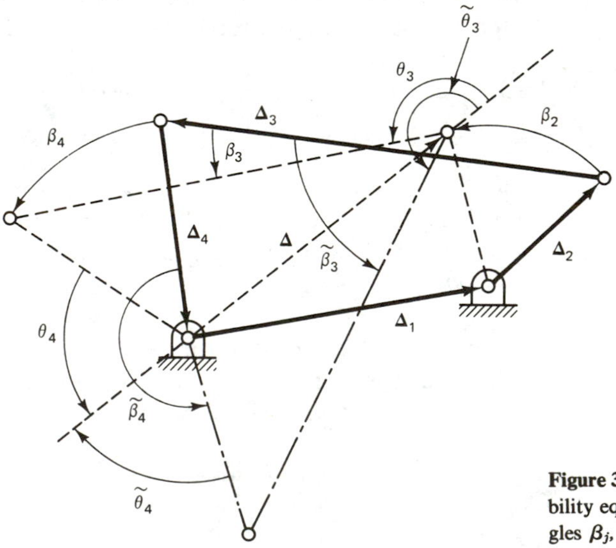


**Figure 3.3** Geometric solution of the compatibility equation [Eq. (3.6)] for the unknown angles $\beta_j$, $j = 3,4$.

imparting a rotational displacement to $\Delta_2$, shown in dashed lines in its new position. The corresponding displaced positions of $\Delta_3$ and $\Delta_4$ also appear with dashed lines. However, Eq. (3.6) can also be satisfied by the dot-dashed positions of $\Delta_3$ and $\Delta_4$, characterized by the respective rotations $\tilde{\beta}_3$, $\tilde{\beta}_4$. Thus, in general, for each assumed value of $\beta_2$, there will be *two sets of values for* $\beta_j$, $j = 3, 4$.

The range within which $\beta_2$ may be assumed is determined by the limits of mobility of the compatibility linkage, found either graphically according to Fig. 3.3, or analytically (see Secs. 3.1 and 3.3 of Vol. 1). Analytical expressions for the computation of $\beta_3$, $\beta_4$, $\tilde{\beta}_3$, and $\tilde{\beta}_4$ for a given value of $\beta_2$ are given in Table 3.1. Any two $\beta$'s from either of two sets, $\beta_2$, $\beta_3$, $\beta_4$ and $\beta_2$, $\tilde{\beta}_3$, $\tilde{\beta}_4$, may be inserted into two of the three standard-form equations [Eq. (3.3)]. Then Cramer's rule or any other method of solving sets of linear equations may be used to solve for **Z** and **W**, from which the circle point

$$\mathbf{k}_1 = \mathbf{R}_1 - \mathbf{Z} \tag{3.10}$$

and the center point

$$\mathbf{m} = \mathbf{k}_1 - \mathbf{W} \tag{3.11}$$

may be found.

If $\beta_2$ is varied in steps from 0 to $2\pi$, each value of $\beta_2$ will yield (when the compatibility linkage closes) two sets of *Burmester point pairs* (BPPs), each consisting of a circle point $k_1$ and a center point $m$. Note that the circle point $k_1$ is a point of the first prescribed position of the moving plane. A plot of the center points for each value of $\beta_2$ will sweep out two branches of the *center-point curve,* one branch associated with $\beta_3$ and $\beta_4$, the other with $\tilde{\beta}_3$ and $\tilde{\beta}_4$. If the "compatibility linkage" of Fig. 3.3 allows complete rotation of $\Delta_2$, these two branches will meet. The circle points may also be plotted similarly to yield the *circle-point curve.* A portion of a typical center- and circle-point curve is shown in Fig. 3.4 (see Sec. 3.5, which covers another technique for generating Burmester curves). Every point on the center-point curve represents a possible ground pivot. This ground pivot can be linked with its conjugate, the circle point in the particular BPP, and with the first prescribed path point. This will form a dyad with ground pivot $m$, crank **W**, pin joint $k_1$, floating link **Z**, and terminal point $P_1$ (Fig. 3.2). This dyad can serve as one-half of the four-bar motion generator and it may be combined with any other similarly formed dyad to complete a four-bar linkage.

In examining the motion of a solution linkage formed by two dyads, it may be observed that, although the moving body will assume each of the four prescribed positions as the input crank rotates in one direction through its range of motion, they may not be reached in the prescribed order, unless certain conditions are fulfilled in choosing the BPP along the two curves [302,303,304,116,110]. Also, the resulting four-bar may have other undesirable characteristics (poor ground and moving pivot locations, low transmission angles, branching, etc.). Some techniques are available for transmission angle control [114,115,306] and branching [301–303]. The infinite number of solutions may be surveyed for the "best" solution by using these techniques in conjunction with computer graphics described next.

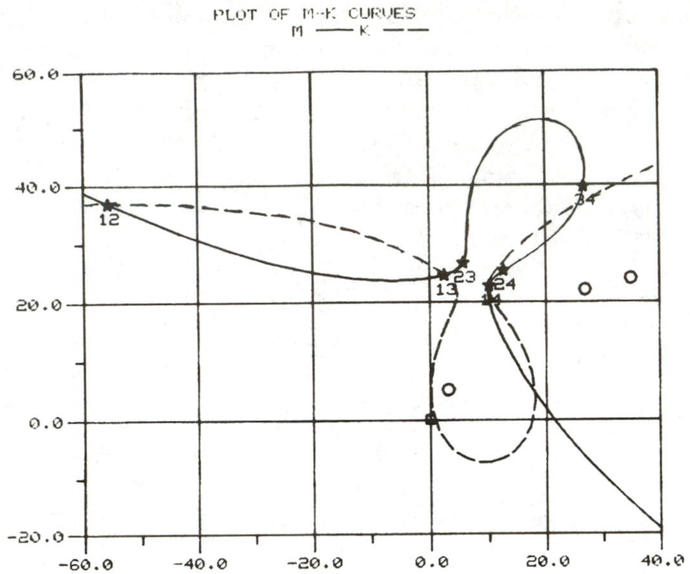


**Figure 3.4** Typical computer-generated Burmester curves. Precision points: open square, $P_1$; open circles, $P_2$, $P_3$, and $P_4$. Poles: $P_{1,2}$ to $P_{3,4}$. Solid line, locus of *m*, ground pivots or centerpoint curve; dashed line, locus of $k_1$, moving pivots or circlepoint curve.

## 3.4 COMPUTER PROGRAM FOR FOUR PRESCRIBED PRECISION POINTS

The solution of the standard-form equations for four prescribed, finitely separated positions has been fully described in Sec. 3.3, which should enable the reader to program it for automatic computation. However, it has been programmed and is part of the LINCAGES* package [78,83,84,218,270] (available from the second author). Rotation $\beta_2$ is used as an independent parameter to generate solutions to the synthesis equations [Eq. (3.3)] resulting in the circle-point and center-point curves.

The four-precision-point option of the LINCAGES package is introduced by way of an example in the appendix to this chapter. Many interactive subroutines are available with graphical output to help the linkage designer to survey numerous possibilities in order to help arrive at an optimal solution.

The following example demonstrates how sensitive the Burmester curves can be to a slight change in input data ($\delta_2$, $\delta_3$, $\delta_4$, $\alpha_2$, $\alpha_3$, $\alpha_4$).

* Copyright, University of Minnesota, 1979. This computer graphics package, developed at the University of Minnesota, contains a Teletype, LSI-11 (programmed on a TERAK microprocessor) and a Tektronix 4010 Series version. Other versions are now available on other mainframe computers and on some turn-key systems. Any planar linkage that can be composed of dyads can be synthesized for either three, four, or five finitely separated precision points for motion, path, or function generation or for their combinations.

**Example 3.1 [318,83]**

If the boom of a front loader is pivoted about a fixed axis, as it is raised above the horizontal position, the bucket tends to arc back in the direction of the loader cab. If the bucket must be lifted fairly high, the forward reach is reduced and if there is any spillback, it might occur near the operator. One manufacturer has overcome this problem by guiding the boom with pivoted links so that the boom is connected to the coupler link of a four-bar linkage. The boom is designed to move in such a manner that the bucket does not arc back toward the operator.*

In Fig. 3.5, the length of boom $k_1P_1$ and crank length $mk_1$ as well as the loader frame were roughly proportioned from a photograph of the above-mentioned commercially available loader. As boom $k_1P_1$ is pivoted clockwise to raise the bucket, link $mk_1$ is pivoted counterclockwise to give four positions of point $P$ that lie approximately on a straight line that slopes outward from the vertical, away from the loader. The angular positions of line $kP$ for the four positions assumed were accurately determined from the drawing (with respect to the horizontal x axis) to be

$$\theta_1 = 191.26°, \qquad \theta_3 = 164.77°$$

$$\theta_2 = 175.68°, \qquad \theta_4 = 152.04°$$

Portions of the center-point and circle-point curves for this example, corresponding to the four positions of the line $kP$ ($\boldsymbol{\delta}_j = \mathbf{P}_j - \mathbf{P}_1$ and $\alpha_j = \theta_j - \theta_1$, $j = 2, 3, 4$) are shown in Fig. 3.5. One $mk_1$ combination that locates the ground pivot $m$ in an ideal position is shown. The only feasible design observed from the Burmester curves would be one in which the fixed pivot for the other dyad would be on a bracket extending backward and upward from the top rear of the cab and the corresponding moving

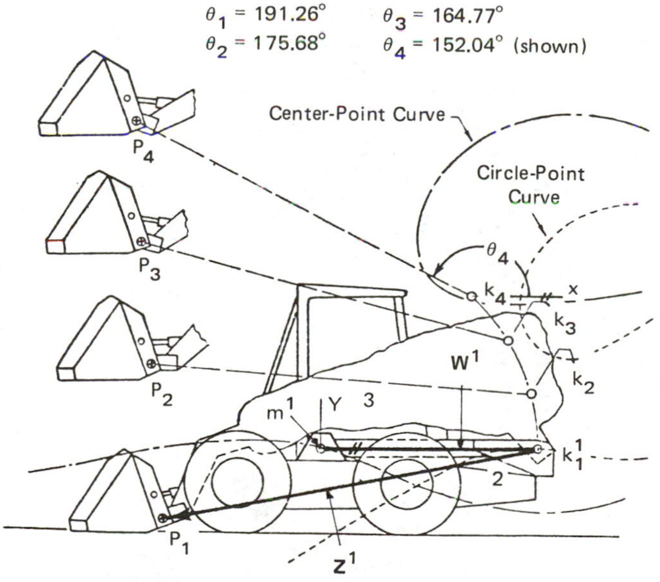


**Figure 3.5** Four assumed positions of front-loader-boom $kP$ and portions of the corresponding Burmester curves. One dyad, $\mathbf{W}^1, \mathbf{Z}^1$, has been chosen from the Burmester curves. No other suitable dyad is found.

* The bucket angle is controlled by a piston pinned between the boom and the bucket.

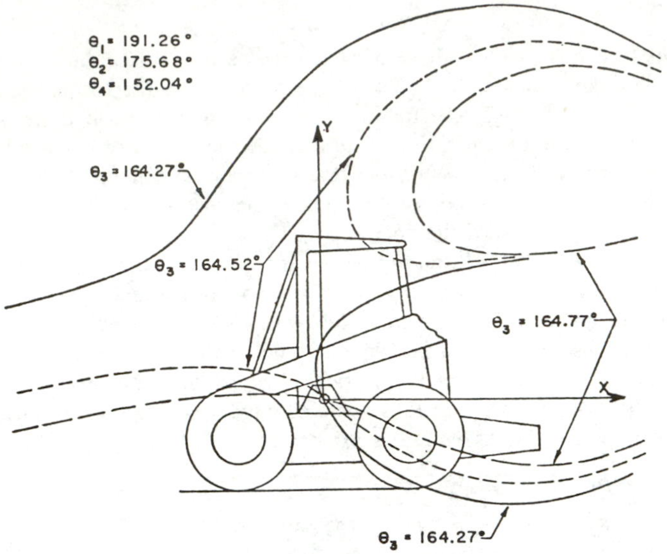


**Figure 3.6** Centerpoint curves for three slightly different angular positions $\theta_3$ of plane $kP$ in position 3.

joint would be on the circle-point curve almost vertically above $k_1$. Assuming that no major extensions are to be built onto the existing cab structure shown, such a design would not be satisfactory. However, it is known that the Burmester curves may shift considerably with small angular changes of the line determining the four positions of the plane. Small angular changes of line $k_1P_1$ would not appreciably affect the general direction in which the bucket travels. This shift of the center-point curve is illustrated in Fig. 3.6, in which three center-point curves are shown which share the same first,

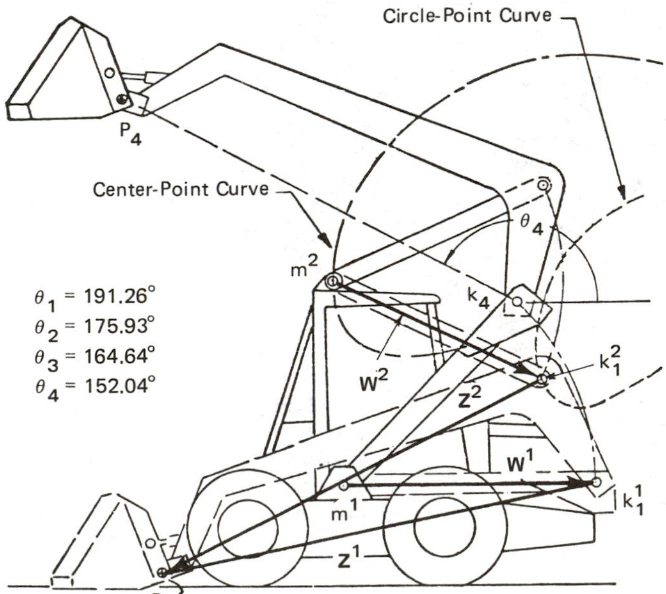


**Figure 3.7** Burmester curves for the selected design positions of plane $kP$. Two dyads, $\mathbf{Z}^1,\mathbf{W}^1$ and $\mathbf{Z}^2,\mathbf{W}^2$, make up the final four-bar solution.

second, and fourth angular positions of line $k_1P_1$. The prescribed path points are also identical for each curve but the third angular position, $\theta_3$, has different values of 164.77°, 164.52°, and 164.27°. With the latter value of $\theta_3$, the center-point curve changes from a one-part curve to a two-part curve. Additional runs were made varying $\theta_2$ and $\theta_4$ by small amounts. From the results of the several runs made, it was apparent that

$$\theta_1 = 191.26°, \qquad \theta_3 = 164.64°$$

$$\theta_2 = 175.93°, \qquad \theta_4 = 152.04°$$

might permit the choice of a fixed pivot near the front top of the cab, where it is located on the commercially available loader previously mentioned, and at the same time give a good location for the moving pivot. The linkage resulting from this choice of angles is shown in Fig. 3.7. The actual center-point and circle-point loci for this case [318,83] are shown in Fig. 3.8.

Notice, that with each change in $\boldsymbol{\delta}_j$ or $\alpha_j$, an infinity of solutions is surveyed. The portions of the ground and moving pivot curves that are too far away are cut off by the "window" chosen on the graphics screen. The designer is now synthesizing

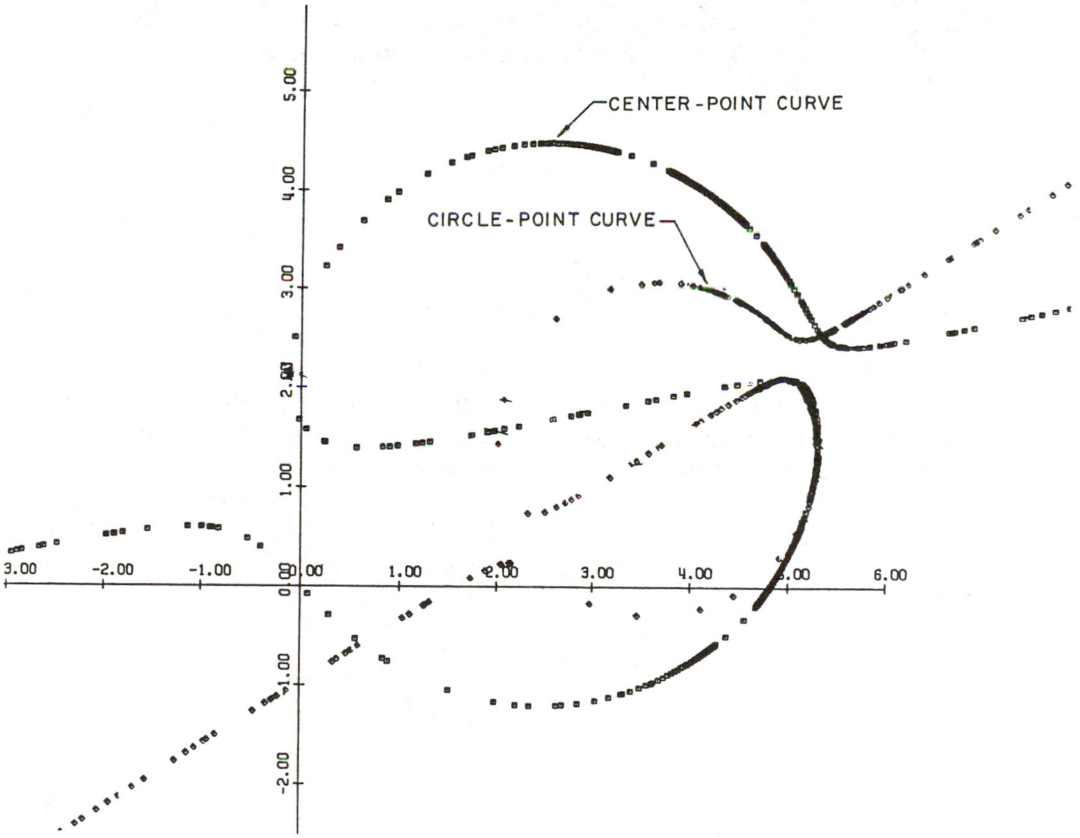


**Figure 3.8** Portion of the plot of the final Burmester curves for the selected design positions for the front loader of Example 3.1 [318].

*infinities at one time* rather than one mechanism at a time as is done graphically or with cardboard and thumb tacks or even by computational methods without the aid of the graphics screen.

## 3.5* MOTION-GENERATION WITH FOUR FINITELY SEPARATED PRESCRIBED POSITIONS: SUPERPOSITION OF TWO THREE-PRECISION-POINT CASES

The four-position motion dyad synthesis problem with $\boldsymbol{\delta}_2$, $\boldsymbol{\delta}_3$, $\boldsymbol{\delta}_4$, $\alpha_2$, $\alpha_3$, and $\alpha_4$ prescribed can be developed from a different viewpoint [170]. Four prescribed positions can be considered as a superposition of two three-position motion-generation subproblems, with ($\boldsymbol{\delta}_2$, $\boldsymbol{\delta}_3$, $\alpha_2$, $\alpha_3$) prescribed in one and ($\boldsymbol{\delta}_2$, $\boldsymbol{\delta}_4$, $\alpha_2$, $\alpha_4$) prescribed in the other. Recall that Sec. 2.20 presented the circle-point circles for the three-position case. For a selected value of the rotation of the ground-link (**W**) (see Fig. 3.2), $\beta_2$, a pair of $M$ and $K$ circles can be drawn for both three-precision-point subproblems; intersections, of such two $M$-circles and $K$-circles, respectively, define points on the $m$ and $k$ curves that satisfy both subproblems simultaneously.

For example, suppose that the following dyad-motion precision positions (Fig. 3.2) were desired:

Problem 1:

$$\boldsymbol{\delta}_2 = 2 + 2i, \qquad \alpha_2 = 60°$$

$$\boldsymbol{\delta}_3 = 5 + 2i, \qquad \alpha_3 = 120°$$

Problem 2:

$$\boldsymbol{\delta}_2 = 2 + 2i, \qquad \alpha_2 = 60°$$

$$\boldsymbol{\delta}_4 = 4 + 3i, \qquad \alpha_4 = 180°$$

(where $\boldsymbol{\delta}_2$ and $\alpha_2$ are the same in both problems). The intersections of the $M$ circles and $K$ circles in both three-point problems (i.e., for each $\beta_2$), as described above, define points on the "circle-point" and "center-point" curves.

Following the graphical construction procedure of Sec. 2.20, we begin by finding the poles $P_{12}$, $P_{13}$, $P_{14}$, $P_{23}$, $P_{24}$, $P'_{23}$, $P'_{24}$. There will be two sets of $M$ and $K$ circles, labeled with superscripts 1 and 2 ($M^1$ corresponds to problem 1). The centers of circles $M^1$, $M^2$, $K^1$, $K^2$ lie on the bisectors of $\overline{P_{13}P_{23}}$, $\overline{P_{14}P_{24}}$, $\overline{P_{13}P'_{23}}$, and $\overline{P_{14}P'_{24}}$, respectively, labeled as "$M^1$ axis," "$K^1$ axis," and so on, in Fig. 3.9. Note that the following length equalities prevail:

$$|\overline{P_{13}P_{23}}| = |\overline{P_{13}P'_{23}}|$$

and

$$|\overline{P_{14}P_{24}}| = |\overline{P_{14}P'_{24}}|$$

* This section may be skipped without loss of continuity.

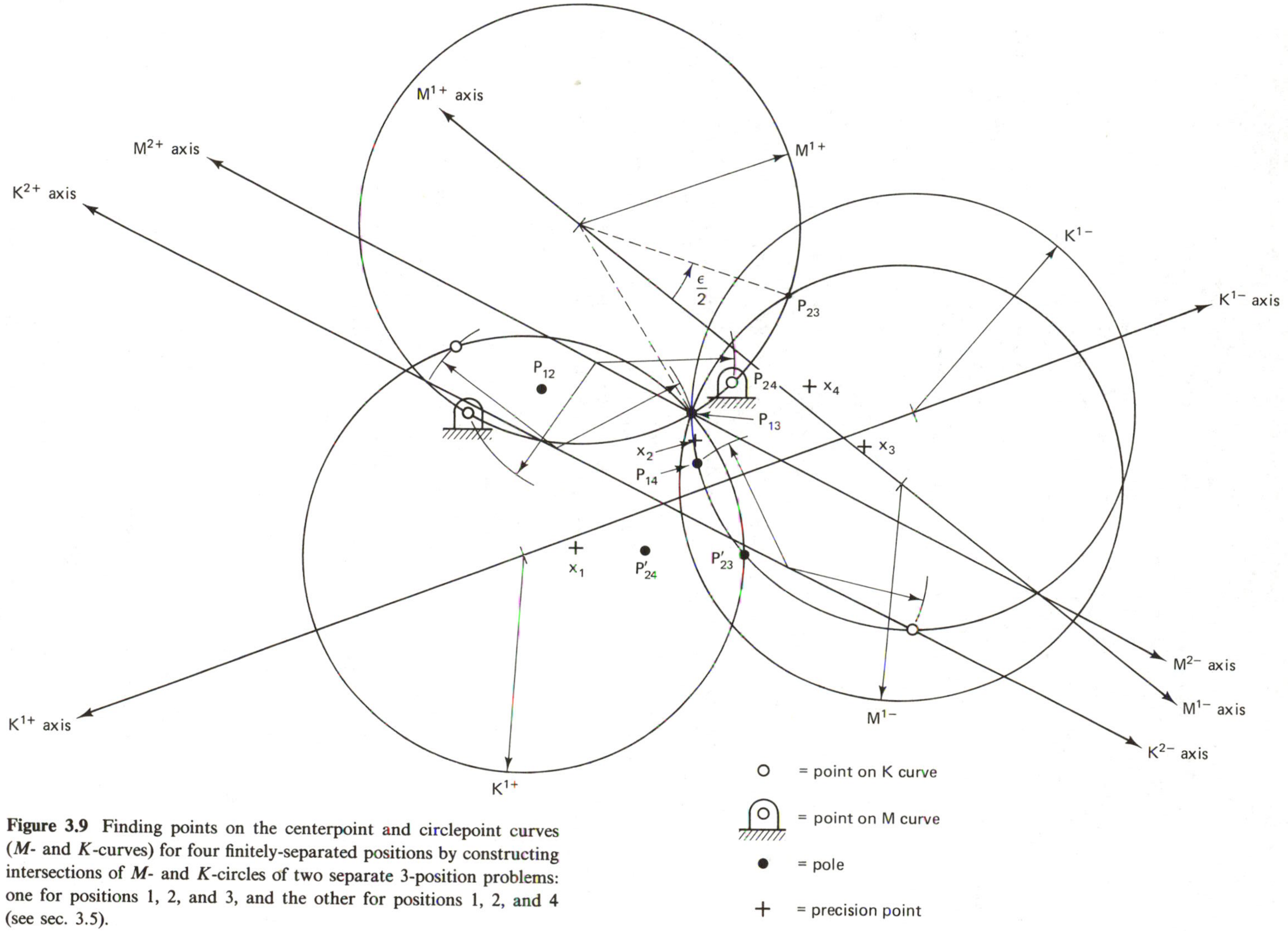


**Figure 3.9** Finding points on the centerpoint and circlepoint curves (*M*- and *K*-curves) for four finitely-separated positions by constructing intersections of *M*- and *K*-circles of two separate 3-position problems: one for positions 1, 2, and 3, and the other for positions 1, 2, and 4 (see sec. 3.5).

Up to four $m$- and $k$-curve points can be found for each value of angle $\epsilon$ of Fig. 3.9 (eight including its supplement, as was done here for ease of construction), the positive angle subtended by the respective pole distances at the corresponding circle centers. For example, for a chosen value of $\epsilon$, strike an arc with radius $\frac{1}{2}|\overline{P_{13}P_{23}}|\csc(\epsilon/2)$ from $P_{13}$ to intersect the $M^{1+}$ and $M^{1-}$ axes.* This locates the centers of one $M^{1+}$ and one $M^{1-}$ circle with this radius. Next, *with this* same value of $\epsilon$, repeat the construction using $|\overline{P_{14}P_{24}}|$ and draw one $M^{2+}$ and one $M^{2-}$ circle. The construction for the $K^1$ and $K^2$ circles is similar, using $|\overline{P_{13}P'_{23}}|$ and $|\overline{P_{14}P'_{24}}|$. Intersections of $K^{1+}$, $K^{1-}$, $M^{1+}$, and $M^{1-}$ with $K^{2+}$, $K^{2-}$, $M^{2+}$, and $M^{2-}$, respectively, are possible moving pivots $k$ and ground pivots $m$ for the four-position problem, and thus are points on the circle-point and center-point curves. If $\epsilon = 40°$, the angles $\beta_2$ corresponding to each intersection pair are (two circles may intersect at two points)

$$M^+: \quad \beta_2 = -\epsilon = -40°$$

$$K^+: \quad \beta_2 = \alpha_2 - \epsilon = 60° - 40° = 20°$$

$$K^-: \quad \beta_2 = \alpha_2 - (-\epsilon) = 60° + 40° = 100°$$

The $K^+$ and the $K^-$ intersections apply for $\beta_2$ less than $\alpha_2$, or for $\beta_2$ greater than $\alpha_2$, respectively. Note, for example, that if one desired the moving pivots for $\beta_2 = 40°$, they would be the $K^+$ intersections for $\epsilon = \alpha_2 - \beta_2 = 20° > 0$.

This procedure, derived here from the three-position $M$ and $K$ circles, is the same as the classical geometric Burmester curve construction based on the *opposite pole quadrilateral* [125]. *Opposite poles* are defined as two poles carrying four different subscripts. For this example there are three pairs of opposite poles: ($P_{13}$, $P_{24}$), ($P_{23}$, $P_{14}$), and ($P_{12}$, $P_{34}$). An opposite pole quadrilateral has its diagonals connecting two opposite poles. Here we are using the opposite pole quadrilateral ($P_{13}$, $P_{23}$, $P_{14}$, $P_{24}$) for the construction of points on the $M$ curve.

Classically, the $M$ circles are not presented as intersections of three-point-solution loci, but as loci of points that subtend equal or supplementary angles at the side of the opposite-pole quadrilateral representing the chord of the circle. The intersections of circles through opposite-pole pairs whose peripheral points subtend equal or supplementary angles at their respective chords are constructed; these points satisfy the theorem of Burmester, which states that the points on the $M$ curve subtend equal or supplemental angles at opposite sides of the opposite-pole quadrilateral. Points on the $K$ curve were generated classically by inverting the motion and repeating the same construction.

To summarize: Points on the $M/K$ curve (center-point curve)/(circle-point curve) for four finitely separated positions may be generated as intersections of the $M/K$ circles of two three-precision-point subproblems, for which the procedure was laid out in Sec. 2.20.

* The plus or minus sign in the superscripts of the axes indicates one direction or the other toward infinity.

## 3.6 SPECIAL CASES OF FOUR-POSITION SYNTHESIS

### The Slider Point or Finite Ball's Point

In Chap. 4 we will discuss "Ball's point," that point of the moving plane $\pi$ whose path has third-order contact with its path tangent. In other words, the path has four-point contact with its tangent in the vicinity of the position under study. In still other words, Ball's point moves in a straight line through four infinitesimally close positions, or through four "infinitesimally separated positions" (ISPs).

The counterpart of Ball's point for four finitely separated positions (FSPs) is the slider point, which we might as well call "finite Ball's point." It is that point of the moving plane whose four corresponding positions fall on a straight line. It is a special circle point, whose conjugate center point is at infinity, and hence is located in the direction of the asymptote of the center-point curve.

Kaufman [148] has shown how the slider point can be formed as a singular solution of the compatibility equation, Eq. (3.5), by setting $\beta_j = 0$, $j = 2, 3, 4$ (see Fig. 3.2). We propose the following intuitive proof of Kaufman's approach. As seen from Eq. (3.4), the vector **W**, which connects the unknown stationary center point $m$ to the unknown moving circle point $k_1$, does not rotate ($\beta_j = 0$) as $k_1$ moves by finite displacements to $k_2$, $k_3$, and $k_4$. This is possible only if **W** is of infinite length. Since all displacements of $k_1$ (to $k_2$, $k_3$, and $k_4$) must be along paths that are always perpendicular to the momentary position of **W**, that path must be a straight line. Therefore, this $k_1$ is the finite Ball's point sought. To signify this we attach the superscript $s$ (for "slider point"), thus: $\mathbf{W}^s$ and $k_1^s$.

However, $\beta_j = 0$ makes the coefficient matrix $M$ singular, and therefore Eqs. (3.5) and (3.4) cannot be solved in their original form to locate $k_1^s$. Kaufman suggests a way around this, illustrated in Fig. 3.10. Here $P_1$ and $P_j$ ($j = 2, 3, 4$) are arbitrarily prescribed finitely separated positions of point $P$ of the moving plane $\pi$, shown in positions $\pi_1$ and $\pi_j$. Rotations of $\pi$, $\alpha_j$, are also prescribed. $\mathbf{Z}^s$ is an unknown vector embedded in $\pi$, defined in position $\pi_1$, which connects the unknown finite Ball's point $K_1^s$ to $P_1$. From positions 1 to $j$, $k_1^s$ moves along the unknown line of the unknown vector **S**, measured from an unknown point $Q$. The unknown distance of such (straight-line sliding motion) is designated by the unknown stretch ratio $\rho_j$ of the vector **S**. With these we write the following equation of closure for the polygon $Qk_j^sP_jP_1k_1^sQ$:

$$\mathbf{S}(\rho_j - 1) + \mathbf{Z}^s(e^{i\alpha_j} - 1) = \boldsymbol{\delta}_j, \qquad j = 2, 3, 4 \tag{3.12}$$

where the unknown $\rho_j$'s are scalars, such that $\rho_j \neq 1$, $\rho_j \neq \rho_k$. The compatibility equation for this system is

$$\begin{vmatrix} \rho_2 - 1 & e^{i\alpha_2} - 1 & \boldsymbol{\delta}_2 \\ \rho_3 - 1 & e^{i\alpha_3} - 1 & \boldsymbol{\delta}_3 \\ \rho_4 - 1 & e^{i\alpha_4} - 1 & \boldsymbol{\delta}_4 \end{vmatrix} = 0 \tag{3.13}$$

Assuming an arbitrary value for $\rho_2 \neq 1$ and expanding about the first column, we obtain

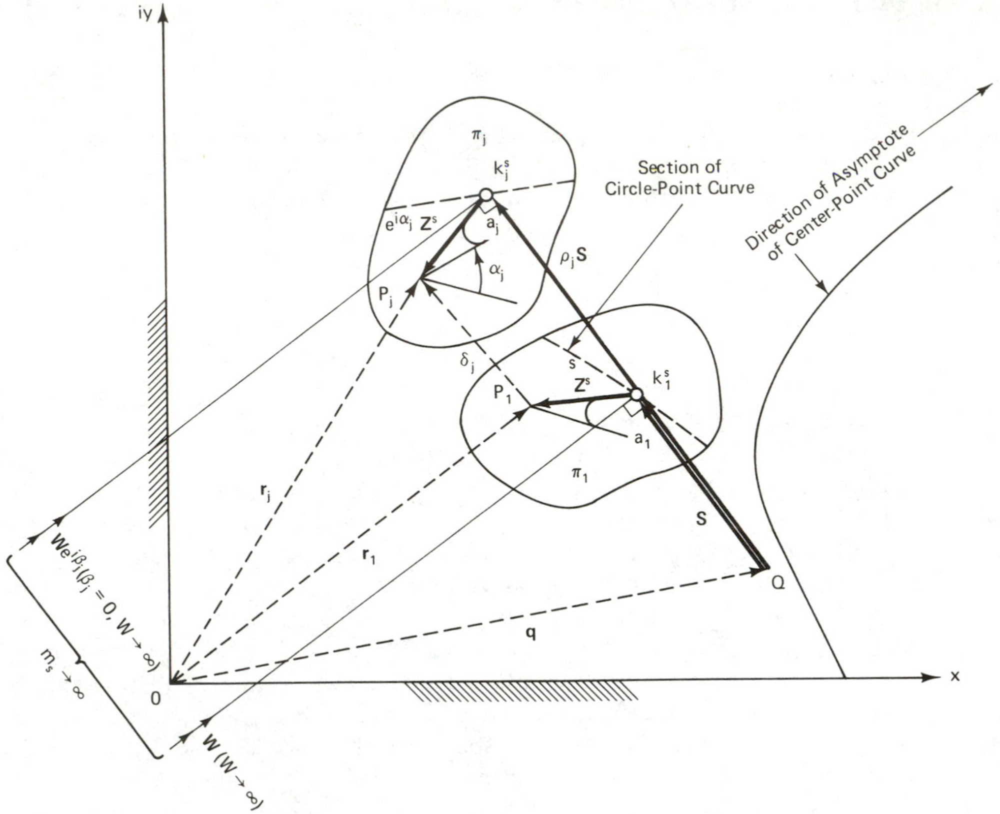


**Figure 3.10** Kaufman's scheme for finding the "finite Ball's point" $k_1^s$ for four finitely-separated positions of a plane (see sec. 3.6).

$$\boldsymbol{\Delta}_3\rho_3 + \boldsymbol{\Delta}_4\rho_4 = -\boldsymbol{\Delta}_1 - \boldsymbol{\Delta}_2\rho_2 \tag{3.14}$$

where the $\boldsymbol{\Delta}_j$ are defined as in Eqs. (3.7) and (3.8). To solve for $\rho_3$ and $\rho_4$, we separate real and imaginary parts and form the real system

$$\begin{bmatrix} \Delta_{3x} & \Delta_{4x} \\ \Delta_{3y} & \Delta_{4y} \end{bmatrix} \begin{bmatrix} \rho_3 \\ \rho_4 \end{bmatrix} = \begin{bmatrix} -\Delta_{1x} - \rho_2\Delta_{2x} \\ -\Delta_{1y} - \rho_2\Delta_{2y} \end{bmatrix} \tag{3.15}$$

where $\boldsymbol{\Delta}_j = \Delta_{jx} + i\Delta_{jy}$, and from which

$$\rho_3 = \frac{\begin{vmatrix} -\Delta_{1x} - \rho_2\Delta_{2x} & \Delta_{4x} \\ -\Delta_{1y} - \rho_2\Delta_{2y} & \Delta_{4y} \end{vmatrix}}{\begin{vmatrix} \Delta_{3x} & \Delta_{4x} \\ \Delta_{3y} & \Delta_{4y} \end{vmatrix}} \tag{3.16}$$

and

$$\rho_4 = \frac{\begin{vmatrix} \Delta_{3x} & -\Delta_{1x} - \rho_2\Delta_{2x} \\ \Delta_{3y} & -\Delta_{1y} - \rho_2\Delta_{2y} \end{vmatrix}}{\begin{vmatrix} \Delta_{3x} & \Delta_{4x} \\ \Delta_{3y} & \Delta_{4y} \end{vmatrix}} \tag{3.17}$$

Substituting back the set of $\rho_j$ ($j = 2, 3, 4$) thus obtained into system (3.12) and solving any two of its equations simultaneously for $\mathbf{S}$ and $\mathbf{Z}^s$, we obtain a slider dyad which guides the plane $\pi$ through its four prescribed positions. Combining this with a pinned dyad forms a slider-crank mechanism, whose coupler goes through the four prescribed positions; a suitable pinned dyad is obtained by assuming a value for $\beta_2 \neq 0$, $\beta_2 \neq \alpha_2$, and solving for $\mathbf{Z}$ and $\mathbf{W}$ as discussed in Sec. 3.3. Figure 3.11 is a sketch of such a slider-crank mechanism: slide guide $Q$, slide $\mathbf{S}$, crank $(m^{(1)}k_1^1) = \mathbf{W}^1$, and coupler $k_1^s P_1 k_1^1$, with sides $\mathbf{Z}^s$ and $\mathbf{Z}^1$.

The slider-crank of Fig. 3.11, while it can assume the positions shown, it would jam in between. To avoid this, we need to seek another $m^1 k_1^1$ pair. The question arises: how many such slider-crank mechanisms can we find for one arbitrary set of four prescribed body positions? It can be shown that no matter what value we pick for $\rho_2 \neq 1$, the solution for $\mathbf{Z}^s$ is always the same, and that $\mathbf{S}$, $\rho_3$, and $\rho_4$ are

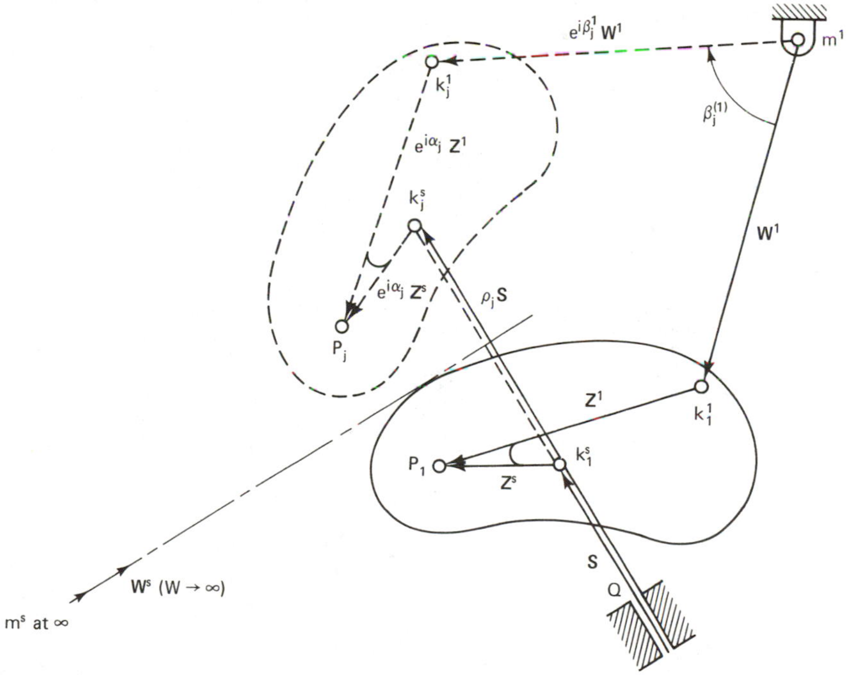


**Figure 3.11** Finding a Burmester Point Pair, $k_1^1$, $m^1$, to form a slider-crank with Kaufman's finite Ball's point $k_1^s$.

such that the resulting $k_j^s$ are the same. Thus it is seen that there is only one slider dyad, which agrees with the fact that there is only one Ball's point for a set of four ISP, and thus only one finite Ball's point for a set of four FSP. However, the number of pinned dyads is infinite. Thus there will be a *single infinity of solutions for the four-position motion generator slider-crank.* Each of these has a cognate (see Sec. 3.9) which is a *four-position path generator slider-crank with prescribed timing,* shown in Fig. 3.12.

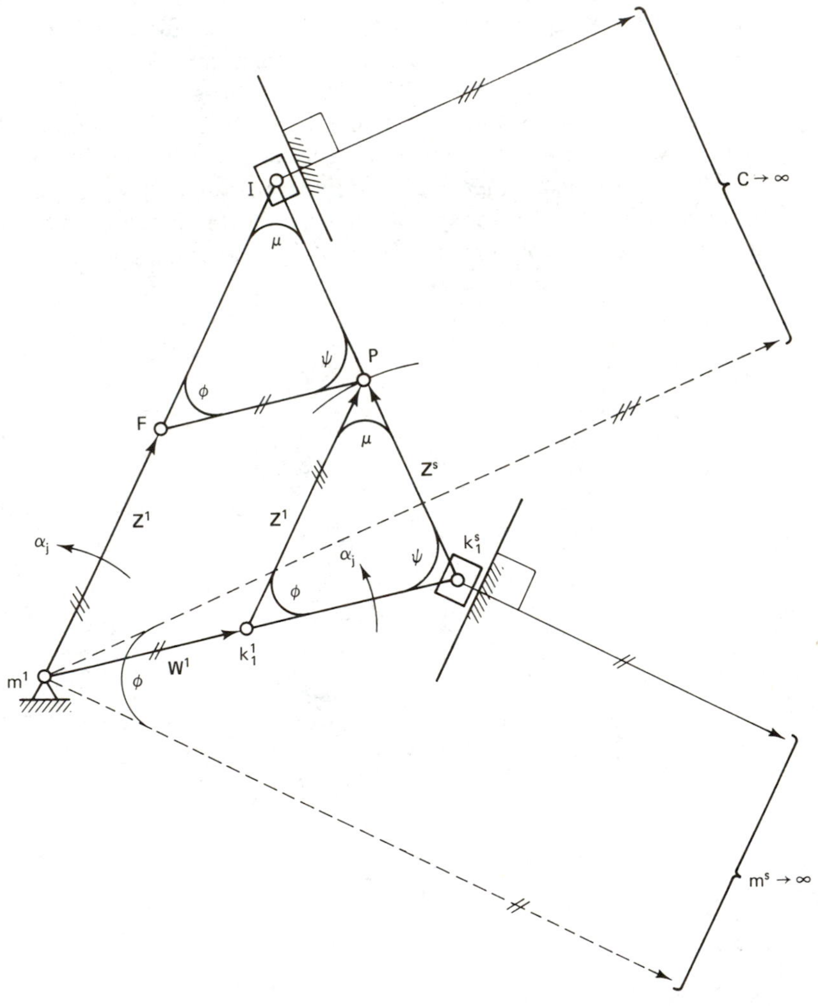


**Figure 3.12** Slider-crank mechanism $m^1k_1^1k_1^sP$ and its cognate $m^1FIP$. Observe that coupler rotations $\alpha_j$ of the first mechanism are the crank rotations in the cognate.

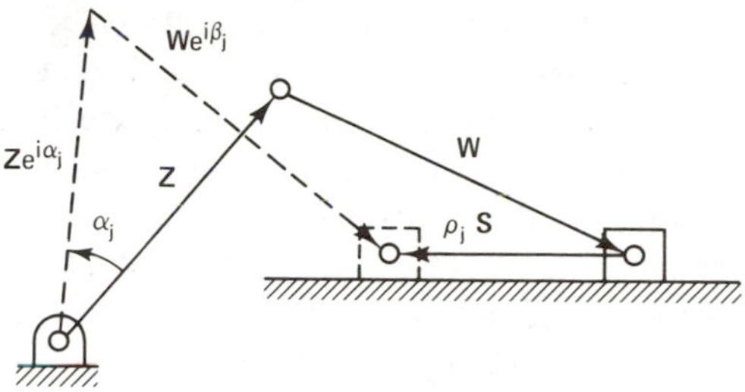


**Figure 3.13** Function-generation synthesis of the slider-crank mechanism for four FSP [see Eqs. (3.18) to (3.21)].

## Four-Point Function Generation with the Slider-Crank

Referring to Fig. 3.13, we write the standard-form equation of closure as follows:

$$\mathbf{W}(e^{i\beta_j}-1)+\mathbf{Z}(e^{i\alpha_j}-1)=\rho_j\mathbf{S} \tag{3.18}$$

The compatibility equation for this system is

$$\begin{vmatrix} e^{i\beta_2}-1 & e^{i\alpha_2}-1 & \rho_2\mathbf{S} \\ e^{i\beta_3}-1 & e^{i\alpha_3}-1 & \rho_3\mathbf{S} \\ e^{i\beta_4}-1 & e^{i\alpha_4}-1 & \rho_4\mathbf{S} \end{vmatrix}=0 \tag{3.19}$$

Here $\alpha_j$, $\rho_j$ are prescribed, correlating crank rotations and slider translations in a functional relationship. **S** is arbitrary, because it determines only the scale and orientation of the linkage. Expanding (3.19), we obtain

$$\mathbf{\Delta}_3 e^{i\beta_3}\mathbf{\Delta}_4 e^{i\beta_4}=-\mathbf{\Delta}_1-\mathbf{\Delta}_2 e^{i\beta_2} \tag{3.20}$$

where

$$\mathbf{\Delta}_2=\begin{vmatrix} e^{i\alpha_3}-1 & \rho_3\mathbf{S} \\ e^{i\alpha_4}-1 & \rho_4\mathbf{S} \end{vmatrix}, \qquad \mathbf{\Delta}_3=-\begin{vmatrix} e^{i\alpha_2}-1 & \rho_2\mathbf{S} \\ e^{i\alpha_4}-1 & \rho_4\mathbf{S} \end{vmatrix}$$
$$\mathbf{\Delta}_4=\begin{vmatrix} e^{i\alpha_2}-1 & \rho_2\mathbf{S} \\ e^{i\alpha_3}-1 & \rho_3\mathbf{S} \end{vmatrix}, \qquad \mathbf{\Delta}_1=-\mathbf{\Delta}_2-\mathbf{\Delta}_3-\mathbf{\Delta}_4 \tag{3.21}$$

The rest of the solution follows the procedure of Table 3.1 to obtain compatible sets of $\beta_j$ and then solving any two of Eqs. (3.18) for **W** and **Z.**

## Four-Position Motion-Generator Turning-Block Mechanism: The Concurrency Point

Equation (3.5) is trivially satisfied when $\alpha_j = \beta_j$, $j = 2, 3, 4$. This means that both **Z** and **W** rotate with the moving body $\pi$. Since **W** rotates about the unknown fixed center point $m$, while **Z** is embedded in and moves with $\pi$, and because **Z** and **W** concur at the unknown circle point $k_1$, a little thought will show that both must be of infinite length and opposite in direction. Since the tip of **W** and the tail of **Z** meet at the circle point $k_1$, the latter must be at infinity, in the direction of the asymptote of the circle-point curve. This is illustrated in Fig. 3.14, where $\pi_1$, $\pi_j$, $P_1$, $P_j$, and $\boldsymbol{\delta}_j$ have the same meaning as before. Along with body rotations $\alpha_j$, they are prescribed. The unknown vectors **Z** and **W** are always parallel, and they are of infinite length. The unknown circle point $k_1$ is at infinity in the unknown direction of **Z** and **W** in the first position. The effect of the infinite lengths of **W** and **Z** is this: The connection between the unknown center point $m$ and the moving body $\pi$ becomes a turn-slide. This enforces the line of the unknown vector **D**, which is embedded in $\pi$, to turn and slide through $m$. Hence $m$ is a *finite concurrency point.* (See also Chap. 4, Path Curvature Theory, for *infinitesimal concurrency point.*) Therefore, we distinguish it with the superscript $c$, and do likewise for **Z** and **W**. Finally, the tip of the unknown **D**, fixed in the body $\pi$, is connected to $P$ by the

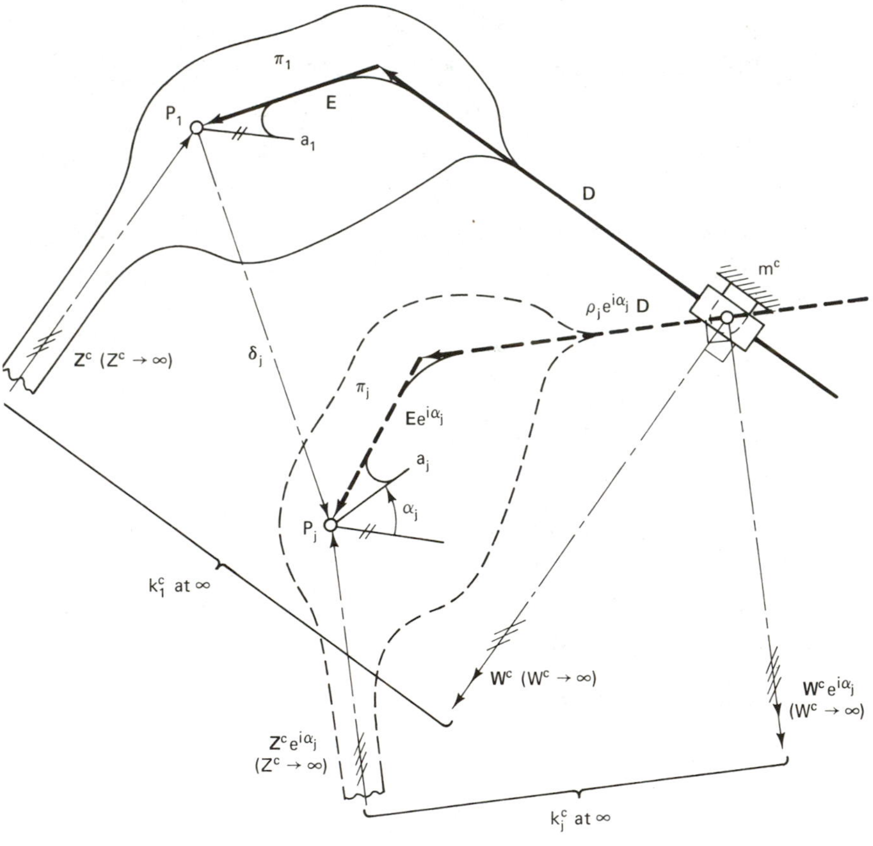


**Figure 3.14** The unique turn-slide dyad for four FSP of plane $\pi$ forms one half of a four-link turn-slide mechanism. The other half is one of a single infinity of pivoted dyads (see Fig. 3.15).

unknown embedded vector **E**. Observe that, while the rotations of both **D** and **E** are the prescribed $\alpha_j$, **E** does not change in length, but **D** does, by the unknown stretch ratios $\rho_j$. With these, the equation of closure in standard form becomes

$$\mathbf{D}(\rho_j e^{i\alpha_j} - 1) + \mathbf{E}(e^{i\alpha_j} - 1) = \boldsymbol{\delta}_j \tag{3.22}$$

where $\alpha_j$ and $\boldsymbol{\delta}_j$ are prescribed. The compatibility equation takes the form

$$\begin{vmatrix} \rho_2 e^{i\alpha_2} - 1 & e^{i\alpha_2} - 1 & \boldsymbol{\delta}_2 \\ \rho_3 e^{i\alpha_3} - 1 & e^{i\alpha_3} - 1 & \boldsymbol{\delta}_3 \\ \rho_4 e^{i\alpha_4} - 1 & e^{i\alpha_4} - 1 & \boldsymbol{\delta}_4 \end{vmatrix} = 0 \tag{3.23}$$

which expands to

$$\rho_2 e^{i\alpha_2}\boldsymbol{\Delta}_2 + \rho_3 e^{i\alpha_3}\boldsymbol{\Delta}_3 + \rho_4 e^{i\alpha_4}\boldsymbol{\Delta}_4 + \boldsymbol{\Delta}_1 = 0 \tag{3.24}$$

where the $\boldsymbol{\Delta}_j$ are the same as those defined in Eqs. (3.7) and (3.8). We now assume an arbitrary real value for $\rho_2$, separate the real and imaginary parts of Eq. (3.24), and solve for $\rho_3$ and $\rho_4$, obtaining

$$\rho_3 = \frac{\begin{vmatrix} \mathscr{R}(-\boldsymbol{\Delta}_1 - \rho_2 e^{i\alpha_2}\boldsymbol{\Delta}_2) & \mathscr{R}(e^{i\alpha_4}\boldsymbol{\Delta}_4) \\ \mathscr{I}(-\boldsymbol{\Delta}_1 - \rho_2 e^{i\alpha_2}\boldsymbol{\Delta}_2) & \mathscr{I}(e^{i\alpha_4}\boldsymbol{\Delta}_4) \end{vmatrix}}{\begin{vmatrix} \mathscr{R}(e^{i\alpha_3}\boldsymbol{\Delta}_3) & \mathscr{R}(e^{i\alpha_4}\boldsymbol{\Delta}_4) \\ \mathscr{I}(e^{i\alpha_3}\boldsymbol{\Delta}_3) & \mathscr{I}(e^{i\alpha_4}\boldsymbol{\Delta}_4) \end{vmatrix}} \tag{3.25}$$

and

$$\rho_4 = \frac{\begin{vmatrix} \mathscr{R}(e^{i\alpha_3}\boldsymbol{\Delta}_3) & \mathscr{R}(-\boldsymbol{\Delta}_1 - \rho_2 e^{i\alpha_2}\boldsymbol{\Delta}_2) \\ \mathscr{I}(e^{i\alpha_3}\boldsymbol{\Delta}_3) & \mathscr{I}(-\boldsymbol{\Delta}_1 - \rho_2 e^{i\alpha_2}\boldsymbol{\Delta}_2) \end{vmatrix}}{\begin{vmatrix} \mathscr{R}(e^{i\alpha_3}\boldsymbol{\Delta}_3) & \mathscr{R}(e^{i\alpha_4}\boldsymbol{\Delta}_4) \\ \mathscr{I}(e^{i\alpha_3}\boldsymbol{\Delta}_3) & \mathscr{I}(e^{i\alpha_4}\boldsymbol{\Delta}_4) \end{vmatrix}} \tag{3.26}$$

where $\mathscr{R}(\cdot)$ signifies the real part of $(\cdot)$ and $\mathscr{I}(\cdot)$ signifies the imaginary part of $(\cdot)$. The resulting compatible set of $\rho_j$, $j = 2, 3, 4$, can now be substituted back into any two of Eqs. (3.22) and these solved simultaneously for **D** and **E**, thus locating $m^c$ and completing the solution.

It is to be noted that the concurrency point $m^c$ thus found is unique: It is the unique slider point or finite Ball's point for the inverted motion in which the fixed plane of reference and the moving plane $\pi$ exchange roles. Thus it is seen that regardless of the arbitrary choice for the value of $\rho_2$, the end result is the same: *There is only one turn-slide dyad* that can guide $\pi$ through the four prescribed finite positions. To complete a single-degree-of-freedom four-link mechanism we find a pivoted dyad (one of the infinite number available for four finite positions), thus obtaining the mechanism shown in Fig. 3.15, one of the single infinity of such *four-position motion-generator turn-slide four-link mechanisms.* Furthermore, if design conditions make it desirable to change it to a *tumbling block mechanism* (Fig. 3.16), the same vector representation and derivations would apply.

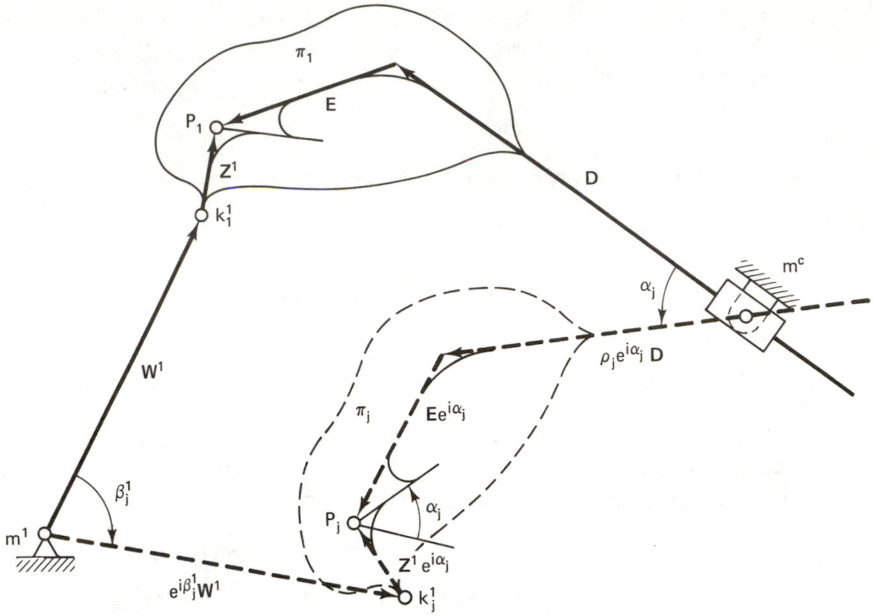


**Figure 3.15** One of Kaufman's singular solutions yields the turn-slide dyad in this four-link turn-slide mechanism [see Eqs. (3.22) to (3.26) and Fig. 3.14].

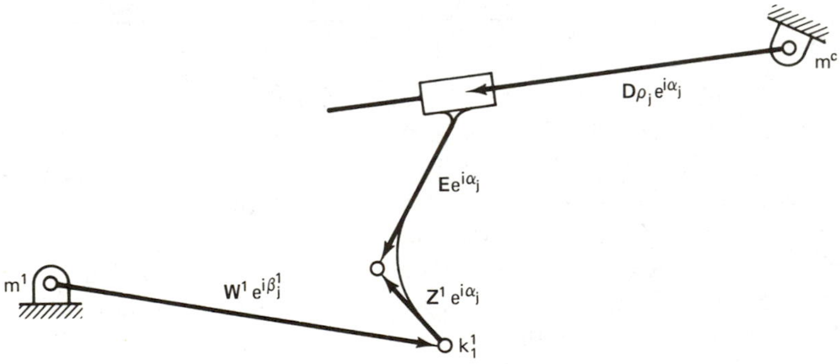


**Figure 3.16** Equations (3.22) to (3.26) can also be used to find the tumbling-block dyad in this four-FSP motion generator four-link mechanism.

## Finding the Poles in Motion-Generation Synthesis

Any pole $P_{jk}$, $j$, $k = 1, 2, 3, 4$, $j \neq k$, can be found as follows (see Fig. 3.16a):

$$(e^{i(\alpha_k - \alpha_j)} - 1)\mathbf{p}_j = \boldsymbol{\delta}_k - \boldsymbol{\delta}_j \tag{3.27}$$

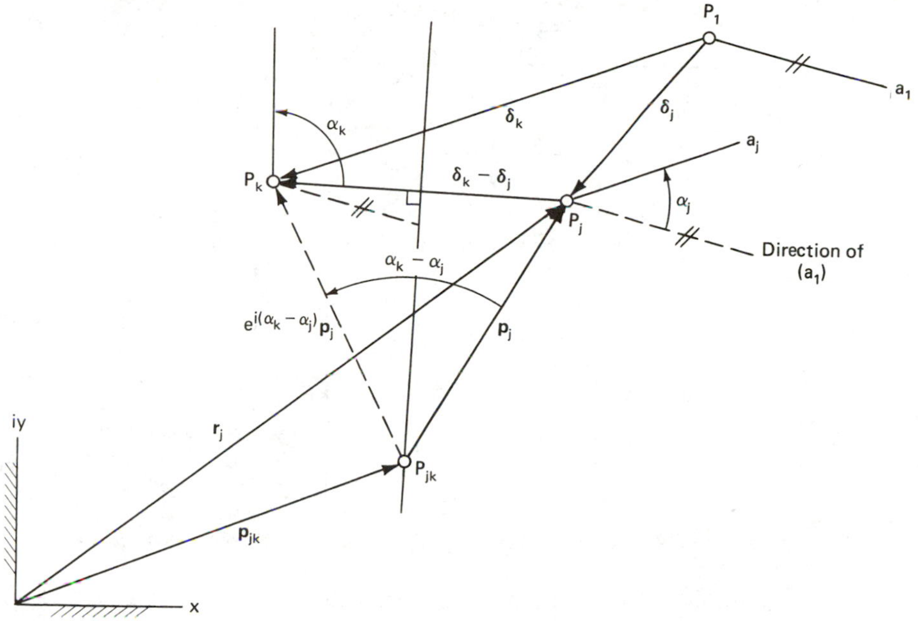


**Figure 3.16.a** Derivation of Eqs. (3.27) to (3.29) for finding the poles in planar motion-generation synthesis.

$$\mathbf{p}_j = \frac{\boldsymbol{\delta}_k - \boldsymbol{\delta}_j}{e^{i(\alpha_k - \alpha_j)} - 1} \tag{3.28}$$

$$\mathbf{p}_{jk} = \mathbf{r}_j - \mathbf{p}_j \tag{3.29}$$

## 3.7 MOTION GENERATION: FIVE POSITIONS

In previous sections we observed that a four-bar linkage could be synthesized for four prescribed positions as a motion generator. Thanks to Burmester and those who continued his work, we know that there are ideally an infinity of pivoted dyads for any four arbitrarily prescribed positions, and that any two of these can form a four-bar mechanism whose coupler will match these prescribed positions. But can we synthesize four-bar motion generators for *five* arbitrary positions?

Our first hint toward the answer to this question came when the tabular formulation was developed (Table 2.1). Although the table shows that there are no free choices for five prescribed positions, the number of real equations and the number of real unknowns are equal, indicating that these equations can be solved.

The second hint will come from further examination of the four-position case and the resulting center- and circle-point curves. Suppose, as in Fig. 3.4 that the circle- and center-point curves are plotted for prescribed positions 1, 2, 3, and 4. In addition, a new set of curves for the same first three positions plus a fifth position

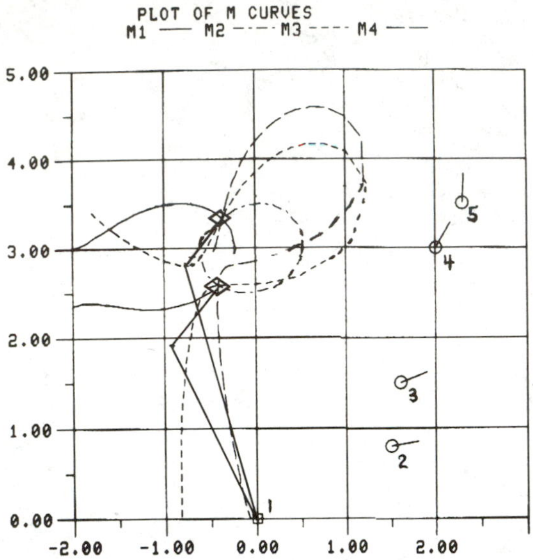


**Figure 3.17** Four combinations of center-point curves for $\boldsymbol{\delta}_2 = 1.5 + 0.8i$, $\boldsymbol{\delta}_3 = 1.6 + 1.5i$, $\boldsymbol{\delta}_4 = 2.0 + 3.0i$, and $\boldsymbol{\delta}_5 = 2.3 + 3.5i$; and $\alpha_2 = 10°$, $\alpha_3 = 20°$, $\alpha_4 = 60°$, and $\alpha_5 = 90°$. $M1$ signifies the centerpoint curve for precision points 1, 2, 3, and 4; $M2 =$ 1, 2, 3, and 5; $M3 =$ 1, 2, 4, and 5; and $M4 =$ 1, 3, 4, and 5. This example was programmed on the LINCAGES package by M. Richardson [218]. Two existing five-point dyads are drawn in. Figure 3.18 shows the resulting four-bar five-point motion-generation solution.

are superimposed over the first set. If these curves *intersect,* a common solution exists and a Burmester pair (or dyad) has been found that will be able to guide a plane through all five prescribed positions. Figure 3.17 shows an overlay of all four combinations of four-precision-point motion-generation cases whose intersections locate the center-point solutions for the combined five-position problem. Figure 3.18 shows the four bar solution for the five precision points of Fig.3.17 in its first position.

The circle- and center-point curves can be shown to be cubics [107,108], so there are a maximum of nine intersections. There are two imaginary intersections at infinity and, discounting the intersections marking the poles $P_{12}$, $P_{13}$, and $P_{23}$, there is a maximum number of four usable real intersections. [$P_{12}$, $P_{13}$, or $P_{23}$ could also be used, but this would be tantamount to the "point-position reduction method" (see Sec. 2.9 and Ref. 169) with its frequent accompanying difficulties of retrograde crank rotation.] Since usable real intersections will come in pairs, we can expect either zero, two, or four solutions for any five arbitrarily prescribed precision points. Let us see whether this geometric concept can be verified by mathematical methods.

Referring again to Fig. 3.2, the same standard form for the equations may be written (four equations in this case):

$$\mathbf{W}(e^{i\beta_j} - 1) + \mathbf{Z}(e^{i\alpha_j} - 1) = \boldsymbol{\delta}_j, \qquad j = 2, 3, 4, 5 \tag{3.30}$$

The augmented matrix of this system is

$$M = \begin{bmatrix} e^{i\beta_2} - 1 & e^{i\alpha_2} - 1 & \boldsymbol{\delta}_2 \\ e^{i\beta_3} - 1 & e^{i\alpha_3} - 1 & \boldsymbol{\delta}_3 \\ e^{i\beta_4} - 1 & e^{i\alpha_4} - 1 & \boldsymbol{\delta}_4 \\ e^{i\beta_5} - 1 & e^{i\alpha_5} - 1 & \boldsymbol{\delta}_5 \end{bmatrix} \tag{3.31}$$

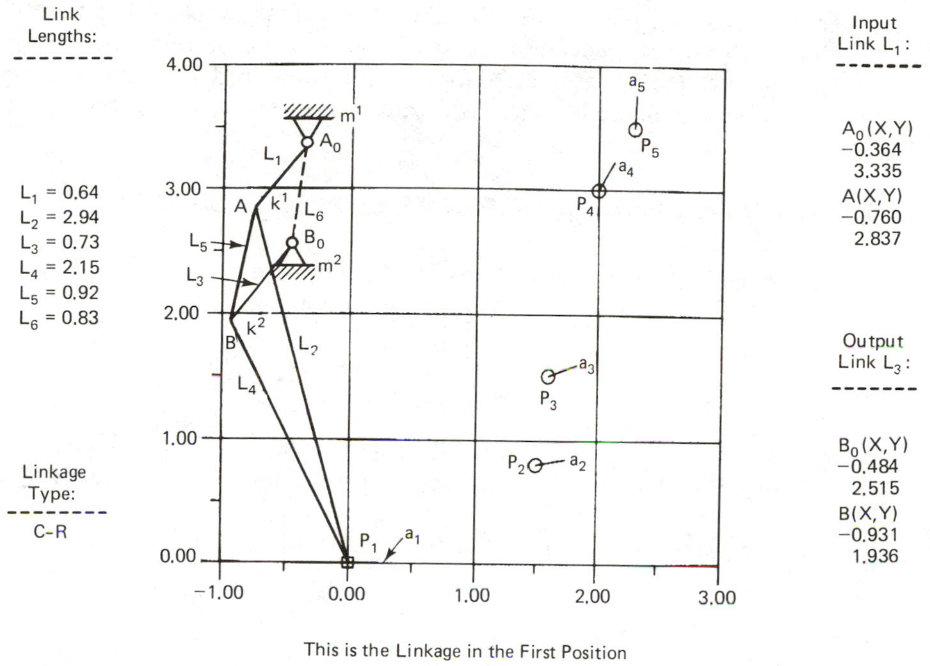


**Figure 3.18** For the five precision points given in Fig. 3.17, two Burmester pairs exist and are shown in the figure. The resulting four-bar is shown in its first position. (For four-bar notation, see Fig. 2.59.) Other precision points are signified by $P_j$ and the prescribed angles by line $P_j a_j$.

For system 3.30 to have simultaneous solutions for the dyad vectors **W** and **Z,** $M$ must be of rank 2. Thus there are two compatibility equations to be satisfied simultaneously for five prescribed positions:

$$\begin{vmatrix} e^{i\beta_2}-1 & e^{i\alpha_2}-1 & \boldsymbol{\delta}_2 \\ e^{i\beta_3}-1 & e^{i\alpha_3}-1 & \boldsymbol{\delta}_3 \\ e^{i\beta_4}-1 & e^{i\alpha_4}-1 & \boldsymbol{\delta}_4 \end{vmatrix} = 0 \tag{3.32}$$

and

$$\begin{vmatrix} e^{i\beta_2}-1 & e^{i\alpha_2}-1 & \boldsymbol{\delta}_2 \\ e^{i\beta_3}-1 & e^{i\alpha_3}-1 & \boldsymbol{\delta}_3 \\ e^{i\beta_5}-1 & e^{i\alpha_5}-1 & \boldsymbol{\delta}_5 \end{vmatrix} = 0 \tag{3.33}$$

Since the second and third columns of determinants (3.32) and (3.33) contain prescribed data, the only unknown (unprescribed) reals are $\beta_j$, $j = 2, 3, 4$, and 5 in the first column. Thus there are no free choices here (Table 2.1) to help solve these two complex (or four real) equations, which are nonlinear and transcendental in $\beta_j$ ($j = 2, 3, 4, 5$). If the solutions for $\beta_j$ are real, then with these values of $\beta_j$, Eqs.

(3.30) are compatible. Then there are up to four usable real intersections of the Burmester curves. This means that there will be up to four dyads that can be used to construct motion generators for the five prescribed positions, for which **Z** and **W** can then be found from any two equations of the system of Eq. (3.30).

## 3.8 SOLUTION PROCEDURE FOR FIVE PRESCRIBED POSITIONS

The determinants [Eqs. (3.32) and (3.33)] will be expanded about their elements in the first column (where the unknown $\beta_j$-*s* are):

$$\mathbf{\Delta}_2 e^{i\beta_2} + \mathbf{\Delta}_3 e^{i\beta_3} + \mathbf{\Delta}_4 e^{i\beta_4} - \mathbf{\Delta}_1 = 0 \tag{3.34}$$

$$\mathbf{\Delta}_2' e^{i\beta_2} + \mathbf{\Delta}_3' e^{i\beta_3} + \mathbf{\Delta}_4 e^{i\beta_5} - \mathbf{\Delta}_1' = 0 \tag{3.35}$$

where the $\mathbf{\Delta}_j$ ($j$ = 1, 2, 3, 4) are the same as before [Eq. (3.8)] and the $\mathbf{\Delta}_j'$ ($j$ = 1, 2, 3) are the same as the $\mathbf{\Delta}_j$ except that each subscript 4 is replaced by a subscript 5.

The complex conjugates* of these compatibility equations also hold true:

$$\overline{\mathbf{\Delta}}_2 e^{-i\beta_2} + \overline{\mathbf{\Delta}}_3 e^{-i\beta_3} + \overline{\mathbf{\Delta}}_4 e^{-i\beta_4} - \overline{\mathbf{\Delta}}_1 = 0 \tag{3.36}$$

$$\overline{\mathbf{\Delta}}_2' e^{-i\beta_2} + \overline{\mathbf{\Delta}}_3' e^{-i\beta_3} + \overline{\mathbf{\Delta}}_4 e^{-i\beta_5} - \overline{\mathbf{\Delta}}_1' = 0 \tag{3.37}$$

We can eliminate $\beta_4$ and $\beta_5$ from Eqs. (3.34) to (3.37) as follows. Equation (3.34) is multiplied by Eq. (3.36) and Eq. (3.35) by Eq. (3.37).† Equations (3.34) and (3.36) yield

$$\begin{aligned}\mathbf{\Delta}_4\overline{\mathbf{\Delta}}_4 = {} & \mathbf{\Delta}_1\overline{\mathbf{\Delta}}_1 - \mathbf{\Delta}_1\overline{\mathbf{\Delta}}_2 e^{-i\beta_2} - \mathbf{\Delta}_1\overline{\mathbf{\Delta}}_3 e^{-i\beta_3} - \mathbf{\Delta}_2\overline{\mathbf{\Delta}}_1 e^{i\beta_2} \\ & + \mathbf{\Delta}_2\overline{\mathbf{\Delta}}_2 + \mathbf{\Delta}_2\overline{\mathbf{\Delta}}_3 e^{i\beta_2} e^{-i\beta_3} - \mathbf{\Delta}_3\overline{\mathbf{\Delta}}_1 e^{i\beta_3} + \mathbf{\Delta}_3\overline{\mathbf{\Delta}}_2 e^{i\beta_3} e^{-i\beta_2} \\ & + \mathbf{\Delta}_3\overline{\mathbf{\Delta}}_3\end{aligned} \tag{3.38}$$

A more compact form of (3.38) is

$$\mathbf{C}_1 e^{i\beta_3} + d_1 + \overline{\mathbf{C}}_1 e^{-i\beta_3} = 0 \tag{3.39}$$

where

$$\mathbf{C}_1 = \mathbf{\Delta}_3(-\overline{\mathbf{\Delta}}_1 + \overline{\mathbf{\Delta}}_2 e^{-i\beta_2})$$

and

$$d_1 = -\overline{\mathbf{\Delta}}_1\mathbf{\Delta}_2 e^{i\beta_2} - \mathbf{\Delta}_1\overline{\mathbf{\Delta}}_2 e^{-i\beta_2} - \mathbf{\Delta}_4\overline{\mathbf{\Delta}}_4 + \Sigma\, \mathbf{\Delta}_j\overline{\mathbf{\Delta}}_j$$

$$j = 1, 2, 3$$

Similarly, Eqs. (3.35) and (3.37) will yield

$$\mathbf{C}_2 e^{i\beta_3} + d_2 + \overline{\mathbf{C}}_2 e^{-i\beta_3} = 0 \tag{3.40}$$

* The superior bar indicates complex conjugates.

† After the $\mathbf{\Delta}_4$ terms are put on the other side of the equals sign.

where $\mathbf{C}_2$ and $d_2$ are the same as $\mathbf{C}_1$ and $d_1$, but with primes on the $\boldsymbol{\Delta}_j$, $j = 1, 2, 3$. Notice that (3.39) and (3.40) can be regarded as homogeneous nonlinear equations in $e^{i\beta_2}$ and $e^{i\beta_3}$, containing their first, zeroth, and minus first powers, with known coefficients. Elimination of the powers containing $\beta_3$ is accomplished through the use of *Sylvester's* dyalitic eliminant.* This is begun by multiplying Eqs. (3.39) and (3.40) by $e^{i\beta_3}$, creating two additional valid equations:

$$\mathbf{C}_1 e^{i2\beta_3} + d_1 e^{i\beta_3} + \overline{\mathbf{C}}_1 = 0 \tag{3.41}$$

$$\mathbf{C}_2 e^{i2\beta_3} + d_2 e^{i\beta_3} + \overline{\mathbf{C}}_2 = 0 \tag{3.42}$$

If $e^{i2\beta_3}$, $e^{i1\beta_3}$, $e^{i0\beta_3}$ and $e^{i(-1)\beta_3}$ are considered as separate "unknowns," Eqs. (3.39) to (3.42) can be regarded as four homogeneous equations, linear in these four unknowns. Since these equations have zeros on the right-hand side, the determinant of the coefficients must be zero for the system to yield simultaneous solutions for the four "unknowns." Therefore, the "eliminant" determinant is

$$E = \begin{vmatrix} 0 & \mathbf{C}_1 & d_1 & \overline{\mathbf{C}}_1 \\ 0 & \mathbf{C}_2 & d_2 & \overline{\mathbf{C}}_2 \\ \mathbf{C}_1 & d_1 & \overline{\mathbf{C}}_1 & 0 \\ \mathbf{C}_2 & d_2 & \overline{\mathbf{C}}_2 & 0 \end{vmatrix} = 0 \tag{3.43}$$

Note that we have successfully eliminated powers of $e^{i\beta_3}$ from the eliminant. Expanding the determinant, a polynomial in $e^{i\beta_2}$ is obtained:

$$\Sigma\, \mathbf{a}_m e^{im\beta 2} = 0 \tag{3.44}$$

where $m = -3, -2, -1, 0, 1, 2, 3$, and where all the coefficients $\mathbf{a}_m$ are deterministic functions of the prescribed quantities $\boldsymbol{\delta}_j$ and $\alpha_j$ ($j = 2, 3, 4, 5$). Also note that $\mathbf{a}_{-k}$ and $\mathbf{a}_k$ ($k = 1, 2, 3$) are each other's complex conjugates, and that $a_0$ is real. Thus the expansion shows that $E$ is real, so that its imaginary part vanishes identically. Therefore, only the real part of the eliminant is of interest. It has the form

$$\sum_m [p_m \cos (m\beta_2) + q_m \sin (m\beta_2)] = 0, \qquad m = 0,1,2,3 \tag{3.45}$$

where $p_m$ and $q_m$ are known reals. By way of trigonometric identities, Eq. (3.45) can be transformed into a form containing powers of sines and cosines of $\beta_2$ (up to the third power), and then by further identities changed to a sixth-degree polynomial in a single variable, $\tau = \tan (\beta_2/2)$, having the form

$$\sum_{0,\,1,\,\ldots}^{6} A_n \tau^n = 0 \tag{3.46}$$

We know that $\beta_2 = 0$ is a trivial solution, which makes $\tau = 0$ a trivial root. Thus Eq. (3.46) can be reduced to a fifth-degree polynomial. Also, from the determinant form of Eqs. (3.32) and (3.33), it is clear that the set of $\beta_j = \alpha_j$, $j = 2, 3, 4, 5$ (here $\beta_2 = \alpha_2$), is another trivial solution. Thus, after dividing the root factor

* Nineteenth-century English mathematician and kinematician.

$[\tau - \tan(\alpha_2/2)]$ out of the remaining fifth-degree polynomial equation, a quartic remains in $\tau$:

$$\tau^4 + \lambda_3\tau^3 + \lambda_2\tau^2 + \lambda_1\tau + \lambda_0 = 0 \tag{3.47}$$

Equation (3.47) will have zero, two, or four real roots. Each real root yields a value for $\beta_2$, which can be substituted back into either Eq. (3.41) or (3.42) to find $\beta_3$, and then into (3.34) and (3.35) to obtain $\beta_4$ and $\beta_5$. Then any two equations of (3.30) can be solved for **Z** and **W**, yielding *up to* four Burmester point pairs for this motion-generation case: $m^{(n)}$, $k_1^{(n)}$, $n = 1, 2, 3, 4$. These, together with the starting position $P_1$ of the prescribed path tracer point $P$, define the four dyads $\mathbf{W}^{(n)}$, $\mathbf{Z}^{(n)}$. $\mathbf{Z}^{(n)}$ exhibit the same set of prescribed rotations $\alpha_j$, $j = 2, 3, 4, 5$, while the four different vectors $\mathbf{W}^{(n)}$ have four different sets of rotations $\beta_j^{(n)}$, $n = 1, 2, 3, 4$, $j = 2, 3, 4, 5$. The example in Figs. 3.17 and 3.18 has only two real roots and thus only two BPPs ($n = 1, 2$) which can be used to form one four-bar linkage which will guide its coupler plane through the five prescribed coplanar positions. In general, combining each of four BPPs with every other, we can obtain up to six different such four-bar motion-generator linkages. (See Appendix A3.2 for computation schedule.)

## 3.9 EXTENSIONS OF BURMESTER POINT THEORY: PATH GENERATION WITH PRESCRIBED TIMING AND FUNCTION GENERATION

Burmester theory was derived above for obtaining dyads suitable for a four-bar motion generator. What about path generation with prescribed timing and function generation with the four-bar? Also, can this theory be extended to other linkage types? Chapter 2 demonstrated that the dyadic standard form equation, Eq. (3.2), was usable in numerous cases. The Roberts–Chebyshev theorem will add more insight to the broad applicability of the dyad form and of the Burmester theory.

### Roberts–Chebyshev Theorem

An extremely useful property of planar four-bar linkages is revealed in the Roberts–Chebyshev theorem [125], which states that one point of each of *three different* but related planar four-bar linkages will trace identical coupler curves. This means that there will be two additional four-bar linkages associated with each "parent" four-bar linkage which will trace the same path as the parent (although the coupler rotations will not be the same). These two additional linkages are called "Roberts–Chebyshev cognates" after their two independent English and Russian discoverers. We can form these cognates geometrically by building on the four-bar linkage shown in full lines in Fig. 3.19 as follows:

1. Complete the parallelograms of $\mathbf{Z}^1$ and $\mathbf{W}^1$ and $\mathbf{Z}^2$ and $\mathbf{W}^2$.
2. Find the third fixed point, $C$, of the Roberts configuration by making triangle

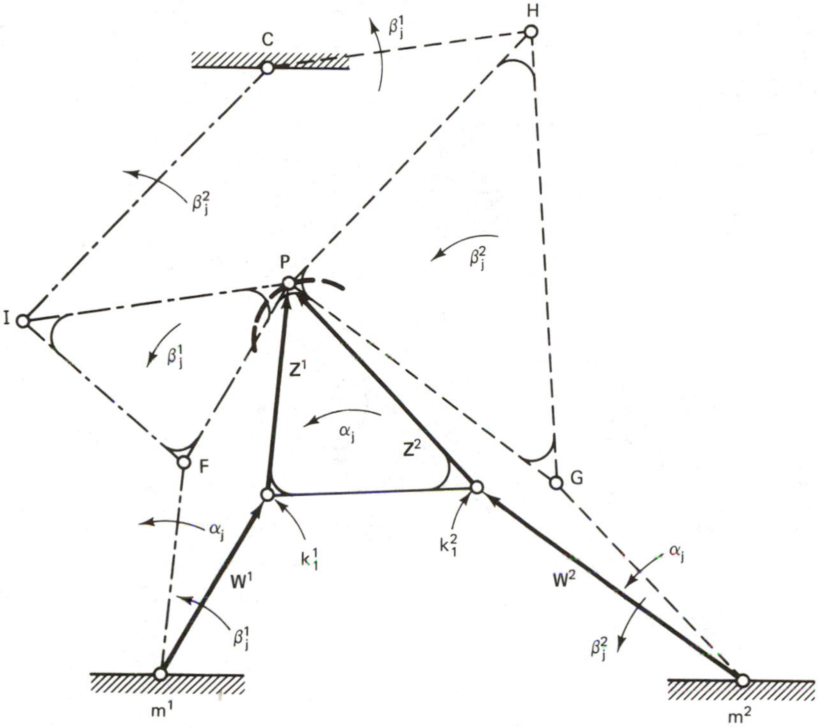


**Figure 3.19** The Roberts-Chebishev Configuration consists of basic motion-generator four-bar mechanism consisting of two (out of the possible maximum of four) Burmester Point Pair (BPP) dyads, $\mathbf{W}^1\mathbf{Z}^1$ and $\mathbf{W}^2\mathbf{Z}^2$. The dashed and dot-dashed four-bar mechanisms are the cognates of the basic one.

$m^1Cm^2$ similar to the coupler triangle $k_1^1Pk_1^2$ such that $C$ corresponds to $P$, $m^1$ to $k_1^1$, and so on.

3. Find one cognate coupler triangle by making $\Delta GPH$ similar to $\Delta m^2m^1C$ or $\Delta k_1^2k_1^1P$. $HC$ is the follower link of the dashed "right cognate" of Fig. 3.19.
4. Find the other cognate by making coupler triangle $FIP$ similar to $\Delta m^1Cm^2$ and/or $\Delta k_1^1Pk_1^2$ and connecting $IC$ to form the follower link of the second cognate, shown in dot-dashed lines. Note that $ICHP$ is a parallelogram. Due to the three parallelograms that concur at $P$, Fig. 3.20 shows that the coupler curves traced by $P$ as a point of the initial four-bar or as a point of either one of its cognates are one and the same curve.

This property, that every four-bar linkage has two cognate linkages which trace the same path as the parent four-bar, is extremely useful to designers. The cognates are different linkages, even though they share one ground-pivot location with one another. A designer may find that although a particular linkage may trace a desired

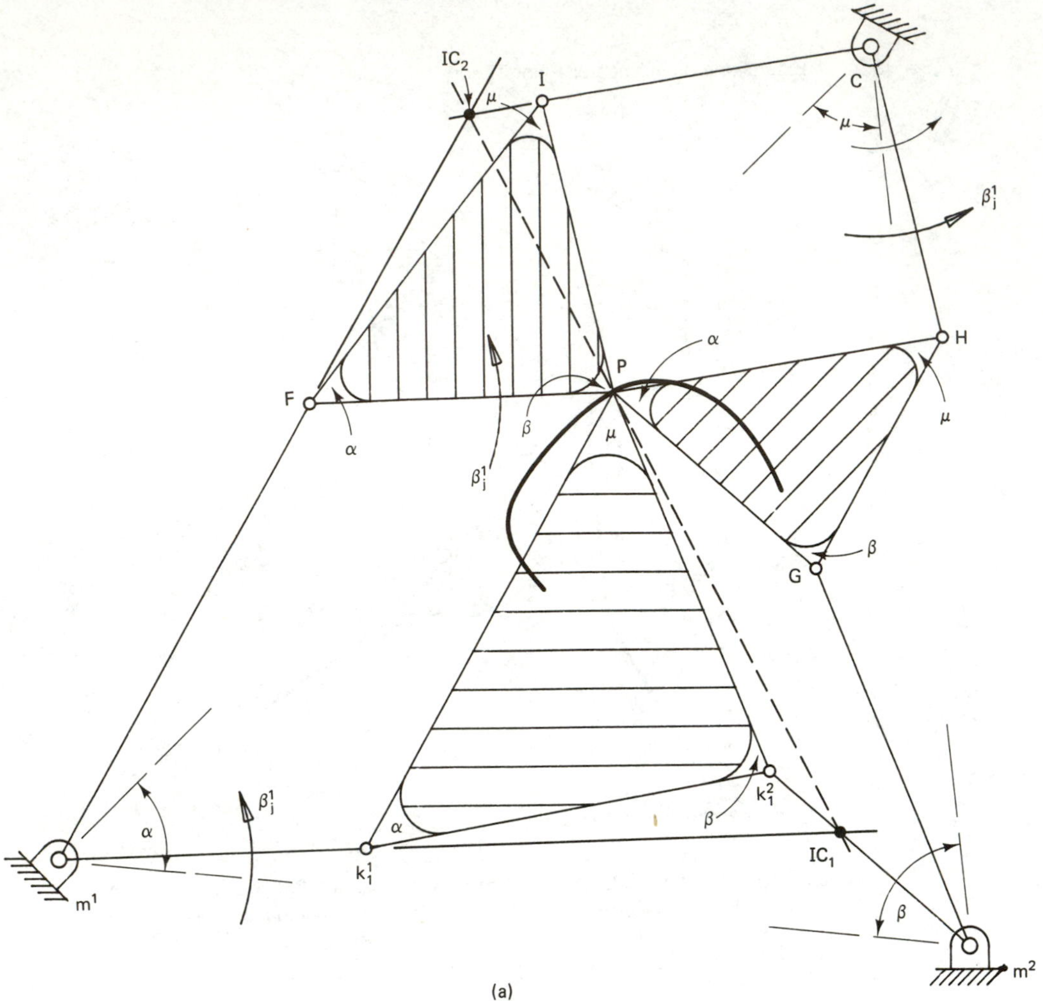
IC$_2$
I
C
$\mu$
$\mu$
$\beta_j^1$
$\alpha$
P
H
F
$\beta$
$\mu$
$\mu$
$\alpha$
$\beta_j^1$
$\beta$
G
$\beta_j^1$
$\alpha$
$k_1^2$
$\alpha$
$\beta$
IC$_1$
$k_1^1$
m$^1$
$\beta$
m$^2$

(a)

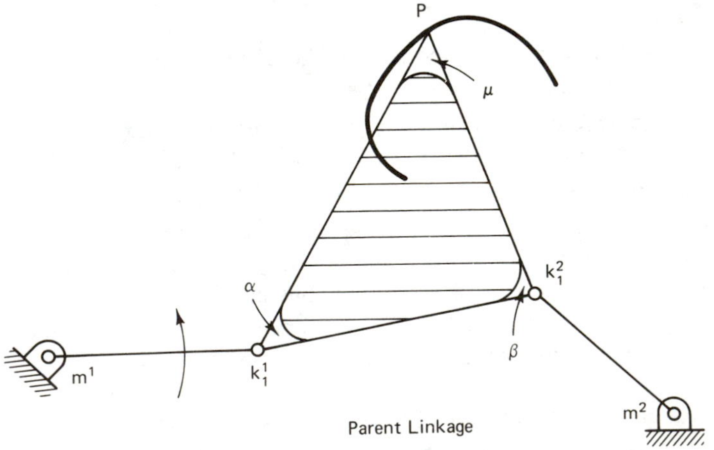
P
$\mu$
$\alpha$
$k_1^2$
$\beta$
m$^1$
$k_1^1$
m$^2$
Parent Linkage

(b)

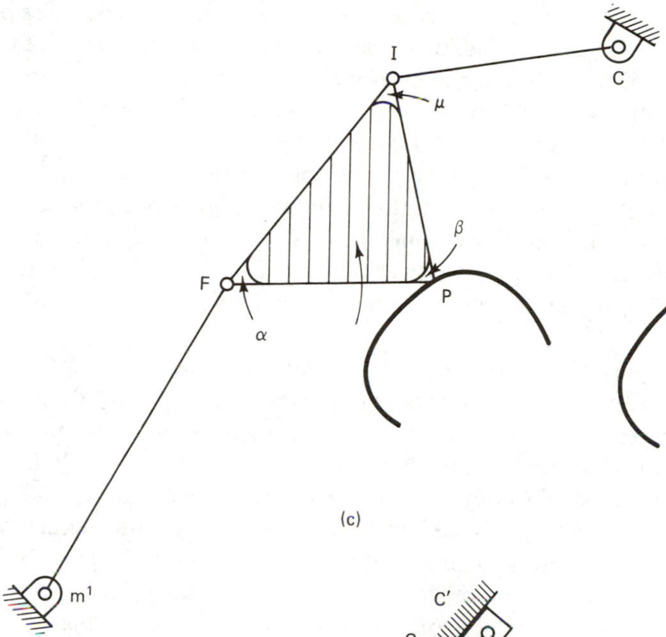


(c)

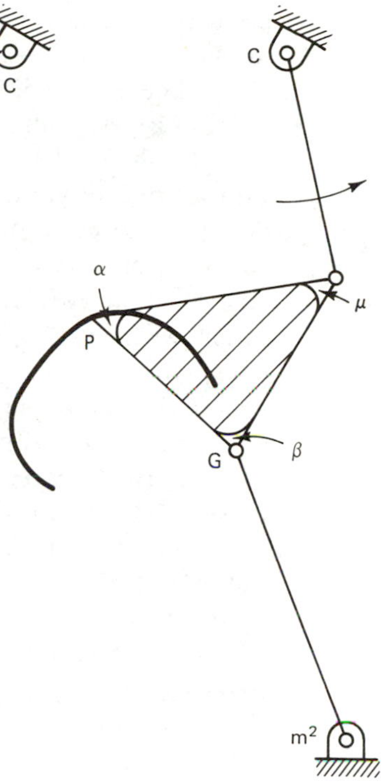


(d)

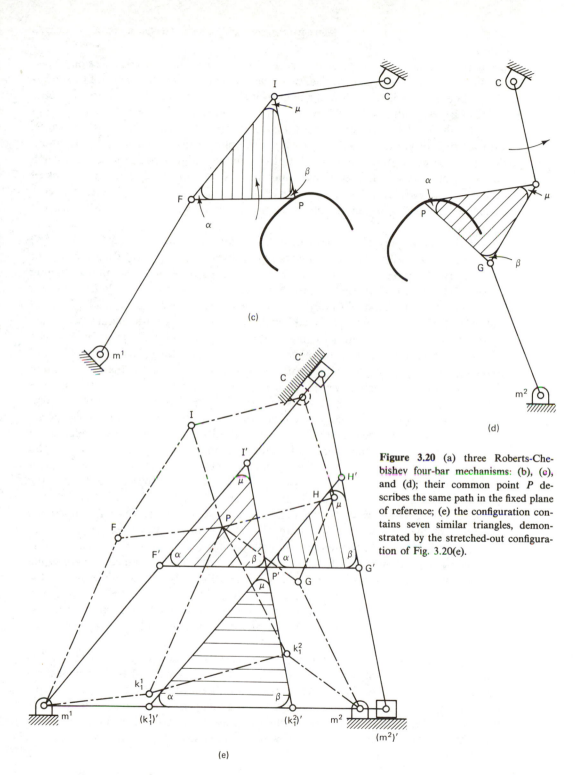

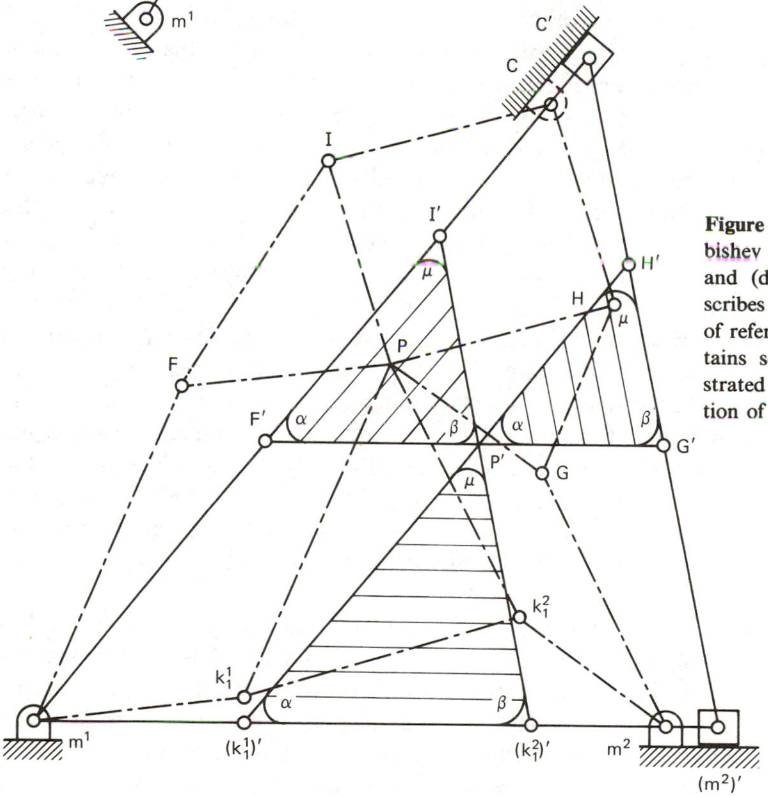


(e)

**Figure 3.20** (a) three Roberts-Chebishev four-bar mechanisms: (b), (c), and (d); their common point $P$ describes the same path in the fixed plane of reference; (e) the configuration contains seven similar triangles, demonstrated by the stretched-out configuration of Fig. 3.20(e).

path, that linkage may not satisfy space requirements, or the transmission angle, mechanical advantage, velocity, and/or acceleration characteristics may not be acceptable. There are, however, two cognates available which, while they trace the same path, in general will display different kinematic and dynamic characteristics.

It should be mentioned here that cognates are not *equivalent* linkages. Equivalent linkages are usually employed to duplicate instantaneously the position, velocity, and perhaps acceleration of a direct-contact (higher-pair) mechanism (such as a cam or a noncircular gear) by a linkage (say, a four-bar). The dimensions of equivalent linkages are different at various positions of such higher-pair parent mechanisms, whereas the link lengths of cognates remain the same for any position of the parent linkage.

Other properties of cognates include the one developed by Cayley [36]: The common coupler–path tracer point and the three instant centers of the three concurrent couplers (with respect to ground) are collinear at all times and the line containing these points is normal to the coupler curve in every position of the linkage system (see points $IC_1$, $IC_2$, and $IC_3$ in Fig. 3.20a). Another observation is: Each grounded link of any of the three FBLs will exhibit the same angular rotations (and will rotate at the same angular velocity) as one of the grounded links of one of its cognate FBLs and the coupler link of its other cognate FBL, as shown in Fig. 3.19 (we will make use of this property shortly). Still another noteworthy fact is that, if the parent linkage and its two cognates were pinned together to form a movable 10-bar linkage, Gruebler's equation (Sec. 1.6) predicts that this linkage has $-1$ degrees of freedom. This is an example of an *overclosed linkage* that has mobility due only to its special geometric properties. Yet another property of the Roberts–Chebyshev configuration is this: In addition to the four rigid similar triangles (the three coupler triangles and the triangle formed by the three ground pivots) there are also three variable-size triangles, all of which remain always similar to the coupler triangles in the course of motion of the linkage. These are: $\Delta m^1 k_1^2 I$, $\Delta k_1^1 m^2 H$, and $\Delta FGC$. The proof may be started as shown in Fig. 3.20e; move $m^2$ away from $m^1$ along the extension of their connecting line until all three links between them are stretched out in a straight line. Proceed similarly with $C$ with respect to $m^1$. In the resulting stretched-out configuration, in which all link lengths have retained their original lengths, the above-mentioned seven triangles are all clearly similar. The rest of the proof is left to the reader as an exercise. [*Hint:* Move $C'$ and $(m^2)'$ toward $m^1$, keeping their triangle similar.] Then, by complex numbers and appropriate rotation operators, show the similarity of the above-mentioned variable triangles with the other four.

Four-bar linkages are not the only linkages that have cognates. The slider-crank (a special case of a four-bar; see Fig. 3.12), five-bar, six-bar, and in fact all planar linkages have cognates. A complex-number proof of the existence of the four-bar cognates, using complex numbers and appropriate rotational operators, can be based on Fig. 3.20e. This is left to the reader as an exercise. For further development of the above-mentioned properties of cognates and a historical note, refer to Ref. 125.

### Four-Bar Path Generator Mechanisms (Four and Five Precision Points)

In addition to the usefulness of cognates just mentioned, an important computational advantage may also be derived. By employing the Roberts–Chebyshev theorem, path generators with prescribed timing may be obtained from motion-generator four bars. Let us look again at the geometric cognates of Fig. 3.19. Suppose that the parent four-bar ($m^1k_1^1Pk_1^2m^2$) is a motion generator that has been synthesized by either the four- or five-precision-point technique. The rotations of the coupler link $\alpha_j$ and the displacements of tracer point $P$ have been prescribed. Input link $m^1k_1^1$ rotates by $\beta_j^1$ (beta rotations of the ground link in the first solution pair or dyad) while the output link $m^2k_1^2$ rotates by $\beta_j^2$ (second dyad).

According to the Roberts–Chebyshev development, all three cognates trace the identical coupler curves with their common tracer point. What do the individual links of the two other cognates rotate by? Noticing that $m^1F$ is always parallel to $k_1^1P$ and $FP$ is always parallel to $m^1k_1^1$, it is clear that $m^1F$ rotates by $\alpha_j$ while $FP$ rotates by $\beta_j^1$. Furthermore, $m^2G$ rotates by $\alpha_j$, $PGH$ and $IC$ rotate by $\beta_j^2$, and $CH$ rotates by $\beta_j^1$. Since the originally prescribed rotations $\alpha_j$ have been transferred to the grounded links in the cognates, *the cognates of a motion generator are path generators with prescribed timing.* For every four-bar motion generator there will be two such four-bar path generators. This development may be utilized to simplify both the four- and five-precision-point synthesis methods, so that the synthesis equations need only be solved *once* for both tasks: motion generation and path generation with prescribed timing. In the second case, the cognates may also be derived via the computer from the parent motion generator.

In the five-precision-point case, how many path generators with prescribed timing might we expect? Since there are either 0, 2, or 4 real roots of the quartic [Eq. (3.47)], there will be 0, 2, or 12 path generators with prescribed timing for each data set.

### Four-Bar Function Generator Mechanism (Four and Five Precision Points)

In Sec. 2.16 it was demonstrated that the four-bar function generator could be synthesized in the "standard form" by treating it like a path generator with prescribed timing where the path was specified along a circular arc (see Fig. 2.60). Furthermore, the tracer points along the arc were chosen so that as a link from the center of the arc to the tracer point rotated from one precision point to the next, this link would rotate by the prescribed output angles $\psi_j$. Correlation of this procedure with the Roberts–Chebyshev configuration can be observed by referring to points $m^1$, $m^2$, $F$ and $P$ in both Figs. 3.20a and 3.21.

How many function generators might we expect from the five-precision-point case? One might initially guess that there will be a maximum of 12 function generators if all the roots of the quartic Eq. (3.47) were real, because we are treating the function

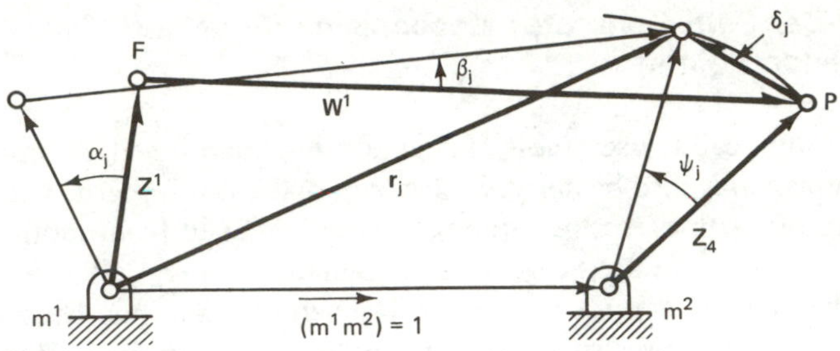


**Figure 3.21** Four-bar function-generator synthesis. Given: $\alpha_j$, $\mathbf{r}_1 = 1 + \mathbf{Z}_4$, and $\mathbf{r}_j = 1 + \mathbf{Z}_4 e^{i\psi_j}$, $j = 2, 3, 4, 5$, where $\mathbf{Z}_4$ is arbitrary and $\alpha_j$ and $\psi_j$ are analogs of the independent and dependent variables of the function to be generated. With these, the standard form of the dyadic synthesis equation becomes: $Z\,(e^{i\alpha_j} - 1) + \mathbf{W}\,(e^{i\beta_j} - 1) = \delta_j$ where $\delta_j = \mathbf{r}_j - \mathbf{r}_1$.

generator as a path generator with prescribed timing. This, however, is not the case. First, only half of the four-bar path generator is required to form a function generator, the dyad $m^1FP$, Figs. 3.20a and 3.21. Furthermore, since there are only four different dyads that make up the 12 path generators, there are only four different dyad solutions available. Also, there will always be one trivial solution since a circular arc, centered at $m^2$, is being specified as the path of $P$. This solution will contain zero length for the grounded link $\mathbf{Z}$ of the dyad and a coupler link $\mathbf{W}$ identical with the specified output link $\mathbf{Z}_4$ (with rotations $\psi_j$). Therefore, there is a maximum of only three different four-bar solutions; but at least one solution is guaranteed because complex roots [for $\tau = \tan(\beta_2/2)$] come in conjugate pairs and there should always be a trivial solution. Table 3.2 summarizes the number of solutions that can be expected for zero, two, or four real roots to the quartic.

**TABLE 3.2** NUMBER OF POSSIBLE SOLUTIONS FOR FOUR-BAR SYNTHESIS WITH FIVE FINITELY SEPARATED PRECISION POINTS BY WAY OF THE STANDARD DYAD FORM [EQ. (3.30)].

| | Number of different four bar solutions expected | | |
|---|---|---|---|
| Number of real roots of the quartic | Motion generation | Path generation with prescribed timing | Function generation (See Fig. 3.21.) |
| 0 | 0 | 0 | 0 |
| 2 | 1 | 2 | 1 |
| 4 | 6 | 12 | 3 |

## 3.10 FURTHER EXTENSIONS OF BURMESTER THEORY

### Geared Five-Bar and Parallelogram Six-Bar Cognates

In the preceding section it was shown that the Burmester dyads can be arranged together to yield four-bar mechanisms for motion, function, and path generation with prescribed timing. Some other useful linkages, with more than four links, can be synthesized from these same dyads using simple construction procedures. Suppose that one wishes to obtain a path generator with prescribed timing directly without computing the cognate of the motion generator. (Perhaps the motion generator ground-pivot locations are acceptable but the cognates exhibit an undesirable ground-pivot location.) Then either the geared five-bar or parallelogram six-bar path generator (with prescribed timing) may be useful.

Referring to Fig. 3.22, complete the vector parallelograms of the original dyads ($\mathbf{W}^1$ and $\mathbf{Z}^1$, $\mathbf{W}^2$ and $\mathbf{Z}^2$) of the four-bar motion generator, shown in dashed lines of Fig. 3.19. Disregarding the parent four-bar, connect the grounded links $\mathbf{Z}^1$ and $\mathbf{Z}^2$ with each other by means of one-to-one gearing (using an idler to assure that $\mathbf{Z}^1$ and $\mathbf{Z}^2$ perform identical rotations). Thus a single-degree-of-freedom geared five-bar linkage $m^1FPGm^2$ is obtained which will trace the prescribed path of $P$ with corresponding prescribed input-crank rotations $\alpha_j$. For each motion generator there

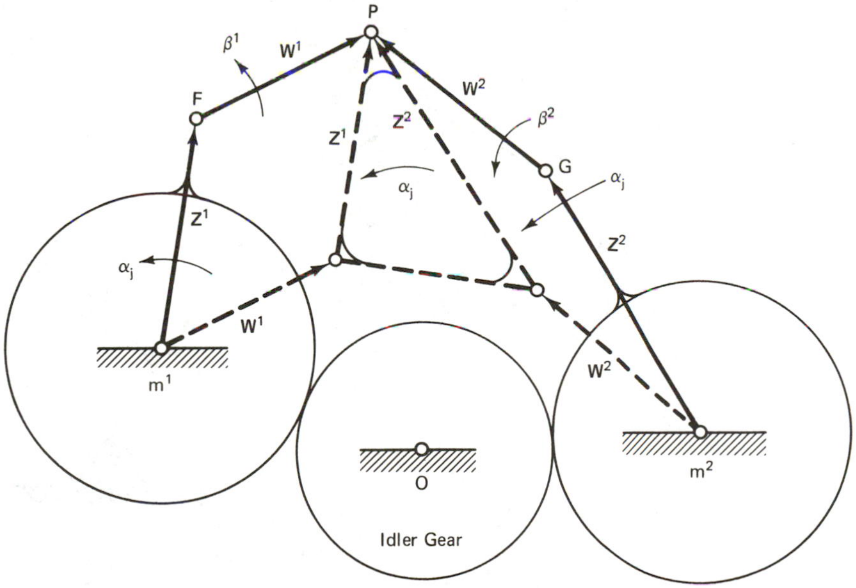


**Figure 3.22** The 1:1 gear ratio between $\mathbf{Z}^1$ and $\mathbf{Z}^2$ of this single-degree-of-freedom geared five-bar assures that point $P$, the joint of $\mathbf{W}^1$ and $\mathbf{W}^2$, will trace the same path as the parent linkage, the four-bar motion generator shown in dashed lines.

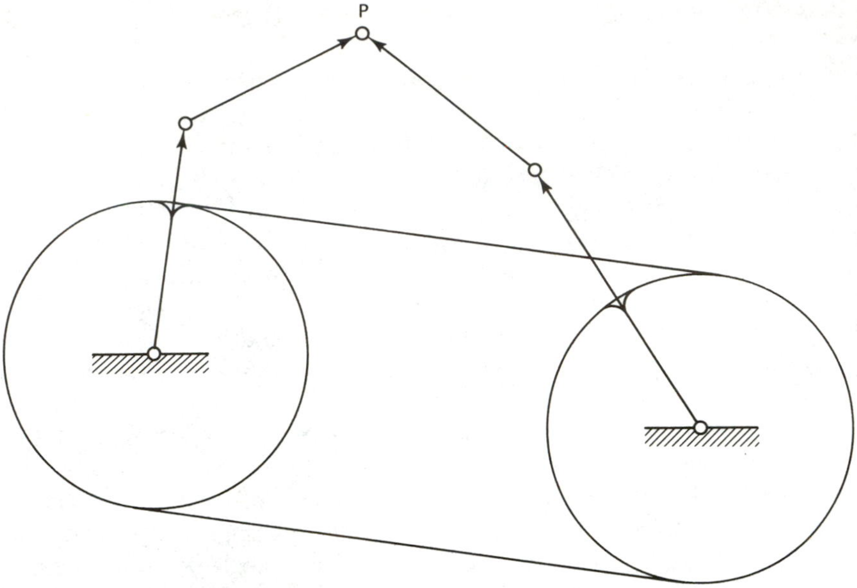


**Figure 3.23** Same as the geared five-bar configuration of Fig. 3.22, but the gears have been replaced by chain and sprockets.

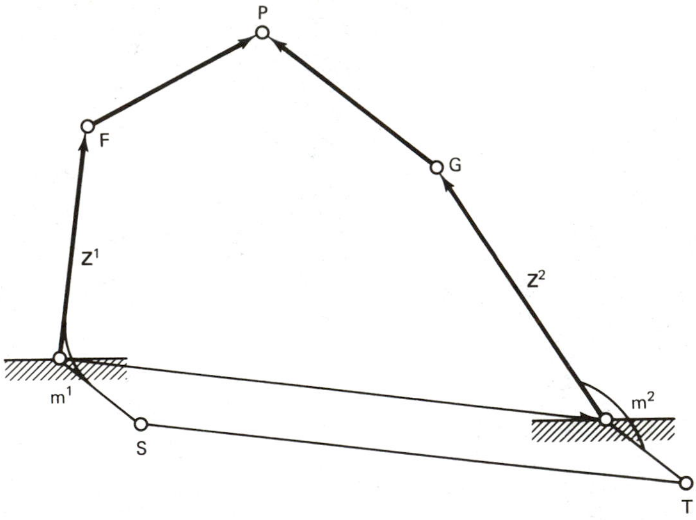


**Figure 3.24** Parallelogram linkage $m^1m^2TS$ assures 1 : 1 velocity ratio between links $\mathbf{Z}^1$ and $\mathbf{Z}^2$ of the five-bar path generator. See Fig. 3.22 for the parent linkage, a four-bar motion generator.

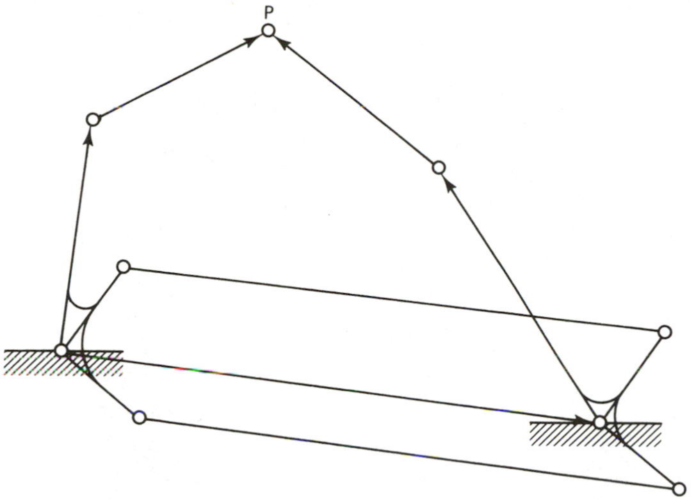


**Figure 3.25** Same five-bar path generator as in Fig. 3.24 with an added parallelogram linkage to avoid branching or binding at dead-center of the original parallelogram.

will be one geared five-bar, in which either grounded link can serve as the input. Another way to design this linkage is shown in Fig. 3.23, where the gearing is replaced by a chain or timing belt and two equal sprockets.

There is yet another way for converting the two-degree-of-freedom five-bar of Fig. 3.22 to a single degree of freedom besides using gears or sprockets with chains or timing belts. The same objective can be accomplished by adding a parallelogram linkage to the five-bar as shown in Fig. 3.24. Notice that the parallelogram connected to $\mathbf{Z}^1$ and $\mathbf{Z}^2$ is not unique. In fact, two parallelograms may be connected together to avoid the dead-center problem (Fig. 3.25). (This is another *overclosed* linkage whose mobility is assured by its link proportions.)

The five-bar $m^1FPGm^2$ may be connected by gears of other than 1:1 ratio, but this would require combining two separate dyad solutions—both with the same $\boldsymbol{\delta}_j$ but with different $\alpha_j$, where one $\alpha_j$ would be proportional to the other.

## Six-Bar Parallel Motion Generator

An extremely useful linkage is one that will trace a coupler curve while the coupler link undergoes no rotation—a parallel motion (curvilinear translation) generator. One can easily observe that this is an inappropriate task for a four-bar linkage except in the trivial case of a circular coupler curve of a parallelogram linkage. The following extension of the Roberts–Chebyshev construction yields a six-bar linkage with one link performing curvilinear translation.

Begin by drawing the initial motion generator four-bar linkage $m^1k_1^1Pk_1^2m^2$,

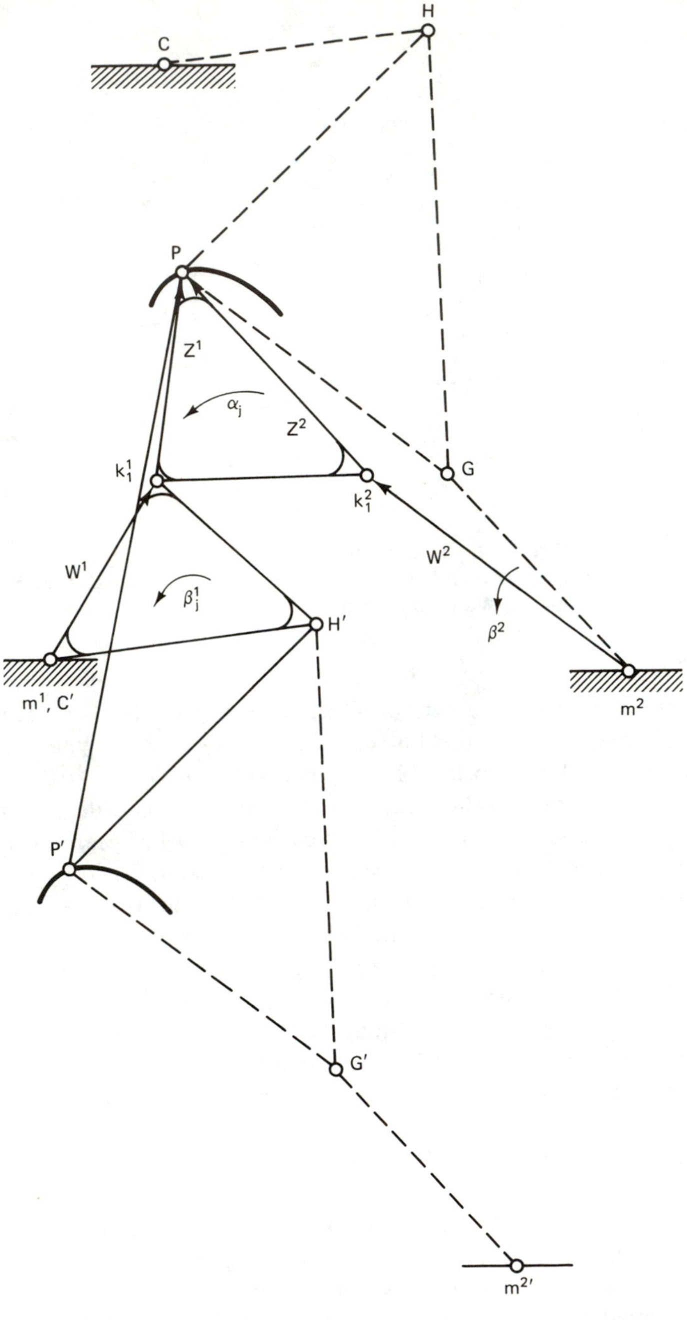


**Figure 3.26** Six-bar parallel-motion generator (solid lines) derived from the parent linkage (four-bar motion generator $m^1k_1^1Pk_1^2m^2$) and its right cognate ($m^2$GPHC). $P$ and $P'$ describe identical paths. If link $m^{2\prime}G'$ is added, a seven-bar (overclosed) parallel-motion generator with prescribed timing results because link $m^{2\prime}G'$ performs the prescribed rotations $\alpha_j$.

whose point $P$ traces the prescribed path (Fig. 3.19). Next, draw one of its cognates, say the right dashed cognate $m^2GPHC$. Now duplicate this cognate by moving it parallel to itself so that point $C$ is coincident with $m^1$ (Fig. 3.26). This yields the four-bar linkage $C'H'P'G'$ and $m^{2'}$. From past discussions we know that link $CH$ rotates by $\beta_j^1$, as does $m^1k_1^1$. Since $C'H'$ (same rotations as $CH$) and $m^1k_1^1$ both rotate identically, triangle $m^1k_1^1H'$ may be rigidly connected. Notice that both $P$ and $P'$ trace the same path—thus $PP'$ may be connected by a rigid link. This link will move parallel to itself while both $P$ and $P'$ trace out the prescribed path. Since we were able to form triangle $m^1k_1^1H'$, a single-degree-of-freedom linkage exists without the need for links $G'P'$, $G'H'$, and $m^{2'}G'$ (shown dashed in Fig. 3.26). The six-bar parallel motion generator is composed of the initial four-bar $m^1k_1^1Pk_1^2m^2$ plus $m^1H'P'P$.

Instead of using the right cognate as shown in Fig. 3.26, we could have used the left cognate. This would yield another six-bar parallel motion generator following a similar construction. Thus, *for every motion generator, there are two six-bar parallel motion generators.*

A closer look at these six-bars may yield some disappointment. The prescribed rotations $\alpha_j$ are seemingly lost—neither the coupler link $PP'$ nor the grounded links rotate by the originally prescribed angles $\alpha_j$. However, by adding the dashed portion $H'G'P'M^{2'}$ to the solid links in Fig. 3.26, we obtain a *seven-bar parallel motion generator with prescribed timing* (an overclosed mechanism) where $M^{2'}G'$ rotates by the prescribed $\alpha_j$ rotations.

There are three additional useful observations to be made about Fig. 3.26:

1. The parallel motion linkages derived from a four-bar motion generator have a special quality: One can prescribe a combined "moving function generator" and parallel motion generator. This can be seen by observing that in Fig. 3.26 link $P\,P'$ does not rotate, while $PK_1^1k_1^2$ rotates by the prescribed angles $\alpha_j$. Thus the relative rotations between $PP'$ and $k_1^1Pk_1^2$ are prescribed. One of the applications of this would be a flying shear, where the object to be cut moves along the prescribed path and is supported by link $PP'$. Meanwhile a blade connected to $k_1^1Pk_1^2$ cuts the object "on the fly."

2. Since the prescribed rotations $\alpha_j$ are not a factor in the parallel-motion-generating quality of the six-bar, one could find more six-bars by varying the choices for $\alpha_j$. Therefore, there are an infinite number of six-bar parallel motion generators that will hit the prescribed precision points along the path of $P$.

3. Observe that the prescribed rotations $\alpha_j$ are the rotations of one of the grounded links in each cognate of the original motion generator. To make use of these prescribed rotations, the six-bar parallel motion construction technique should be applied to the cognates rather than the parent motion generator. This will yield two different six-bars per cognate for a total of four *six-bar parallel motion generators with prescribed timing* for every motion-generator four-bar.

Table 3.3 indicates the number of solutions expected for all the extensions of the Roberts–Chebyshev constructions described above. One can see that one motion generator four-bar breeds numerous useful offspring.

**TABLE 3.3** MECHANISMS SYNTHESIZED BY WAY OF BURMESTER POINT PAIRS OBTAINED USING THE DYADIC STANDARD-FORM EQUATION AND EXTENDING THE RESULTS ON THE BASIS OF THE ROBERTS–CHEBYSHEV THEOREM.

| Number of real roots of the quartic equation (3.47) | Four-bar motion generators (Fig. 3.19) (The Parent Linkage) | Four-bar path generators with prescribed timing (Fig. 3.19) (Cognates) | Four-bar function generators (Fig. 3.21) | Geared five-bar path generators with prescribed timing[a] (Fig. 3.22) |
|---|---|---|---|---|
| 0 | 0 | 0 | — | 0 |
| 2 | 1 | 2 | 1 | 1 |
| 4 | 6 | 12 | 3 | 6 |

[a] Several configurations possible (see Figs. 3.22 and 3.23).

## 3.11 SYNTHESIS OF MULTILOOP LINKAGE MECHANISMS

The dyad synthesis approach may be used to synthesize virtually all planar mechanisms. This was suggested in Chap. 2, where the standard-form equations were derived for the Stephenson III six-bar (see Fig. 2.66).

Why try to design a multiloop linkage such as the Stephenson III six-bar when the four-bar can do so much? If only *three* positions are required, the answer to this question is that in most cases we do not need more than the four-bar chain. In the three-position case all coefficients in the dyad synthesis equations are either specified or picked arbitrarily, so that even motion generation with prescribed timing is possible. Also, there are two infinities of solutions to inspect—usually more than enough to find a "good" solution if the motion requirements are "well behaved."*

In the *four*-position case, multiloop linkages become more attractive for several reasons: (1) motion generation with prescribed timing is no longer possible with the four-bar linkage; (2) even an infinite number of solutions which one can expect from four prescribed positions may not produce a suitable four-bar, especially if enough requirements are imposed on the final linkage (e.g., ground-pivot locations); and (3) multiloop linkages can exhibit more complex motion than four-bar linkages since coupler plane motions are no longer restricted by the requirement of two points to follow either circular arcs or straight lines (except the Watt II six-bar, which consists of two four-bar chains). In the *five*-prescribed-position case, multiloop linkages present a valuable alternative since at best there are only a finite number of four-bar solutions.

Kinematic loops consisting of five, six, seven or more bars may be synthesized for more than five prescribed positions. Recall that in the Stephenson III six-bar of Fig. 2.66, the loop containing $\mathbf{Z}_5$, $\mathbf{Z}_4$, and $\mathbf{Z}_3$ has a loop closure equation $\mathbf{Z}_5(e^{i\psi_j} - 1) + \mathbf{Z}_4(e^{i\beta_j} - 1) - \mathbf{Z}_3(e^{i\gamma_j} - 1) = \boldsymbol{\delta}_j$ [Eq. (2.31)]. This loop was

* No sudden changes in direction, velocity, and acceleration.

| Six-bar parallelogram path generators with prescribed timing[b] (Fig. 3.24) | Geared five-bar combined path and constant-velocity-ratio generation[c] (Fig. 3.22, but $n \neq 1$) | Six-bar parallel motion generators with prescribed timing | Six-bar parallel motion generation[d] (Fig. 3.26) | Seven-bar parallel motion generator with prescribed timing (Fig. 3.26) |
|---|---|---|---|---|
| 0<br>1<br>6 | 0<br>4<br>16 | 0<br>4<br>24 | 0<br>2<br>12 | 0<br>2<br>12 |

[b] The basic five-bar of Fig. 3.24 ($m^1FPGm^2$) remaining the same but any parallelogram $m^1STm^2$ may be added to complete the linkage. (See Fig. 3.25 for a seven-bar version which avoids the dead-center problem.)

[c] Requires two runs of the program, both prescribing the same path displacement vectors $\boldsymbol{\delta}_j$ but different $\alpha_j$ such that $\alpha'_j = n(\alpha_j)$, where $n$ is any rational number. (The number of solutions in each row assumes that both quartics have the same number of real roots.)

[d] By varying the prescribed $\alpha_j$, an infinite number of sets of 2 or 12 new solutions may be generated.

analyzed in Table 2.2, which showed that these vectors could be synthesized for a maximum of seven positions. Since the rest of the mechanism could be designed for only five positions, there is little reason to require seven from just one loop. In the great majority of design situations five precision points are sufficient. Of more importance perhaps is to make proper use for other purposes of the extra free choices that occur in multiloop linkage synthesis.

In order to synthesize the triad $\mathbf{Z}_3\mathbf{Z}_4\mathbf{Z}_5$ in the Stephenson III six-bar of Fig. 2.66a for five precision points, Table 2.2 tells us that we must make free choices of two unknown reals. If we choose the vector $\mathbf{Z}_3$, the standard form will be achieved:

$$\mathbf{Z}_5(e^{i\psi_j} - 1) + \mathbf{Z}_4(e^{i\beta_j} - 1) = \boldsymbol{\delta}'_j \tag{3.48}$$

where

$$\boldsymbol{\delta}'_j = \boldsymbol{\delta}_j + \mathbf{Z}_3(e^{i\gamma_j} - 1)$$

The other loops were described by (see Fig. 2.66b)

$$\mathbf{Z}_1(e^{i\phi_j} - 1) + \mathbf{Z}_2(e^{i\gamma_j} - 1) = \boldsymbol{\delta}_j \tag{3.49}$$

$$\mathbf{Z}_6(e^{i\theta_j} - 1) + \mathbf{Z}_7(e^{i\beta_j} - 1) = \boldsymbol{\delta}'_j \tag{3.50}$$

What tasks can we ask of this linkage? Two major tasks are evident after inspection of the three equations above (recalling that the $\boldsymbol{\delta}$'s and one set of the rotations are prescribed for the standard form): (1) *combined path and function generation* (the path of point $P$ and the rotations $\phi_j$ and $\psi_j$ or $\theta_j$), and (2) *motion generation with prescribed timing* (the path of point $P$ and the rotations $\gamma_j$ and $\psi_j$ or $\theta_j$).

Besides the greater usefulness of this linkage for the designer, another "nice" by-product of this procedure for design is the free choice of $\mathbf{Z}_3$. This vector, which forms the rest of the coupler plane once $\mathbf{Z}_1$ and $\mathbf{Z}_2$ are synthesized, can be used advantageously in three ways: (1) the shape of the coupler plane may be picked by the linkage designer; (2) by choosing $\mathbf{Z}_3$ in a certain direction, the designer may influence the form of the rest of the linkage: $\mathbf{Z}_5$, $\mathbf{Z}_4$, $\mathbf{Z}_6$, and $\mathbf{Z}_7$ (e.g., influence the resulting ground-pivot locations); or (3) by simply varying $\mathbf{Z}_3$, the designer may generate a larger number of solutions. With regard to the last observation, how many possible solutions for five prescribed positions could we expect from either of the two combined tasks mentioned above? Equation (3.49), for $j = 2, 3, 4, 5$, yields up to four solutions; $\mathbf{Z}_3$ may be varied between $-\infty$ and $+\infty$ in both $x$ and $y$ directions, and the four-bar left over (which includes $\mathbf{Z}_5$, $\mathbf{Z}_6$, $\mathbf{Z}_4$, and $\mathbf{Z}_7$) has up to 12 solutions for each value of $\mathbf{Z}_3$. Thus there are many infinities of solutions for five-precision-point synthesis of the Stephenson III six-bar mechanism of Fig. 2.66.

**Example 3.2 [213]**

The Stephenson III type of six-bar of Fig. 2.66a is to be synthesized as a combined approximate path and function generator for five prescribed positions. The path is an approximate straight line while the function to be approximated is $y = x^2$, $1 \le x \le 3$, with the range in input $\Delta\phi = 40°$ and output $\Delta\psi = 90°$. The prescribed precision points are

$$\boldsymbol{\delta}_2 = 0.7 - 0.5i, \qquad \boldsymbol{\delta}_4 = 2.55 - 1.8i$$

$$\boldsymbol{\delta}_3 = 1.5 - 1.1i, \qquad \boldsymbol{\delta}_5 = 3.6 - 2.6i$$

$$\phi_2 = 10°, \qquad \psi_2 = 14.06°$$

$$\phi_3 = 20°, \qquad \psi_3 = 33.75°$$

$$\phi_4 = 30°, \qquad \psi_4 = 59.06°$$

$$\phi_5 = 40°, \qquad \psi_5 = 90.00°$$

Figure 3.27 shows one of the resulting linkage mechanisms in its first and fifth prescribed positions, and Fig. 3.28 shows the second, third, and fourth positions of this same mechanism.

## 3.12 APPLICATIONS OF DUAL-PURPOSE MULTILOOP MECHANISMS

Multiloop mechanisms have numerous applications in assembly line operations. For example, in a soap-bar-wrapping process, where a piece of thin cardboard must be fed between rollers which initiate the wrapping operation, an eight-link mechanism is employed such as that shown in Fig. 3.29.* This mechanism is a combined function and motion generator with prescribed timing. The motion of link $\mathbf{Z}_3$ is prescribed in order to pick up one card from a gravitation feeder (the suction cups mounted on the coupler must approach and depart from the card in the vertical direction) and insert the card between the rollers (the card is fed in a horizontal direction). The input timing is prescribed in such a fashion that the cups pick up the card

* This application was brought to the authors' attention by Delbert Tesar of the University of Florida.

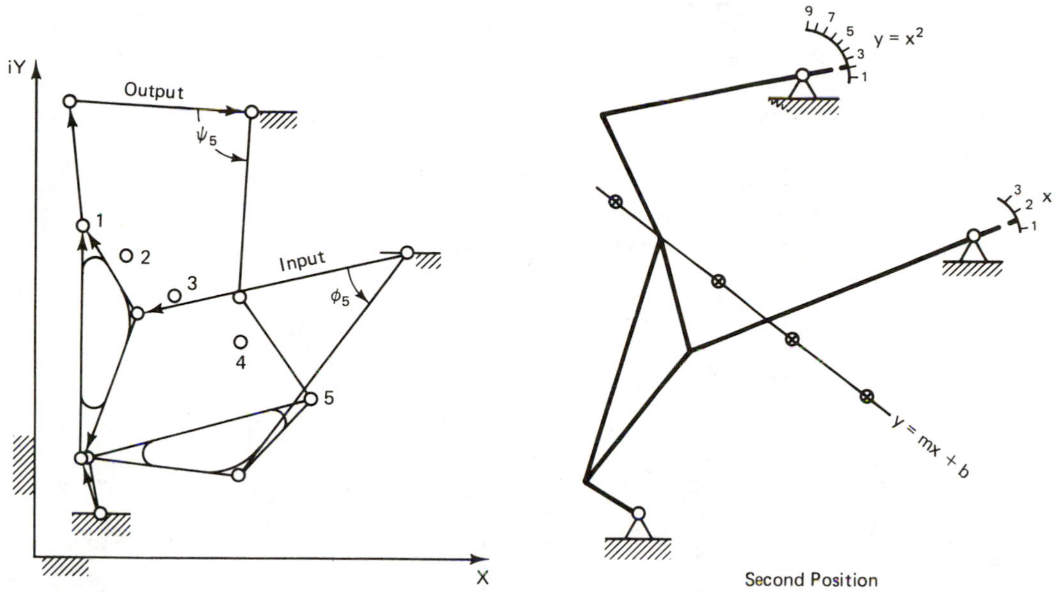


**Figure 3.27** Synthesized six-bar simultaneous path and function generator of Ex. 3.2 in initial and final prescribed positions.

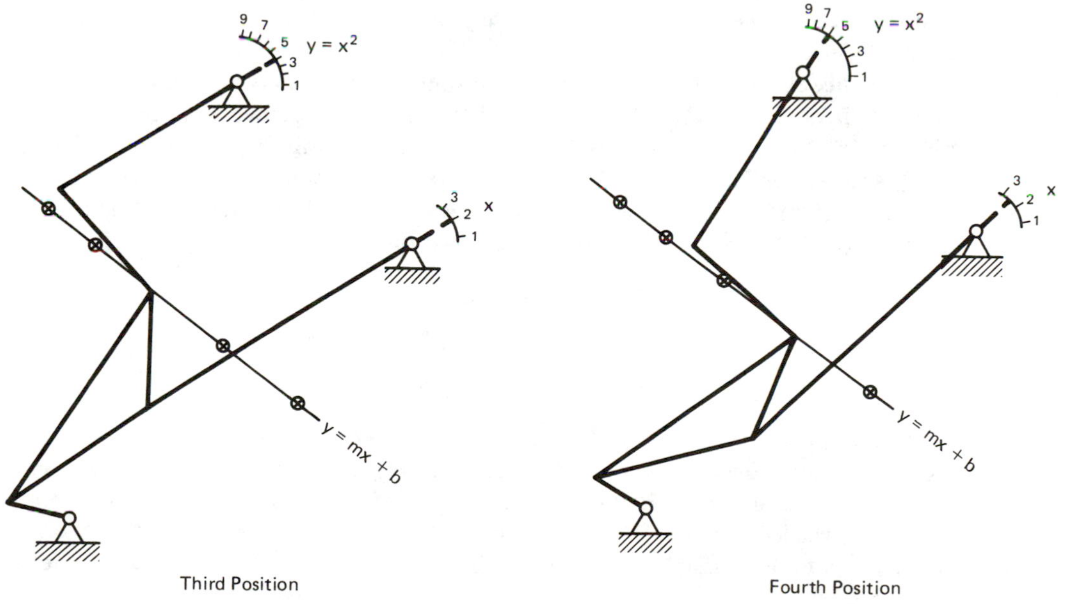


**Figure 3.28** Intermediate positions of the mechanism of Fig. 3.27.

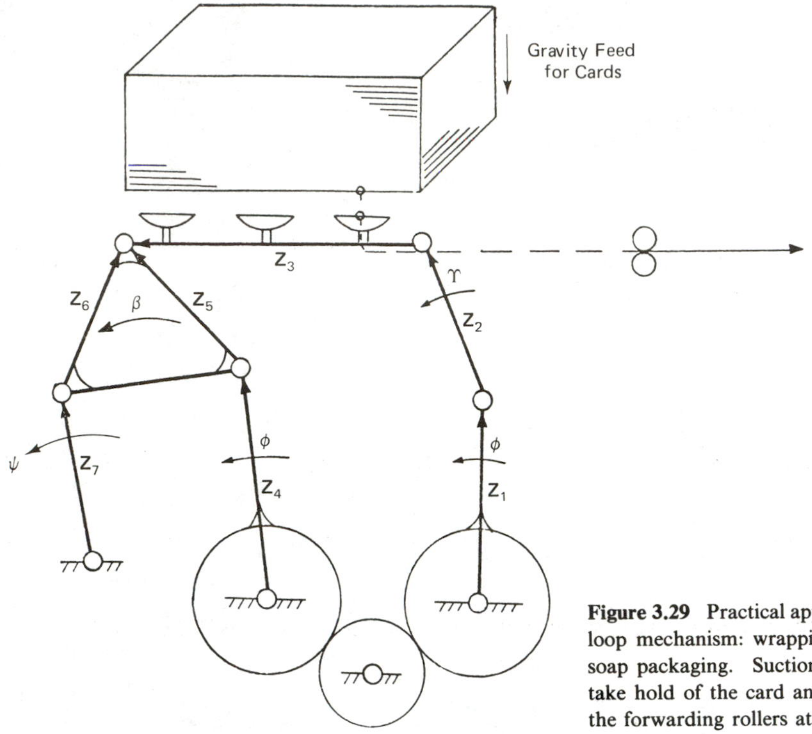


**Figure 3.29** Practical application of multi-loop mechanism: wrapping-card feeder in soap packaging. Suction cups on link $\mathbf{Z}_3$ take hold of the card and feed it between the forwarding rollers at the right.

during a dwell period and release the card in a position and at a velocity that assures that it is fed into the rollers at approximately the same speed as the tangential velocity of the rollers.

In this example, there is a one-to-one functional relationship between the rotations of links $\mathbf{Z}_1$ and $\mathbf{Z}_4$ since they are shown geared together with gears of equal pitch radii. The motion of link $\mathbf{Z}_3$ is prescribed by assigning the path of the tips of $\mathbf{Z}_2$ and $\mathbf{Z}_5$ (or $\mathbf{Z}_6$). Since $\mathbf{Z}_3$ is a rigid link, the distance between the tips of $\mathbf{Z}_2$ and $\mathbf{Z}_5$ must remain the same. In fact, we are free to choose the length and initial orientation of $\mathbf{Z}_3$. Thus the loop-closure equations are

$$\mathbf{Z}_1(e^{i\phi_j} - 1) + \mathbf{Z}_2(e^{i\gamma_j} - 1) = \boldsymbol{\delta}_j \tag{3.51}$$

$$\mathbf{Z}_4(e^{i\phi_j} - 1) + \mathbf{Z}_5(e^{i\beta_j} - 1) = \boldsymbol{\delta}_j \tag{3.52}$$

$$\mathbf{Z}_7(e^{i\psi_j} - 1) + \mathbf{Z}_6(e^{i\beta_j} - 1) = \boldsymbol{\delta}_j \tag{3.53}$$

This mechanism may be synthesized for five prescribed positions in two steps. First synthesize the right-side dyad by utilizing Eq. (3.51) (up to four possible solutions). Second, utilizing the four-bar generator option of the LINCAGES program, synthesize the left four-bar [Eqs. (3.52) and (3.53)] for path generation with prescribed timing (up to 12 possible solutions). Thus, in general, we may expect up to 48 solutions.

## Watt II Examples

One way to illustrate the tremendous number of multiloop linkage applications is to point out several instances in which Watt II six-bars are employed. The standard-form equations for this linkage are easily found by recognizing that the Watt II is just two four-bars connected together. Therefore, one simply synthesizes two four-bar function generators, making sure that the output link of the first and the input link of the second rotate by the same angles (see Ex. 2.5). For example, Hain [119] describes the need for a six-link mechanism for large angular output oscillation. He states: "It would be very difficult to solve this problem with one four-bar linkage, because it is difficult to design a four-bar linkage having such a large oscillation of a crank without running into problems of poor transmission-angle characteristics; it might be possible to use linkages in combinations with gears, but this would make the mechanism more expensive, less efficient, and probably noisier." This statement of Hain's provides strong motivation for developing the kinematic synthesis of gearless multilink mechanisms. Figure 3.30 shows an agitator mechanism used in certain washing machines. Certainly, Hain's advice would be seconded by the designer of this device. This Watt II six-bar has approximately 150° of rotation on the output link.

Hain also cites another application which could be very nicely fulfilled by a Watt II six-link mechanism. A feeding mechanism (see Fig. 3.31) is required to transfer cylindrical parts from a hopper to a chute for further machining. A combined path and function generator will be an ideal solution to this problem. The Watt II six-link mechanism may be synthesized for this task. A schematic configuration of this linkage is shown in Fig. 3.31. Link 6 provides the rotating cupped platform

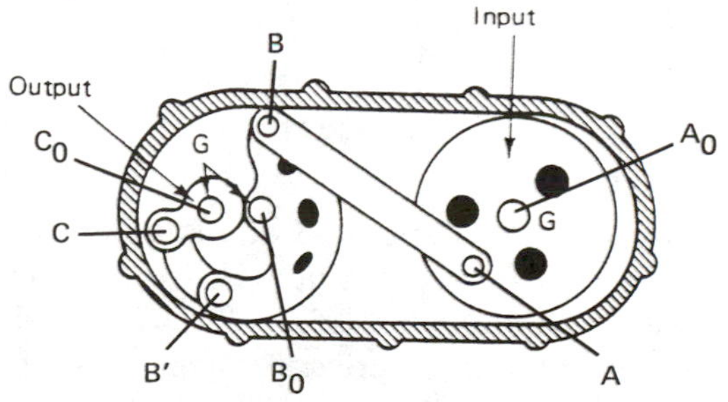


**Figure 3.30** Watt II six-link washing-machine agitator mechanism with crank $A_oA$, coupler no. 1 $AB$, bellcrank $BB_oB'$, coupler no. 2 $B'C$ and rocker $C_oC$. The latter oscillates 150°.

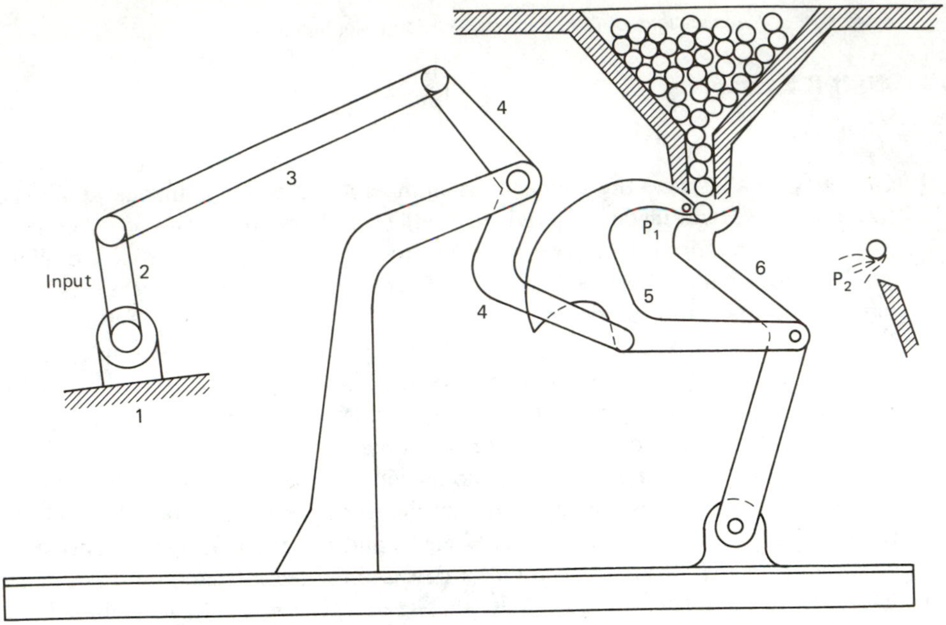


**Figure 3.31** Schematic drawing of feeder mechanism (a Watt II six-bar simultaneous path and function generator with prescribed timing).

(whose rotations are prescribed functions with respect to the input link) which transfers the cylinder from the hopper to the chute while the prescribed path of the coupler (point $P$) positions the cylinder on the platform and then pushes the cylinder into the hopper (see points $P_1$ and $P_2$ in Fig. 3.31).

The following three examples were described by Kramer and Sandor [164]. They suggested the design of an automobile throttle linkage whereby the following angular positions must be coordinated:

| Gas pedal movement | Throttle opening |
|---|---|
| 0° | Closed |
| 5° | 14° |
| 10° | 28° |
| 15° | 44° |
| 20° (to the floor) | (Wide open) 60° |

A six-bar is to be used instead of a four-bar, due to the space that the engine takes up in the engine compartment. The required positions of the fixed pivots for this Watt II linkage are shown in Fig. 3.32. The four-bar function generator sublinkages are first synthesized for prescribed input and output angles and then stretch-rotated so that their ground pivots match these locations. Since the rotation of the intermediate bell crank is of minor importance, it is arbitrarily chosen to be along the function $y = 1.4x$. For this choice, a final solution is shown in Fig. 3.32, while an analysis confirmed that the transmission angles vary from 56° to 90° and

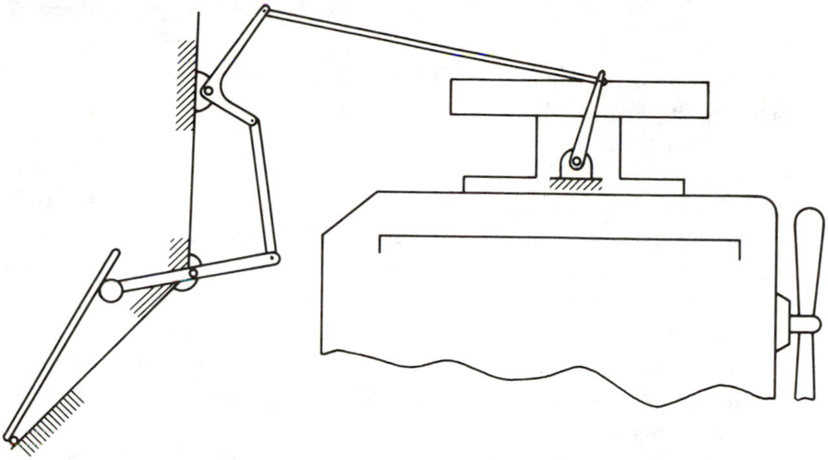

**Figure 3.32** In this example, the rotations of the accelerator pedal are to be coordinated with those of the carburetor throttle valve. The motion of the intermediate crank is not of primary importance; its arbitrary choice can be used to influence the design as to proportions and transmission angles.

the maximum error of bell-crank rotations between precision points is 1.001°. If a more accurate solution is desired, another choice for the location of the fixed pivots and/or the rotation of the intermediate crank would be suggested.

In designing a linkage for the IBM Selectric typewriter, the printer element needs to be tilted a specified amount and the velocity and acceleration are required to be about the same in the vicinity of each precision point. The "tilt tape system" (Fig. 3.33) transforms a linear pull of the tilt latch to a rotation of the tilt bell

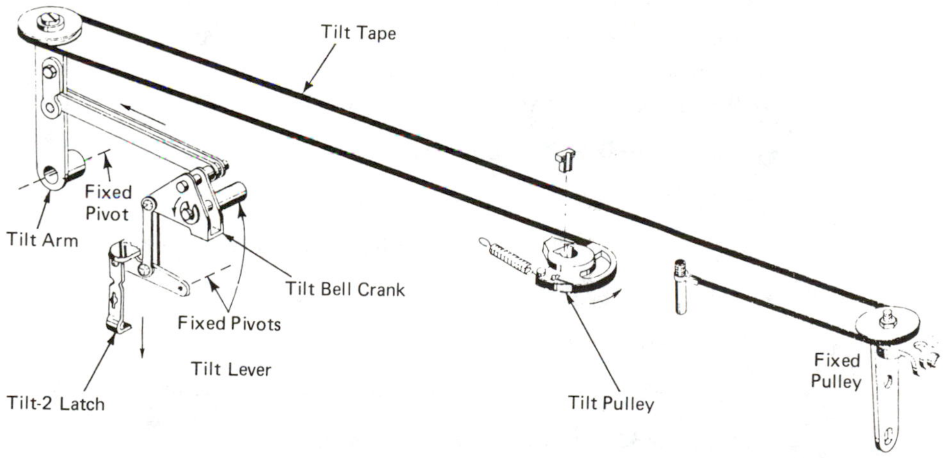


**Figure 3.33** In the Watt II six-bar tilting mechanism for the type ball of the IBM Selectric typewriter, by pulling the tilt-2 latch, the tilt bell crank is made to rotate counterclockwise. The tilt arm is forced to oscillate to the left, thus rotating the tilt pulley. (From Ref. 164.)

crank. The bell crank in turn rotates the tilt arm, which is connected to the printer element by way of the tilt tape and tilt pulley. The tilt latch is to pull the tilt lever down from the starting position: 2.5, 5.0, 7.5, and 10.0°. The rotation of the bell crank may be arbitrarily chosen so that a solution results in acceptable velocity and acceleration values as well as linkage dimensions. The tilt pulley is connected to the tilt arm in such a way that the arm must rotate: 3.5, 7.1, 10.0, and 14.0°. After several unsuccessful passes with the computer-aided design and analysis programs, in a third run the function $y = x^3$ was prescribed for the bell crank. The transmission angles were excellent and the maximum error of bell-crank rotations between precision points was 0.005°. A larger but similar mechanism is used to rotate the printer element, and the two linkages are used in conjunction to give the typehead over 80 print positions.

The control of the heating ducts in a compact car by way of a Watt II six-bar mechanism is shown in Fig. 3.34. The input crank $\mathbf{Z}'_2$ controlled by the driver must rotate enough to open the valve $\mathbf{Z}_2$ on top while closing the flap $\mathbf{Z}'_4$ on the bottom (right), whereby the circulation of air is directed through the tubes leading into the cabin instead of through the exit passages at the rear of the vehicle. For proper control of the air flow we must coordinate the following rotations:

| Top valve | Driver control | Rear flap |
|---|---|---|
| (Closed) | (Rest position) | (Open) |
| 11° | 12° | 10° |
| 23° | 24° | 10° |
| 34° | 36° | 30° |
| 45° (open) | 50° | 40° (closed) |

After synthesizing the two four-bar linkage mechanisms for function generation, they are stretched and rotated to match the fixed links: $\mathbf{z}_1 = 2.800 - 9.000i$ and $\mathbf{z}'_1 = 9.000 - 1.500i$. The resulting Watt II linkage is, for the first four-bar:

$$\mathbf{z}_2 = +2.025 - 0.517i = \text{top valve link}$$

$$\mathbf{z}_3 = +2.580 - 8.671i = \text{coupler link}$$

$$\mathbf{z}_4 = -1.805 + 0.188i = \text{driver-control link (lower branch of bell crank)}$$

$$\mathbf{z}_1 = +2.800 - 9.000i - \text{fixed link}$$

and for the second four-bar:

$$\mathbf{z}'_2 = +0.368 + 1.420i = \text{input link (upper branch of bell crank)}$$

$$\mathbf{z}'_3 = 9.432 - 1.409i = \text{coupler link}$$

$$\mathbf{z}'_4 = 0.800 - 1.511i = \text{output link}$$

$$\mathbf{z}'_1 = +9.000 - 1.500i = \text{fixed link}$$

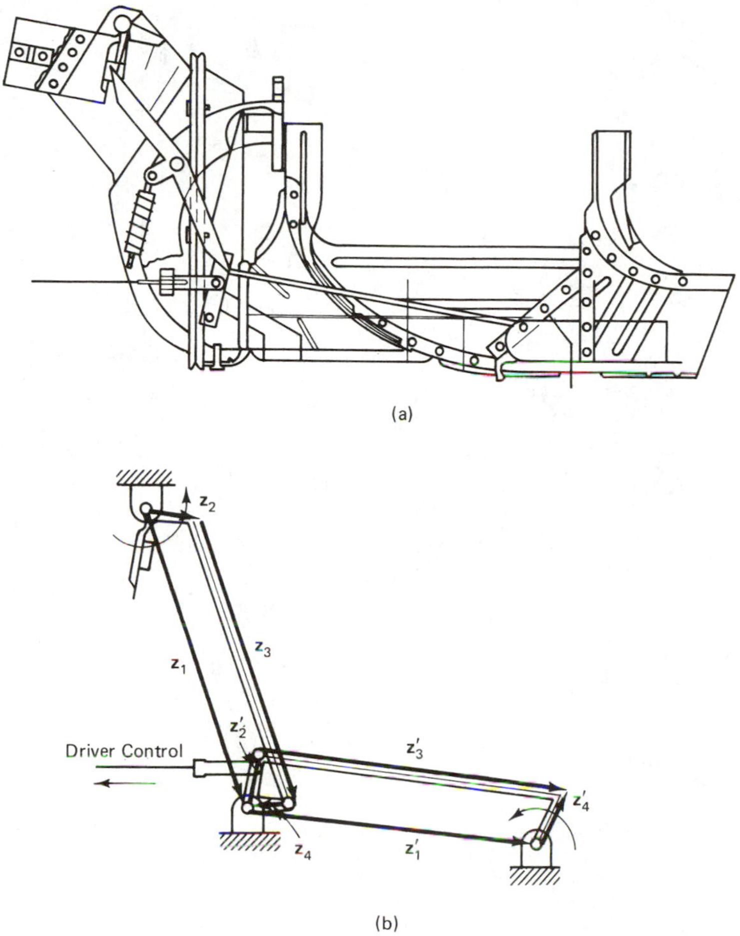


**Figure 3.34** Watt II six-bar mechanism controls the airflow in the heating system of an imported car: (a) original design, (b) proposed Watt II design. (From Ref. 164.)

The foregoing design examples illustrate a few of the many applications of the Watt II mechanism.

### Case Study: Application of the Five-Precision-Point Synthesis in an Industrial Situation [91]

A linkage synthesis problem arose in building a machine for the assembly of a connector (which is used in the installation of telephones) shown in Fig. 3.35. Five metal clips are to be automatically inserted into the five slots in the plastic base of the connector. The first attempt at building a production machine for this project used

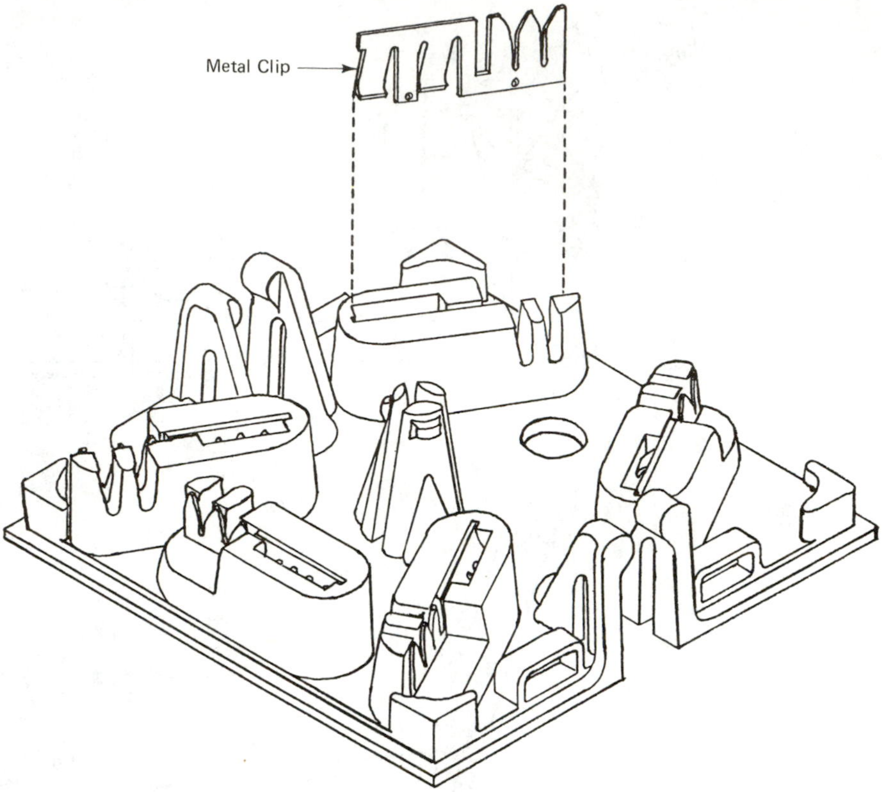


**Figure 3.35** Telephone connector and metal clips. Five metal clips are to be inserted in the five slots of the plastic base. (*Courtesy of the 3M Company.*)

five clip insertion heads which were fed by five separate feed bowls (vibratory feeders; see Sec. 5.6) positioned in the same configuration as the slots. The five heads inserted all clips simultaneously into the base, which was fixed to ground. Because of unreliable performance of the insertion heads, considerable downtime resulted when any one of the insertion heads malfunctioned and the entire machine had to be shut down for repair.

Rather than fixing the base to ground, a mechanism was sought to reposition the base in each of the five desired positions under a single insertion head. The telephone connector would be indexed through the five positions necessary for one head to insert all clips. Because of the simplicity of linkages as compared to other types of mechanisms for motion generation, a linkage was sought to move the telephone connector.

The first step in designing the motion generation linkage was to determine the number of links needed to solve the problem. The four-bar was the first linkage that was synthesized. All six solution linkages were obtained based on four real roots to Eq. (3.47). When these linkages were evaluated, however, it was found that none of them was the crank-and-rocker or double-rocker type which could easily be built into a production machine.

Thus it became evident that a more complex linkage was needed to solve the problem. Inasmuch as a multiloop mechanism would be sought, it was decided to place two additional requirements on the linkage: (1) the input crank angles corresponding to the precision positions of the motion generator link would be equally spaced and, (2) the crank should be capable of 360° rotation. These requirements would simplify the indexing mechanism needed to drive this mechanism.

The eight-bar linkage type shown in Fig. 3.36 was chosen to solve this problem, although many six-bars or seven-bars could also have been considered.

The vectors describing the linkage in position 1, $\mathbf{Z}_1$ through $\mathbf{Z}_{11}$, are shown in Fig. 3.37. Of these vectors $\mathbf{Z}_3$, $\mathbf{Z}_8$, and $\mathbf{Z}_{11}$ are chosen arbitrarily by the designer.

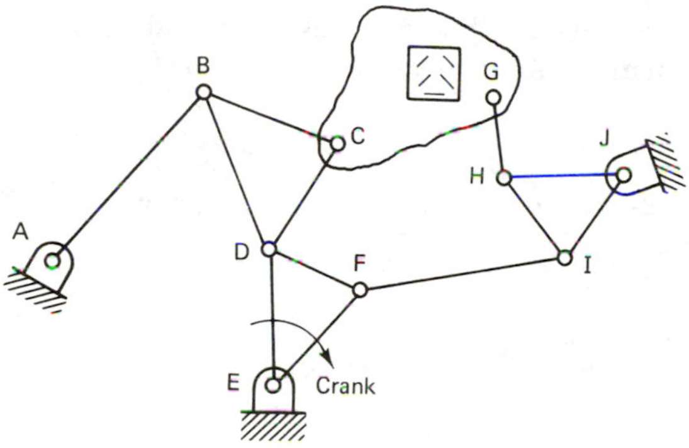


**Figure 3.36** Eight-bar linkage chosen to successively move the plastic base of Fig. 3.35 to the five positions, placing the five slots in sequence under one metal-clip-inserting head.

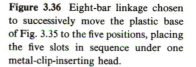

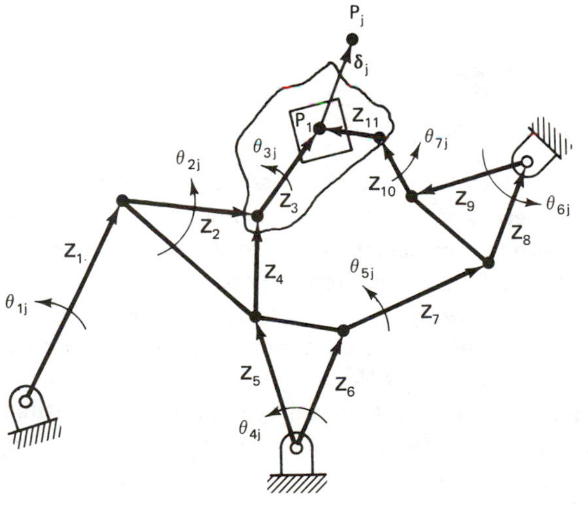


**Figure 3.37** Vector representation of the linkage of Fig. 3.36 in position 1 showing rotations to position *j*. Displacements $\delta_j$ and rotations $\theta_{3j}$ are prescribed (see Table 3.4).

When the crank rotates from position 1 to position $j$, the links rotate through angles $\theta_{kj}$, $k = 1$ through 7. The standard-form equations for the eight-bar linkage are

$$\mathbf{Z}_1(e^{i\theta_{1j}} - 1) + \mathbf{Z}_2(e^{i\theta_{2j}} - 1) = \boldsymbol{\delta}_j - \mathbf{Z}_3(e^{i\theta_{3j}} - 1) = \boldsymbol{\Delta}_j \tag{3.54}$$

$$\mathbf{Z}_5(e^{i\theta_{4j}} - 1) + \mathbf{Z}_4(e^{i\theta_{2j}} - 1) = \boldsymbol{\delta}_j - \mathbf{Z}_3(e^{i\theta_{3j}} - 1) = \boldsymbol{\Delta}_j \tag{3.55}$$

$$\mathbf{Z}_9(e^{i\theta_{6j}} - 1) + \mathbf{Z}_{10}(e^{i\theta_{7j}} - 1) = \boldsymbol{\delta}_j - \mathbf{Z}_{11}(e^{i\theta_{3j}} - 1) = \boldsymbol{\Delta}'_j \tag{3.56}$$

$$\mathbf{Z}_6(e^{i\theta_{4j}} - 1) + \mathbf{Z}_7(e^{i\theta_{5j}} - 1) = -\mathbf{Z}_8(e^{i\theta_{6j}} - 1) = \boldsymbol{\Delta}''_j \tag{3.57}$$

where $j = 2, 3, 4, 5$. Equation systems (3.54) to (3.57) form the set of synthesis equations which are to be solved in order to obtain the desired linkage. The given (or known from having solved a previous equation), unknown, and designer-specified quantities for the synthesis equations are tabulated in Table 3.4.

**TABLE 3.4** STRATEGY FOR SOLVING EQUATION SYSTEMS 3.54 TO 3.57 IN SEQUENCE.

| Equation system number | Given or known from previously solved systems | Unknown | Designer-specified |
|---|---|---|---|
| (3.54) | $\boldsymbol{\delta}_j$, $\theta_{3j}$, $\theta_{1j}$ | $\mathbf{Z}_1$, $\mathbf{Z}_2$, $\theta_{2j}$ | $\mathbf{Z}_3$ |
| (3.55) | $\boldsymbol{\delta}_j$, $\theta_{3j}$, $\theta_{2j}$ | $\mathbf{Z}_4$, $\mathbf{Z}_5$, $\theta_{4j}$ | $\mathbf{Z}_3$ |
| (3.56) | $\boldsymbol{\delta}_j$, $\theta_{3j}$, $\theta_{6j}$ or $\theta_{7j}$ | $\mathbf{Z}_9$, $\mathbf{Z}_{10}$, $\theta_{7j}$ or $\theta_{6j}$ | $\mathbf{Z}_{11}$ |
| (3.57) | $\theta_{4j}$, $\theta_{6j}$ | $\mathbf{Z}_6$, $\mathbf{Z}_7$, $\theta_{5j}$ | $\mathbf{Z}_8$ |
| | $j = 2, 3, 4, 5$ | $j = 2, 3, 4, 5$ | |

One slight problem exists with this procedure, however. When the last loops ($\mathbf{Z}_6$ and $\mathbf{Z}_7$) are synthesized, there is no guarantee that the ground pivots of the $\mathbf{Z}_6$, $\mathbf{Z}_7$, $\mathbf{Z}_8$ loop will match up with the ground pivots already fixed by the base of $\mathbf{Z}_5$ and $\mathbf{Z}_9$. Fortunately, since $\mathbf{Z}_6$, $\mathbf{Z}_7$, $\mathbf{Z}_8$ is a function-generating loop, it can be stretched and/or rotated without affecting the rotations $\theta_{4j}$, $\theta_{5j}$, $\theta_{6j}$. Therefore, the synthesized vectors $\mathbf{Z}_6$, $\mathbf{Z}_7$ as well as $\mathbf{Z}_8$ are simply stretched uniformly and rotated to fit the gap between the base of $\mathbf{Z}_5$ and the base of $\mathbf{Z}_9$.

After an eight-bar linkage has been synthesized by the procedure outlined above, it must be analyzed to determine if it has acceptable transmission angles throughout its entire crank rotation. At the time when this project was accomplished an analysis program was written for this purpose which analytically stepped the crank through incremental rotations and printed out the transmission angles at each step. Unfortunately, no interactive graphics hardware was available, so the synthesis–analysis process was time consuming.

A linkage thus designed is shown schematically in Fig. 3.38 together with the telephone connector. Although this linkage has not been optimized, it is certainly operational. Its minimum transmission angle is 27° and its maximum link-length ratio is 9.5. Once the necessary programming had been done, about 150 hours of trial-and-error design effort were used to obtain this linkage. A model of the linkage was built to demonstrate its motion. The final design is shown in Fig. 3.39. Interactive computer graphics displays of a synthesized linkage with appropriate analysis options would have significantly reduced the time of the trial-and-error design steps. A nonanalytical solution for this problem would have been very difficult.

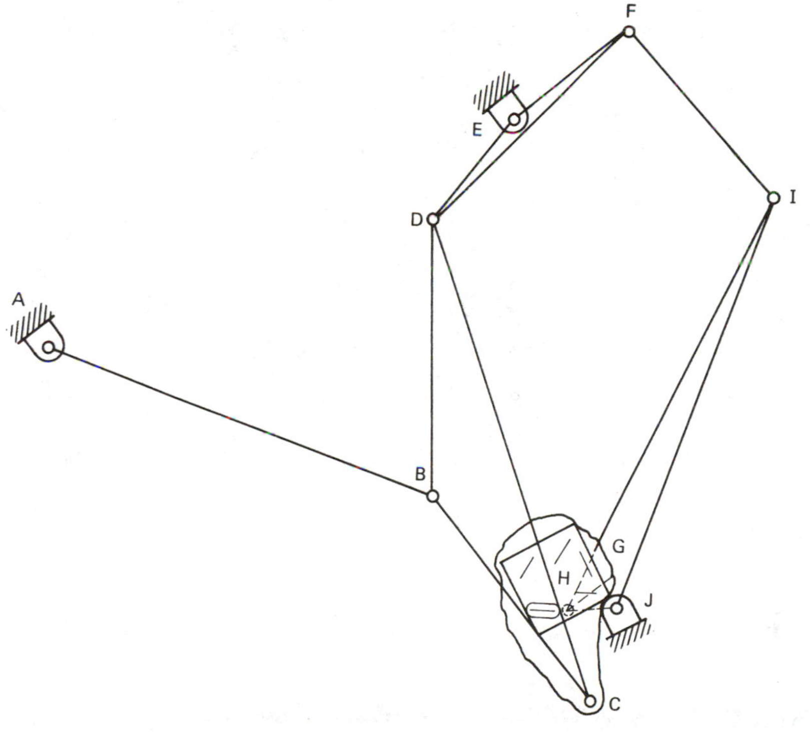


**Figure 3.38** Schematic of an eight-link mechanism synthesized according to Fig. 3.37, Eqs. (3.54) to (3.57), and Table 3.4. Bell crank EFD is input.

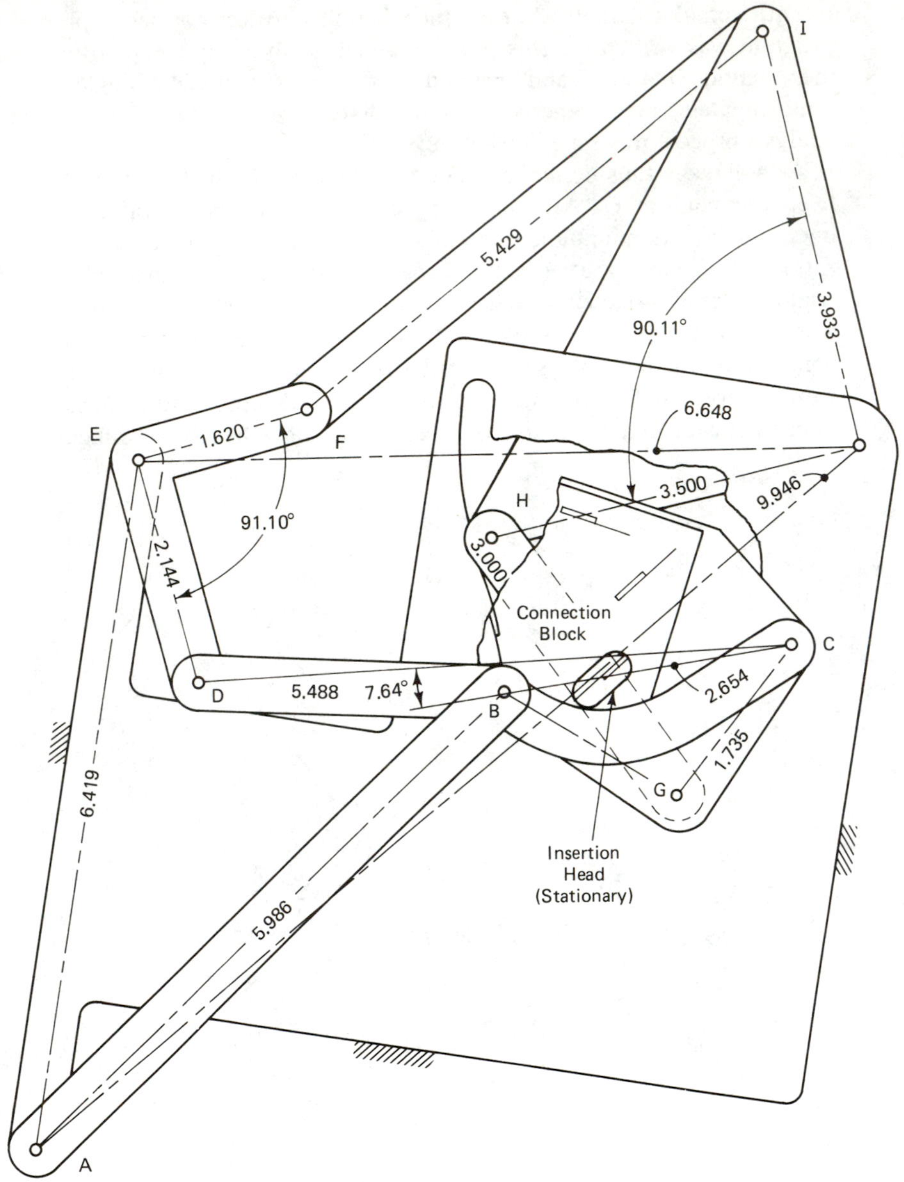


**Figure 3.39** Scale drawing of solution linkage synthesized according to Fig. 3.37, Eq. systems (3.54) to (3.57), and Table 3.4.

## 3.13 KINEMATIC SYNTHESIS OF GEARED LINKAGES

Planar geared linkages readily lend themselves to function, path, and motion generation. Function generation includes any problems in which rotations or sliding motion of input and output elements (either links, racks, or gears) must be correlated.

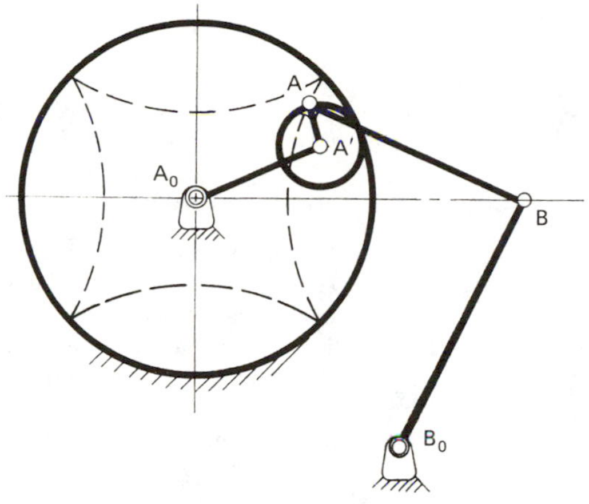


**Figure 3.40** Geared five-bar dwell mechanism.

For example, a frequently encountered industrial problem involves the generation of intermittent or nonuniform motions. Simple linkages, such as the four-bar drag-link mechanisms, are usually the least complicated and most satisfactory devices for such tasks. However, when the desired motion is too complex to produce with a four-bar linkage or a slider-crank mechanism, a geared linkage can often be designed to fulfill the design requirements economically.

In a packaging machine, it may be necessary to connect an input shaft and an output shaft so that the output shaft oscillates with a prescribed dwell period and timing while the input shaft rotates continuously. A simple geared five-bar mechanism, such as that shown in Fig. 3.40, can be readily designed to produce such a dwell period [45].*

In the case shown, the gear ratios of the *cycloidal crank* $A_0A'A$ have been chosen to produce a hypocycloid of four cusps (an astroid). The path of point $A$ on the pitch circle of the planet gear between cusps is approximately an arc of a circle, and the length of the coupler $AB$ is equal to the radius of that arc. The follower link $B_0B$ has been arranged so that in its extreme right position its moving pin $B$ coincides with the center of the arc. If desired, the coupler could be attached to an output slider rather than to a rocker. Still different motion characteristics could be obtained through the use of a three- or five-cusp hypocycloid.

## Degrees of Freedom

Mobility of a planar geared linkage can be studied through the use of Eq. (1.3) (see Table 1.1):

$$F = 3(n-1) - 2f_1 - 1f_2$$

where $F$ is the number of degrees of freedom of the linkage, $n$ is the number of links, $f_1$ is the number of joints that constrain two degrees of freedom (revolute and slider joints), and $f_2$ is the number of joints that constrain one degree of freedom

* See also Prob. 3.49.

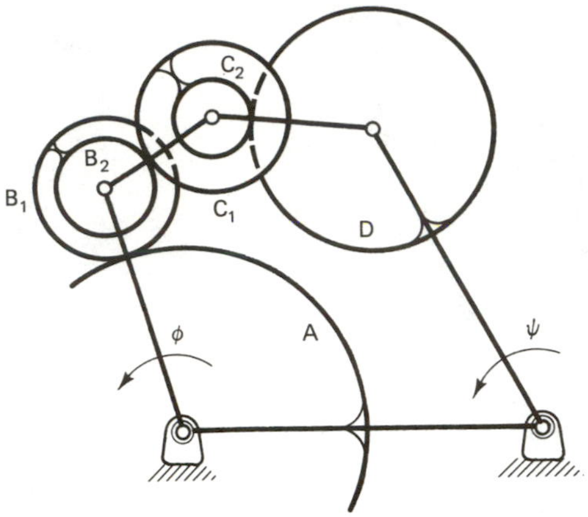


Gear ratios:

$$\frac{T_A}{T_{B_1}} = r_2, \quad \frac{T_{B_2}}{T_{C_1}} = r_3, \quad \frac{T_{C_1}}{T_D} = r_4$$

where $T_A$ is the number of teeth in gear A, and so on.

**Figure 3.41** Geared five-bar function generator linkage. Gear A is stationary, compound gears B and C are pivoted at link joints, and gear D is rigidly attached to the output link. Gear ratios: $T_A/T_{B_1} = r_2$, $T_{B_2}/T_{C_1} = r_3$, $T_{C_1}/T_D = r_4$, where $T_A$ is the number of teeth in gear A, and so on.

(gear meshes in this case). If the gears are attached rigidly to $p$ links, then, in general, $f_2 = p/2$. For instance, the mechanism of Fig. 3.41 has five bars and a train of four gears. The gears are fixed at two points, to the frame and to the output crank. Thus $n = 5$, $p = 2$, $f_1 = 5$, and $f_2 = 1$, so the mechanism has a single degree of freedom. Here we tacitly recognized that gears $B$ and $C$ are idlers and therefore do not participate in the degrees-of-freedom computation when the quantity $f_2$ is defined as given above. However, we could also regard the idlers as separate links, in which case $n = 7$, $f_1 = 7$, and the number of gear meshes $f_2 = 3$. Since each $f_2$ subtracts one freedom of rotation, the result is again a single degree of freedom for the mechanism.

## Synthesis Equations in Complex Numbers

The method of complex numbers is particularly well suited to the synthesis of geared linkages because links and gear ratios between links are readily represented and manipulated mathematically. When a limited number of precision conditions is imposed on a linkage, the method provides synthesis equations which are linear in the unknown link vectors describing the mechanism in its starting configuration. In function generation with a single-loop mechanism, a linear solution can be obtained for one fewer precision conditions than the number of bars in the loop. For example, in the case of a geared five-bar (such as that shown in Fig. 3.41), a linear solution can be obtained for up to four first-order (finitely separated) precision points or for, say, two first-order and one second-order precision point (the latter equivalent to two infinitesimally close precision points) (see Sec. 2.24). One can prescribe more than four precision conditions for this mechanism, but the solution is made more difficult because some of the coefficients of the link vectors must then be treated as unknowns. Nonlinear compatibility equations must then be solved.

For finite synthesis, vector loop equations and displacement equations written in terms of complex numbers form the system of synthesis equations. In higher-order synthesis involving prescribed derivatives, velocities, accelerations, and higher accelerations, derivatives of the loop equations are taken with respect to a reference variable or time. In both finite and infinitesimal synthesis, the system of synthesis equations can be made linear in the unknown link vectors. The coefficients would contain the prescribed performance parameters and the gearing velocity ratios. These can be arbitrarily assigned convenient values by the mechanism designer. By varying these arbitrary choices the designer can obtain an infinite spectrum of solutions from which to select a suitable mechanism. All of these solutions will satisfy the given precision conditions. Selection of the best available solution can be based on such optimization criteria as most favorable transmission angles, best gear ratios, or ratio between longest and shortest link lengths closest to unity. These criteria can be used either singly or in weighted combinations.

## Geared Five-Bar Example [248]

Suppose that one wishes to synthesize a geared linkage of the type shown in Fig. 3.41 to generate a function $y = f(x)$ over some given range. Let the rotation of the input crank ($\phi$) be the linear analog of $x$ and the rotation of the output link ($\psi$) be the linear analog of $y$.

One can represent this mechanism by a closed vector pentagon (Fig. 3.42). In the reference position of the mechanism, vectors $\mathbf{Z}_1$, $\mathbf{Z}_2$, $\mathbf{Z}_3$, $\mathbf{Z}_4$, and $\mathbf{Z}_5$ define the orientation and length of links 1, 2, 3, 4, and 5. In a general displaced position of the mechanism, say the $j$th position, at which requirements on the motion have been prescribed, the mechanism is defined by these vectors multiplied by their appropriate rotation operators:

$$
\begin{aligned}
\mathbf{Z}'_1 &= \mathbf{Z}_1 \\
\mathbf{Z}'_2 &= e^{i\phi_j}\mathbf{Z}_2 \\
\mathbf{Z}'_3 &= e^{i\gamma_j}\mathbf{Z}_3 \\
\mathbf{Z}'_4 &= e^{i\mu_j}\mathbf{Z}_4 \\
\mathbf{Z}'_5 &= e^{i\psi_j}\mathbf{Z}_5
\end{aligned}
\tag{3.58}
$$

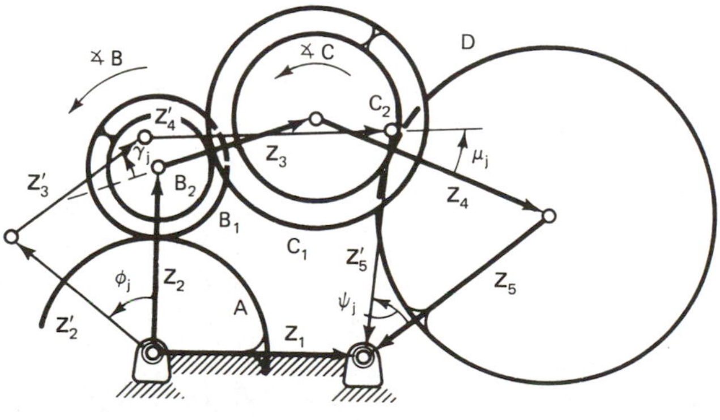


**Figure 3.42** Vector schematic of geared five-bar mechanism of Fig. 3.41. Note that gear A is fixed to vector 1, which represents the frame, and gear D is fixed to vector 5, which represents the output link. Compound gears B and C are free to rotate.

A mathematical relationship must be developed to express the relative, and finally the absolute, rotations of the various links. In other words, for a specified rotation of the input crank (link 2), can an expression for the rotation of some other link be found in terms of this input rotation and the gear ratios? Obviously, there must be some such relationship since the mechanism has but one degree of freedom.

### Determining the Effect of Gear Ratios on the Rotation of the Links

In reference to Fig. 3.43, the following general relationship (see Chap. 6 of Vol. 1) will prove very helpful in solving this rotation problem:

$$\phi_{j(k+1)} = \phi_{jk} + [\phi_{jk} - \phi_{j(k-1)}]\frac{T_{k-1}}{T_{k+1}} \tag{3.59}$$

where, for example, $\phi_{jk}$ is a finite rotation of the $k$th link from the first position of that link to its $j$th position. $T_k$ refers to the number of teeth on the gear rigidly attached to the $k$th link. For the linkage in Fig. 3.42 let $\measuredangle B$ *and* $\measuredangle C$ be the absolute rotations of gears $B$ and $C$ for any given input rotation $\phi_j$, measured from the starting position of $\mathbf{Z}_2$. $\gamma$ and $\mu$ are absolute rotations of links $\mathbf{Z}_3$ and $\mathbf{Z}_4$ from their respective starting positions. Similarly, $\psi$ is the rotation of link $\mathbf{Z}_5$, the output. Considering now each gear-pair separately, from Fig. 3.44, omitting the position subscript $j$, we have,

$$\measuredangle B = \frac{T_A}{T_{B_1}}\phi + \phi$$

or, letting $T_A/T_B = r_2$,

$$\measuredangle B = (r_2 + 1)\phi \tag{3.60}$$

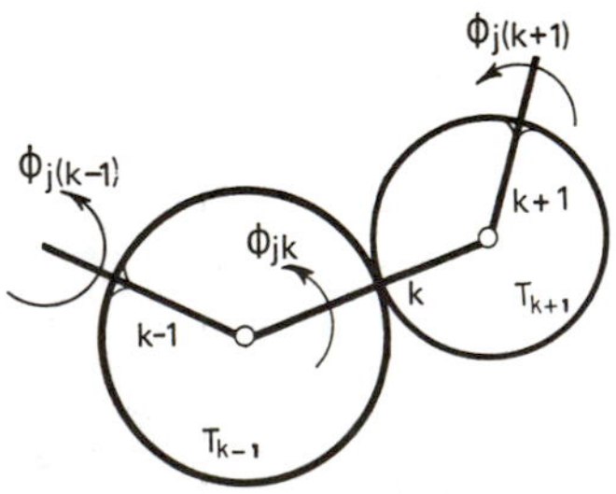


**Figure 3.43** General geared pair showing notation for the link rotations.

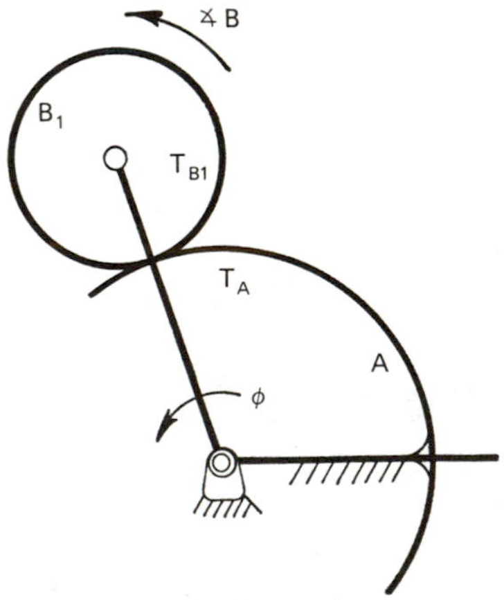


**Figure 3.44** Input side of the geared five-bar of Fig. 3.41.

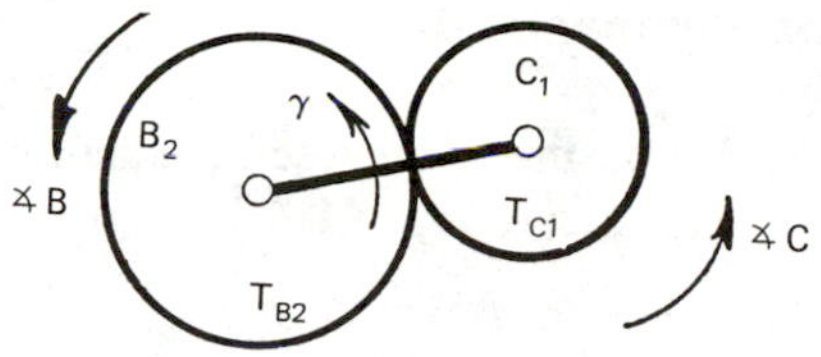


**Figure 3.45** Intermediate gears and floating link $\mathbf{Z}_3$ in the geared five-bar of Fig. 3.41.

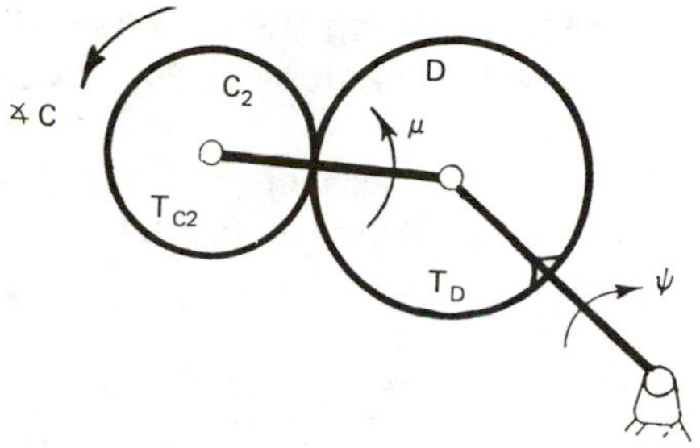


**Figure 3.46** Output side of the geared five-bar of Fig. 3.41.

Continuing around the loop of the mechanism of Fig. 3.42, applying the general relationship to Fig. 3.45 yields

$$\measuredangle C = \gamma + (\gamma - \measuredangle B) r_3 \tag{3.61}$$

, where $r_3 = T_{B2}/T_{C1}$

Finally, applying the general relationship to Fig. 3.46, we obtain

$$\psi = \mu + (\mu - \measuredangle C) r_4 \tag{3.62}$$

, where $r_4 = T_{C2}/T_D$

Substituting Eq. (3.60) into Eq. (3.61) yields

$$\measuredangle C = \gamma + [\gamma - (r_2 + 1)\phi] r_3$$

$$\measuredangle C = \gamma + \gamma r_3 - r_2 r_3 \phi - r_3 \phi \tag{3.63}$$

and substituting Eq. (3.63) in Eq. (3.62) gives

$$\psi = \mu + [\mu - (\gamma + \gamma r_3 - r_2 r_3 \phi - r_3 \phi)] r_4$$
$$= \mu + \mu r_4 - \gamma r_4 - \gamma r_3 r_4 + \phi r_2 r_3 r_4 + \phi r_3 r_4$$

and

$$\psi = \mu(1 + r_4) - \gamma(r_4 + r_3 r_4) + \phi(r_2 r_3 r_4 + r_3 r_4) \tag{3.64}$$

So it can be seen that there is a direct relationship between $\psi$, $\phi$, $\mu$, $\gamma$ and the gear ratios. As expected, the absolute rotations $\measuredangle B$ and $\measuredangle C$ of the idlers do not appear in this expression, but the gear ratios do.

Since $\phi$ and $\psi$ are prescribed in accordance with the function to be generated, if values of $\gamma$ are assumed, then corresponding values of $\mu$ are defined by Eq. (3.64).

For convenience, let

$$Q = 1 + r_4, \qquad R = r_4 + r_3 r_4, \qquad S = r_2 r_3 r_4 + r_3 r_4 \tag{3.65}$$

Therefore,

$$\psi = \mu Q - \gamma R + \phi S \tag{3.66}$$

$$\mu = \frac{1}{Q}(\psi + \gamma R - \phi S) \tag{3.67}$$

## Determining the Number of Precision Positions for Which the Mechanism of Fig. 3.42 Can be Synthesized

For the mechanism of Fig. 3.42 in the $j$th position, the vector equation of closure can be written as follows:

$$\mathbf{Z}_2 e^{i\theta_j} + \mathbf{Z}_3 e^{i\gamma_j} + \mathbf{Z}_4 e^{i\mu_j} + \mathbf{Z}_5 e^{i\psi_j} = \mathbf{Z}_1 = 1 \tag{3.68}$$

Note that $\mathbf{Z}_1$ is assigned the value 1 for convenience. This is permissible because only angular relationships are of interest (see Sec. 2.23).

Recall that the $\phi$'s and $\psi$'s are prescribed and some of the $\gamma$'s may be chosen arbitrarily, in which case the $\mu$'s are known through Eq. (3.67).

Table 3.5 can now be constructed.

**TABLE 3.5** NUMBER OF POSITIONS FOR WHICH THE GEARED FIVE-BAR MECHANISM OF FIG. 3.42 CAN BE SYNTHESIZED FOR PRESCRIBED INPUT AND OUTPUT ROTATIONS, ONCE THE GEAR RATIOS ARE SPECIFIED

| Number of prescribed positions | Number of independent real equations [excluding Eq. (3.67)] | Number of independent unknown reals [excluding $\mu$ owing to Eq. (3.67)] | | Arbitrary choices of reals | Number of solutions |
|---|---|---|---|---|---|
| 1 | 2 | $\mathbf{Z}_2, \mathbf{Z}_3, \mathbf{Z}_4, \mathbf{Z}_5$ | (8) | 6, say $\mathbf{Z}_{2,3,4}$ | $O(\infty)^6$ |
| 2 | 4 | $\mathbf{Z}_2, \mathbf{Z}_3, \mathbf{Z}_4, \mathbf{Z}_5, \gamma_2$ | (9) | 5, say $\mathbf{Z}_{2,3}, \gamma_2$ | $O(\infty)^5$ |
| 3 | 6 | $\mathbf{Z}_2, \mathbf{Z}_3, \mathbf{Z}_4, \mathbf{Z}_5, \gamma_2, \gamma_3$ | (10) | 4, say $\mathbf{Z}_2, \gamma_{2,3}$ | $O(\infty)^4$ |
| 4 | 8 | $\mathbf{Z}_2, \mathbf{Z}_3, \mathbf{Z}_4, \mathbf{Z}_5, \gamma_2, \gamma_3, \gamma_4$ | (11) | 3, say $\gamma_{2,3,4}$ | $O(\infty)^3$ |
| 5 | 10 | $\mathbf{Z}_2, \mathbf{Z}_3, \mathbf{Z}_4, \mathbf{Z}_5, \gamma_2, \gamma_3, \gamma_4, \gamma_5$ | (12) | 2, say, $\gamma_{2,3}$ | $O(\infty)^2$ |
| 6 | 12 | $\mathbf{Z}_2, \mathbf{Z}_3, \mathbf{Z}_4, \mathbf{Z}_5, \gamma_2$ to $\gamma_6$ | (13) | 1, say, $\gamma_2$ | $O(\infty)^1$ |
| 7 | 14 | $\mathbf{Z}_2, \mathbf{Z}_3, \mathbf{Z}_4, \mathbf{Z}_5, \gamma_2$ to $\gamma_7$ | (14) | 0 | Finite |

Up to four precision positions, the designer can pick all the $\gamma$'s arbitrarily. For five, six, or seven positions, only some or none of the $\gamma$'s can be picked, and nonlinear compatibility relationships must be solved for the remaining unknown $\gamma$'s. Thus it can be seen that the limiting number of precision positions, beyond which nonlinear compatibility equations become necessary in the solution, is four. For predetermined gear ratios and scale factors, the mechanism may be synthesized for up to seven positions. However, if the gear ratios and scale factors are also regarded as unknowns, the number of attainable precision points can be further increased.

For four finitely separated first-order (finitely separated) precision points the synthesis equations for this mechanism, written in matrix form, are

$$\begin{bmatrix} 1 & 1 & 1 & 1 \\ e^{i\phi_2} & e^{i\gamma_2} & e^{i\mu_2} & e^{i\psi_2} \\ e^{i\phi_3} & e^{i\gamma_3} & e^{i\mu_3} & e^{i\psi_3} \\ e^{i\phi_4} & e^{i\gamma_4} & e^{i\mu_4} & e^{i\psi_4} \end{bmatrix} \begin{bmatrix} \mathbf{Z}_2 \\ \mathbf{Z}_3 \\ \mathbf{Z}_4 \\ \mathbf{Z}_5 \end{bmatrix} = \begin{bmatrix} 1 \\ 1 \\ 1 \\ 1 \end{bmatrix} \tag{3.69}$$

As all the quantities in the coefficient matrix are either prescribed or arbitrarily assumed, one can easily solve this linear system of complex equations for the four complex unknowns, $\mathbf{Z}_2$, $\mathbf{Z}_3$, $\mathbf{Z}_4$, and $\mathbf{Z}_5$. Varying the arbitrary values $\gamma_2$, $\gamma_3$, and $\gamma_4$ and varying the choice of gear ratios and scale factors allows one to obtain an infinite spectrum of solutions.

**Example 3.3**

Suppose that the function to be generated is $y = \tan(x)$, $0 \le x \le 45°$. For four accuracy points with Chebyshev spacing, values of $x$ and $y$ can be found (see Sec. 2.2):

$$x_1 = 1.71°, \qquad y_1 = 0.03$$
$$x_2 = 13.89°, \qquad y_2 = 0.25$$
$$x_3 = 31.11°, \qquad y_3 = 0.60$$
$$x_4 = 43.29°, \qquad y_4 = 0.94$$

Let $\Delta\phi$ = range of input crank = 90° and $\Delta\psi$ = range of output = 90°. With these

$$\phi_2 = 24.36°, \qquad \psi_2 = 19.80°$$
$$\phi_3 = 58.80°, \qquad \psi_3 = 51.30°$$
$$\phi_4 = 83.16°, \qquad \psi_4 = 81.90°$$

Figures 3.47, 3.48, and 3.49 and Table 3.6 show some typical computer-synthesized linkages for generating the function $y = \tan(x)$. Note that the gears are not shown, but their effect is clearly in evidence: the gears transfer rotary motion directly, transmission angles are of no interest between geared links.

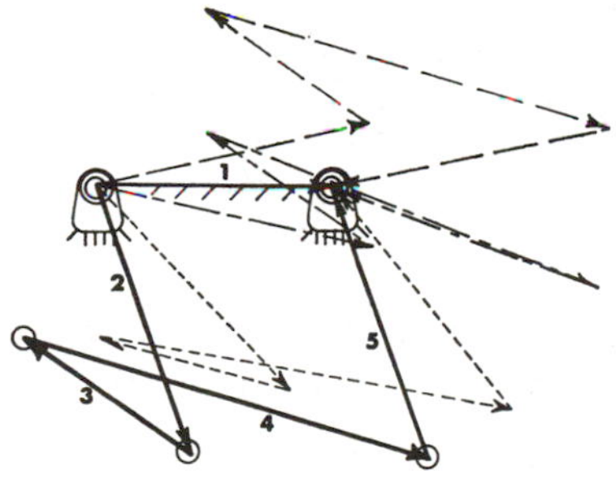


**Figure 3.47** Synthesized geared five-bar generating the tangent function. Example A of Table 3.6 is shown in its first (solid), second (short dashed), third (uneven dashed), and fourth (long dashed) precision positions. Gears are not shown.

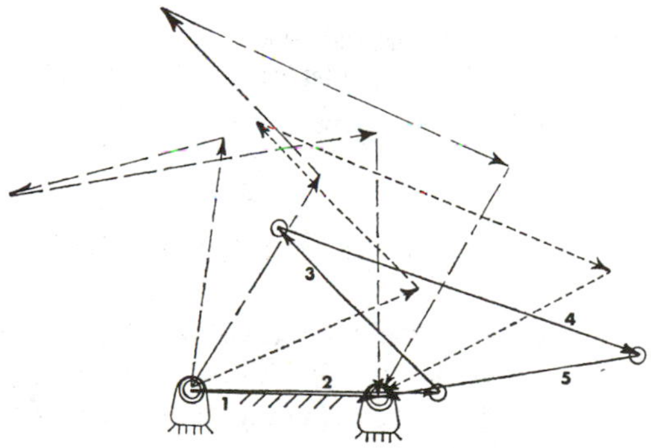


**Figure 3.48** Example B of Table 3.6 is shown in its four precision positions (same notation as Fig. 3.47). Gears are not shown.

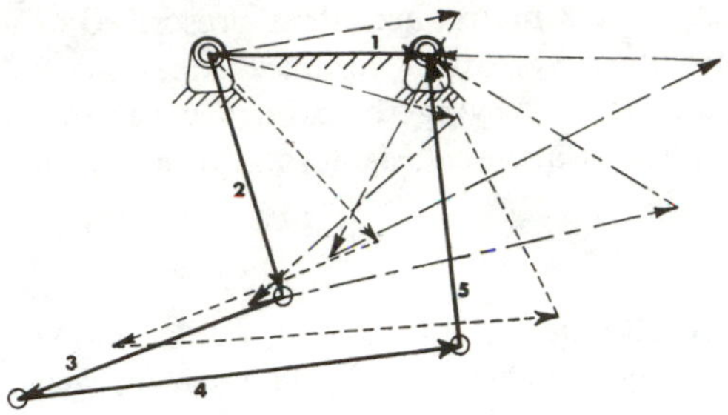


**Figure 3.49** Example C of Table 3.6 is shown in its four precision positions (same notation as Fig. 3.47). Gears are not shown.

**TABLE 3.6** THREE DIFFERENT GEARED FIVE-BAR DESIGNS SYNTHESIZED FOR FOUR-PRECISION-POINT FUNCTION GENERATOR (See Fig. 3.42 and Table 3.5.)

| | Example A | Example B | Example C |
|---|---|---|---|
| Figure | 3.47 | 3.48 | 3.49 |
| Function | $y = \tan x$ | $y = \tan x$ | $y = \tan x$ |
| Range | $0 \leq x \leq 45°$ | $0 \leq x \leq 45°$ | $0 \leq x \leq 45°$ |
| Scale factors | $\Delta\phi = 90°$ | $\Delta\phi = 90°$ | $\Delta\phi = 90°$ |
| | $\Delta\psi = 90°$ | $\Delta\psi = 90°$ | $\Delta\psi = 90°$ |
| Gear ratios $r_2$ | 3 | 3 | 3 |
| $r_3$ | 0.5 | 0.5 | 0.5 |
| $r_4$ | 0.5 | 0.5 | 0.5 |
| Link vectors | | | |
| $\mathbf{Z}_1$ | 1.000, +0.000$i$ | 1.000, +0.000$i$ | 1.000, +0.000$i$ |
| $\mathbf{Z}_2$ | 0.402, −1.115$i$ | 1.335, +0.027$i$ | 0.333, −1.126$i$ |
| $\mathbf{Z}_3$ | −0.709, +0.475$i$ | −0.888, +0.846$i$ | −1.225, −0.482$i$ |
| $\mathbf{Z}_4$ | 1.714, −0.4688$i$ | 1.919, −0.611$i$ | 2.029, +0.254$i$ |
| $\mathbf{Z}_5$ | −0.407, +1.109$i$ | −1.366, −0.262$i$ | −0.137, +1.354$i$ |
| $\mathbf{Z}_6$ | — — | — — | — |
| Arbitrary link rotations | $\gamma_2 = 20°$ | $\gamma_2 = 0°$ | $\gamma_2 = 0°$ |
| | $\gamma_3 = 0°$ | $\gamma_3 = 0°$ | $\gamma_3 = 20°$ |
| | $\gamma_4 = 0°$ | $\gamma_4 = 60°$ | $\gamma_4 = 40°$ |

## Geared-Linkage Compatibility Equations

The system of equations for five-point function-generation synthesis of the geared five-bar of Fig. 3.42 is obtained by adding another equation in system (3.69). This will yield the compatibility equation

$$\begin{vmatrix} 1 & 1 & 1 & 1 & 1 \\ e^{i\phi_2} & e^{i\gamma_2} & e^{i\mu_2} & e^{i\psi_2} & 1 \\ e^{i\phi_3} & e^{i\gamma_3} & e^{i\mu_3} & e^{i\psi_3} & 1 \\ e^{i\phi_4} & e^{i\gamma_4} & e^{i\mu_4} & e^{i\psi_4} & 1 \\ e^{i\phi_5} & e^{i\gamma_5} & e^{i\mu_5} & e^{i\psi_5} & 1 \end{vmatrix} = 0 \tag{3.70}$$

With $\gamma_2$ and $\gamma_5$ assumed arbitrarily, this expands to

$$\mathbf{\Delta}_3 e^{i\gamma_3} + \mathbf{\Delta}_4 e^{i\gamma_4} = \mathbf{\Delta} \tag{3.71}$$

where the $\mathbf{\Delta}$'s are known. Equation (3.71) can therefore be solved geometrically for $\gamma_3$ and $\gamma_4$, as shown in Fig. 3.3 and Table 3.1 for $\beta_3$ and $\beta_4$. Then, with compatible sets of $\gamma_j$, $j = 2, 3, 4, 5$, any four of the five equations such as Eq. (3.69) can be solved simultaneously for $\mathbf{Z}_k$, $k = 2, 3, 4, 5$.

In case of six-point synthesis, the five-column augmented matrix will have six rows and therefore yield two compatibility equations, say, one consisting of the first four and the sixth rows, and another of the first three plus the last two rows. After assuming an arbitrary value for $\gamma_6$, these will expand to

$$\mathbf{\Delta}_2 e^{i\gamma_2} + \mathbf{\Delta}_3 e^{i\gamma_3} + \mathbf{\Delta}_4 e^{i\gamma_4} = \mathbf{\Delta} \tag{3.72}$$

and

$$\mathbf{\Delta}'_2 e^{i\gamma_2} + \mathbf{\Delta}'_3 e^{i\gamma_3} + \mathbf{\Delta}_4 e^{i\gamma_5} = \mathbf{\Delta}' \tag{3.73}$$

where all the $\Delta$'s are known. These are of the same form as Eqs. (3.34) and (3.35), and can be solved the same way for $\gamma_j$, $j = 2, 3, 4, 5$. Thus, with the assumed value of $\gamma_6$, we will have compatible sets of values for all the $\gamma_j$, $j = 2, 3, 4, 5, 6$. Any such set can be substituted back into our set of synthesis equations of the form of Eq. (3.68) with $j = 1, 2, \ldots, 6$ (note that all angles for $j = 1$ are zero), and solve any four of these simultaneously for $\mathbf{Z}_k$, $k = 2, 3, 4, 5$, thus yielding a solution for a *six-precision point function generator geared five-bar.*

## 3.14 DISCUSSION OF MULTIPLY SEPARATED POSITION SYNTHESIS

Section 2.24 introduced the concept of prescribing precision points where the generated function or path is to have higher-order contact with an ideal curve. By taking derivatives of displacement equations, contacts through two, three, and so on, infinitesimally separated positions can be obtained. When such infinitesimally separated positions (contained in a higher-order precision point) are specified in addition to finitely separated first-order (single-point match) precision points, we have "multiply separated position synthesis." Depending on the total number of prescribed positions and derivatives and the number of unknowns in the synthesis equations (see Table 2.6), the solution procedure may involve either linear or nonlinear methods. Two examples of a nonlinear method follow.

### Synthesis of a Fifth-Order Path Generator

In industrial practice, problems are frequently encountered that require the design of mechanisms that will generate a prescribed path in a plane. The problem is further complicated if the velocity, acceleration, and higher accelerations of the motion are critical, as might be the case where possible damage to the mechanism or to the objects handled may result from large accelerations or rates of change of acceleration and the resultant shock. This example [250] presents an analytical closed-form

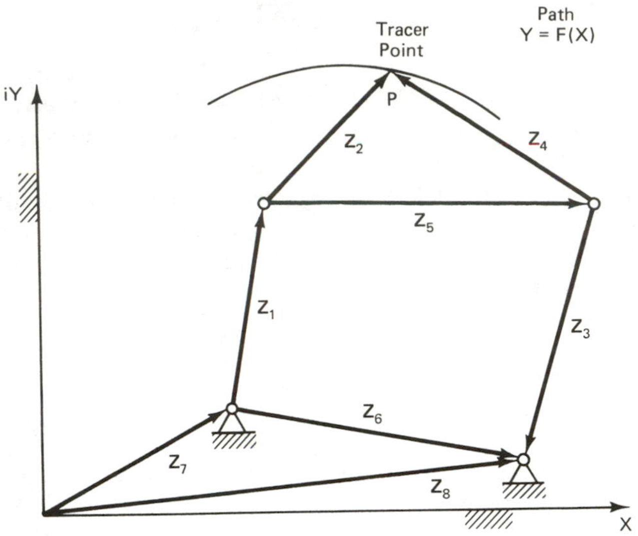


**Figure 3.50** Four-bar path generator for higher-order path generation.

solution for the synthesis of a four-bar linkage that will give fifth-order path approximation in the vicinity of a single precision point. Recall from Table 2.6 that this is the maximum order that can be prescribed for a four-bar. At the single precision point, derivatives of the path-point position vector up to the fourth are specified, which, when taken with respect to time, can be interpreted as the velocity, acceleration, shock, and third acceleration of the path-tracer coupler point. The closed-form solution to be derived in this section will yield a maximum of 12 different linkages for each data set.

The problem is to synthesize a four-bar linkage with the notation shown in Fig. 3.50 that will generate a path given by $y = f(x)$ with prescribed timing, i.e., with prescribed input-crank motion. First, as before, the input side (dyad) of the four-bar shown in Fig. 3.51 will be synthesized. Then, for each possible solution for the input side, the corresponding output side solutions will be found.

In Fig. 3.51, the vector $\mathbf{R}$ locates the precision point on the prescribed ideal path; vector $\mathbf{Z}_7$ locates the unknown fixed pivot; $\mathbf{Z}_1$ represents the unknown input link, and vector $\mathbf{Z}_2$ represents one side of the unknown floating or coupler link. The following loop equation can now be written for the input side at the precision point:

$$\mathbf{Z}_7 + e^{i\phi}\mathbf{Z}_1 + e^{i\gamma}\mathbf{Z}_2 = \mathbf{R} \tag{3.74}$$

where $\phi$ and $\gamma$ are rotations measured from some reference position shown in dashed lines.

In order to achieve fifth-order path generation, Eq. (3.74) must be successively differentiated up to the fourth derivative with respect to time:

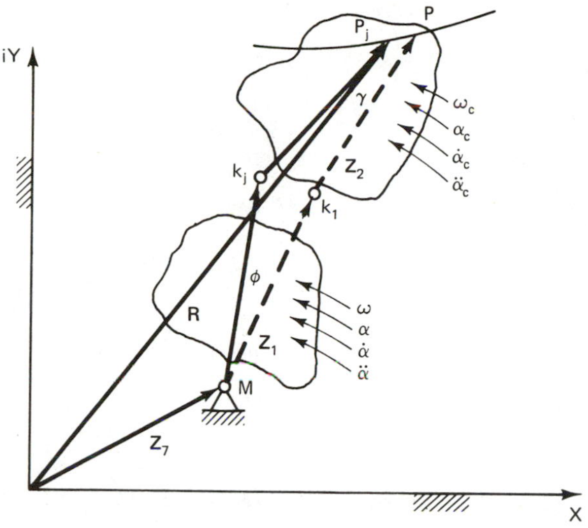


**Figure 3.51** Input side of the four-bar path generator of Fig. 3.50. Observe the *M,k,* "fifth-order Burmester Point Pair."

$$i\omega e^{i\phi}\mathbf{Z}_1 + i\omega_c e^{i\gamma}\mathbf{Z}_2 = \overset{1}{\mathbf{R}} \tag{3.75}$$

$$(i\alpha - \omega^2)\mathbf{Z}_1 e^{i\phi} + (i\alpha_c - \omega_c^2)\mathbf{Z}_2 e^{i\gamma} = \overset{2}{\mathbf{R}} \tag{3.75a}$$

$$(i\dot{\alpha} - 3\alpha\omega - i\omega^3)\mathbf{Z}_1 e^{i\phi} + (i\dot{\alpha}_c - 3\alpha_c\omega_c - i\omega_c^3)\mathbf{Z}_2 e^{i\gamma} = \overset{3}{\mathbf{R}} \tag{3.76}$$

$$\begin{aligned}[][i(\ddot{\alpha} - 6\alpha\omega^2) + \omega^4 - 4\dot{\alpha}\omega - 3\alpha^2]\mathbf{Z}_1 e^{i\phi} \\ + [i(\ddot{\alpha}_c) - 6\alpha_c\omega_c^2) + \omega_c^4 - 4\dot{\alpha}_c\omega_c - 3\alpha_c^2]\mathbf{Z}_2 e^{i\gamma} = \overset{4}{\mathbf{R}}\end{aligned} \tag{3.77}$$

where $\overset{1}{\mathbf{R}}, \overset{2}{\mathbf{R}}, \overset{3}{\mathbf{R}}, \overset{4}{\mathbf{R}}$ are the successive derivatives of $\mathbf{R}$ with respect to $t$, and

$$\omega = \frac{d\phi}{dt}, \qquad \dot{\alpha} = \frac{d\alpha}{dt}$$

$$\omega_c = \frac{d\gamma}{dt}, \qquad \dot{\alpha}_c = \frac{d\alpha_c}{dt}$$

$$\alpha = \frac{d\omega}{dt}, \qquad \ddot{\alpha} = \frac{d\dot{\alpha}}{dt}$$

$$\alpha_c = \frac{d\omega_c}{dt}, \qquad \ddot{\alpha}_c = \frac{d\dot{\alpha}_c}{dt}$$

Simplifications of Eqs. (3.74) to (3.77) are possible since a single precision point is being considered, and $P$ coincides with $P_j$ in Fig. 3.51. Therefore,

$$\phi = \gamma = 0$$

and

$$e^{i\phi} = e^{i\gamma} = 1$$

With these Eqs. (3.74) to (3.77) become

$$R = \mathbf{Z}_7 + \mathbf{Z}_1 + \mathbf{Z}_2 \tag{3.78}$$

$$\overset{1}{\mathbf{R}} = i\omega\mathbf{Z}_1 + i\omega_c\mathbf{Z}_2 \tag{3.79}$$

$$\overset{2}{\mathbf{R}} = (-\omega^2 + i\alpha)\mathbf{Z}_1 + (-\omega_c^2 + i\alpha_c)\mathbf{Z}_2 \tag{3.80}$$

$$\overset{3}{\mathbf{R}} = [-3\alpha\omega + i(\dot{\alpha} - \omega^3)]\mathbf{Z}_1 + [-3\alpha_c\,\omega_c + i(\dot{\alpha}_c - \omega_c^3)]\mathbf{Z}_2 \tag{3.81}$$

$$\begin{aligned}\overset{4}{\mathbf{R}} = [\omega^4 - 4\dot{\alpha}\omega - 3\alpha^2 + i(\ddot{\alpha} - 6\alpha\omega^2)]\mathbf{Z}_1 \\ + [\omega_c^4 - 4\dot{\alpha}_c\omega_c - 3\alpha_c^2 + i(\ddot{\alpha}_c - 6\alpha_c\omega_c^2)]\mathbf{Z}_2\end{aligned} \tag{3.82}$$

The prescribed quantities in Eqs. (3.78) to (3.82) are the position vector $\mathbf{R}$ with its time derivatives $\overset{1}{\mathbf{R}}$, $\overset{2}{\mathbf{R}}$, $\overset{3}{\mathbf{R}}$, and $\overset{4}{\mathbf{R}}$, plus $\omega$, $\alpha$, $\dot{\alpha}$, $\ddot{\alpha}$, which are the angular velocity, angular acceleration, angular shock, and angular third acceleration of the input link. The unknown quantities are the complex vectors defining the input side of the mechanism in its starting position $\mathbf{Z}_7$, $\mathbf{Z}_1$, and $\mathbf{Z}_2$ plus the unprescribed quantities $\omega_c$, $\alpha_c$, $\dot{\alpha}_c$, and $\ddot{\alpha}_c$, which are the angular velocity, angular acceleration, angular shock, and angular third acceleration of the coupler link.

The path function $y = f(x)$ is introduced into the equations by the position vector $\mathbf{R}$ and its derivatives. The first derivative of $\mathbf{R}$ is defined as

$$\overset{1}{\mathbf{R}} = \frac{d\mathbf{R}}{dt} = \frac{d\mathbf{R}}{dS}\frac{dS}{dt} \tag{3.83}$$

where $S$ represents the scalar arc length along the path, measured from some reference point on the path. The term $d\mathbf{R}/dS$ is a unit vector tangent to the path at the precision point and the term $dS/dt$ is the speed of the tracer point along the path, which is a scalar. The second through fourth derivatives of $\mathbf{R}$ are

$$\overset{2}{\mathbf{R}} = \frac{d^2\mathbf{R}}{dS^2}\left(\frac{dS}{dt}\right)^2 + \frac{d\mathbf{R}}{dS}\frac{d^2S}{dt^2} \tag{3.84}$$

$$\overset{3}{\mathbf{R}} = \frac{d^3\mathbf{R}}{dS^3}\left(\frac{dS}{dt}\right)^3 + 3\left(\frac{d^2\mathbf{R}}{dS^2}\right)\frac{dS}{dt}\frac{d^2S}{dt^2} + \frac{d\mathbf{R}}{dS}\frac{d^3S}{dt^3} \tag{3.85}$$

$$\begin{aligned}\overset{4}{\mathbf{R}} = \frac{d^4\mathbf{R}}{dS^4}\left(\frac{dS}{dt}\right)^4 + 6\left(\frac{d^3\mathbf{R}}{dS^3}\right)\left(\frac{dS}{dt}\right)^2\frac{d^2S}{dt^2} \\ + 4\left(\frac{d^2\mathbf{R}}{dS^2}\right)\frac{dS}{dt}\frac{d^3S}{dt^3} + 3\left(\frac{d^2\mathbf{R}}{dS^2}\right)\left(\frac{d^2S}{dt^2}\right)^2 + \frac{d\mathbf{R}}{dS}\frac{d^4S}{dt^4}\end{aligned} \tag{3.86}$$

The solution of the synthesis equation (3.78) to (3.82) will be accomplished by first solving Eqs. (3.79) to (3.82) for the unknown link vectors $\mathbf{Z}_1$ and $\mathbf{Z}_2$ and then returning to Eq. (3.78) for solving it for the unknown vector $\mathbf{Z}_7$.

The solution of $\mathbf{Z}_1$ and $\mathbf{Z}_2$ requires the simultaneous solution of four equations with complex coefficients linear in two complex unknowns. In order for simultaneous solutions to exist for Eqs. (3.79) to (3.82), the augmented matrix of the coefficients must be of rank 2. The augmented matrix is

$$M = \begin{bmatrix} i\omega & i\omega_c & \overset{1}{\mathbf{R}} \\ -\omega^2 + \alpha i & -\omega_c^2 + \alpha_c i & \overset{2}{\mathbf{R}} \\ -3\alpha\omega + (\dot{\alpha} - \omega^3)i & -3\alpha_c\omega_c + (\dot{\alpha}_c - \omega_c^3)i & \overset{3}{\mathbf{R}} \\ \omega^4 - 4\dot{\alpha}\omega - 3\alpha^2 + (\ddot{\alpha} - 6\alpha\omega^2)i & \omega_c^4 - 4\dot{\alpha}_c\omega_c - 3\alpha_c^2 + (\ddot{\alpha}_c - 6\alpha_c\omega_c^2)i & \overset{4}{\mathbf{R}} \end{bmatrix} \tag{3.87}$$

This will be assured by the vanishing of the following two determinants:

$$\mathbf{D}_1 = \begin{vmatrix} i\omega & i\omega_c & \overset{1}{\mathbf{R}} \\ -\omega^2 + \alpha i & -\omega_c^2 + \alpha_c i & \overset{2}{\mathbf{R}} \\ -3\alpha\omega + (\dot{\alpha} - \omega^3)i & -3\alpha_c\omega_c + (\dot{\alpha}_c - \omega_c^3)i & \overset{3}{\mathbf{R}} \end{vmatrix} = 0 \tag{3.88}$$

$$\mathbf{D}_2 = \begin{vmatrix} i\omega & i\omega_c & \overset{1}{\mathbf{R}} \\ -\omega^2 + i\alpha & -\omega_c^2 + i\alpha_c & \overset{2}{\mathbf{R}} \\ \omega^4 - 4\dot{\alpha}\omega - 3\alpha^2 + (\ddot{\alpha} - 6\alpha\omega^2)i & \omega_c^4 - 4\dot{\alpha}_c\omega_c - 3\alpha_c^2 + (\ddot{\alpha}_c - 6\alpha_c\omega_c^2)i & \overset{4}{\mathbf{R}} \end{vmatrix} = 0 \tag{3.89}$$

The two complex "compatibility equations" (3.88) and (3.89) may be solved for the four unknown reals, $\omega_c$, $\alpha_c$, $\dot{\alpha}_c$ and $\ddot{\alpha}_c$. Expanding the determinants $\mathbf{D}_1$ and $\mathbf{D}_2$ according to the elements of the second column and their cofactors, separating real and imaginary parts and employing Sylvester's dyalitic eliminant results in a sixth-degree polynomial in $\omega_c$ with real coefficients and with no constant term, which can be written as follows:

$$H\omega_c^6 + J\omega_c^5 + K\omega_c^4 + L\omega_c^3 + M\omega_c^2 + N\omega_c = 0 \tag{3.90}$$

where the coefficients $H$ to $N$ are deterministic functions of the prescribed quantities in the first and third columns of the determinants in Eqs. (3.88) and (3.89).

Factoring out the zero root results in

$$H\omega_c^5 + J\omega_c^4 + K\omega_c^3 + L\omega_c^2 + M\omega_c + N = 0 \tag{3.91}$$

The solution of (3.91) gives five roots for $\omega_c$. An examination of Eqs. (3.88) and (3.89) shows that $\omega_c = \omega$ is also a trivial root. Dividing out the $\omega_c = \omega$ root results in

$$A_4\omega_c^4 + A_3\omega_c^3 + A_2\omega_c^2 + A_1\omega_c + A_0 = 0 \tag{3.92}$$

where $A_j$, $j = 0, 1, \ldots 4$, are known real coefficients. This leaves four possible roots remaining as solutions. These four remaining roots are real roots and/or complex pairs. Only the real roots are possible solutions for $\omega_c$. Each real value of $\omega_c$ is substituted back into the real and imaginary parts of Eqs. (3.88) and (3.89) to solve any three of the resulting four real equations simultaneously for the corresponding values of $\alpha_c$, $\dot{\alpha}_c$, and $\ddot{\alpha}_c$ associated with each of the four $\omega_c$ values. Any one of these sets of $\omega_c$, $\alpha_c$, $\dot{\alpha}_c$, and $\ddot{\alpha}_c$ values are then substituted into any two of the original synthesis equations (3.79) to (3.82), from which $\mathbf{Z}_1$ and $\mathbf{Z}_2$ are determined. Then $\mathbf{Z}_7$ may be obtained by solving Eq. (3.78). Since four is the maximum number of possible real roots of the polynomial (3.92), there may exist four possible sets of input-side vectors $\mathbf{Z}_7$, $\mathbf{Z}_1$, and $\mathbf{Z}_2$. The output-side dyad, $\mathbf{Z}_3$ and $\mathbf{Z}_4$, with $\mathbf{Z}_8$ locating its ground pivot, is synthesized by the same method as outlined above, using the same $\omega_c$, $\alpha_c$, $\dot{\alpha}_c$ and $\ddot{\alpha}_c$ set as the prescribed rotation of link $\mathbf{Z}_4$, solving the compatibility equations for the rotation of $\mathbf{Z}_3$, say $\omega_3$, $\alpha_3$, $\ddot{\alpha}_3$, and $\ddot{\alpha}_3$, and then going back to solve for the dyad and ground-pivot vectors. It can be shown that this procedure will yield the same set of values for these vectors as those found for the input-side dyad. Thus, since each of the four such dyads can be combined with any of the other three to form the path generator FBL, there will be 12 such FBLs: up to 12 possible solutions for this higher-order path generation synthesis with prescribed timing, just as for the five-finitely-separated-precision-point case described in Sec. 3.9.

**Example 3.4**

The solution of the synthesis equations and the analysis of the synthesized linkages were carried out on the IBM 360 digital computer by programs based on the foregoing equations. An example of a solution is given in Table 3.7 and Fig. 3.52.

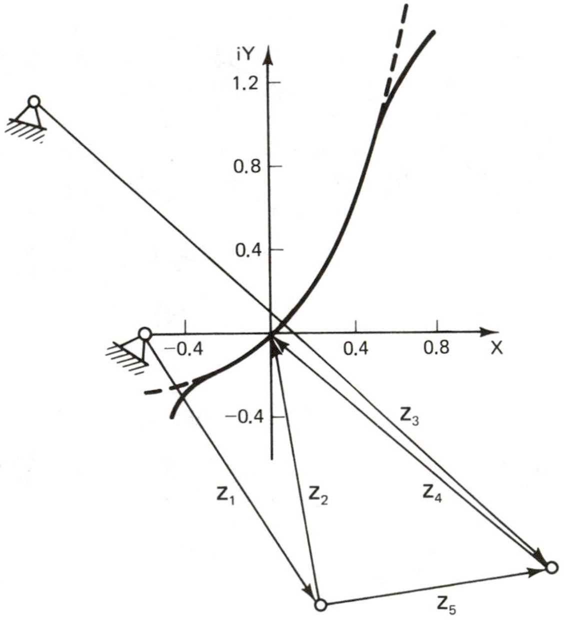


**Figure 3.52** Example of Table 3.7. A four-bar higher-order path generator synthesized for the path $y = xe^x$. The ideal path is shown dashed and the generated path is shown solid.

**TABLE 3.7** THE MECHANISM SHOWN IN FIG. 3.52[a], SYNTHESIZED FOR PATH GENERATION WITH FIVE INFINITESIMALLY CLOSE POSITIONS (FIFTH-ORDER APPROXIMATION WITH PRESCRIBED TIMING)

Path: $y = xe^x$

Precision point: $\mathbf{R} = (0.000, 0.000)$

$\omega = 1, \quad \alpha = \dot{\alpha} = \ddot{\alpha} = 0$ (constant-velocity input crank)

$$\frac{ds}{dt} = 1, \qquad \frac{d^2s}{dt^2} = 1, \qquad \frac{d^3s}{dt^3} = \frac{d^4s}{dt^4} = 0$$

Link vectors:

| | | |
|---|---|---|
| $\mathbf{Z}_1 =$ | 0.81691 | −1.33976*i* |
| $\mathbf{Z}_2 =$ | −0.23055 | 1.32842*i* |
| $\mathbf{Z}_4 =$ | −1.34553 | 1.15162*i* |
| $\mathbf{Z}_3 =$ | 2.43881 | −2.27172*i* |

[a] In Fig. 3.52 the generated path appears to depart from the ideal path on the same side at both ends of the curve. However, in reality, the generated curve does depart the ideal curve on opposite sides, which would indicate an even number of (infinitesimally close) precision points. This, however, is not the case, because in the positive *x* and positive *y* quadrant the departure on the positive *y* side is so slight that it is detectable only in the numerical values of the computer output.

If the motion $\omega_c$, $\alpha_c$, $\dot{\alpha}_c$, and $\ddot{\alpha}_c$ of link $\mathbf{Z}_2$ is prescribed together with $\overset{m}{\mathbf{R}}$, $m = 0, 1, 2, 3, 4$, the preceding synthesis would be motion generation synthesis of the $\mathbf{Z}_1$, $\mathbf{Z}_2$ dyad with five infinitesimally close prescribed positions. Figure 3.51 shows one of the up-to-four *M, $k_1$ Fifth-Order Burmester Point Pairs* associated with such a dyad. Using one other of the up-to-four such dyads together with the first, a four-bar higher-order motion generator is obtained. There are up to six such four-bar mechanisms, with a total of 12 cognates. The cognates are higher-order path generators with prescribed timing. The geared, parallelogram-connected and tape- or chain-connected five-bar path generators, as well as the parallel-motion generator discussed for five finitely separated (discrete) prescribed positions, can all be adapted for higher-order synthesis with the method of this section, as can the many multiloop linkage mechanisms presented in the preceding sections.

## Position-Velocity Synthesis of a Geared Five-Bar Linkage

The geared-five bar linkage of Fig. 3.41 was synthesized for four finitely separated positions of function generation. According to Table 3.5, this linkage can be designed for seven total positions but the five-, six-, and seven-position cases will involve compatibility equations.

Following the logic laid out in Sec. 2.24, the number of prescribed positions in column 1 of Table 3.5 may be either finitely or infinitesimally separated. Note, however, that an acceleration involving three infinitesimally close positions may not be prescribed without the position and velocity (two infinitesimally close positions)

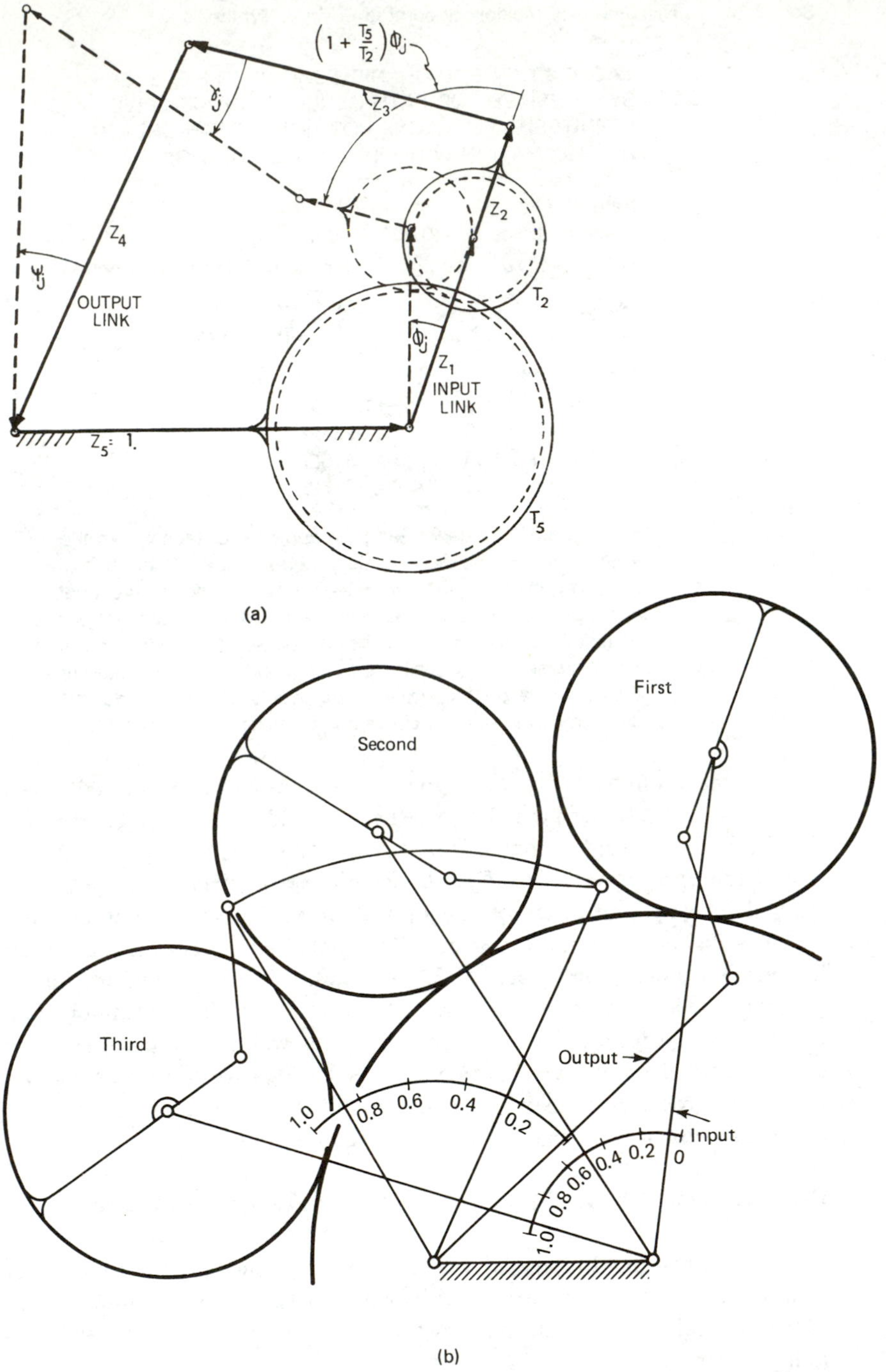


**Figure 3.53** (a) geared five-bar function generator is shown in the first precision position (solid). The $j$th position (dashed) corresponds to a rotation of the input link $\mathbf{Z}_1$ by $\phi_j$; (b) scale drawing of Ex. 1, Table 3.8 in its three Chebyshev-spaced second-order precision positions.

at that location also being prescribed. Figure 3.53a shows another form of a geared five-bar with $F = 1$, since there is a geared constraint between link 1 and link 2, forming a *cycloidal crank*. Let us write the equations and synthesize this geared five-bar for six mixedly separated prescribed positions of function generation—three pairs of prescribed corresponding positions and three prescribed corresponding velocity pairs for the input and output links [85].

We will use the following notation for the rotational operators:

$$\text{Given:} \quad \begin{cases} \boldsymbol{\lambda}_j = e^{i\phi_j} \\ \boldsymbol{\epsilon}_j = e^{i(1+(T_5/T_2)\phi_j} \\ \boldsymbol{\mu}_j = e^{i\psi_j} \end{cases}$$

$$\text{Unknown:} \quad \boldsymbol{\nu}_j = e^{i\gamma_j}$$

The equation of closure for this mechanism may be written as follows:

$$\boldsymbol{\lambda}_j \mathbf{Z}_1 + \boldsymbol{\epsilon}_j \mathbf{Z}_2 + \boldsymbol{\nu}_j \mathbf{Z}_3 + \boldsymbol{\mu}_j \mathbf{Z}_4 = -1, \qquad j = 1, 2, 3 \tag{3.93}$$

The first derivative loop equation

$$\dot{\boldsymbol{\lambda}}_j \mathbf{Z}_1 + \dot{\boldsymbol{\epsilon}}_j \mathbf{Z}_2 + \dot{\boldsymbol{\nu}}_j \mathbf{Z}_3 + \dot{\boldsymbol{\mu}}_j \mathbf{Z}_4 = 0, \qquad j = 1, 2, 3 \tag{3.94}$$

where the superior dot represents differentiation with respect to the input-crank rotation. Here

$$\dot{\boldsymbol{\lambda}}_j = i\left(\frac{d\phi}{d\phi}\right)_j \boldsymbol{\lambda}_j = i\boldsymbol{\lambda}_j$$

$$\dot{\boldsymbol{\epsilon}}_j = i\left(1 + \frac{T_5}{T_2}\right)\boldsymbol{\epsilon}_j$$

$$\dot{\boldsymbol{\mu}}_j = i\left(\frac{d\psi}{d\phi}\right)_j \boldsymbol{\mu}_j$$

are known from prescribed data for $j = 1, 2, 3$, and

$$\dot{\boldsymbol{\nu}}_j = i\left(\frac{d\gamma}{d\phi}\right)_j \boldsymbol{\nu}_j$$

are unknown. Note that for $j = 1$, $\phi_1 = \psi_1 = \gamma_1 = 0$, and therefore $\lambda_1 = \epsilon_1 = \nu_1 = \mu_1 = 1$. The augmented matrix of systems (3.93) and (3.94) in full array is as follows:

$$M = \begin{bmatrix} 1 & 1 & 1 & 1 & -1 \\ \boldsymbol{\lambda}_2 & \boldsymbol{\epsilon}_2 & \boldsymbol{\nu}_2 & \boldsymbol{\mu}_2 & -1 \\ \boldsymbol{\lambda}_3 & \boldsymbol{\epsilon}_3 & \boldsymbol{\nu}_3 & \boldsymbol{\mu}_3 & -1 \\ i & \dot{\boldsymbol{\epsilon}}_1 & i\dot{\gamma}_1 & \dot{\boldsymbol{\mu}}_1 & 0 \\ i\boldsymbol{\lambda}_2 & \dot{\boldsymbol{\epsilon}}_2 & i\dot{\gamma}_2\boldsymbol{\nu}_2 & \dot{\boldsymbol{\mu}}_2 & 0 \\ i\boldsymbol{\lambda}_3 & \dot{\boldsymbol{\epsilon}}_3 & i\dot{\gamma}_3\boldsymbol{\nu}_3 & \dot{\boldsymbol{\mu}}_3 & 0 \end{bmatrix} \tag{3.95}$$

This matrix must be of rank 4 to ensure that system (3.93)–(3.94) yields simultaneous solutions for $\mathbf{Z}_k$. Thus if we say that

$$D_1 = \det(M)_{1,2,3,4,5} = 0 \tag{3.96}$$

$$D_2 = \det(M)_{1,2,3,4,6} = 0 \tag{3.97}$$

where the subscripts designate row numbers, we can regard system (3.96)–(3.97) as the compatibility system. This system, containing four real equations, can be solved for four real unknowns. Elements in the first, second, and fourth columns of Eq. (3.95) are known from prescribed data. Column 3, however contains five unprescribed reals. If we assume $\dot{\gamma}_1$ arbitrarily, this leaves the four unknown reals $\gamma_2$, $\gamma_3$, $\dot{\gamma}_2$, and $\dot{\gamma}_3$. Expanding $D_1$ and $D_2$, each according to the elements of the third column and their cofactors, we get

$$D_1 = \mathbf{\Delta}_1 + \mathbf{\Delta}_2\boldsymbol{\nu}_2 + \mathbf{\Delta}_3\boldsymbol{\nu}_3 + \mathbf{\Delta}_4 i\dot{\gamma}_1 + i\mathbf{\Delta}_5\dot{\gamma}_2\boldsymbol{\nu}_2 = 0 \tag{3.98}$$

$$D_2 = \mathbf{\Delta}'_1 + \mathbf{\Delta}'_2\boldsymbol{\nu}_2 + \mathbf{\Delta}'_3\boldsymbol{\nu}_3 + \mathbf{\Delta}'_4 i\dot{\gamma}_1 + i\mathbf{\Delta}_5\dot{\gamma}_3\boldsymbol{\nu}_3 = 0 \tag{3.99}$$

where the $\mathbf{\Delta}$'s are the appropriate cofactors, known from prescribed data. Divide Eq. (3.98) by $i\mathbf{\Delta}_5\boldsymbol{\nu}_2$ and Eq. (3.99) by $i\mathbf{\Delta}_5\boldsymbol{\nu}_3$, obtaining

$$\frac{\mathbf{\Delta}_1}{i\mathbf{\Delta}_5}\boldsymbol{\nu}_2^{-1} + \frac{\mathbf{\Delta}_2}{i\mathbf{\Delta}_5} + \frac{\mathbf{\Delta}_3}{i\mathbf{\Delta}_5}\boldsymbol{\nu}_2^{-1}\boldsymbol{\nu}_3 + \frac{\mathbf{\Delta}_4 i\dot{\gamma}_1}{i\mathbf{\Delta}_5}\boldsymbol{\nu}_2^{-1} + \dot{\gamma}_2 = 0 \tag{3.100}$$

$$\frac{\mathbf{\Delta}'_1}{i\mathbf{\Delta}_5}\boldsymbol{\nu}_3^{-1} + \frac{\mathbf{\Delta}'_2}{i\mathbf{\Delta}_5}\boldsymbol{\nu}_2\boldsymbol{\nu}_3^{-1} + \frac{\mathbf{\Delta}'_3}{i\mathbf{\Delta}_5} + \frac{\mathbf{\Delta}'_4 i\dot{\gamma}_1}{i\mathbf{\Delta}_5}\boldsymbol{\nu}_3^{-1} + \dot{\gamma}_3 = 0 \tag{3.101}$$

Simplifying the notation of known quantities, we rewrite Eq. (3.100):

$$\mathbf{a}_1 + \mathbf{a}_2\boldsymbol{\nu}_2^{-1} + \mathbf{a}_3\boldsymbol{\nu}_2^{-1}\boldsymbol{\nu}_3 + \dot{\gamma}_2 = 0 \tag{3.102}$$

The complex conjugate of Eq. (3.102) is

$$\bar{\mathbf{a}}_1 + \bar{\mathbf{a}}_2\boldsymbol{\nu}_2 + \bar{\mathbf{a}}_3\boldsymbol{\nu}_2\boldsymbol{\nu}_3^{-1} + \dot{\gamma}_2 = 0 \tag{3.103}$$

Subtracting Eq. (3.102) from Eq. (3.103) and dividing by $2i$ we obtain*

$$a_{1y} + a_{2y}\cos\gamma_2 - a_{2x}\sin\gamma_2 + a_{3y}\cos(\gamma_3 - \gamma_2) - a_{3x}\sin(\gamma_3 - \gamma_2) = 0 \tag{3.104}$$

Similarly, combining Eq. (3.101) with its complex conjugate and using primed symbols for the known factors yields

$$a'_{1y} + a'_{2y}\cos\gamma_3 - a'_{2x}\sin\gamma_3 + a'_{3y}\cos(\gamma_3 - \gamma_2) + a'_{3x}\sin(\gamma_3 - \gamma_2) = 0 \tag{3.105}$$

In Eqs. (3.104) and (3.105) all $a$ and $a'$ values are real deterministic functions of the known coefficients in columns 1, 2, and 4 of the matrix $M$. To simplify Eqs. (3.104) and (3.105) we use the following identities:

$$\cos(\gamma_3 - \gamma_2) = \cos\gamma_2\cos\gamma_3 + \sin\gamma_2\sin\gamma_3$$

* Alternatively, we can say that, since $\dot{\gamma}_2$ is real, the imaginary part of Eq. (3.102) is zero, which also leads to Eq. (3.104).

$$\sin(\gamma_3 - \gamma_2) = \sin\gamma_3 \cos\gamma_2 - \cos\gamma_3 \sin\gamma_2$$

$$\cos\gamma_j = \frac{1-\tau_j^2}{1+\tau_j^2}, \qquad \sin\gamma = \frac{2\tau_j}{1+\tau_j^2}$$

where

$$\tau_j = \tan\frac{\gamma_j}{2}$$

With these we rewrite Eq. (3.104):

$$a_{1y} + a_{2y}\frac{1-\tau_2^2}{1+\tau_2^2} - a_{2x}\frac{2\tau_2}{1+\tau_2^2} + a_{3y}\left[\frac{1-\tau_2^2}{1+\tau_2^2}\,\frac{1-\tau_3^2}{1+\tau_3^2} + \frac{4\tau_2\tau_3}{(1+\tau_2^2)(1+\tau_3^2)}\right] - a_{3x}\frac{2\tau_3(1-\tau_2^2) - 2\tau_2(1-\tau_3^2)}{(1+\tau_2^2)(1+\tau_3^2)} = 0 \tag{3.106}$$

Multiply by $(1 + \tau_2^2)(1 + \tau_3^2)$:*

$$a_{1y}(1+\tau_2^2)(1+\tau_3^2) + a_{2y}(1-\tau_2^2)(1+\tau_3^2) - a_{2x}2\tau_2(1+\tau_3^2) + a_{3y}[(1-\tau_2^2)(1-\tau_3^2) + 4\tau_2\tau_3] - 2a_{3x}[\tau_3(1-\tau_2^2) - \tau_2(1-\tau_3^2)] = 0 \tag{3.107}$$

Expanding, we get

$$a_{1y}(1+\tau_2^2+\tau_3^2+\tau_2^2\tau_3^2) + a_{2y}(1-\tau_2^2+\tau_3^2-\tau_2^2\tau_3^2) - 2a_{2x}(\tau_2+\tau_2\tau_3^2) + a_{3y}[(1-\tau_2^2-\tau_3^2+\tau_2^2\tau_3^2) + 4\tau_2\tau_3] - 2a_{3x}(\tau_3-\tau_2^2\tau_3-\tau_2+\tau_2\tau_3^2) = 0 \tag{3.108}$$

Arrange in descending powers of $\tau_3$:

$$\tau_3^2[a_{1y}(1+\tau_2^2) + a_{2y}(1-\tau_2^2) - 2a_{2x}\tau_2 - a_{3y}(1-\tau_2^2) - 2a_{3x}\tau_2] + \tau_3[4a_{3y}\tau_2 - 2a_{3x}(1-\tau_2^2)] + \tau_3^0[a_{1y}(1+\tau_2^2) + a_{2y}(1-\tau_2^2) - 2a_{2x}\tau_2 + a_{3y}(1-\tau_2^2) + 2a_{3x}\tau_2] = 0 \tag{3.109}$$

This is in the form

$$\tau_3^2P_2 + \tau_3P_1 + P_0 = 0 \tag{3.110}$$

where $P_j$ ($j = 0, 1, 2$) denote second-degree polynomials in $\tau_2$ with real coefficients.

Similarly, from Eq. (3.105) we obtain

$$\tau_3^2\pi_2 + \tau_3\pi_1 + \pi_0 = 0 \tag{3.111}$$

where $\pi_j$ ($j=0, 1, 2$) also denote second-degree polynomials in $\tau_2$ with real coefficients.

We eliminate $\tau_3$ by writing Sylvester's dyalitic eliminant:†

* Multiplication through by $(1 + \tau_2^2)(1 + \tau_3^2)$ introduces the extraneous roots of $\pm i$ for $\tau_2$ and $\tau_3$.

† Multiply Eqs. (3.110) and (3.111) by $\tau_3$, yielding two more equations, which are third-degree polynomials in $\tau_3$. The resulting four equations will have simultaneous solutions for $\tau_3^{(n)}$, $n = 0, 1, 2, 3$, if the augmented matrix of the coefficients is of rank 4, which leads to Eq. (3.112).

$$S_1 = \begin{vmatrix} 0 & P_2 & P_1 & P_0 \\ 0 & \pi_2 & \pi_1 & \pi_0 \\ P_2 & P_1 & P_0 & 0 \\ \pi_2 & \pi_1 & \pi_0 & 0 \end{vmatrix} = 0 \tag{3.112}$$

Expanding, we obtain an eighth-degree polynomial in $\tau_2$ (since every element of $S_1$ is a quadratic in $\tau_2$) with two trivial roots ($\pm i$).

Similarly, by eliminating $\tau_2$ from the system of Eqs. (3.104) and (3.105) we obtain an octic in $\tau_3$, whose solutions also include four trivial roots: $\pm i$ and

$$\tau_3 = \tan\left(\frac{\phi_3}{2}\right) \qquad \text{and} \qquad \tau_3 = \tan\left(\frac{\psi_3}{2}\right)$$

Corresponding simultaneous values of the nontrivial roots of $\tau_2$ and $\tau_3$ can be identified by direct substitution in systems (3.110) and (3.111) of one $\tau_2$ value at a time. This will yield two roots for $\tau_3$ from Eq. (3.110) and two roots for $\tau_3$ from equation (3.111), one of which will be a common root satisfying the system (3.110)–(3.111) and identical with one of the nontrivial roots of the octic in $\tau_3$. Indeed, the procedure of the foregoing paragraph need not be performed to find these roots. Instead, any two different nontrivial real roots of Eq. (3.112) can be used as $\tau_2$ and $\tau_3$. The maximum number of simultaneous nontrivial $\tau_2$, $\tau_3$ pairs is six.

Thus having found up to six pairs of values for $\gamma_2$ and $\gamma_3$, each such pair can be substituted back in Eqs. (3.108) and (3.109) to find corresponding values for $\dot{\gamma}_2$ and $\dot{\gamma}_3$.

Any one of such sets of solutions for $\gamma_2$, $\gamma_3$, $\dot{\gamma}_2$, and $\dot{\gamma}_3$ can then be substituted back into the original system of Eqs. (3.93) and (3.94), any four of which can then be solved simultaneously for the mechanism dimensions in the starting position, namely $\mathbf{Z}_k$, $k = 1, 2, 3, 4$, yielding up to *six different* designs.

**Example 3.5**

Table 3.8 lists several results of the geared five-bar function generator synthesized for three second-order precision points. Although respacing for optimal error has not been employed, the maximum error in several examples (2, 4, 5, 6) is considerably smaller than the optimum four-bar synthesized by Freudenstein [104] for the identical set of parameters (i.e., the function, range, and scale factors). In some cases (1, 3) the output provides a "near fit" to the ideal function for a sector of the range. An extra first-order precision point has contributed to the accuracy of the function generator in examples (1, 4).

Specifying a second-order precision point and receiving a third-order precision point does not seem to be unlikely. Example 3 has one second-order and two third-order precision points.

Since a second-order precision point is actually two precision points infinitesimally close to one another, the actual curve approaches and departs from the ideal curve without crossing it. Second-order precision points have applications wherever the first derivative of a function (tangent to a curve) must be reproduced exactly. As suggested by McLarnan [181], if all the precision points are second order, the maximum error can be halved by shifting the ideal curve by one-half the maximum error. Examples 2a, 5a, and 6a each have only second-order precision points which are shifted in parts b to halve the maximum error. The negative gear ratios, examples 3 and 4 denote that the gears lie on the same side of their common tangent (a hypocycloidal-crank mechanism) (see Fig. 3.40).

Example 1 is shown in Fig. 3.53b in its three Chebyshev-spaced second-order precision positions. The output generates $x^2$, corresponding to an input of $x$ for $0 \leq x \leq 1$. The range of both the input and output is 90°. An arbitrary gear ratio of 2 and $\dot{\gamma}_1$ of 1.5 gave an extra precision point at $x = 0.2299$. This extra precision point causes the error for nearly 60% of the range to be less than 0.0176°.

Example 4 is a special case where $\mathbf{Z}_2$ is for all practical purposes zero. Thus we have a four-bar mechanism which, because of unprescribed precision points at $x = 0.0666$ and $x = 0.2157$, has eight precision conditions (one third-order, one first-order, and two second-order precision points). Note that the maximum error over the entire range is 0.0584°, while over 95% of the range the error is less than one-tenth of the maximum error of the optimal four-bar linkage of Ref. 104.

## APPENDIX: A3.1 The LINCAGES Package*

It is not the purpose here to fully explain all the options of the interactive subroutines of the LINCAGES package [78,83,84,218,270]. Some of the subroutines will be illustrated, however, by way of an example. Since there are two solutions for each choice of $\beta_2$ (Fig. 3.2) (for which the compatibility linkage closes (Fig. 3.3), each synthesized dyad is designated by a $\beta_2$ value (0° to 360°) and a *set number* (1 or 2) indicating whether it is the first or second of the two available solutions for $\beta_3$ and $\beta_4$ for the particular $\beta_2$.

**Example 3.6**

The assembly of a filter product begins by forming the filtration material into what is known as a filter blank. Next the filter blank is placed by hand onto a mandrel. This mandrel is part of a machine that completes the assembly of the filter. The objective of this problem is to design a four-bar linkage mechanism for removing the filter blanks from the hopper and transferring them to the mandrel.

Figure 3.54 diagrams the design objective. A gravity-feed hopper holds the semi-cylindrical filter blanks with the diametral plane surface initially at a 27° angle from vertical. The blank must be rotated until this diametral plane is horizontal on the mandrel. The position of the hopper can be located within the sector indicated, although the angle must remain at 27°.

At the beginning of the "pick and place" cycle, it is desirable to pull the blank in a direction approximately perpendicular to the face of the hopper. To prevent folding the filter blank on the mandrel, it is necessary to have the rotation of the blank completed at a position of approximately 2 cm above the mandrel and then translate without rotation onto the mandrel. The motion of the linkage should then reverse to remove the completed filter from the mandrel and eject it onto a conveyor belt. After this, the linkage should return to the hopper and pick another blank. Because of the requirement of both forward and reverse motions over the same path, a crank-rocker would have no real advantage over a double-rocker linkage.

An acceptable linkage solution (a four-bar chain is desirable here) must have

* Available from the second author.

**TABLE 3.8** GEARED FIVE-BAR FUNCTION GENERATOR[a] OF FIG. 3.53, EX. 3.5.

| Mech-anism number | Function $y^{b}$ | Range $x_0 \le x \le x_4$ | Scale factors[a] $\Delta\phi$, $\Delta\psi$ | Gear ratio | Angular velocity[a] $\dot{\gamma}_1$ | **Z** vectors in initial position[a] of geared five-bar | Precision points of geared five-bar at $x =$ | Geared five-bar maximum error | Four-bar maximum error (Freudenstein) | Remarks on geared five-bar performance | Shape of error curve of geared five-bar |
|---|---|---|---|---|---|---|---|---|---|---|---|
| 1 | $x^2$ | $0 \le x \le 1$ | 90, 90 | 2 | 1.5 | $0.2759 + 2.291i$<br>$-0.1451 - 0.3883i$<br>$0.2270 - 0.638i$<br>$-1.358 - 1.264i$ | 0.0666[c], 0.2299[d], 0.5000[c], 0.9333[c], | 0.0802° | 0.0673° | Extra precision point at $x = 0.2299$; for $0.03 \le x \le 6$, error $< 0.0176°$ | 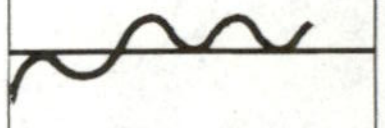 |
| 2a | $x^2$ | $0 \le x \le 1$ | 90, 90 | 2 | 1.5 | $-3.340 + 2.131i$<br>$-0.2015 + 0.1525i$<br>$2.638 - 1.671i$<br>$-0.0968 - 0.6122i$ | 0.0666[c], 0.5000[c], 0.9333[c] | 0.0708° | 0.0673° | All precision points are second-order |  |
| 2b | $x^2$ | $0 \le x \le 1$ | 90, 90 | 2 | 1.5 | Same | 0.0169[e], 0.1517[e], 0.3611[e], 0.6409[e], 0.8403[e], 0.9788[e] | 0.0354° | 0.0673° | Shifting precision points of example 2a yields 1/2 the maximum error; $y_{2b} = y_{2a} + 1/2\,\lvert\theta_{max}\rvert(2a)$ |  |
| 3 | $x^2$ | $0 \le x \le 1$ | 90, 90 | −2 | 1.0 | $0.2946 + 1.396i$<br>$-0.0148 - 0.1921i$<br>$-0.0918 - 0.7607i$<br>$-1.188 - 0.4433i$ | 0.0666[c, f]<br>0.5000[c, f]<br>0.9333[c] | 0.3816° | 0.0673° | For $0.33 < x \le 1.0$, error $< .011$; for $0.40 < x < 0.56$ error $< 0.001$ |  |
| 4 | $x^{1.5}$ | $0 \le x \le 1$ | 90, 90 | −2 | 1.5 | $-3.803 + 4.852i$<br>$0.0071 - 0.0052i$<br>$2.424 - 2.659i$<br>$0.3712 - 2.187i$ | 0.0666[c, f]<br>0.2157[d], 0.5000[c]<br>0.9333[c] | 0.0584° | 0.146° | Extra precision point at 0.2157; for $0.05 \le x \le 1.0$, error $< 0.0138°$ |  |
| 5a | $x^{2.5}$ | $0 \le x \le 1$ | 90, 90 | 2 | 1.5 | $-3.511 + 1.435i$<br>$-0.3032 + 0.2282i$<br>$2.951 - 1.405i$<br>$-0.1368 - 0.2573i$ | 0.0666[c]<br>0.5000[c]<br>0.9333[c] | 0.0951° | 0.412° | All precision points are second-order |  |

| | | | | | | | | | | | |
|---|---|---|---|---|---|---|---|---|---|---|---|
| 5b | $x^{2.5}$ | $0 \le x \le 1$ | 90, 90 | 2 | 1.5 | Same | 0.0142[e], 0.1540[e], 0.3763[e], 0.6152[e], 0.8582[e], 0.9829[e] | 0.0476° | 0.412° | Shifting (5a) yields ½ maximum error |  |
| 6a | $x^3$ | $0 \le x \le 1$ | 90, 90 | 2 | 1.5 | $-3.615+1.251i$<br>$-0.3235+0.2889i$<br>$3.059-1.411i$<br>$-0.1193-0.1292i$ | 0.0666[c], 0.5000[c], 0.9333[c] | 0.1281° | 0.566° | All precision points are second-order | 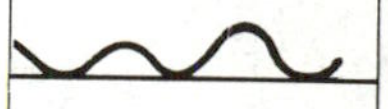 |
| 6b | $x^3$ | $0 \le x \le 1$ | 90, 90 | 2 | 1.5 | Same | 0.6338[d]<br>0.8359[d] | 0.0641° | 0.566° | Shifting (6a) yields ½ maximum error |  |

[a] The units of the scale factors are in degrees per unit of $x$. The velocity $\dot{\gamma}_1$ is in radians per second. The **Z** vectors are (from the top down) $\mathbf{Z}_2$, $\mathbf{Z}_3$, $\mathbf{Z}_4$, and $\mathbf{Z}_5$, where $\mathbf{Z}_1 = 1$.

[b] $x^p$ function generator also good for $x^{1/p}$.

[c] Precision point derived from Chebyshev spacing (second-order precision points).

[d] There is an extra, unprescribed first-order precision point present.

[e] These are precision points obtained by shifting the error curve of example 2a upward by ½ of the maximum error of that example.

[f] These turned out to be gratuitous third-order precision points.

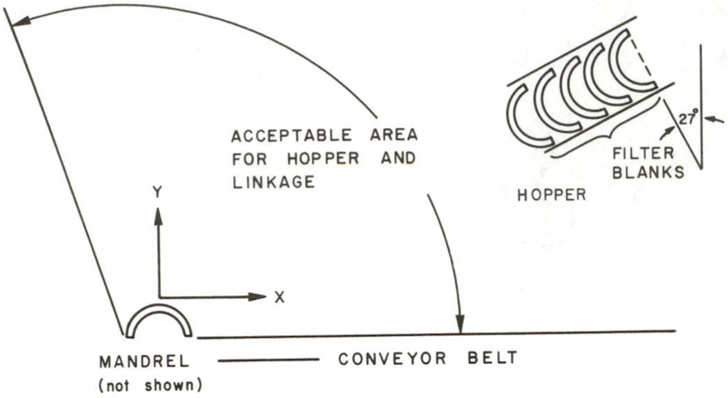


**Figure 3.54** The prescribed task of motion generation for Ex. 3.6: Filter blanks are to be taken from the hopper and placed on the mandrel.

ground pivots and linkage motions within areas that do not interfere with the hopper and the filter assembler. Also, since the resulting linkage may be driven by an added dyad (to provide a fully rotating input), the total angular travel of the input link of the four-bar synthesized here should be minimized so as to obtain acceptable transmission angles for the entire mechanism, including the driving crank and connecting link, formed by the added dyad, which actuates the input of the motion-generating four-bar linkage.

This example is a typical challenging problem that often faces linkage designers in practice. Some of the constraints are firm, whereas others can vary within some specified range. This means (mathematically) a number of infinities of solution possibilities. The computer graphics screen is an ideal tool to help survey a large number of possible solutions.

## Method of Solution

The problem clearly requires motion-generation synthesis (or rigid-body guidance), in which the position and angle of the filter blank is specified at different precision points. Four points along a specified path and four corresponding angular positions were chosen.

The first set of precision points chosen are shown in Table 3.9. The mandrel

**TABLE 3.9** FIRST ATTEMPT AT SYNTHESIZING A FOUR-BAR MOTION GENERATOR WITH FOUR PRECISION POINTS (EX. 3.6, FIG. 3.54).

| Position | *X* coordinate (cm) | *Y* coordinate (cm) | Rotation (deg) |
|---|---|---|---|
| 1 | 0 | 0 | 0 |
| 2 | 1 | 7 | 0 |
| 3 | 17 | 18 | 60 |
| 4 | 38 | 21 | 117 |

was designated position 1, while the second position was picked above the mandrel with no rotation (to prevent folding the filter blank). The third position was chosen to be about halfway between the second and fourth positions with approximately half the required rotation. The fourth position corresponded to the angle and position of the hopper.

A portion of the $M$–$K$ curves (center- and circle-point curves) is shown in Fig. 3.55. The solid and short-dashed lines represent the portions of the center-point curve from sets 1 and 2 of the $\beta_j$ ($j = 2, 3, 4$), respectively, while the long-dashed and dash-dotted lines are the circle-point curves from these sets. Figure 3.56 shows an $M$–$K$ curve for set 1 solutions only, where the $\beta_2$ values are correlated to their $m$–$k_1$ positions on the curves by letters corresponding to the table along the left-hand side of the figure. For example, letter $B$ represents $m$ and $k_1$ points for $\beta_2 = 30°$. The results of interactively locating ground pivots and moving pivots (by using the crosshairs on the graphics screen which can be positioned by the operator to indicate his choice of a point on the $M$ or $K$ curve), corresponding to $\beta_2 = 330°$ and $\beta_2 = 30°$, are also shown in Fig. 3.56. As the designer locates a ground pivot with the crosshairs, the computer finds the moving pivot of the dyad and draws lines on the screen to represent the dyad. These two dyads formed what looked to be an acceptable four-bar solution to be further analyzed. Coupler curves of four-bars picked are generated by another subroutine. Figure 3.57 shows that the coupler curve shifts to the left between points 1 and 2, approaches precision point 4 vertically, and has a cusp at point 3. These are all unacceptabe characteristics. Although this linkage is not useful, what about the others which also satisfy the same set of prescribed precision points?

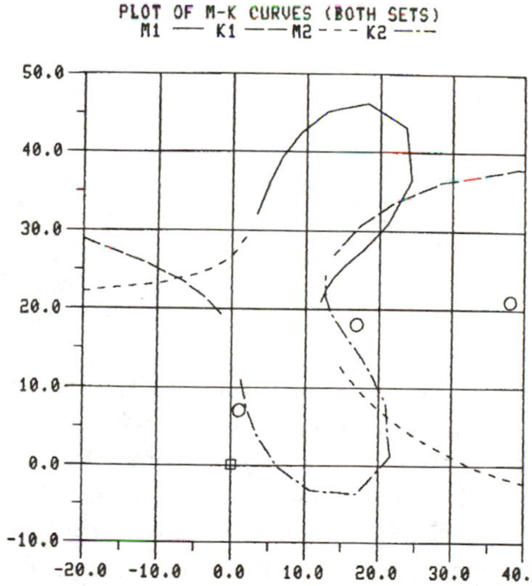


**Figure 3.55** Plot of $M$–$K$ curves (both sets), Ex. 3.6, Fig. 3.54, for the four precision points specified in Table 3.9.

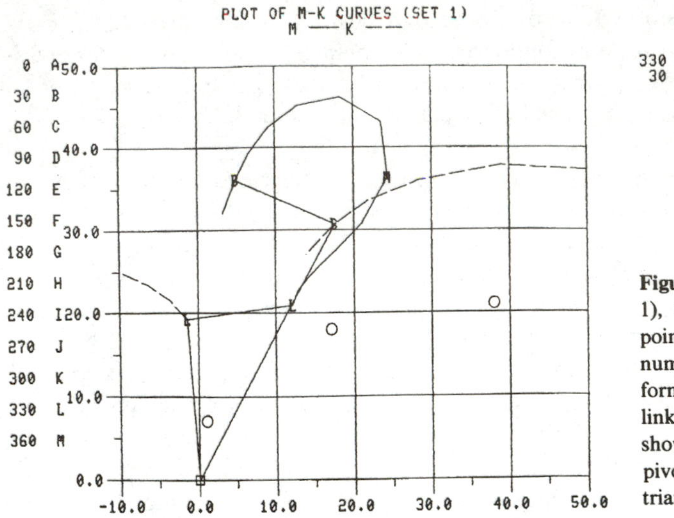


**Figure 3.56** Plot of *M–K* curves (set 1), Ex. 3.6, Fig. 3.54, for precision points of Table 3.9. One of an infinite number of four-bar linkage mechanisms formed from two *M–K* dyads: the fixed link is LB (nearly vertical, line not shown). BB and LL are ground-pivoted links and $LP_1B$ is the coupler triangle, with coupler LB not shown.

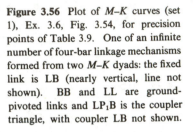

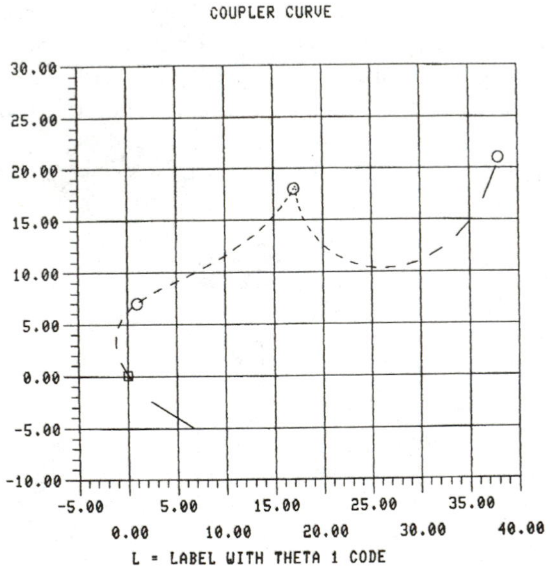


**Figure 3.57** Coupler curve of the four-bar mechanism of Fig. 3.56 (Table 3.9).

Figure 3.58 shows a copy of one of the other options available (the so-called "BETAS") in the LINCAGES package, which is helpful in surveying possible dyads. Grounded link (**W**) rotations $\beta_3$ and $\beta_4$ from sets 1 and 2 (BTA31, BTA21, BTA41, and BTA42) are plotted here against $\beta_2$. Notice that there are no values of $\beta_3$, $\beta_4$ between $30° < \beta_2 < 330°$. This shows that the "compatibility linkage" does not close for this range of $\beta_2$. Another useful design characteristic of this plot is the ability now to pick dyads in which link **W** exhibits constant directional rotation between precision points 1 to 4. For example, the solution corresponding to $\beta_2 =$

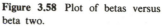

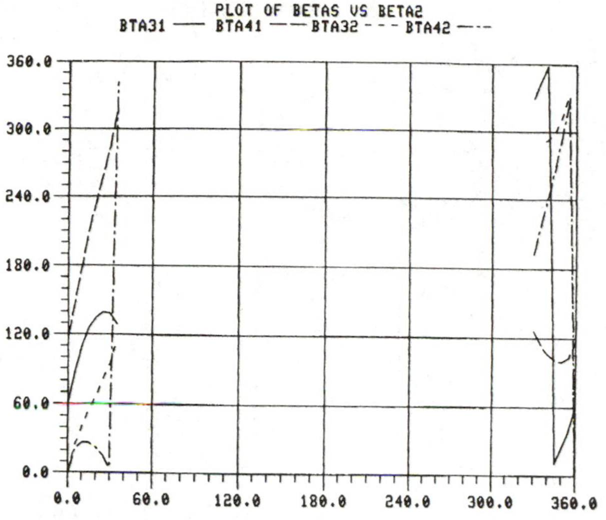


**Figure 3.58** Plot of betas versus beta two.

10° from set 1 has constant directional rotations with $\beta_3 = 110°$ and $\beta_4 = 180°$ (see Fig. 3.58).

After investigating a representative number of other four-bar solutions without successfully improving the coupler path trajectories, another set of precision points was picked. An attempt was made to position the hopper above and closer to the mandrel (Table 3.10).

The "TABLE" subroutine of the "LINCAGES" package helps to survey a large number of four-bar combinations (in sort of a "shotgun" manner) as to whether they branch between precision points and what the resulting maximum link-length ratio is. Figure 3.59 shows a sample TABLE from the second choice of precision points. Twelve dyads corresponding to user specified $\beta_2$'s from set 1 or set 2 form the vertical and horizontal axes of the table (both dyad sets are from set 2 of the $\beta_2$'s in Fig. 3.59). Each of the 36 boxes of the matrix lists a measure of the mobility (see Sec. 3.1 of Vol. 1) and the link-length ratio of the resulting four-bar mechanism made up of these dyads. Table 3.11 lists the mobility abbreviations generated by this table. Regions that showed promise were expanded, with an eye toward the

**TABLE 3.10** SECOND SET OF PRECISION POINTS FOR THE FOUR-BAR MOTION GENERATOR OF EX. 3.6, SELECTED AFTER THE HOPPER (FIG. 3.54) WAS MOVED CLOSER TO AND DIRECTLY ABOVE THE MANDREL.

| Position | *X* coordinate (cm) | *Y* coordinate (cm) | Rotation (deg) |
|---|---|---|---|
| 1 | 0 | 0 | 0 |
| 2 | 1 | 7 | 0 |
| 3 | 9 | 17 | 60 |
| 4 | 17 | 22 | 117 |

TABLE OF LINKAGE PARAMETERS

MINIMUM TRANSMISSION ANGLE — N.NN
N.NN — MAXIMUM LINK LENGTH RATIO

| SET 2, $\beta_2$ ↓ \ $\beta_2$ → | 300. | 312. | 324. | 336. | 348. | 360. |
|---|---|---|---|---|---|---|
| 60. | T-RR 2.1 | TOG 2.1 | TOG 3.0 | TOG 10.9 | TOG 14.0 | T-RR **** |
| 50. | R-C 1.9 | T-RR 1.6 | T-RR 3.0 | TOG 11.0 | TOG 14.1 | T-RR **** |
| 40. | T-RR 1.8 | T-RR 1.5 | T-RR 3.1 | T-RR 11.3 | TOG 14.4 | T-RR **** |
| 30. | TOG 1.9 | T-RR 1.7 | T-RR 3.3 | T-RR 11.9 | TOG 15.0 | T-RR **** |
| 20. | TOG 2.9 | T-RR 2.4 | T-RR 4.1 | T-RR 13.8 | T-RR 16.4 | T-RR **** |
| 10. | TOG 5.7 | TOG 4.7 | BRAN 7.1 | T-RR 21.2 | T-RR 21.8 | T-RR **** |

SET 2

**Figure 3.59** Table of linkage parameters of four-bar mechanisms formed from two dyads. For example, the first box in the top row refers to the mechanism formed out of a dyad obtained with $\beta_2 = 60°$, and $\beta_3, \beta_4$ taken from set 2, plus the dyad obtained with $\beta_2 = 300°$, also with $\beta_3, \beta_4$ from set 2.

**TABLE 3.11** ABBREVIATIONS FOR FIG. 3.59.

| | |
|---|---|
| R-C | Rocker-crank mechanism (input side is rocker) |
| ROC | Double-rocker does not toggle between precision points |
| BRAN | Linkage branches |
| T-RR | Linkage passes through toggle position between precision points when input is driven but does not toggle if follower is driven |
| RR-T | Double-rocker will toggle if follower is driven |
| TOG | Double-rocker will toggle when either side is driven |
| **** | Link ratio greater than 99.9:1 |
| ---- | No solution for one of the dyads |
| | (Blank) no linkage—essentially identical dyads |

BETAS output to ensure that the total required input angle rotation ($\beta_4$) was not too large. For example, the four-bar formed by the $\beta_2 = 50°$ and $\beta_2 = 300°$ solutions (both from set 2 in this case) looks promising, with the maximum link-length ratio = 1.9. When driven from the $\beta_2 = 50°$ side, it would be a rocker-rocker linkage; while driven from the $\beta_2 = 300°$ side, it would be a crank-rocker. Unfortunately, no acceptable solutions were found from this search.

One further attempt was made at specifying the precision points. Precision

**TABLE 3.12** THIRD SET OF PRECISION POINTS FOR THE FOUR-BAR MOTION GENERATOR OF EX. 3.6.

| Position | *X* coordinate (cm) | *Y* coordinate (cm) | Rotation (deg) |
|---|---|---|---|
| 1 | 0 | 0 | 0 |
| 2 | 3 | 5 | 5 |
| 3 | 27 | 22 | 90 |
| 4 | 35 | 24 | 117 |

point 2 was given at a slight angle (see Table 3.12); also, it was moved downward, closer to, and over to the right of precision point 1.

After a search through several possible solutions from this set of precision points, a final solution was found. The dyads of this linkage are shown in Fig. 3.60 and the coupler curve in Fig. 3.61. The total angular travel of the input link is only 113°, which is small enough to drive with another dyad. The link-length ratio (see Table 3.13) for the entire linkage is 2.51 and the transmission angles are satisfactory over the range of motion. The "choose" option of Table 3.13 allows the user to put two dyads together to form a four-bar mechanism. The "side 1" and "side 2" columns give specific numerical data of interest for the four-bar; the "minimum transmission angles" row indicates that with either side driven the linkage would be a rocker-rocker linkage. The coupler angles refer to the angles *PAB* and *PBA* (see Fig. 3.60). Finally, the linkage fits within the physical constraints required (Fig. 3.62 displays the mechanism in its four prescribed positions).

This example represents a typical design situation in which there are numerous constraints that would be difficult to make part of the mathematical model. With

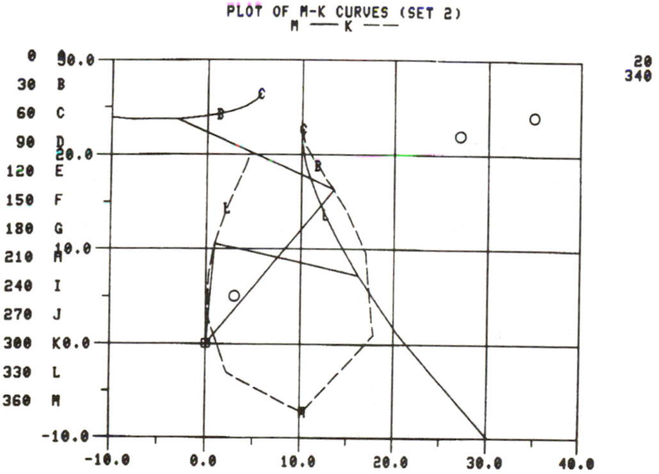


**Figure 3.60** Plot of *M–K* curves (set 2). Final solution of the four-bar motion-generator of Ex. 3.6 with precision points listed in Table 3.12.

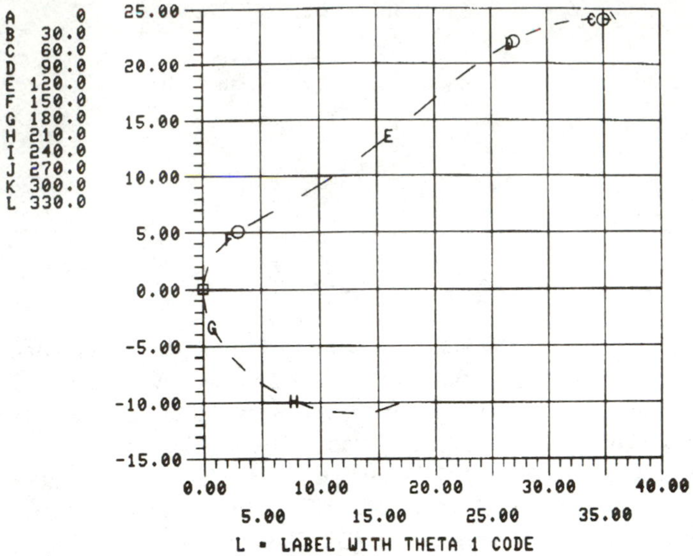


**Figure 3.61** Coupler curve of the final solution, a four-bar motion-generator of Ex. 3.6.

**TABLE 3.13** "CHOOSE" SUBROUTINE—FOUR-BAR MOTION GENERATOR SYNTHESIZED WITH THE INTERACTIVE "LINCAGES" PACKAGE (EX. 3.6).

| | SIDE 1 | | SIDE 2 |
|---|---|---|---|
| SOLUTION SET | 2 | | 2 |
| BETA 2 | 340.00 | | 18.00 |
| BETA 3 | 281.35 | | 60.93 |
| BETA 4 | 247.13 | | 40.06 |
| MX | 16.16 | | −4.66 |
| MY | 7.17 | | 23.63 |
| KX | .85 | | 13.98 |
| KY | 10.54 | | 15.51 |
| INPUT LENGTHS (LINKS 1 AND 3) | 15.68 | | 20.34 |
| COUPLER SIDE LENGTHS (LINKS 2 AND 4) | 10.58 | | 20.88 |
| ********** TOGGLES IF SIDE TWO DRIVEN ********** | | | |
| MINIMUM TRANSMISSION ANGLES | ROCKER | | ROCKER |
| COUPLER ANGLES (PAB AND PBA) | 115.31 | | 27.25 |
| COUPLER LENGTH (LINK 5) | | 14.04 | |
| GROUND LENGTH (LINK 6) | | 26.54 | |
| MAXIMUM LINK LENGTH RATIOS | | | |
| TOTAL (LINKS 1–6) | | 2.51 | |
| FOUR BAR (LINKS 1,5,3,6) | | 1.89 | |
| COUPLER (LINKS 2,4,5) | | 1.97 | |

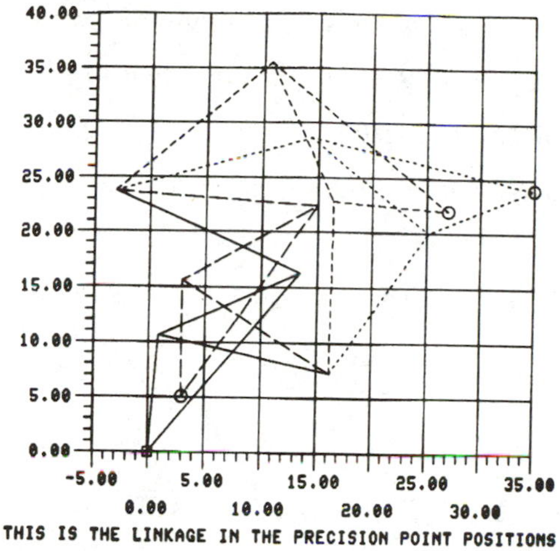


**Figure 3.62** Linkage of Fig. 3.60 in the precision-point positions.

the designer "in the loop," nonspecified constraints may be considered as well, especially when there is a manageable total number of solutions and a visual method of surveying many solutions at once, as is available with the LINCAGES package.

## APPENDIX A3.2

Computation schedule for standard form synthesis of motion-generator dyad for five finitely separated positions of the moving plane—Burmester Point Pairs.* See Fig. 3.2 for notation.

Prescribed quantities:

$$\mathbf{R}_j, \qquad j = 1, 2, 3, 4, 5$$

$$\alpha_j, \qquad j = 2, 3, 4, 5$$

Compute:

$$\boldsymbol{\delta}_j = \mathbf{R}_j - \mathbf{R}_1, \qquad j = 2, 3, 4, 5$$

Compute $\boldsymbol{\Delta}_k$, $\quad k = 2, 3, 4$ according to Eq. (3.8)

$$\boldsymbol{\Delta}_1 = -\sum_{k=2}^{4} \boldsymbol{\Delta}_k$$

Compute:

$$\boldsymbol{\Delta}_2' = \begin{vmatrix} (e^{i\alpha_3} - 1) & \boldsymbol{\delta}_3 \\ (e^{i\alpha_5} - 1) & \boldsymbol{\delta}_5 \end{vmatrix}, \qquad \boldsymbol{\Delta}_3' = \begin{vmatrix} \boldsymbol{\delta}_2 & (e^{i\alpha_2} - 1) \\ \boldsymbol{\delta}_5 & (e^{i\alpha_5} - 1) \end{vmatrix}$$

$$\boldsymbol{\Delta}_1' = -\boldsymbol{\Delta}_2' - \boldsymbol{\Delta}_3' - \boldsymbol{\Delta}_4$$

$$\Delta_k = |\boldsymbol{\Delta}_k|, \qquad k = 1, 2, 3, 4; \qquad \Delta_k' = |\boldsymbol{\Delta}_k'|, \qquad k = 1, 2, 3$$

* Enables the reader to write a program for five precision positions.

Using the above-obtained (now known) quantities, compute the real coefficients of a fifth-degree polynomial in the unknown $\tau = \tan \dfrac{\beta_2}{2}$ according to the following sequence (superior bar signifies complex conjugate):

$$\mathbf{a}' = \overline{\boldsymbol{\Delta}_1}\overline{\boldsymbol{\Delta}'_1}, \quad \mathbf{a}'' = \boldsymbol{\Delta}'_2\boldsymbol{\Delta}_3, \quad \mathbf{a}''' = \boldsymbol{\Delta}_2\boldsymbol{\Delta}'_3, \quad \mathbf{a}'''' = \mathbf{a}'' - \mathbf{a}'''$$

$$\mathbf{a} = \mathbf{a}'\mathbf{a}''''$$

$$\mathbf{b}' = \overline{\boldsymbol{\Delta}'_1}\,\overline{\boldsymbol{\Delta}_2}, \quad \mathbf{b}'' = \overline{\boldsymbol{\Delta}_1}\,\overline{\boldsymbol{\Delta}'_2}, \quad \mathbf{b}''' = \overline{\boldsymbol{\Delta}_1}\,\boldsymbol{\Delta}_3, \quad \mathbf{b}'''' = \overline{\boldsymbol{\Delta}'_1}\,\boldsymbol{\Delta}'_3$$

$$n = \sum_{k=1}^{3} \Delta_k^2 - \Delta_4^2, \quad n' = -\Delta_4^2 + \sum_{k=1}^{3} (\Delta'_k)^2$$

$$\mathbf{b} = \mathbf{b}'\mathbf{a}'' - \mathbf{b}''\mathbf{a}''' + \mathbf{b}'''\,n' - \mathbf{b}''''\,n$$

$$\mathbf{c}' = \boldsymbol{\Delta}'_1\,\boldsymbol{\Delta}_3, \quad \mathbf{c}'' = \boldsymbol{\Delta}_1\,\boldsymbol{\Delta}'_3, \quad \mathbf{c}''' = \overline{\boldsymbol{\Delta}_2}\,\boldsymbol{\Delta}_3, \quad \mathbf{c}'''' = \overline{\boldsymbol{\Delta}'_2}\,\boldsymbol{\Delta}'_3$$

$$\mathbf{c} = \mathbf{b}''\mathbf{c}' - \mathbf{b}'\mathbf{c}'' + \mathbf{c}'''\,n' - \mathbf{c}''''\,n$$

$$\mathbf{d}' = \overline{\boldsymbol{\Delta}_2}\,\overline{\boldsymbol{\Delta}'_2}, \quad \mathbf{d}'' = \mathbf{c}' - \mathbf{c}'', \quad \mathbf{d} = \mathbf{d}'\mathbf{d}''$$

$$\mathbf{u} = \boldsymbol{\Delta}_3\overline{\boldsymbol{\Delta}'_3}, \quad \mathbf{f}' = \overline{\boldsymbol{\Delta}_1}\,\boldsymbol{\Delta}'_2, \quad \mathbf{f} = \mathbf{f}'\mathbf{u}$$

$$\mathbf{h}' = \boldsymbol{\Delta}'_1\,\overline{\boldsymbol{\Delta}_2}, \quad \mathbf{h} = \mathbf{h}'\mathbf{u}, \quad \mathbf{k} = \mathbf{f} - \bar{\mathbf{h}}, \quad k = |\mathbf{k}|$$

$$\mathbf{g}' = \boldsymbol{\Delta}'_1\,\overline{\boldsymbol{\Delta}_1}, \quad \mathbf{g}'' = \boldsymbol{\Delta}'_2\,\overline{\boldsymbol{\Delta}_2}, \quad \mathbf{g}''' = \mathbf{g}' + \mathbf{g}''$$

$$g_y = u_x g_y''' + u_y g_x''', \quad \mathbf{v} = i g_y(4\mathbf{k})$$

$$m = -4g_y^2 - 2k^2, \quad \mathbf{p} = \mathbf{a}\bar{\mathbf{d}}$$

$$\mathbf{q} = \mathbf{a}\bar{\mathbf{c}} + \mathbf{b}\bar{\mathbf{d}} + \mathbf{k}^2, \quad \mathbf{s} = \mathbf{a}\bar{\mathbf{b}} + \mathbf{b}\bar{\mathbf{c}} + \mathbf{c}\bar{\mathbf{d}} + \mathbf{v}$$

$$t = \frac{1}{2}(a^2 + b^2 + c^2 + d^2 + m), \quad \text{where } a = |\mathbf{a}|, \text{ etc.}$$

$$A_1 = -6p_y - 4q_y - 2s_y$$

$$A_2 = -15p_x - 5q_x + s_x + 3t$$

$$A_3 = 20p_y - 4s_y, \quad A_4 = 15p_x - 5q_x - s_x + 3t$$

$$A_5 = -6p_y + 4q_y - 2s_y, \quad A_6 = -p_x + q_x - s_x + t$$

Check: $A_0 = p_x + q_x + s_x + t = 0$

$$a_j = A_{j+1}/A_6, \quad j = 0, 1, 2, 3, 4$$

Solve the following fifth-degree polynomial equation having real coefficients by means of any polynomial-solver routine for all five roots, both real and complex, for the unknown $\tau$:

$$\tau^5 + a_4\tau^4 + a_3\tau^3 + a_2\tau^2 + a_1\tau + a_0 = 0$$

Check: one of the roots should be:

$$\tau_0 = \frac{1 - \cos \alpha_2}{\sin \alpha_2}$$

This is a trivial root. Discard it and all complex roots. Keep the remaining real roots:

$$\tau_1, \tau_2 \quad \text{or} \quad \tau_1, \tau_2, \tau_3 \text{ and } \tau_4$$

(If no real roots remain, no solution exists. Go back to use different prescribed quantities.)

Using the real roots, compute $\beta_2$ as follows:

$$\beta_2 = \arg(1 - \tau_u^2 + i2\tau_u) \qquad \text{(up-to-four different values)}$$

where

$$u = 1 \text{ or } 2 \text{ when there are 2 real roots } (\tau_1 \text{ and } \tau_2)$$

and

$$u = 1 \text{ or } 2 \text{ or } 3 \text{ or } 4 \text{ when there are 4 real roots } (\tau_1, \tau_2, \tau_3 \text{ and } \tau_4)$$

Take one of the $\beta_2$ values and use Table 3.1 to solve for $\beta_3$ and $\beta_4$. Then use any two equations of the system Eq. (3.3) to compute **W** and **Z**. Repeat this for all (up-to-four) $\beta_2$ values.

This completes the computation of the (up-to-four) Burmester Point Pairs.

## PROBLEMS

**3.1.** Several different four-bar linkages are shown in Fig. P3.1.
- **(a)** Find the two four-bar cognates for the specified four-bar(s).
- **(b)** Compare the coupler curves of the cognates with the parent four-bar(s).
- **(c)** Construct the single degree of freedom geared five-bar path generator mechanism (with the same path).

**3.2.** Figure P3.2 shows four slider-crank mechanisms. Find a cognate and compare coupler curves for the slider crank in:
- **(a)** Figure P3.2a
- **(b)** Figure P3.2b
- **(c)** Figure P3.2c
- **(d)** Figure P3.2d

**3.3.** Figure P3.3 shows several four-bar linkages that should form the base of a six-bar linkage that generates the same path as point $P$ of the four-bar, but does so without rotating the coupler link of the six-bar. Construct those six-bars and compare paths of the four-bar and six-bar.

**3.4.** We wish to synthesize a six-bar motion generator to move stereo equipment from a shelf to closed storage when not in use. Rotation of the coupler link is not permitted, to avoid tipping the turntable or having the equipment slide off the moving platform. Figure P3.4 shows a four-bar that was synthesized such that its ground pivots are con-

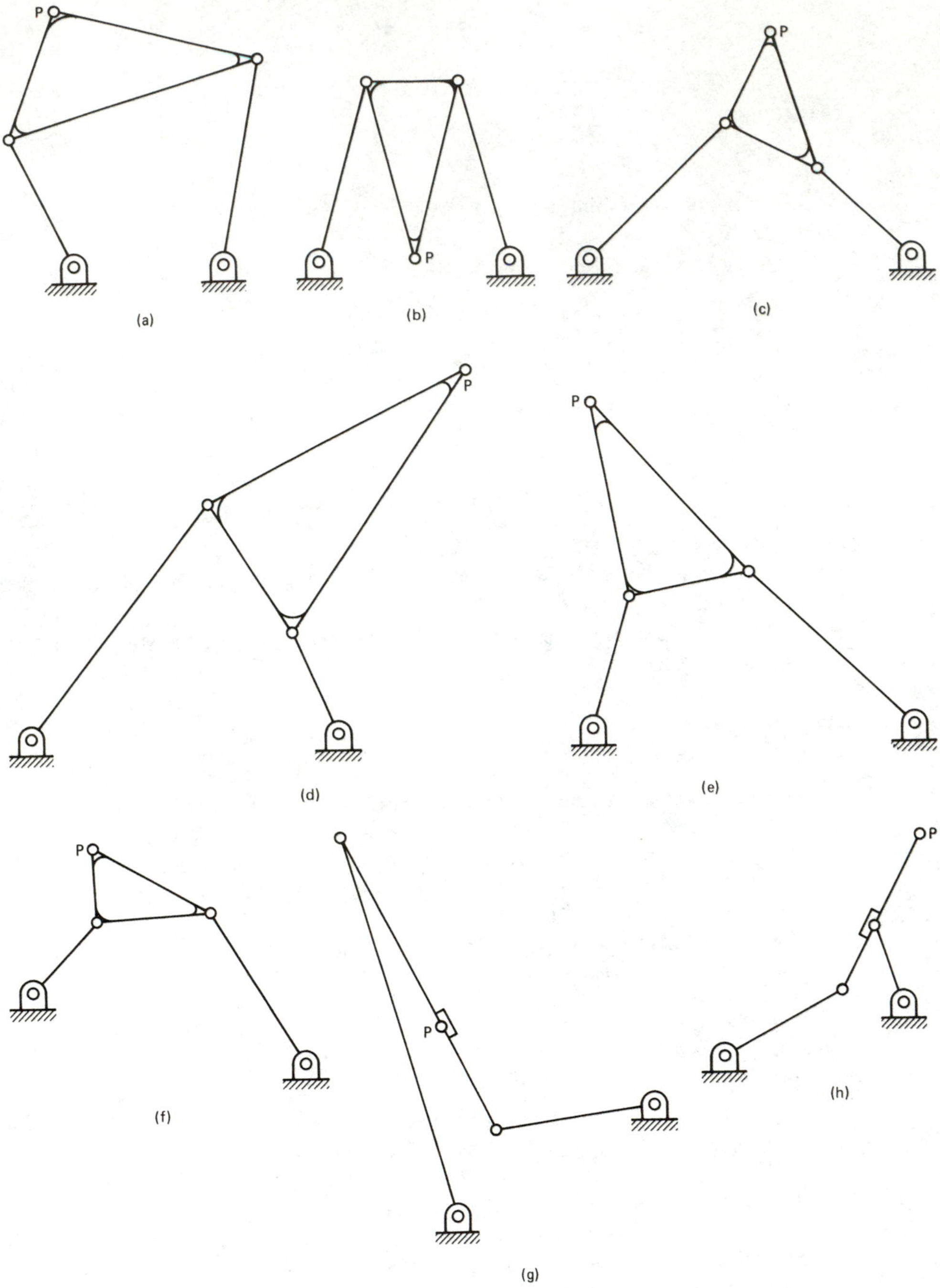


**Figure P3.1**

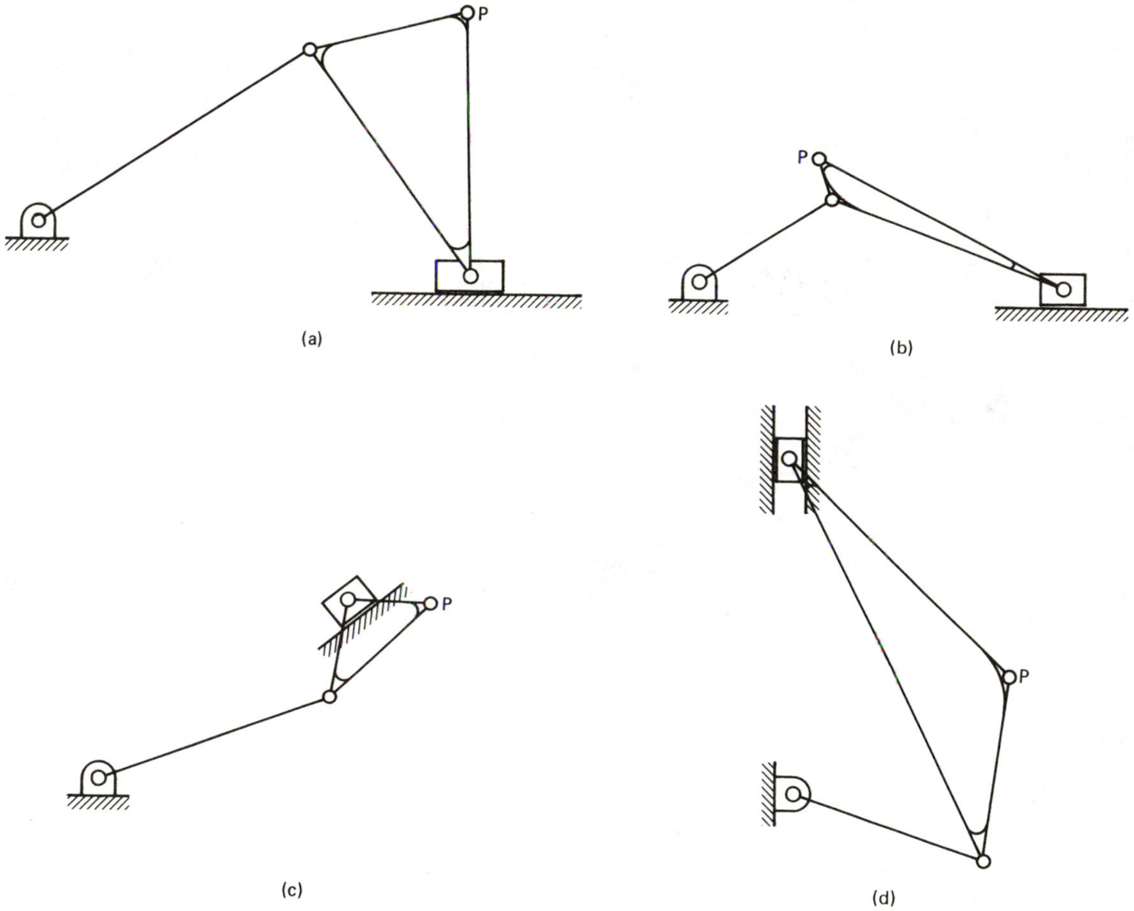


Figure P3.2

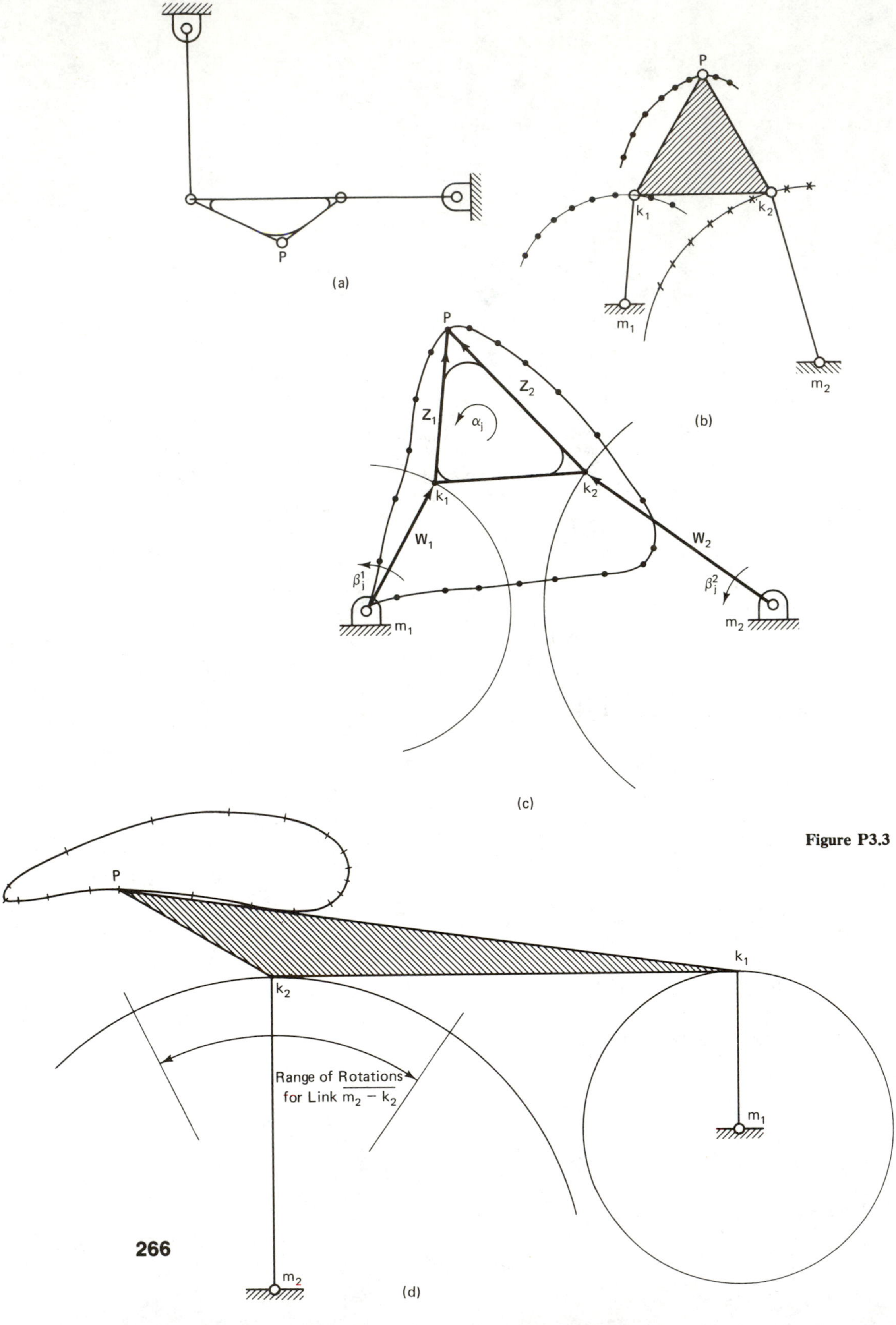
P
(a)
P
$k_1$
$k_2$
$m_1$
$m_2$
(b)
P
$Z_1$
$Z_2$
$\alpha_j$
$k_1$
$k_2$
$W_1$
$W_2$
$\beta_j^1$
$\beta_j^2$
$m_1$
$m_2$
(c)
P
$k_1$
$k_2$
Range of Rotations
for Link $\overline{m_2 - k_2}$
$m_1$
$m_2$
(d)

Figure P3.3

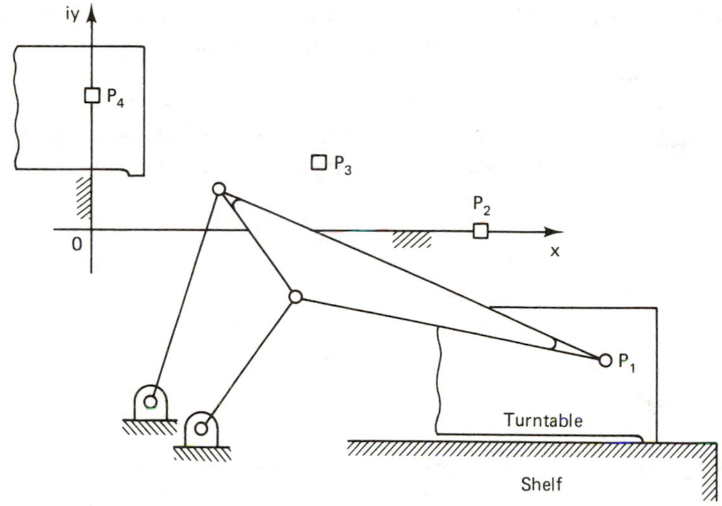


Figure P3.4

strained with the area from 1.5 to 4.0 in the $x$ direction and from −5.0 to −7.0 in the $y$ direction. The precision points prescribed were:

| Number | Precision points $x$ | Precision points $y$ | Coupler angle |
|---|---|---|---|
| 1 | 16 | −4 | 0° |
| 2 | 12 | 0 | 20° |
| 3 | 7 | 2 | 50° |
| 4 | 0 | 4 | 95° |

Design a six-bar linkage based on this acceptable four-bar such that the coupler of the six-bar does not rotate the turntable.

**3.5.** A six-bar linkage is to be designed to raise a portable bench grinder from an initial position on the bench to a final position resting against the garage wall. Figure P3.5

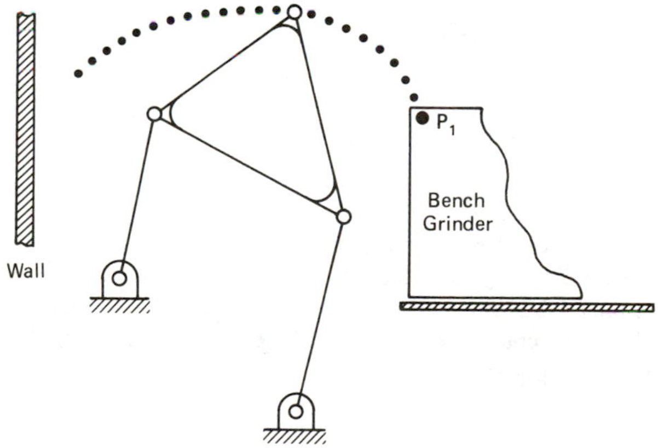


Figure P3.5

shows a synthesized four-bar in an intermediate position along the required path. Based on the cognate development in this chapter, design a six-bar to carry the grinder along the path without rotating the grinder.

**3.6.** Design a six-bar parallel motion generator to lift a small boat off the top of a car to a rack attached to the rafters of a garage. Figure P3.6 shows an acceptable four-bar that was synthesized for the following prescribed positions of motion generation:

| Number | Precision points | | Coupler angle |
|---|---|---|---|
| | $x$ | $y$ | |
| 1 | 0 | 0 | 0° |
| 2 | 1.8 | 2.9 | 36° |
| 3 | 3.7 | 3.3 | 48° |
| 4 | 6.0 | 3.0 | 60° |

Use this four-bar in the development of your solution.

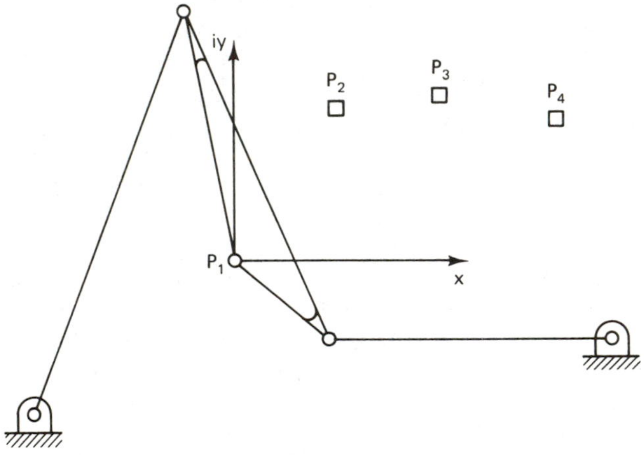


**Figure P3.6**

**3.7.** Design a six-link, approximate double-dwell mechanism (to replace a cam linkage) using all revolute pairs based on the synthesized four-bar of Fig. P3.7a. Notice that the four-bar traces a symmetric coupler curve [125] which has two circular arc sections. The output link should have a change of angle of 15° (see Fig. P3.7b) while the minimum transmission angle at the output is 65°.

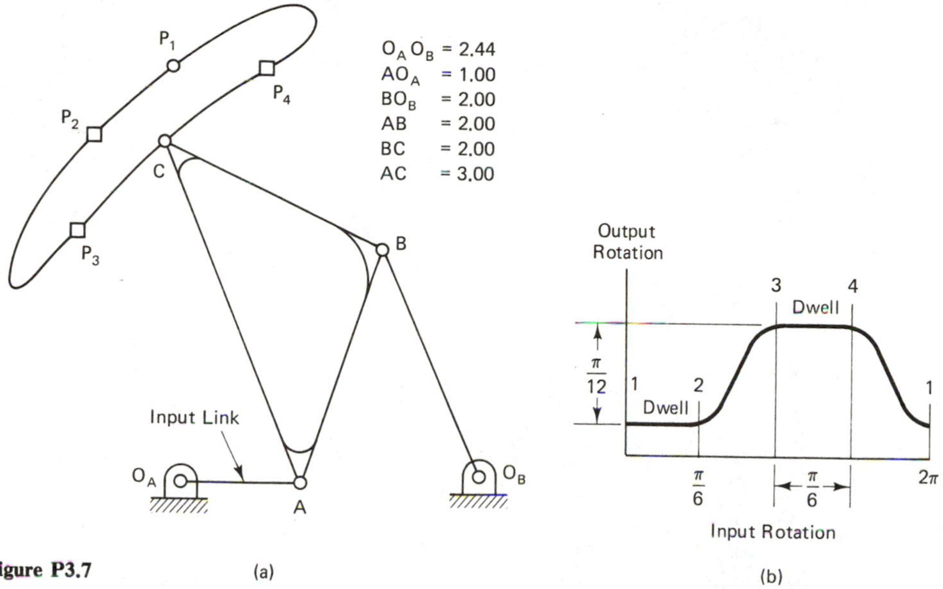


**Figure P3.7**

**3.8.** Design a six-bar linkage with revolute joints to replace a cam double-dwell mechanism. The output should oscillate 20° while the two approximate dwell periods should be 100° and 35° of input link rotation. Figure P3.8 shows the prescribed path of the primary four-bar linkage (path generation with prescribed timing) together with the five precision points and the input timing:

| Number | Precision points | | Input crank rotation |
|---|---|---|---|
| | $x$ | $y$ | |
| 1 | 0 | 0 | 0° |
| 2 | −26.4 | −5.0 | 100° |
| 3 | −21.0 | −7.2 | 220° |
| 4 | −7.0 | −4.0 | 240° |
| 5 | 0.0 | −4.25 | 255° |

**(a)** Synthesize a four-bar path generator with prescribed timing for the precision points. (If you wish to synthesize for only four precision points, leave out the fourth point.)

**(b)** Design the rest of the dwell mechanism such that the output link will swing only 20°.

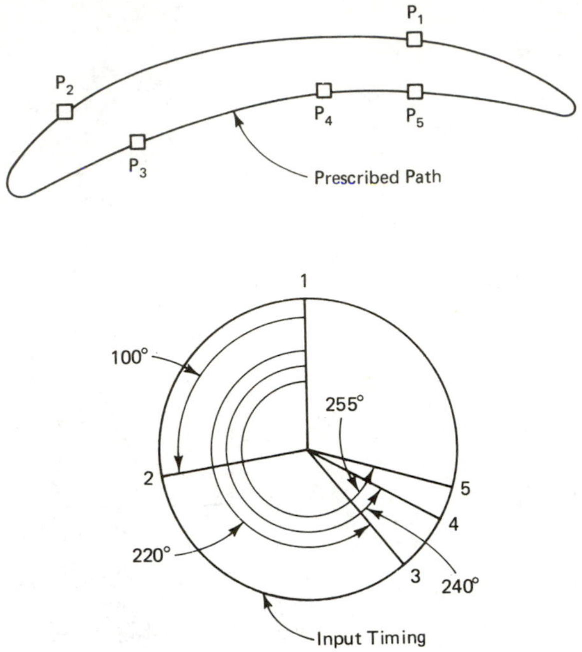


**Figure P3.8**

**3.9.** A small autoclave is to be used to sterilize medical instruments. The door must be stored on the inside of the autoclave when it is open. The door must be closed by a mechanism from the inside to form a seal with a gasket that allows the steam pressure to reach 15 psi on the inside of the vessel, forcing the door to stay closed. Figure P3.9 shows the space limitations of the door during its movement as well as its initial and final positions. Synthesize a four-bar mechanism that will open and close the auto-

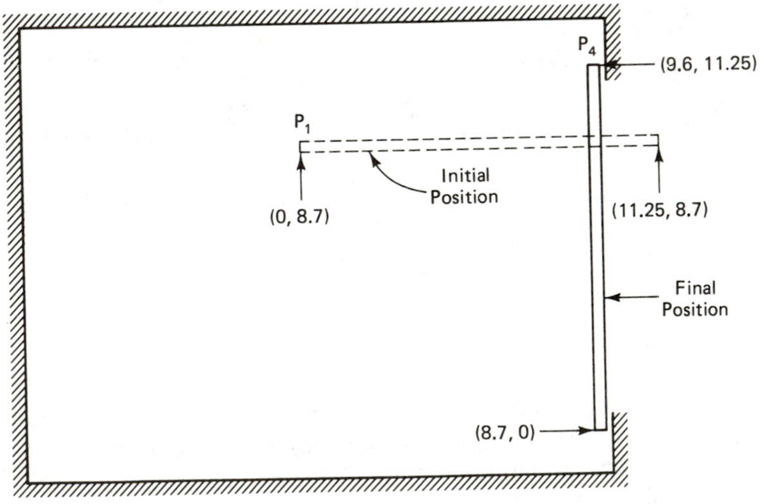


**Figure P3.9** Side view of autoclave. Door is 11.25 units; opening is 10.5 units. Ignore the thickness of the door.

clave door. The suggested precision points are:

| Number | Precision points | | Coupler angle |
|---|---|---|---|
| | $x$ | $y$ | |
| 1 | 0 | 8.7 | 0° |
| 2 | 2.8 | 11.125 | −40.62° |
| 3 | 7.5 | 11.6 | −79.24° |
| 4 | 9.6 | 11.25 | −90.00° |

**3.10.** Pick a fifth precision point for the linkage in Prob. 3.9 and compare the best linkage synthesized with all five points to the linkage synthesized with four points. Suggested additional precision point: $x = 0.8135$, $y = 9.5640$, angle $= -16.8842°$.

**3.11.** Synthesize a four-bar linkage that will move a small door from a vertical position in front of an automobile headlamp to a horizontal position above the headlamp. Figure P3.10 shows the door in five precision positions during its travel. The linkage you synthesize must fit into the space available (within the rectangle). Although the precision points are along a straight-line path, a slider is to be avoided if possible due to the wear that will occur in long-term use of this mechanism. Use the following precision points (skip precision points 3 or 4 if only prescribing four points):

| Number | Precision points | | Coupler angle |
|---|---|---|---|
| | $x$ | $y$ | |
| 1 | 2.0 | 8.0 | 0° |
| 2 | 4.0 | 8.0 | 30° |
| 3 | 6.0 | 8.0 | 50° |
| 4 | 8.0 | 8.0 | 70° |
| 5 | 10.0 | 8.0 | 90° |

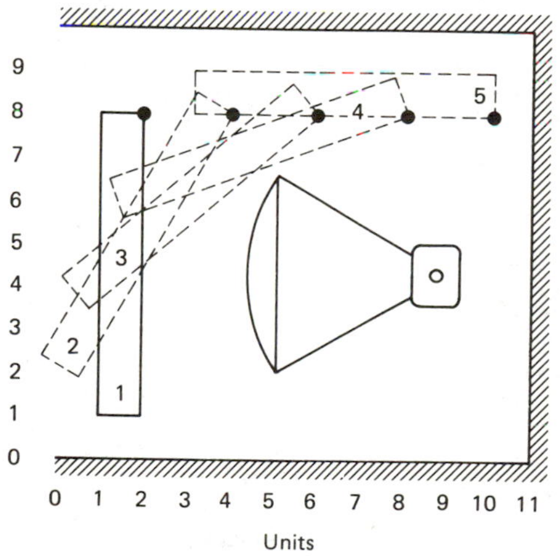


**Figure P3.10** Rotations are 0°, 30°, 50°, 70°, 90°cw.

**3.12.** Figure P3.11 shows a small bucket that is to be dumped into a large container and a desired path for the center of the bucket. Synthesize a four-bar mechanism that will lift the bucket and dump its contents into the container. Ground pivots are attached to the container. Synthesize for either four precision points (leaving out precision point 3) or for all five precision points.

| Number | Precision points $x$ | Precision points $y$ | Coupler angle |
|---|---|---|---|
| 1 | 0 | 0 | 0° |
| 2 | −0.5 | 4 | 5° |
| 3 | −1.5 | 5 | 5° |
| 4 | −2.0 | 5.5 | 60° |
| 5 | −2.5 | 5 | 120° |

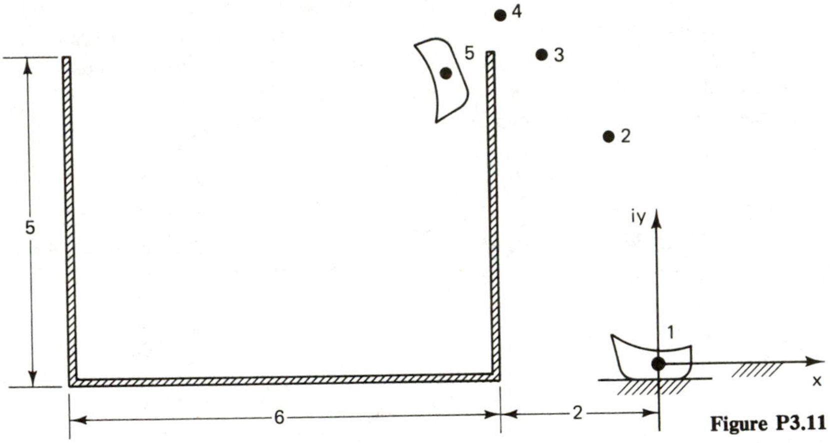


**Figure P3.11**

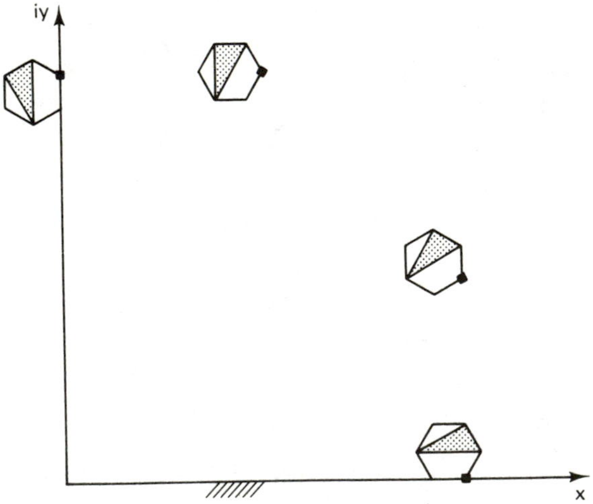


**Figure P3.12**

**3.13.** Synthesize a linkage to pick an object off the ground at (2, 0) and smoothly translate it, rotate it, and put it down at (0, 2) (see Fig. P3.12). Choose (2, 1) and (1, 2) as additional precision points and rotate the coupler 0°, 30°, 60°, and 90°, respectively. Restrict the ground pivots to within the triangle formed by the origin (0, 0) and the initial (2, 0) and final points (0, 2).

**3.14.** Figure P3.13 shows a square 20 x 20 units with the origin at the center. Inside the square is a circle of radius 5 units centered at the origin. Design a four-bar motion generator that will contain the arrow within the coupler triangle and have it point at the lower right corner. Contain the ground pivots within the square. The four prescribed precision points are:

| Number | Precision points | | Coupler angle |
|---|---|---|---|
| | $x$ | $y$ | |
| 1 | 5 | 0 | 0° |
| 2 | 0 | 5 | 7.12° |
| 3 | −5 | 0 | 29.74° |
| 4 | 0 | −5 | 36.86° |

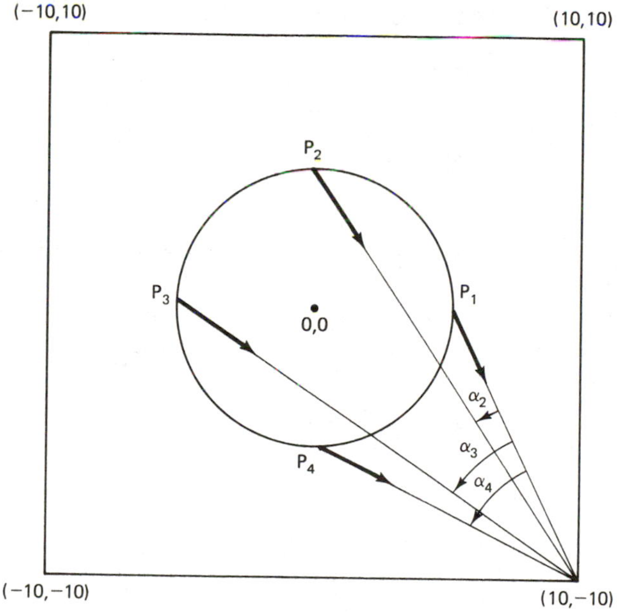


**Figure P3.13**

**3.15.** Choose a fifth precision point for Prob. 3.14 and find a linkage that gives similar results [we suggest (−2.38, 4.02) and a coupler angle of 13.44°].

**3.16.** An obstacle is blocking the path of an object as shown in Fig. P3.14. Synthesize a four-bar linkage using the given precision points to move the object over the obstacle.

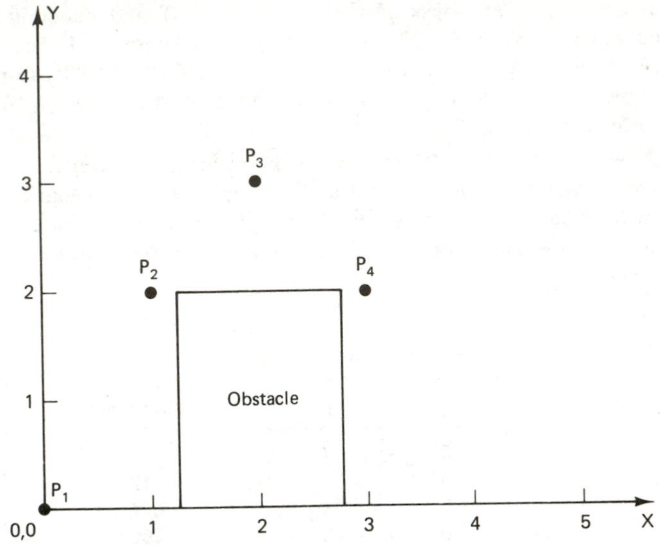


Figure P3.14

| Number | Precision points | | Coupler angle |
|---|---|---|---|
| | $x$ | $y$ | |
| 1 | 0 | 0 | 0° |
| 2 | 1 | 2 | 45° |
| 3 | 2 | 3 | 0° |
| 4 | 3 | 2 | 315° |

**3.17.** Add a fifth precision point to those used in Prob. 3.16 and find a four-bar linkage solution. The suggested point is (0.799, 1.42) and a coupler angle of 59°.

**3.18.** Synthesize a four-bar linkage that can be used to open the garage door in Fig. P3.15. The following precision points and door angular positions are given:

| Number | Precision points | | Angle of door |
|---|---|---|---|
| | $x$ | $y$ | |
| 1 | 0 | 6 | 90° |
| 2 | 0.5 | 6.5 | 60° |
| 3 | 1 | 7 | 30° |
| 4 | 1.5 | 7.5 | 0° |

(Notice that these points lie on a straight line.)

**3.19.** Figure P3.16 [(a) front view and (b) side view] is taken from U.S. patent 4,084,411 (A. B. Mayfield). This device is a radial misalignment coupling that transmits constant angular velocity between shafts. The two shafts (12 and 13) are shiftable during operation, and the linkage system remains dynamically balanced.

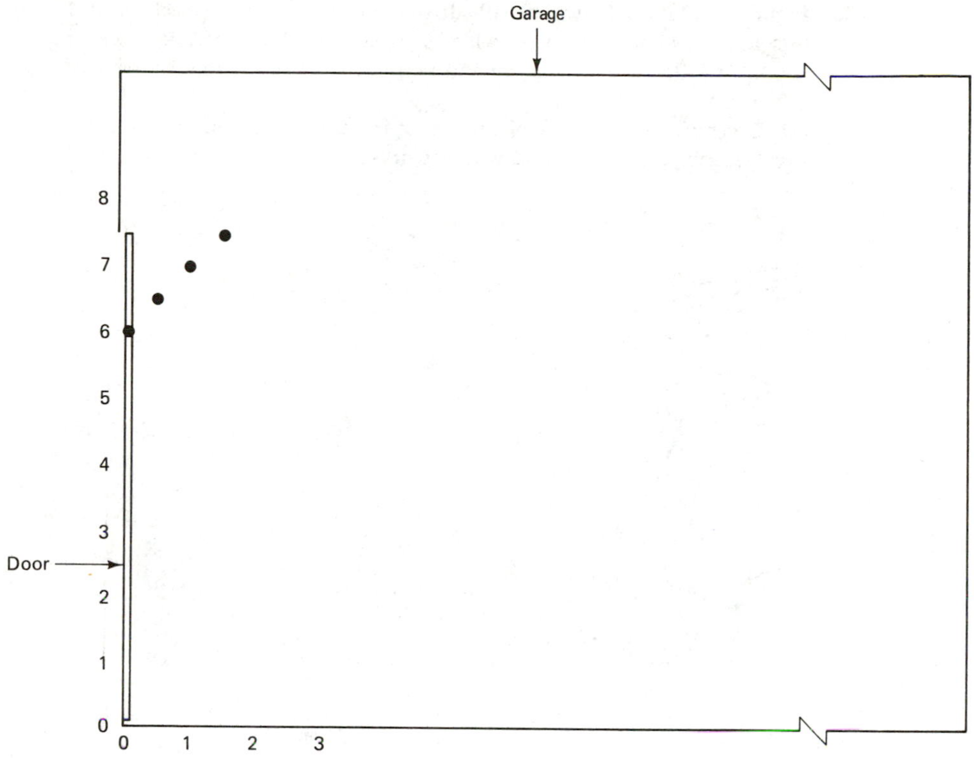


Figure P3.15

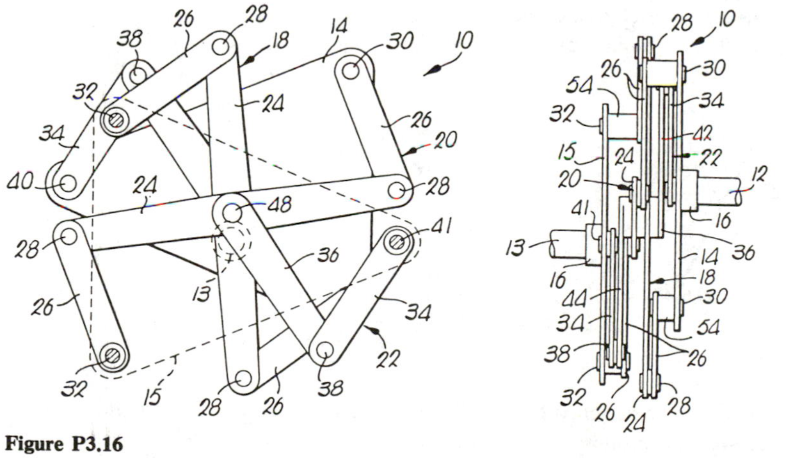


Figure P3.16

**(a)** Determine the degrees of freedom of this device.
**(b)** Describe briefly how this mechanism works.
**(c)** Why is it designed the way it appears? Could you make some improvements?

**3.20.** Figure P3.17 shows a hydraulically driven industrial lift-table mechanism. The mechanism lifts a table (4 ft × 8 ft) on which a weight of 2500 lbf (max.) can be placed to a height of 4 ft. The mechanism and table will collapse into a box-shaped region 4 ft × 8 ft × 1 ft.

**(a)** Determine the number of degrees of freedom of this mechanism.

**(b)** Describe briefly how and why it works.

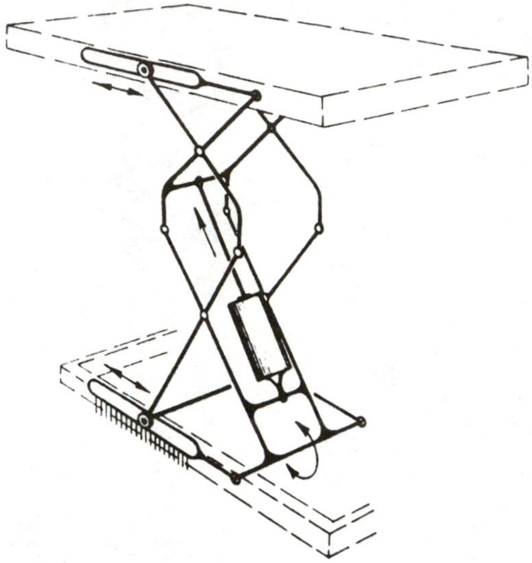

**Figure P3.17**

**3.21.** Figure P3.18 shows a version of a lazy-tongs linkage.

**(a)** Determine the number of degrees of freedom of this mechanism.

**(b)** Describe briefly how and why it works.

**(c)** Can you design a different mechanism for this task?

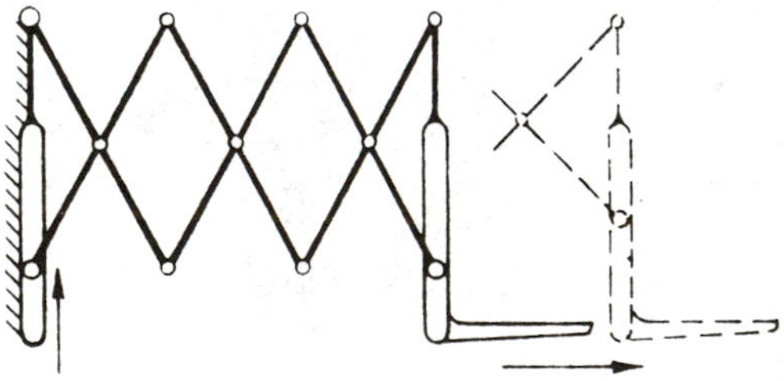

**Figure P3.18**

**3.22.** A hinge is to be designed to be entirely inside a container when the lid is closed. Figure P3.19 shows a proposed six-bar design in the form of parallelograms.*

**(a)** What type of six-bar is this?

**(b)** Write the standard-form synthesis equations for a motion-generator task of moving the lid with respect to the container.

* Suggested by T. Carlson.

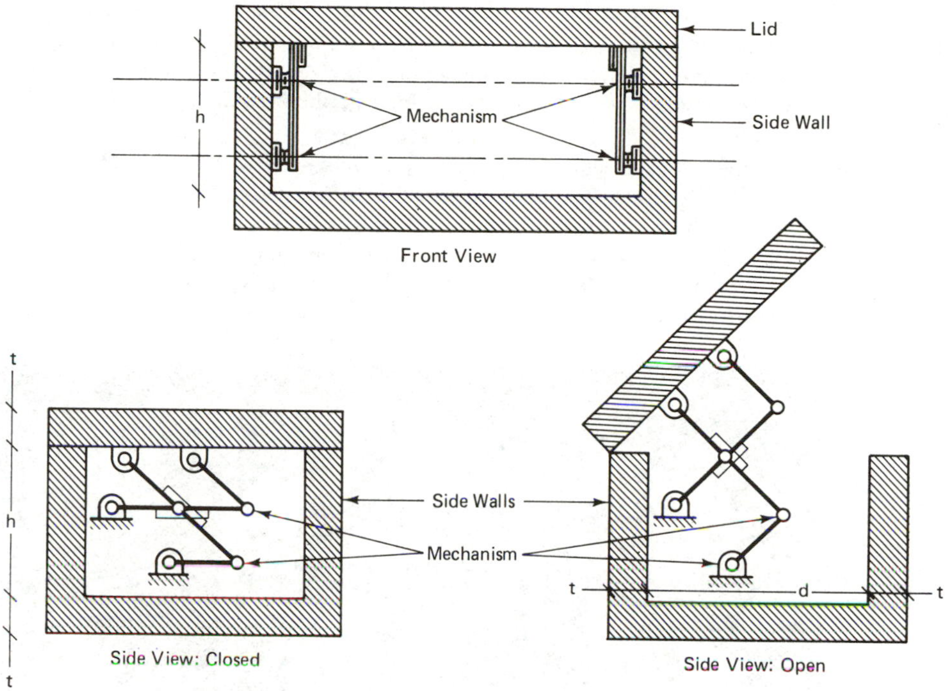


**Figure P3.19**

(c) For a set of four precision positions of your choice, design a six-bar to go through your specific set of positions.

**3.23.** Every golfer realizes the necessity of a good drive (or first shot) and also the important role that consistency plays in a good game. With these needs in mind, an automatic tee-reset mechanism for use at a driving range was conceived.* This machine would help in practicing driving by automatically replacing another golf ball on the tee, and it would aid consistency because the golfer would not need to change his or her stance. The machine's task is to take one ball from a ball reservoir, gently place it on a tee, and retract out of the way without knocking the ball off (see Fig. P3.20). A crank-

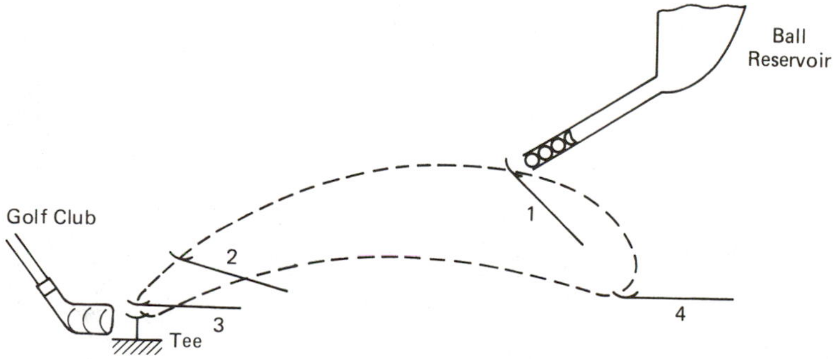


**Figure P3.20**

* This problem was originated by J. Peters, S. Yassin, and J. Arnold.

rocker motion-generator four-bar is desired. Measure the precision points from the figure and synthesize a four-bar for this task. Ground pivots are allowed to be below ground since a permanent placement for this mechanism is assumed. The entire mechanism should be well out of the way of the golfer's swing in the rest position (position 4).

**3.24.** An interesting mechanism is used by cabinetmakers—the Soss [254] concealed hinge (see Fig. P3.21). The entire mechanism is very compact and is embedded into the wood wall and door of the cabinet. Figure P3.21b shows that it will open 180°.

**(a)** What type of linkage is the Soss hinge (see Chap. 1)?

**(b)** Write the standard-form equation for the synthesis of this mechanism.

**(c)** How does this design compare with the type shown in Fig. 3.19?

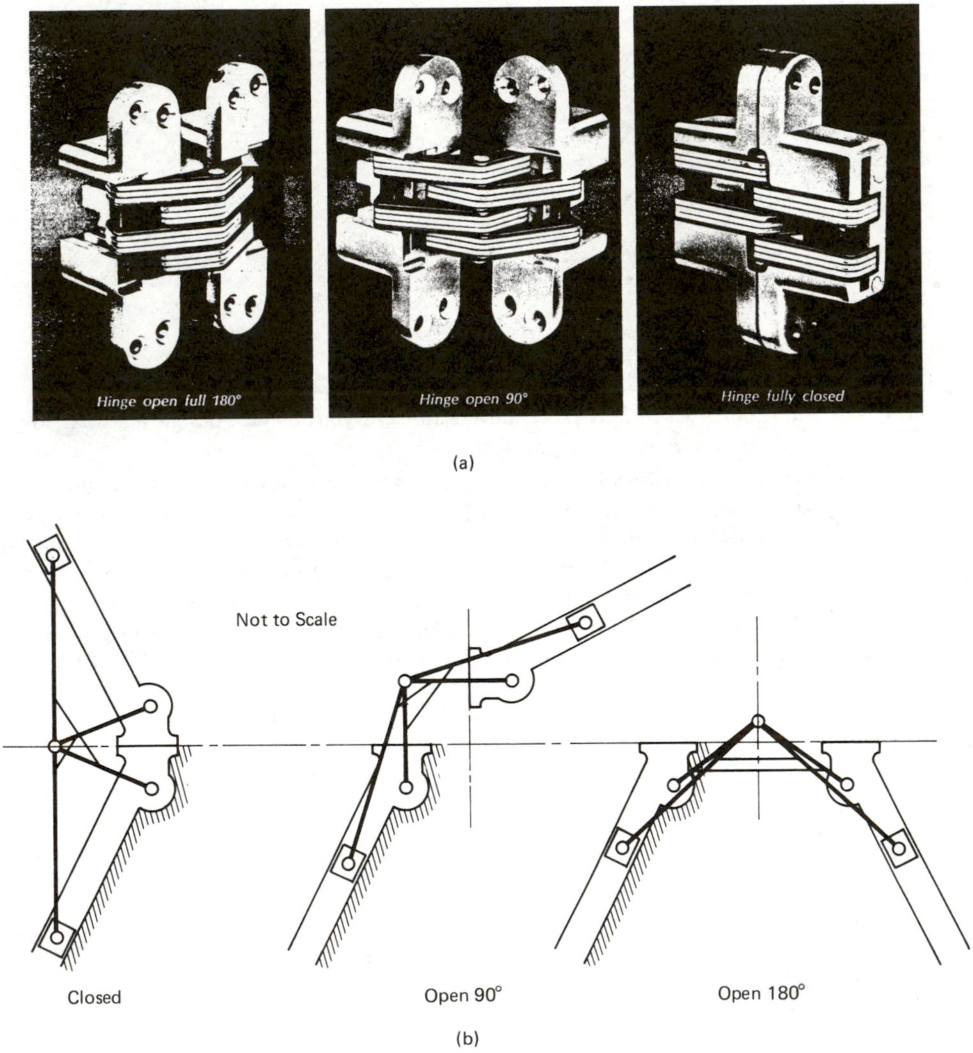


**Figure P3.21**

**3.25.** Figure P3.22 shows a type of mechanism used as an automobile window guidance linkage [275]. To enter the door cavity properly, the window should have minimal rotation while the path of $p$ follows a prescribed trajectory.

**(a)** Verify that this mechanism has a single degree of freedom.

**(b)** What type of linkage is this?

**(c)** Write the synthesis equations for the linkage in the standard form.

**(d)** Synthesize a mechanism of this type that satisfies a path and space constraints of your choice.

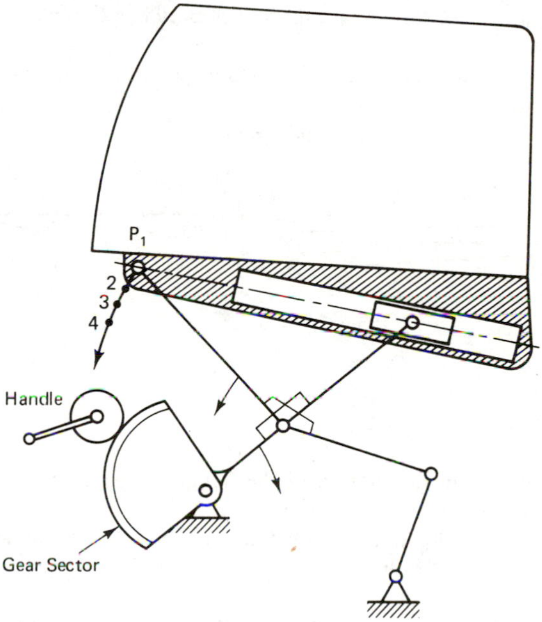


**Figure P3.22**

**3.26.** A lock mechanism for a window must be designed such that the key will turn the input crank of a slider-crank mechanism while the slider (the bolt) will travel a total of 0.5 in. (see Fig. P3.23). The restricted space for the mechanism is such that the maximum distances are $H = 0.65$ in. at $\phi = \phi_{max}$ and $L = 1.75$ in. at $\phi = 0$. The mechanical advantage in the first position of the mechanism must be maximized while the deviation angle $\delta$, is minimized at $\phi = \phi_{max}$. Design the slider-crank ($\ell_1$, $\ell_2$, $L$) for this objective.

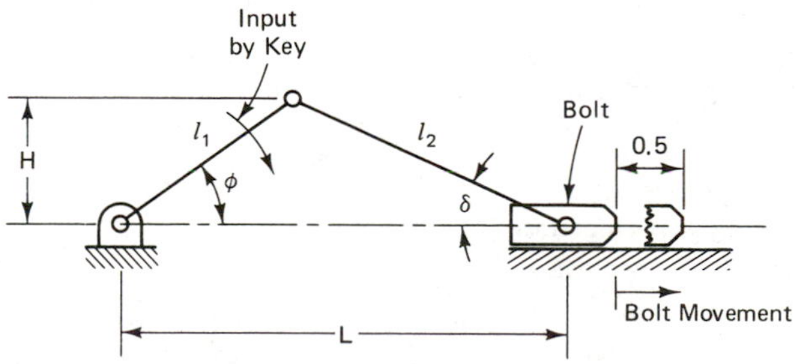


**Figure P3.23** First position schematic.

**3.27.** It is proposed to design a Fowler wing-flap mechanism of the type shown in Fig. P3.24.* The objective is to avoid sliding contacts which are employed in present designs. The motion specified is a linear translation along the mean chamber for 15% of the wing width, followed by a 40° rotation downward. Further constraints are that the linkage fits in all its positions inside a 10° angle between the top and bottom of the wing profile, and approximate the motion specified above between the precision points.

**(a)** Determine the number of degrees of freedom of the linkage in the figure.

**(b)** Check graphically to see that the mechanism shown accomplishes the design objectives.

**(c)** Write the synthesis equations for this linkage for four prescribed positions in standard form.

**(d)** Describe how the synthesis will proceed. How many possible solutions are there for this objective?

**(e)** Design a mechanism of the type shown in the figure.

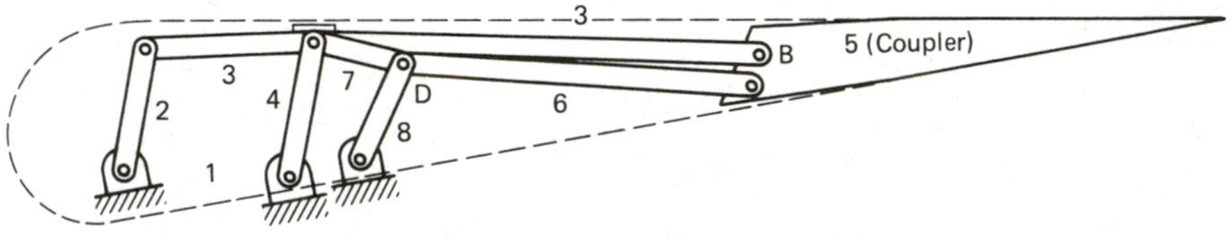


**Figure P3.24**

**3.28.** A linkage is required† to duplicate the motion of the human finger from the knuckle to the tip of the finger (Fig. P3.25). After careful study, a Watt I six-bar linkage (see Fig. 1.9) was chosen as the most likely to match four prescribed positions and to be narrow enough to match the size of a finger. The positions of $P$ and the rotations of link 6 are:

*First position:* finger fully extended, parallel with the back of the hand:

$$\boldsymbol{\delta}_1 = 0, 0i; \qquad \theta_1 = 0°$$

*Second position:* finger slightly bent, as if one were holding a medium-sized glass:

$$\boldsymbol{\delta}_2 = 1.475 - 5.650i; \qquad \theta_2 = 68°$$

*Third position:* finger and thumb touching as if one were holding a piece of paper:

$$\boldsymbol{\delta}_3 = 5.350 - 8.100i; \qquad \theta_3 = 124°$$

*Fourth position:* fingers almost forming a closed fist, such as when grasping a steering wheel:

$$\boldsymbol{\delta}_4 = 10.350 - 6.650i; \qquad \theta_4 = 196°$$

* This problem was suggested by J. Boomgaarden [22].

† This problem was suggested by Kevin J. Olson.

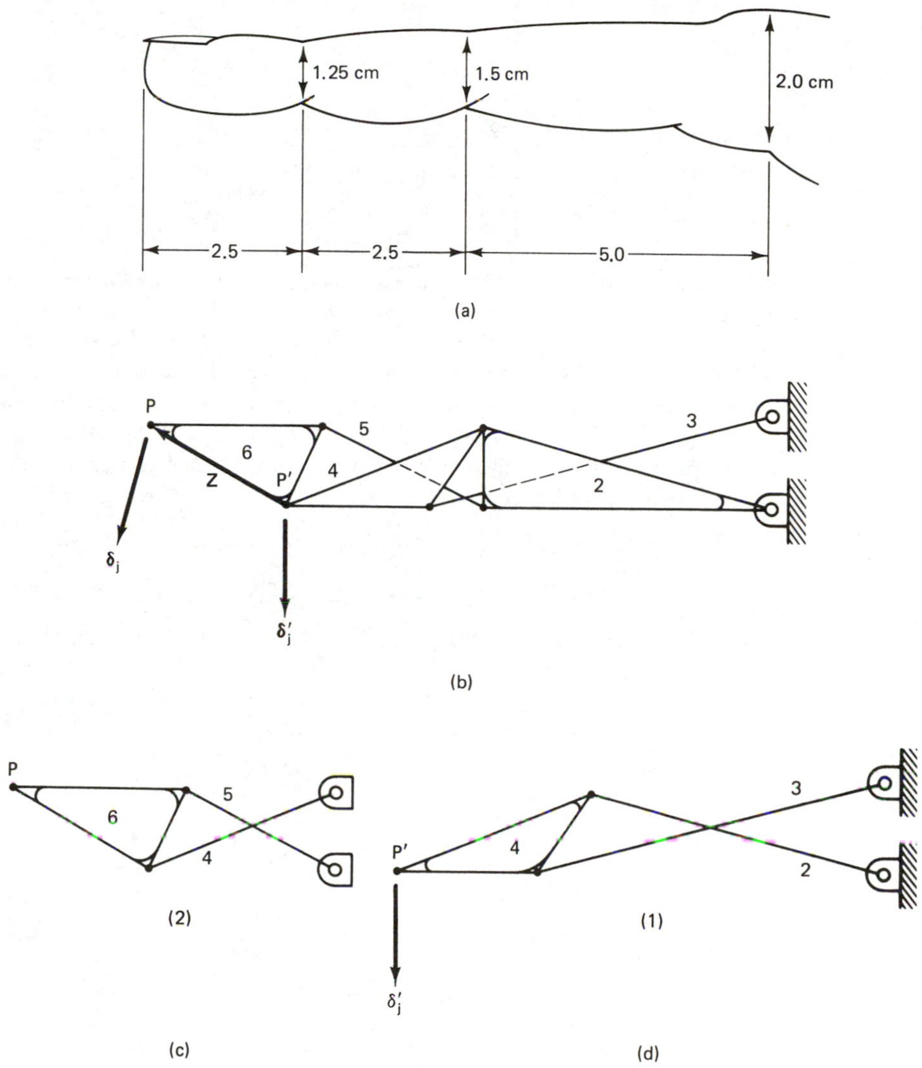


**Figure P3.25**

The four-bar labeled 1 (Fig. P3.25d) must be synthesized first. But to do this, one must relate the $\boldsymbol{\delta}_j$ vectors to the $\boldsymbol{\delta}'_j$ vectors. This can be done by choosing the vector **Z** and solving for the following vector equation.

$$\boldsymbol{\delta}'_j = \boldsymbol{\delta}_j - \mathbf{Z}(e^{i\theta_j} - 1), \qquad j = 1, 2, 3, 4$$

choosing $\mathbf{Z} = -1.90,\ 1.00i$.

The calculated positions for point $P'$ and rotations for coupler 4 become:

| | $X$ | $Y$ | $\alpha°$ |
|---|---|---|---|
| $\delta'_1$ | 0 | 0 | 0 |
| $\delta'_2$ | 1.214 | −3.263 | 44.5 |
| $\delta'_3$ | 3.217 | −4.966 | 80 |
| $\delta'_4$ | 6.348 | −5.212 | 133 |

Synthesize a Watt I six-bar for this task. The most challenging aspect of synthesizing this linkage is designing the mechanism to fit the constraints of a human finger. The mechanism should be approximately 10 cm long, 2.5 cm from the tip to the first joint, 2.5 cm between the first and second joints, and 5.0 cm between the second joint and the knuckle. The height of the first joint should not exceed 1.25 cm, the second should be no more than 1.5 cm, and the knuckle should not be greater than 2.0 cm.

**3.29.** Figure P3.26 shows four different schematics of bucket loaders seen on work sites.

**(a)** For each bucket loader:

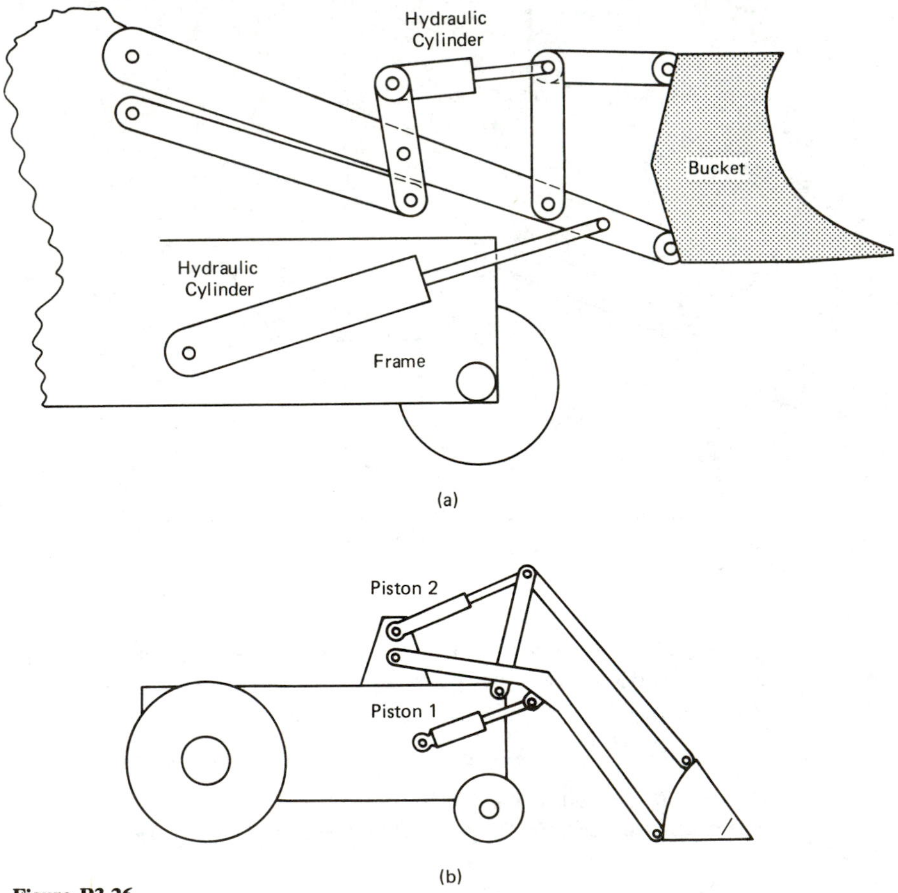


**Figure P3.26**

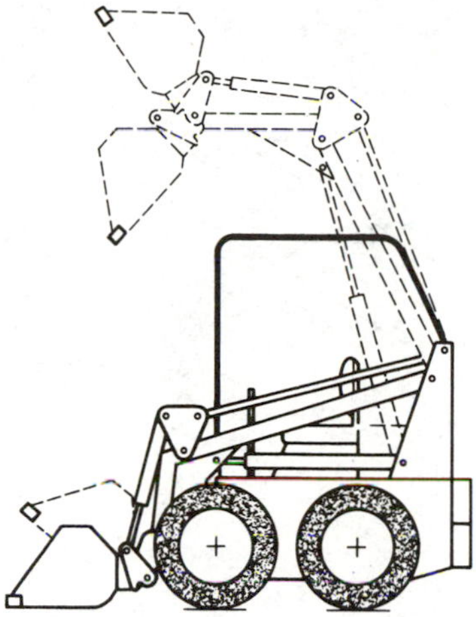

(c)

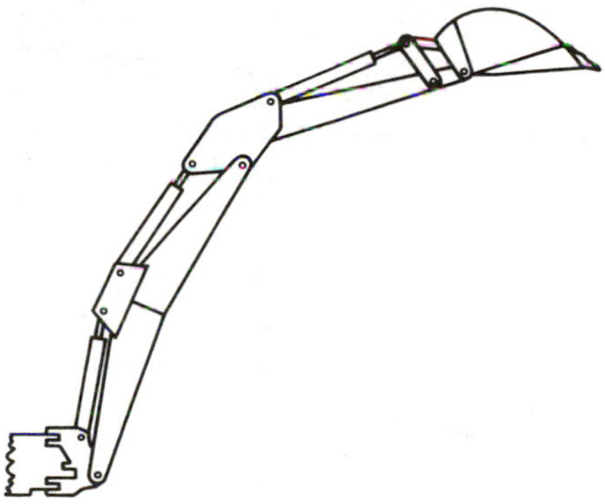

(d)

**Figure P3.26 (cont.)**

(1) Draw an unscaled kinematic diagram.
(2) What kind of mechanism is this? What task does it perform?
(3) Determine the degrees of freedom of the mechanism.
(4) How would you synthesize this mechanism in the standard form?

**(b)** Compare the performance of each design based on intuition and the knowledge gained in part (a).

**3.30.** Figure P3.27 shows schematics of two alternative designs for desk lamp mechanisms.

**(a)** For each desk lamp mechanism:

(1) Draw an unscaled kinematic diagram.
(2) What kind of mechanism is this? What task does it perform?
(3) Determine the degrees of freedom of the mechanism.
(4) How would you synthesize this mechanism in the standard form?

**(b)** Compare the performance of each design based on intuition and the knowledge gained in part (a).

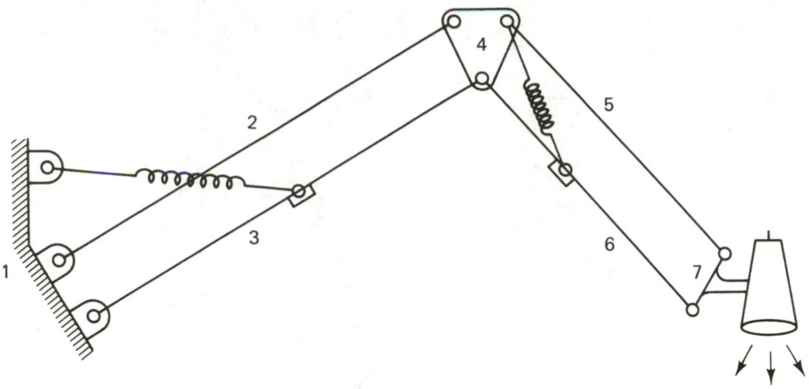


(a)

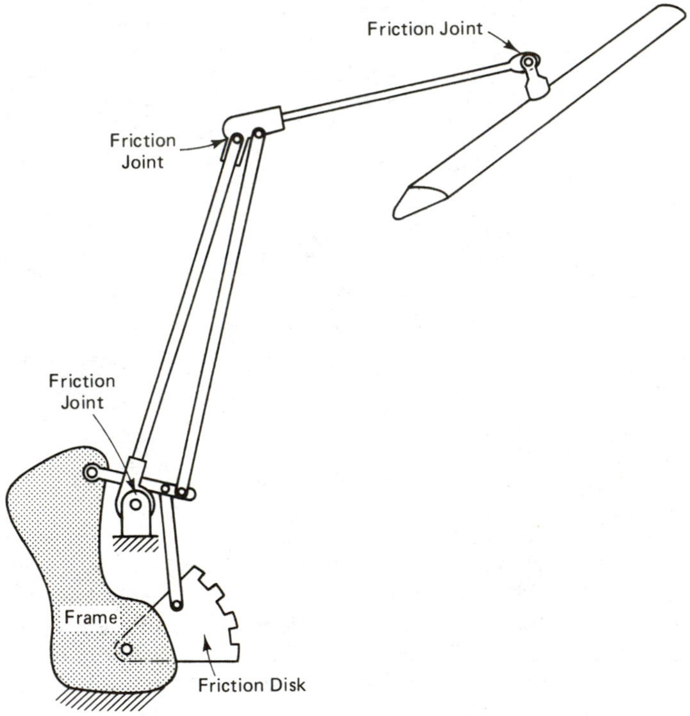


(b)

**Figure P3.27**

**3.31.** Figure P3.28 shows a schematic of a fill valve mechanism for a toilet tank.

**(a)** Draw an unscaled kinematic diagram of this device.

**(b)** What kind of mechanism is this? What task does it perform?

**(c)** Determine the degrees of freedom.

**(d)** How would you synthesize this mechanism in the standard form?

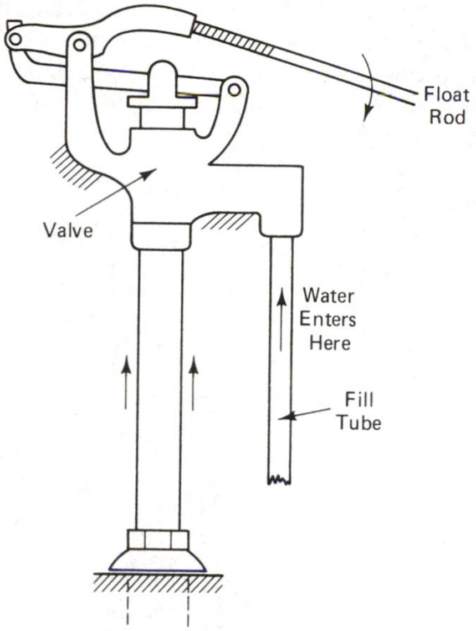


**Figure P3.28**

**3.32.** Figure P3.29 shows an automobile hood mechanism [different from the two in Chap. 1 (Figs. 1.2c and P1.24)].

**(a)** Draw an unscaled kinematic diagram of this device.

**(b)** What kind of mechanism is this? What task does it perform?

**(c)** Determine the degrees of freedom of this mechanism.

**(d)** How would you synthesize the mechanism in the standard form?

**(e)** Compare this linkage's performance to that of the other two hood linkages based on intuition and knowledge gained in parts **(a)** to **(d)**.

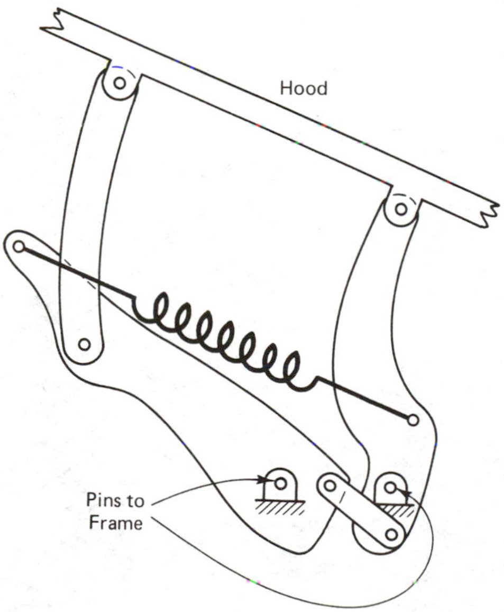


**Figure P3.29**

3.33. An industrial robot manipulator designed for extracting formed articles (such as castings) is shown in Fig. P3.30 (designed by C. A. Burton, U.S. patent 3,765, 474, courtesy of Rimrock Corporation, Columbus, Ohio). This machine was designed so that the linkage could move out of the way of the die-casting process and have a nearly straight line

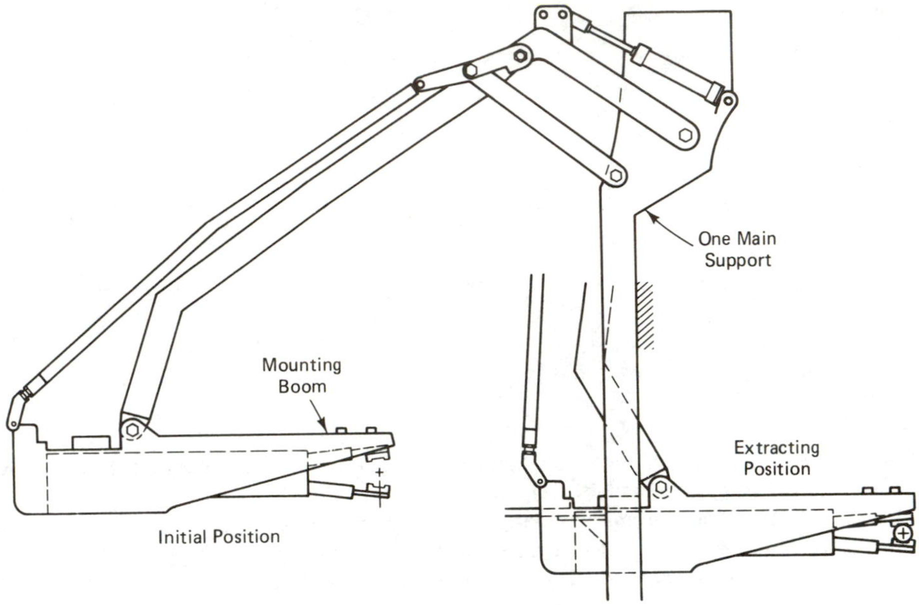


(a)

(b)

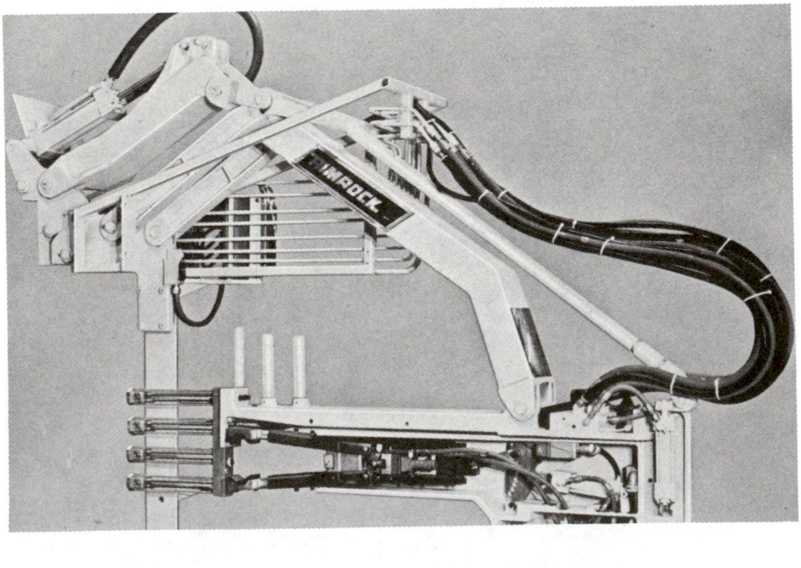


Figure P3.30

motion for up to a 70-in. stroke. The mechanism that closes the extractor is not shown here.

**(a)** Determine the degrees of freedom of this linkage.
**(b)** Determine the type of this mechanism.
**(c)** Write the standard form equations for the synthesis of this device.
**(d)** Determine the length and accuracy of the straight-line path.
**(e)** Try to design a linkage with better straight-line characteristics.

**3.34.** Figure P3.31 shows one of the early designs for typewriters. A multiloop linkage transfers the finger movement of the typist to the magnified movement of the type bar.
**(a)** Determine the degrees of freedom of this linkage.
**(b)** What type of mechanism is this?
**(c)** Write the equations for this mechanism in the standard form.
**(d)** Design a typewriter mechanism with your own set of four or five precision points.

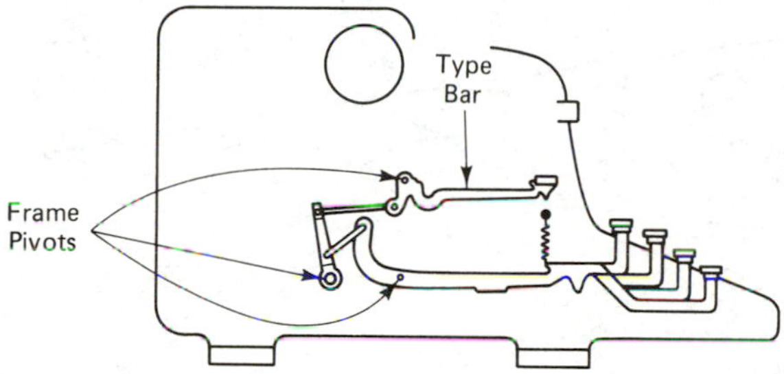


**Figure P3.31**

**3.35.** Figure P3.32 shows the Garrard Zero 100 "zero-tracking" error mechanism for "elimination of distortion" in playback.* The articulating arm is designed to constantly decrease

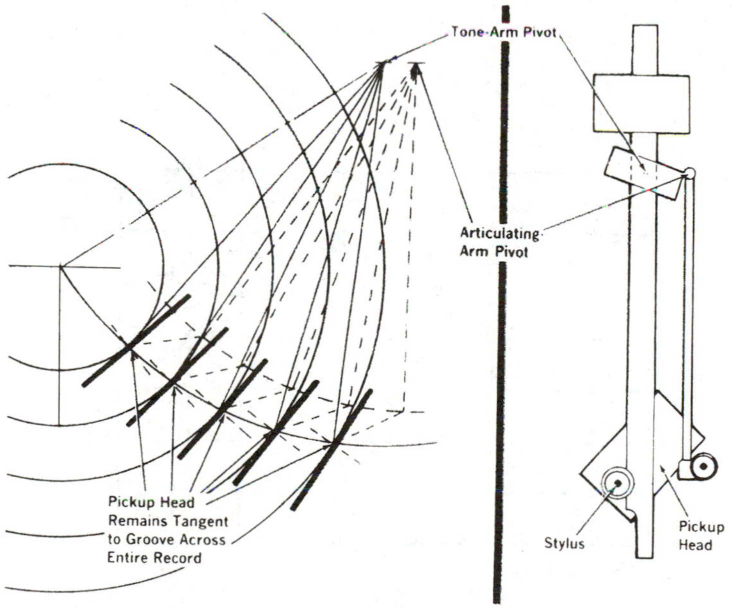


**Figure P3.32**

* *Popular Science,* November 1971, pp. 94–96.

the pickup head/tone arm angle so the pickup head forms a tangent with the groove being played.

**(a)** What task does this four-bar satisfy?

**(b)** Write the standard-form synthesis equations for this task.

**(c)** Design your own tone arm mechanism using your choice of four or five precision points.

**3.36.** Interactive computer graphics were used to design a wing-flap mechanism [201] (see Fig. P3.33). Some desired positions were digitized and displayed on the computer screen (Fig. P3.33a). Four such positions yielded the *m–k* curves shown in Fig. P3.33b. A final linkage was picked interactively as shown in Fig. P3.33c and d.

**(a)** Pick four or five of your own design positions and synthesize your own linkage for this task.

**(b)** Include in this synthesis the trailing flap that is shown in Fig. P3.33d.

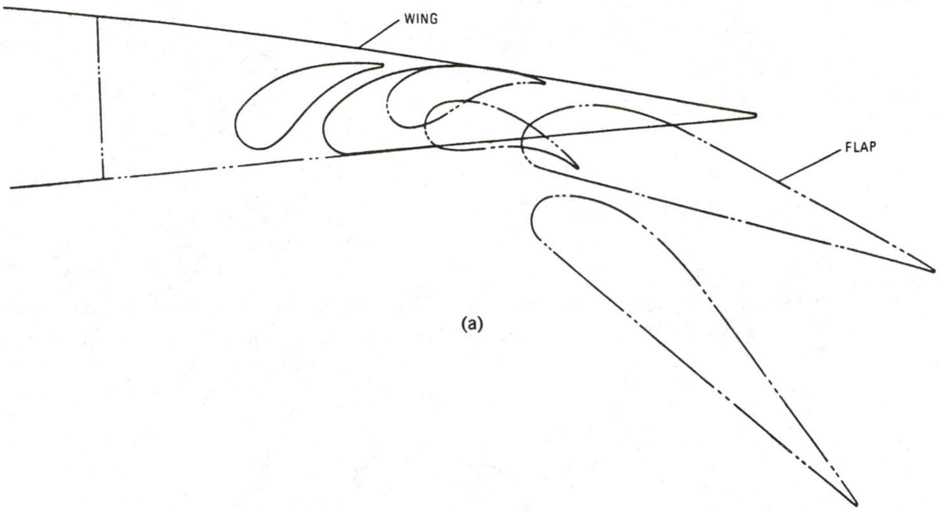


(a)

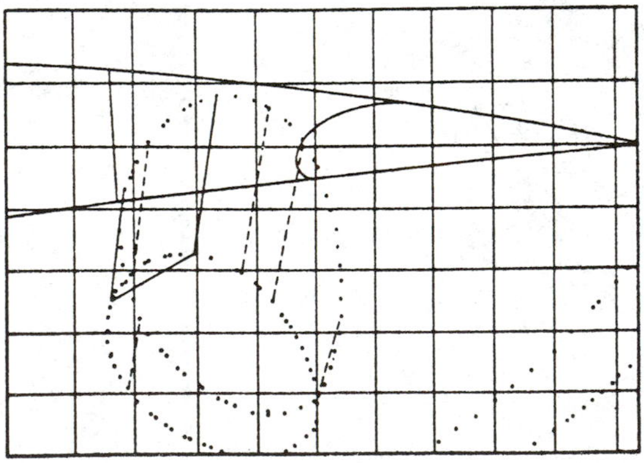

(b)

**Figure P3.33**

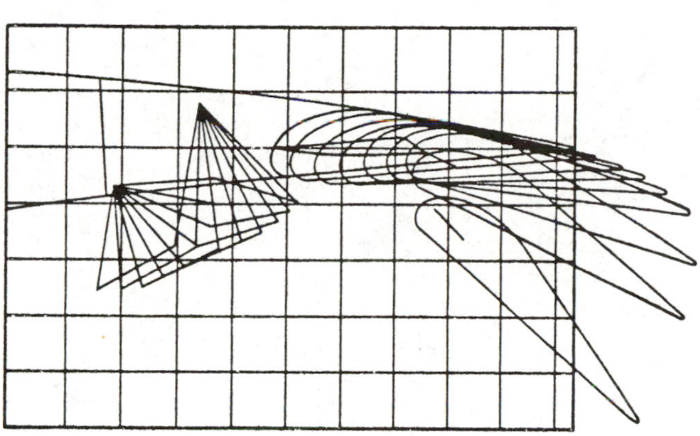

(c)

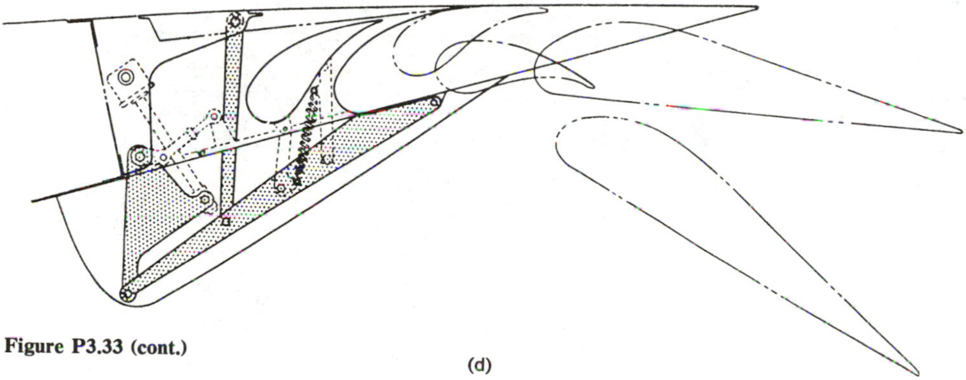

**Figure P3.33 (cont.)**

(d)

**3.37.** Figure P3.34 shows a proposed design of an aircraft spoiler assist device [201]. (A spoiler is a device that "spoils" the airflow around a wing to decrease lift. Spoilers are used in landing and for roll control.) The "$q$ pot" is to act as a balancing device to the spoiler. With the assistance of the $q$ pot, no power, other than that exerted by the pilot, is required to actuate the spoilers. A linkage is desired to balance the $q$ pot and the spoiler throughout 60° rotation of the spoiler. After the entire dynamic system was modeled, a governing equation was derived. Using Chebyshev spacing for four input positions within a 90° range results in the following values of $\theta$:

$$\theta_1 = 138.4254°, \qquad \theta_3 = 197.2215°$$

$$\theta_2 = 162.7794°, \qquad \theta_4 = 221.5746°$$

For an approximate 60° spoiler rotation, the following values of $\phi$ result:

$$\phi_1 = 10.7342°, \qquad \phi_3 = 51.5953°$$

$$\phi_2 = 33.0073°, \qquad \phi_4 = 60.3387°$$

Four coordinated positions of each crank and the locations of the fixed pivots were specified at the interactive graphics console, after which the computer displayed the $M$–$K$ curves shown in Fig. P3.34b. After selecting several linkages from the curves,

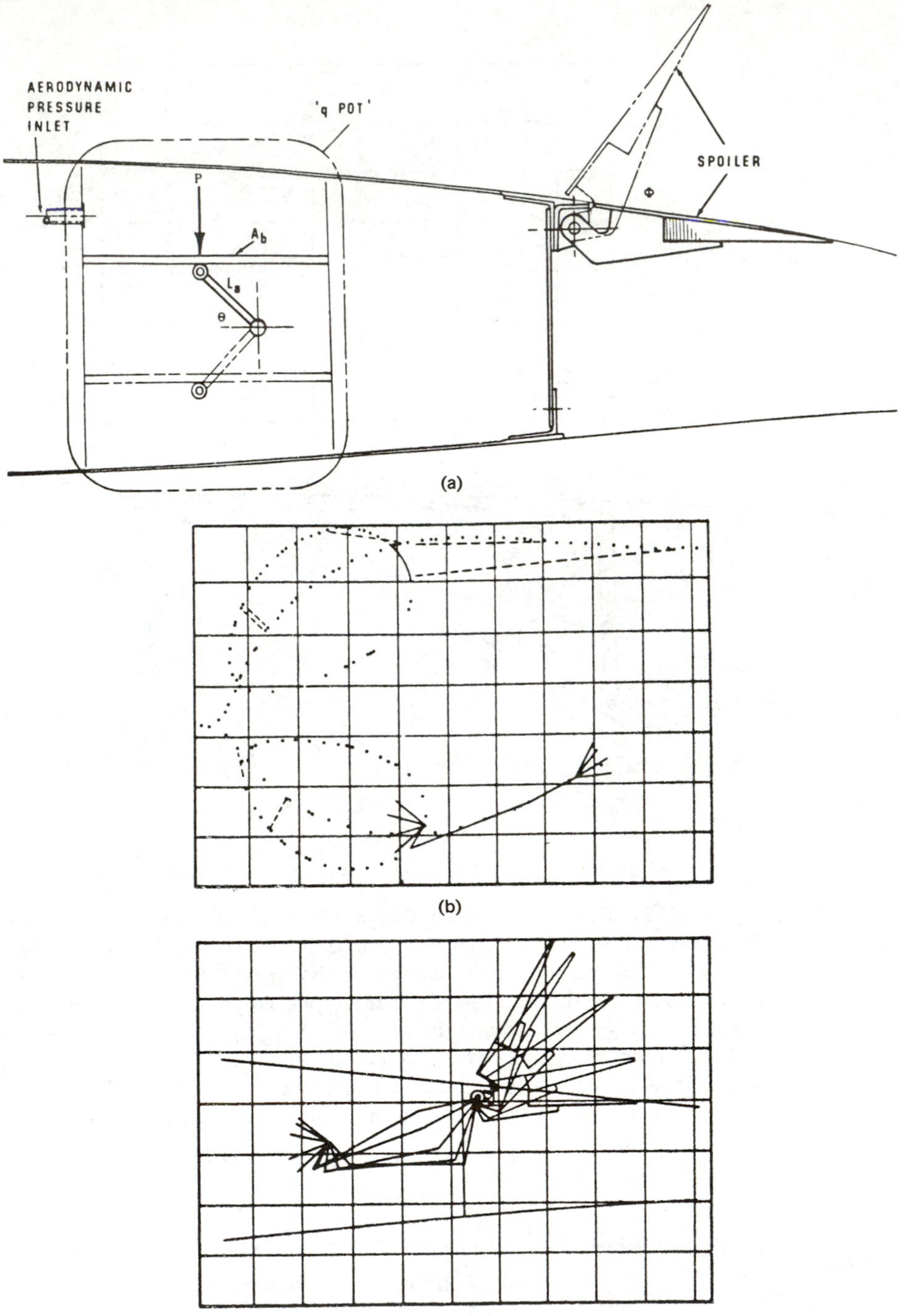


**Figure P3.34** (d) Final design; pilot has only to break linkage past dead-center position and q pot will then assist motion.

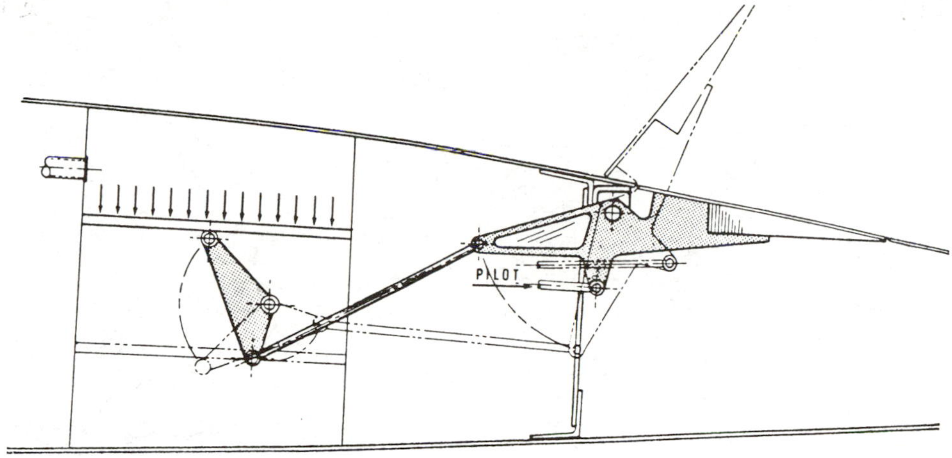


(d)

**Figure P3.34 (cont.)**

the linkage shown in Fig. P3.34 c and d was chosen because it best fits within the space requirements.

**(a)** Using the precision points above, see if you can find the same or a better mechanism for this task.

**(b)** How sensitive is the final choice to small variations in input data (i.e., if you truncate the decimals on the angles, what happens?)

**3.38.** A helicopter skid is to be retracted to clear a large rotating antenna (Fig. P3.35). A mechanism was designed for this task as shown in Fig. P3.35 [201].

**(a)** Determine the degrees of freedom of this mechanism.

**(b)** Write the synthesis equation for this mechanism in the standard form.

**(c)** Pick four or five positions from the figure and design your own retracting linkage.

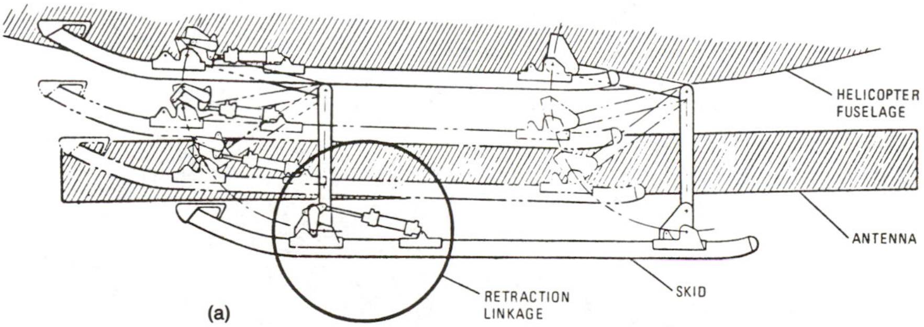


**Figure P3.35** Helicopter skid retraction mechanism.

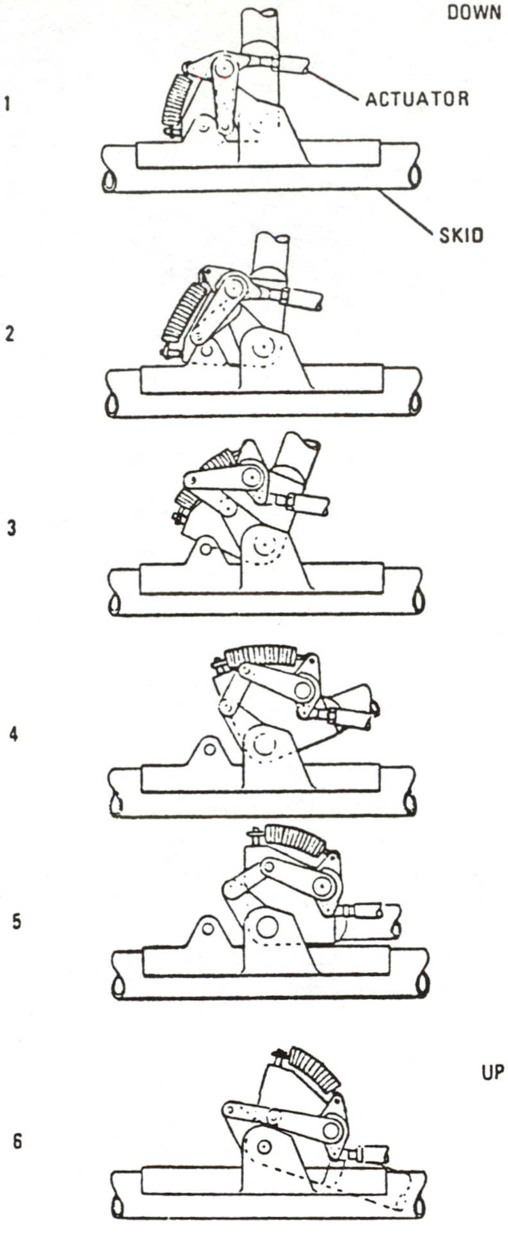


(b)

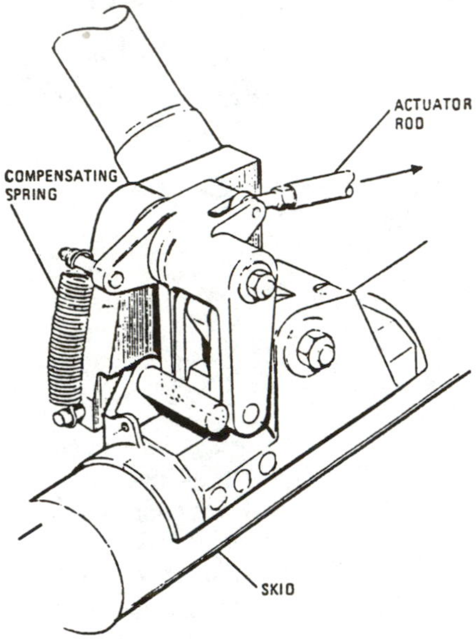


(c)

Figure P3.35 (cont.)

**3.39.** Figure P3.36 shows several suggested mechanisms that have been designed [208,209] to replace bulky, noncosmetic, steel slide joints with linkages located entirely below the stump for the through-knee amputee. These mechanisms exhibit instant centers and fixed centrode (Chap. 4) that pass through the femoral condyle (upper portion of the knee) for stability reasons. These mechanisms are all different than the one shown in Fig. 1.16. For one or more of these designs:

**(a)** Draw the kinematic diagram of the mechanism, and check the degrees of freedom.

**(b)** What type of linkage is this?

**(c)** How would you synthesize such a mechanism for this task?

**(d)** Pick four or five prescribed positions and design your own through-knee prosthesis.

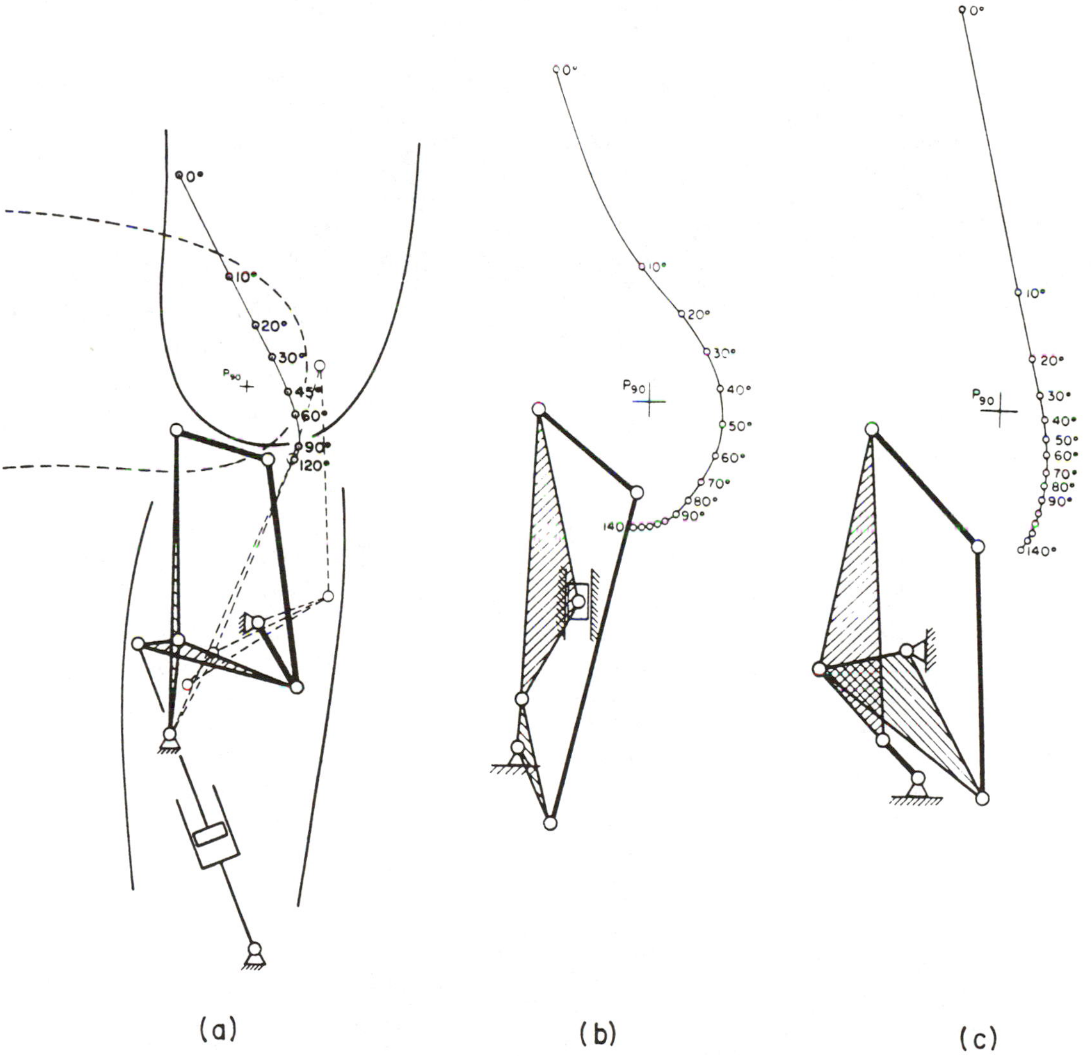


**Figure P3.36**

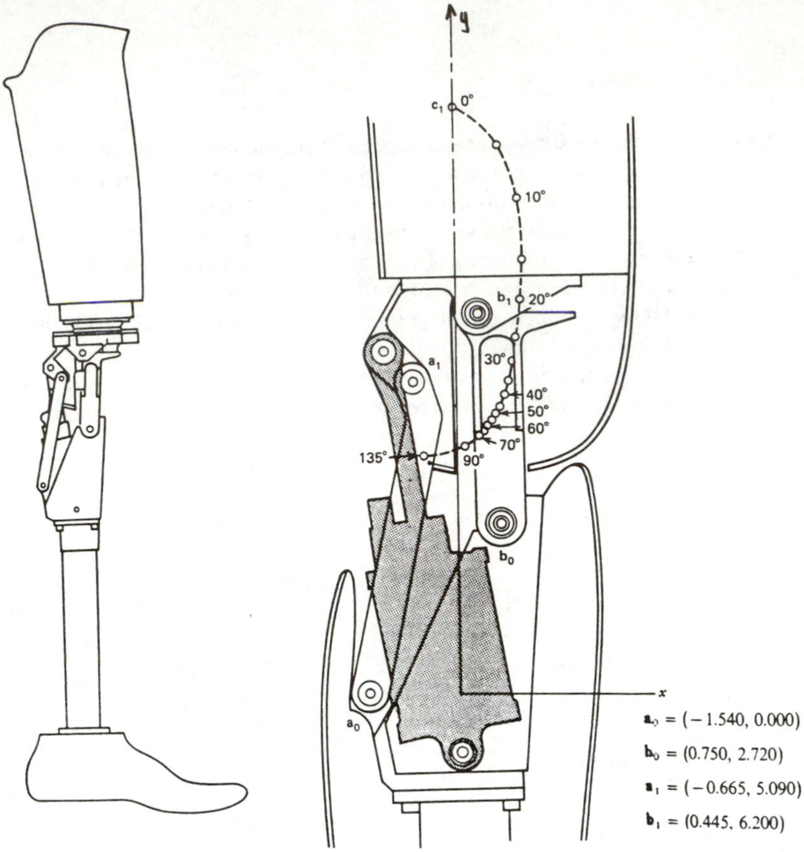


(d)

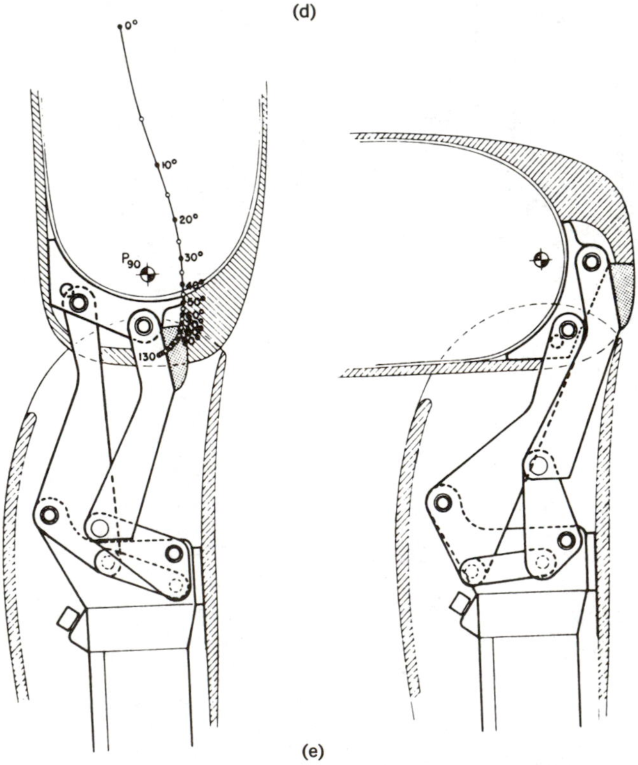


(e)

**Figure P3.36 (cont.)**

**3.40.** A six-link mechanism of Fig. 1.13 was designed for generating the five given path precision positions of the coupler point of link 5:

$$P_1(X_1, Y_1) = (5.0, 6.0)$$

$$P_2(X_2, Y_2) = (3.9, 5.71)$$

$$P_3(X_3, Y_3) = (3.06, 5.202)$$

$$P_4(X_4, Y_4) = (2.716, 4.429)$$

$$P_5(X_5, Y_5) = (3.386, 4.936)$$

Figure P3.37 shows the designed mechanism. The calculated values of the coordinates of $C_0$ and $C_1$ are

$$C_0 = (X, Y) = (2.347846, 2.916081)$$

$$C_1 = (X, Y) = (0.045052, 6.231479)$$

See if you can duplicate these results.

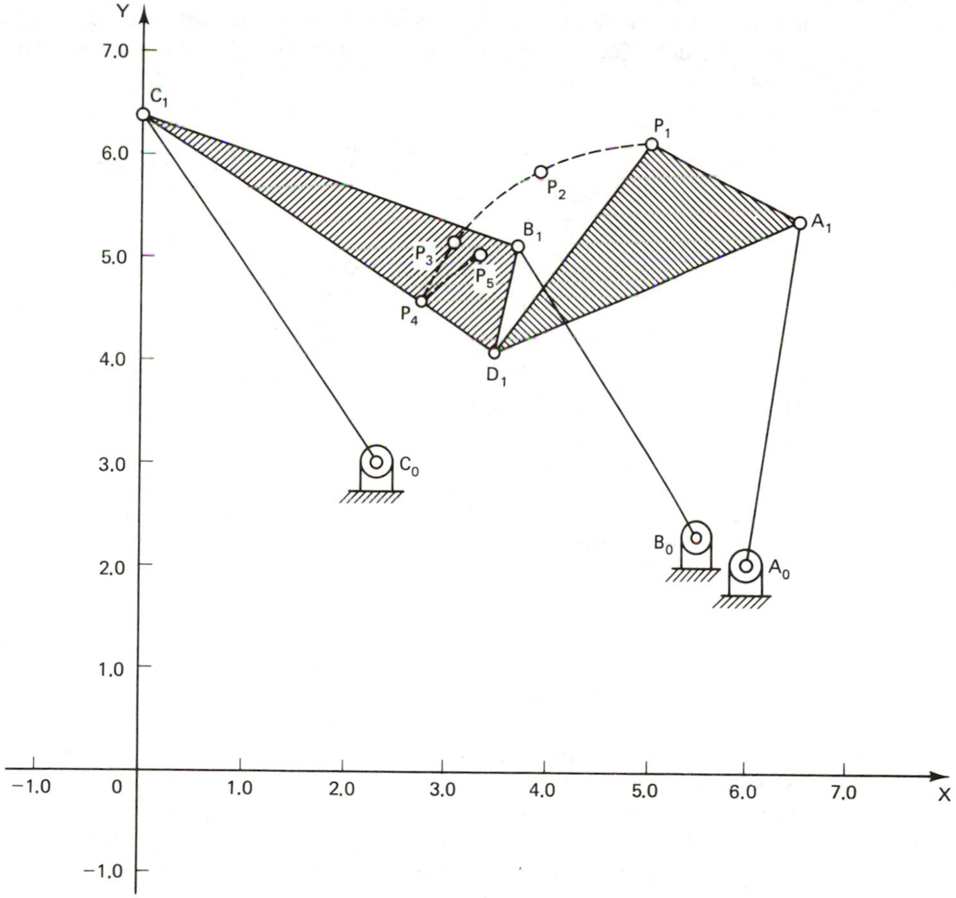


Figure P3.37

**3.41.** Prove that Eq. (3.38) contains only real terms.

**3.42.** Design a four-bar path generator with prescribed timing such that the path is an approximate straight line traveling through the following precision points along a straight line:

| Number | Precision points | | Input angle $\theta°$ |
|---|---|---|---|
| | $X$ | $Y$ | |
| 1 | 0.0 | −0.25 | 0.0 |
| 2 | 1.0 | −0.25 | −16.093 |
| 3 | 1.4 | −0.25 | −21.423 |
| 4 | 1.79 | −0.25 | −26.404 |
| 5 | 2.16 | −0.25 | −31.006 |

**3.43.** Could the approximate straight-line four-bar mechanism of Prob. 3.42 be used to form a dwell mechanism? How?

**3.44.** A Stephenson III six-bar linkage (Figs. 1.13 and 2.66a) is to be designed so that the coupler bodies rotate in opposite directions so as to be used as a "flying shear" or a crimping tool.* The following precision points are to be used for the initial dyad (links 5 and 6):

| Number | Precision points | | Coupler rotation $\alpha°$ |
|---|---|---|---|
| | $X$ | $Y$ | |
| 1 | 0.0 | 0.0 | 0 |
| 2 | −0.625 | −0.22 | 10 |
| 3 | −1.0 | −0.625 | 20 |
| 4 | −1.25 | −1.25 | 37 |
| 5 | −1.0625 | −1.50 | 58 |

Once an acceptable dyad is chosen, design the rest of the linkage (we suggest specifying $\mathbf{Z}_3 = -1.007 - 1.092i$).

| Number | Precision points | | Coupler rotation $\alpha°$ |
|---|---|---|---|
| | $X$ | $Y$ | |
| 1 | 0.0 | 0.0 | 0 |
| 2 | −0.4172 | −0.3809 | −10 |
| 3 | −0.5658 | −0.9036 | −20 |
| 4 | −0.3900 | −1.636 | −32 |
| 5 | −0.3369 | −1.8407 | −45 |

**3.45.** Dry powder ingredients for forming ceramic tile are contained in the hopper. At the proper time of the press cycle, the gate pivots open to dispense the "dust" into a transfer slide which transports the dust to the die cavity on the next stroke.† The hopper and gate are existing. It is desired to use a 2-in.-stroke air cylinder to open and close the gate. It is further desired to have an adjustable gate opening to meter

* This problem contributed by A. S. Adams.

† Suggested by M. Nelson.

the amount of dust. The gate opening is to be variable by 2° increments from 10° to 18° using the constant 2-in. air cylinder stroke. Design a mechanism for this task.

**3.46.** A four-bar path generator with prescribed timing is to be synthesized to generate a sausage-like curve to be used to make a double-dwell mechanism [210]. The five precision points are:

| Number | Precision points (polar form) | | Input angle |
|---|---|---|---|
| | $R$ | $\theta°$ | $\beta°$ |
| 1 | 1.0 | 0.0 | 0.0 |
| 2 | 1.740 | −29.50 | 117.0 |
| 3 | 1.740 | −10.70 | 150.0 |
| 4 | 1.740 | 10.30 | 191.0 |
| 5 | 1.740 | 25.90 | 228.0 |

**(a)** Design an acceptable four-bar for this task.

**(b)** Complete the design for a double-dwell mechanism with good transmission-angle characteristics.

**3.47.** Design a four-bar motion generator for five prescribed positions [210]:

| Number | Precision points (polar form) | | Coupler angle |
|---|---|---|---|
| | $R$ | $\theta°$ | $\alpha°$ |
| 1 | 1.5 | 0.0 | 0.0 |
| 2 | 1.275 | 33.7 | 12.0 |
| 3 | 1.0 | 90.0 | 24.0 |
| 4 | 1.275 | 146.3 | 36.0 |
| 5 | 1.5 | 180.0 | 48.0 |

**3.48.** The table lists three examples of four-bar function generation [210]. The first example is identical with Freudenstein's optimum four-bar function generator based on Chebyshev spacing [104].

| | (A) | (B) | (C) |
|---|---|---|---|
| Function | $X^2$ | $-\frac{X}{8}(X+2)$ | $X + \sin X$ |
| Interval of $X$ | $0 \leq x \leq 1$ | $0 \leq X \leq 6$ | $0 \leq X \leq 1$ |
| Range of $\phi$ (deg) | 90.0 | 100.0 | 85.0 |
| Range of $\psi$(deg) | 90.0 | 60.0 | 60.0 |
| Precision points: | | | |
| $X_1$ | 0.033689272 | 0.1468304511 | 0.02447174185 |
| $X_2$ | 0.24917564 | 1.1236644243 | 0.2061073739 |
| $X_3$ | 0.54280174 | 3.00000000 | 0.50000000 |
| $X_4$ | 0.81636273 | 4.763355757 | 0.7938926261 |
| $X_5$ | 0.9786319 | 5.8316954 | 0.9755282581 |

Design one or more of these five-point function generators.

**3.49.*** In the design of machinery, it is often necessary to use a mechanism to convert uniform input rotational motion into nonuniform output rotation or reciprocation. Mechanisms designed for such purposes are almost invariably based on four-bar linkages. Such linkages produce a sinusoidal output that can be modified to yield a variety of motions.

Four-bar linkages cannot produce dwells of useful duration. A further limitation of four-bar linkages is that only a few types have efficient force-transmission capabilities. Nevertheless, the designer may not choose to use a cam when a dwell is desired and accept the inherent speed restrictions and vibration associated with cams. Therefore, he/she goes back to linkages.

One way to increase the variety of output motions of a four-bar linkage, and obtain longer dwells and better force transmissions, is to add a link and a gear set.

Figure P3.38a shows a practical geared five-bar configuration including paired

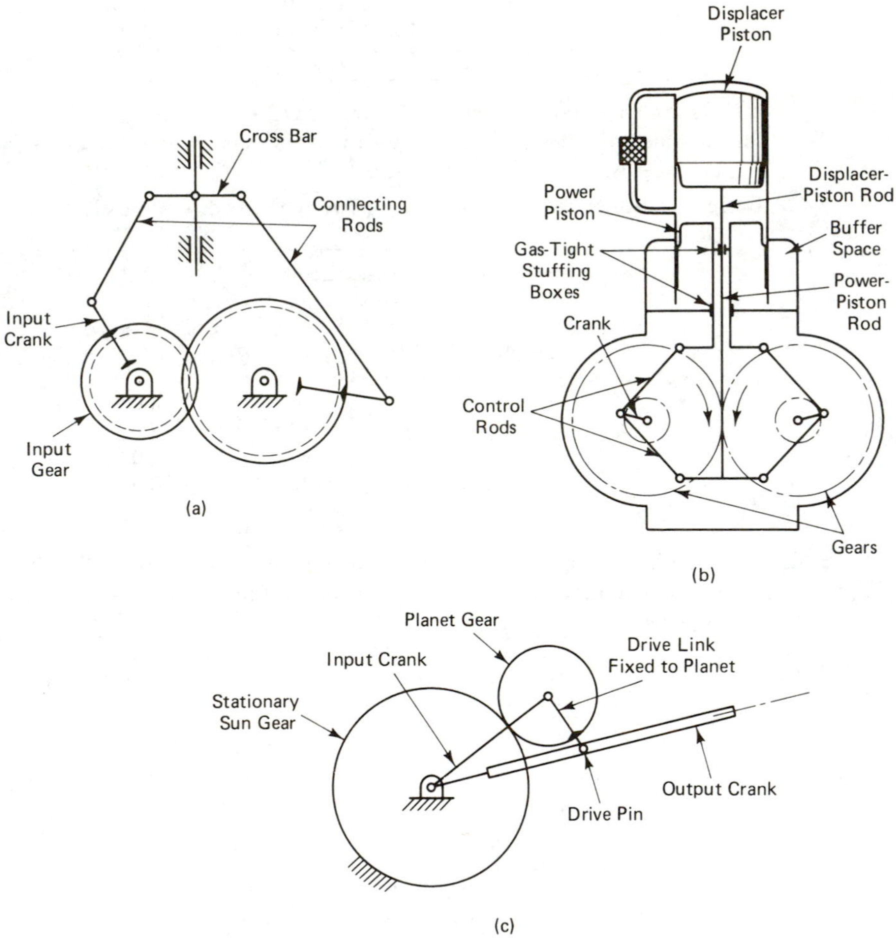


**Figure P3.38** (a) fixed-crank external gear system; (b) stirling engine system; (c) external planetary gear system.

* This problem adapted from Ref. 28.

external gears pinned rotatably to ground. The coupler link (cross bar) is pinned to a slider. The system has been successful in high-speed machines, where it transforms rotary motion into high-impact linear motion. A similar system (Fig. P3.38b) is used in a Stirling engine.

**(a)** Verify the degrees of freedom of geared linkages in Fig. P3.38a and b.

**(b)** Draw all the inversions of the geared five-bar in Fig. P3.38a.

**(c)** Figure P3.38c shows a different type of geared dwell mechanism using a slotted output crank.
Verify the degrees of freedom of this mechanism.

**3.50.** A multiloop dwell linkage has been designed* as a combination punching and indexing device. The principle used is based on synthesizing a nearly circular portion of a four-bar coupler curve. While the four-bar traces that portion of the curve, a dyad pinned to the tracer point at one end and to ground at the other (the intermediate point being located at the center of curvature of the path) will exhibit a near dwell in the dyad segment that is pivoted to ground. Figure P3.39a and b show photographs of the drive in two positions. A computer-generated animation of the motion of the dwell mechanism at 20° increments of the input crank angle is displayed in Fig. P3.39c. The complex-number method was used to design a portion of this linkage.

**(a)** Determine the degrees of freedom of this mechanism.

**(b)** Describe the function of each loop of the dwell mechanism.

**(c)** Describe how this mechanism could be synthesized using the standard-form approach.

(a)

(b)

**Figure P3.39**

* By W. Farrell, D. Johnson, and M. Popjoy under the direction of A. Midha at Pennsylvania State University, September 1980 and described in Ref. [185].

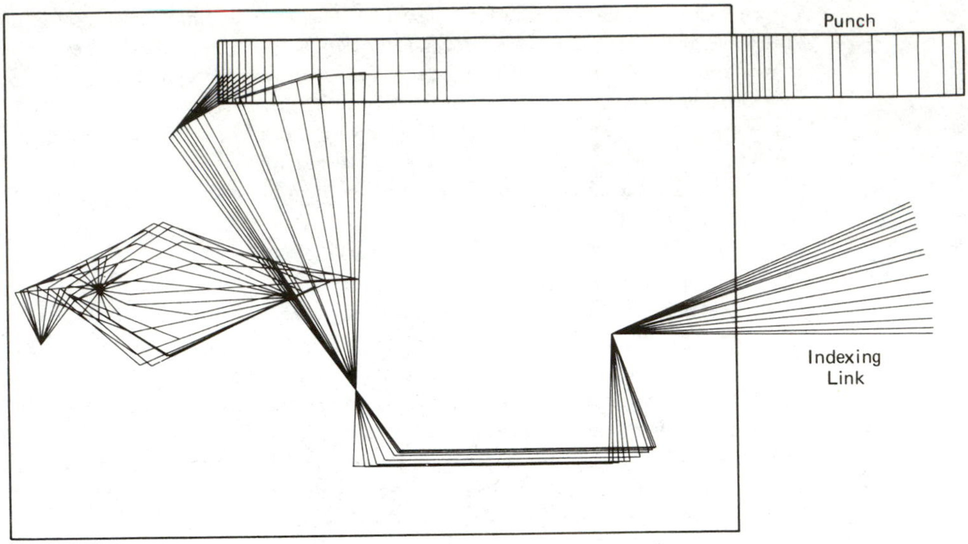


(c)

**Figure P3.39 (cont.)**

# 4 Path Curvature Theory

## 4.1 INTRODUCTION

In the synthesis or analysis of linkages it is often of interest to find points in the moving plane which exhibit certain path characteristics, such as to trace a curve of a certain radius or trace a straight line. The techniques presented here are useful in both analyzing the path-curvature characteristics of points on a given linkage and synthesizing new designs based on path curvature considerations. The finite kinematic synthesis methods presented in this text are nicely complemented by the following treatment, which is based on infinitesimal displacements.

## 4.2 FIXED AND MOVING CENTRODES

Before proceeding to the study of the path curvature of tracer points on a moving link, an important concept must be introduced, that of the fixed and moving centrodes. As a linkage moves, a typical floating link (one without a fixed pivot) will move in planar motion with respect to ground in a combination of translational and rotational motion. When only two finitely separated positions are involved, a single link rigidly attached to the moving link and pinned to ground will guide it through these two positions. In this case, a pure rotational motion will take the moving link through the two prescribed positions. Such a ground pivot is called a *pole.*

Two positions of a floating link are shown in Fig. 4.1 together with the procedure for finding the pole $P_{12}$. The procedure for locating the pole is to connect point $A_1$ to $A_2$ and point $B_1$ to $B_2$ by straight lines. The midnormals (perpendicular

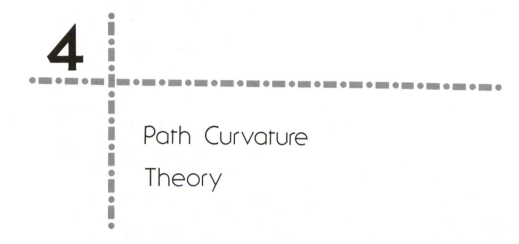

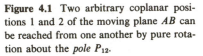

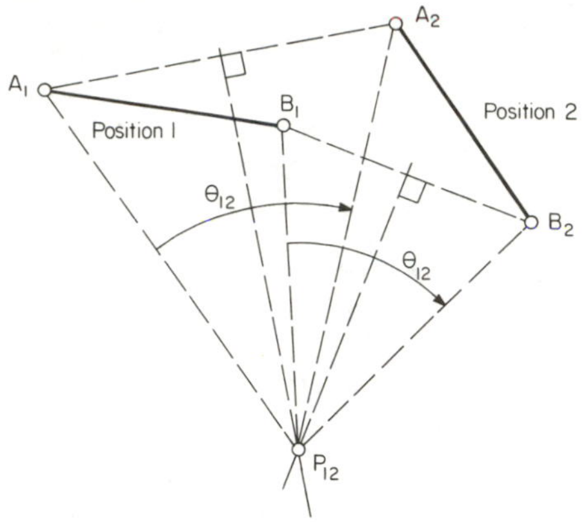


**Figure 4.1** Two arbitrary coplanar positions 1 and 2 of the moving plane $AB$ can be reached from one another by pure rotation about the *pole* $P_{12}$.

bisectors) to the lines $A_1$–$A_2$ and $B_1$–$B_2$ are constructed as in Fig. 4.1. Their intersection is the pole $P_{12}$.

The pole is not the same as an instant center. The "instant center" is a momentary center of rotation of the floating link. Rotation about this center can match the instantaneous velocity of every point on the link with that in the actual link motion, while the pole deals with two finitely separated positions of the moving plane. The instant center can be thought of as the limiting case of a pole as the angle $\theta_{12}$ of Fig. 4.1 goes to zero.

An important use of the pole concept is observed when more than two positions of the moving plane are considered. The two positions of the floating link in Fig. 4.1 are redrawn in Fig. 4.2 together with a third and a fourth position.

$P_{12}$ is the pole for link positions 1 and 2, $P_{23}$ is the pole for positions 2 and 3, and so on. Thus there are three consecutive poles for the four link positions, and in general if $n$ consecutive positions of the floating link are considered, there will be $n - 1$ consecutive poles, one associated with each incremental motion between successive positions.* Adjacent consecutive poles are connected by lines in Fig. 4.2. If a large number of positions of the coupler link is considered and each coupler† location is infinitesimally close to the adjacent position, then in general, the poles will form a smooth curve instead of a polygon. This smooth curve is the *fixed centrode.*

A schematic drawing of a fixed centrode and four positions of the moving link are shown in Fig. 4.3, together with poles $P_{12}$, $P_{23}$, and $P_{34}$. Since the instant center of the moving link with respect to ground is the limiting case of the pole as the displacement of the coupler link goes to zero, it is clear that such instant centers will always lie on the fixed centrode. In fact, the fixed centrode is the locus of all instant centers for the motion of the floating link with respect to ground.

* These are not the only poles. Any two positions, $j$ and $k$, consecutive or not, have a pole $P_{jk}$ associated with them. Note also that $P_{kj} \equiv P_{jk}$.

† *Coupler* is used here interchangeably with *floating link,* since both words can apply to the connecting link of a four-bar mechanism.

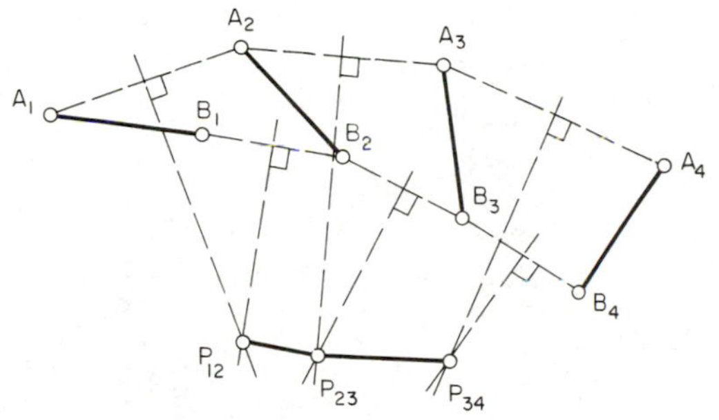


**Figure 4.2** The *fixed centrode* is the limiting case of the *pole-polygon* as the number of intermediate positions is increased and adjacent consecutive poles move infinitesimally close together (see Fig. 4.3).

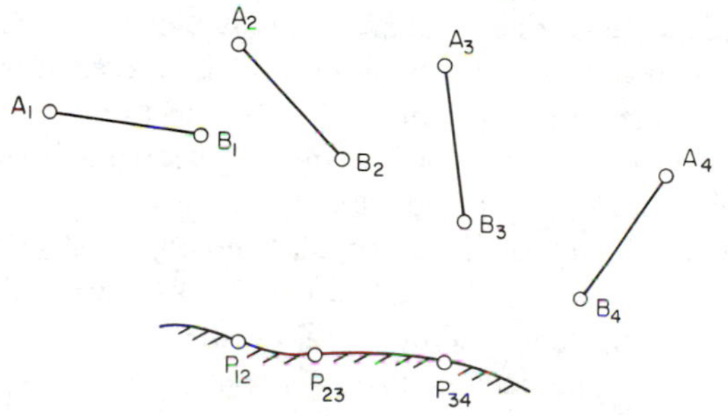


**Figure 4.3** The fixed centrode is the locus of all instant centers for the motion of link *AB*.

Now consider three corresponding link positions, as shown in Fig. 4.4. The motion will be inverted as follows. Poles $P_{12}$ and $P_{23}$ are connected by a line, and this line is rotated by an angle $-\theta_{12}$ to yield the point $M_{23}$. A line is now drawn to connect $P_{12}$ and $M_{23}$. $M_{23}$ is that point of the moving plane shown in position 1, which moves into coincidence with $P_{23}$ in position 2. We can also define point $M_{12}$ of the coupler plane, coincident with $P_{12}$ in position 1. Additional $M$ points can be constructed as follows. To find $M_{34}$, measure $\phi = \measuredangle P_{12} P_{23} P_{34}$ and rotate the extension of line $M_{12}$–$M_{23}$ about $M_{23}$ by $\phi$ in the same sense as $\phi = \measuredangle P_{12}P_{23}P_{34}$, as shown in Fig. 4.5. Then rotate this new line about $M_{23}$ by $-\theta_{23}$. Locate $M_{34}$ on this line at the same distance from $M_{23}$ as $P_{34}$ is from $P_{23}$. A similar procedure can be repeated for each successive position of the coupler.

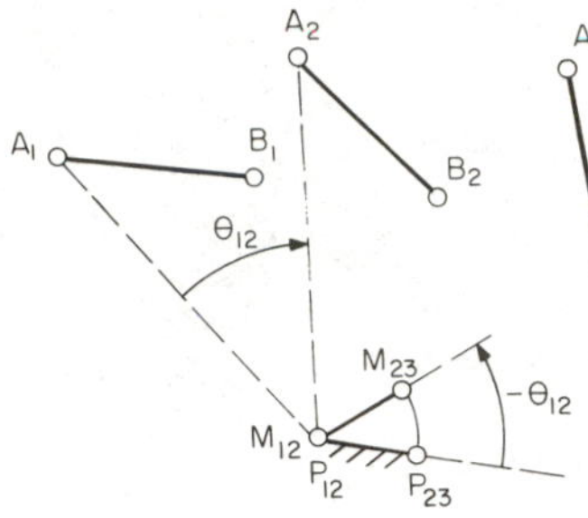


**Figure 4.4** $M_{23}$ is that point of the moving plane in the first position $A_1B_1$ which becomes $P_{23}$ in the second position $A_2B_2$.

To emphasize the fact that $A_1$, $B_1$, $M_{12}$, $M_{23}$, and $M_{34}$ are all points embedded in the moving link, we connected the $M$'s rigidly to $A_1B_1$ in Fig. 4.5. Now, as this five-point system rotates about $P_{12}$ until $M_{23}$ reaches $P_{23}$, $A_1B_1$ will have moved into coincidence with $A_2B_2$. Then, as the five-point system rotates about $P_{23}$ until $M_{34}$ reaches $P_{23}$, $A_1B_1$ will have come into coincidence with $A_3B_3$. In other words, rolling the "moving pole polygon" on the "fixed pole polygon," the link $AB$ will go through the successive corresponding positions $A_jB_j$, $j = 1, 2, 3, 4, \ldots$.

If successive poles are taken infinitely close to each other on the fixed centrode, successive $M$ points will also be infinitely close to each other and thus form a smooth curve. This is the moving centrode, shown in Fig. 4.6. Note that the fixed centrode is fixedly attached to ground. The moving centrode is rigidly attached to the coupler and is shown in Fig. 4.6 in position 1.

The importance of the fixed and moving centrodes (also called "polodes" by some authors) is that the actual motion of a floating link with respect to the fixed link can be duplicated by simply rolling the moving centrode on the fixed centrode without sliding. The fixed centrode is connected to ground and is always stationary, while the moving centrode can be thought of as being rigidly attached to the moving link. So-called "relative centrodes" can be constructed for the motion of any two links with respect to each other, not just for motion of a floating link, such as the coupler of a four-bar with respect to ground, but also for the follower link with respect to the input link, for example.

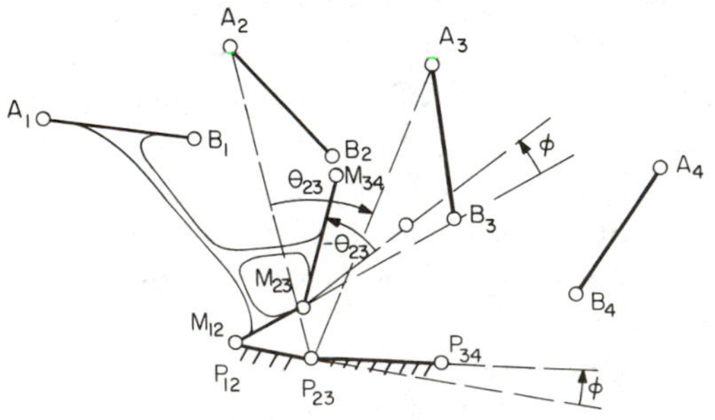


**Figure 4.5** As the *moving pole-polygon* $M_{12}M_{23}M_{34}$ rolls on *fixed pole-polygon* $P_{12}P_{23}P_{34}$, the plane $AB$ moves through positions $A_1B_1$, $A_2B_2$, $A_3B_3$, $A_4B_4$.

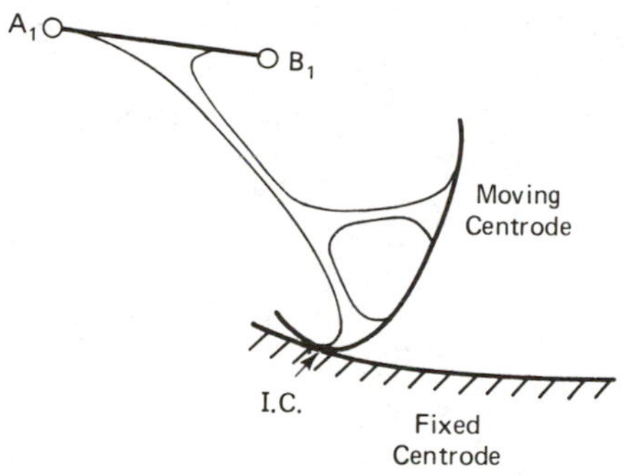


**Figure 4.6** As intermediate consecutive positions increase in number, moving and fixed poles are closer and closer. In the limit, their polygons become the *moving* and *fixed centrodes*. The point of tangency of the rolling centrodes is the *instant center*, that point of the moving plane AB which is momentarily stationary, that is, it has zero velocity.

As shown in Fig. 4.6, as the moving centrode rolls on the fixed centrode, there is generally a single point of contact between them; this point is the instant center of the moving link with respect to ground. This is also true for relative centrodes, in which case their point of contact is the relative instant center.

## 4.3 VELOCITIES

Consider the fixed and moving centrodes, $p$ and $\pi$, for the planar motion of some arbitrary plane with respect to the fixed plane of reference, as shown in Fig. 4.7. Although some authors call these "polodes," poles are centers of finite rotations, while centrodes are loci of instant centers. Therefore, we use the term "centrodes."

Now let us define the term "osculating circle." Consider a general curve $C$ (Fig. 4.8) and a point $A_0$ on it. For finding the radius of curvature of $C$ at point $A_0$, we take two points $A_{-1}$ and $A_1$ on either side of $A_0$. The perpendicular bisectors of the secants $A_{-1}A_0$ and $A_0A_1$ intersect at a point $O$. With $O$ as the center and $OA_0$ as the radius ($OA_{-1} = OA_0 = OA_1$) we can always draw a circle passing through the three points $A_{-1}$, $A_0$, and $A_1$. Now let the points $A_{-1}$ and $A_1$ approach point $A_0$ (i.e., the arc lengths $\Delta S \to 0$; see Fig. 4.8). In the limiting case, the circle passing through the three points as $\Delta S \to 0$ is called the "osculating circle." The radius of the osculating circle, ρ, is the radius of curvature of curve $C$ at point $A_0$ and the center of the osculating circle, $O$, is the center of curvature. Thus the osculating circle has contact with curve $C$ at three infinitesimally close positions. The vector $\overrightarrow{OA_0}$ is called the "radius vector of curvature at $A_0$," and is designated by $\boldsymbol{\rho}_{A_0}$.

In Fig. 4.7 the osculating circle radii $\boldsymbol{\rho}_p$ and $\boldsymbol{\rho}_\pi$, which are the respective radius vectors of curvature of the fixed and moving centrodes, are shown. These circles have three-point contact with the respective centrodes at their point of tangency, which is the instant center $I$ of the fixed and moving planes. As the motion progresses, $\pi$ rolls on $p$ without slipping. Therefore, the osculating circle of radius $\rho_\pi$ rolls without slipping on the osculating circle of radius $\rho_p$ through three infinitesimally close positions.

Let us define the fixed-coordinate system embedded in the fixed plane at $O_p, x, iy$. Also, let $A$ be a point of the $\pi$ plane. The first (present) position of $A$ is $A_1$,

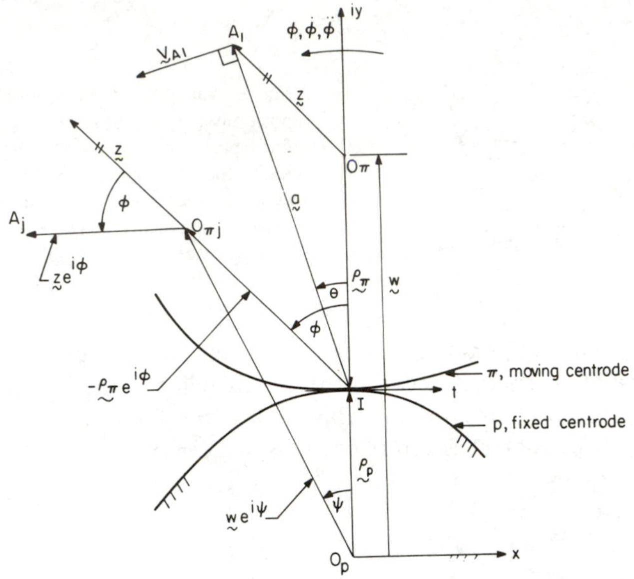


Figure 4.7 Moving centrode $\pi$ is embedded in the moving plane; fixed centrode $p$ is embedded in the fixed plane. The *osculating circles* (see Fig. 4.8) of $\pi$ and $p$ (radii $\boldsymbol{\rho}_\pi$ and $\boldsymbol{\rho}_p$) roll on one another through three infinitesimally close points at the instant center $I$.

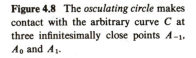

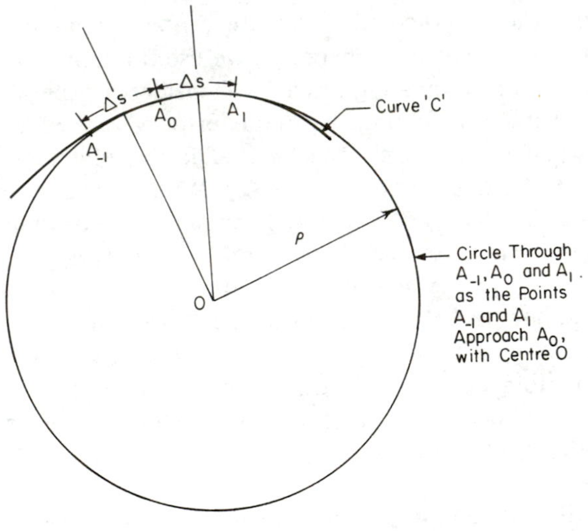


**Figure 4.8** The *osculating circle* makes contact with the arbitrary curve $C$ at three infinitesimally close points $A_{-1}$, $A_0$ and $A_1$.

located by the following vectors, as shown in Fig. 4.7:

$$\overrightarrow{O_pA_1} = \mathbf{w} + \mathbf{z} \tag{4.1}$$

where

$$\mathbf{w} = \boldsymbol{\rho}_p - \boldsymbol{\rho}_\pi \tag{4.2}$$

$$\mathbf{z} = \boldsymbol{\rho}_\pi + \mathbf{a} \tag{4.3}$$

and **a** is the vector $\overrightarrow{IA_1}$ on the "ray" emanating from $I$ in the direction $\theta$ with respect to the $iy$ axis, the common normal of the centrodes at $I$.

After an infinitesimally small displacement of point $A_1$ to point $A_j$, its position vector becomes

$$\overrightarrow{O_pA_j} = \mathbf{w}e^{i\psi} + \mathbf{z}e^{i\phi}$$

where $\psi$ and $\phi$ are infinitesimally small rotations of the vectors **w** and **z**, respectively.

Since the osculating circles contact the centrodes for three infinitesimally close positions of the moving plane, the centrodes can be represented by their osculating circles for up to and including the second derivative of displacement (i.e., for the purpose of velocity and acceleration analyses). Thus the vectors **w** and **z** can be considered constant through three infinitesimally close positions, or two consecutive infinitesimally close displacements, the first of which is from position 1 to $j$.

Now let us focus on the center of the osculating circle of the moving centrode, $O_\pi$. After an infinitesimally small displacement to $O_{\pi j}$, its location becomes

$$\overrightarrow{O_pO_{\pi j}} = (\boldsymbol{\rho}_p - \boldsymbol{\rho}_\pi)e^{i\psi} \qquad \text{or} \qquad \overrightarrow{IO_{\pi j}} = -\boldsymbol{\rho}_\pi e^{i\phi}$$

Therefore, we can express the velocity of $O_\pi$ in two ways:

$$\mathbf{V}_{O\pi} = \dot{\overrightarrow{O_pO_{\pi j}}} = \dot{\overrightarrow{IO_{\pi j}}}$$

or

$$\mathbf{V}_{O\pi} = (\boldsymbol{\rho}_p - \boldsymbol{\rho}_\pi)i\dot{\psi}e^{i\psi} = -\boldsymbol{\rho}_\pi i\dot{\phi}e^{i\phi} \tag{4.4}$$

and

$$(\boldsymbol{\rho}_p - \boldsymbol{\rho}_\pi)\dot{\psi}e^{i\psi} = -\boldsymbol{\rho}_\pi\dot{\phi}e^{i\phi}$$

from which

$$\dot{\psi} = \frac{-\boldsymbol{\rho}_\pi}{\boldsymbol{\rho}_p - \boldsymbol{\rho}_\pi}\frac{e^{i\phi}}{e^{i\psi}}\dot{\phi} \tag{4.5}$$

In the starting position $\phi = \psi = 0$, and

$$\dot{\psi} = \frac{-\boldsymbol{\rho}_\pi}{\boldsymbol{\rho}_p - \boldsymbol{\rho}_\pi}\dot{\phi} \tag{4.6}$$

or by defining

$$\boldsymbol{\rho} = \frac{-\boldsymbol{\rho}_\pi}{\boldsymbol{\rho}_p - \boldsymbol{\rho}_\pi} \tag{4.7}$$

$$\dot{\psi} = \boldsymbol{\rho}\dot{\phi} \tag{4.8}$$

Let us examine the nature of the vector $\boldsymbol{\rho}$. When the osculating circles are on opposite sides of their common tangent $t$, as shown in Fig. 4.7,

$$\boldsymbol{\rho} = \frac{-\rho_\pi e^{-i\pi/2}}{\rho_p e^{i\pi/2} - \rho_\pi e^{-i\pi/2}} = \frac{\rho_\pi e^{i\pi/2}}{\rho_p e^{i\pi/2} + \rho_\pi e^{i\pi/2}} = \frac{\rho_\pi}{\rho_p + \rho_\pi}\begin{cases} > 0 \\ < 1 \end{cases} \tag{4.7a}$$

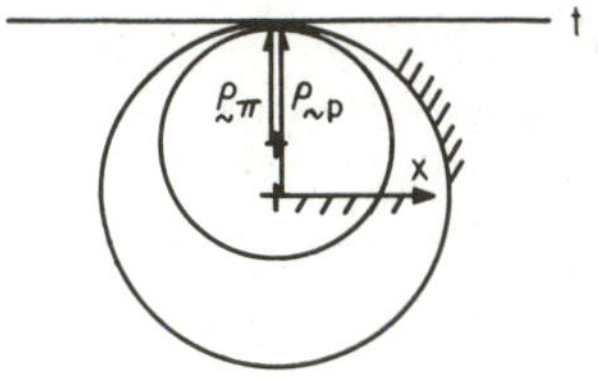


(a)

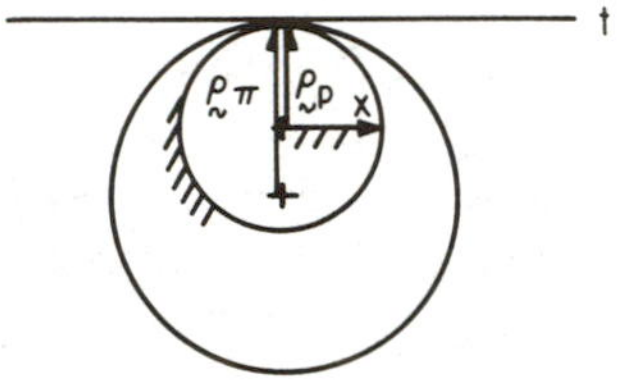


(b)

**Figure 4.9** Value of the vector

$$\boldsymbol{\rho} = \frac{-\boldsymbol{\rho}_\pi}{\boldsymbol{\rho}_p - \boldsymbol{\rho}_\pi}:$$

in (a) $\boldsymbol{\rho}$ is a negative real; in (b) $\boldsymbol{\rho}$ is a positive real, always greater than 1 [see Eq. (4.7b) and (c)].

So for this case $\boldsymbol{\rho}$ is a positive real. However, if the osculating circles are on the same side of their common tangent and $\rho_p > \rho_\pi$ (Fig. 4.9a),

$$\boldsymbol{\rho} = \frac{-\rho_\pi e^{i\pi/2}}{\rho_p e^{i\pi/2} - \rho_\pi e^{i\pi/2}} = \frac{-\rho_\pi}{\rho_p - \rho_\pi} < 0 \tag{4.7b}$$

Thus here $\boldsymbol{\rho}$ is a negative real. But if $\rho_p < \rho_\pi$, as shown in Fig. 4.9b, then

$$\boldsymbol{\rho} = \frac{-\rho_\pi e^{i\pi/2}}{\rho_p e^{i\pi/2} - \rho_\pi e^{i\pi/2}} = \frac{-\rho_\pi}{\rho_p - \rho_\pi} = \frac{\rho_\pi}{\rho_\pi - \rho_p} > 1 \tag{4.7c}$$

Here $\boldsymbol{\rho}$ is a positive real greater than 1.

These values of $\boldsymbol{\rho}$ can be verified by considering $\psi$ and $\phi$ for a small rolling displacement of the $\pi$ circle on the $p$ circle in Figs. 4.7 and 4.9a, as shown in Fig. 4.10.

In Fig. 4.10a, the osculating circles are on opposite sides of their common tangent, and it is clear that, for a small positive (CCW) displacement, $\psi < \phi$ and both have the same sense. Therefore, $\dot{\psi} < \dot{\phi}$ and they have the same sense. This verifies that, for this case, in Eq. (4.8), $0 < \boldsymbol{\rho} < 1$, and is real, as shown above analytically.

In Fig. 4.10b, the osculating circles are on the same side of $t$, and $\rho_p > \rho_\pi$. Here, $\psi$ and $\phi$ are of opposite sense, and $\left|\psi\right| \gtreqless \left|\phi\right|$. This verifies that, for this case, $\boldsymbol{\rho} < 0$, and is real, as derived above.

Finally, in Fig. 4.10c, the osculating circles are on the same side of $t$, but now $\rho_p < \rho_\pi$, and, while $\psi$ and $\phi$ have the same sense, $\psi$ is always greater than $\phi$. Hence the fact that, for this case, $\boldsymbol{\rho} > 1$, and is real, derived above analytically, is now verified by geometry.

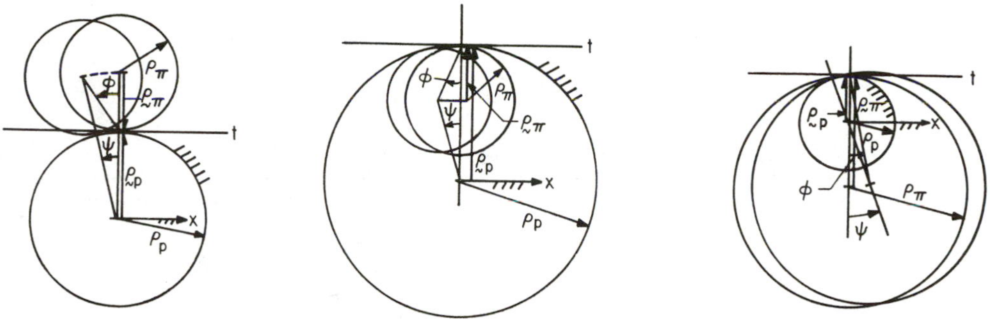


**Figure 4.10** Value of the vector $\boldsymbol{\rho}$ of Eq. 4.7 verified geometrically by way of visualizing Eq. (4.8): $\dot{\psi} = \boldsymbol{\rho}\dot{\phi}$. For the three configurations of this figure: (a) $\boldsymbol{\rho}$ is a positive real, less than 1; (b) $\boldsymbol{\rho}$ is a negative real; (c) $\boldsymbol{\rho}$ is a positive real, greater than 1.

Instead of the vector notation of Eq. (4.7), we could define $\boldsymbol{\rho}$ by the scalar notation shown at the end of Eqs. (4.7a), (4.7b), and (4.7c), using only the magnitudes $\rho_p$ and $\rho_\pi$. But then it would be necessary to set up a "sign convention" for these magnitudes, which destroys the generality of the vector-form equation (4.7), which is the same for all cases. Since the complex vector form is easily programmed in FORTRAN, this work will use that form and dispense with the traditional sign conventions. The practical advantage of the generality of the use of complex numbers, which makes the derived expressions universally valid for all geometric configurations, relieving us of traditional, hard-to-program sign conventions, will be evident many more times in this treatment of path-curvature theory. The relationship of the geometric configuration of the osculating circles and the value of $\boldsymbol{\rho}$ is pictured in Fig. 4.11.

It is also noteworthy that, although for convenience in derivation we used a coordinate system with its real and imaginary axes aligned with the common tangent and common normal of the roll curves (centrodes) $p$ and $\pi$, the results of Eqs. (4.7a), (4.7b), and (4.7c) are independent of the choice of coordinate system. This is true because $\boldsymbol{\rho}_p$ and $\boldsymbol{\rho}_\pi$ are always collinear, and therefore their unit vectors cancel out, leaving only their absolute values, $\rho_p$ and $\rho_\pi$ in the expression, *with the appropriate algebraic signs.* For example, take Eq. (4.7a), and instead of $\pi/2$, put any arbitrary angle $\phi$ in the exponentials. It is clear that the result is the same real quantity as before. This independence of the choice of coordinate systems applies to all derivations with complex numbers. Therefore, in all derivations and also in solving problems, we are free to choose any convenient coordinate system.

To find the velocity of point $A$, we start by taking the derivative of the position vector of $A_j$ (see Fig. 4.7):

$$\mathbf{V}_{AJ} = \dot{\overrightarrow{O_pA_j}} = \mathbf{w}i\dot{\psi}e^{i\psi} + \mathbf{z}i\dot{\phi}e^{i\phi} \tag{4.9}$$

When referred back to position 1, where $\psi = \phi = 0$, Eq. (4.9) becomes

$$\mathbf{V}_{A1} = i\dot{\psi}\mathbf{w} + i\dot{\phi}\mathbf{z} \tag{4.10}$$

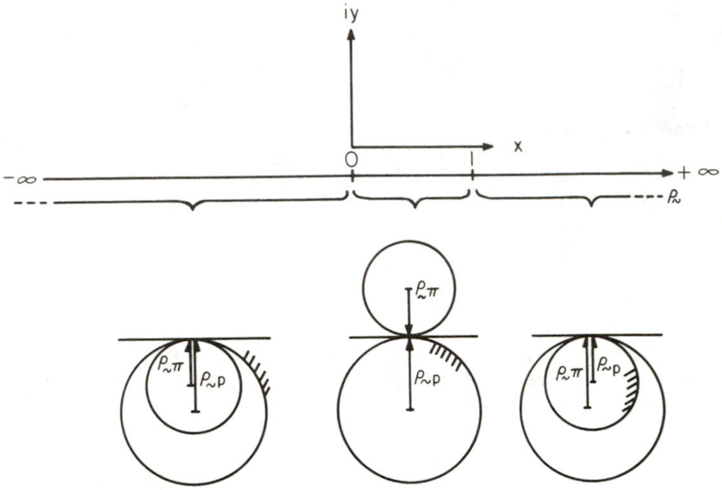

**Figure 4.11** Correlation of the values of the vector $\boldsymbol{\rho}$ with various geometric configurations of roll-curve osculating circles.

After substituting Eqs. (4.2), (4.3), and (4.8) into Eq. (4.10), we obtain

$$\mathbf{V}_{A1} = i\boldsymbol{\rho}\dot{\phi}\mathbf{w} + i\dot{\phi}(\mathbf{a} + \boldsymbol{\rho}_\pi) \tag{4.11}$$

or

$$\mathbf{V}_{A1} = i\dot{\phi}(\mathbf{a} + \boldsymbol{\rho}\mathbf{w} + \boldsymbol{\rho}_\pi) \tag{4.12}$$

But

$$\boldsymbol{\rho}\mathbf{w} + \boldsymbol{\rho}_\pi = \frac{-\boldsymbol{\rho}_\pi}{\boldsymbol{\rho}_p - \boldsymbol{\rho}_\pi}(\boldsymbol{\rho}_p - \boldsymbol{\rho}_\pi) + \boldsymbol{\rho}_\pi = 0 \tag{4.13}$$

Therefore,

$$\mathbf{V}_{A1} = i\dot{\phi}\mathbf{a} \tag{4.14}$$

which verifies the well-known fact that the absolute velocity of a point of a moving plane is equal to the velocity difference between that point and the instant center of that plane with respect to the fixed plane (see Chap. 3 of Vol. 1). Example 4.1 illustrates this point.

**Example 4.1**

Given the four-bar linkage $O_AABO_B$ of Fig. 4.12 and the angular velocity of its coupler, find the absolute velocity of coupler point $C$.

**Solution** Use Eq. (4.14):

$$\mathbf{V}_C = i\dot{\phi}\mathbf{a} \tag{4.15}$$

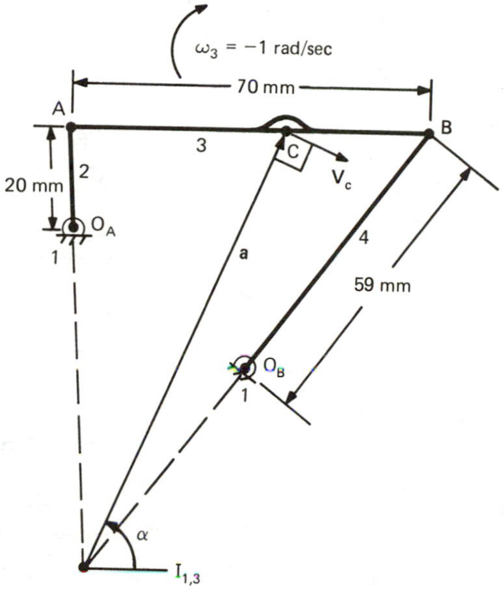


**Figure 4.12** Calculating the velocity of the coupler point $C$ in the four-bar mechanism $O_AABO_B$ by way of the instant center $I_{1,3}$ (see Ex. 4.1).

First, note that in Eq. (4.15), $\dot{\phi} = \omega_3 = -1$ rad/sec. To find **a**, construct the instant center $I_{1,3}$ and draw the vector $\overrightarrow{I_{1,3}C}$. Then $\mathbf{V}_C = i\dot{\phi}\mathbf{a} = i(-1)ae^{i\alpha}$, where $a$ is measured to be 9.3 cm and $\alpha$ is the argument of **a**, also measured from the drawing. Thus, $\mathbf{V}_C = -(9.3)ie^{i\alpha}$ cm/sec, as shown in the drawing to a scale different from that of **a.**

So far in Ex. 4.1 we used a combined graphical–analytical approach, reminiscent of the instant center method of velocity analysis discussed in Chap. 3 of Vol. 1. Applying this method to the four-bar mechanism of Fig. 4.13, we measure $|\mathbf{c}| = |\overrightarrow{IC}| = 105$ mm, and $\mathbf{V}_c = -105$ (mm/sec) $ie^{i\arg(\mathbf{c})}$. Now we will show how this solution can be obtained by automatic computation. To this end we designate each moving link of the given FBL as a vector (see Fig. 4.13). These are $\mathbf{z}_2$, $\mathbf{z}_3$, and $\mathbf{z}_4$. We specify these with reference to any convenient coordinate system, say, letting $O_A$ be the origin and the $iy$ axis taken in the direction and sense of $\mathbf{z}_2$. It will be clear from this example that the results of all our derivations so far are independent of the choice of the coordinate system. The one chosen for the derivations was most convenient for that purpose, but it did not impair the generality of the outcome.

In the chosen coordinate system, $\mathbf{z}_2 = 0 + i(23)$, $\mathbf{z}_3 = 78 - i8$, $\mathbf{z}_4 = 47.5 + i(54)$, where all lengths are millimeters, $\lambda_c = 0.6$, and $\dot{\phi}_3 = -1$ rad/sec. The computation flowchart is shown in Fig. 4.14. The first task is to find the instant center $I$:

$$\mathbf{I} = \overrightarrow{O_AI} = \lambda_2\mathbf{z}_2 = \mathbf{z}_2 + \mathbf{z}_3 - \mathbf{z}_4(1 + \lambda_4)$$

or

$$\mathbf{z}_2(1 - \lambda_2) + \mathbf{z}_3 - \mathbf{z}_4(1 + \lambda_4) = 0 \tag{4.16}$$

$$\mathbf{I} = \lambda_2\mathbf{z}_2 \tag{4.17}$$

$$\overrightarrow{IC} = \mathbf{c} = -\mathbf{I} + \mathbf{z}_2 + \lambda_c\mathbf{z}_3 \tag{4.18}$$

$$\mathbf{V}_C = i\dot{\phi}_3\mathbf{c} \tag{4.19}$$

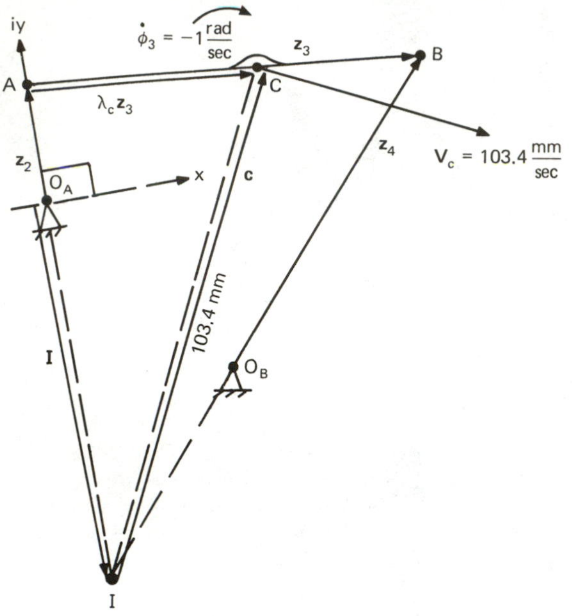


**Figure 4.13** Example 4.1: Complex-number calculation of the velocity of a coupler point.

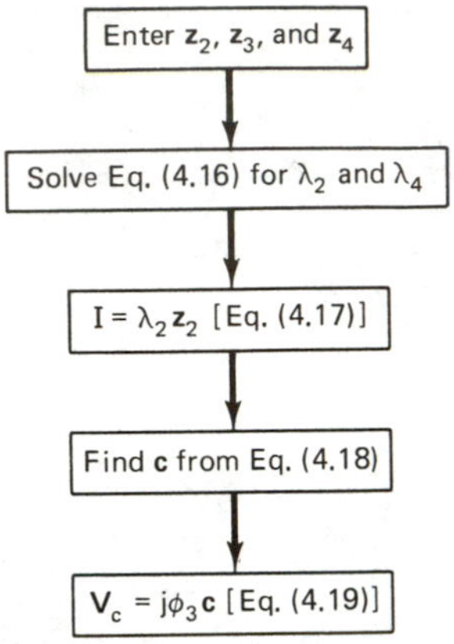


**Figure 4.14** Flowchart for computing velocity of a point on the coupler link of a four-bar mechanism by means of complex numbers.

Plugging in the numerical values in (4.16), we obtain

$$[0 + i(23)](1 - \lambda_2) + 78 - i8 - [47.5 + i(54)](1 + \lambda_4) = 0$$

$$i(23) - i(23)\lambda_2 + (78) - i8 - (47.5) - (47.5)\lambda_4 - i(54) - i(54)\lambda_4 = 0$$

$$\left.\begin{aligned} (0)\lambda_2 - (47.5)\lambda_4 &= -30.5 \\ -(23)\lambda_2 - (54.0)\lambda_4 &= \ \ 39 \end{aligned}\right\} \Rightarrow \lambda_4 = \frac{\begin{vmatrix} 0 & -30.5 \\ -23 & 39 \end{vmatrix}}{\begin{vmatrix} 0 & -47.5 \\ -23 & -54 \end{vmatrix}} = \frac{-30.5}{-47.5} = 0.642$$

$$\lambda_2 = \frac{\begin{vmatrix} -30.5 & -47.5 \\ 39 & -54 \end{vmatrix}}{-47.5(23)} = -3.2$$

$$\mathbf{I} = [0 + i(23)](-3.2) = 0 - i74$$

$$\mathbf{c} = [0 + i(74)] + [0 + i(23) + 0.6(78 - i8)]$$

$$= 0.6(78) + i[74 + 23 - 0.6(8)] = 46.8 + i(92.2) = 103.4 \text{ mm}e^{i63.09^\circ}$$

$$\mathbf{V}_C = i\dot{\phi}_3\mathbf{c} = i(-1)[46.8 + i(92.2)] = (92.2) - i(46.8) = 103.4 \frac{\text{mm}}{\text{sec}} e^{i(-26.91^\circ)}$$

which agrees with the result of the combined geometric–analytical method within the accuracy of graphical construction.

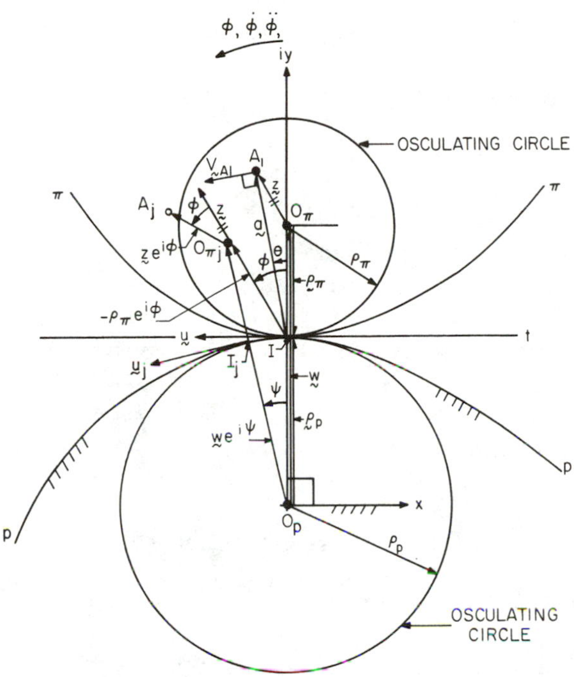


**Figure 4.15** Complex-number vector notation for defining and deriving **u**, the "Instant Center Transfer Velocity," the vector-rate of change of position of the point of tangency of the rollcurves $\pi$ and $p$ along $p$.

## 4.4 ACCELERATIONS

To find the acceleration of point $A$ at position $A_1$ in Fig. 4.15, once again we start with the velocity of $A_j$ [Eq. (4.9)] and take its time derivative:

$$\mathbf{A}_{Aj} = \dot{\mathbf{V}}_{Aj} = \mathbf{w}\ddot{\psi}ie^{i\psi} - \mathbf{w}\dot{\psi}^2e^{i\psi} + \mathbf{z}\ddot{\phi}ie^{i\phi} - \mathbf{z}\dot{\phi}^2e^{i\phi} \tag{4.20}$$

or

$$\mathbf{A}_{Aj} = (i\ddot{\psi} - \dot{\psi}^2)\mathbf{w}e^{i\psi} + (i\ddot{\phi} - \dot{\phi}^2)\mathbf{z}e^{i\phi} \tag{4.21}$$

In position $A_1$, where $\psi = \phi = 0$, this reduces to

$$\mathbf{A}_{A1} = (i\ddot{\psi} - \dot{\psi}^2)\mathbf{w} + (i\ddot{\phi} - \dot{\phi}^2)\mathbf{z} \tag{4.22}$$

Taking the time derivative of Eq. (4.8) yields

$$\ddot{\psi} = \boldsymbol{\rho}\ddot{\phi}$$

Using this and Eqs. (4.2), (4.3), and (4.7), Eq. (4.22) becomes

$$\mathbf{A}_{A1} = \left[i\frac{-\boldsymbol{\rho}_\pi}{\boldsymbol{\rho}_p - \boldsymbol{\rho}_\pi}\ddot{\phi} - \left(\frac{-\boldsymbol{\rho}_\pi}{\boldsymbol{\rho}_p - \boldsymbol{\rho}_\pi}\dot{\phi}\right)^2\right](\boldsymbol{\rho}_p - \boldsymbol{\rho}_\pi) + (i\ddot{\phi} - \dot{\phi}^2)(\mathbf{a} + \boldsymbol{\rho}_\pi)$$

or

$$\mathbf{A}_{A1} = -i\boldsymbol{\rho}_\pi\ddot{\phi} - \frac{\boldsymbol{\rho}_\pi^2}{\boldsymbol{\rho}_p - \boldsymbol{\rho}_\pi}\dot{\phi}^2 + i\ddot{\phi}\mathbf{a} + i\ddot{\phi}\boldsymbol{\rho}_\pi - \dot{\phi}^2\mathbf{a} - \dot{\phi}^2\boldsymbol{\rho}_\pi$$

Regrouping terms, we have

$$\mathbf{A}_{A1} = i\ddot{\phi}(-\boldsymbol{\rho}_\pi + \mathbf{a} + \boldsymbol{\rho}_\pi) - \dot{\phi}^2\left(\frac{\boldsymbol{\rho}_\pi^{\,2}}{\boldsymbol{\rho}_p - \boldsymbol{\rho}_\pi} + \mathbf{a} + \boldsymbol{\rho}_\pi\right)$$

which reduces to

$$\mathbf{A}_{A1} = i\ddot{\phi}\mathbf{a} - \dot{\phi}^2\mathbf{a} - \dot{\phi}^2\boldsymbol{\rho}_\pi(1 - \boldsymbol{\rho}) \tag{4.23}$$

Here the first term is the pathwise tangential acceleration component, the second term is the centripetal component, and the third term is a pure imaginary, because $\boldsymbol{\rho}_\pi$ is defined along the imaginary axis (Fig. 4.15) and $\boldsymbol{\rho}$ is a positive or negative real. To examine this term, let us express $\boldsymbol{\rho}$ in terms of a kinematic quantity, the "instant center transfer velocity" (ICTV), $\mathbf{u}$, which is defined as the time rate of change of position along $p$ of the instant center $I$ as $\pi$ rolls on $p$.

Referring to Fig. 4.15, the position vector of $I_j$ is

$$\overrightarrow{O_pI_j} = \boldsymbol{\rho}_p e^{i\psi}$$

and the ICTV at $I_j$ is $\mathbf{u}_j = \boldsymbol{\rho}_p\dot{\psi}ie^{i\psi}$. For $I$, where $\psi = 0$, this reduces to

$$\mathbf{u} = i\dot{\psi}\boldsymbol{\rho}_p = i\dot{\phi}\boldsymbol{\rho}_p\boldsymbol{\rho} = i\dot{\phi}\frac{-\boldsymbol{\rho}_p\boldsymbol{\rho}_\pi}{\boldsymbol{\rho}_p - \boldsymbol{\rho}_\pi} \tag{4.24}$$

Equation (4.24) shows that $\mathbf{u}$ is invariant under kinematic inversion: Both $\dot{\phi}$ and $-\boldsymbol{\rho}_p\boldsymbol{\rho}_\pi/(\boldsymbol{\rho}_p - \boldsymbol{\rho}_\pi)$ change sign but maintain their magnitudes unchanged.

Let us now define the vector $\boldsymbol{\delta}$ as follows:

$$\boldsymbol{\delta} \equiv -i\frac{\mathbf{u}}{\dot{\phi}} = \frac{-\boldsymbol{\rho}_p\boldsymbol{\rho}_\pi}{\boldsymbol{\rho}_p - \boldsymbol{\rho}_\pi} = \boldsymbol{\rho}_p\boldsymbol{\rho}; \qquad \text{then} \quad \mathbf{u} = i\dot{\phi}\boldsymbol{\delta} \tag{4.24a}$$

From this

$$\frac{\boldsymbol{\delta}}{\boldsymbol{\rho}_\pi} = \frac{-\boldsymbol{\rho}_p}{\boldsymbol{\rho}_p - \boldsymbol{\rho}_\pi} = \frac{-\boldsymbol{\rho}_p}{\boldsymbol{\rho}_p - \boldsymbol{\rho}_\pi} + \frac{\boldsymbol{\rho}_\pi}{\boldsymbol{\rho}_p - \boldsymbol{\rho}_\pi} + \frac{-\boldsymbol{\rho}_\pi}{\boldsymbol{\rho}_p - \boldsymbol{\rho}_\pi} = -1 + \boldsymbol{\rho}$$

and

$$\boldsymbol{\rho} = 1 + \frac{\boldsymbol{\delta}}{\boldsymbol{\rho}_\pi} \tag{4.25}$$

Note that, for the case of Fig. 4.10a, $0 < 1 + \boldsymbol{\delta}/\boldsymbol{\rho}_\pi < 1$, just like $0 < \boldsymbol{\rho} < 1$, as shown before (see Exer. 4.1). Also see Exers. 4.2 and 4.3 for the cases of Fig. 4.10b and c, which verify the complete generality of Eq. (4.25), provided that it is written in complex vector form.

With the use of Eqs. (4.25) and (4.24a), the last term of Eq. (4.23) becomes

$$-\dot{\phi}^2\boldsymbol{\rho}_\pi\left(1 - 1 - \frac{\boldsymbol{\delta}}{\boldsymbol{\rho}_\pi}\right) = \dot{\phi}^2\boldsymbol{\delta} = \dot{\phi}^2\left(-i\frac{\mathbf{u}}{\dot{\phi}}\right) = -i\dot{\phi}\mathbf{u} \tag{4.26}$$

Substituting this into Eq. (4.23), we have

$$\mathbf{A}_{A1} = i\ddot{\phi}\mathbf{a} - \dot{\phi}^2\mathbf{a} - i\dot{\phi}\mathbf{u} \tag{4.27}$$

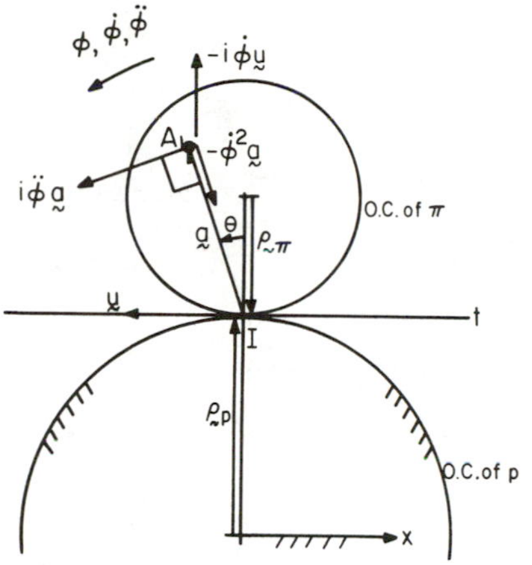


**Figure 4.16** Tangential, centripetal and "invariant" components of the acceleration of point $A$. The invariant acceleration component is the same for all points of $\pi$. including that point of $\pi$ which momentarily coincides with instant center $I$, namely $I_\pi$. Thus, $A_{I\pi} = -i\dot{\phi}\mathbf{u}$.

It is clear that the last term of $\mathbf{A}_{A1}$ is a pure imaginary. All three terms of $\mathbf{A}_{A1}$ are shown in Fig. 4.16 for the case of Fig. 4.10a. However, Eq. (4.27) is in complex vector form, is quite generally applicable, and from it the correct directions and magnitudes are easily drawn or computed once $\boldsymbol{\rho}_p$, $\boldsymbol{\rho}_\pi$, $\mathbf{a}$, $\dot{\phi}$, and $\ddot{\phi}$ are known. The nature of the third term is revealed by observing that it is invariant for all points of the $\pi$ plane. For example, if we let $A_1$ approach $I$ by shrinking $\mathbf{a}$ to zero, we get the acceleration of the point of plane $\pi$, $I_\pi$, momentarily coincident with the instant center:

$$\mathbf{A}_{I\pi} = i\dot{\phi}\mathbf{u} \qquad (4.28)$$

Some authors call $\mathbf{A}_{I\pi}$ the "pole acceleration." This is confusing to many students, because the "pole," or more correctly the instant center, is transferring along the fixed centrode $p$. We therfore propose the term "invariant acceleration component" (IAC) because it is the same for all points of the $\pi$ plane.

## 4.5 INFLECTION POINTS AND THE INFLECTION CIRCLE

Let us resolve the third term of Eq. (4.27) into a tangential and a normal component by projecting it onto the path tangent $i\mathbf{a}$ and path normal $\mathbf{a}$. To this end we will now define the "projection operator."

The *projection operator* rotates a vector into a direction parallel with the line of action of another and multiplies its magnitude by the cosine of the angle between them (Fig. 4.17). Here $\mathbf{e}_1 = e^{i\beta_1}$ and $\mathbf{e}_2 = e^{i\beta_2}$ are the respective unit vectors. Let the projection operator, which is to project $\mathbf{z}_1$ onto $\mathbf{z}_2$, be defined as

$$\mathbf{p}_{12} \equiv \cos(\beta_2 - \beta_1)e^{i(\beta_2-\beta_1)} \qquad (4.29)$$

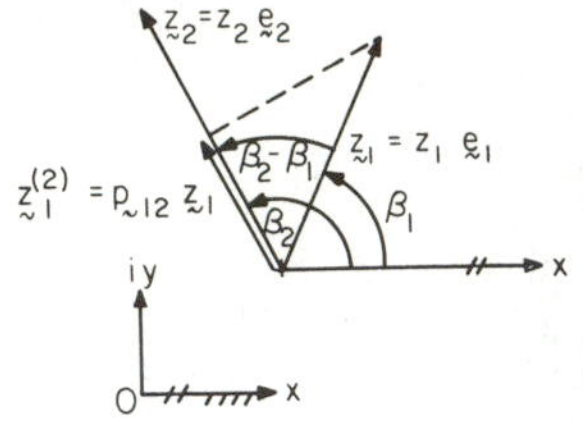


**Figure 4.17** The "projection operator" $\mathbf{p}_{12}$ of Eq. (4.29), applied to $\mathbf{z}_1$ gives the component of $\mathbf{z}_1$ in the direction of $\mathbf{z}_2$.

With this, the projection of $\mathbf{z}_1$ onto $\mathbf{z}_2$ becomes

$$\begin{aligned}\mathbf{z}_1^{(2)} = \mathbf{p}_{12}\mathbf{z}_1 &= \cos(\beta_2 - \beta_1)e^{i(\beta_2-\beta_1)}z_1 e^{i\beta_1} \\ &= \cos(\beta_2 - \beta_1)z_1 e^{i\beta_2}\end{aligned} \tag{4.30}$$

Applying this to the invariant acceleration component and using the projection operator from the $-i\mathbf{u}$ direction onto the $i\mathbf{a}$ direction (Fig. 4.18), $\mathbf{p}_t$, is

$$\mathbf{p}_t = \cos\left(\theta + \frac{\pi}{2}\right)e^{i[\theta+(\pi/2)]} = -(\sin\theta)e^{i[\theta+(\pi/2)]}$$

With this, the tangential component of the third term in Eq. (4.27) becomes

$$\mathbf{p}_t\mathbf{A}_{I\pi} = -(\sin\theta)e^{i[\theta+(\pi/2)]}(-i\dot{\phi}\mathbf{u}) = \mathbf{A}_{I\pi}^t$$

The projection operator from $-i\mathbf{u}$ to $\mathbf{a}$ is

$$\mathbf{p}_n = (\cos\theta)e^{i\theta}$$

With this

$$\mathbf{p}_n\mathbf{A}_{I\pi} = (\cos\theta)e^{i\theta}(-i\dot{\phi}\mathbf{u}) = \mathbf{A}_{I\pi}^n$$

and

$$\mathbf{A}_{I\pi} = -i\dot{\phi}\mathbf{u} = -(\sin\theta)e^{i[\theta+(\pi/2)]}(-i\dot{\phi}\mathbf{u}) + (\cos\theta)e^{i\theta}(-i\dot{\phi}\mathbf{u}) \tag{4.31}$$

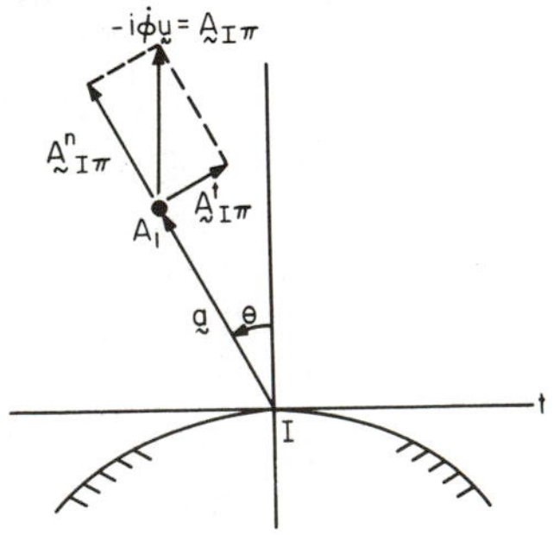


**Figure 4.18** Resolution of the invariant acceleration component, $-i\dot{\phi}\mathbf{u}$, into its tangential and normal components.

Fig. 4.18 shows these components. Substituting Eq. (4.31) into Eq. (4.27) and separating the tangential and normal components yields

$$\left.\begin{aligned}\mathbf{A}^t_{A1} &= i\ddot{\phi}\mathbf{a} - (\sin\theta)e^{i[\theta+\pi/2)]}(-i\dot{\phi}\mathbf{u})\\ \mathbf{A}^n_{A1} &= -\dot{\phi}^2\mathbf{a} + (\cos\theta)e^{i\theta}(-i\dot{\phi}\mathbf{u})\end{aligned}\right\} \mathbf{A}_{A1} = \mathbf{A}^t_{A1} + \mathbf{A}^n_{A1} \tag{4.32}$$

We now define an *inflection point* to be a point of the plane $\pi$ which at the moment is going through a point of inflection of its path with respect to the fixed plane $p$. Such points will have zero acceleration normal to their paths (i.e., $\mathbf{A}^n_{A1} = 0$). Thus on the $\theta$ ray (the extended line $IA_1$) we find such a point by solving the following equation for **a:**

$$-\dot{\phi}^2\mathbf{a} + (\cos\theta)e^{i\theta}(-i\dot{\phi}\mathbf{u}) = 0 \tag{4.33}$$

from which

$$\mathbf{a} = (\cos\theta)e^{i\theta}\left[-i\frac{\mathbf{u}}{\dot{\phi}}\right] = (\cos\theta)e^{i\theta}\boldsymbol{\delta} \tag{4.34}$$

where we have used the definition of $\boldsymbol{\delta}$ from Eq. (4.24a). Equation (4.24a) also shows that $\boldsymbol{\delta}$ is a purely geometric property of the rolling centrodes and is collinear with their common normal. Observe that, on the right side of Eq. 4.34, $(\cos\theta)e^{i\theta}$ is a projection operator which projects $\boldsymbol{\delta}$ onto the $\theta$ ray. Pictured in Fig. 4.19, this shows that, as $\theta$ is allowed to vary, **a** traces the locus of all inflection points of the moving plane $\pi$: a circle of diameter $\boldsymbol{\delta}$, touching the common tangent $t$ at $I$. The tip of the vector $\boldsymbol{\delta}$, marked $J$, is called the *inflection pole* and the locus of $J_\theta$ for all $\theta$ values is called the *inflection circle.* Points at the tips of the vectors **a** are on the inflection circle. They are marked $J_\theta$ and are called *inflection points.*

The inflection circle is embedded in the moving plane $\pi$. However, its existence is only momentary, because the instant center $I$, the common tangent $t$, the common normal, and the diameter vector $\boldsymbol{\delta}$ all change as the rolling of the centrodes progresses. Thus the inflection circle must be constructed separately for each position of the plane $\pi$. This is shown schematically in Fig. 4.20.

The path of every inflection point as a point of the moving plane $\pi$ has second-order contact (three infinitesimally close point contacts) with its "circle of curvature" or "path osculating circle," which is of zero curvature (i.e., of infinite radius), in other words, it is a straight line. This property is often used by designers to design

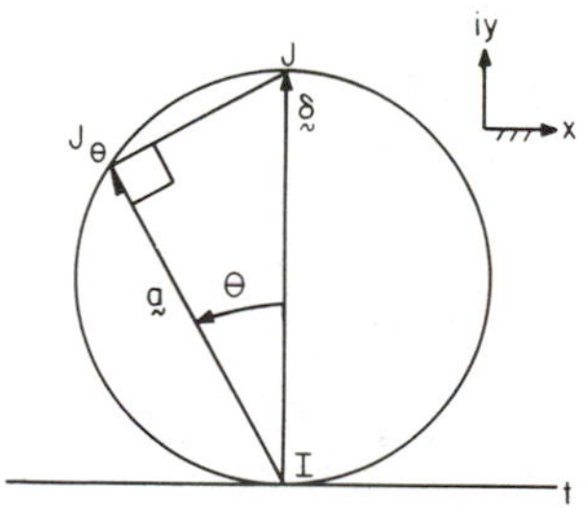


**Figure 4.19** The "inflection circle" is the locus of projections of the "inflection pole" $J$ onto the $\theta$ rays.

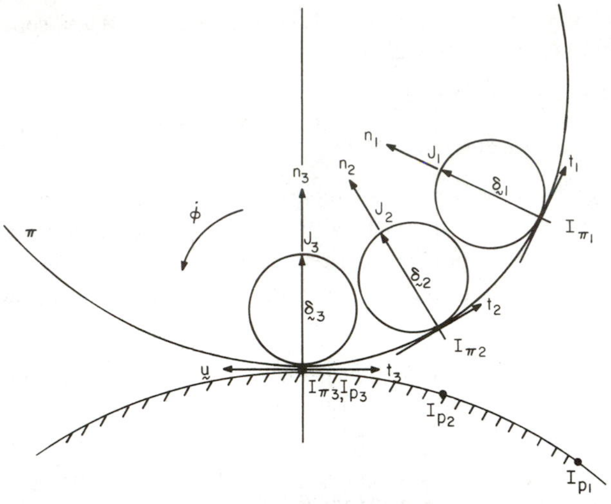


**Figure 4.20** The inflection circle exists only momentarily. In general, it changes as $\pi$ rolls on $p$.

path generators for tracing an approximate straight line with three infinitesimally close precision points. While the resulting linkage often traces an acceptable approximation of a straight line for an appreciable distance, there is no assurance that this will be the case in a particular instance.

To trace an exact straight line, the tracer point must coincide with an inflection point in every position of the moving plane; that is, it must always be on the inflection circle. One way to accomplish this has been shown by Cardano in the *Cardan mechanism,* a planetary gear train in which the planet gear rolls on the inside of the sun gear, which is twice as large as the planet gear. This mechanism is shown in Fig. 4.21, and it is treated analytically in Ex. 4.2.

**Example 4.2**

Let

$$\boldsymbol{\rho}_p = 0 - i2$$
$$\boldsymbol{\rho}_\pi = 0 - i1$$

Then, from Eq. (4.7),

$$\boldsymbol{\rho} = \frac{-\boldsymbol{\rho}_\pi}{\boldsymbol{\rho}_p - \boldsymbol{\rho}_\pi} = \frac{i1}{-i2 + i1} = \frac{1}{-1} = -1$$

From Eq. (4.24a),

$$\boldsymbol{\delta} = \frac{-\boldsymbol{\rho}_p \boldsymbol{\rho}_\pi}{\boldsymbol{\rho}_p - \boldsymbol{\rho}_\pi} = \boldsymbol{\rho}_p \boldsymbol{\rho} = -i2(-1) = i2$$

Therefore, $J \equiv O_p$.

Let $\dot{\phi} = 1$ rad/sec. Then $\mathbf{u} = i\dot{\phi}\boldsymbol{\delta} = i(1)i2 = -2$. Let $\theta = \pi/4$ radian, $a = 1$; then $\mathbf{a} = 1e^{i(3\pi/4)}$, and from Eq. (4.14),

$$\mathbf{V}_A = i\dot{\phi}\mathbf{a} = i(1)1e^{i(3\pi/4)} \tag{4.14}$$
$$\mathbf{V}_A = 1e^{i(5\pi/4)}$$

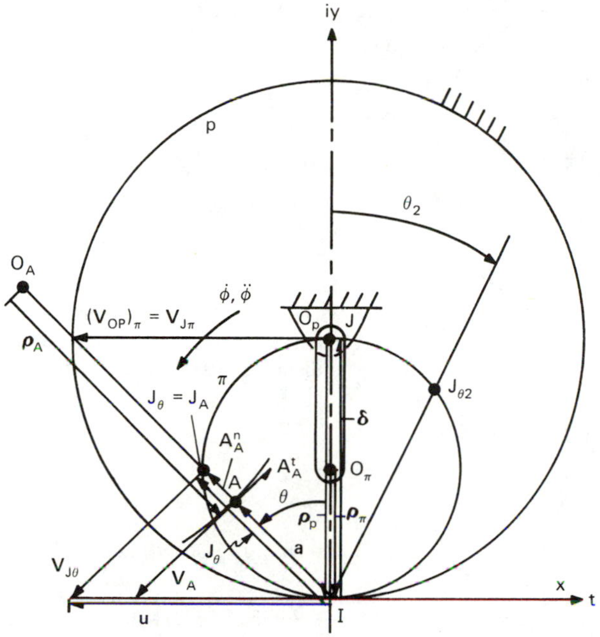


**Figure 4.21** Cardano's mechanism: gear $\pi$ rolling inside the sun gear $p$, twice as large as $\pi$ (see Ex. 4.2).

From Eq. (4.34),

$$\mathbf{J}_\theta = \overrightarrow{IJ_\theta} = (\cos\theta)e^{i\theta}\boldsymbol{\delta} = \frac{\sqrt{2}}{2}\, e^{i(\pi/4)} i2 = \sqrt{2}\, e^{i(3\pi/4)}$$

Let $\ddot{\phi} = 1$ rad/sec². Then, from Eq. (4.32),

$$\mathbf{A}^t_A = i\ddot{\phi}\mathbf{a} - (\sin\theta)e^{i[\theta+(\pi/2)]}(-i\dot{\phi}\mathbf{u}) = i1\,[1e^{i(3\pi/4)}] - \frac{\sqrt{2}}{2}\, e^{i(3\pi/4)}[-i(1)(-2)]$$

$$\mathbf{A}^t_A = e^{i(5\pi/4)} - \sqrt{2}\, e^{i(5\pi/4)} = (1-\sqrt{2})e^{i(5\pi/4)} = (\sqrt{2}-1)e^{i(\pi/4)}$$

$$\mathbf{A}^n_A = -\dot{\phi}^2\mathbf{a} + (\cos\theta)e^{i\theta}(-i\dot{\phi}\mathbf{u}) = -1^2 1e^{i(3\pi/4)} + \frac{\sqrt{2}}{2}\, e^{i(\pi/4)}\,[-i(1)(-2)]$$

$$\mathbf{A}^n_A = -e^{i3\pi/4} + i\sqrt{2}\, e^{i(\pi/4)} = (\sqrt{2}-1)e^{i(3\pi/4)}$$

$$\mathbf{V}_{J\theta} = i\dot{\phi}\mathbf{J}_\theta = i(1)\sqrt{2}\, e^{i(3\pi/4)} = \sqrt{2}\, e^{i(5\pi/4)}$$

$$A^t_{J\theta} = i\ddot{\phi}\mathbf{J}_\theta - (\sin\theta)e^{i[\theta+(\pi/2)]}(-i\dot{\phi}\mathbf{u}) = i(1)\sqrt{2}\, e^{i(3\pi/4)} - \frac{\sqrt{2}}{2}\, e^{i(3\pi/4)}(i2) = 0$$

$$A^n_{J\theta} = -\dot{\phi}^2\mathbf{J}_\theta + (\cos\theta)e^{i\theta}(-i\dot{\phi}\mathbf{u}) = -1^2\sqrt{2}\, e^{i(3\pi/4)} + \frac{\sqrt{2}}{2}\, e^{i(\pi/4)}(i2) = 0$$

The velocity of $J \equiv O_p$ as a point of the moving (planet) plane $\pi$:

$$(\mathbf{V}_{J\pi}) = (\mathbf{V}_{OP})_\pi = i\dot{\phi}\boldsymbol{\delta} = i1(i2) = -2$$

$$(\mathbf{A}_{OP})^t_\pi = i\ddot{\phi}\boldsymbol{\delta} - (\sin\theta)e^{i[\theta+(\pi/2)]}(-i\dot{\phi}\mathbf{u}) = i1(i2) - 0 = -2$$

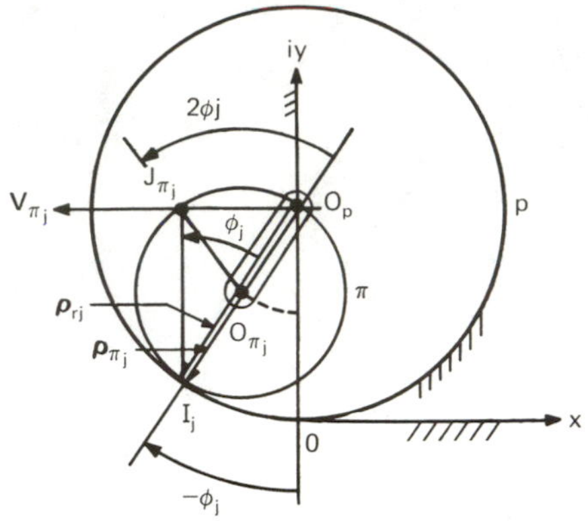


**Figure 4.22** Kinematic proof that a point on the periphery of a Cardano planet traces a diameter of the Cardano sun gear.

Next let $\ddot{\phi} = 0$. Calculate $\mathbf{A}_A$ and $\mathbf{A}_{J\theta}$.

$$\mathbf{A}_A^t = -\frac{\sqrt{2}}{2} e^{i(3\pi/4)} (i2) = \sqrt{2}\, e^{i(\pi/4)},$$

$$\mathbf{A}_A^n = -1^2 1 e^{i(3\pi/4)} + \sqrt{2}\, e^{i(3\pi/4)} = (\sqrt{2} - 1) e^{i(3\pi/4)}$$

$$\mathbf{A}_{J\theta}^t = 0 - \sqrt{2}\, e^{i(\pi/4)}, \qquad \mathbf{A}_{J\theta}^n = -1^2 \sqrt{2}\, e^{i(3\pi/4)} + \frac{\sqrt{2}}{2} e^{i(\pi/4)}(i2) = 0$$

Now let us find the velocity of $J$ as a point of $\pi$, ($J_\pi$) after it has moved away from $O_p$ in the course of the rolling of the planet $\pi$ by the arbitrary angle $\phi_j$. The position of $J_{\pi j}$ is shown in Fig. 4.22.

Now, it is easy to show in Fig. 4.22 by way of the velocity of $O_\pi$ as a point of $\pi$ and also of the planet support arm, that the angular velocity of the latter is $-\dot{\phi}$. Therefore, in Fig. 4.22 the angle $OO_pI_j$ is $-\phi_j$. Since the circumference of $p$ is twice that of $\pi$, the angle $O_pO_{\pi j}J_{\pi j}$ is $2\phi_j$, and therefore the ray $I_jJ_{\pi j}$ is vertical in Fig. 4.22. The angle of this ray with respect to $I_jO_p$ must be $\phi_j$. The velocity of $J_{\pi j}$ is perpendicular to $I_jJ_{\pi j}$, which makes it horizontal in Fig. 4.22. Thus, since $\phi_j$ is arbitrary, the direction of the velocity of $J_\pi$ never changes and its path is a straight line, the diameter of $p$. Because of the geometry of the Cardan planetary, at some time every point of the circle $\pi$ becomes the inflection pole $J$ and thus describes a diameter of $p$. Here we have a purely kinematic proof of a well-known property of the Cardan gear pair.

## 4.6 THE EULER–SAVARY EQUATION [245]

In Fig. 4.23, consider the following four points on the $\theta$ ray:

1. $I$, the instant center
2. $A$, an arbitrary point
3. $J_A = J_\theta$, the inflection point on the $\theta$ ray (the ray of $A$)
4. $O_A$, the center of curvature of the path of $A$ described in $p$

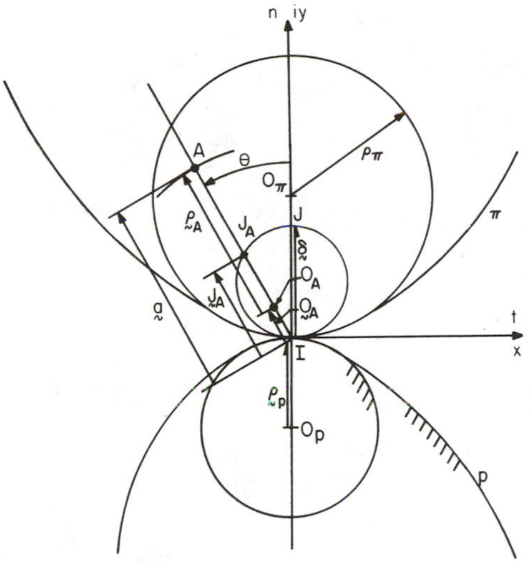


**Figure 4.23** The complex-number forms of the Euler-Savary equation [Eqs. (4.40, 4.41, 4.43) and (4.45)] are explicit expressions of the vectors **a**, **p**, $\mathbf{J}_A$ and $\mathbf{O}_A$, each in terms of the others, suitable for digital computer programming.

Also define the following four vectors, all collinear with the $\theta$ ray:

1. $\mathbf{a} = \overrightarrow{IA}$
2. $\mathbf{J}_A = \overrightarrow{IJ_A}$
3. $\mathbf{O}_A = \overrightarrow{IO_A}$
4. $\boldsymbol{\rho}_A = \overrightarrow{O_AA}$, which is the vector radius of path curvature

The Euler–Savary equation (ESE) correlates these vectors and thus provides a way to find any one of the four points $I$, $A$, $J_A$, and $O_A$ if the other three are known. Although the ESE is a purely geometric concept, the authors believe that an easily visualizable kinematic, complex-number derivation of the various forms of the ESE is most instructive. Furthermore it leads to explicit complex-vector formulas well adapted for digital computation.

As a preliminary to our derivation, consider any point $A$ describing some arbitrary path $\sigma$ in the fixed plane of reference $\Sigma$ (Fig. 4.24). It is well known from elementary-particle kinematics that

$$\mathbf{A}_A^n = -(\dot{\beta})^2 \boldsymbol{\rho}_A \tag{4.35}$$

where $\boldsymbol{\rho}_A$ is the vector radius of curvature of $\sigma$ at $A$. Observe that the center of path curvature, $O_A$, is fixed in $\Sigma$. Therefore, in Fig. 4.23 consider $O_A$ fixed in $p$, and observe that the angular velocity (say, $\dot{\gamma}$) of the vector radius of curvature $\boldsymbol{\rho}_A$ pivoted at $O_A$ is

$$\dot{\gamma} = \dot{\phi}\,\frac{\mathbf{a}}{\boldsymbol{\rho}_A} \tag{4.36}$$

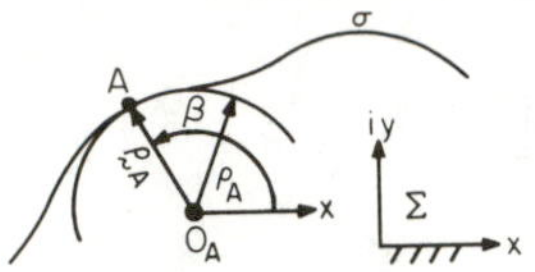


**Figure 4.24** Definition of the radius of curvature $\boldsymbol{\rho}_A$ of the path of point $A$.

Note that $\mathbf{a}/\boldsymbol{\rho}_A$ is a positive or negative real, according as $\mathbf{a}$ and $\boldsymbol{\rho}_A$ have the same or opposite sense. Therefore, applying Eqs. (4.35) and (4.36) to Fig. 4.23, we get

$$\mathbf{A}_A^n = -(\dot{\gamma})^2\boldsymbol{\rho}_A = -(\dot{\phi})^2\frac{a^2}{\rho_A^2}\boldsymbol{\rho}_A \tag{4.37}$$

Combining Eq. (4.37) with Eq. (4.32) and noting that using Eq. (4.24a)

$$-i\dot{\phi}\mathbf{u} = -i\dot{\phi}i\dot{\phi}\boldsymbol{\delta} = (\dot{\phi})^2\boldsymbol{\delta} \tag{4.38}$$

we obtain

$$-(\dot{\phi})^2\frac{a^2}{\rho_A^2}\boldsymbol{\rho}_A = -\dot{\phi}^2\mathbf{a} + (\cos\theta)e^{i\theta}\dot{\phi}^2\boldsymbol{\delta}$$

and using the fact that $(\cos\theta)e^{i\theta}\boldsymbol{\delta} = \mathbf{J}_A$, after canceling $\dot{\phi}^2$, we get

$$-\frac{a^2}{\rho_A^2}\boldsymbol{\rho}_A = \mathbf{J}_A - \mathbf{a} \tag{4.39}$$

or

$$-\frac{\boldsymbol{\rho}_A}{\rho_A^2} = \frac{\mathbf{J}_A - \mathbf{a}}{a^2} \tag{4.39a}$$

Now for the left side of Eq. (4.39a),

$$-\frac{\boldsymbol{\rho}_A}{\rho_A^2} = -\frac{1}{\rho_A}e^{i\arg\boldsymbol{\rho}_A} = \frac{1}{\rho_A}(-e^{i\arg\boldsymbol{\rho}_A})$$

and, on the right side of Eq. (4.39a),

$$\mathbf{J}_A - \mathbf{a} = |\mathbf{J}_A - \mathbf{a}|e^{i\arg(\mathbf{J}_A-\mathbf{a})} = |J_A - a|(-e^{i\arg(\mathbf{a}-\mathbf{J}_A)})$$

With these, taking the reciprocal and then the absolute value of Eq. (4.39a), we get

$$\rho_A = \frac{a^2}{|\mathbf{a} - \mathbf{J}_A|}$$

and equating the arguments of both sides of Eq. (4.39a):

$$\arg\boldsymbol{\rho}_A = \arg(\mathbf{a} - \mathbf{J}_A)$$

or

$$\boldsymbol{\rho}_A = \frac{a^2}{|\mathbf{a} - \mathbf{J}_A|}e^{i\arg(\mathbf{a}-\mathbf{J}_A)} \qquad \text{ESE-1} \tag{4.40}$$

We may regard Eq. (4.40) as the first form of the Euler–Savary equation, applicable when points $I$, $A$, and $J_A$ are known, and $O_A$ is sought.

Let us apply ESE-1 to Ex. 4.2, Fig. 4.21:

$$\boldsymbol{\rho}_A = \frac{a^2}{|\mathbf{a}-\mathbf{J}_A|} e^{i\arg(\mathbf{a}-\mathbf{J}_A)} = \frac{(1)^2}{|1e^{i(3\pi/4)} - |\sqrt{2}|e^{i(3\pi/4)}|}[-e^{i(3\pi/4)}]$$

$$= \frac{1}{|1-|\sqrt{2}||}[-e^{i(3\pi/4)}] = \frac{1}{|-(|\sqrt{2}|-1)|}[-e^{i(3\pi/4)}] = \frac{1}{|\sqrt{2}|-1}[-e^{i(3\pi/4)}]$$

$$= (1+|\sqrt{2}|)[-e^{i(3\pi/4)}]$$

This is shown in Fig. 4.21. We have given every algebraic step in the computation to demonstrate how to avoid certain typical errors. Of course, when programmed for digital computation using complex arithmetic, these steps take care of themselves.

Equation (4.40) is the first form of the Euler–Savary equation in "sign-proof" complex-number notation. With this, when $\mathbf{a}$ and $\mathbf{J}_A$, that is, points $I$, $A$, and $J_A$ are known, the vector radius of path curvature of point $A$ can be calculated without sign ambiguity and the center of curvature $O_A$ can thus be found. The argument part of ESE-1 also shows that the path of $A$ is always concave toward $J_A$, which the traditional scalar forms of ESE fail to point out. In other words, this part of the equation reveals the fact that $O_A$ is always on the same side of $A$ as $J_A$, which in conventional scalar ESE calculations must be remembered and separately applied. Thus it is seen that this form of ESE-1 is well adapted for automatic computation.

In the absolute value part of ESE-1, if $\mathbf{a} \to \mathbf{J}_A$ (i.e., when $A$ becomes the inflection point on its ray), then $\rho_A \to \infty$, as it should at the inflection point. It is also instructive to observe what happens when $A$ approaches $I$. To this end, let $a = \epsilon$ in the first part of ESE-1, where $\epsilon$ is a small number, and let $\epsilon \to 0$:

$$\lim_{a\to 0} \rho_A = \lim_{a\to 0} \frac{a^2}{|\mathbf{a}-\mathbf{J}_A|} = \lim_{a\to 0} \frac{2a}{J_A} = 0$$

where we applied l'Hospital's rule in the limiting process.

So we see that, as $A$ approaches $I$, $\rho_A \to 0$, and the path becomes cusped. As for the direction of the path tangent in the vicinity of the cusp, let us observe what happens to the second part of ESE-1 as $a \to 0$ becomes very small:

$$\lim_{a\to 0} \arg \boldsymbol{\rho}_A = \arg(-\mathbf{J}_A)$$

which shows that the tangent to the path in the vicinity of the cusp is collinear with the ray of $A$. However, at the cusp, when $a = 0$, this no longer holds because a zero vector has no argument, and the tangent does not exist.

A second form of ESE is readily obtained by solving Eq. (4.39) for $\mathbf{J}_A$:

$$\mathbf{J}_A = \mathbf{a} - \frac{a^2}{\rho_A^2}\boldsymbol{\rho}_A \qquad \text{ESE-2} \tag{4.41}$$

for the case when points $I$, $A$, and $O_A$ are known, and $J_A$ is sought.

Now let us solve the ESE for **a**. To this end we proceed as follows:

$$\frac{1}{\mathbf{J}_A} = \frac{1}{\mathbf{a} - (a^2/\rho_A^2)\,\boldsymbol{\rho}_A} = \frac{1}{\mathbf{a} - (a^2/\rho_A^2)\,\rho_A e^{i\arg \boldsymbol{\rho}_A}}$$

$$= \frac{1}{\mathbf{a} - (a^2 e^{i2\arg \mathbf{a}}/\rho_A^2\, e^{i2\arg \mathbf{a}})\,\rho_A e^{i\arg \boldsymbol{\rho}_A}}$$

Consider $e^{i\arg \boldsymbol{\rho}_A}/e^{i2\arg \mathbf{a}}$. Now

$$e^{i\arg \boldsymbol{\rho}_A} = \begin{cases} +e^{i\arg \mathbf{a}} \\ \text{or} \\ -e^{i\arg \mathbf{a}} \end{cases}$$

Take the + sign first. Then

$$\frac{e^{i\arg \boldsymbol{\rho}_A}}{e^{i2\arg \mathbf{a}}} = \frac{e^{i\arg \boldsymbol{\rho}_A}/e^{i\arg \mathbf{a}}}{e^{i\arg \mathbf{a}}} = \frac{1}{e^{i\arg \boldsymbol{\rho}_A}}$$

Now take the − sign:

$$\frac{e^{i\arg \boldsymbol{\rho}_A}}{e^{i2\arg \mathbf{a}}} = \frac{-1}{-e^{i\arg \boldsymbol{\rho}_A}} = \frac{1}{e^{i\arg \boldsymbol{\rho}_A}}$$

Therefore, we can write

$$\frac{1}{\mathbf{J}_A} = \frac{1}{\mathbf{a} - (\mathbf{a}^2/\rho_A e^{i\arg \boldsymbol{\rho}_A})} = \frac{1}{\mathbf{a} - \mathbf{a}^2/\boldsymbol{\rho}_A} = \frac{\boldsymbol{\rho}_A}{\boldsymbol{\rho}_A \mathbf{a} - \mathbf{a}^2} = \frac{\boldsymbol{\rho}_A}{\mathbf{a}(\boldsymbol{\rho}_A - \mathbf{a})}$$

$$\frac{-\boldsymbol{\rho}_A}{\mathbf{a}(\mathbf{a} - \boldsymbol{\rho}_A)} = \frac{\mathbf{a} - \mathbf{a} - \boldsymbol{\rho}_A}{\mathbf{a}(\mathbf{a} - \boldsymbol{\rho}_A)} = \frac{\mathbf{a} - \boldsymbol{\rho}_A}{\mathbf{a}(\mathbf{a} - \boldsymbol{\rho}_A)} - \frac{\mathbf{a}}{\mathbf{a}(\mathbf{a} - \boldsymbol{\rho}_A)} = \frac{1}{\mathbf{a}} - \frac{1}{\mathbf{a} - \boldsymbol{\rho}_A}$$

But referring to Fig. 4.23, we see that $\mathbf{a} - \boldsymbol{\rho}_A = \mathbf{O}_A$. With this

$$\frac{1}{\mathbf{J}_A} = \frac{1}{\mathbf{a}} - \frac{1}{\mathbf{O}_A} \tag{4.42}$$

which is similar to a traditional scalar form of ESE, except that Eq. (4.42) is in vector form, and therefore contains information about the sense of the vectors without the need for traditional sign conventions, which is an important advantage in automatic computation. Solving for **a** yields

$$\frac{1}{\mathbf{a}} = \frac{1}{\mathbf{J}_A} + \frac{1}{\mathbf{O}_A} \qquad \text{or} \qquad \frac{1}{\mathbf{a}} = \frac{\mathbf{O}_A + \mathbf{J}_A}{\mathbf{J}_A O_A}$$

and

$$\mathbf{a} = \frac{\mathbf{J}_A \mathbf{O}_A}{\mathbf{O}_A + \mathbf{J}_A} \qquad \text{ESE-3} \tag{4.43}$$

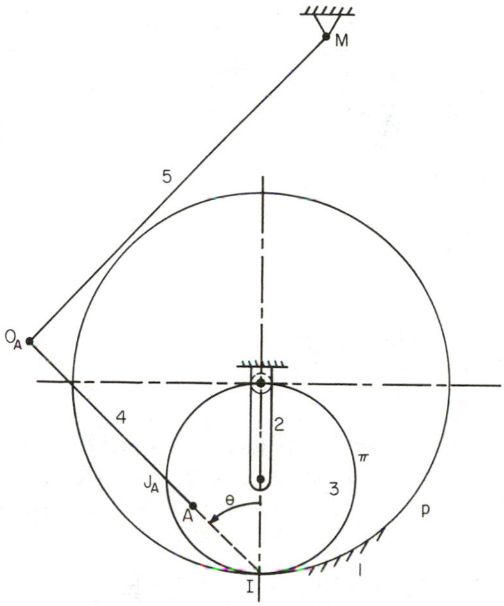


**Figure 4.25** Dwell mechanism based on the Cardano planetary gears (a hypo-cycloidal crank) and the path curvature of point $A$ on the planet. $A$ momentarily describes an approximate circular arc with center $O_A$. Thus link 5 "dwells" momentarily (see Ex. 4.2).

In ESE-3, we have a form of the Euler–Savary equation which allows locating the tracer point $A$ when the locations of points $I$, $O_A$, and $J_A$ are known on a certain ray. Again, we have a new complex-number form solved explicitly for the radius vector $\mathbf{a}$ in terms of the vectors $\mathbf{O}_A$ and $\mathbf{J}_A$, without needing to use the traditional sign conventions and therefore well suited for automatic computation.

Suppose, in the Cardan mechanism of Fig. 4.21 (Ex. 4.2), that we wish to find a point on gear $\pi$ to which we could attach a dyad to form an approximate dwell linkage in the vicinity of the position shown, with link 5 as the output (Fig. 4.25). Note that in this case the pivot $M$ and the direction of the output link in the dwell position would be assumed arbitrarily and $O_A$ then found by forming the right triangle $MO_AI$ with the dwell position of link 5. As a numerical example for this task, take the known values of $\mathbf{O}_A$ and $\mathbf{J}_A$ from Ex. 4.2. Here

$$\mathbf{O}_A = (2+\sqrt{2})e^{i(3\pi/4)} \qquad \text{and} \qquad \mathbf{J}_A = \sqrt{2}\, e^{i(3\pi/4)}$$

from which using Eq. (4.43)

$$\mathbf{a} = \frac{\sqrt{2}(2+\sqrt{2})e^{i(3\pi/2)}}{(2+2\sqrt{2})e^{i(3\pi/4)}} = 1e^{i(3\pi/4)}$$

which completes the design of the dwell mechanism.

Let us take stock of our results so far. ESE-1 lets us locate the point $O_A$ if points $I$, $A$, and $J_A$ are known on a ray. ESE-2 helps find $J_A$ when points $I$, $A$, and $O_A$ are known. ESE-3 locates $A$ when $I$, $J_A$, and $O_A$ are known. One more case remains: Can we find a form of ESE which will locate $I$ when $A$, $O_A$, and $J_A$ are given? $A$ and $O_A$ determine the vector $\boldsymbol{\rho}_A$, but all three other vectors, $\mathbf{O}_A$, $\mathbf{J}_A$, and **a**, are unknown. Finding any one of them will locate *I*. Let us start with ESE-2:

$$\mathbf{J}_A = \mathbf{a} - \frac{a^2}{\rho_A^2}\boldsymbol{\rho}_A \qquad \text{ESE-2}$$

$\mathbf{J}_A - \mathbf{a} = \overrightarrow{AJ_A}$ is known. Therefore,

$$\overrightarrow{AJ_A} = -\frac{a^2}{\rho_A^2}\boldsymbol{\rho}_A \qquad \text{or} \qquad \overrightarrow{J_AA} = \frac{a^2}{\rho_A^2}\boldsymbol{\rho}_A$$

from which

$$a^2 = (\overrightarrow{J_AA})\frac{\rho_A^2}{\boldsymbol{\rho}_A} \qquad \text{ESE-4} \tag{4.44}$$

This is the fourth form of the Euler–Savary equation. Now, ESE-1 has shown that $J_A$ and $O_A$ are always on the same side of $A$ along the ray. Therefore, the arguments of $\overrightarrow{J_AA}$ and $\overrightarrow{O_AA} = \boldsymbol{\rho}_A$ are the same. As a consequence, the right side of Eq. (4.44) is always positive, and its square root is real:

$$a = \pm\sqrt{(\overrightarrow{J_AA})\frac{\rho_A^2}{\boldsymbol{\rho}_A}} = \pm\rho_A\sqrt{\frac{\overrightarrow{J_AA}}{\boldsymbol{\rho}_A}} = \pm\rho_A\sqrt{\frac{|\overrightarrow{J_AA}|}{\rho_A}} = \pm\sqrt{|\overrightarrow{J_AA}|\rho_A} \tag{4.44a}$$

Note that Eq. (4.44) can also be derived more directly from the first equation of ESE-1 (see Exer. 4.6). How to choose between the positive and the negative signs of this square root? Since $a$ is an absolute value, it must be positive. But what about its argument? In other words, what is arg **a**? Does **a** have the same sense as one of the other known vectors, $\overrightarrow{J_AA}$, $\boldsymbol{\rho}_A$, or $\overrightarrow{O_AJ_A}$? The answer is that either sense of the vector **a** is a valid solution. Putting it another way, one can assume the sense of **a** in either direction along the ray, but then one must use both the positive and negative values of the square root in Eq. (4.44a) and thus obtain the two possible locations of $I$ for the given locations of $A$, $O_A$, and $J_A$. Thus ESE-4 can be written in the following explicit vector form:

$$\mathbf{a} = |(|\overrightarrow{J_AA}|\rho_A)^{1/2}|(\pm e^{i\arg\boldsymbol{\rho}_A}) \qquad \text{ESE-4} \tag{4.45}$$

which is easily programmed for computation. The choice between the (+) or (−) signs for the unit vector ($e^{i\arg\boldsymbol{\rho}_A}$) can then be programmed to be determined from other information or requirements contained in the particular problem.

This fact is also born out by the well-known Bobillier construction. This can be shown by observing that Eq. (4.44), owing to the equality of arg $\boldsymbol{\rho}_A$ and arg $(\overrightarrow{J_AA})$, can just as well be written in scalar form:

$$a^2 = |\overrightarrow{J_AA}|\rho_A \qquad \text{or} \qquad \frac{\rho_A}{a} = \frac{a}{|\overrightarrow{J_AA}|} \tag{4.46}$$

which is the basis for Bobillier's construction for locating one of the four points $I$, $A$, $O_A$, or $J_A$ if the other three are given. When $I$ is one of the three given points, this graphical procedure yields a unique answer for the unknown fourth point.

However, if $I$ is the unknown, there are two possible locations for $I$, owing to its presence only in $a$, which appears as $a^2$ in Eq. (4.44), yielding a positive and a negative root for $a$. So let us describe the *Bobillier constructions* (BCs).

## 4.7 BOBILLIER'S CONSTRUCTIONS

### Case 1

Given $I$, $A$, and $O_A$ on a ray, find $J_A$ by $BC$ (Fig. 4.26).

**Procedure:**

Draw arbitrary line 1 through $O_A$.

Draw arbitrary line 2 through $I$.

Draw line 3 through $A$ and through the intersection of lines 1 and 2 (point 12).

Draw line 4 through $I$ parallel to line 1.

Through intersection 34 draw line 5 parallel with line 2. Line 5 cuts the ray $IO_AA$ at $J_A$.

For proof of this construction and the uniqueness of $J_A$, see Exer. 4.7.

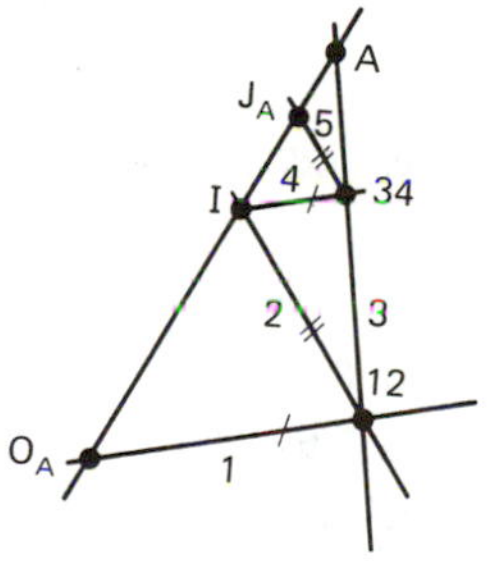


**Figure 4.26** Bobillier's construction for finding the inflection point $J_A$ when $I$, $A$, and $O_A$ are known on the ray $IA$. Single-digit numbers indicate the sequence of drawing the lines of construction. Double-digit numbers signify from which two lines the intersections were found. Note parallelism of lines 1,4 and 2,5.

### Case 2

Given $I$, $A$, and $J_A$ on a ray, find $O_A$ by Bobillier's construction (Fig. 4.27).

**Procedure** (Fig. 4.27):

Draw arbitrary line 1 through $I$.

Draw arbitrary line 2 through $J_A$ to form intersection 12.

Connect $A$ and 12 with line 3.

Draw line 4 through $I$, parallel with line 2 to intersect line 3 at 34.

Through 34 draw line 5 parallel to line 1 to intersect the ray $AJ_AI$. This is $O_A$.

For proof of this construction and of its uniqueness, see Exer. 4.8.

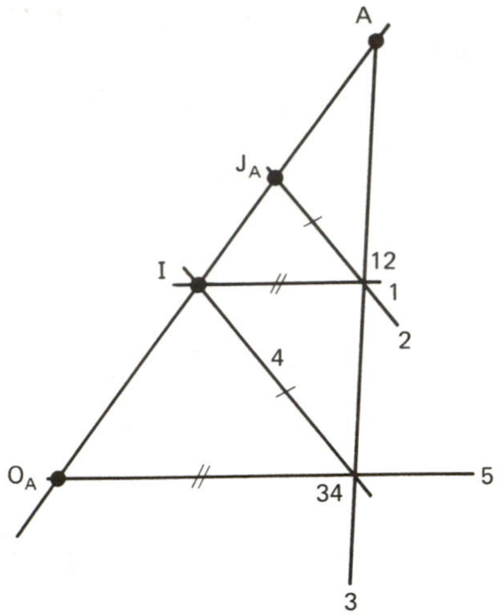


**Figure 4.27** Bobillier's construction to find $O_A$ when $I$, $A$, and $J_A$ are known (also see caption of Fig. 4.26).

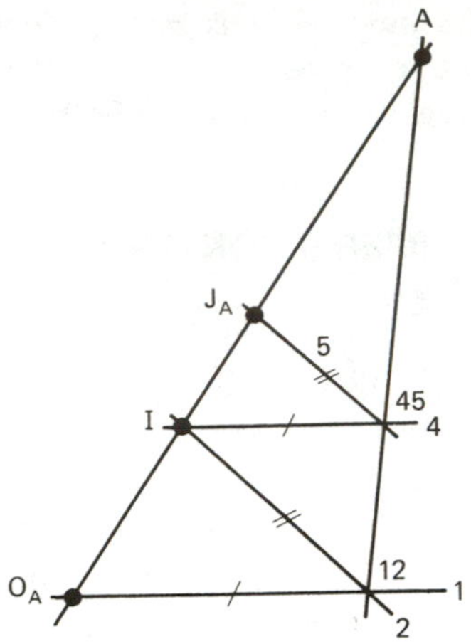


**Figure 4.28** Bobillier's construction for finding $A$ when $I$, $J_A$, and $O_A$ are given (see also caption of Fig. 4.26).

**Case 3**

Given $I$, $O_A$, and $J_A$ on a ray, find $A$ by Bobillier's Construction.

**Procedure** (Fig. 4.28):

Draw arbitrary line 1 through $O_A$.
Draw arbitrary line 2 through $I$ to intersect 1. Call this point 12.
Draw line 4 through $I$ parallel with line 1.
Draw line 5 through $J_A$ parallel with line 2 until it intersects line 4 at point 45.
Draw a line through intersections 12 and 45 and extend it to intersect the ray. This locates point $A$.

The agreement of the constructions of cases 2 and 3 with Eq. (4.46) is easily verified by way of the appropriate similar triangles (see Exers. 4.8 and 4.9).

**Case 4**

In the remaining case, let $O_A$, $A$, and $J_A$ be given on a ray. To find $I$, proceed as follows (see Fig. 4.29):

**Procedure** (Fig. 4.29, *First Solution*):

Draw arbitrary line 1 through $A$.
Draw arbitrary trial line 2′ through $J_A$ and mark its intersection with line 1: 12′.

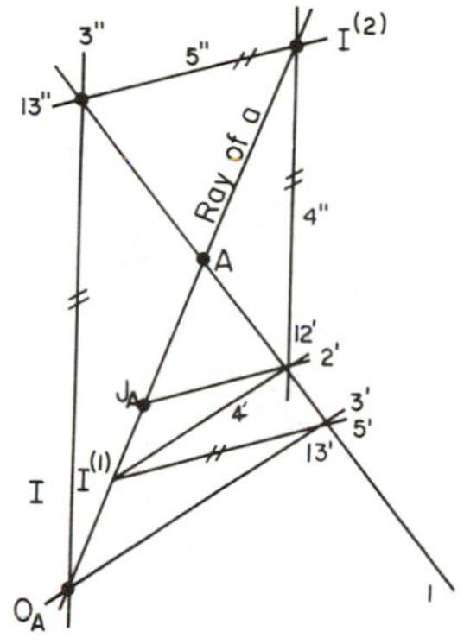


**Figure 4.29** Bobillier's construction for finding $I$ when points $A$, $J_A$ and $O_A$ are given yields 2 solutions for the location of I, equidistant on either side of A on the ray $AJ_AO_A$. This agrees with ESE-4, Eqs. (4.45) and (4.46).

Draw arbitrary trial line 3′ through $O_A$ and mark its intersection with line 1: 13′.

Through 12′ draw line 4′ parallel with 3′.

Through 13′ draw line 5′ parallel with 2′.

If lines 4′ and 5′ intersect on the ray, the intersection is the first solution for the instant center ($I^{(1)}$).

If lines 4′ and 5′ do not intersect on the ray, then iterate from the second step until they do.

In this first solution, both lines 2′ and 3′ through $J_A$ and $O_A$, respectively, were drawn to intersect line 1 on the same side of $A$. A second solution is found if either line 2′ or 3′ is drawn to intersect line 1 on the other side of $A$. This is illustrated in Fig. 4.29 by line 3″ through $O_A$ in the following second solution:

**Procedure** (Fig. 4.29, *Second Solution*):

Draw arbitrary line 1 through $A$.

Draw arbitrary line 2′ through $J_A$ and mark its intersection with line 1: (12′).

Draw arbitrary line 3″ through $O_A$ to intersect line 1 at 13″ on opposite side of $A$ from intersection 12′.

Through 12′ draw line 4″ parallel with 3″.

Through 13″ draw line 5″ parallel with 2′.

If lines 4″ and 5″ intersect on the ray, the intersection is $I^{(2)}$, the second solution for the instant center.

If not, then iterate from the second step until they do.

It is easy to show by similar triangles (see Exer. 4.10) that both $I^{(1)}$ and $I^{(2)}$ satisfy equation ESE-4 (Eq. 4.46), and that the magnitudes of the distances $I^{(1)}A$ and $I^{(2)}A$ are equal. Thus, in the foregoing Bobillier constructions we have geometric verification of the fact that, when points $O_A$, $A$, and $J_A$ [vectors $\boldsymbol{\rho}_A$ and $(\overrightarrow{J_AA})$] are given on a ray, then, as Eq. (4.45) shows, there are two solutions for $a = IA$, equal in magnitude, but opposite in sign. Thus $(\pm \mid a \mid e^{i\arg \boldsymbol{\rho}_A})$ are both solutions for the vector **a**.

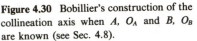

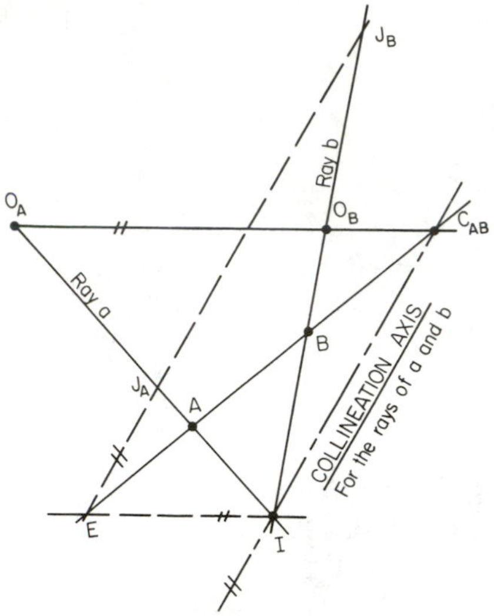


**Figure 4.30** Bobillier's construction of the collineation axis when $A$, $O_A$ and $B$, $O_B$ are known (see Sec. 4.8).

## 4.8 THE COLLINEATION AXIS

Suppose that we know the location of $I$, $A$, $O_A$, $B$, and $O_B$ as shown in Fig. 4.30, and nothing else. If we draw a line through $A$ and $B$, and intersect it with a line through $O_A$ and $O_B$, we obtain a point $C_{AB}$ on the "collineation axis" for the rays of $A$ and $B$, which goes through the instant center $I$. Bobillier has shown how the collineation axis can be used to find the inflection points $J_A$ and $J_B$:

> Draw a line through $I$, parallel with $O_AO_B$, to intersect $AB$ at $E$.
>
> Draw a line through $E$, parallel with the collineation axis, to intersect the rays of $A$ and $B$ at $J_A$ and $J_B$.

We can prove the validity of this construction, using our vector notation, as follows. From similar triangles

$$\Delta AO_AC_{AB} \approx \Delta AIE$$

we have

$$\frac{\mathbf{a}}{\boldsymbol{\rho}_A} = \frac{\overrightarrow{IE}}{\overrightarrow{O_AC_{AB}}} \tag{4.47}$$

and from similar triangles

$$\Delta IJ_AE \approx \Delta O_AIC_{AB}$$

$$\frac{\overrightarrow{IE}}{\overrightarrow{O_AC_{AB}}} = \frac{\mathbf{J}_A}{-\mathbf{O}_A} \tag{4.48}$$

and, combining Eqs. (4.47) and (4.48), we obtain

$$\frac{1}{\mathbf{a}} = \frac{-\mathbf{O}_A}{\boldsymbol{\rho}_A \mathbf{J}_A} = \frac{\boldsymbol{\rho}_A - \mathbf{a}}{\boldsymbol{\rho}_A \mathbf{J}_A} = \frac{1}{\mathbf{J}_A} - \frac{\mathbf{a}}{\boldsymbol{\rho}_A \mathbf{J}_A} \tag{4.49}$$

but

$$\frac{1}{\mathbf{a}} = \frac{-\mathbf{O}_A}{\boldsymbol{\rho}_A \mathbf{J}_A} \quad \text{can be written as} \quad \frac{\mathbf{a}}{\boldsymbol{\rho}_A \mathbf{J}_A} = -\frac{1}{\mathbf{O}_A}$$

and

$$\frac{1}{\mathbf{a}} = \frac{1}{\mathbf{J}_A} + \frac{1}{\mathbf{O}_A} \quad \text{or} \quad \mathbf{a} = \frac{\mathbf{J}_A \mathbf{O}_A}{\mathbf{O}_A + \mathbf{J}_A}$$

which is ESE-3, Eq. (4.43). Q.E.D.

We can similarly show that $I$, $B$, $O_B$, and $J_B$ satisfy ESE (see Exer. 4.11). In fact, any ray through $I$ with a tracer point, say $F$, *collinear* with $A$ and $B$, has $O_F$ *collinear* with $O_A$ and $O_B$, and has $J_F$ *collinear* with $J_A$ and $J_B$ (see Exer. 4.12), hence the name for $IC_{AB}$: the "collineation axis." Now, any *two* rays, such as those of $A$, $B$, or $F$, can be used to determine an inflection circle by way of $I$, $J_A$, and $J_B$, or $I$, $J_A$, and $J_F$, and so on, thereby establishing a "pole tangent" and "pole normal" associated with each pair of rays. It is easy to show that, if $\mathbf{O}_F = \mathbf{J}_F$, then $\mathbf{f} = \mathbf{J}_F/2$ (see Exer. 4.13).

**Lemma.** Any two rays $a$ and $b$ (Fig. 4.31) through $I$, with $J_A$ and $J_B$ given on them, uniquely determine their collineation axis through $I$, which is parallel with $J_AJ_B$. For a proof of this lemma, see Exer. 4.14.

## 4.9 BOBILLIER'S THEOREM [29]

"The path normals (pole rays) of two points of a moving body make equal angles with their collineation axis and with the pole tangent."* The bisector of the angle between two rays also bisects the angle between their collineation axis and the common tangent of the centrodes. We prove this by way of Fig. 4.32, where $C_{AB}$ is any point on the collineation axis ab and where $J_A$ and $J_B$ are inflection points. Here $\alpha$ is the angle between the secant $\mathbf{J}_A$ of the inflection circle and the tangent at one end of this secant. Therefore, according to a well-known fact from elementary geometry, $\alpha$ equals any peripheral angle subtended by the secant, such as $\beta$. Furthermore, $\beta$ and $\gamma$ have one common side and their other sides are parallel. Thus $\beta = \gamma$, and therefore $\alpha = \gamma$. This completes our proof of Bobillier's theorem, which has many applications in kinematic synthesis and analysis (see the exercises at the end of the chapter).

* As quoted by Rudolph A. Beyer [24].

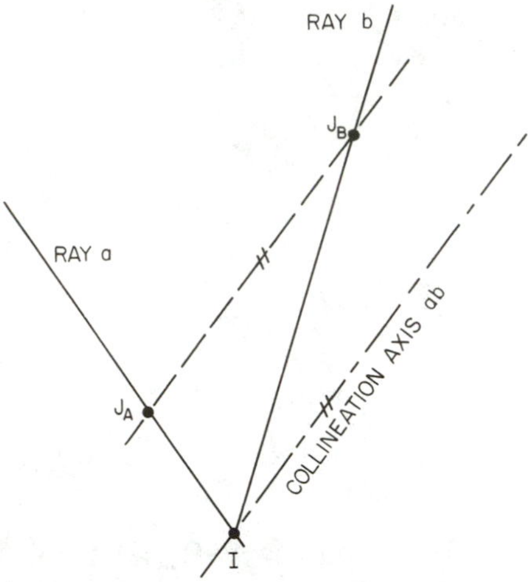


**Figure 4.31** The collineation axis for rays $a$ and $b$ is parallel with the line connecting the inflection points on these two rays.

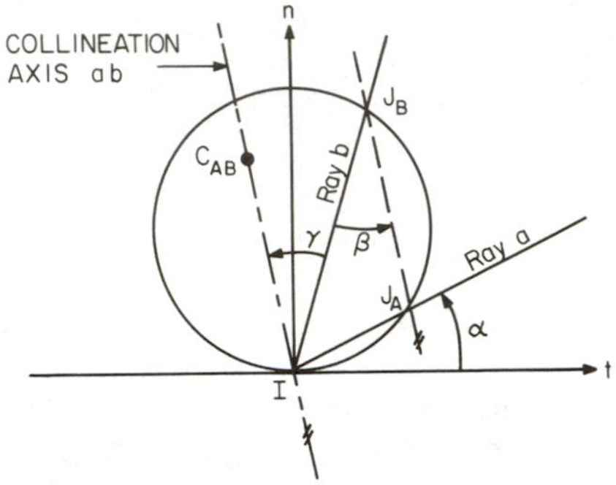


**Figure 4.32** Illustruction of Bobillier's theorem (see Sec. 4.9): the bisector of the angle between two rays $a$ and $b$ also bisects the angle between their collineation axis and the common tangent of the centrodes.

The following equations can easily be deduced from Fig. 4.32:

$$\arg \mathbf{J}_A + \arg \mathbf{J}_B = \arg \mathbf{C}_{AB} \tag{4.50}$$

where the $+x$ axis coincides with $t$. From this

$$\theta_a + \theta_b + \frac{\pi}{2} = \theta_{cab} \tag{4.51}$$

where the $\theta$'s are measured from $n$, the common normal of the centrodes, positive ccw. Equations (4.50) and (4.51) are programmable statements of Bobillier's theorem.

## 4.10 HARTMANN'S CONSTRUCTION [126]

The original Hartmann method starts with given common tangent and common normal of the centrodes, the velocity of point $A$ of the moving plane, $\mathbf{v}_A$, and the transfer velocity $\mathbf{u}$ of the instant center $I$. He then seeks the center of path curvature for $A$, namely $O_A$, using a combined graphical–analytical procedure, as follows.

With $\mathbf{v}_A$ and $\mathbf{u}$ drawn to scale in Fig. 4.33, he constructs the component of $\mathbf{u}$ parallel to $\mathbf{v}_A$, namely $\mathbf{u}_{at}$ (subscript $at$ = tangential for points on ray $a$). He now draws a line through the tips of the vectors $\mathbf{v}_A$ and $\mathbf{u}_{at}$ to intersect ray $a$. This is $O_A$.

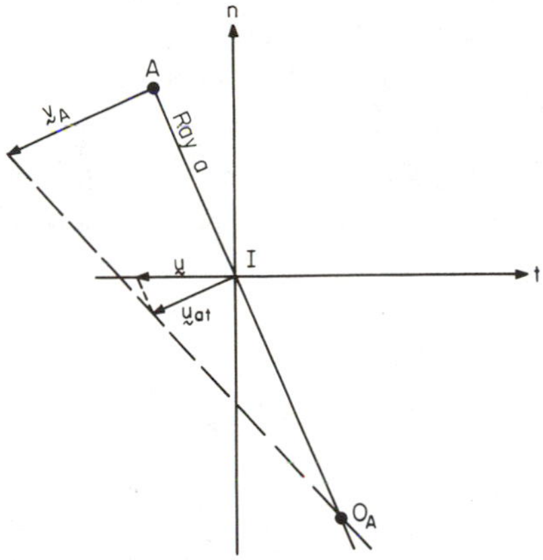


**Figure 4.33** Hartman's construction for finding $O_A$ when $t, n, A$ $\mathbf{V}_A$ and $\mathbf{u}$ are known can be visualized by way of a small rotation of a ficticious link pivoted at the unknown $O_A$ and guiding the point A in its path (see Sec. 4.10).

We can verify this construction as follows. Imagine a fictitious rigid link, connecting the point $A$ to the fixed pivot $O_A$. As this link swings with $A$ moving at the velocity $\mathbf{v}_A$, the point of the link momentarily coincident with $I$ will have the velocity $\mathbf{u}_{at}$, because the transfer velocity of $I$ itself is $\mathbf{u}$ along the common tangent. Thus it is clear that our fictitious link guides the point $A$ momentarily along its actual path, and therefore $O_A$ must be the center of path curvature.

Hartmann's original construction, just described, requires that $\mathbf{u}$ be known. For certain special cases, such as a circle rolling on a straight line, $\mathbf{u}$ is easily found once the angular velocity of rolling is determined from the given velocity of the moving point $A$. In fact, this is the usual example described in the literature. It is, however, possible to generalize Hartmann's method for any combination of centrode curvatures and, at the same time, adapt it for automatic computation by way of complex numbers. Here is how.

Like Hartmann, we start with the knowledge of $t$, $n$, $A$, and $\mathbf{v}_A$, but in general, $\mathbf{u}$ will not be known. If the centrode curvatures are known, we find $\mathbf{u}$ as follows (see Fig. 4.34). From Eq. (4.14),

$$\dot{\phi} = \frac{\mathbf{v}_A}{i\mathbf{a}}$$

and from Eq. (4.24),

$$\mathbf{u} = i\dot{\phi}\boldsymbol{\rho}_p\boldsymbol{\rho}$$

where

$$\boldsymbol{\rho} = \frac{-\boldsymbol{\rho}_\pi}{\boldsymbol{\rho}_p - \boldsymbol{\rho}_\pi}$$

as defined in Eq. (4.7). With $\mathbf{u}$ thus determined, $\mathbf{u}_{at}$ is obtained by using the projection operator $(\cos\theta)e^{i\theta}$:

$$\mathbf{u}_{at} = (\cos\theta)e^{i\theta}\mathbf{u} \qquad (4.52)$$

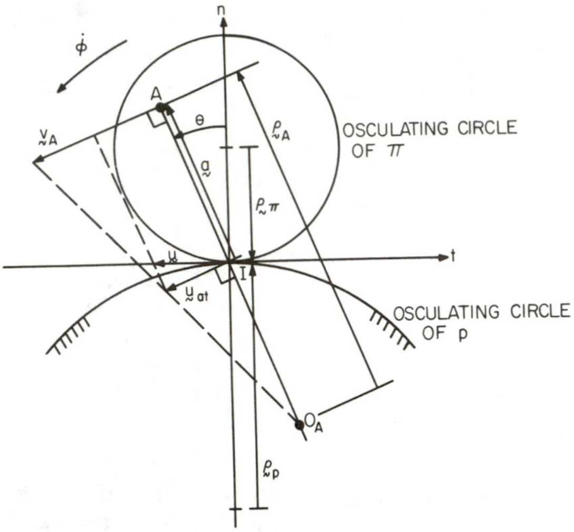


**Figure 4.34** Notation for the complex-number form of Hartman's construction for finding $O_A$ when the osculating circles of $p$ and $\pi$, the angular velocity $\dot{\phi}$ of $\pi$ rolling on $p$ and point $A$ are given [see Eqs. (4.24, 4.52, 4.53) and (4.54)].

Then from similar triangles,

$$\frac{\boldsymbol{\rho}_A}{\mathbf{a}} = \frac{\mathbf{v}_A}{\mathbf{v}_A - \mathbf{u}_{at}} \quad \text{and} \quad \boldsymbol{\rho}_A = \frac{\mathbf{a}\mathbf{v}_A}{\mathbf{v}_A - \mathbf{u}_{at}} \tag{4.53}$$

We can now proceed to find the vector diameter of the inflection circle $\boldsymbol{\delta}$ with Hartmann's method. From Eq. (4.24a),

$$\boldsymbol{\delta} = -i\frac{\mathbf{u}}{\dot{\phi}} = -i\frac{\mathbf{u}_{at}(1/\cos\theta)e^{-i\theta}}{-i(\mathbf{v}_A/\mathbf{a})} = \frac{\mathbf{u}_{at}\mathbf{a}e^{-i\theta}}{\mathbf{v}_A\cos\theta} = -\frac{\mathbf{O}_A}{\boldsymbol{\rho}_A}\mathbf{a}\frac{e^{-i\theta}}{\cos\theta} \tag{4.54}$$

where we used Eqs. (4.14) and (4.52) and similar triangles in Fig. 4.34.

Note that the resulting expression for $\boldsymbol{\delta}$ in Eq. (4.54) is purely geometric and is independent of $\mathbf{v}_A$ or $\dot{\phi}$. Looking at it another way, it is equivalent to assuming that $\dot{\phi} = 1$ rad/sec, which makes $v_A = a$. This leads to a simple, purely geometric construction of $\boldsymbol{\delta}$ when $A$, $O_A$, $I$, and the common normal $n$ are known. The procedure goes like this (Fig. 4.35):

1. At $A$ erect a ⊥ to **a** of length a.
2. From the end of this ⊥, draw a line to $O_A$.
3. At $I$ erect another ⊥ to **a** extending to the line just drawn and swing its length over onto **a**.
4. At the point on **a** just found, erect still another ⊥ to **a**. This ⊥ intersects $n$ at the tip of $\boldsymbol{\delta}$.

For verification of this construction, see Exer. 4.15.

Note that this form of Hartmann's construction requires knowing the direction of $n$. To obtain this knowledge for the coupler motion of a four-bar linkage, we

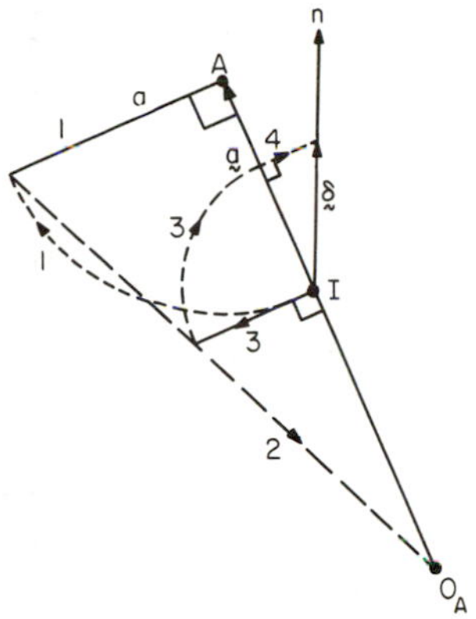


**Figure 4.35** Hartman's construction for finding the vector-diameter $\boldsymbol{\delta}$ of the inflection circle when $A$, $O_A$, $I$ and the common normal $n$ are known.

use another form of Hartmann's construction for the transfer velocity $\mathbf{u}$ for the instant center of the coupler motion, which gives the direction of the common tangent of the centrodes, $t$, and thus also of the common normal, $n$. The procedure, pictured in Fig. 4.36, goes like this:

1. Find $I$ of coupler $AB$ and fixed link $O_AO_B$ of the four-bar linkage in Fig. 4.36 (see construction lines for step 1 labeled with the number 1, and similarly for subsequent steps).
2. Assume $\mathbf{v}_A$.
3. Draw a line from the tip of $\mathbf{v}_A$ through $O_A$ to intersect the $\perp$ to $AO_A$ erected at $I$. This gives $\mathbf{u}_{at}$.
4. Using the fact that the projections of $\mathbf{v}_A$ and $\mathbf{v}_B$ onto the link $AB$ are equal, construct $\mathbf{v}_B$.
5. Repeat step 3 for $B$ to find $\mathbf{u}_{bt}$.
6. The $\perp$'s erected from the tips of $\mathbf{u}_{at}$ and $\mathbf{u}_{bt}$ intersect at the tip of $\mathbf{u}$. This determines the direction of the common tangent, $t$, collinear with but opposed in sense to $\mathbf{u}$.

For verification of this construction, see Exer. 4.16.

To program this procedure for digital computation, we proceed as follows:

1. Choose a convenient coordinate system, say $x \equiv \overrightarrow{O_AO_B}$, $O \equiv O_A$, and thus $iy$ will point upward from $O_A$ in Fig. 4.36 (not shown).
2. Let $\overrightarrow{O_AA} = \boldsymbol{\rho}_A$, $\overrightarrow{O_BB} = \boldsymbol{\rho}_B$, and $\overrightarrow{BA} = \boldsymbol{\rho}_A - \overrightarrow{O_AO_B} - \boldsymbol{\rho}_B$.
3. Assume the magnitude of $\mathbf{v}_A$. Then $\mathbf{v}_A = v_A i e^{i \arg \boldsymbol{\rho}_A}$.
4. $\mathbf{v}_B = \{\cos(\arg \mathbf{v}_A - \arg \overrightarrow{BA})e^{-i(\arg \mathbf{v}_A - \arg \overrightarrow{BA})} \sec[\arg \boldsymbol{\rho}_B + (\pi/2) - \arg \overrightarrow{BA}]e^{i(\arg \boldsymbol{\rho}_B + (\pi/2) - \arg \overrightarrow{BA})}\}\mathbf{v}_A$.
5. Solve $\lambda_a \boldsymbol{\rho}_A = \overrightarrow{A_AO_B} + \lambda_b \boldsymbol{\rho}_B$ for $\lambda_a$ and $\lambda_b$.
6. $\mathbf{I} = \lambda_a \boldsymbol{\rho}_A$.

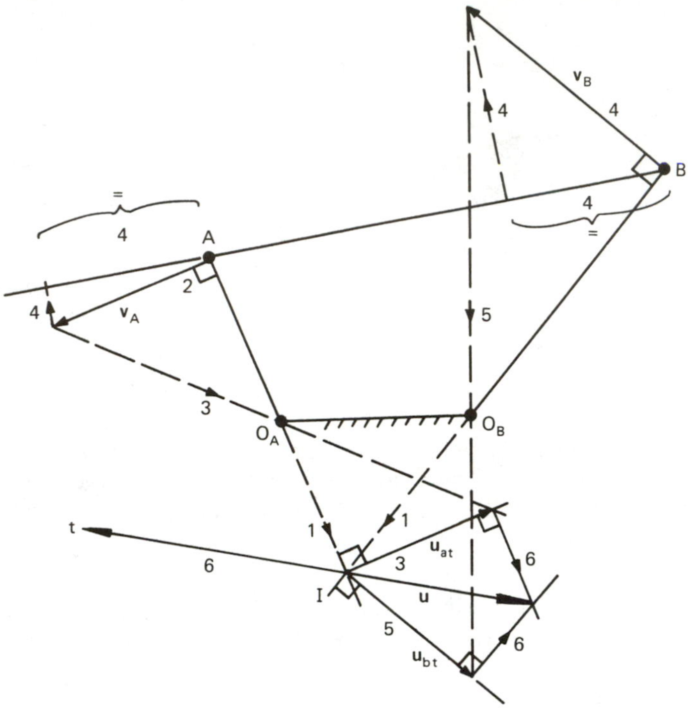


**Figure 4.36** Hartman's construction for finding the common tangent $t$ and the instant-center-transfer velocity **u** for a four-bar mechanism when the velocity of a moving joint is known. Numbers indicate sequence of drawing the construction lines.

7. $\mathbf{u}_{at} = (\mathbf{I}/\boldsymbol{\rho}_A)\,\mathbf{v}_A = \lambda_a \mathbf{v}_A,\ \mathbf{u}_{bt} = \lambda_b \mathbf{v}_B.$
8. Solve $\lambda_{bt}\boldsymbol{\rho}_B + \mathbf{u}_{bt} = \lambda_{at}\boldsymbol{\rho}_A + \mathbf{u}_{at}$ for $\lambda_{at}$ and $\lambda_{bt}$.
9. $\mathbf{u} = \mathbf{u}_{at} + \lambda_{at}\boldsymbol{\rho}_A$. Check: $\mathbf{u} = \mathbf{u}_{bt} + \lambda_{bt}\,\boldsymbol{\rho}_B$.

## 4.11 THE BRESSE CIRCLE

Let us now find the locus of all points in the rolling plane $\pi$ whose tangential acceleration is zero in the position shown. To this end, we refer to Eq. (4.32) and equate $\mathbf{A}^t_{A1}$ with zero:

$$\mathbf{A}^t_{A1} = i\ddot{\phi}\mathbf{a} - (\sin\theta)e^{i[\theta+(\pi/2)]}(-i\dot{\phi}\mathbf{u}) = 0$$

or

$$i\ddot{\phi}\mathbf{a} = (\sin\theta)e^{i[\theta+(\pi/2)]}(-i\dot{\phi}\mathbf{u})$$

Recalling that $-i\dot{\phi}\mathbf{u} = \dot{\phi}^2\boldsymbol{\delta}$ [Eq. (4.26)], we have

$$\mathbf{a} = (\sin\theta)\frac{e^{i[\theta+(\pi/2)]}}{i}\frac{\dot{\phi}^2}{\ddot{\phi}}\boldsymbol{\delta} = (\sin\theta)e^{i\theta}\frac{\dot{\phi}^2}{\ddot{\phi}}\boldsymbol{\delta} \tag{4.55}$$

Multiply and divide by $i$:

$$\mathbf{a} = (\sin\theta)e^{i[\theta-(\pi/2)]}\left(i\frac{\dot{\phi}^2}{\ddot{\phi}}\boldsymbol{\delta}\right) \tag{4.56}$$

Figure 4.37 is a geometric representation of Eq. (4.56), showing that this is the equation of the Bresse circle, the locus of all points of the $\pi$ plane with zero tangential acceleration. Its intersection with the inflection circle is the *acceleration pole, $A_p$,* which is that point of the $\pi$ plane that has zero total acceleration in the position shown.

The vector $(\overrightarrow{IA_p})$, which locates $A_p$, is found by equating the right sides of Eqs. (4.34) and (4.55) of the inflection circle and the Bresse circle:

$$(\cos\theta)e^{i\theta}\boldsymbol{\delta} = (\sin\theta)e^{i\theta}\frac{\dot{\phi}^2}{\ddot{\phi}}\boldsymbol{\delta} \Rightarrow \tan\theta_{AP} = \frac{\ddot{\phi}}{\dot{\phi}^2} \tag{4.57}$$

To resolve the ambiguity due to the double-valued nature of the $\tan^{-1}(\cdot)$ function, we observe from Fig. 4.37 that

$$-\frac{\pi}{2} < \theta_{AP} < \frac{\pi}{2} \tag{4.58}$$

It will also be shown [Eq. (4.65)] that

$$\theta_{AP} = \arg(\dot{\phi}^2 + i\ddot{\phi}) \tag{4.59}$$

$$\mathbf{a}_{AP} = (\cos\theta_{AP})e^{i\theta_{AP}}\boldsymbol{\delta} \tag{4.60}$$

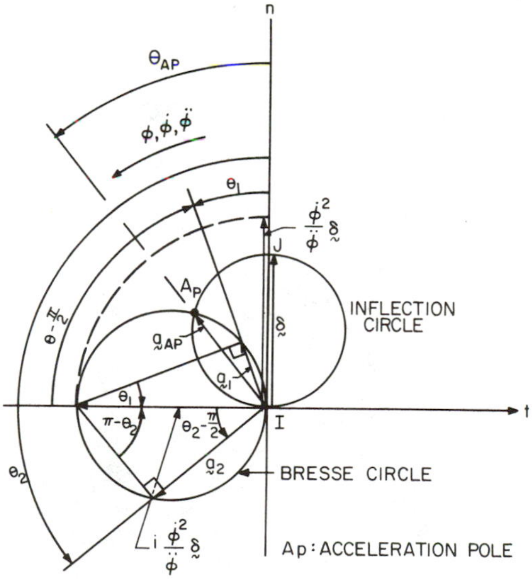


**Figure 4.37** The Bresse Circle is the locus of all points of the moving plane $\pi$ whose pathwise (tangential) acceleration is momentarily zero. Its intersection with the inflection circle (locus of all points of $\pi$ whose path-normal acceleration is momentarily zero) is the *acceleration pole $A_p$*: that point of $\pi$ which momentarily has zero acceleration.

What happens when $\ddot{\phi} = 0$? It can be shown (see Exer. 4.5) that: (1) the Bresse circle degenerates into a straight line, namely the normal $n$ to the stationary and moving centrodes $p$ and $\pi$ at $I$, and (2) that $A_p = J$.

Figure 4.37 shows the Bresse circle and the acceleration pole for the case when $\ddot{\phi} > 0$. What if $\ddot{\phi} < 0$? Equation (4.56) shows that, in that case, the diameter of the Bresse circle $[i(\dot{\phi}^2/\ddot{\phi})\boldsymbol{\delta}]$ changes sign, and the Bresse circle would appear on the right side of the centrode normal $n$ in Fig. 4.37. It is also clear from Eq. (4.57) that if $\ddot{\phi}$ is very small, the Bresse circle becomes very large. This agrees with the fact (see Exer. 4.17) that as $\ddot{\phi}$ approaches zero from either the positive or the negative side, the Bresse circle degenerates into the normal $n$.

## 4.12 THE ACCELERATION FIELD

It can be shown that the acceleration of any point of the moving plane $\pi$ can be expressed as

$$\mathbf{A}_A = c_a e^{i(-\theta_{AP})}(\mathbf{a}_{AP} - \mathbf{a}_A) \tag{4.61}$$

as shown in Fig. 4.38, where $c_a$ is a scalar constant. In words: The acceleration of any point $A$ of the $\pi$ plane can be obtained by taking the vector $\overrightarrow{AA_p}$ and stretch-rotating it by the operator $[c_a e^{i(-\theta_{AP})}]$. This operator is the same for all points of the moving plane. To prove this, and find the value of $c_a$, we proceed as follows.

Observe that, since $\mathbf{A}_{AP} = 0$, the total acceleration of any point $A$ is equal to the acceleration difference between the points $A_{AP}$ and $A$, or

$$\mathbf{A}_A = \mathbf{A}_{AP} + \mathbf{A}_{A,AP} = (-\dot{\phi}^2 + i\ddot{\phi})(\mathbf{a}_A - \mathbf{a}_{AP}) \tag{4.62}$$

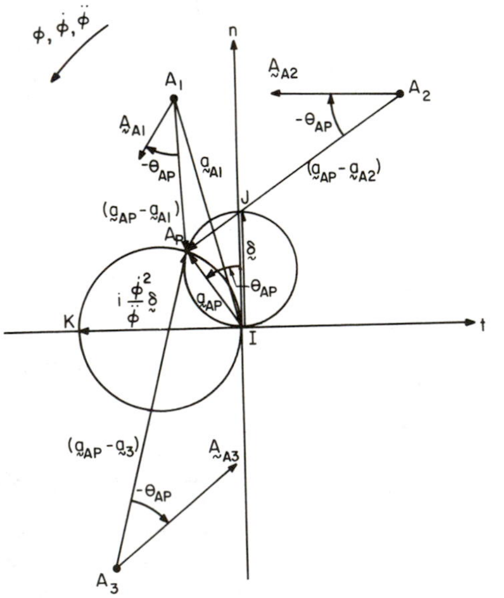


**Figure 4.38** Complex-vector illustration of the acceleration field equation, Eq. (4.69), which covers all points of the moving plane $\pi$.

Combining Eqs. (4.61) and (4.62), the stretch-rotation operator is obtained as

$$\frac{\mathbf{A}_A}{\mathbf{a}_{AP} - \mathbf{a}_A} = \frac{\mathbf{A}_A}{-(\mathbf{a}_A - \mathbf{a}_{AP})} = \dot{\phi}^2 - i\ddot{\phi} = c_a e^{i(-\theta_{AP})} \tag{4.63}$$

which is invariant for all $A$'s, and can be expressed in polar form as follows:

$$\dot{\phi}^2 - i\ddot{\phi} = (\dot{\phi}^4 + \ddot{\phi}^2)^{1/2} e^{i\arg(\dot{\phi}^2 - i\ddot{\phi})} \tag{4.64}$$

from which

$$\arg(\dot{\phi}^2 - i\ddot{\phi}) = -\theta_{AP} \tag{4.65}$$

which proves Eq. (4.59) and

$$c_a = (\dot{\phi}^4 + \ddot{\phi}^2)^{1/2} \tag{4.66}$$

which indeed are constant for any point of the $\pi$ plane.

Let us now put our result to the test by applying it to the point whose total acceleration is the invariant acceleration component, $-i\dot{\phi}\mathbf{u}$, namely the point $I_\pi$, which is that point of the moving plane $\pi$ which momentarily coincides with the instant center $I$. We therefore have to show that

$$-i\dot{\phi}\mathbf{u} = (\dot{\phi}^4 + \ddot{\phi}^2)^{1/2} e^{i(-\theta_{AP})}(-\mathbf{a}_{AP}) \tag{4.67}$$

To this end, let us express both $\mathbf{u}$ and $\mathbf{a}_{AP}$ in terms of $\boldsymbol{\delta}$, the vector diameter of the inflection circle. Recall from Eq. (4.26) that $-i\dot{\phi}\mathbf{u} = \dot{\phi}^2\boldsymbol{\delta}$, and from Eq. (4.60) that $\mathbf{a}_{AP} = (\cos\theta_{AP})e^{i\theta_{AP}}\boldsymbol{\delta}$. Substituting these in Eq. (4.67), we get

$$\dot{\phi}^2\boldsymbol{\delta} = (\dot{\phi}^4 + \ddot{\phi}^2)^{1/2} e^{i(-\theta_{AP})}(-\cos\theta_{AP})e^{i\theta_{AP}}\boldsymbol{\delta}$$

Canceling $\boldsymbol{\delta}$, and observing that $e^{i(-\theta_{AP})}e^{i\theta_{AP}} = 1$, we have

$$\dot{\phi}^2 = (-\cos\theta_{AP})(\dot{\phi}^4 + \ddot{\phi}^2)^{1/2}$$

or that

$$(\cos\theta_{AP})(\dot{\phi}^4 + \ddot{\phi}^2)^{1/2} = -\dot{\phi}^2 \tag{4.68}$$

But

$$(\cos\theta_{AP})(\dot{\phi}^4 + \ddot{\phi}^2)^{1/2} = \mathscr{R}(-\dot{\phi}^2 + i\ddot{\phi}) = -\dot{\phi}^2$$

Thus we have reached an identity, and our invariant stretch-rotation operator of the acceleration field is verified. It can be shown that Eqs. (4.23) and (4.62) yield the same value for $\mathbf{A}_A$ (see Exer. 4.18).

Let us summarize our results, now verified. To obtain the acceleration field of all points of the moving plane $\pi$, proceed as follows.

1. Find the acceleration pole $A_p$ as the intersection of the inflection circle and the Bresse circle (other than $I$):

$$\mathbf{a}_{AP} = \boldsymbol{\delta}\,(\cos\theta_{AP})e^{i\theta_{AP}}$$

where

$$\theta_{AP} = \arg(\dot{\phi}^2 + i\ddot{\phi}) \tag{4.59}$$

2. To find the acceleration of any point $A$ of $\pi$, obtain the vector $\overrightarrow{AA_p} = \mathbf{a}_{AP} - \mathbf{a}_A$ and multiply it by the invariant stretch-rotation operator of the acceleration field:

$$\mathbf{A}_A = (\mathbf{a}_{AP} - \mathbf{a}_A)(\dot{\phi}^4 + \ddot{\phi}^2)^{1/2} e^{i(-\theta_{AP})} \tag{4.69}$$

This procedure is illustrated in Fig. 4.38.

Note that in all acceleration field calculations it is preferable to use $\theta_{AP} = \arg(-\dot{\phi}^2 + i\ddot{\phi})$ rather than $\theta_{AP} = \tan^{-1}(\ddot{\phi}/\dot{\phi}^2)$, because the "arg" function eliminates the 180° ambiguity of the "$\tan^{-1}$" function. Therefore, on hand calculators use the rectangular-to-polar (R → P) transformation key and in FORTRAN IV or WATIV use the "ATAN2" statement. Thus the significance of the sense of $\ddot{\phi}$ will not be lost due to the "$\tan^{-1}$" ambiguity. It is also beneficial to sketch $(-\dot{\phi}^2 + i\ddot{\phi})$ as a complex vector in order to visually verify the sense and quadrant of $\theta_{AP}$.

## 4.13 THE RETURN CIRCLE

Named by literal translation of the German "Rückkehrtskreis," the return circle is the locus of those points of the fixed plane $p$ through which a straight line embedded in the moving plane $\pi$ momentarily slide-rotates. An instantaneously equivalent physical embodiment of such a point $Q$ is the tumbling block of Fig. 4.39, showing three successive positions of the line $l$ and the point $A$ on it. By letting these positions approach infinitesimally close together, we have pictured the motion described above.

The name "return circle" (referred to as the "cuspidal circle" by some authors) comes from the fact that the envelope in the $p$ plane of the line of the $\pi$ plane, momentarily sliding through and rotating around a point on this circle, forms a cusp as it approaches to and recedes from this "return point" of the fixed plane. Note, however, that the point of the moving plane $\pi$ momentarily coincident with the "return point" describes a continuous path without a break, not a cusp, as it traverses the return point (see successive positions of point $A$: $A_1$, $A_2$, and so on, in Fig. 4.39).

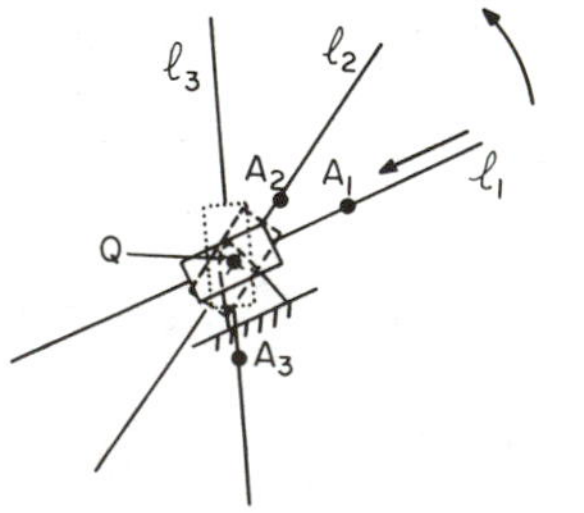


**Figure 4.39** If a line $l$ of the moving plane $\pi$ slide-rotates through point $Q$ of the fixed plane $p$, then $Q$ is a point of the *return circle,* the locus of all such points embedded in $p$.

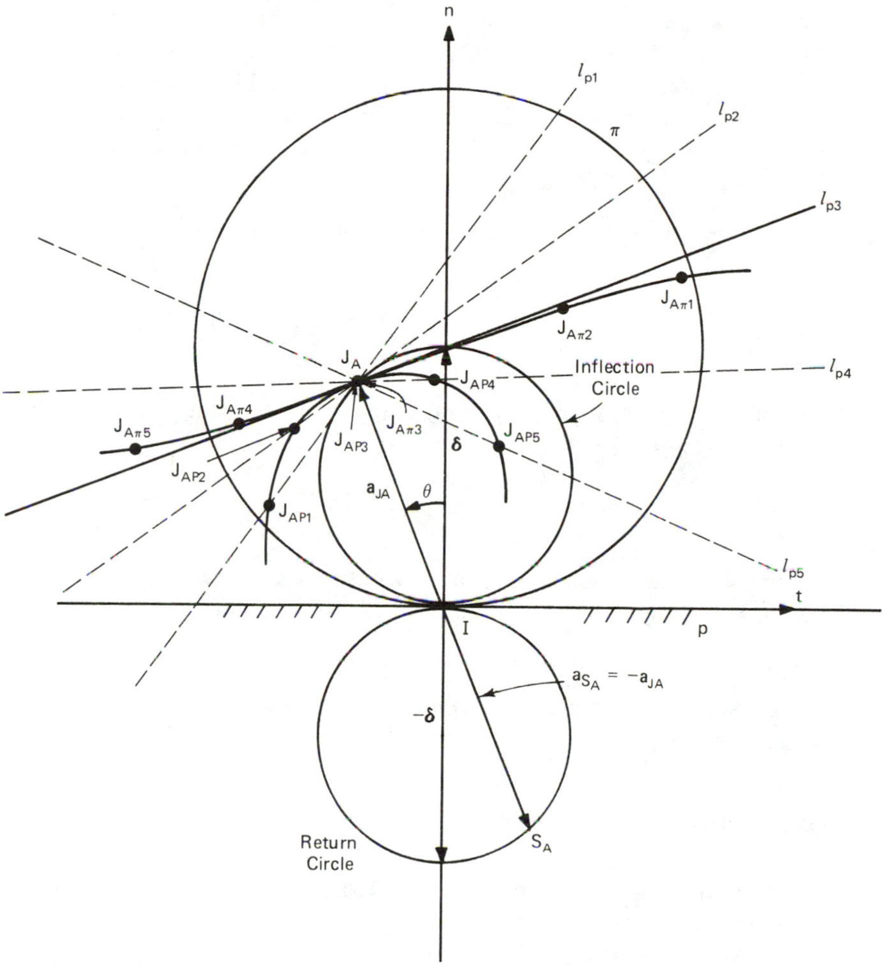


**Figure 4.40** As $\pi$ roles on $p$, joint $J_{A\pi}$ of $\pi$ describes a path in $p$ with an inflection at $J_A$ on the inflection circle. The path tangent at $J_A$ is $l_{p3}$, a line embedded in $p$. In the inversion, when $\pi$ is fixed, line $l_p$ slide-rotates through $J_{A\pi3}$, now fixed. Thus, in the inversion, the inflection circle becomes the *return circle*.

How can we find such points in the fixed plane $p$? We can find them by kinematic inversion. Consider the inflection point $J_A$ as a point of the moving plane $\pi$, $J_{A\pi}$, and observe its path in the vicinity of the position shown in Fig. 4.40. Following its path through positions $J_{A\pi1}$ to $J_{A\pi5}$, we see that, in the vicinity of $J_A = J_{A\pi3}$, it momentarily moves along the path tangent $l_p$, a line embedded in the fixed plane $p$, which touches the path at the inflection point $J_A$.

Now consider the kinematic inversion when $\pi$ is fixed and $p$ rolls on it. Then the line $l_p$, carrying with it the point $J_{AP}$, will slide-rotate through the now fixed point $J_{A\pi3}$. Thus we see that by the inversion above, the inflection circle of the $\pi$ plane becomes the return circle for the $p$ plane. Recognizing this, we obtain the equation of the return circle by interchanging $\boldsymbol{\rho}_\pi$ and $\boldsymbol{\rho}_p$ in the equation of the inflection circle as follows:

$$\mathbf{a}_{J_A} = (\cos\theta)e^{i\theta}\boldsymbol{\delta} \qquad \text{(from the equation of the inflection circle)} \tag{4.34}$$

Recall that

$$\boldsymbol{\delta} = \frac{-\boldsymbol{\rho}_p\boldsymbol{\rho}_\pi}{\boldsymbol{\rho}_p - \boldsymbol{\rho}_\pi} \tag{4.24a}$$

and interchange $\boldsymbol{\rho}_p$ and $\boldsymbol{\rho}_\pi$ to obtain the expression for the vector diameter of the return circle, $\boldsymbol{\sigma}$:

$$\boldsymbol{\sigma} = \frac{-\boldsymbol{\rho}_\pi\boldsymbol{\rho}_p}{\boldsymbol{\rho}_\pi - \boldsymbol{\rho}_p} = -\boldsymbol{\delta} \tag{4.70}$$

Designating the general point of the return circle $S_A$, in view of Eqs. (4.34) and (4.70) we can write the equation of the return circle as

$$\mathbf{a}_{S_A} = (\cos\theta)e^{i\theta}(-\boldsymbol{\delta}) = -(\cos\theta)e^{i\theta}\boldsymbol{\delta} = -\mathbf{a}_{J_A} \tag{4.71}$$

which shows that the return circle and the inflection circle are symmetric about the instant center $I$.

## 4.14 CUSP POINTS

A cusp point of the moving plane $\pi$ describes a cusp in its path in the fixed plane $p$ at the position considered. What is a cusp? It is a point where the path has zero radius of curvature. Also, the tracer point approaches to and recedes from the cusp along the same tangent in the immediate vicinity of the cusp. As an example, consider point $I_\pi$, a point of the $\pi$ plane momentarily coincident with the instant center $I$. Its velocity is therefore zero and its total acceleration is the invariant component of accelerations:

$$\mathbf{A}_{I\pi} = -i\dot{\phi}\mathbf{u} \tag{4.31}$$

But

$$\mathbf{u} = i\dot{\phi}\boldsymbol{\delta} \tag{4.24a}$$

and therefore

$$\mathbf{A}_{I\pi} = -i\dot{\phi}i\dot{\phi}\boldsymbol{\delta} = \dot{\phi}^2\boldsymbol{\delta} \tag{4.72}$$

which shows that the total acceleration of $I_\pi$ in coincidence with $I$ is in the direction of the common normal of the centrodes, in the sense of $\boldsymbol{\delta}$. This means that $I_\pi$

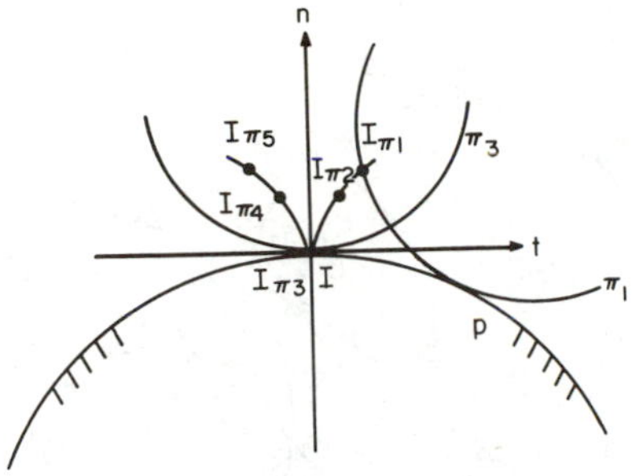


**Figure 4.41** The rollcurve $\pi$ is the locus of all *cusp points* of the $\pi$ plane.

approaches to and recedes from $I$ in a direction perpendicular to the common tangent. This is shown in Fig. 4.41, where the moving centrode is shown in the $\pi_1$ position as it rolls toward the $\pi_3$ position, in which the point $I_\pi$ assumes coincidence with $I$. Thus it is seen that the locus of all cusp points of the $\pi$ plane is the moving centrode $\pi$, since every point on it eventually becomes momentarily coincident with the instant center. Furthermore, the path tangent in the vicinity of the cusp is perpendicular to the fixed centrode.

## 4.15 CRUNODE POINTS

A crunode is a double point of the path where it recrosses itself forming a loop. To find crunode points, extend the $\theta$ ray to the side of the moving centrode $\pi$ opposite the inflection circle (Fig. 4.42). On this ray extension, locate a point $C_\pi$. Now freeze this ray in the moving plane $\pi$ and follow its motion through several positions preceding and following the central position under study. Here, successive positions of the moving centrode are labeled $\pi_1, \pi_2, \ldots, \pi_6$ and those of $C_\pi$ are marked $C_{\pi1}, C_{\pi2}, \ldots$. Note that $C_{\pi2}$ and $C_{\pi5}$ coincide at the path crunode, where the path has two tangents: $t_{C\pi2}$ and $t_{C\pi5}$, each perpendicular to the ray emanating from the corresponding instant center $I_2$ and $I_5$. Thus it is seen that there is a "doubly infinite" number of crunode points, all on the fixed-polode side of the moving polode.

## 4.16 THE ρ CURVE [5,6]

The ρ curve is the locus of points of the moving plane $\pi$ whose radius vectors of path curvature have equal magnitudes, say $\rho_g$. To find this locus, we start by rearranging Eq. (4.39) taken from the derivation of ESE-1.

$$-\frac{\boldsymbol{\rho}_A}{\rho_A^2}\,a^2 + \mathbf{a} - \mathbf{J}_A = 0 \tag{4.40a}$$

or, using Eq. (4.34),

$$-\frac{\boldsymbol{\rho}_A}{\rho_A^2}\,a^2 + \mathbf{a} - \delta e^{i\theta}\cos\theta = 0 \tag{4.73}$$

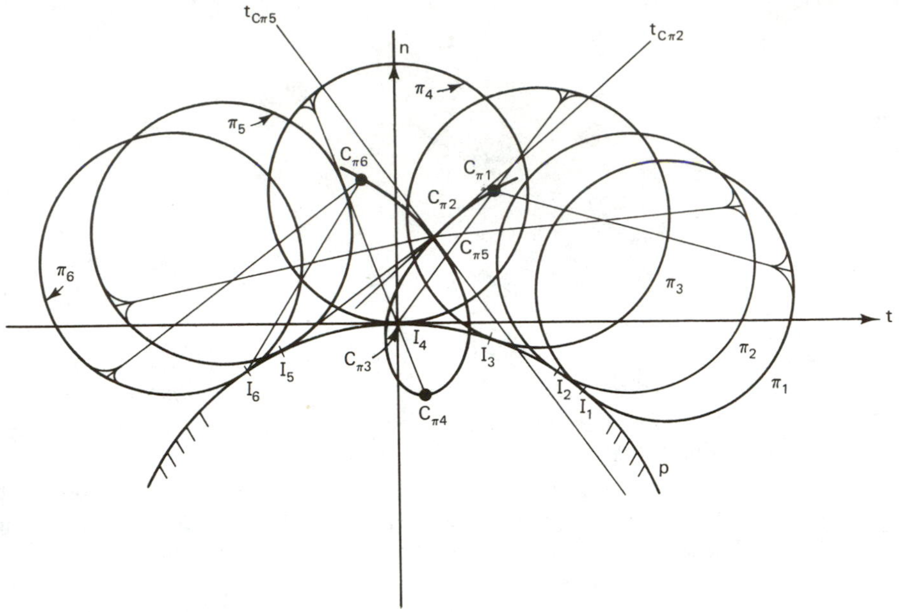


**Figure 4.42** A *crunode point* of the $\pi$ plane describes a looped path in the fixed plane. The crossing or double point of the loop is the *crunode* (see Sec. 4.15).

Now let $e^{i\arg a} = \mathbf{u}_a$, $e^{i\arg \delta} = \mathbf{u}_\delta$, and $e^{i\arg \rho} = \mathbf{u}_\rho$. With these, and by dividing through by $\mathbf{u}_a$, we get

$$\frac{-\rho_A \mathbf{u}_\rho}{\rho_A^2 \mathbf{u}_a} a^2 + a - (\cos\theta)\delta \frac{\mathbf{u}_\delta e^{i\theta}}{\mathbf{u}_a} = 0 \tag{4.74}$$

Note that $\mathbf{u}_\delta e^{i\theta} = \mathbf{u}_a$, and that $\mathbf{u}_\rho/\mathbf{u}_a = \pm 1$, according as $\boldsymbol{\rho}_A$ has the same or opposite sense to $\mathbf{a}$ on ray $a$. Therefore,

$$a^2 - \rho_A \frac{\mathbf{u}_a}{\mathbf{u}_\rho} a + \frac{\mathbf{u}_a}{\mathbf{u}_\rho} \rho_A \delta \cos\theta$$

which leads to the two quadratics in $a$:

$$\left.\begin{aligned} a^2 - \rho_A a + \rho_A \delta \cos\theta = 0 \Rightarrow a_1, a_2 \\ a^2 + \rho_A a - \rho_A \delta \cos\theta = 0 \Rightarrow a_3, a_4 \end{aligned}\right\} \Rightarrow \mathbf{a}_j = a_j e^{i(\theta + \arg \delta)} \qquad \begin{aligned}(4.75)\\(4.76)\end{aligned}$$

where $j = 1, 2, 3, 4$. Although only real values of $a_j$ are meaningful, either positive or negative reals are acceptable, because a sign change of $a_j$ merely reverses its direction along the ray $a$.

With a constant value of $\rho_A$, say $\rho_g$, Eqs. (4.75) and (4.76) correlate the values of $a$ and $\theta$, and therefore represent a set of parametric equations in $a$ with $\theta$ as the parameter. In other words, they are a set of equations of the $\rho$ curve.

Let us now test the validity of our results by applying it to Ex. 4.2, Fig. 4.21. From previous calculations we know that when $\theta = \pi/4$ and $a = 1$, then $\rho_A = (1+\sqrt{2})e^{-i\theta}$. Therefore, let $\rho_A = \rho_g = 1+\sqrt{2}$, $\theta = \pi/4$, and noting that $\delta = 2$, calculate all four roots $a_{1,2,3,4}$ and see if one of them is really equal to 1. Solving (4.75), we have

$$a_{1,2} = \frac{\rho_A \pm \sqrt{\rho_A^2 - 4\rho_A\delta\cos\theta}}{2} = \frac{1+\sqrt{2} \pm \sqrt{(1+\sqrt{2})^2 - 4(1+\sqrt{2})\,2\,(1/\sqrt{2})}}{2}$$

$$= 1.21 \pm i1.40$$

These roots, being complex, have no physical meaning for the scalar $a$. Solving (4.76) yields

$$a_{3,4} = \frac{-\rho_A \pm \sqrt{\rho_A^2 + 4\rho_A\delta\cos\theta}}{2} = \frac{-1-\sqrt{2} \pm \sqrt{(1+\sqrt{2})^2 + 4(1+\sqrt{2})2\dfrac{1}{\sqrt{2}}}}{2}$$

$$= -3.41, +1$$

Thus $a_4 = 1$ is one of the roots, and our set of equations is verified.

The $\rho$ curve is easily plotted by digital computation using $\theta$ as the parameter, solving Eqs. (4.75) and (4.76) for all real values of $a$ and letting $\mathbf{a} = ae^{i[\theta+(\pi/2)]}$ in the coordinate system in which $\boldsymbol{\delta} = |\delta|e^{i(\pi/2)}$.

The Cartesian form of the equation of the $\rho$ curve in this coordinate system (see Exer. 4.22), with

$$\mathbf{a} = x + iy \tag{4.77}$$

is

$$(x^2+y^2)^3 - \rho_A^2(x^2+y^2-\delta y)^2 = 0 \tag{4.78}$$

For plotting by digital computation on the basis of Eq. (4.78), see Exer. 4.23.

For digital plotting of the locus in the fixed plane $p$ of the centers of path curvature for points of the moving plane $\pi$ on the $\rho$ curve, the "$\rho_m$ curve," see Exer. 4.24.

## 4.17 THE CUBIC OF STATIONARY CURVATURE OR BURMESTER'S CIRCLE-POINT AND CENTER-POINT CURVES FOR FOUR INFINITESIMALLY CLOSE POSITIONS OF THE MOVING PLANE [244]

So far we have considered only three infinitesimally close positions of the $\pi$ plane as the $\pi$ curve rolls over the $p$ curve of the fixed plane. This was sufficient to derive all relationships involving velocities, accelerations, path tangents, path curvature, and so on. Now we wish to extend our studies to include the rate of change of curvature in the immediate vicinity of the position considered. This requires four infinitesimally close positions.

The fixed and rolling centrodes make contact with their respective osculating circles in three infinitesimally close positions. These osculating circles define the radius of curvature. This is clear, because three points define a circle. Let us number these positions 1, 2, and 3. Now add a fourth infinitesimally close position. Positions 2, 3, and 4 now define another circle, infinitesimally different from the first. This is easy to see by considering the four positions to be finitely separated, and then going to the limit by letting them approach infinitesimally close to one and all to the other. The two infinitesimally close osculating circles define the rate of change of the length of the radius of curvature of the centrodes:

$$\frac{d\boldsymbol{\rho}_p}{ds_p} = \boldsymbol{\rho}'_p \qquad \text{and} \qquad \frac{d\boldsymbol{\rho}_\pi}{ds_\pi} = \boldsymbol{\rho}'_\pi \tag{4.79}$$

where $\boldsymbol{\rho}'_p$ is the rate of change of $\boldsymbol{\rho}_p$ with respect to the arc length $s_p$ of the fixed centrode and $\boldsymbol{\rho}'_\pi$ is that of $\boldsymbol{\rho}_\pi$ with respect to $s_\pi$. The *cubic of stationary curvature* (CSC) is the locus of all points of the moving plane $\pi$ whose radius of path curvature remains the same length through four infinitesimally close positions. This means that the rate of change of the magnitude of the path curvature is momentarily zero:

$$\frac{d\boldsymbol{\rho}_A}{ds_A} = \boldsymbol{\rho}'_A = 0 \tag{4.80}$$

where $s_A$ is the arc length along the path of the point $A$.

We begin our derivation of the CSC starting with ESE-1 [Eq. (4.40)] in the following form:

$$\boldsymbol{\rho}_A = \left[\frac{a^2}{\mathbf{a} - \mathbf{J}_A} e^{i[\theta+(\pi/2)]}\right] e^{i[\theta+(\pi/2)]} \qquad \text{ESE-1} \tag{4.81}$$

where the first (bracketed) quantity on the right side is $\pm\rho_A$, according as $\mathbf{a} - \mathbf{J}_A$ has the same or opposite sense as $e^{i[\theta+(\pi/2)]}$ (see Exer. 4.25).

Let us consider the rate of change of $\boldsymbol{\rho}_A$ with respect to the arc length along the path of $A$, $s_A$:

$$\frac{d\boldsymbol{\rho}_A}{ds_A} = \frac{d\boldsymbol{\rho}_A}{ds_\pi}\frac{ds_\pi}{ds_A} \tag{4.82}$$

Note that, in general, $ds_\pi/ds_A \neq 0$. Therefore, on the CSC,

$$\frac{d[\rho_A]}{ds_\pi} = 0$$

where $[\rho_A]$ is the first quantity on the right side of Eq. (4.81). The equation of the CSC is, therefore,

$$\frac{d[\rho_A]}{ds_\pi} = d\left[\frac{a^2}{\mathbf{a} - \mathbf{J}_A} e^{i[\theta + (\pi/2)]}\right] \Big/ ds_\pi = 0 \tag{4.83}$$

The bracketed expression on the right side of Eq. (4.83) will now be expanded in preparation for taking the derivative as indicated. To this end we first list the expansion of each quantity on the right side of the equation:

$$a^2 = a_x^2 + a_y^2 \qquad \text{or, for short,} \qquad a^2 = x^2 + y^2 \tag{4.83a}$$

where $x$ and $y$ are the coordinates of the point $A$ in a Cartesian system embedded in the moving plane $\pi$, where the common tangent of the centrodes coincides at the moment with the real axis and the common normal with the imaginary axis. Call this coordinate system $x_\pi I_\pi y_\pi$, where $I_\pi$ is its origin. Furthermore, $\mathbf{J}_A = e^{i\theta}\boldsymbol{\delta} \cos\theta$, and $\boldsymbol{\delta} = \delta e^{i(\pi/2)}$ in this system. Substituting, we obtain

$$\left[\frac{a^2}{\mathbf{a} - \mathbf{J}_A} e^{i[\theta+(\pi/2)]}\right] = \frac{a^2}{[a]e^{i[\theta+(\pi/2)]} - \cos\theta\, e^{i\theta}[\delta]e^{i[\pi/2)}} e^{i[\theta+(\pi/2)]}$$
$$= \frac{a^2}{[a] - \cos\theta[\delta]} \tag{4.84}$$

where

$$[\delta] = \frac{-\rho_p \rho_\pi}{\rho_p - \rho_\pi} e^{-i(\pi/2)} = \pm|\delta| \qquad \text{[from Eq. (4.24a)]},$$

and where

$$[a] = \mathbf{a}e^{-i[\theta+(\pi/2)]} = \pm|a| \tag{4.84a}$$

Substituting in Eq. (4.83), we obtain

$$\frac{d[\rho_A]}{ds_\pi} = \frac{2[a]\dfrac{d[a]}{ds_\pi}([a] - [\delta]\cos\theta) - a^2\left(\dfrac{d[a]}{ds_\pi} + [\delta](\sin\theta)\dfrac{d\theta}{ds_\pi} - \dfrac{d[\delta]}{ds_\pi}\cos\theta\right)}{([a] - \cos\theta[\delta])^2} \tag{4.85}$$

From here on in this derivation, we omit the brackets from $[a]$ and $[\delta]$, but keep in mind that they are positive or negative reals determined from Eqs. (4.24a) and (4.84a). To evaluate $da/ds_\pi$ and $d\theta/ds_\pi$, we proceed as follows. Let the unit vector in the negative $x_\pi$ direction be

$$\boldsymbol{\tau} = \frac{\mathbf{u}}{|u|} \tag{4.86}$$

where $\mathbf{u}$ is the "instant center transfer velocity" [Eq. (4.24)]. Then

$$\mathbf{a} = a\boldsymbol{\tau}e^{-i[(\pi/2)-\theta]} = a\boldsymbol{\tau}e^{i[\theta-(\pi/2)]} \tag{4.87}$$

Recall that, with respect to the $x_\pi I_\pi y_\pi$ system, $\mathbf{I} = \overrightarrow{I_\pi I}$ and that $\mathbf{a} = \overrightarrow{I_\pi A}$. Therefore,

$$\overrightarrow{IA} = \overrightarrow{I_\pi A} - \overrightarrow{I_\pi I} = a\boldsymbol{\tau}e^{i[\theta-(\pi/2)]} \tag{4.88}$$

Taking the derivative of Eq. (4.88) with respect to $s_\pi$ yields

$$\frac{d(\overrightarrow{I_\pi A})}{ds_\pi} - \frac{d(\overrightarrow{I_\pi I})}{ds_\pi} = \frac{da}{ds_\pi} \boldsymbol{\tau} e^{i[\theta-(\pi/2)]} + a\frac{d\boldsymbol{\tau}}{ds_\pi} e^{i[\theta-(\pi/2)]} + ia\boldsymbol{\tau}\frac{d\theta}{ds_\pi} e^{i[\theta-(\pi/2)]} \tag{4.89}$$

Since $x_\pi I_\pi y_\pi$ moves with the plane $\pi$ and so does $A$,

$$\frac{d(\overrightarrow{I_\pi A})}{ds_\pi} = 0 \tag{4.90}$$

Also note that the time rate of change of $\overrightarrow{I_\pi I}$ is equal to the pole transfer velocity $\mathbf{u}$:

$$\frac{d(\overrightarrow{I_\pi I})}{dt} = \mathbf{u}, \qquad \text{and} \qquad \frac{d(\overrightarrow{I_\pi I})}{ds_\pi} = \frac{d(\overrightarrow{I_\pi I})}{dt}\frac{dt}{ds_\pi} \tag{4.91}$$

But

$$\frac{dt}{ds_\pi} = \frac{1}{ds_\pi/dt} = \frac{1}{|\mathbf{u}|} \tag{4.92}$$

Therefore,

$$\frac{d(\overrightarrow{I_\pi I})}{ds_\pi} = \frac{\mathbf{u}}{|\mathbf{u}|} = \boldsymbol{\tau} \tag{4.93}$$

Furthermore, $d\boldsymbol{\tau}/ds_\pi$ can be expressed in terms of itself and of the radius vector of curvature of the moving centrode (Fig. 4.43). Referring to the figure, observing that $d\boldsymbol{\tau} = i\boldsymbol{\tau}\, d\gamma$ and $ds_\pi = [\boldsymbol{\rho}_\pi/e^{i(\pi/2)}]\, d\gamma$, where $d\gamma$ is the angle swept in the $\pi$ plane by $\boldsymbol{\rho}_\pi$ as its tip follows the point of contact $I$ through an infinitesimal roll, and that $ds_\pi$ is positive, we see that

$$\frac{d\boldsymbol{\tau}}{ds_\pi} = \frac{i\boldsymbol{\tau}\, d\gamma}{[\boldsymbol{\rho}_\pi/e^{i(\pi/2)}]\, d\gamma} = i\boldsymbol{\tau}\frac{1}{\boldsymbol{\rho}_\pi} e^{i(\pi/2)} \tag{4.94}$$

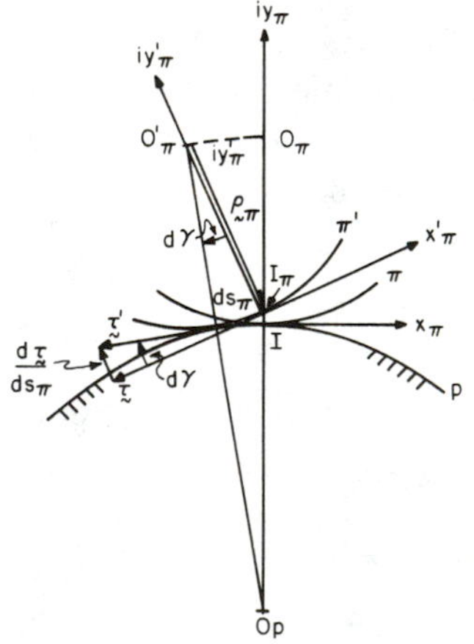


**Figure 4.43** Notation for the derivation of the *cubic of stationary curvature* (*Burmester's circlepoint curve*) for four infinitesimally close positions of a moving plane using complex numbers as vectors and differential geometry (see Sec. 4.17).

where $\boldsymbol{\rho}_\pi$ is expressed in the $x_\pi I_\pi y_\pi$ coordinate system. Substituting Eqs. (4.90) to (4.94) into Eq. (4.89), we obtain

$$0 - \boldsymbol{\tau} = \left(\frac{da}{ds_\pi} + \frac{ia}{\boldsymbol{\rho}_\pi} e^{i(\pi/2)} + ia\frac{d\theta}{ds_\pi}\right) \boldsymbol{\tau} e^{i[\theta-\pi/2)]} \tag{4.95}$$

and dividing by $\boldsymbol{\tau} e^{i[\theta-\pi/2)]}$ yields

$$-e^{-i[\theta-(\pi/2)]} = \frac{da}{ds_\pi} + ia\left[\left(\frac{e^{i(\pi/2)}}{\boldsymbol{\rho}_\pi}\right) + \frac{d\theta}{ds_\pi}\right]$$

or

$$-\sin\theta - i\cos\theta = \frac{da}{ds_\pi} + ia\left[\left(\frac{e^{i(\pi/2)}}{\boldsymbol{\rho}_\pi}\right) + \frac{d\theta}{ds_\pi}\right] \tag{4.96}$$

where

$$\frac{e^{i(\pi/2)}}{\boldsymbol{\rho}_\pi} = \pm\frac{1}{|\boldsymbol{\rho}_\pi|} \tag{4.97}$$

which is a positive or negative real, according as the moving centrode is convex or concave toward the side of the common tangent on which the common normal (the $+iy_\pi$ axis) is located. Equating real and imaginary parts from both sides of Eq. (4.96), we obtain

$$\frac{da}{ds_\pi} = -\sin\theta, \qquad \frac{d\theta}{ds_\pi} = -\frac{\cos\theta}{a} - \frac{e^{i(\pi/2)}}{\boldsymbol{\rho}_\pi} \tag{4.98}$$

Substituting Eq. (4.98) in Eq. (4.85) yields (see Exer. 4.26)

$$\frac{d[\boldsymbol{\rho}_A]}{ds_\pi} = \frac{3a^2\delta\sin\theta\cos\theta}{(a-\delta\cos\theta)^2}\left[\frac{1}{\cos\theta}\frac{1}{3}\left(\frac{e^{i(\pi/2)}}{\boldsymbol{\rho}_\pi} - \frac{1}{\delta}\right) + \frac{1}{\sin\theta}\frac{1}{3}\left(\frac{d\delta}{ds_\pi}\frac{1}{\delta}\right) + \frac{1}{a}\right] = 0 \tag{4.99}$$

This equation of the CSC is satisfied if the bracketed expression on the right side vanishes. If either $(\sin\theta)$ or $(\cos\theta)$ is zero, the value of the right side becomes indeterminate and must be evaluated by a limiting process using L'Hospital's rule. Since $3\delta/(a - \delta\cos\theta)^2$ cannot be zero, it can be factored out and ignored. It is customary to designate the other groupings of generally given quantities by

$$\frac{1}{M} = \frac{1}{3}\left(\frac{e^{i(\pi/2)}}{\boldsymbol{\rho}_\pi} - \frac{1}{\delta}\right) \quad \text{and} \quad \frac{1}{N} = \frac{1}{3}\left(\frac{d\delta}{ds_\pi}\frac{1}{\delta}\right) \tag{4.100}$$

With these, the equation of the CSC in polar coordinates becomes

$$a^2\sin\theta\cos\theta\left(\frac{1}{M\cos\theta} + \frac{1}{N\sin\theta} + \frac{1}{a}\right) = 0 \tag{4.101}$$

Points on the CSC will have the radius vector $\mathbf{a} = ae^{i[\theta+(\pi/2)]}$, where $a$ and $\theta$ are pairs of values satisfying Eq. (4.101). Thus, once $M$ and $N$ are known, the CSC can easily be generated by using $\theta$ as the parameter, solving for $a$. Note that both $a \gtrless 0$ are acceptable: If $a < 0$, then $\mathbf{a}$ will have the opposite sense from $e^{i(\theta+\pi/2)}$ on the $\theta$ ray.

If the radius vectors of curvature of the fixed and moving centrodes and their rates of change are known, $M$ can be evaluated as follows:

$$\frac{1}{M}=\frac{1}{3}\left(\frac{e^{i(\pi/2)}}{\boldsymbol{\rho}_\pi}-\frac{\boldsymbol{\rho}_p-\boldsymbol{\rho}_\pi}{-\boldsymbol{\rho}_p\boldsymbol{\rho}_\pi}e^{i(\pi/2)}\right)=\frac{e^{i(\pi/2)}}{3}\left(\frac{1}{\boldsymbol{\rho}_\pi}+\frac{\boldsymbol{\rho}_p-\boldsymbol{\rho}_\pi}{\boldsymbol{\rho}_p\boldsymbol{\rho}_\pi}\right)$$
$$=\frac{e^{i(\pi/2)}}{3}\left(\frac{2}{\boldsymbol{\rho}_\pi}-\frac{1}{\boldsymbol{\rho}_p}\right) \tag{4.102}$$

where we have used Eq. (4.24a) for $\boldsymbol{\delta}$. Since we referred all the vectors to the $x_\pi I_\pi y_\pi$ coordinate system, the factor $e^{i(\pi/2)}$ makes the right side of Eq. (4.102) real and assigns the proper sign to it.

Similarly,

$$\frac{1}{N}=\frac{1}{3}\frac{d\boldsymbol{\delta}}{ds_\pi}\frac{1}{\boldsymbol{\delta}}=\frac{1}{3}\frac{d}{ds_\pi}\left(\frac{-\boldsymbol{\rho}_p\boldsymbol{\rho}_\pi}{\boldsymbol{\rho}_p-\boldsymbol{\rho}_\pi}\right)\frac{\boldsymbol{\rho}_p-\boldsymbol{\rho}_\pi}{-\boldsymbol{\rho}_p\boldsymbol{\rho}_\pi} \tag{4.103}$$

Since $\boldsymbol{\delta}$ is a pure imaginary in the $x_\pi I_\pi y_\pi$ coordinate system, letting $d\boldsymbol{\delta}/ds_\pi = \boldsymbol{\delta}'$, $\boldsymbol{\delta}'/\boldsymbol{\delta}$ is real and will have the correct sign. We derive $\boldsymbol{\delta}'$ in terms of $\boldsymbol{\rho}_p$, $\boldsymbol{\rho}_\pi$ and their derivatives as follows:

$$\boldsymbol{\delta}'=\frac{d}{ds_\pi}\left(\frac{-\boldsymbol{\rho}_p\boldsymbol{\rho}_\pi}{\boldsymbol{\rho}_p-\boldsymbol{\rho}_\pi}\right)=\frac{-(\boldsymbol{\rho}_p'\boldsymbol{\rho}_\pi+\boldsymbol{\rho}_p\boldsymbol{\rho}_\pi')(\boldsymbol{\rho}_p-\boldsymbol{\rho}_\pi)-(-\boldsymbol{\rho}_p\boldsymbol{\rho}_\pi)(\boldsymbol{\rho}_p'-\boldsymbol{\rho}_\pi')}{(\boldsymbol{\rho}_p-\boldsymbol{\rho}_\pi)(\boldsymbol{\rho}_p-\boldsymbol{\rho}_\pi)}$$
$$=\frac{\boldsymbol{\rho}_p\boldsymbol{\rho}_\pi\boldsymbol{\rho}_p'-\boldsymbol{\rho}_p\boldsymbol{\rho}_\pi\boldsymbol{\rho}_\pi'-\boldsymbol{\rho}_p'\boldsymbol{\rho}_\pi\boldsymbol{\rho}_p+\boldsymbol{\rho}_p'\boldsymbol{\rho}_\pi^2-\boldsymbol{\rho}_p^2\boldsymbol{\rho}_\pi'+\boldsymbol{\rho}_p\boldsymbol{\rho}_\pi'\boldsymbol{\rho}_\pi}{\boldsymbol{\rho}_p^2-2\boldsymbol{\rho}_p\boldsymbol{\rho}_\pi+\boldsymbol{\rho}_\pi^2} \tag{4.104}$$
$$=\frac{\boldsymbol{\rho}_p'\boldsymbol{\rho}_\pi^2-\boldsymbol{\rho}_p^2\boldsymbol{\rho}_\pi'}{\boldsymbol{\rho}_p^2-2\boldsymbol{\rho}_p\boldsymbol{\rho}_\pi+\boldsymbol{\rho}_\pi^2}$$

where

$$\boldsymbol{\rho}_p'=\frac{d\boldsymbol{\rho}_p}{ds_\pi}=\frac{d\boldsymbol{\rho}_p}{ds_p} \quad\text{and}\quad \boldsymbol{\rho}_\pi'=\frac{d\boldsymbol{\rho}_\pi}{ds_\pi} \tag{4.105}$$

Substituting (4.104) in (4.103), we obtain

$$\frac{1}{N}=\frac{1}{3}\frac{(\boldsymbol{\rho}_p'\boldsymbol{\rho}_\pi^2-\boldsymbol{\rho}_p^2\boldsymbol{\rho}_\pi')(\boldsymbol{\rho}_p-\boldsymbol{\rho}_\pi)}{(\boldsymbol{\rho}_p-\boldsymbol{\rho}_\pi)^2(-\boldsymbol{\rho}_p\boldsymbol{\rho}_\pi)}$$

or

$$\frac{1}{N}=\frac{1}{3}\frac{1}{\boldsymbol{\rho}_p-\boldsymbol{\rho}_\pi}\left(\frac{\boldsymbol{\rho}_\pi'\boldsymbol{\rho}_p}{\boldsymbol{\rho}_\pi}-\frac{\boldsymbol{\rho}_p'\boldsymbol{\rho}_\pi}{\boldsymbol{\rho}_p}\right) \tag{4.106}$$

Now let us transform Eq. (4.101) to Cartesian coordinates in the $x_\pi I_\pi y_\pi$ system. We substitute

$$a = (x^2 + y^2)^{1/2}, \qquad \sin\theta = -\frac{x}{(x^2 + y^2)^{1/2}}$$

$$\cos\theta = \frac{y}{(x^2 + y^2)^{1/2}} \tag{4.107}$$

With these, Eq. (4.101) becomes

$$(x^2 + y^2)\frac{-xy}{(x^2 + y^2)}\left[\frac{1}{M}\frac{(x^2 + y^2)^{1/2}}{y} - \frac{1}{N}\frac{(x^2 + y^2)^{1/2}}{x} + \frac{1}{(x^2 + y^2)^{1/2}}\right] = 0$$

$$\frac{1}{M}\frac{(x^2 + y^2)(-xy)}{y} - \frac{1}{N}\frac{(x^2 + y^2)(-xy)}{x} + (-xy) = 0 \tag{4.108}$$

$$(x^2 + y^2)\left(\frac{y}{N} - \frac{x}{M}\right) - xy = 0$$

The factor $(x^2 + y^2)$ in Eq. (4.108) shows that the CSC is a circular algebraic curve, and, together with the adjacent factor, that it is a cubic. Thus it is a "circular cubic."

**Example 4.3**

Determine the equation and plot the CSC for a planet gear of 2 in. radius on a stationary sungear of 3 in. radius (Fig. 4.44).

**Solution**

$$\frac{1}{M} = \frac{e^{i(\pi/2)}}{3}\left(\frac{2}{\rho_\pi} - \frac{1}{\rho_p}\right)$$

$$= \frac{i}{3}\left(\frac{2}{-2i} - \frac{1}{3i}\right) \tag{4.102}$$

Therefore, $M = -2.25$. Since the centrodes are circles, Eq. (4.106) shows that $1/N = 0$. The equation of the CSC becomes [see Eq. (4.101)]

$$a^2 \sin\theta \cos\theta\left(\frac{1}{-2.25\cos\theta} + \frac{1}{a}\right) = 0$$

or

$$\frac{\sin\theta}{-2.25}a^2 + \sin\theta\cos\theta\, a = 0$$

or the line on which $\sin\theta = 0$, and therefore $\theta = 0$, which is the path normal, and

$$a^2 - (2.25)\cos\theta\, a = 0 \qquad \text{or} \qquad a = 0$$

is a point of the CSC, and $a = (2.25)\cos\theta$, which is a circle of 2.25 in. diameter with its center at $(0, (1.125)i)$ in the $x_\pi I_\pi y_\pi$ system (see point $O_{CSC}$ in Fig. 4.44). Thus, in this case, the CSC has degenerated into a circle (a quadratic) and a straight line, which preserves the status of a cubic.

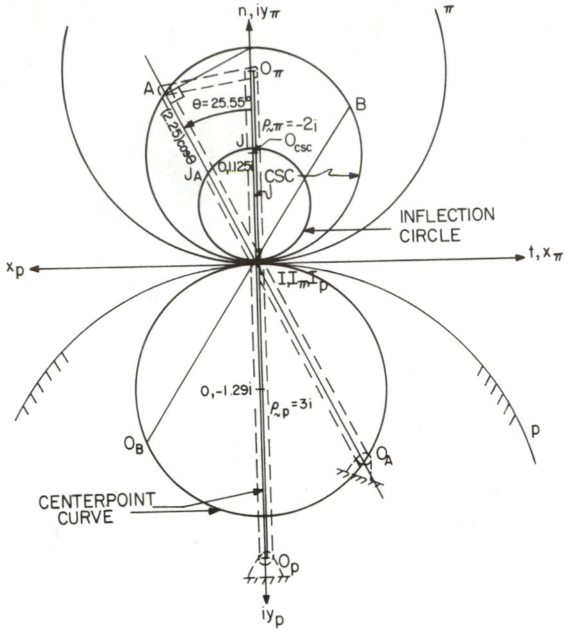


**Figure 4.44** The *cubic of stationary curvature* (*Burmester's circlepoint curve*) for a planet gear of 2″ diameter rolling on a fixed sun gear of 3″ diameter degenerates into a circle and a line, namely the common normal. Its conjugate, obtained by kinematic inversion, is the *centerpoint curve.* Conjugate points on these can be connected by a rigid link to form a linkage equivalent to the rollcurve mechanism (see Sec. 4.18).

## 4.18 THE CIRCLE-POINT CURVE AND CENTER-POINT CURVE FOR FOUR ISPs

Some authors call the CSC the "circling-point curve," which is a literal translation of the German "Kreisungspunktkurve." However, since it is the locus of all points of the moving plane momentarily describing circular paths through four "infinitesimally separated positions," we might as well call it "Burmester's circle-point curve for four ISPs."

What is the "center-point curve" associated with this circle-point curve, and how do we find conjugate center-point–circle-point pairs?

Well, the center-point curve is the locus of all points in the fixed plane which are centers of the momentary circular paths of circle points. It is easy to see that in kinematic inversion these two curves exchange roles. This gives us a clue how to find the equation of the center-point curve: It is the CSC of the kinematically inverted mechanism, in which the former moving plane becomes fixed and the former fixed plane is mobilized. Its equation will thus be expressed in a coordinate system embedded in the fixed plane, with origin at $I_p$, positive real axis coinciding with the former negative real axis, and positive imaginary axis coinciding with the former negative imaginary axis ($x_p I_p y_p$ in Fig. 4.45). It should be easy to transform this equation into the $x_\pi I_\pi y_\pi$ system.

How do we find conjugate center-point–circle-point pairs? Easy! We pick a circle point $A$ on the CSC and find its center of path curvature by way of ESE. Also, if both the CSC and its conjugate have been plotted, any ray through the instant center will intersect these curves in conjugate center point–circle point pairs.

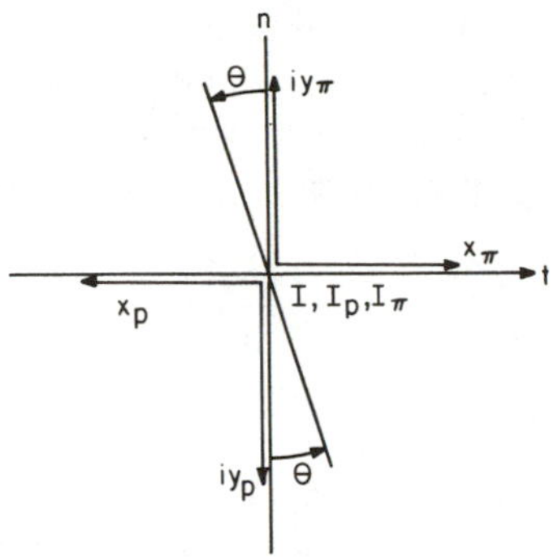


**Figure 4.45** Coordinate systems embedded in the moving plane $\pi$ and in the fixed plane $p$.

So let us derive all these relationships. We start to find the equation of the center-point curve, expressed in the $x_p I_p y_p$ system, by writing the equation of the CSC with the subscripts $p$ and $\pi$ interchanged.

Referring to Eq. (4.101) and Fig. 4.45, we observe that $\theta$ remains unchanged, but is now measured from the negative branch of the common normal $n$, still ccw. However, for locating a point on the $\theta$ ray, positive $a$ now means the sense opposite from before. To find $M$ for this case, we kinematically invert Eq. (4.102) by interchanging the subscripts $p$ and $\pi$:

$$\frac{1}{M_p} = \frac{e^{i(\pi/2)}}{3}\left(\frac{1}{\boldsymbol{\rho}_p} + \frac{\boldsymbol{\rho}_\pi - \boldsymbol{\rho}_p}{\boldsymbol{\rho}_\pi \boldsymbol{\rho}_p}\right) = \frac{i}{3}\left(\frac{2}{\boldsymbol{\rho}_p} - \frac{1}{\boldsymbol{\rho}_\pi}\right) \tag{4.109}$$

(here all vectors are to be expressed in the $x_p I_p y_p$ system). Similarly, for $N$ [from Eq. (4.106)]:

$$\frac{1}{N_p} = \frac{1}{3(\boldsymbol{\rho}_\pi - \boldsymbol{\rho}_p)}\left(\frac{\boldsymbol{\rho}_p' \boldsymbol{\rho}_\pi}{\boldsymbol{\rho}_p} - \frac{\boldsymbol{\rho}_\pi' \boldsymbol{\rho}_p}{\boldsymbol{\rho}_\pi}\right) = \frac{1}{N} \tag{4.110}$$

Thus we see that $M_p \neq M$, but $N_p = N$.

With these, let us now find the centerpoint curve for Ex. 4.3. Here, $\boldsymbol{\rho}_\pi = 2i$ and $\boldsymbol{\rho}_p = -3i$ in the $x_p I_p y_p$ system.

$$\frac{1}{N_p} = 0, \qquad \frac{1}{M_p} = \frac{i}{3}\left(\frac{2}{-3i} - \frac{1}{2i}\right) = \frac{-7}{18}$$

$$M_p = \frac{-18}{7} = -2.257 \tag{4.111}$$

Therefore, the equation of the center-point curve in the $x_p I_p y_p$ system is

$$a^2 \sin\theta \cos\theta\left(\frac{1}{-2.57\cos\theta} + \frac{1}{a}\right) = 0$$

or

$$\sin\theta\left(\frac{a^2}{-2.57} + a\cos\theta\right) = 0$$

The first factor, $\sin\theta = 0$, is the common normal, and the second factor, after canceling $a$, becomes $a = (2.57)\cos\theta$, which is the circle of 2.57 in. diameter with its center at $(0, (1.29)i)$ in the $x_pI_py_p$ system (Fig. 4.45), or $(0, (-1.29)i)$ in the $x_\pi I_\pi y_\pi$ system. For example, for $\theta = 25.55°$, $a_A = 2.03$ in., $a_{OA} = 2.32$ in. As Fig. 4.45 shows,

$$\boldsymbol{\rho}_A = \overrightarrow{O_AA} = (2.03 + 2.32)e^{i[(25.55\pi/180)+(\pi/2)}$$

$$= (4.35)e^{i(2.02)}$$

Let us check this result by ESE in the $x_\pi O_\pi y_\pi$ system. To start this, we must find $J_A$ on the inflection circle. The vector diameter of the inflection circle $\boldsymbol{\delta}$, is:

$$\boldsymbol{\delta} = \frac{-\boldsymbol{\rho}_p\boldsymbol{\rho}_\pi}{\boldsymbol{\rho}_p - \boldsymbol{\rho}_\pi} = \frac{-3i(-2i)}{3i-(-2i)} = \frac{-6}{5i} = (1.2)i$$

$$\mathbf{J}_A = \boldsymbol{\delta}\cos\theta\, e^{i\theta} = i(1.2)\cos\theta\, e^{i\theta} = (108)e^{i[\theta+(\pi/2)]}$$

$$\boldsymbol{\rho}_A = \left[\frac{a^2}{\mathbf{a}-\mathbf{J}_A}e^{i[\theta+(\pi/2)]}\right]e^{i[\theta+(\pi/2)]} = \left[\frac{(2.03)^2e^{i[\theta+(\pi/2)]}}{(2.03-1.08)e^{i[\theta+(\pi/2)]}}\right]e^{i[\theta+(\pi/2)]}$$

$$= 4.35e^{i[\theta+(\pi/2)]}$$

which checks with our previous result.

It is interesting to note that, within the vicinity of the position shown in Fig. 4.44, the motion of the planet $\pi$ with respect to the sun gear $p$ can be closely approximated by the coupler $O_\pi A$ of a four-bar linkage with $O_pO_A$ the fixed link, $O_pO_\pi$ the input link, and $O_AA$ the follower link. This linkage is shown in phantom in Fig. 4.44. In other words, the link $O_AA$ could be pin-connected to the sun and planet gears at $O_A$, respectively $A$, and the gears would still roll freely through a limited range straddling the position shown. In fact, any conjugate point pair on the CSC and the center-point curve, say $BO_B$, located by drawing another ray through $I$, could serve the same purpose. Even if $AO_A$ and $BO_B$ are used simultaneously as redundant links connecting the two gears, the mechanism still would not lock within a limited range around the position shown.

## 4.19 BALL'S POINT

All points of the moving plane on the inflection circle have momentary paths making contact with their path tangents through three infinitesimally close positions (ISPs). On the other hand, all points of the moving plane on the CSC have momentary paths making contact with their osculating circles through four ISPs.

Question: Is there any point in the moving plane whose path is making four-ISP contact with its path tangent? A little thought will show that, if there is such a point, it must be the intersection of the CSC and the inflection circle. Indeed, in general there is such a point, and it is called Ball's point.

In Ex. 4.3 (Fig. 4.44) Ball's point happens to be the inflection pole $J$, where the inflection circle is intersected by the straight-line branch of the CSC, which branch coincides with the common normal of the centrodes.

In general, Ball's point satisfies both the equations of the CSC and of the inflection circle. If we let the Ball's point be at $a_B e^{i[\theta_B+(\pi/2)]}$ in the $x_\pi I_\pi y_\pi$ coordinate system,

$$a_B^2 \cos\theta_B \sin\theta_B \left(\frac{1}{M\cos\theta_B}+\frac{1}{N\sin\theta_B}+\frac{1}{a_B}\right)=0 \tag{4.112}$$

$$a_B=\delta\cos\theta_B \qquad \text{or} \qquad \cos\theta_B=\frac{a_B}{\delta} \tag{4.113}$$

and (see Fig. 4.19)

$$\sin\theta_B=\frac{(\delta^2-a_B^2)^{1/2}}{\delta} \tag{4.114}$$

Substituting (4.114) and (4.113) in (4.112), we get

$$a_B^2\frac{a_B}{\delta}\frac{(\delta^2-a_B^2)^{1/2}}{\delta}\left(\frac{\delta}{Ma_B}+\frac{\delta}{N(\delta^2-a_B^2)^{1/2}}+\frac{1}{a_B}\right)=0 \tag{4.115}$$

Solving Eq. (4.115) for $a_B$ and then Eqs. (4.113) and (4.114) for $\cos\theta_B$ and $\sin\theta_B$ gives the polar coordinates of the Ball's point. Computer-plotting the CSC and the inflection circle and locating their intersection will give excellent first approximations, which can then be refined numerically. We can proceed similarly in Cartesian coordinates:

$$(x_B^2+y_B^2)\left(\frac{y_B}{N}-\frac{x_B}{M}\right)-x_B y_B=0 \tag{4.116}$$

and

$$(x_B^2+y_B^2)^{1/2}=\delta\cos\theta_B=\delta\frac{y_B}{(x_B^2+y_B^2)^{1/2}} \tag{4.117}$$

or

$$x_B^2+y_B^2=\delta y_B \tag{4.118}$$

$$x_B^2=\delta y_B-y_B^2 \tag{4.119}$$

Substituting Eqs. (4.118) and (4.119) in (4.116), we have

$$\delta y_B\left(\frac{y_B}{N}-\frac{(\delta y_B-y_B^2)^{1/2}}{M}\right)-y_B(\delta y_B-y_B^2)^{1/2}=0 \tag{4.120}$$

Again, solving Eq. (4.120) for $y_B$ and then Eq. (4.119) for $x_B$ yields the answer. Also, after computer-plotting (4.116) and the inflection circle, their intersection will give excellent first approximations to be refined by Eq. (4.120) for $y_B$ and solving (4.119) for $x_B$.

The "Ball's point curve" is the locus of points of the moving plane which successively become Ball's points as the $\pi$ curve rolls on the $p$ curve. This curve can be computer-plotted as follows. Roll the $\pi$ curve by a small increment around the $p$ curve, and define a new $x_\pi I_\pi y_\pi$ coordinate system coincident with the new common tangent and common normal to the centrodes, as before. Find the Ball's point for this new position. Now determine the transformation from the present $x_\pi I_\pi y_\pi$ system to that of the original position and transform the radius vector locating the Ball's point and plot it in the original system. For instance, in Ex. 4.3, Fig. 4.44, the Ball's point curve is a circle centered at $O_\pi$ and having radius $\overline{O_\pi J}$ (not shown).

## PROBLEMS*

**4.1.** For the linkage (identified below) construct fixed and moving centrodes for coupler link 3. Include the common tangent and common normal of the centrodes. What kind of motion occurs as the moving centrode rolls on the fixed centrode? What is the path of *C*?
**(a)** Figure P4.1
**(b)** Figure P4.2
Use graphical methods and check the results analytically.

**4.2.** In the following problems construct and label the fixed and moving centrodes for the motion of coupler link 3 for the linkages identified. Include the tangent and common

* Contribution of Lee Hunt in developing these problems is acknowledged.

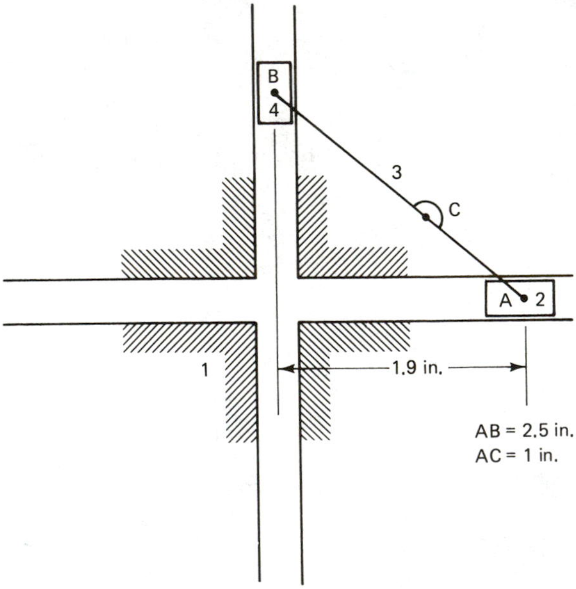


**Figure P4.1**

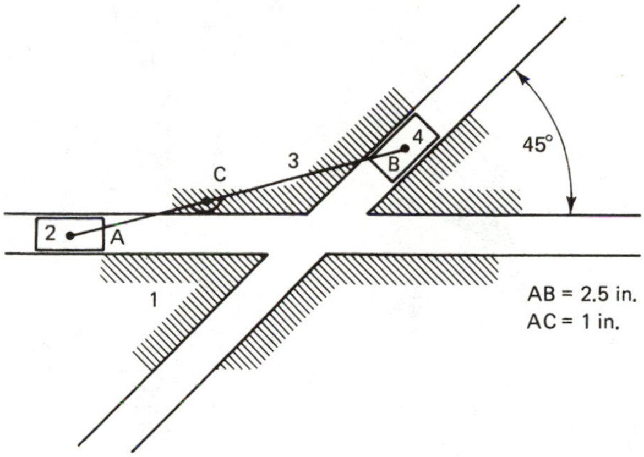


Figure P4.2

normal of the centrodes. Use graphical methods, spot-check the results analytically, program for digital computation, and plot both centrodes.

**(a)** Figure P4.3

**(b)** Figure P4.4

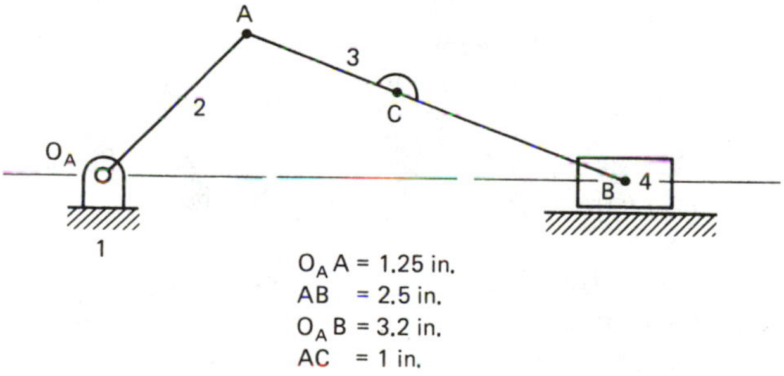


Figure P4.3

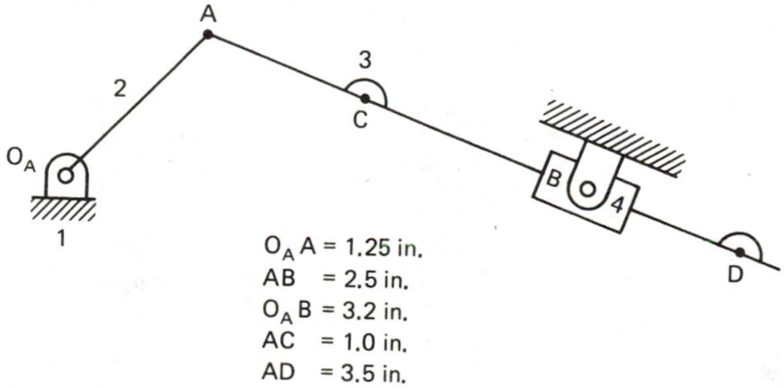


Figure P4.4

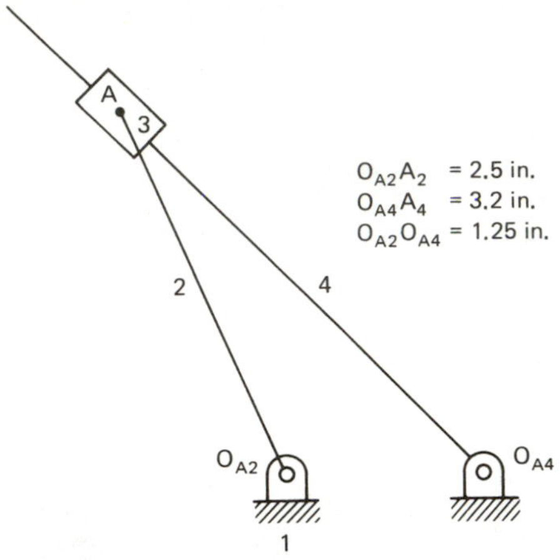


**Figure P4.5**

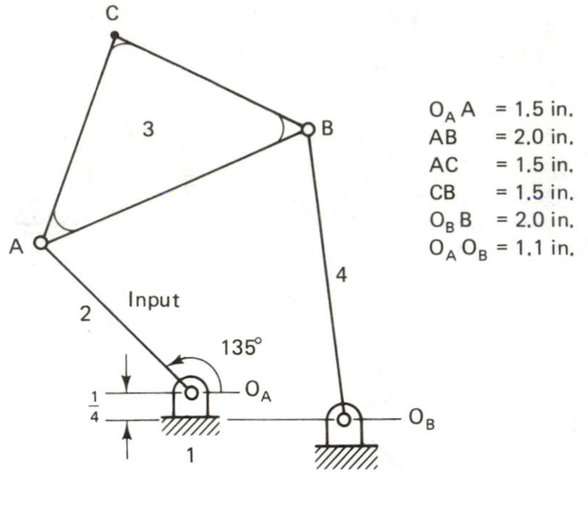


**Figure P4.6**

(c) Figure P4.5

(d) Figure P4.6

**4.3.** For the linkage of Fig. P4.1, find the velocity of point $C$ (direction and magnitude) using both the graphical and the complex-number method. The velocity of the slider, link 2, is 3 cm/sec to the right.*

**4.4.** Find the velocity (direction and magnitude) of coupler point $C$ for the linkage of Fig. P4.2 both graphically and by the complex-number method. The velocity of link 2, a slider, is 1.5 cm/sec to the right.*

**4.5.** For the slider-crank mechanism of Fig. P4.3, find the velocity of coupler point $C$ (direction and magnitude). Use both graphical and complex-number methods. The input angular velocity ($\omega_2$) is 1 rad/sec ccw.*

**4.6.** Find the velocity of coupler point $C$ (direction and magnitude) for the inverted slider-crank mechanism of Fig. P4.4. Use both graphical and analytical methods. The input angular velocity ($\omega_2$) is 1.5 rad/sec cw.*

**4.7.** For the four-bar linkage of Fig. P4.6, find the velocity (direction and magnitude) of coupler point $C$ using both graphical and complex-number methods. The input angular velocity ($\omega_2$) is 1 rad/sec cw.*

**4.8.** For the six-bar linkage of Fig. P4.7, find the velocity (direction and magnitude) of points $C$ and $D$. Use both graphical and analytical methods. Assume that the input angular velocity ($\omega_2$) is 1.5 rad/sec cw. Use the methods of this chapter.

**4.9.** For the linkage of Fig. P4.1, construct the inflection circle for the given position. Then find the center of path curvature of coupler point $C$. Use both graphical and analytic methods.

**4.10.** For the linkage of Fig. P4.2, construct the inflection circle for the given position. Then use the Euler–Savary equation to find the center of path curvature of coupler point $C$. Check this with the Hartmann construction.

* Use the results of Problems 1 and 2.

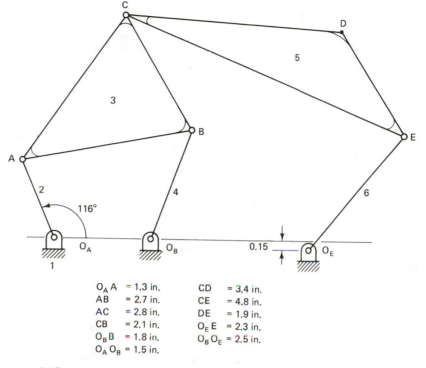

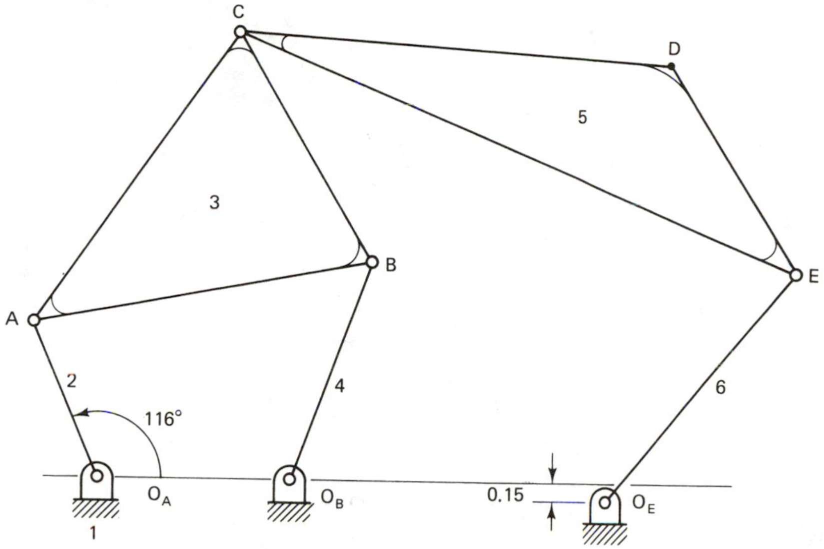


| | | | |
|---|---|---|---|
| $O_A A$ | = 1.3 in. | CD | = 3.4 in. |
| AB | = 2.7 in. | CE | = 4.8 in. |
| AC | = 2.8 in. | DE | = 1.9 in. |
| CB | = 2.1 in. | $O_E E$ | = 2.3 in. |
| $O_B B$ | = 1.8 in. | $O_B O_E$ | = 2.5 in. |
| $O_A O_B$ | = 1.5 in. | | |

**Figure P4.7**

**4.11.** For the linkage of Fig. P4.3, construct the inflection circle and the Bresse circle. Identify the inflection pole and the acceleration pole. Find the center of path curvature of coupler point *C*. The angular velocity ($\omega_3$) of link 3 is 2 rad/sec ccw and the angular acceleration ($\alpha_3$) of link 3 is 1 rad/sec² ccw. Use both graphical and analytical methods.

**4.12.** For the linkage of Fig. P4.4, construct the inflection circle and the Bresse circle, and identify the inflection pole and acceleration pole. Then find the centers of path curvature for coupler points *C* and *D*. The angular velocity ($\omega_3$) of link 3 is 1 rad/sec cw and the angular acceleration ($\alpha_3$) of link 3 is 1.7 rad/sec² cw. Use both graphical and analytical methods.

**4.13.** Use Bobillier's construction to construct the inflection circle for the slider-crank mechanism of Fig. P4.3. Check analytically.

**4.14.** Use Bobillier's construction to construct the inflection circle for the four-bar mechanism of Fig. P4.6. Then using the same method, find the center of path curvature of coupler point *C*. Check analytically.

**4.15.** Construct the inflection and Bresse circles for the coupler links 3 and 5 of the six-bar linkage of Fig. P4.7. The input angular velocity ($\omega_2$) is 2 rad/sec cw. Then find the centers of path curvature for coupler points *C* and *D*.

**4.16.** Verify the results of Prob. 4.15 using Bobillier's construction.

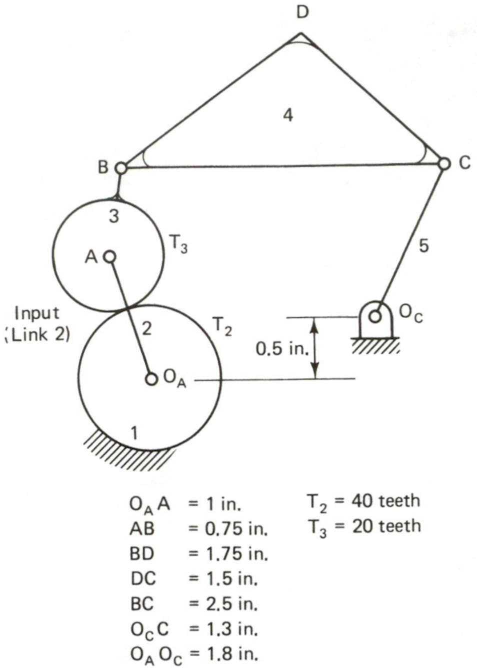


**Figure P4.8**

**4.17.** For the geared five-bar linkage of Fig. P4.8, construct the inflection circle for the motion of link 4 with respect to link 1 and then use the Euler–Savary equation to find the center of path curvature of coupler point *D*.

**4.18.** For the slider-crank mechanism of Fig. P4.3, the constant angular velocity of link 2 ($\omega_2$) is 1 rad/sec ccw. Find the acceleration of coupler point *C* (direction and magnitude) using the methods in this chapter. Use the instant center method to find $\omega_3$ and find $\alpha_3$ using both the acceleration-difference and the acceleration-field methods.

**4.19.** For the slider-crank inversion shown in Fig. P4.4, find the acceleration, in direction and magnitude, of coupler points *C* and *D*. The input angular velocity ($\omega_2$) is 1 rad/sec cw. Use the methods of Prob. 4.18.

**4.20.** Find the acceleration of point *C* in the four-bar linkage of Fig. P4.6. The constant angular velocity of the input link (2) is 1.5 rad/sec cw. Use the methods of this chapter.

**4.21.** For the six-bar linkage of Fig. P4.7, use the methods of this chapter to find the acceleration (in direction and magnitude) for points *C* and *D*. The constant angular velocity of the input link ($\omega_2$) is 1.5 rad/sec cw.

**4.22.** For the four-bar of Fig. P4.9:

**(a)** Construct the inflection circle of the coupler with respect to ground.

**(b)** Generate the cubic of stationary curvature.

**(c)** Check the actual path curvature of six path tracer points that coincide with the cubic of stationary curvature.

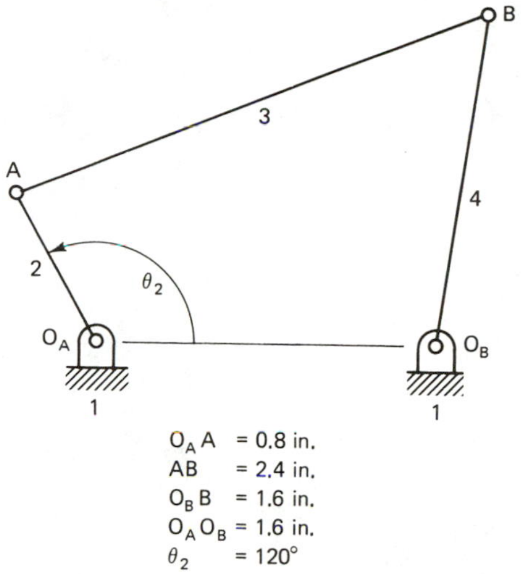


Figure P4.9

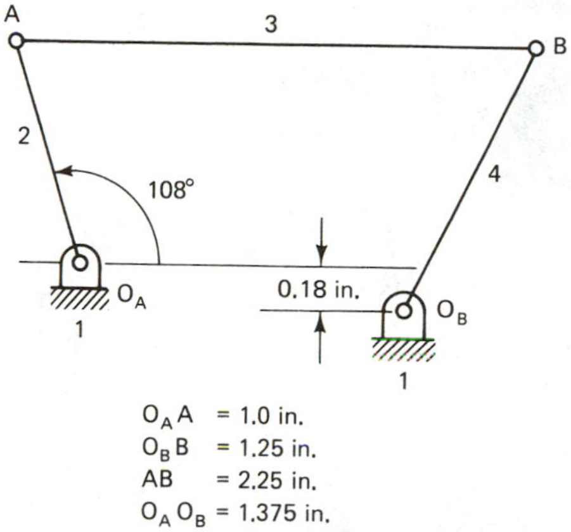


Figure P4.10

**4.23.** For the four-bar of Fig. P4.10:

**(a)** Construct the inflection circle of the coupler with respect to ground.

**(b)** Generate the cubic of stationary curvature.

**(c)** Check the actual path curvature of six path tracer points that coincide with the cubic of stationary curvature.

**4.24.** For the four-bar of Fig. P4.11:

**(a)** Construct the inflection circle of the coupler with respect to ground.

**(b)** Generate the cubic of stationary curvature.

**(c)** Check the actual path curvature of six path tracer points that coincide with the cubic of stationary curvature.

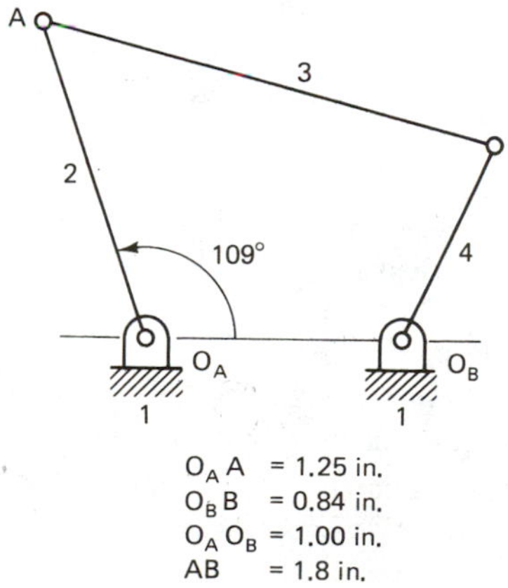


Figure P4.11

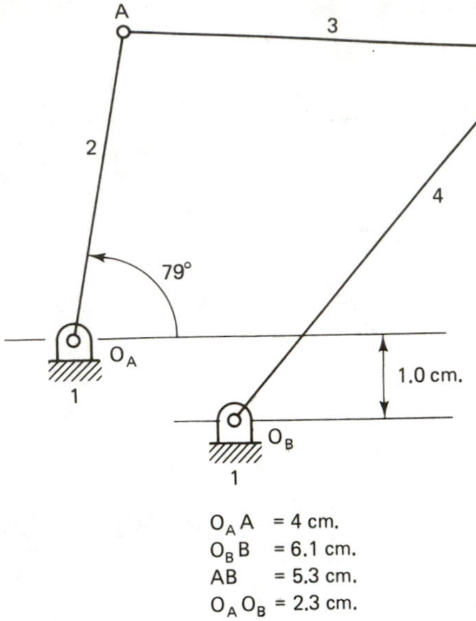


**Figure P4.12**

**4.25.** For the four-bar of Fig. P4.12:

**(a)** Construct the inflection circle of the coupler with respect to ground.

**(b)** Generate the cubic of stationary curvature.

**(c)** Check the actual path curvature of six path tracer points that coincide with the cubic of stationary curvature.

## EXERCISES

**4.1.** Using Eqs. (4.24a) and (4.25), show that, for the case of Fig. 4.10a, $0 < 1 + (\delta/\rho_\pi) < 1$, verifying that $0 < \rho < 1$, and that $\rho$ is real.

**4.2.** Use the procedure of Exer. 4.1 to verify that, for the case of Fig. 4.10b, $\rho$ is real and negative.

**4.3.** Similarly to Exers. 4.1 and 4.2, verify that for the case of Fig. 4.10c, $\rho$ is real and is greater than 1.

**4.4.** Show analytically and geometrically that:

**(a)** $\rho \to 1$ when $\rho_\pi \to \infty$ (i.e., when $\pi$ is a straight line or has an inflection at $I$)

**(b)** $\rho \to 0$ when $\rho_p \to \infty$ (i.e., when $p$ is a straight line or has an inflection at $I$)

**(c)** $\rho \to -\infty$ when $\rho_p > \rho_\pi$ and $\rho_\pi \to \rho_p$

**(d)** $\rho \to \infty$ when $\rho_p < \rho_\pi$ and $\rho_p \to \rho_\pi$

**4.5.** Substitute appropriate values from Ex. 4.2 into the right side of equations ESE-1 and ESE-2 and thus verify these equations by actual calculations.

**4.6.** Using argument manipulations similar to those in the derivation of ESE-3 [Eq. 4.43)], start with ESE-1 [Eq. (4.40)] and derive Eq. (4.44) from it directly, without going through ESE-2.

**4.7.** Using similar triangles from Fig. 4.26, show that the Bobillier construction (BC) satisfies Eq. (4.45). Then start the construction over again by drawing another arbitrary line 1, now extending on the other side of the ray, and show that it leads to the same point, *A*.

**4.8.** Same as Exer. 4.7, but based on Fig. 4.27.

**4.9.** Same as Exers. 4.7 and 4.8, but based on Fig. 4.28.

**4.10.** Using appropriate similar triangles in Fig. 4.29, show that both $I^{(1)}$ and $I^{(2)}$ satisfy Eq. (4.45) and that therefore Bobillier's construction is a geometric verification of ESE-4.

**4.11.** In Fig. 4.30, show that $I$, $B$, $O_B$, and $J_B$, with $J_B$ constructed by way of the collineation axis, satisfies ESE-3, Eq. (4.43). (*Hint:* Use the method of similar triangles as demonstrated in the text.)

**4.12.** Draw an arbitrary ray through $I$ in Fig. 4.30 and on it mark $F$ collinear with $A$ and $B$, $O_F$ collinear with $O_A$ and $O_B$, and $J_F$ collinear with $J_A$ and $J_B$. Then repeat Exer. 4.11 for $I$, $F$, $O_F$, and $J_F$.

**4.13.** In Fig. 4.30, draw a ray through $I$ and the intersection of $O_AO_B$ and $J_AJ_B$; mark the resulting three-way intersection $O_G$ and also $J_G$, and also mark the intersection of this ray with $AB$ as $F$. Then show geometrically and analytically that $\mathbf{f} = \mathbf{J}_G/2 = \mathbf{O}_G/2$. (*Hint:* Use the parallelogram $IEJ_GC_{AB}$ and ESE-3 [Eq. (4.43)].)

**4.14.** Prove the lemma associated with Fig. 4.31. [*Hints:* (1) Choose an arbitrary point on the collineation axis and call it $C_{AB}$. (2) Draw an arbitrary line $l$ through $C_{AB}$ and call its intersection with ray $a$, $A$; with ray $b$, $B$; and with $J_AJ_B$, $E$. (3) Through $C_{AB}$ draw a line parallel with $IE$; call its intersection with ray $a$, $O_A$, and with ray $b$, $O_B$. (4) By similar triangles (as in Exer. 4.11), show that $I$, $A$, $O_A$, and $J_A$, as well as $I$, $B$, $O_B$, and $J_B$, satisfy ESE-3 [Eq. (4.43)]. (5) Use the fact of the arbitrary nature of your choice of $C_{AB}$ and the line $l$, together with the method of constructing the collineation axis as described in connection with Fig. 4.30, to assert the lemma associated with Fig. 4.31.]

**4.15.** Verify the purely geometric version of Hartmann's construction for determining $\boldsymbol{\delta}$ when $A$, $I$, $O_A$, and $n$ are known (see Fig. 4.35). [*Hint:* Assume that $\dot{\phi} = 1$ rad/sec and use the final form of Eq. (4.54).]

**4.16.** Verify Hartmann's construction for finding $\mathbf{u}$ in Fig. 4.36. (*Hint:* Compare with the construction for $O_A$ in Fig. 4.34.)

**4.17.** Prove the validity of statements 1 and 2 that follow Eq. (4.60). [*Hint:* In Eq. (4.57), let $\ddot{\phi} \rightarrow 0$.]

**4.18.** Expressing $\mathbf{a}_{AP}$ in terms of $\boldsymbol{\delta}$, and $\boldsymbol{\delta}$ in terms of $\boldsymbol{\rho}_\pi$ and $\boldsymbol{\rho}_p$, show that equating the right sides of Eqs. (4.23) and (4.62) leads to an identity, and thus prove that both equations yield the same value for the total acceleration of a point $A$ of the $\pi$ plane.

**4.19.** Applying Eq. (4.69) to the inflection pole $J$, show that the total acceleration of the point $J$ as a point of the moving plane $\pi$, $\mathbf{A}_{J\pi}$, is always parallel with the common tangent $t$ of the fixed and moving centrodes, and thus verify the fact that $J_\pi$ has no normal acceleration.

**4.20.** Applying Eq. (4.69) to the point $K$ of the Bresse circle (Fig. 4.38), show that the total acceleration of the point $K$ as a point of the moving plane $\pi$, $\mathbf{A}_{K\pi}$ of $K_\pi$, is collinear with the common tangent $t$ of the fixed and moving centrodes, and thus verify that the tangential acceleration of $K_\pi$ is zero.

**4.21.** Show that the total acceleration of any point $A$ of the $\pi$ plane located anywhere along the ray of $A_p$ or its extensions can be expressed as $b\boldsymbol{\delta}$, where $b$ is a scalar. Derive the expression for $b$ in terms of $\dot{\phi}$, $\ddot{\phi}$, and $\overrightarrow{AA_p}$.

**4.22.** Derive Eq. (4.78), the Cartesian form of the ρ curve. [*Hints:* Use Eq. (4.77) referred to the coordinate system $x = t$, $O = I$, and $iy = n$; express $\cos\theta$ and the arguments of $\boldsymbol{\delta}$, $\mathbf{a}$, and $\boldsymbol{\rho}$ in this system. Then substitute these in Eq. (4.73) and simplify. Finally, square the resulting expression, thus including both Eqs. (4.75) and (4.76) in one sixth-degree equation.]

**4.23.** Program Eq. (4.78) for digital plotting of the ρ curve. (*Hints:* Expand and regroup terms to form a sextic in $x$ with coefficients which are deterministic functions of $\rho_A$, $\delta$, and $y$; reduce this sextic to a cubic by the change of variable $z = x^2$; using $y$ as the parameter, vary it from $0$ along the $+iy$ axis and then along the $-iy$ axis until all three solutions for $z$ become complex, imaginary, or negative. For each positive real value of $z$, find $\pm x$ on the ρ curve and plot.)

**4.24.** Plot the $\rho_m$ curve by digital computation. [*Hints:* For each value of $\mathbf{a}$ on the ρ curve found by solving Eqs. (4.75) and (4.76), find $\mathbf{J}_A$ by Eq. (4.34) and then $\boldsymbol{\rho}_A$ by Eq. (4.40).]

**4.25.** Show that Eq. (4.81) gives the same result as ESE-1 [Eq. (4.40) or Eq. (4.40a)]. [The latter appears just preceding Eq. (4.73).]

**4.26.** Starting with Eq. (4.85) and using Eqs. (4.98), derive Eq. (4.99) by appropriate algebraic manipulations.

**4.27.** For the motion of the coupler with respect to the fixed link of the four-bar linkage in the position shown in Fig. X4.1, using complex-number analytical methods, proceed as follows and check your results by both Bobillier's and Hartmann's constructions:

**(a)** Find the instant center $I$ (Fig. X4.2).

**(b)** Find the inflection points $J_A$ and $J_B$.

**(c)** Find the center and the vector diameter of the inflection circle (see Fig. X4.3 for method of construction and computation.)

**(d)** Find the path tangent and the path normal, establish the $x_\pi I_\pi y_\pi$ system, and transform all vectors to be expressed in that system.

**(e)** Using the fact that points $A$ and $B$ describe circular paths, and are therefore points of the CSC, write the equation of the CSC twice, once for $A$ and once for $B$, and solve the resulting system of two real equations, linear in $1/M$ and $1/N$, simultaneously for $M$ and $N$.

**(f)** Use the values of $M$ and $\boldsymbol{\delta}$ to find $\boldsymbol{\rho}_\pi$ and then $\boldsymbol{\rho}_p$.

**(g)** Use the values of $1/N$ and $\boldsymbol{\delta}$ to find $\delta'$.

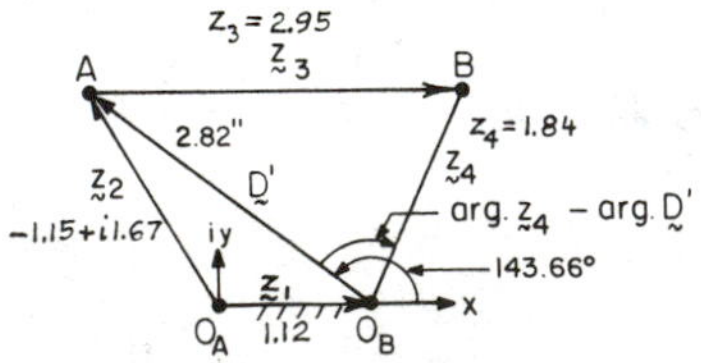


**Figure X4.1**

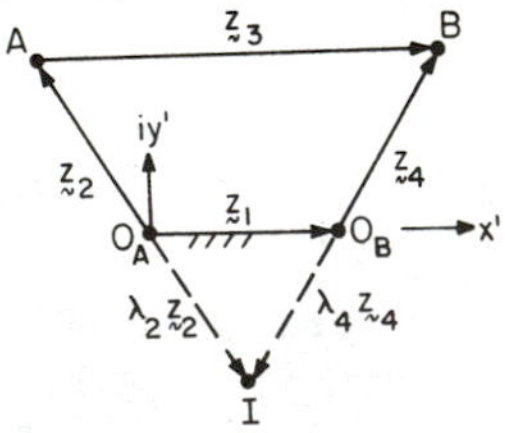


**Figure X4.2**

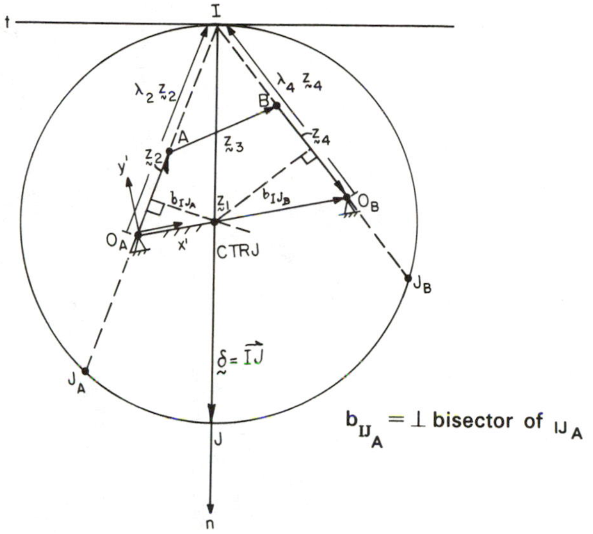


**Figure X4.3**

**(h)** Find the Bresse circle.

**(i)** Find the acceleration pole.

**(j)** Write the equation of the acceleration field.

**(k)** Choose a general point of the coupler plane and determine its acceleration in two ways: (1) by way of the acceleration field equation, and (2) using the principle of acceleration difference. Do the two results agree? (They should!)

**(l)** Find the return circle.

**(m)** Write the equation of the CSC.

**(n)** Plot a few points of the CSC in the vicinity of its intersection with the inflection circle to find the approximate location of Ball's point.

**(o)** Plot a few points on the path of Ball's point in the vicinity of the present position on either side. How long is the approximate straight-line path within a 10% error? [% = 100(error normal to path tangent)/range.]

**(p)** Write the equation of the center-point curve by writing the CSC for the inverted mechanism and transforming it into the original $x_\pi I_\pi y_\pi = xIy$ coordinate system.

**(q)** By way of conjugate point pairs on the CSC and the center-point curve, construct an alternate FBL to match the coupler motion of the original one through four ISPs.

**(r)** For the coupler motion of this new FBL, find the instant center $I$, the inflection circle, acceleration pole and acceleration field, and the CSC. How do these agree with those of the original FBL? (They should agree!)

# 5

# Dynamics of Mechanisms: Advanced Concepts

## 5.1 INTRODUCTION

Chapter 5 of Vol. 1 introduced the basic principles related to dynamics of mechanisms. Dynamics can be divided into subcategories as defined in Table 5.1. The degree of difficulty of a dynamic analysis rests on what is known and unknown in a specific problem and what assumptions can be made about the unknowns.

Methods of static and kinetostatic mechanism analysis were described in Vol. 1. In these cases, the independent variable is either an angular position or a location of a link of the mechanism. Thus the "dynamic" equations are algebraic and easily solvable.

In this chapter we pursue more complex problems, starting with time-response problems and ending with high-speed mechanism analysis. The latter includes non-rigid mechanism members. (See Table 2.1 of Vol. 1 for a summary of assumptions versus analysis classifications.)

In both of the kinetostatic methods introduced in Chap. 5 of Vol. 1, the motion of the linkage was prescribed: the position, velocity, and acceleration of the input link were given. In kinetostatics, we seek the joint forces that result and the input torque that would be required to drive the linkage with these prescribed kinematic conditions. Unfortunately, there may be no means of providing the torque versus input angle relationship derived in the position by position results of kinetostatic analysis.

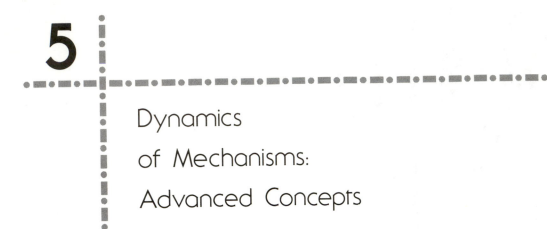

**TABLE 5.1** CLASSIFICATION OF METHODS OF STATIC AND DYNAMIC ANALYSIS OF MECHANISMS[a]

| | | Method | | |
|---|---|---|---|---|
| | | Statics (mechanical advantage) | Kinetostatics | Dynamics (time response) |
| Input information, assumptions (given) | Masses | Zero[b] | Specified | Specified |
| | Loading | Specified or parameterized, as in "input/output" ratio | Specified at each position | Specified in terms of position, velocity, and/or time |
| | Motion | Positions specified | Position, velocity, and acceleration specified | Unknown |
| Output information (sought) | | Required input force to balance load<br>Mechanical advantage at each position<br>Reactions in joints | Required input force to sustain assumed motion<br>Joint reactions | Position, velocity, and acceleration of each member as a function of time: that is, the *actual* motion |
| Required analytical tools | | Statics, linear algebra | D'Alembert's principle, statics, linear algebra | Writing differential equations of motion, solution by computer |

[a] Arranged in order of increasing complexity from left to right.

[b] The weight of the links may play a part in the analysis but not the inertia.

Source: Ref. 88.

## 5.2 REVIEW OF KINETOSTATIC ANALYSIS USING THE MATRIX METHOD

A standard approach to the kinetostatic analysis of a linkage is the matrix method. This section reviews the approach as presented in Sec. 5.6 of Vol. 1. The advantage of the matrix method is that the equations of motion are quickly derived. The disadvantage is the need for matrix manipulation in order to solve the equations. This technique will be applied to the four-bar of Fig. 5.1a. $\mathbf{A}_{gi}$, $\mathbf{F}_{Oi}$, $F_{ji}$, $F_{Oix}$, $g_i$, $T_s$, $\omega_i$, and $\alpha_i$ are the acceleration of the center of mass of link $i$, the inertia force of link $i$ acting through the center of mass, the force of link $j$ on link $i$, the $x$ component of the inertia force of link $i$, the center of mass of link $i$, the driving torque on the linkage, the angular velocity of link $i$, and the angular acceleration of link $i$, respectively. The $x$ and $y$ components of the inertia forces are modeled

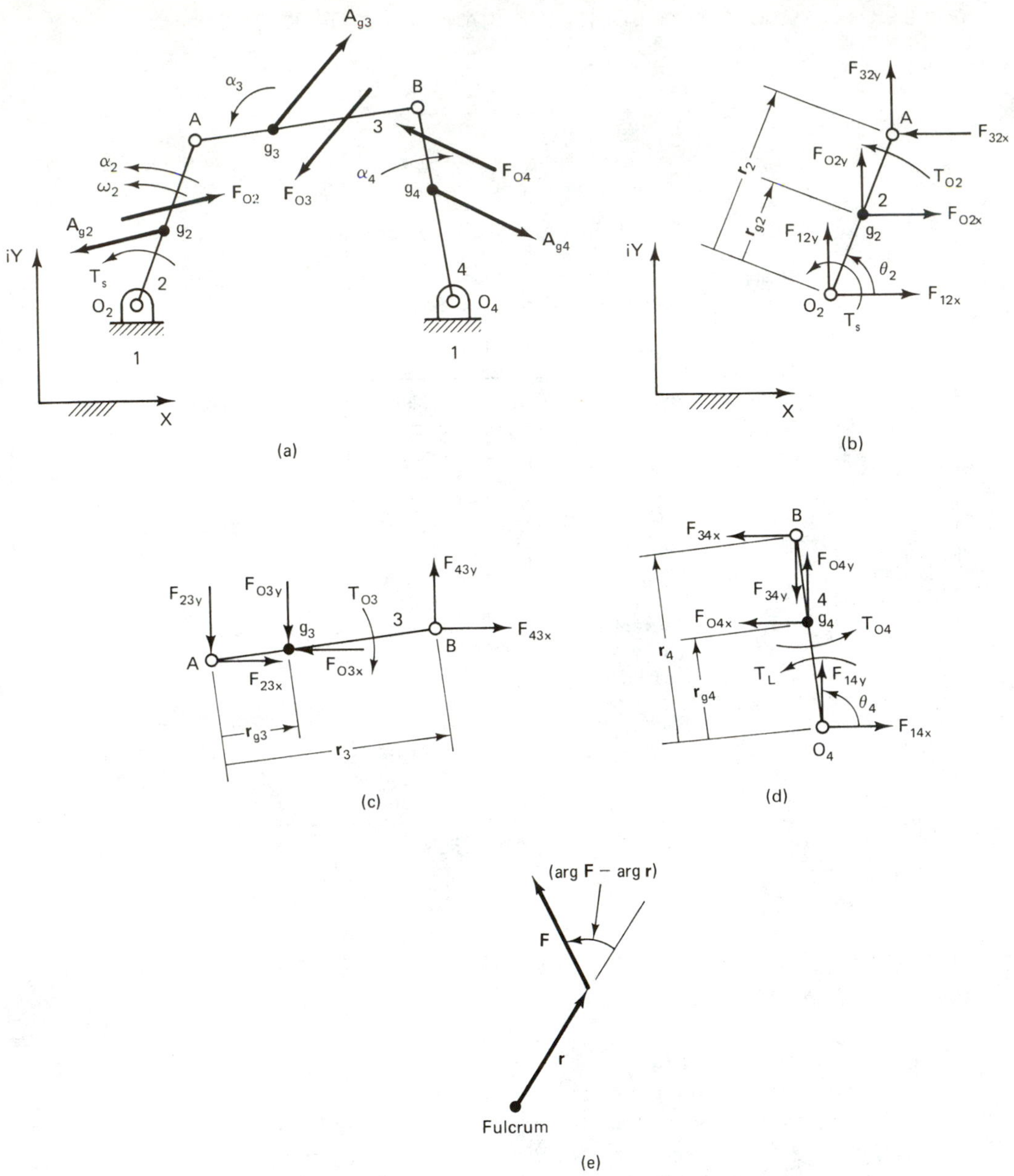


**Figure 5.1** (a) notation for kinetostatic analysis of a four-bar mechanism; (b) free-body diagram (FBD) of link 2; (c) FBD of link 3; (d) FBD of link 4; (e) the vector equation $\mathbf{M} = \mathbf{r} \times \mathbf{F}$ of the moment of **F** about the fulcrum leads to the scalar equation (eq. 5.3) in two-dimensions, compatible with complex-number notation; (f) free-body diagram (FBD) of link 2 for time-response Eqs. [5.18(1), 5.18(2), and 5.18(3)]. (g) FBD of link 3 for time-response Eqs. [5.18(4), 5.18(5), and 5.18(6)]. (h) FBD of link 4 for time-response Eqs. [5.18(7), 5.18(8), and 5.18(9)].

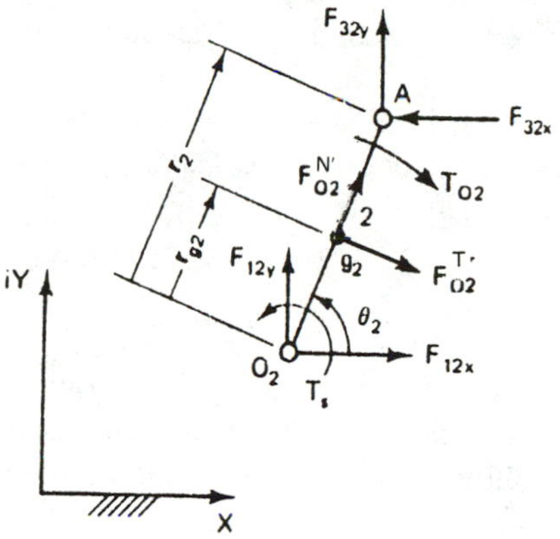


(f)

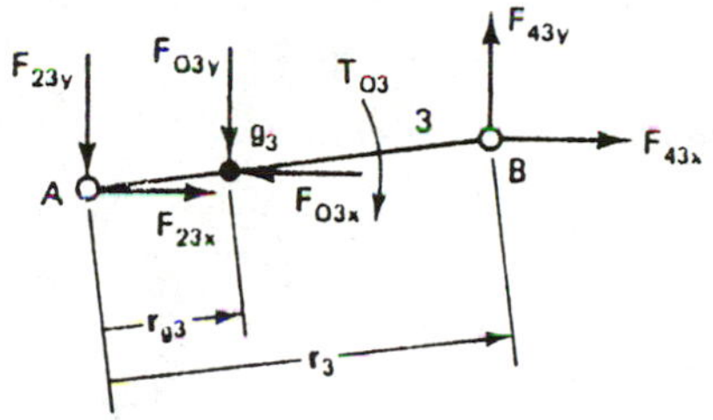


(g)

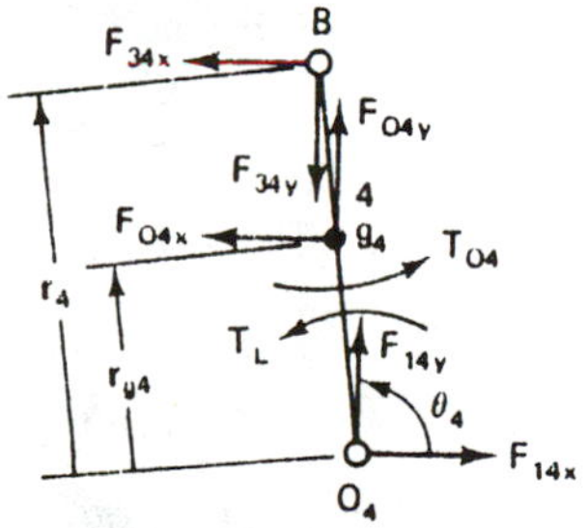


(h)

**Figure 5.1 (Continued)**

as intersecting the center of mass of each moving link so that an inertia torque is added to each free-body diagram (to account for angular acceleration). Referring to the three free-body diagrams of Fig. 5.1b, c, and d, three static equilibrium equations are written for each link: $\Sigma F_x = 0$, $\Sigma F_y = 0$, and $\Sigma T_g = 0$, where the last equation is the sum of moments about the center of gravity, $g_i$.

*Link 2* (Fig. 5.1b):

$$F_{12x} + F_{32x} + F_{O2x} = 0 \tag{5.1}$$

$$F_{12y} + F_{32y} + F_{O2y} = 0 \tag{5.2}$$

In order to write the moment-equilibrium equations in a uniform, programmable manner, without the need for visual determination of the sense of each moment, we will use the following expression of the moment of a force F around a fulcrum:

$$M_F = r_x F_y - r_y F_x \tag{5.3}$$

which comes from the determinant form of the vector cross product $\mathbf{M}_F = \mathbf{r} \times \mathbf{F}$ and where $\mathbf{r}$ is a vector from the fulcrum to the point of application of the force (Fig. 5.1e). With this, for example in Fig. 5.1b, the moment of $\mathbf{F}_{12}$ about $g_2$ is

$$\begin{aligned} M_{F\,12/g} &= (-\mathbf{r}_{g2}) \times \mathbf{F}_{12} = (-r_{g2x})F_{12y} - (-r_{g2y})F_{12x} \\ &= F_{12x} r_{g2} \sin\theta_2 - F_{12y} r_{g2} \cos\theta_2 \end{aligned}$$

Accordingly, the moment equation for link 2 becomes

$$\begin{aligned} T_{O2} + T_s + F_{12x} r_{g2} \sin\theta_2 - F_{12y} r_{g2} \cos\theta_2 - F_{32x}(r_2 - r_{g2}) \sin\theta_2 \\ + F_{32y}(r_2 - r_{g2}) \cos\theta_2 = 0 \end{aligned} \tag{5.4}$$

*Link 3* (Fig. 5.1c):

$$F_{23x} + F_{43x} + F_{O3x} = 0 \tag{5.5}$$

$$F_{23y} + F_{43y} + F_{O3y} = 0 \tag{5.6}$$

$$\begin{aligned} T_{O3} + F_{23x} r_{g3} \sin\theta_3 - F_{23y} r_{g3} \cos\theta_3 \\ - F_{43x}(r_3 - r_{g3}) \sin\theta_3 + F_{43y}(r_3 - r_{g3}) \cos\theta_3 = 0 \end{aligned} \tag{5.7}$$

where we used "external" forces $F_{23}$ and $F_{43}$ acting on link 3.

*Link 4:*

$$F_{34x} + F_{14x} + F_{O4x} = 0 \tag{5.8}$$

$$F_{34y} + F_{14y} + F_{O4y} = 0 \tag{5.9}$$

$$\begin{aligned} T_{O4} - F_{34x}(r_4 - r_{g4}) \sin\theta_4 + F_{34y}(r_4 - r_{g4}) \cos\theta_4 \\ + T_L + F_{14x} r_{g4} \sin\theta_4 - F_{14y} r_{g4} \cos\theta_4 = 0 \end{aligned} \tag{5.10}$$

where $T_L$ represents the torque acting on link 4 due to the external loading on the unit. Equations (5.1) to (5.10) (except 5.3) represent a system of nine equations linear in nine unknowns which describe the instantaneous dynamic force and torque equilibrium on each moving link of the four-bar linkage of Fig. 5.1a. Assuming the load torque $T_L$ and the motion characteristics $\omega_i$ and $\alpha_i$ are given, the nine unknowns are $F_{12x}$, $F_{12y}$, $F_{32x}$, $F_{32y}$, $F_{34x}$, $F_{34y}$, $F_{14x}$, $F_{14y}$, and $T_s$. Note that $F_{jkx} = -F_{kjx}$, and so on. Notice also that the real and imaginary components of the inertia forces, $\mathbf{F}_{Oi}$ and $T_{Oi}$ are obtained from

$$\mathbf{F}_{Oi} = -m_i \mathbf{A}_{gi} = (m_i A_{gi}) e^{i(\beta_i + \pi)} \qquad \text{and} \qquad T_{Oi} = -\alpha_i I_i, \tag{5.11}$$

where $m_i$ is the mass, $\mathbf{A}_{gi}$ the acceleration, $\beta_i$ the argument of $\mathbf{A}_{gi}$, $\alpha_i$ the angular acceleration and $I_i$ the mass moment of inertia about the center of mass $g$ of link $i$. The nine equations are now rewritten so that the unknown terms are shifted to the right side:

$$
\begin{aligned}
F_{O2x} &= -F_{12x} + F_{23x} \\
F_{O2y} &= -F_{12y} + F_{23y} \\
T_{O2} &= -F_{12x}(r_{g2} \sin \theta_2) + F_{12y}(r_{g2} \cos \theta_2) - T_s - F_{23x}(r_2 - r_{g2}) \sin \theta_2 \\
&\quad + F_{23y}(r_2 - r_{g2}) \cos \theta_2 \\
F_{O3x} &= -F_{23x} + F_{34x} \\
F_{O3y} &= -F_{23y} + F_{34y} \\
T_{O3} &= -F_{23x} r_{g3} \sin \theta_3 + F_{23y} r_{g3} \cos \theta_3 \\
&\quad - F_{34x}(r_3 - r_{g3}) \sin \theta_3 + F_{34y}(r_3 - r_{g3}) \cos \theta_3 \\
F_{O4x} &= -F_{34x} - F_{14x} \\
F_{O4y} &= -F_{34y} - F_{14y} \\
T_{O4} &= F_{34x}(r_4 - r_{g4}) \sin \theta_4 - F_{34y}(r_4 - r_{g4}) \cos \theta_4 \\
&\quad - F_{14x} r_{g4} \sin \theta_4 + F_{14y} r_{g4} \cos \theta_4 - T_L
\end{aligned}
\tag{5.12}
$$

This system is linear in the unknowns (the pin forces plus $T_s$) since the sines and cosines are known in this kinetostatic analysis for each position of the linkage. If we define

$$
\begin{aligned}
R_2 &\equiv r_2 - r_{g2} \\
R_3 &\equiv r_3 - r_{g3} \\
R_4 &\equiv r_4 - r_{g4}
\end{aligned}
$$

the system of Eqs. (5.12) may be expressed in matrix form, as shown in Eq. (5.13) on p. 372, or in symbolic form

$$[F_I] = [L][F_B] \tag{5.14}$$

where $[F_I]$ = column matrix of known inertia forces and torques plus the load torque

$[L]$ = square matrix of known linkage and position parameters

$[F_B]$ = column matrix of unknown joint forces and input torque

$$\begin{bmatrix} F_{O2x} \\ F_{O2y} \\ T_{O2} \\ F_{O3x} \\ F_{O3y} \\ T_{O3} \\ F_{O4x} \\ F_{O4y} \\ T_{O4} + T_L \end{bmatrix} = \begin{bmatrix} -1 & 0 & 0 & 1 & 0 & 0 & 0 & 0 & 0 \\ 0 & -1 & 0 & 0 & 1 & 0 & 0 & 0 & 0 \\ -r_{g2}\sin\theta_2 & r_{g2}\cos\theta_2 & -1 & -R_2\sin\theta_2 & R_2\cos\theta_2 & 0 & 0 & 0 & 0 \\ 0 & 0 & 0 & -1 & 0 & 1 & 0 & 0 & 0 \\ 0 & 0 & 0 & 0 & -1 & 0 & 1 & 0 & 0 \\ 0 & 0 & 0 & -r_{g3}\sin\theta_3 & r_{g3}\cos\theta_3 & -R_3\sin\theta_3 & R_3\cos\theta_3 & 0 & 0 \\ 0 & 0 & 0 & 0 & 0 & -1 & 0 & -1 & 0 \\ 0 & 0 & 0 & 0 & 0 & 0 & -1 & 0 & -1 \\ 0 & 0 & 0 & 0 & 0 & R_4\sin\theta_4 & -R_4\cos\theta_4 & -r_{g4}\sin\theta_4 & r_{g4}\cos\theta_4 \end{bmatrix} \begin{bmatrix} F_{12x} \\ F_{12y} \\ T_s \\ F_{23x} \\ F_{23y} \\ F_{34x} \\ F_{34y} \\ F_{14x} \\ F_{14y} \end{bmatrix} \tag{5.13}$$

The system may be solved for the unknowns by matrix computation techniques as follows. The right and left sides of Eq. (5.14) are premultiplied by $[L]^{-1}$ (the inverse of $[L]$).

$$[L]^{-1}[F_I] = [L]^{-1}[L][F_B] \tag{5.15}$$

since

$$[L]^{-1}[L] = [I] \qquad \text{the identity matrix}$$

$$[F_B] = [L]^{-1}[F_I] \tag{5.16}$$

The solution of Eq. (5.16) requires calculation of the inverse of the linkage parameter matrix and the multiplication of the two matrices on the right-hand side of Eq. (5.16). In most instances, these matrix operations are available as standard functions on computer systems.

It should be noted that taking the inverse in Eq. (5.15) can be a time-consuming step on the computer. If minimizing computer time is important, a large reduction in the number of calculations can be accomplished by using the Gauss–Jordan reduction to put the array of coefficients of Eq. (5.13) into upper-right-triangle form and then solving by direct back substitution [118]. This generally requires $[(3m)^3/3 + (3m)^2/2]$ multiplications (where $m = 3$ is the number of moving links), a reduction of two-thirds from the inverse method. Partial pivoting of the array is usually performed to avoid dividing by small coefficients and to retain accuracy [50].

### Design Example: Kinetostatic Analysis of a Card Punch [30]

Tabulating card punches and readers are common types of peripheral equipment used in digital computer systems. The card feeder mechanisms, which feed cards from the input hopper into the machine, have to be capable to feed cards at a high rate. Typical rates are 300 to 1000 cards per minute. Techniques for card feeding, which are simpler and cheaper but reliable, are desired.

Figure 5.2 shows one proposed design concept for a card feeder involving a four-bar linkage. A crank is driven by a suitable driving mechanism to produce oscillating motion of a picker knife assembly. When the mechanism starts from rest, the picker knives engage a 0.007- to 0.009-in.-thick card. When the picker knives have reached maximum velocity, the card is caught by the feeder wheels, which are driven by another source. The card is then fed along its path through the machine. Finally, the mechanism is brought to rest back at its original position ready for the next card.

While the picker knives engage the card during the initial part of the cycle, the linkage has to overcome the card-to-card and the card-to-steel drag forces while supporting part of the weight of the card stack. After the card being fed is engaged by the feed wheels, only the card-to-steel drag force is acting. This drag force is small compared to the card-to-card drag force.

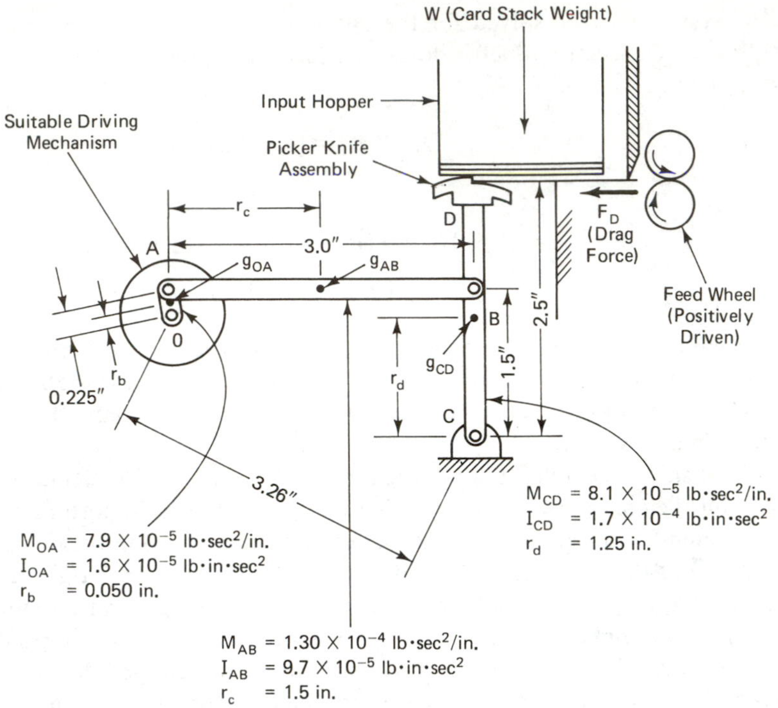


**Figure 5.2** Schematic of card feeder mechanism using the four-bar mechanism *OABC*.

Figures 5.3 to 5.5 show the kinematic characteristics and the loading assumed for the mechanism. From Fig. 5.3 the angular position, velocity, and acceleration with respect to time are derived from the drive motor characteristics:

When $t \leq 0.01234$ sec:

$$\theta_2 = -\frac{21{,}610.2}{2} t^2 + 3.6183 \qquad \text{rad}$$

$$\dot{\theta}_2 = -21{,}610.2t \qquad \text{rad/sec}$$

$$\ddot{\theta}_2 = -21{,}610.2 \qquad \text{rad/sec}^2$$

When $t \geq 0.01234$ sec:

$$\theta_2 = -266.67(t - 0.01234) + \frac{7667.2}{2}(t - 0.01234)^2 + 1.9729 \qquad \text{rad}$$

$$\dot{\theta}_2 = -266.67 + 7667.2(t - 0.01234) \qquad \text{rad/sec}$$

$$\ddot{\theta}_2 = 7667.2 \qquad \text{rad/sec}^2$$

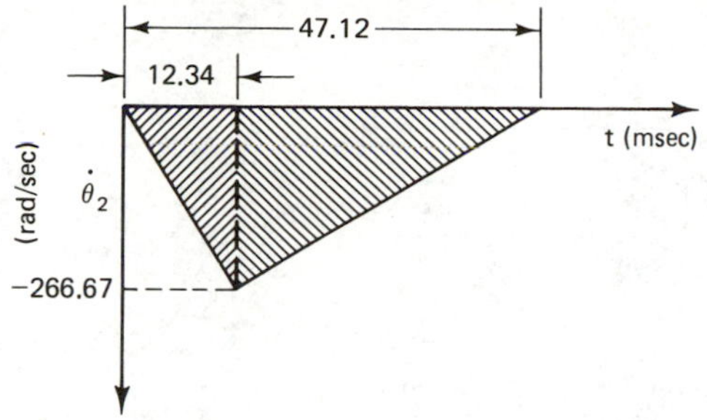


**Figure 5.3** Angular velocity input to driving crank of sample four-bar linkage.

The area under the velocity curve in Fig. 5.3 represents one revolution or $2\pi$ radians of input crank rotation.

Figures 5.4 and 5.5 depict the loads imposed in this application, which are $W_C$, the weight of the stack of cards in the hopper, and $F_D$, the drag force on the card. Assuming that $W_C$ acts along link 4 and $F_D$ acts perpendicular to it, which

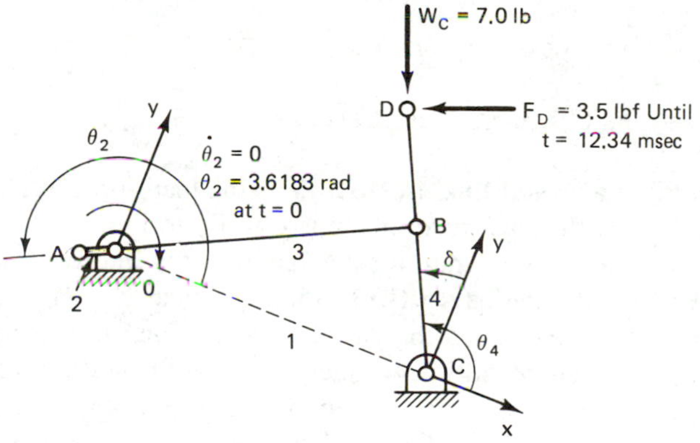


**Figure 5.4** Definition of loads and angular coordinates at time zero.

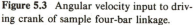

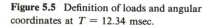

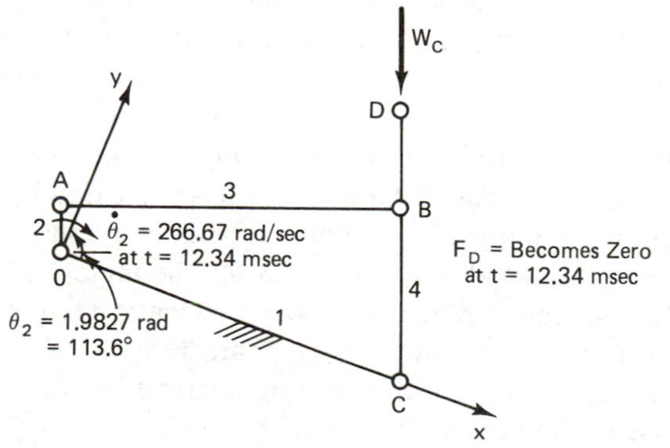


**Figure 5.5** Definition of loads and angular coordinates at $T = 12.34$ msec.

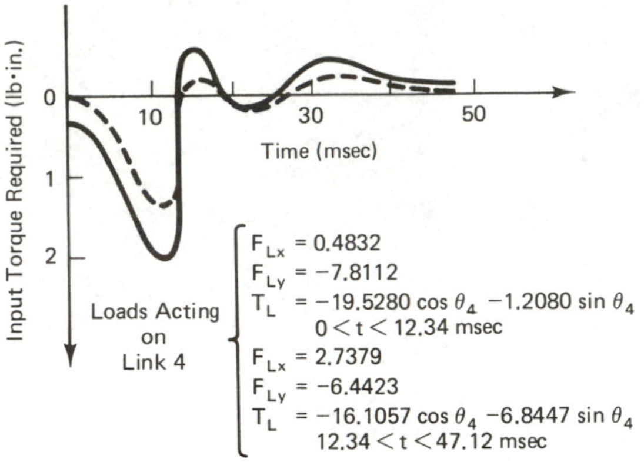


**Figure 5.6** Torque required to maintain equilibrium, given the load acting on link 4 and the kinematic input (see Fig. 5.3).

is approximately true throughout the motion cycle,

$$F_{L_x} = W_C \sin\delta - F_D \cos\delta$$

$$F_{L_y} = -W_C \cos\delta - F_D \sin\delta$$

$$T_L = -F_{L_x}\overline{CD}\sin\theta_4 + F_{L_y}\overline{CD}\cos\theta_4$$

where $\delta$ is the angle between the $y$ axis and link 4 ($DC$), $\mathbf{F}_L$ is the load force applied to the linkage at point $D$ and $T_L$ is the load torque applied by $\mathbf{F}_L$ to link 4.

The solid line in Fig. 5.6 shows the input torque required for one complete revolution of the crank (link 2) based on Eqs. (5.12) to (5.16). Since $F_{L_x}$, $F_{L_y}$, and $T_L$ are not functions of $\dot{\theta}_2$, $\ddot{\theta}_2$, or $t$, we can see the effects of inertia on input torque by setting $\dot{\theta}_2$ and $\ddot{\theta}_2$ equal to zero. The resulting torque should be the static torque required to maintain equilibrium. The dashed line in Fig. 5.6 shows the static torque required. The difference between the solid and dashed line may be thought of as being proportional to the equivalent inertia of the four-bar linkage referred to link 2. If a variable inertia load equal to the equivalent inertia were attached to the input link (without the rest of the linkage), the same variable input torque would be required. It should be noticed that the equivalent inertia is not constant during the cycle.

Although the kinematic characteristics, as shown in Fig. 5.3, may be desirable, in practice, they would be *difficult to achieve.* One would have to carefully measure the torque–speed characteristics of a motor while loading it with the presumed varying inertia of the linkage in order to obtain the true Fig. 5.3 for the motor. Therefore, the problem remains whether we can achieve the strict kinematic requirements of feeding cards without knowing the exact input torque characteristics of the driving mechanism. If more precision is required, a kinetostatic analysis is not appropriate and the time-response method of the next section must be used.

## 5.3 TIME RESPONSE

The design example of the preceding section illustrates one instance where a kinetostatic analysis was not appropriate, because the kinematic input is uncertain. The analysis objective must be turned around in order to produce a more exact technique. In *time-response* analysis (refer to Table 5.1), we are given a mechanism of known geometry, mass, and inertia that is subjected to known external loads and known driving forces or torques (such as a torque–speed relationship). Time response analysis produces kinematic information about the linkage positions, velocities, and accelerations as functions of *time,* as well as the joint forces, if desired, also as functions of time.

Since the independent variable is now time rather than a position parameter and since we seek the unknown motion of the linkage, differential equations of motion must be solved. These equations have coefficients, such as equivalent inertia and damping at the input, that are functions of position, and therefore require numerical integration techniques.

### Description of the Runge–Kutta Method

Among the many available numerical integration techniques, one of the most widely used is the Runge–Kutta fourth-order numerical analysis routine [33,50,259]. This method will be demonstrated by two examples, one involving first-order and the other second-order differential equations.

Runge–Kutta numerical integration provides an algorithm by which the integral of a function may be evaluated at specific values of the independent variable. For the problem at hand, it could be restated as an algorithm by which the velocity may be determined from the acceleration at specific values of time. The Runge–Kutta method was developed from consideration of Taylor-series expansion of a function about a point and is given here without derivation. The value of the dependent variable $y(x)$ at point $n + 1$ is determined in terms of the dependent variable $y$ at point $n$, and the derivative expression at time $n$:

$$y_{n+1} = y_n + \tfrac{1}{6}(k_1 + 2 * k_2 + 2 * k_3 + k_4) \qquad (5.17)$$

where $*$ is the symbol for multiplication.

$$k_1 = h * f(x_n, y_n)$$

$$k_2 = h * f\left(x_n + \frac{h}{2}, y_n + \frac{k_1}{2}\right)$$

$$k_3 = h * f\left(x_n + \frac{h}{2}, y_n + \frac{k_2}{2}\right)$$

$$k_4 = h * f(x_n + h, y_n + k_3)$$

$f(x_n, y_n) = y'_n$ value of the derivative of the dependent variable $y$ with respect to the independent variable $x$, the function to be integrated in order to determine $y(x)$, evaluated at $x_n$, $y_n$

$h$ = step size specified for the independent variable $x$

$n$ = step number (i.e., 0, 1, 2, 3, . . .)

$x_n$ = value of independent variable at step $n$

**Example 5.1**

To illustrate the method, let us assume that we are asked to solve the first-order differential equation

$$\frac{dy(t)}{dt} = y' = t^2$$

for step size of $\Delta t = h = 1$ and with the initial condition

$$t = 1, \qquad y = \tfrac{1}{3}$$

Note that here $y$ is the dependent and $t$ is the independent variable, and, knowing the time derivative $y'(t)$, we seek $y(t)$.

**Solution** This particular example can be solved analytically by separating variables and integrating both sides:

$$\frac{dy}{dt} = t^2, \qquad dy = t^2\,dt, \qquad \int dy = \int t^2\,dt$$

which gives us

$$y = \frac{t^3}{3} + C$$

Evaluating the constant by applying the initial condition $[y_{t=1} = \frac{1}{3}]$, we obtain

$$\tfrac{1}{3} = \tfrac{1}{3} + C \qquad \text{or} \qquad C = 0$$

with the result that

$$y(t) = \frac{t^3}{3}$$

We can now evaluate the dependent variable $y$ at the intervals of $t$ for which results are required. This will provide a basis of comparison for which to appraise the effectiveness of the Runge–Kutta method. Since we have a solution at $t = 1$ (the initial condition), begin by applying Runge–Kutta at $t = 2$. Before proceeding, it should be noted that the general form of Runge–Kutta assumes that the derivative is a function of both the independent variable and the integral or dependent variable [in the equation $y' = f(t, y)$], but for this example, the derivative is a function of the independent variable alone.

First evaluate the $k$'s:

$$k_1 = h * t^2 = 1.00$$

$$k_2 = 1 * (1 + \tfrac{1}{2})^2 = 2.25$$

$$k_3 = 1 * (1 + \tfrac{1}{2})^2 = 2.25$$

$$k_4 = 1 * (1 + 1)^2 = 4.00$$

Substituting into Eq. (5.17), where $y_1 = \frac{1}{3}$, we have

$$y_2 = \tfrac{1}{3} + \tfrac{1}{6}(1 + 4.50 + 4.50 + 4.00) = 2.667$$

We repeat the procedure for the next interval.

Evaluating the $k$'s, we get

$$k_1 = 1 * 2^2 = 4.000$$

$$k_2 = 1 * (2 + \tfrac{1}{2})^2 = 6.250$$

$$k_3 = 1 * (2 + \tfrac{1}{2})^2 = 6.250$$

$$k_4 = 1 * (2 + 1)^2 = 9.000$$

and substituting into Eq. (5.17), we have

$$y_3 = 2.667 + \tfrac{1}{6}(4.000 + 12.500 + 12.500 + 9.000)$$
$$= 9.000$$

Note that the analytic expressions for the same intervals yield identical solutions. In general, however, the error in Runge–Kutta will depend on such things as step size and computer round-off, but with proper control of step size, accuracy to several decimal places can reasonably be expected.

We have seen thus far that Runge–Kutta works well with a first-order differential equation, but in kinematics problems, where we start with known accelerations and seek displacements, second-order differential equations must be solved.

To apply Runge–Kutta to higher-order systems, transform the higher-order equation into a series of first-order equations and solve them one at a time, always using the results of the previous integration as the input for the next integration. Thus when there is an equation involving known acceleration (second time derivative of displacement) and the position is desired, first solve for velocity and use this velocity as input to a second Runge–Kutta process which will integrate from velocity to obtain displacement, and thus position. An example will demonstrate this procedure.

**Example 5.2**

A second-order differential equation is to be solved:

$$y'' = 2t$$

with the following stated initial conditions at $t = 1$:

$$y = \tfrac{1}{3} \qquad \text{and} \qquad y' = 1$$

As before, we wish to solve in step sizes of $\Delta t = h = 1$.

**Solution** Observe that the second-order equation was generated by differentiating the equation presented in the first example, and the initial conditions are such that both constants of integration disappear. The analytical solution to this equation is

$$y' = t^2 \qquad \text{and} \qquad y = \frac{t^3}{3}$$

The Runge–Kutta analysis is begun much the same as before with one exception: The first derivative will be evaluated at every half-step (the reason will become clear soon).

First, evaluate the $k$'s for the half-step $h = \frac{1}{2}$:

$$k_1 = \tfrac{1}{2} * 2 * 1 = 1.0000$$

$$k_2 = \tfrac{1}{2} * 2 * 1.25 = 1.2500$$

$$k_3 = \tfrac{1}{2} * 2 * 1.25 = 1.2500$$

$$k_4 = \tfrac{1}{2} * 2 * 1.5 = 1.5000$$

Substituting into Eq. (5.17), we get

$$y'_{1.5} = 1 + \tfrac{1}{6}(1 + 2.5 + 2.5 + 1.5) = 2.2500$$

Now evaluate the $k$'s for $h = 1$:

$$k_1 = 1 * 2 * 1 = 2.0000$$

$$k_2 = 1 * 2 * 1.5 = 3.0000$$

$$k_3 = 1 * 2 * 1.5 = 3.0000$$

$$k_4 = 1 * 2 * 2 = 4.0000$$

Substituting into Eq. (5.17) yields

$$y'_2 = 1 + \tfrac{1}{6}(2 + 6 + 6 + 4) = 4.0000$$

This is the solution for the first equation of the second-order system.

Evaluating the $k$'s for the second (velocity) equation, we get

$$k_1 = 1 * y'_1 = 1.0000$$

$$k_2 = 1 * y'_{1.5} = 2.2500$$

$$k_3 = 1 * y'_{1.5} = 2.2500$$

$$k_4 = 1 * y'_2 = 4.0000$$

and substituting into Eq. (5.17) yields

$$y_2 = 0.3333 + \tfrac{1}{6}(1 + 4.5 + 4.5 + 4) = 2.6667$$

Repeating the entire procedure for the next interval, the $k$'s are evaluated for the half-interval:

$$k_1 = \tfrac{1}{2} * 2 * 2 = 2.0000$$

$$k_2 = \tfrac{1}{2} * 2 * 2.25 = 2.2500$$

$$k_3 = \tfrac{1}{2} * 2 * 2.25 = 2.2500$$

$$k_4 = \tfrac{1}{2} * 2 * 2.5 = 2.5000$$

Substituting into Eq. (5.17), we get

$$y'_{2.5} = 4 + \tfrac{1}{6}(2 + 4.5 + 4.5 + 2.5) = 6.2500$$

Evaluating the $k$'s for the full interval yields

$$k_1 = 1 * 2 * 2 = 4.0000$$

$$k_2 = 1 * 2 * 2.5 = 5.0000$$

$$k_3 = 1 * 2 * 2.5 = 5.0000$$

$$k_4 = 1 * 2 * 3 = 6.0000$$

Substituting into Eq. (5.17), we obtain

$$y'_3 = 4 + \tfrac{1}{6}(4 + 10 + 10 + 6) = 9.0000$$

Evaluating the $k$'s for the second equation yields

$$k_1 = 1 * y'_2 = 4.0000$$

$$k_2 = 1 * y'_{2.5} = 6.2500$$

$$k_3 = 1 * y'_{2.5} = 6.2500$$

$$k_4 = 1 * y'_3 = 9.0000$$

Substituting into Eq. (5.17), we have

$$y_3 = 2.6667 + \tfrac{1}{6}(4 + 12.5 + 12.5 + 9) = 9.0000$$

There is one last point to be made regarding the form of the differential equation. In both examples, the right side of the differential equation had the form of a mathematical expression. This is *not* a requirement for Runge–Kutta solution. It is only necessary to be able to *evaluate* the derivative at the specific point. The evaluation may come from a simple expression as in our examples, or it may be generated numerically by finite-difference methods, as is the case in a mechanism problem in the next section.

## Equations of Motion: Time Response of a Four-Bar Linkage

The same equations of motion that were used for the kinetostatic analysis of the four-bar linkage of Fig. 5.1 are valid for performing a time-response analysis. The form of the matrix [Eq. (5.13)] must be modified, however. In the kinetostatic analysis the motion of the input link was completely known or specified (i.e., $\theta_2$, $\dot{\theta}_2$, and $\ddot{\theta}_2$ were given). In a time-response problem the input torque $T_s$ is known, and the initial position and velocity are known at the start of the time step in question, say at $t_1$. Thus $\ddot{\theta}_2$ becomes the unknown, to be determined at $t_1$ through a modified form of Eq. (5.13), in which the input torque and output angular acceleration terms must exchange places: $T_s$ appears in the vector of known quantities, $\ddot{\theta}_2$ in the vector of unknowns. Portions of the acceleration terms of the coupler and output links that are functions of $\ddot{\theta}_2$ must also be modified so that the matrix of linkage parameters contains only known quantities. To this end, the free-body equilibrium equations are rearranged as follows.

For link 2 (see Fig. 5.1f):

$$\Sigma \mathbf{F} = 0, \qquad \mathbf{F}_{12} + \mathbf{F}_{32} + \mathbf{F}_{02}^N + \mathbf{F}_{02}^T = 0$$

$$\mathbf{F}_{02}^N = M_2 \mathbf{r}_{g2} \dot{\theta}_2^2 = M_2 (r_{g2x} + i r_{g2y}) \dot{\theta}_2^2$$

$$\mathbf{F}_{02}^T = -M_2 \mathbf{r}_{g2i} \ddot{\theta}_2 = -M_2 (i r_{g2x} - r_{g2y}) \ddot{\theta}_2$$

$$\Sigma F_x = 0, \qquad M_2 \dot{\theta}_2^2 r_{g2} \cos\theta_2 = -F_{12x} + F_{23x} - M_2 \ddot{\theta}_2 r_{g2} \sin\theta_2 \qquad [5.18(1)]$$

$$\Sigma F_y = 0, \qquad M_2 \dot{\theta}_2^2 r_{g2} \sin\theta_2 = -F_{12y} + F_{23y} + M_2 \ddot{\theta}_2 r_{g2} \cos\theta_2 \qquad [5.18(2)]$$

$$\Sigma M = 0, \qquad T_s = -F_{12x} r_{g2} \sin\theta_2 + F_{12y} r_{g2} \cos\theta_2 - F_{23x}(r_2 - r_{g2}) \sin\theta_2 + F_{23y}(r_2 - r_{g2}) \cos\theta_2 + I_{g2} \ddot{\theta}_2 \qquad [5.18(3)]$$

For link 3 (see Fig. 5.1g):

$$\ddot{\theta}_3 = \frac{1}{r_3 \sin(\theta_3 - \theta_4)} \{[-\dot{\theta}_2^2 r_2 \cos(\theta_2 - \theta_4) - \dot{\theta}_3^2 r_3 \cos(\theta_3 - \theta_4) + \dot{\theta}_4^2 r_4] - [\ddot{\theta}_2 r_2 \sin(\theta_2 - \theta_4)]\}$$

$$\mathbf{A}_{g3} = \mathbf{A}_A + \mathbf{r}_{g3} i \ddot{\theta}_3 - \mathbf{r}_{g3} \dot{\theta}_3^2$$
$$= \mathbf{r}_2 i \ddot{\theta}_2 - \mathbf{r}_2 \dot{\theta}_2^2 + \mathbf{r}_{g3} i \ddot{\theta}_3 - \mathbf{r}_{g3} \dot{\theta}_3^2$$

$$\Sigma \mathbf{F} = 0, \qquad \mathbf{F}_{23} + \mathbf{F}_{43} + \mathbf{F}_{03} = 0$$

$$\mathbf{F}_{23} + \mathbf{F}_{43} - M_3(\mathbf{r}_2 i \ddot{\theta}_2 - \mathbf{r}_2 \dot{\theta}_2^2 + \mathbf{r}_{g3} i \ddot{\theta}_3 - \mathbf{r}_{g3} \dot{\theta}_3^2) = 0$$

$$\Sigma F_x = 0 \qquad -F_{23x} - F_{34x} + M_3 \dot{\theta}_2^2 r_2 \cos\theta_2 + M_3 \dot{\theta}_3^2 r_{g3} \cos\theta_3 + M_3 \ddot{\theta}_2 r_2 \sin\theta_2$$
$$+ M_3 r_{g3} \sin\theta_3 \frac{1}{r_3 \sin(\theta_3 - \theta_4)} \{[-\dot{\theta}_2^2 r_2 \cos(\theta_2 - \theta_4) - \dot{\theta}_3^2 r_3 \cos(\theta_3 - \theta_4) + \dot{\theta}_4^2 r_4] - \ddot{\theta}_2 r_2 \sin(\theta_2 - \theta_4)\} = 0$$

$$\Rightarrow \quad M_3 \left\{ \dot{\theta}_2^2 r_2 \cos\theta_2 + \dot{\theta}_3^2 r_{g3} \cos\theta_3 - \frac{r_{g3} \sin\theta_3}{r_3 \sin(\theta_3 - \theta_4)} [\dot{\theta}_2^2 r_2 \cos(\theta_2 - \theta_4) + \dot{\theta}_3^2 r_3 \cos(\theta_3 - \theta_4) - \dot{\theta}_4^2 r_4] \right\}$$

$$= F_{23x} + F_{34x} - M_3 \ddot{\theta}_2 r_2 \left[ \sin\theta_2 - \frac{r_{g3} \sin\theta_3}{r_3 \sin(\theta_3 - \theta_4)} \sin(\theta_2 - \theta_4) \right] \qquad [5.18(4)]$$

$$\Sigma F_y = 0, \qquad M_3 \left\{ \dot\theta_2^2 r_2 \sin\theta_2 + \dot\theta_3^2 r_{g3} \sin\theta_3 + \frac{r_{g3}\cos\theta_3}{r_3\sin(\theta_3-\theta_4)} \right.$$

$$\left. [\dot\theta_2^2 r_2 \cos(\theta_2-\theta_4) + \dot\theta_3^2 r_3 \cos(\theta_3-\theta_4) - \dot\theta_4^2 r_4] \right\}$$

$$= F_{23y} + F_{34y} + M_3\ddot\theta_2 r_2 \left[ \cos\theta_2 - \frac{r_{g3}\cos\theta_3}{r_3\sin(\theta_3-\theta_4)} \sin(\theta_2-\theta_4) \right] \qquad [5.18(5)]$$

$$\Sigma M = 0, \qquad T_{03} = -F_{23x} r_{g3} \sin\theta_3 + F_{23y} r_{g3}\cos\theta_3$$

$$+ F_{34x}(r_3 - r_{g3})\sin\theta_3 - F_{34y}(r_3 - r_{g3})\cos\theta_3 \qquad [5.18(6)]$$

For link 4 (see Fig. 5.1h):

$$\ddot\theta_4 = \frac{1}{r_4\sin(\theta_3-\theta_4)} \{[-\dot\theta_2^2 r_2 \cos(\theta_3-\theta_2)$$

$$- \dot\theta_3^2 r_3 + \dot\theta_4^2 r_4 \cos(\theta_3-\theta_4)] + \ddot\theta_2 r_2 \sin(\theta_3-\theta_2)\}$$

$$\Sigma \mathbf{F} = 0, \qquad \mathbf{F}_{14} + \mathbf{F}_{34} + \mathbf{F}_{04}^N + \mathbf{F}_{04}^T = 0$$

$$\mathbf{F}_{04}^N = M_4 \mathbf{r}_{g4} \dot\theta_4^2 = M_4(r_{g4x} + i r_{g4y})\dot\theta_4^2$$

$$\mathbf{F}_{04}^T = -M_4 \mathbf{r}_{g4} i \ddot\theta_4 = -M_4(i r_{g4x} - r_{g4y})\ddot\theta_4$$

$$\Sigma F_x = 0, \qquad F_{14x} - F_{34x} + M_4\dot\theta_4^2 r_{g4}\cos\theta_4$$

$$+ M_4 r_{g4}\sin\theta_4 \frac{1}{r_4\sin(\theta_3-\theta_4)} \{[-\dot\theta_2^2 r_2\cos(\theta_3-\theta_2) - \dot\theta_3^2 r_3$$

$$+ \dot\theta_4^2 r_4 \cos(\theta_3-\theta_4)] + \ddot\theta_2 r_2 \sin(\theta_3-\theta_2)\} = 0$$

$$\Rightarrow \quad M_4 r_{g4} \left\{ \dot\theta_4^2 \cos\theta_4 - \frac{\sin\theta_4}{r_4\sin(\theta_3-\theta_4)} [\dot\theta_2^2 r_2 \cos(\theta_3-\theta_2) \right.$$

$$\left. + \dot\theta_3^2 r_3 - \dot\theta_4^2 r_4 \cos(\theta_3-\theta_4)] \right\}$$

$$= F_{34x} - F_{14x} - M_4\ddot\theta_2 \frac{r_{g4}\sin\theta_4}{r_4\sin(\theta_3-\theta_4)} r_2 \sin(\theta_3-\theta_2) \qquad [5.18(7)]$$

$$\begin{bmatrix} M_2\dot{\theta}_2^2 r_{g2}\cos\theta_2 \\ M_2\dot{\theta}_2^2 r_{g2}\sin\theta_2 \\ T_s \\ F_{k03x}{}^* \\ F_{k03y}{}^* \\ T_{03} \\ F_{k04x}{}^* \\ F_{k04y}{}^* \\ T_{04}-T_L \end{bmatrix} = \begin{bmatrix} -1 & 0 & -M_2 r_{g2}\sin\theta_2 & 1 & 0 & 0 & 0 & 0 & 0 \\ 0 & -1 & M_2 r_{g2}\cos\theta_2 & 0 & 1 & 0 & 0 & 0 & 0 \\ -r_{g2}\sin\theta_2 & r_{g2}\cos\theta_2 & I_{g2} & -(r_2-r_{g2})\sin\theta_2 & (r_2-r_{g2})\cos\theta_2 & 0 & 0 & 0 & 0 \\ 0 & 0 & F_{c03x}{}^* & 1 & 0 & 1 & 0 & 0 & 0 \\ 0 & 0 & F_{c03y}{}^* & 0 & 1 & 0 & 1 & 0 & 0 \\ 0 & 0 & 0 & -r_{g3}\sin\theta_3 & r_{g3}\cos\theta_3 & (r_3-r_{g3})\sin\theta_3 & -(r_3-r_{g3})\cos\theta_3 & 0 & 0 \\ 0 & 0 & F_{c04x}{}^* & 0 & 0 & 1 & 0 & -1 & 0 \\ 0 & 0 & F_{c04y}{}^* & 0 & 0 & 0 & 1 & 0 & -1 \\ 0 & 0 & 0 & 0 & 0 & (r_4-r_{g4})\sin\theta_4 & -(r_4-r_{g4})\cos\theta_4 & r_{g4}\sin\theta_4 & -r_{g4}\cos\theta_4 \end{bmatrix} \begin{bmatrix} F_{12x} \\ F_{12y} \\ \ddot{\theta}_2 \\ F_{23x} \\ F_{23y} \\ F_{34x} \\ F_{34y} \\ F_{14x} \\ F_{14y} \end{bmatrix} \quad (5.18)$$

* Complete expressions for the starred quantities are as follows:

The known parts of the inertia forces in the left column matrix [Eq. (5.18)]are:

$$F_{k03x} = M_3\left\{\dot{\theta}_2^2 r_2\cos\theta_2 + \dot{\theta}_3^2 r_{g3}\cos\theta_3 - \frac{r_{g3}\sin\theta_3}{r_3\sin(\theta_3-\theta_4)}[\dot{\theta}_2^2 r_2\cos(\theta_2-\theta_4) + \dot{\theta}_3^2 r_3\cos(\theta_3-\theta_4) - \dot{\theta}_4^2 r_4]\right\}$$

$$F_{k03y} = M_3\left\{\dot{\theta}_2^2 r_2\sin\theta_2 + \dot{\theta}_3^2 r_{g3}\sin\theta_3 + \frac{r_{g3}\cos\theta_3}{r_3\sin(\theta_3-\theta_4)}[\dot{\theta}_2^2 r_2\cos(\theta_2-\theta_4) + \dot{\theta}_3^2 r_3\cos(\theta_3-\theta_4) - \dot{\theta}_4^2 r_4]\right\}$$

$$F_{k04x} = M_4 r_{g4}\left\{\dot{\theta}_4^2\cos\theta_4 - \frac{\sin\theta_4}{r_4\sin(\theta_3-\theta_4)}[\dot{\theta}_2^2 r_2\cos(\theta_3-\theta_2) + \dot{\theta}_3^2 r_3 - \dot{\theta}_4^2 r_4\cos(\theta_3-\theta_4)]\right\}$$

$$F_{k04y} = M_4 r_{g4}\left\{\dot{\theta}_4^2\sin\theta_4 + \frac{\cos\theta_4}{r_4\sin(\theta_3-\theta_4)}[\dot{\theta}_2^2 r_2\cos(\theta_3-\theta_2) + \dot{\theta}_3^2 r_3 - \dot{\theta}_4^2 r_4\cos(\theta_3-\theta_4)]\right\}$$

The coefficients of the unknown parts of the inertia forces in the body of the matrix are:

$$F_{c03x} = -M_3 r_2\left[\sin\theta_2 - r_{g3}\sin\theta_3\frac{\sin(\theta_2-\theta_4)}{r_3\sin(\theta_3-\theta_4)}\right]$$

$$F_{c03y} = M_3 r_2\left[\cos\theta_2 - r_{g3}\cos\theta_3\frac{\sin(\theta_2-\theta_4)}{r_3\sin(\theta_3-\theta_4)}\right]$$

$$F_{c04x} = -M_4 r_2 r_{g4}\sin\theta_4\frac{\sin(\theta_3-\theta_2)}{r_4\sin(\theta_3-\theta_4)}$$

$$F_{c04y} = M_4 r_2 r_{g4}\cos\theta_4\frac{\sin(\theta_3-\theta_2)}{r_4\sin(\theta_3-\theta_4)}$$

$$\Sigma F_y = 0, \qquad F_{14y} - F_{34y} + M_4\dot{\theta}_4^2 r_{g4} \sin\theta_4$$

$$- M_4 r_{g4} \cos\theta_4 \frac{1}{r_4 \sin(\theta_3 - \theta_4)} \{[-\dot{\theta}_2^2 r_2 \cos(\theta_3 - \theta_2) - \dot{\theta}_3^2 r_3$$

$$+ \dot{\theta}_4^2 r_4 \cos(\theta_3 - \theta_4)] + \ddot{\theta}_2 r_2 \sin(\theta_3 - \theta_2)\} = 0$$

$$\Rightarrow \quad M_4 r_{g4} \left\{\dot{\theta}_4^2 \sin\theta_4 + \frac{\cos\theta_4}{r_4 \sin(\theta_3 - \theta_4)} [\dot{\theta}_2^2 r_2 \cos(\theta_3 - \theta_2) \right.$$

$$\left. + \dot{\theta}_3^2 r_3 - \dot{\theta}_4^2 r_4 \cos(\theta_3 - \theta_4)]\right\}$$

$$= F_{34y} - F_{14y} + M_4 \ddot{\theta}_2 r_{g4} \cos\theta_4 \frac{r_2 \sin(\theta_3 - \theta_2)}{r_4 \sin(\theta_3 - \theta_4)} \qquad [5.18(8)]$$

$$\Sigma M = 0, \qquad T_{04} - T_L = F_{14x} r_{g4} \sin\theta_4 - F_{14y} r_{g4} \cos\theta_4$$
$$+ F_{34x}(r_4 - r_{g4}) \sin\theta_4 - F_{34y}(r_4 - r_{g4}) \cos\theta_4 \qquad [5.18(9)]$$

Equations [5.18(1)] to [5.18(9)] can now be arranged in matrix form Eq. (5.18) (see p. 384):
In compact form, the matrix equation (5.18) is

$$[C] = [A] \begin{bmatrix} F_{12x} \\ \vdots \\ \ddot{\theta}_2 \end{bmatrix} \qquad (5.19)$$

Here we see that $\ddot{\theta}_2$ has been moved to the last element in the column matrix on the right-hand side of the matrix equation with corresponding changes in $[C]$ and $[A]$. Thus we can solve for $\ddot{\theta}_2$:

$$\ddot{\theta}_2 = \text{last element of the column matrix} = [A^{-1}][C]_{91} \qquad (5.20)$$

where we added the subscript of $[C]$ to indicate that it is a $9 \times 1$ matrix (a "vector").

Since we are given $\theta_2$ and $\dot{\theta}_2$ at $t = t_i$, before we calculate the values of $\theta_2$ and $\dot{\theta}_2$ at $(t_i + \Delta t)$ using the Runge–Kutta method described in the preceding section [Eq. (5.17)], we can determine the value of $\ddot{\theta}_2$ at $t_i$ using the matrix Equation (5.20).

**Design example.** Referring back to the card punch of Fig. 5.2, a servomotor was evaluated to see if it could bring the four-bar linkage up to an angular velocity of $\dot{\theta}_2 = 266.67$ rad/sec at $\theta_2 = 1.9827$ rad, starting from rest at $\theta_2 = 3.6183$ rad, without overheating under various loads. The servomotor has the following characteristics:

$$I_m = \frac{EV}{R} + \frac{K_V}{R} \dot{\theta}_2 \qquad (5.21)$$

and

$$T_i(t) = K_T I_m - 0.1 \tag{5.22}$$

where $EV$ = step change in voltage supplied to motor = 24 V dc

$R$ = total resistance in motor circuit (ohms)

$I_m$ = current in motor circuit (amperes)

$K_V = 0.043$ V·sec/rad

$K_T = 0.3625$ lbf·in./A

0.1 = drag torque (lbf·in.)

In addition, the motor had an internal inertia of $3.125 \times 10^{-5}$ lbf·in./sec² which had to be added to $I_{A0}$, the mass moment of inertia of the input crank about center 0. The motor coils had an internal resistance of 1 Ω. Additional external resistance is added in series with the internal resistance to achieve a minimum current drain from the power supply furnishing the step change in voltage.

In this particular application, the drag force $F_D$ is directly proportional to the weight of the card stack $W_C$. The resistance, $R$, was incremented until the desired output was obtained for a given $W_C$.

Figure 5.7 shows $\theta_2$ and $\dot{\theta}_2$ for a $W_C$ of 16.5 lbf versus time in milliseconds. It can be seen that the desired output has been obtained with a total resistance of 1.52 Ω = 1 Ω internal + 0.52 Ω external.

Figure 5.8 shows the time in milliseconds and resistance required to obtain the required output for various loads.

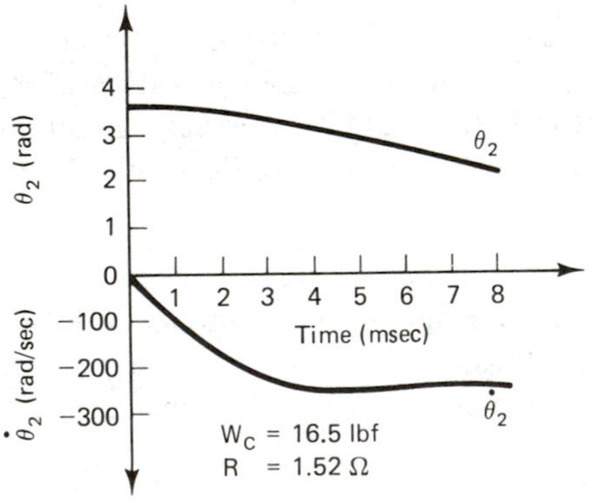


**Figure 5.7** Angular displacement and velocity using dc motor to achieve design requirements.

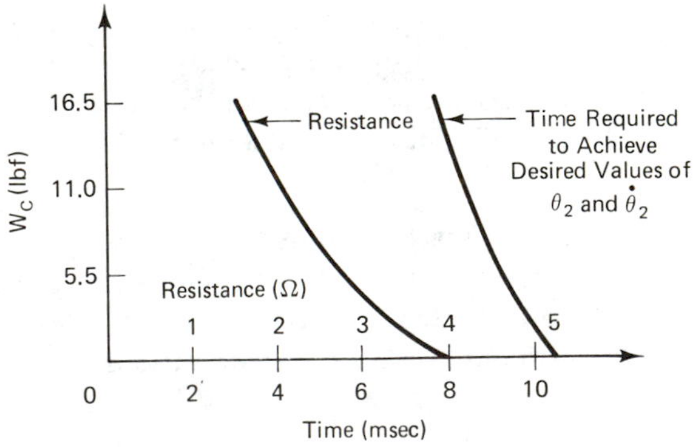


**Figure 5.8** Loads versus required resistance and time using dc motor.

## 5.4 MODIFICATION OF THE TIME RESPONSE OF MECHANISMS

In cases where a linkage has been synthesized kinematically but fails to perform adequately with respect to time, nonkinematic changes can often be made to optimize the time response of the mechanism. The first "fix" that comes to mind is a change to the mass and inertial properties of a linkage. In Sec. 5.11 to 5.18, shifting of mass within a linkage will be used to balance linkage mechanisms.

Sometimes redistribution of moving mass in a linkage will not be sufficient to produce an optimal time response—either due to the large change required or to the unsuitability of the linkage for mass adjustment. In such cases, addition of "dynamic elements" such as springs and/or dampers may be a feasible alternative. Minimizing the input torque to a machine is often a requirement of the total design. Examples of analytical techniques applied to this kind of problem can be found in the literature [15,121,178,179]. Carson [35,121] has correlated analytical and experimental results using his "time systems synthesis" techniques.

Tacheny [277] has devised a four-bar linkage model, constructed primarily of aluminum, designed to study and modify time response of the mechanism (see Figs. 5.9 and 5.10 and Table 5.2). The ground link is a two foot square by $\frac{1}{4}$ in. thick

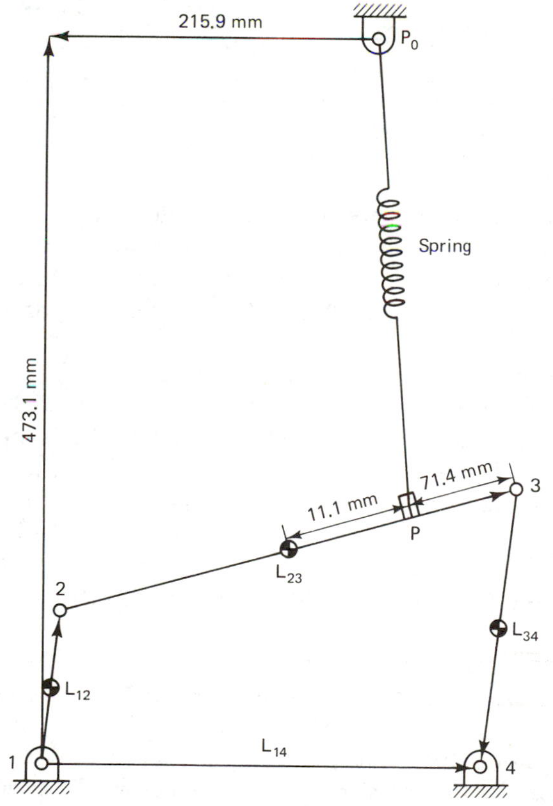


**Figure 5.9** For link dimensions, see Table 5.2.

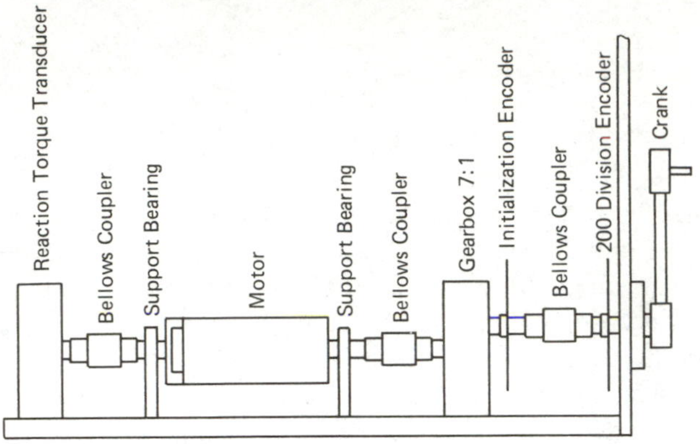


**Figure 5.10**

**TABLE 5.2**[a] LINK DIMENSIONS FOR THE FOUR-BAR MECHANISM OF FIG. 5.9

| | Link length (mm) | Distance to center of gravity (mm) from lower-numbered joint | Mass (kg) | Moment of inertia (m²·kg) |
|---|---|---|---|---|
| $L_{12}$ | 87.6 | 48.7 | 0.0680 | 0.5883 |
| $L_{23}$ | 308.0 | 154 | 0.1134 | $1.533 \times 10^{-3}$ |
| $L_{34}$ | 182.9 | 84.5 | 0.1021 | $6.41 \times 10^{-4}$ |
| $L_{14}$ | 280.4 | — | — | — |

[a] For a schematic of the linkage, see Fig. 5.9.

aluminum plate. The crank and follower are supported on steel shafts mounted in double ball bearings. The coupler is attached with double-ball-bearing pin joints to the crank and follower. The device is designed such that springs can be connected either between links or from a link to ground. In Fig. 5.9 a particular spring connection is shown.

The drive train shown in Fig. 5.10 consists of motor, gearbox, and couplers. The motor is a permanent-magnet dc type with velocity feedback control.

The maximum motor torque measured through the gearbox is 1.15 N·m (162.79 in·oz). The overall dynamic friction associated with the drive train is 0.19 N·m. A constant crank velocity will be maintained when the required torque is between −0.19 and 0.96 N·m. All shafts in the drive train are supported on ball bearings to minimize friction and to maintain alignment. The 7:1 gearbox was included to step down from the motor rpm because of the low operating velocity of the crank (≃5 Hz).

Torque measurements are made with a reaction torque sensor. The motor is supported at both ends on ball bearings and is tied to ground only through the torque sensor.

Because of the low level of output signal, it was necessary to provide shielded cables and shielded amplification for the signal prior to entering the oscilloscope

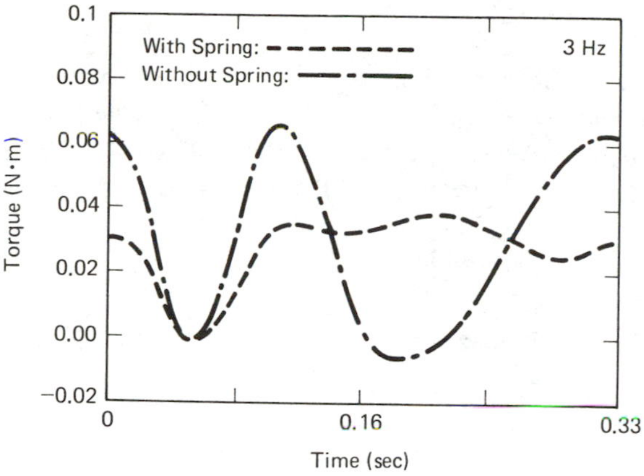


**Figure 5.11** Experimentally observed time response of the four-bar mechanism of Fig. 5.9: reaction-torque variation measured between motor frame (Fig. 5.10) and ground.

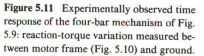

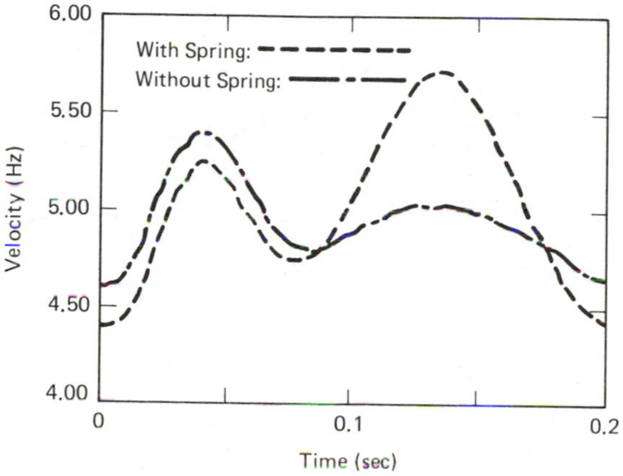


**Figure 5.12** Variation of input angular velocity of the four-bar mechanism of Fig. 5.9 at a nominal frequency of 5Hz (5 rev/sec).

for visual display. It was also necessary to pass the signal through a two-stage, low-pass filter to remove noise from the signal that resulted from mechanical resonance within the transducer.

Time-response curves were generated experimentally with and without a spring attached. The "best" spring location was determined simply by moving the grounded end of the spring (point $P_0$ in Fig. 5.9), keeping it in the plane of the linkage until the mechanism was observed to run more smoothly. Certainly this is not a sophisticated technique and most probably was not the optimal choice of the spring* (spring constant and free length) and placement (since point $P$ on the coupler was not changed), but very quickly an improved time response was observed. Figure 5.11 shows a sample experimental input torque result for one cycle of motion for 3-Hz nominal driving speed. With the spring attached, one can see the smoothing effect on the required motor torque. Figure 5.12 shows an angular velocity comparison for a nominal speed of 5 Hz.

* Analytical determination of spring parameters for torque balancing has been worked out by Matthew and Tesar [178,179]. However, time-response synthesis by analytical means in connection with problems like the one treated experimentally by Tacheny is still to be developed.

## 5.5 VIRTUAL WORK

The concept of virtual work is another useful device for solving both static and quasi-static* force-analysis problems. Recall that real work ($U$) is defined in engineering mechanics as the integral of the scalar product of the force vector (or moment vector) and the path element vector (or rotation element vector):

$$U = \int_a^b \mathbf{F} \cdot d\mathbf{s} = \int_a^b F(\cos\theta)ds \tag{5.23}$$

for a force vector **F**, where in the plane $\theta = \arg \mathbf{F} - \arg d\mathbf{s}$; or

$$U = \int_\alpha^\beta \mathbf{M} \cdot d\boldsymbol{\phi} = \int_\alpha^\beta M(\cos\gamma)d\phi \tag{5.24}$$

for a moment (couple) vector **M**, which in plane motion where **M** and $d\boldsymbol{\phi}$ are collinear, $U = M\,d\phi$, with $M$ and $d\phi$ positive or negative scalars, where

$d\mathbf{s}$ = path element vector

$\theta$ = angle between **F** and $d\mathbf{s} = \arg \mathbf{F} - \arg d\mathbf{s}$

$a, b$ = start and terminus of the path

$d\boldsymbol{\phi}$ = rotation element vector, or $d\phi$, a scalar in plane motion

$\gamma$ = angle between **M** and $d\boldsymbol{\phi}$,

$\alpha, \beta$ = start and terminus of the angle of rotation

Virtual work, however, refers to imagined work—the displacement does not actually occur—it is introduced as an imagined quantity. A virtual displacement is an infinitesimally small movement. Virtual work ($\delta U$) is again a scalar quantity:

$$\delta U = \mathbf{F} \cdot \delta\mathbf{s} = F(\cos\theta)\delta s \tag{5.25}$$

for a force and

$$\delta U = \mathbf{M} \cdot \delta\boldsymbol{\phi} = M(\cos\gamma)\delta\phi \tag{5.26}$$

or $M\,d\phi$ in plane motion for a moment, where

$\delta\mathbf{s}$ = virtual linear displacement

$\delta\boldsymbol{\phi}$ = virtual angular displacement, or $\delta\phi$ in plane motion

A mechanism with rigid components is in a state of static equilibrium if the sum of the virtual work done by all real forces and moments is zero for every virtual displacement consistent with the kinematic constraints. If elastic components are a part of the mechanical system, the total virtual work done by these elastic components is equal to the total virtual work of all real forces and moments (acting on the

* Synonymous to kineto-static.

nonelastic components) for virtual displacements consistent with the constraint. Thus for such a system

$$\sum_P (\mathbf{F}_i \cdot \delta \mathbf{s}_i) + \sum_P (\mathbf{M}_i \cdot \delta \boldsymbol{\phi}_i) + \sum_Q (\mathbf{R}_j \cdot \delta \mathbf{s}_j) + \sum_Q (\mathbf{T}_j \cdot \delta \boldsymbol{\phi}_j) = \sum_S (\mathbf{E}_k \cdot \delta \mathbf{s}_k) + \sum_S (\mathbf{K}_k \cdot \delta \boldsymbol{\phi}_k) \tag{5.27}$$

where $\mathbf{F}_i$ = active force vector at $i$

$\mathbf{M}_i$ = active moment vector at $i$

$i = 1, 2, \ldots P$ is the point of application

$\mathbf{R}_j$ = reactive force vector at $j$

$\mathbf{T}_j$ = reactive moment vector at $j$

$j = 1, 2, \ldots, Q$ is the point of reaction

$\mathbf{E}_k$ = elastic force resultant in resilient connector $k$

$\mathbf{K}_k$ = elastic moment resultant in resilient connector $k$

$k = 1, 2, \ldots, S$ is the point of attachment

$\delta \mathbf{s}_k$, $\delta \boldsymbol{\phi}_k$ = virtual deflections of resilient connectors.

Friction and inertia forces can be easily added as forces and moments in Eq. (5.27).

The following example will be used to demonstrate the application of the principle.

**Example 5.3**

Determine the equilibrium position (angle $\alpha$) of the slider-crank mechanism of Fig. 5.13a given the weight of the input and connecting links $W = 100$ N each, the external force $P = 50$ N, and $b = 40$ m. Friction is assumed negligible in this example.

**Solution** The application of Eq. (5.27) requires the introduction of virtual displacements consistent with the kinematic constraints of the mechanism. In simple cases this can be done semigraphically as shown in Fig. 5.13b and c, where the virtual angle $\delta\alpha$ introduces the virtual displacements $\delta \mathbf{s}_{12}$, $\delta \mathbf{s}_{13}$, $\delta \mathbf{s}_{14}$, $\delta \mathbf{s}_{15}$ which must satisfy the conditions imposed by the pin joints and the slider.

In general, however, it is more convenient to use the analytical approach, dealing with the differentials of the position coordinates of the active forces (forces moving through virtual displacements). Since the slider-crank mechanism of Fig. 5.13a has one degree of freedom, either the angle $\alpha$ or $\beta$, or the length $2a$ can be used as the independent variable. In terms of $\alpha$, the position coordinates of the points of application of the active forces are

$$x_{12} = \frac{c}{2}\cos\alpha, \qquad y_{12} = \frac{c}{2}\sin\alpha$$

$$x_{14} = \frac{3c}{2}\cos\alpha, \qquad y_{14} = \frac{c}{2}\sin\alpha$$

$$x_{15} = 2c\cos\alpha, \qquad y_{15} = 0$$

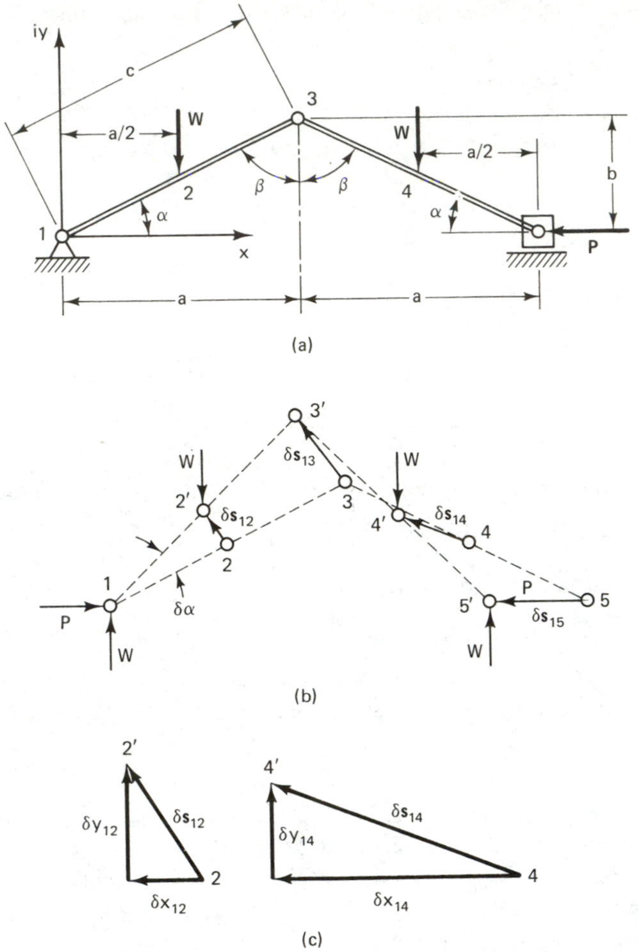


**Figure 5.13** Static equilibrium determination via virtual work: (a) loads on a slider crank; (b) virtual displacements; (c) components of virtual displacements.

The virtual displacements associated with these coordinates are

$$\delta x_{12} = -\frac{c}{2}\sin\alpha\,\delta\alpha, \qquad \delta y_{12} = \frac{c}{2}\cos\alpha\,\delta\alpha, \qquad \delta\mathbf{s}_{12} = \delta x_{12} + i\delta y_{12}$$

$$\delta x_{14} = -\frac{3c}{2}\sin\alpha\,\delta\alpha, \qquad \delta y_{14} = \frac{c}{2}\cos\alpha\,\delta\alpha, \qquad \delta\mathbf{s}_{14} = \delta x_{14} + i\delta y_{14}$$

$$\delta x_{15} = -2c\sin\alpha\,\delta\alpha, \qquad \delta y_{15} = 0, \qquad \delta\mathbf{s}_{15} = \delta x_{15} + i\delta y_{15}$$

By Eq. (5.27),

$$\underbrace{(0)\delta x_{12} + (-W)\delta y_{12}}_{\mathbf{W}\cdot\delta\mathbf{s}_{12}} + \underbrace{(0)\delta x_{14} + (-W)\delta y_{14}}_{\mathbf{W}\cdot\delta\mathbf{s}_{14}} + \underbrace{(-P)\delta x_{15} + (0)\delta y_{15}}_{\mathbf{P}\cdot\delta\mathbf{s}_{15}} = 0$$

where $W$ and $P$ are absolute values.

Substituting from above, we get

$$(-W)\left(\frac{c}{2}\right)\cos\alpha\,\delta\alpha + (-W)\left(\frac{c}{2}\right)\cos\alpha\,\delta\alpha + (-P)(-2c)\sin\alpha\,\delta\alpha = 0$$

or

$$\tan\alpha = \frac{W}{2P}$$

For $W = 100$ N and $P = 50$ N, $\tan\alpha = 1.00$, $\alpha = 45°$, and $a = 40$ m.

As an exercise for the student, add a 10 N frictional force between the slider and ground and determine the new equilibrium position of the linkage.

**Example 5.4**

A rocker-rocker type of four-bar linkage has been designed to remove a box from a conveyor belt, rotate it 90°, and place it on a second conveyor belt (see Fig. 5.14). To assist in lifting the box, a torsion spring is attached to output link 4. In an intermediate position shown in Fig. 5.14, determine the required input torque $T_2$ assuming frictionless pin joints and massless input and output links (i.e., negligible compared to the coupler). The required input angular velocity is $\omega_2 = 1$ rad/sec, and angular acceleration $\alpha_2 = 0$ rad/sec$^2$. The appropriate lengths and angles for this example are shown to scale in Figs. 5.14 and 5.15. The other given data are:

Mass of coupler: $m_3 = 20$ kg

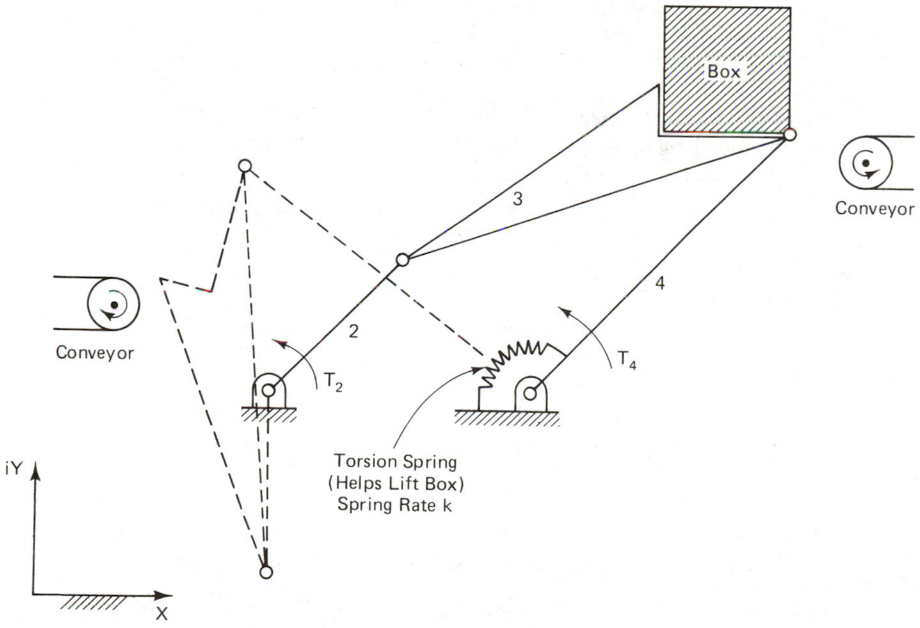


**Figure 5.14** Application of the virtual work method for dynamic equilibrium. (See Ex. 5.4.)

Mass of box: $m_b = 100$ kg

Spring rate: $k = 0.2$ N·m/deg

$\mathbf{g} = (9.81 \text{ m/sec}^2)(-i)$ is the acceleration of gravity as a complex vector in the vertical plane of the mechanism

**Solution** Using the virtual work approach [Eq. (5.23)] the expression for equilibrium is that the total virtual work of all forces and moments acting on the mechanism are zero:

$$\sum_n \mathbf{F}_n \cdot \delta \mathbf{s}_n + \sum_p \mathbf{T}_p \cdot \delta \boldsymbol{\theta}_p = 0.$$

Going to the differential and dividing by *dt:*

$$\sum_n \mathbf{F}_n \cdot \frac{d\mathbf{s}_n}{dt} + \sum_p \mathbf{T}_p \cdot \frac{d\boldsymbol{\theta}_p}{dt} = 0, \qquad \text{or} \qquad \sum_n \mathbf{F}_n \cdot \mathbf{V}_n + \sum_p \mathbf{T}_p \cdot \boldsymbol{\omega}_p = 0.$$

[See also *instantaneous power,* Eq. (5.234.)]

Accordingly:

$$T_2\omega_2 + T_{O3}\omega_3 + T_b\omega_3 + m_3\mathbf{g}\cdot\mathbf{V}_{g3} + m_b\mathbf{g}\cdot\mathbf{V}_{gb} + \mathbf{F}_{O3}\cdot\mathbf{V}_{g3} + \mathbf{F}_{Ob}\cdot\mathbf{V}_{gb} + T_4\omega_4 = 0 \tag{5.28}$$

The required torque is, therefore,

$$T_2 = -\frac{1}{\omega_2}[-I_{g3}\alpha_3\omega_3 - I_{gb}\alpha_3\omega_3 + T_4\omega_4 - m_3\mathbf{A}_{g3}\cdot\mathbf{V}_{g3} - m_b\mathbf{A}_{gb}\cdot\mathbf{V}_{gb} + m_3\mathbf{g}\cdot\mathbf{V}_{g3} + m_b\mathbf{g}\cdot\mathbf{V}_{gb}] \tag{5.29}$$

In order to determine the instantaneous input torque $T_2$ in Eq. (5.29), the velocity and acceleration terms must be determined. According to Eqs. (3.84) and (3.85), derived in Chap. 3 of Vol. 1:

$$\omega_3 = -\frac{r_2}{r_{31}}\omega_2\frac{\sin(\theta_4-\theta_2)}{\sin(\theta_4-\theta_3)} \tag{5.30}$$

and

$$\omega_4 = \frac{r_2}{r_4}\omega_2\frac{\sin(\theta_3-\theta_2)}{\sin(\theta_3-\theta_4)} \tag{5.31}$$

where $r_2 = |\mathbf{R}_2|$, $r_{31} = |\mathbf{R}_{31}|$, and $r_4 = |\mathbf{R}_4|$, and where the angles $\theta_2$, $\theta_3$, and $\theta_4$ are defined in Fig. 5.15. From Eq. (4.21) of Vol. 1 (with $\alpha_2 = 0$),

$$\alpha_3 = \frac{\omega_2^2 r_2\cos(\theta_4-\theta_2) + \omega_3^2 r_{31}\cos(\theta_4-\theta_3) - \omega_4^2 r_4}{r_{31}\sin(\theta_4-\theta_3)} \tag{5.32}$$

Using complex numbers, the position vector of the center of mass of the coupler is

$$\mathbf{Z}_{g3} = \mathbf{R}_2 + \mathbf{R}_{32} = r_2 e^{i\theta_2} + r_{32}e^{i(\theta_3+\gamma_1)} \tag{5.33}$$

The velocity of the center of mass of the coupler is

$$\mathbf{V}_{g3} = \dot{\mathbf{Z}}_{g3} = \omega_2 r_2 i e^{i\theta_2} + \omega_3 r_{32} i e^{i(\theta_3+\gamma_1)} \tag{5.34}$$

while the acceleration is

$$\mathbf{A}_{g3} = \dot{\mathbf{V}}_{g3} = \alpha_3 r_{32} i e^{i(\theta_3+\gamma_1)} - \omega_2^2 r_2 e^{i\theta_2} - \omega_3^2 r_{32} e^{i(\theta_3+\gamma_1)} \tag{5.35}$$

Similarly, for the center of mass of the box,

$$\mathbf{V}_{gb} = \omega_2 r_2 i e^{i\theta_2} + \omega_3 r_{33} i e^{i(\theta_3+\gamma_2)} \tag{5.36}$$

$$\mathbf{A}_{gb} = \alpha_3 r_{33} i e^{i(\theta_3+\gamma_2)} - \omega_2^2 r_2 e^{i\theta_2} - \omega_3^2 r_{33} e^{i(\theta_3+\gamma_2)} \tag{5.37}$$

Measurements from Fig. 5.15 yield the following dimensions:

| | | |
|---|---|---|
| $r_1 = 2.0$ m, | $\theta_1 = 0°$, | $r_{32} = 2.16$ m |
| $r_2 = 1.4$ m, | $\theta_2 = 45°$, | $r_{33} = 2.92$ m |
| $r_{31} = 3.2$ m, | $\theta_3 = 18.4°$, | $\gamma_1 = 7.1°$ |
| $r_4 = 2.8$ m, | $\theta_4 = 45°$, | $\gamma_2 = 12.6°$ |

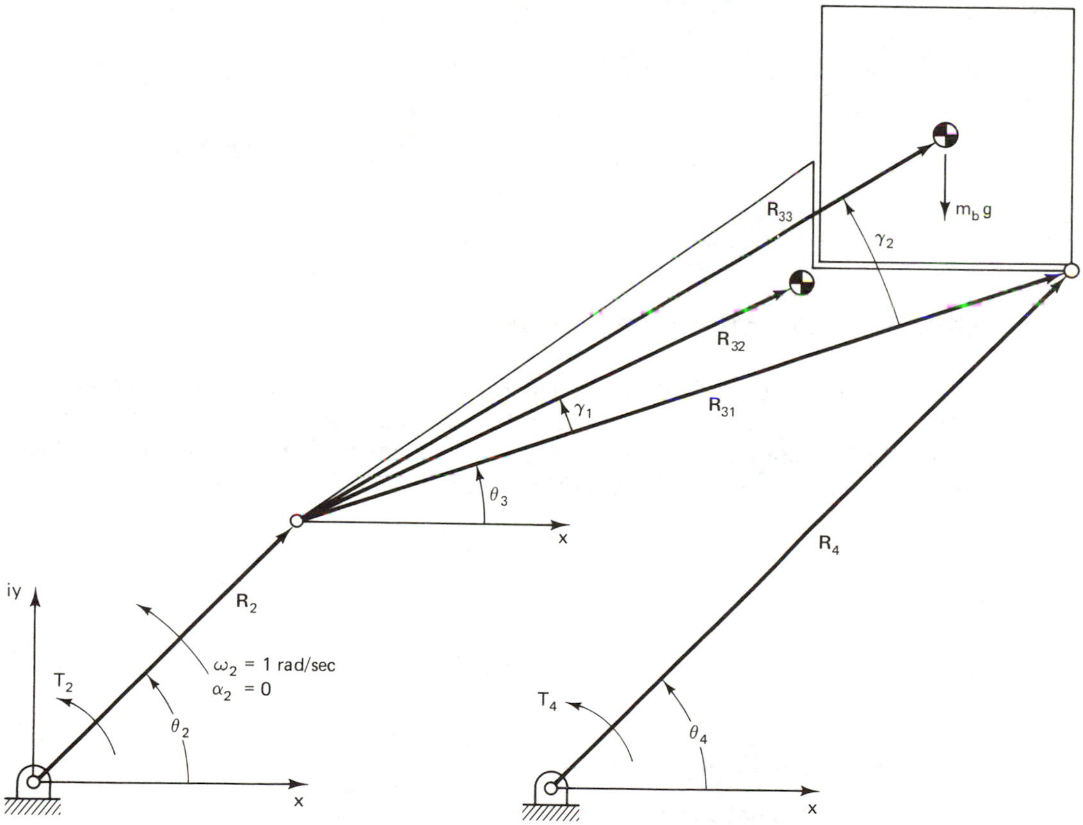


**Figure 5.15** Scaled drawing of the mechanism in Fig. 5.14. (See Ex. 5.4.)

Also given are

$$I_{g3} = 5.4 \text{ kg}\cdot\text{m}^2$$

$$I_{gb} = 16.7 \text{ kg}\cdot\text{m}^2$$

and the relaxed equilibrium (zero torque) position for the torsional spring is at $\theta_4 = 135°$.

Equations (5.30) to (5.37) can now be evaluated:

$$\omega_3 = 0$$

$$\omega_4 = 0.5 \text{ rad/sec}$$

$$\alpha_3 = 0.49 \text{ rad/sec}^2$$

$$\mathbf{V}_{g3} = \mathbf{V}_{gb} = -0.99 + 0.99i \text{ m/sec}$$

$$\mathbf{A}_{g3} = -1.51 + 0.10i \text{ m/sec}^2$$

$$\mathbf{A}_{gb} = -1.83 + 0.42i \text{ m/sec}^2$$

Also,

$$T_4 = 18 \text{ N}\cdot\text{m}$$

From Eq. (5.29) the required input torque is

$$T_2 = -\{0 + 0 + 18(0.5) - 20[-1.51(-0.99) + (0.10)(0.99)] \\ -100[-1.83(-0.99) + (0.42)(0.99)] + 20[0 + (-9.8)(0.99)] \\ + 100[0 + (-9.8)(0.99)]\}$$

$$= 1{,}410 \text{ N}\cdot\text{m ccw}$$

## 5.6 LAGRANGE'S EQUATIONS OF MOTION

Thus far we have used Newton's laws to generate equations of motion of a mechanism. The Lagrange method is also a most useful technique, especially when internal forces and reactions are not of interest. The compact form of the Lagrangian equations are given here without derivation [111,207,86]:

$$\frac{d}{dt}\left(\frac{\partial L}{\partial \dot{q}_r}\right) - \frac{\partial L}{\partial q_r} = F_{qr} \tag{5.38}$$

where

$$L = T - V \tag{5.39}$$

and

$T$ = total kinetic energy of the mechanism

$V$ = total potential energy of the mechanism

$q_r$ = generalized position coordinates

$\dot{q}_r$ = generalized velocities

$Fq_r$ = generalized force corresponding to the coordinate $q_r$

$r$ = subscript enumerating the general coordinates, one for each degree of freedom of the mechanism

Therefore, there are as many Lagrange equations of motion as there are degrees of freedom in the system. The uniform standard form of these equations holds no matter how complicated the relation between the kinematic constraints and the generalized coordinates. In fact, one can strategically choose a set of coordinates to facilitate algebraic manipulation due to the invariance of the form of the equations with respect to the choice of generalized coordinates.

The use of Lagrange's equations will be illustrated by way of an example—a double pendulum.

**Example 5.5**

Figure 5.16 shows a double pendulum in its initial position, in which the first link is displaced from the vertical by $\theta_1$ and the second by $\theta_2$. Assumptions include massless rods of length $l_1$ and $l_2$, frictionless pivots, and point masses $m_1$ and $m_2$. In this displaced initial position, mass 1 has initial velocity $\mathbf{V}_1$, while mass 2 of the second pendulum has initial velocity $\mathbf{V}_2$. The equations of motion of this two-degree-of-freedom system are to be derived.

The kinetic energy of the system is

$$T = \tfrac{1}{2}m_1(V_1)^2 + \tfrac{1}{2}m_2(V_2)^2 \tag{5.40}$$

while the potential energy is

$$\begin{aligned} V = {} & m_1 g l_1(1 - \cos\theta_1) \\ & + m_2 g[l_1(1 - \cos\theta_1) + l_2(1 - \cos\theta_2)] \end{aligned} \tag{5.41}$$

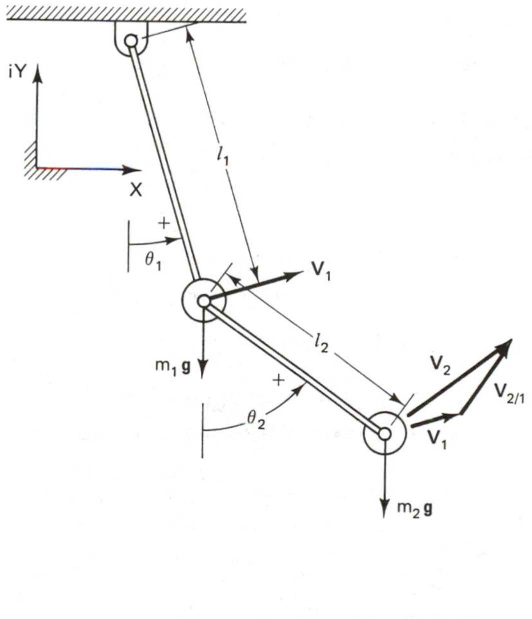


**Figure 5.16** Nomenclature for the double pendulum. (See Ex. 5.5.)

Notice that

$$\mathbf{V}_1 = l_1\dot{\theta}_1 \cos\theta_1 + il_1\dot{\theta}_1 \sin\theta_1 \tag{5.42}$$

and

$$\mathbf{V}_2 = (l_1\dot{\theta}_1 \cos\theta_1 + l_2\dot{\theta}_2 \cos\theta_2) + i(l_1\dot{\theta}_1 \sin\theta_1 + l_2\dot{\theta}_2 \sin\theta_2) \tag{5.43}$$

Replacing the velocities of Eq. (5.40) by Eqs. (5.41) and (5.42) and inserting Eqs. (5.40) and (5.41) into (5.39) yields

$$\begin{aligned} L &= T - V \\ &= \{\tfrac{1}{2}m_1(l_1\dot{\theta}_1)^2 + \tfrac{1}{2}m_2[(l_1\dot{\theta}_1 \cos\theta_1 + l_2\dot{\theta}_2 \cos\theta_2)^2 \\ &\qquad + (l_1\dot{\theta}_1 \sin\theta_1 + l_2\dot{\theta}_2 \sin\theta_2)^2]\} \\ &\qquad - \{m_1 g l_1(1 - \cos\theta_1) + m_2 g[l_1(1 - \cos\theta_1) + l_2(1 - \cos\theta_2)]\} \end{aligned} \tag{5.44}$$

The independent variables in this example are $\theta_1$ and $\theta_2$. Thus there will be two Lagrange equations:

$$\frac{d}{dt}\left(\frac{\partial L}{\partial \dot{\theta}_1}\right) - \frac{\partial L}{\partial \theta_1} = F_{\mathrm{EXT1}} = 0 \tag{5.45}$$

$$\frac{d}{dt}\left(\frac{\partial L}{\partial \dot{\theta}_2}\right) - \frac{\partial L}{\partial \theta_2} = F_{\mathrm{EXT2}} = 0 \tag{5.46}$$

where $F_{\mathrm{EXT1}}$ and $F_{\mathrm{EXT2}}$ are the external forces associated with $\theta_1$ and $\theta_2$, respectively, both zero in this case. After taking derivatives and canceling terms, these equations reduce to

$$\begin{aligned} &(m_1 + m_2)l_1\ddot{\theta}_1 + m_2 l_2 \cos(\theta_2 - \theta_1)\ddot{\theta}_2 \\ &\qquad - m_2 l_2 \sin(\theta_2 - \theta_1)(\dot{\theta}_2)^2 + (m_1 + m_2)g \sin\theta_1 = 0 \end{aligned} \tag{5.47}$$

and

$$m_2 l_2\ddot{\theta}_2 + m_2 l_2 \cos(\theta_2 - \theta_1)\ddot{\theta}_1 + m_2 l_1 \sin(\theta_2 - \theta_1)(\dot{\theta}_1)^2 + m_2 g \sin\theta_2 = 0 \tag{5.48}$$

These two equations of motion can be simplified by using a small-angle assumption:

(1) $\sin\theta \simeq \theta$ (2% error at $\theta = 20°$)
(2) $\theta_2 \simeq \theta_1 \rightarrow \sin(\theta_2 - \theta_1) \simeq 0$
(3) $\cos(\theta_2 - \theta_1) \simeq 1$

With these, Eqs. (5.47) and (5.48) reduce to

$$(m_1 + m_2)l_1\ddot{\theta}_1 + m_2 l_2\ddot{\theta}_2 + (m_1 + m_2)g\theta_1 = 0 \tag{5.49}$$

$$m_2 l_1\ddot{\theta}_1 + m_2 l_2\ddot{\theta}_2 + m_2 g\theta_2 = 0 \tag{5.50}$$

A copy of a flow chart and the coding of a computer program written for this example* is given in Fig. 5.17 and Table 5.3. The Runge–Kutta method (Sec. 5.3) is used to

* Contributed by Wm. Dahlof.

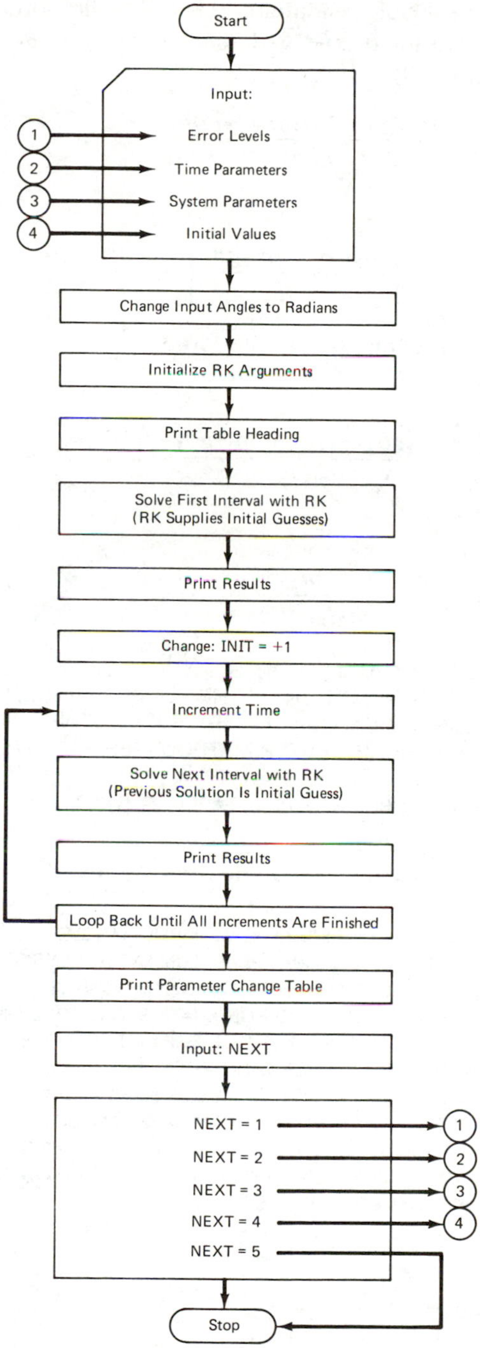


**Figure 5.17** Flowchart for program "RKPEND": Analysis of double pendulum of Fig. 5.16 by the Runge-Kutta method. (See Table 5.3 for listing and Table 5.4 for output.)

solve for the time response of the double pendulum. Therefore, first-order differential equations are required. To solve for $\ddot{\theta}_1$ and $\ddot{\theta}_2$ in succession, first $\ddot{\theta}_2$ is eliminated by subtracting Eq. (5.50) from (5.49). Thus

$$\ddot{\theta}_1 = \frac{m_2 g \theta_2 - (m_1 + m_2) g \theta_1}{m_1 l_1} \tag{5.51}$$

Then, from Eq. (5.50),

$$\ddot{\theta}_2 = -\left[\frac{m_2 l_1 \ddot{\theta}_1 + m_2 g \theta_2}{m_2 l_2}\right] \tag{5.52}$$

Equations (5.51) and (5.52) are programmed in the form

$$\dot{\theta} = \phi_1$$

$$\dot{\phi}_1 = \frac{m_2 g \theta_2 - (m_1 + m_2) g \theta_1}{m_1 l_1}$$

$$\dot{\theta}_2 = \phi_2$$

$$\dot{\phi}_2 = \frac{m_2 l_1 \dot{\phi}_1 + m_2 g \theta_2}{m_2 l_2}$$

**TABLE 5.3** LISTING OF PROGRAM FOR INTEGRATING THE DIFFERENTIAL EQUATIONS OF MOTION OF THE DOUBLE PENDULUM OF FIG. 5.16. THE PROGRAM CALLS A LIBRARY SUBROUTINE "RK" WHICH IS THE STANDARD RUNGE–KUTTA METHOD (SEE SEC. 5.3)

```
      PROGRAM RKPEND (INPUT, OUTPUT, TAPE5=OUTPUT)

C
C*****************************************************************
C
C           RKPEND CALCULATES THE KINEMATIC VARIABLES (ANGLE,
C VELOCITY, & ACCELERATION) OF THE DUAL PENDULUM SYSTEM DESCRIBED
C IN FIG. 5.16 OF THE TEXT.  THE TWO DIFFERENTIAL EQUATIONS OF
C MOTION ARE SOLVED USING A LIBRARY RUNGE-KUTTA SUBROUTINE (RK),
C THE USER MUST SPECIFY: 1) THE ERROR LEVELS (ERR1 & ERR2) FOR
C RK 2) THE TIME INTERVAL & THE NUMBER OF INTERVALS (T=0
C INITIALLY) 3) THE PHYSICAL PARAMETERS (MASS & LENGTH) AND 4)
C THE INITIAL VALUES (THETA1, THETA1 DOT, THETA2, & THETA2 DOT).
C RKPEND WILL THEN PRINT A TABLE OF RESULTS FOR T=0 TO T=(INTERVAL
C * # OF INTERVALS) AND PROVIDE A LOOPING MECHANISM TO ALLOW THE
C USER TO CHANGE ANY PARAMETERS IF DESIRED.
C                                               By Wm. Dahlof—1979
C
C
C
      EXTERNAL PEND, TABLE
      COMMON RM(2), RL(2)
      DIMENSION TH(4), ERR1(4), RTH(4), THPRIME(4), W(16)
C
```

**TABLE 5.3 Continued**

```
C
C INPUT INFORMATION
  10     PRINT, 'INPUT ERROR LEVELS: ERR2 & ERR1(1)'
         READ, ERR2, ERR1(1)
         ERR1(2)=ERR1(3)=ERR1(4)=ERR1(1)
  20     PRINT, 'INPUT TIME INTERVAL & NUMBER OF INTERVALS'
         READ, TINCRE, NINCRE
  30     PRINT, 'INPUT M1, M2, L1, L2'
         READ,RM(1),RM(2),RL(1),RL(2)
  40     PRINT, 'INPUT INITIAL VALUES: TH1, DTH1, TH2, DTH2'
+         '(DEG & DEG/SEC)'
         READ, TH
C
C
C INITIALIZE VARIABLES
         DO 50 I = 1,4
               RTH(I) = TH(I)/57.29578
  50     CONTINUE
         N = 4
         T = 0.0
         TFINAL = TINCRE
         INIT = +1
C
C
C PRINT TABLE HEADING & INITIAL VALUES
         WRITE (5,52)
  52     FORMAT (3/,
+         35H TIME      THETA1       OMEGA1       ALPHA1
+         30H      THETA2       OMEGA2       ALPHA2)
         WRITE (5,54)
  54     FORMAT (
+        35H(SEC)       (DEG)  (RAD/SEC) (RAD/SEC2)
+        30H       (DEG)  (RAD/SEC)(RAD/SEC2) , / )
         WRITE(5,56) T, TH(1), TH(2), T, TH(3), TH(4), T
  56     FORMAT ( F5.2, F10.1, 2(F10.5), F10.1, 2(F10.5) )
C
C
C SOLVE EQUATIONS & PRINT RESULTS FOR T=(0 + TINCRE)
         CALL RK(T,TFINAL,RTH,THPRIME,PEND,N,ERR2,ERR1,INIT,W,STEP)
         CALL TABLE (T,RTH,THPRIME)
C
C
C SOLVE & PRINT RESULTS FOR REMAINING TIME VALUES
         INIT = -1
         DO 60 I = 2,NINCRE
               TFINAL = T + TINCRE
         CALL RK(T,TFINAL,RTH,THPRIME,PEND,N,ERR2,ERR1,INIT,W,STEP)
               CALL TABLE(T,RTH,THPRIME)
  60     CONTINUE
C
C
```

**TABLE 5.3 Continued**

```
C PARAMETER CHANGE COMMANDS
      WRITE(5,70)
   70 FORMAT (3/)
      PRINT,'WHAT NEXT?                          TYPE'
      PRINT,'CHANGE ERROR LEVELS                    1'
      PRINT,'CHANGE TIME PARAMETERS                 2'
      PRINT,'CHANGE SYSTEM PARAMETERS               3'
      PRINT,'CHANGE INITIAL VALUES                  4'
      PRINT,'END                                    5'
      READ,NEXT
      GO TO (10,20,30,40,80)NEXT
C
C
   80 STOP
      END
C
C*****************************************************************
C SUBROUTINE FOR PENDULUM EQUATIONS
C
C
      SUBROUTINE PEND(T,RTH,THPRIME)
      COMMON RM(2), RL(2)
      DIMENSION RTH(4), THPRIME(4)
C
C
      THPRIME(1) = RTH(2)
      THPRIME(2) = 9.8*(RM(2)*(RTH(3) - (RM(1)+(RM(2))*RTH(1))/
     +              (RM(1)*RL(1))
      THPRIME(3) = RTH(4)
      THPRIME(4) = -(RM(2)*RL(1)*THPRIME(2) + RM(2)*9.8*RTH(3))
     +             /(RM(2)*RL(2))
C
      RETURN
      END
C
C*****************************************************************
C SUBROUTINE TO PRINT TABLE
C
C
      SUBROUTINE TABLE(T,RTH,THPRIME)
      DIMENSION RTH(4), THPRIME(4), DTH(4), DTHPRIM(4)
C
C
  CHANGE ANGLES TO DEGREES
      DTH(1) = RTH(1) * 57.296
      DTH(3) = RTH(3) * 57.296
C
      WRITE(5,210) T, DTH(1), RTH(2), THPRIME(2), DTH(3), RTH(4),
     +              THPRIME(4)
  210 FORMAT(F5.2, F10.1, 2(F10.5), F10.1, 2(F10.5) )
      RETURN
      END
```

Program RKPEND calculates the kinematic variables (angle, angular velocity) of the double pendulum shown in Fig. 5.16. The Runge–Kutta subroutine (RK) is called by the program. The user will have to investigate the requirements of the differential-equation-solver subroutine available to determine the requirements of input and output to that subroutine. A flowchart is given in Fig. 5.17. The program listed in Table 5.3 was written for a CDC Cyber 74 in Minnesota FORTRAN. The first 11 lines give user instructions.

Tables 5.4 and 5.5 show two examples. Both input and output data are shown in Table 5.4, but only output is listed in Table 5.5. Input for both is shown in Table 5.6.

**TABLE 5.4**

```
MNF        PROGRAM        RKPEND

 INPUT ERROR LEVELS: ERR2 & ERR1(1)
? 1.0E-5, 1.0E-7
 INPUT TIME INTERVAL & TIME SPAN
? 0.1, 20
 INPUT M1, M2, L1, L2
? 2, 0.01, 1, 0.01
 INPUT INITIAL VALUES: TH1, DTH1, TH2, DTH2
? 10, 0, 10, 0
```

| t TIME (SEC) | $\theta_1$ THETA1 (DEG) | $\dot{\theta}_1$ OMEGA1 (RAD/SEC) | $\ddot{\theta}_1$ ALPHA1 (RAD/SEC2) | $\theta_2$ THETA2 (DEG) | $\dot{\theta}_2$ OMEGA2 (RAD/SEC) |
|---|---|---|---|---|---|
| 0 | 10.0 | 0 | 0 | 10.0 | 0 |
| .10 | 9.5 | −.16825 | −1.62713 | 9.7 | −.16977 |
| .20 | 8.1 | −.32015 | −1.38602 | 8.1 | −.32373 |
| .30 | 5.9 | −.44094 | −1.00986 | 6.1 | −.44484 |
| .40 | 3.1 | −.51886 | −.53588 | 3.1 | −.52478 |
| .50 | .1 | −.54635 | −.00947 | .2 | −.55095 |
| .60 | −3.0 | −.52074 | .51752 | −3.2 | −.52704 |
| .70 | −5.8 | −.44451 | .99454 | −5.8 | −.44774 |
| .80 | −8.0 | −.32508 | 1.37456 | −8.2 | −.32977 |
| .90 | −9.5 | −.17405 | 1.62131 | −9.5 | −.17420 |
| 1.00 | −10.0 | −00611 | 1.71014 | −10.2 | −.00794 |
| 1.10 | −9.5 | .16243 | 1.63309 | −9.5 | .16602 |
| 1.20 | −8.2 | .31519 | 1.39696 | −8.4 | .31622 |
| 1.30 | −6.0 | .43730 | 1.02539 | −6.0 | .44400 |
| 1.40 | −3.2 | .51691 | .55382 | −3.4 | .51962 |
| 1.50 | −.2 | .54628 | .02876 | −.1 | .55444 |
| 1.60 | 2.9 | .52256 | −.49944 | 2.8 | .52497 |
| 1.70 | 5.7 | .44804 | −.97875 | 5.9 | .45556 |
| 1.80 | 8.0 | .32997 | −1.36327 | 8.0 | .33009 |
| 1.90 | 9.4 | .17983 | −1.61495 | 9.6 | .18501 |
| 2.00 | 10.0 | .01221 | −1.71000 | 10.0 | .00878 |

**TABLE 5.5**

| t TIME (SEC) | $\theta_1$ THETA1 (DEG) | $\dot{\theta}_1$ OMEGA1 (RAD/SEC) | $\ddot{\theta}_1$ ALPHA1 (RAD/SEC2) | $\theta_2$ THETA2 (DEG) | $\dot{\theta}_2$ OMEGA2 (RAD/SEC) | $\ddot{\theta}_2$ ALPHA2 (RAD/SEC2) |
|---|---|---|---|---|---|---|
| 0 | 10.0 | 0 | 0 | 10.0 | 0 | 0 |
| .10 | 9.5 | −.16554 | −1.54688 | 10.0 | −.00548 | −.16219 |
| .20 | 8.2 | −.29993 | −1.10287 | 9.9 | −.04131 | −.58676 |
| .30 | 6.2 | −.38089 | −.50379 | 9.4 | −.12612 | −1.10803 |
| .40 | 3.9 | −.40095 | .08672 | 8.3 | −.25906 | −1.51315 |
| .50 | 1.7 | −.36891 | .51715 | 6.4 | −.41850 | −1.61243 |
| .60 | −.3 | −.30600 | .69575 | 3.6 | −.56788 | −1.30526 |
| .70 | −1.8 | −.23842 | .61741 | −.0 | −.66668 | −.61630 |
| .80 | −3.0 | −.18847 | .36255 | −3.9 | −.68321 | .30758 |
| .90 | −4.0 | −.16726 | .06769 | −7.7 | −.60482 | 1.24113 |
| 1.00 | −5.0 | −.17143 | −.12258 | −10.7 | −.44248 | 1.95042 |
| 1.10 | −6.0 | −.18477 | −.10446 | −12.6 | −.22803 | 2.26350 |
| 1.20 | −7.1 | −.18440 | .14841 | −13.3 | −.00502 | 2.12368 |
| 1.30 | −8.1 | −.14924 | .57430 | −12.7 | .18398 | 1.60585 |
| 1.40 | −8.7 | −.06809 | 1.04245 | −11.3 | .30948 | .89010 |
| 1.50 | −8.8 | .05543 | 1.39597 | −9.3 | .36289 | .20140 |
| 1.60 | −8.0 | .20277 | 1.50142 | −7.3 | .35740 | −.26079 |
| 1.70 | −6.4 | .34522 | 1.29499 | −5.3 | .32203 | −.38832 |
| 1.80 | −4.1 | .45218 | .80420 | −3.6 | .29054 | −.19614 |
| 1.90 | −1.4 | .50022 | .14078 | −1.9 | .28925 | .18636 |
| 2.00 | 1.5 | .48002 | −.53175 | −.2 | .32752 | .55976 |
| 2.10 | 4.0 | .39908 | −1.04951 | 1.9 | .39420 | .72525 |
| 2.20 | 6.0 | .27920 | −1.29776 | 4.4 | .46118 | .55297 |
| 2.30 | 7.2 | .14966 | −1.24538 | 7.1 | .49292 | .02879 |
| 2.40 | 7.7 | .03824 | −.95209 | 9.9 | .45877 | −.73711 |
| 2.50 | 7.7 | −.03699 | −.54661 | 12.2 | .34413 | −1.54261 |
| 2.60 | 7.4 | −.07250 | −.18289 | 13.7 | .15667 | −2.15677 |
| 2.70 | 6.9 | −.07929 | .01127 | 13.9 | −.07451 | −2.39314 |
| 2.80 | 6.5 | −.07799 | −.02368 | 12.8 | −.30648 | −2.16968 |
| 2.90 | 6.0 | −.09083 | −.25986 | 10.5 | −.49463 | −1.53576 |
| 3.00 | 5.4 | −.13321 | −.59198 | 7.3 | −.60541 | −.65742 |
| 3.10 | 4.4 | −.20753 | −.87308 | 3.7 | −.62566 | .23412 |
| 3.20 | 3.0 | −.30149 | −.96526 | .3 | −.56563 | .91572 |
| 3.30 | 1.0 | −.39143 | −.78583 | −2.6 | −.45460 | 1.23885 |
| 3.40 | −1.5 | −.44963 | −.33863 | −4.9 | −.33087 | 1.17600 |
| 3.50 | −4.1 | −.45321 | .28486 | −6.5 | −.22914 | .82402 |
| 3.60 | −6.5 | −.39184 | .93182 | −7.6 | −.16955 | .36872 |
| 3.70 | −8.5 | −.27162 | 1.43587 | −8.5 | −.15185 | .01923 |
| 3.80 | −9.6 | −.11385 | 1.66694 | −9.4 | −.15663 | −.06135 |
| 3.90 | −9.7 | .05084 | 1.57338 | −10.3 | −.15305 | .18597 |
| 4.00 | −9.0 | .19129 | 1.19656 | −11.1 | −.11055 | .69615 |
| 4.10 | −7.6 | .28468 | .65691 | −11.4 | −.01063 | 1.30014 |
| 4.20 | −5.9 | .32261 | .11493 | −11.1 | .14532 | 1.78008 |
| 4.30 | −4.0 | .31262 | −.28022 | −9.7 | .33467 | 1.94137 |
| 4.40 | −2.3 | .27458 | −.43767 | −7.3 | .51930 | 1.67827 |
| 4.50 | −.9 | .23328 | −.35226 | −3.9 | .65670 | 1.01111 |
| 4.60 | .4 | .20964 | −.10359 | .1 | .71275 | .08381 |
| 4.70 | 1.6 | .21350 | .17284 | 4.1 | .67225 | −.87917 |
| 4.80 | 2.9 | .24028 | .33272 | 7.6 | .54368 | −1.64102 |
| 4.90 | 4.3 | .27267 | .27351 | 10.2 | .35670 | −2.02621 |
| 5.00 | 5.9 | .28674 | −.03024 | 11.7 | .15322 | −1.97191 |

**TABLE 5.6**

| | Table 5.4 | Table 5.5 |
|---|---|---|
| Time interval | 0.1 sec | 0.1 sec |
| Time span | 20 sec | 50 sec |
| $m_1, m_2$ | 2, 0.01 kg. | 2, 2 kg |
| $\ell_1, \ell_2$ | 1, 0.01 m | 1, 1 m |
| $\theta_1, \dot{\theta}_1$ | 10°, 0 | 10°, 0 |
| $\theta_2, \dot{\theta}_2$ | 10°, 0 | 10°, 0 |

## Case Study: Analysis of Vibratory Feed Bowls [93]

Vibratory feed bowls (or vibratory feeders) feed parts into an assembly station. They are designed with parameters based on past experience. Examples of these parameters include the bowl diameter, base weight, track angle, spring angle, spring constant, and drive frequency. Many hours of effort are usually spent experimentally (and in some cases analytically) developing designs to build a feed bowl for a given part.

This design example presents a mathematical model (based on Lagrange equations) of the feed bowl's suspension system, a spatial mechanism. The model calculates the natural frequencies of a given feed bowl. Once a bowl is designed for a given part on the basis of past experience, this mathematical model may be helpful in selecting parameter values to optimize a feature of the feed bowl. For example, one may need to minimize the motion of the base of the feed bowl or the level of parts inside the bowl. The design of bowls for a given part is not covered here.

The vibratory feed bowl consists of a helical track welded to the inside wall of a bowl (see Figs. 5.18, 5.19, and 5.20). The bowl is supported by inclined leaf

**Figure 5.18** Photograph of the feed bowl used.

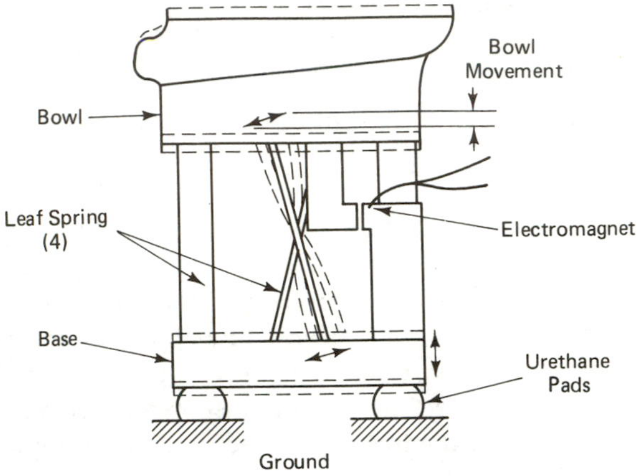


**Figure 5.19** Exaggerated motion of the feed bowl.

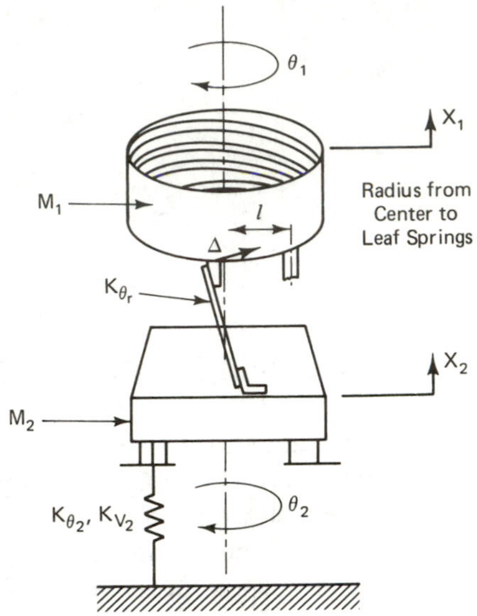


**Figure 5.20** Feed bowl model.

springs attached to a heavy base (see Fig. 5.19). An electromagnet, placed horizontally and attached to the base, periodically pulls against a flat plate attached to the bowl. The result is a cyclic "spiral up/spiral down" oscillating-screw motion. The movements of the track are small compared with the bowl diameter and the length of the leaf springs. As the bowl moves, parts in the bowl are vibrated up the track through selectors. The "spiral-up" motion imparts near uniform velocity to the parts traveling up with the bowl. The "spiral-down" movement is quick enough, however, to leave the parts behind in the up position due to slipping (or parts becoming airborne). The selectors allow only correctly oriented parts to pass out of the bowl. The misoriented parts are forced off the track and drop back into the bottom of the bowl.

The Lagrange method is used to derive the equations of motion of the feed bowl. Figures 5.20 and 5.21 show a model of the feed bowl with displacements (linear and angular) identified by $X_1$, $X_2$, $\theta_r = (\theta_1 - \theta_2)$, $\theta_2$, and $\Delta$. Using the energy method, the following equations are valid.

Kinetic energy:

$$T = \tfrac{1}{2}M_2\dot{X}_2^2 + \tfrac{1}{2}M_1\dot{X}_1^2 + \tfrac{1}{2}I_2\dot{\theta}_2^2 + \tfrac{1}{2}I_1[\dot{\theta}_2 + \dot{\theta}_r]^2 \tag{5.53}$$

Potential energy:

$$V = \tfrac{1}{2}K_{\theta_2}\theta_2^2 + \tfrac{1}{2}K_{V_2}X_2^2 + \tfrac{1}{2}K_{\theta_r}\Delta^2 \tag{5.54}$$

Using the following kinematic constraints derived from Figs. 5.20 and 5.21, $X_1$ and

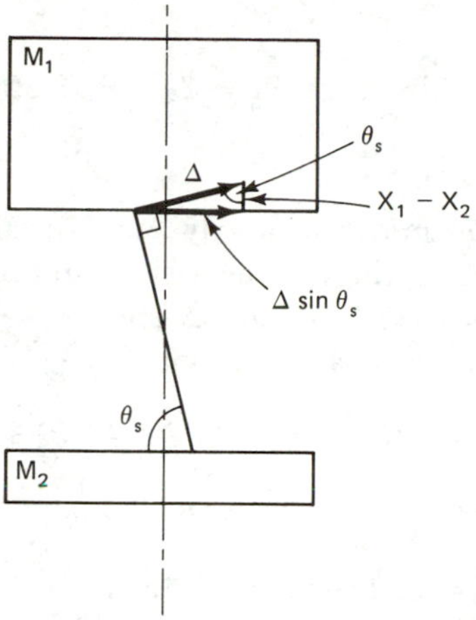


**Figure 5.21** Geometric constraint diagram.

**TABLE 5.7** PHYSICAL PARAMETERS OF THE FEEDBOWL OF FIGS. 5.18, 5.19, 5.20, AND 5.21.

| | |
|---|---|
| Mass of bowl $M_1$ | $= 31.15$ lb-mass |
| Mass of base $M_2$ | $= 95.65$ lb-mass |
| Moment of inertia of bowl $I_1$ | $= 1001.22$ lb·in² |
| Moment of inertia of base $I_2$ | $= 2159.04$ lb·in² |
| Vertical spring constant of urethane pads $K_{v_2}$ | $= 105{,}000$ lbf/in. |
| Tangential spring constant of urethane pads $K_{\theta_2}$ | $= 366{,}000$ lbf/in. |
| Tangential spring constant of leaf springs $K_{\theta_r}$ | $= 627{,}000$ lbf/in. |
| Leaf spring angle $= \theta_s$ | $= 76°$ |
| Radius from center to leaf springs $l$ | $= 3.35$ in. |

$\Delta$ will be eliminated, leaving $X_2$, $\theta_r$, and $\theta_2$ as the primary degrees of freedom of the system (see Table 5.7 for values).

$$X_1 - X_2 = \Delta \cos \theta_s$$

$$\theta_r l = \Delta \sin \theta_s$$

$$\Delta = \frac{\theta_r l}{\sin \theta_s} = \frac{\theta_r(3.35'')}{\sin 76°},$$

where $l$ is the radius from center of bowl to leaf springs

$$X_1 = X_2 + \Delta \cos \theta_s$$

$$X_1 = X_2 + 0.835\theta_r \qquad [\text{because } X_1 - X_2 = \theta_r(3.35) \cot 76°]$$

$$\dot{X}_1 = \dot{X}_2 + 0.835\dot{\theta}_r$$

It is assumed that the bowl moves in nearly pure spiral motion about its center. Also, the twist of the leaf springs is neglected and motion is assumed small. Substituting $(3.35/\sin 76°)\theta_r = (3.45\ \theta_r)$ for $\Delta$ and $(X_2 + 0.835\theta_r)$ for $X_1$ into Eqs. (5.53) and (5.54), the following equations result:

$$T = \tfrac{1}{2}M_2\dot{X}_2^2 + \tfrac{1}{2}M_1(0.835\dot{\theta}_r + \dot{X}_2)^2 + \tfrac{1}{2}I_2\dot{\theta}_2^2 + \tfrac{1}{2}I_1[\dot{\theta}_r + \dot{\theta}_2]^2$$
$$V = \tfrac{1}{2}K_{\theta_2}\theta_2^2 + \tfrac{1}{2}K_{V_2}X_2^2 + \tfrac{1}{2}K_{\theta_r}(3.45\ \theta_r)^2 \qquad (5.55)$$

Assuming a conservative system $L = T - V$, then

$$L = [\tfrac{1}{2}M_2\dot{X}_2^2 + \tfrac{1}{2}M_1(0.698\dot{\theta}_r^2 + 1.67\dot{X}_2\dot{\theta}_r + \dot{X}_2^2) + \tfrac{1}{2}I_2\dot{\theta}_r^2$$
$$+ \tfrac{1}{2}I_1(\dot{\theta}_2^2 + 2\dot{\theta}_2\dot{\theta}_r + \dot{\theta}_r^2)] - [\tfrac{1}{2}K_{\theta_2}\theta_2^2 + \tfrac{1}{2}K_{V_2}X_2^2 + \tfrac{1}{2}K_{\theta_r}(3.45\theta_r)^2] = 0 \qquad (5.56)$$

The partial derivatives are

$$\frac{\partial L}{\partial \dot{X}_2} = M_2\dot{X}_2 + 0.835M_1\dot{\theta}_r + M_1\dot{X}_2 \qquad (5.57)$$

$$\frac{\partial L}{\partial \dot{\theta}_2} = I_1\dot{\theta}_2 + I_1\dot{\theta}_r \qquad (5.58)$$

$$\frac{\partial L}{\partial \dot{\theta}_r} = 0.698M_1\dot{\theta}_r + 0.835M_1\dot{X}_2 + I_1\dot{\theta}_2 + (I_1 + I_2)\dot{\theta}_r \qquad (5.59)$$

$$\frac{\partial L}{\partial X_2} = -K_{V_2}X_2, \qquad \frac{\delta L}{\delta \theta_2} = -K_{\theta_2}\theta_2 \qquad (5.60)$$

$$\frac{\partial L}{\partial \theta_r} = -11.9K_{\theta_r}\theta_r \qquad (5.61)$$

Combining terms and taking time derivatives yields

$$\frac{d}{dt}[(M_1 + M_2)\dot{X}_2 + 0.835M_1\dot{\theta}_r] + K_{V_2}X_2 = 0 \qquad (5.62)$$

$$\frac{d}{dt}[(I_1)\dot{\theta}_2 + I_1\dot{\theta}_r] + K_{\theta_2}\theta_2 = 0 \qquad (5.63)$$

$$\frac{d}{dt}[(0.698M_1 + I_1 + I_2)\dot{\theta}_r + 0.835M_1\dot{X}_2 + I_1\dot{\theta}_2] + 11.9K_{\theta_r}\theta_r = 0 \qquad (5.64)$$

Equations (5.62) to (5.64) may be written

$$\begin{vmatrix} M_1+M_2 & 0 & 0.835M_1 \\ 0 & I_1 & I_1 \\ 0.835M_1 & I_1 & 0.697M_1+I_1+I_2 \end{vmatrix} \begin{vmatrix} \ddot{X}_2 \\ \ddot{\theta}_2 \\ \ddot{\theta}_r \end{vmatrix} + \begin{vmatrix} K_{V2} & 0 & 0 \\ 0 & K_{\theta_2} & 0 \\ 0 & 0 & 11.9K_{\theta_r} \end{vmatrix} \begin{vmatrix} X_2 \\ \theta_2 \\ \theta_r \end{vmatrix} = 0 \tag{5.65}$$

Equation (5.65) is the mathematical model of the feed bowl of Table 5.7.

The displacement coordinates of the feed bowl were so chosen that the spring matrix would be uncoupled, making matrix inversion economical. To find the eigenvalues of this system, the technique outlined in Ref. 123 was used as follows.*

Let the mass matrix $= M$, the stiffness matrix $= K$, and let the damping matrix $B = K^{-1}M$. Then $B\boldsymbol{\nu} = \mu\boldsymbol{\nu}$ where $\boldsymbol{\nu}$ represents the eigenvectors and $\mu$ represents the three eigenvalues of $B$. Also, $\mu = 1/\omega^2$, where $\omega$ represents the natural frequencies.

A computer program was written in FORTRAN which reads the $M$ and $K$ matrices. The program inverts $K$ and premultiples $M$ by $K^{-1}$ to find $B$. It then calls up a library function which finds the three eigenvalues of $B$ and then calculates the natural frequencies using $\omega = \sqrt{1/\mu}$.

* A brief refresher on eigenvalues and eigenvectors is as follows. Let

$$B = \begin{bmatrix} 0 & 1 & 0 \\ 0 & 0 & 1 \\ -8 & -14 & -7 \end{bmatrix}$$

The eigenvalues of $B$ are defined by

$$|B - \mu I| = 0, \text{ or}$$

$$\begin{vmatrix} 0 & 1 & 0 \\ 0 & 0 & 1 \\ -8 & -14 & -7 \end{vmatrix} - \begin{vmatrix} \mu & 0 & 0 \\ 0 & \mu & 0 \\ 0 & 0 & \mu \end{vmatrix} = 0$$

$$\begin{vmatrix} -\mu & 1 & 0 \\ 0 & -\mu & 1 \\ -8 & -14 & -7-\mu \end{vmatrix} = \mu^3 + 7\mu^2 + 14\mu + 8 = 0$$

The eigenvalues are the roots $\mu_1 = -1$, $\mu_2 = -2$, $\mu_3 = -4$. The eigenvectors $\boldsymbol{\nu}_i$, $i = 1, 2, 3$ are defined by $B\boldsymbol{\nu}_i = \mu_i\boldsymbol{\nu}_i$. For $\mu_i = \mu_1 = -1$:

$$\begin{bmatrix} 0 & 1 & 0 \\ 0 & 0 & 1 \\ -8 & -14 & -7 \end{bmatrix} \begin{bmatrix} \nu_{11} \\ \nu_{12} \\ \nu_{13} \end{bmatrix} = \begin{bmatrix} -\nu_{11} \\ -\nu_{12} \\ -\nu_{13} \end{bmatrix} \Rightarrow \begin{array}{l} \nu_{12} = -\nu_{11}, \\ \nu_{13} = -\nu_{12} \\ -8\nu_{11} - 14\nu_{12} - 7\nu_{13} = -\nu_{13} \end{array} \Rightarrow$$

$$\nu_{11} + \nu_{12} = 0$$

$$\nu_{12} + \nu_{13} = 0$$

$$-8\nu_{11} - 14\nu_{12} - 6\nu_{13} = 0$$

This is a homogeneous set of equations linear in $\nu_{1j}$, $j = 1, 2, 3$. Therefore we can assume one $\nu_{1j}$ arbitrarily. Letting $\nu_{11} = 1$, then $\nu_{12} = -1$ and $\nu_{13} = 1$. Substituting these in the third equation checks these values. Thus: $\boldsymbol{\nu}_1 = \{1, -1, 1\}$. Similarly we can find $\boldsymbol{\nu}_2 = \{1, -2, 4\}$ and $\boldsymbol{\nu}_3 = \{1, -4, 16\}$ (Example based on Ref. 57.)

**Results.** The feed bowl used in verifying the mathematical model was a 12-in.-diameter "Moorefeed" feed bowl. It was designed to feed $\frac{1}{4}$-in.$^2$ electrical contacts, each weighing a few grams. The drive unit is an electromagnet powered at line frequency through a Variac which controls the amplitude of the bowl vibration. This bowl can feed about 150 parts per minute. Table 5.7 lists the physical parameters of the feed bowl.

The spring constants were found by attaching a linear voltage displacement transformer (LVDT) to a spring and then statically loading the spring with known weights.

The moment of inertia values were found by placing the base or bowl on a torsional pendulum and comparing the periods with that of a known cylinder using the method of Sec. 5.1 of Vol. 1.

To determine which frequencies would excite this feed bowl and cause parts to move up the track, a frequency generator was used to drive the feed bowl. An electronic digital frequency counter was used to accurately monitor the driving frequency. Table 5.8 shows the frequencies at which the feed bowl fed parts. The table also shows the results from a spectrum analyzer. An accelerometer was excited by repetitive tapping with a hammer, and the response cathode-ray-tube display photographed (see Fig. 5.22).

As shown in Table 5.8, the computer model closely matches the feed bowl experiments. Notice the natural frequency of about 40 Hz not predicted by the mathematical model. This fourth degree of freedom was due to the table on which the feed bowl vibrated during the tests. Based on the close match between analytical and experimental results, the model has the potential for use in an optimization study to both improve transport efficiency and alter the mass of the components and/or the energy consumption of the feeder.

An example would be to add a heavier base ($M_2$) to reduce the energy transmitted through the urethane pads to the machine on which the feed bowl is mounted. A search routine [216] could be employed to more desirably alter bowl components while retaining a stationary value of the governing transport (natural) frequency.

**TABLE 5.8** COMPARISON OF ACTUAL, MEASURED AND COMPUTED FREQUENCIES OF THE FEEDBOWL OF FIGS. 5.18, 5.19, 5.20, AND 5.21.

| Parts feeding frequency (Hz) | Spectrum analyzer | Computer model |
|---|---|---|
| 8–11 | 9 | 10 |
| 26 | 29 | 28 |
| 37 | 40 | — |
| 102.6–104.4 | 104 | 103 |

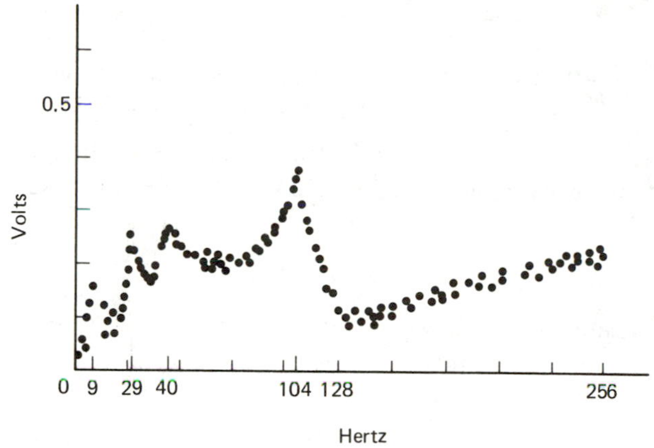


**Figure 5.22** Tracing of the photograph taken of the spectrum analyzer CRT showing the measured natural frequencies listed in Table 5.9.

This approach offers a rapid technique for modifying or designing a feed bowl, reducing prototype cost incurred in achieving desired performance.

## 5.7 FREE VIBRATION OF SYSTEMS WITH ONE DEGREE OF FREEDOM*

Before moving to complex problems of linkage balancing and high-speed elastic mechanisms, we will review the fundamentals of vibrations and rotor balancing. Readers who are already familiar with this material can proceed to Sec. 5.11.

The simplest vibrating system has only one degree of freedom. A well-known example is a mass $M$ connected to a wall by a spring of stiffness $K$ as shown in Fig. 5.23. The force $F$ exerted by the spring is assumed linear with displacement so that $F = -Kx$, where $K$ is the "spring constant" and $x$ is the displacement from equilibrium, positive to the right. Note that $F$ is a "return force," since it is opposite in sign to the displacement.

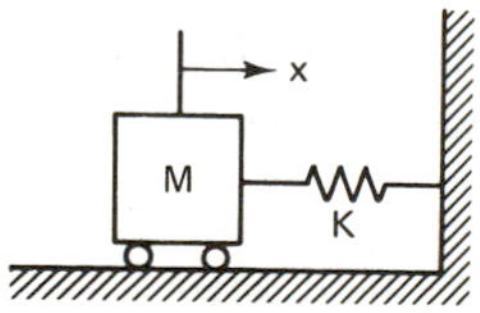


**Figure 5.23** Single-degree-of-freedom undamped vibrating system. Direction of arrow shows positive direction for x and all its time derivatives.

* The next four sections were prepared with the help of Tom Carlson, Exxon Corp.

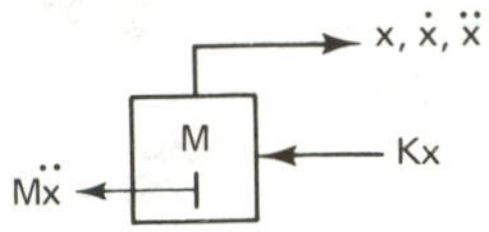


**Figure 5.24** Free-body diagram of the mass in the system of Fig. 5.23.

The only external force on the mass is that of the spring. The equation of motion is derived from the free-body diagram (Fig. 5.24) using Newton's second law:

$$M\ddot{x} = -Kx \qquad \text{or} \qquad M\ddot{x} + Kx = 0 \tag{5.66}$$

This differential equation is readily solved if we assume that $x$ is periodic in time. Let $x(t)$ be given by

$$x = A \sin \omega t \tag{5.67}$$

in which $A$ is a constant amplitude and $\omega$ is the frequency of vibration given in units of radians per unit of time "circular frequency." Substituting Eq. (5.67) into Eq. (5.66), we find that

$$-MA\omega^2 \sin \omega t + kA \sin \omega t = K - M\omega^2 = 0 \tag{5.68}$$

or

$$\omega^2 = \frac{K}{M} \tag{5.69}$$

Thus the "circular frequency" (radians per second) (often referred to briefly as the "frequency") of vibration is uniquely determined by the system's physical constants. This frequency is called the "natural frequency," $\omega_n$, of the system, and is given by

$$\omega_n = \left| \sqrt{\frac{K}{M}} \right| \tag{5.70}$$

where the absolute-value symbols indicate that only the positive value of the radical has physical meaning. Periodic oscillation of this kind, as expressed in Eq. (5.67), is called "simple harmonic motion," and the freely vibrating system of Fig. 5.23 is called a "simple harmonic oscillator."

The motion of this system can also be found using complex exponential notation instead of trigonometric functions. Equation (5.67) can be rewritten in the form $x = R(\mathbf{z})$, $\mathbf{z} = Ae^{i\omega t}$ (see Fig. 5.25). Substitution of this expression into Eq. (5.66) yields the same result for $\omega_n$.

**Example 5.6**

If the weight of the sliding block of Fig. 5.23 is 20 lbf and the spring deflects $\frac{1}{10}$ in. under a 20 lbf load, what is the natural frequency of the spring–mass system?

**Solution** Using Eq. (5.70), we obtain

$$\omega_n = \left| \sqrt{\frac{2400 \text{ lbf/ft}}{0.621 \text{ slug}}} \right| = 62.2 \text{ rad/sec} = 9.89 \text{ Hz}$$

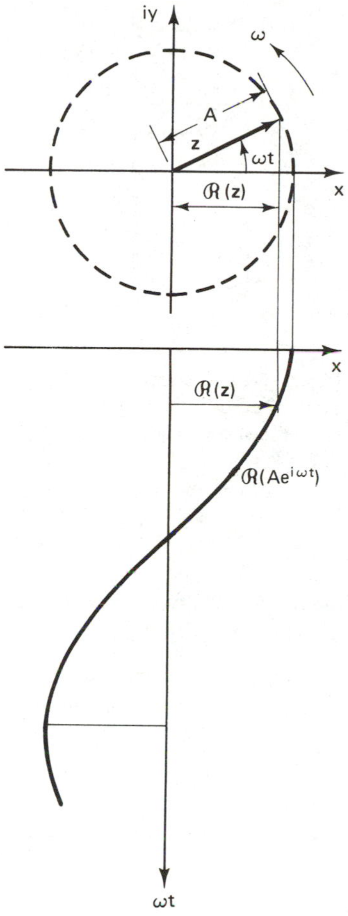


**Figure 5.25** Sinusoidal vibration modeled by the real part of the rotating "*phasor*" $\mathbf{Z} = Ae^{iwt}$.

Many other systems can be characterized by equations of the form of Eq. (5.70). Several examples are presented below. In each case a mass is acted on by a return force proportional to deflection from the equilibrium position. For example, in Fig. 5.26 the equilibrium position is where the spring force just balances the weight of *M*.

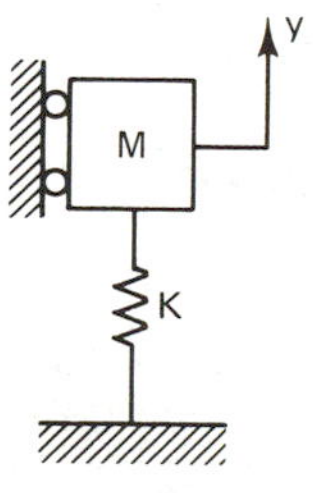


**Figure 5.26** At equilibrium position, the spring force just balances the weight of M.

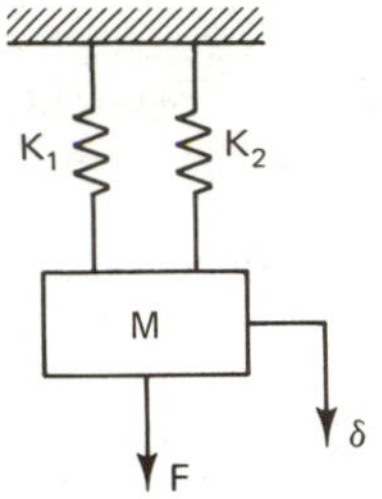


**Figure 5.27** Combined spring rate of side-by-side springs is equal to the sum of their spring rates.

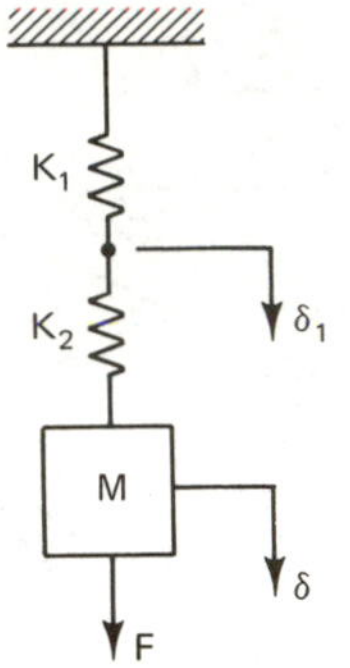


**Figure 5.28** Combined "*compliance*" of in-line springs is the sum of their compliances: $1/K = 1/K_1 + 1/K_2$.

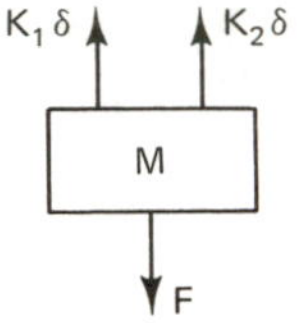


**Figure 5.29** Free-body diagram of M in Fig. 5.27.

**Example 5.7**

Determine the expression for the natural frequency for the case of two springs side by side (Fig. 5.27) and two springs in line (Fig. 5.28).

**Solution** For springs side by side, the free-body diagram of Fig. 5.29 shows that given a force $F$ (due to the weight of the block and/or some external force) that causes deflection $\delta$,

$$F - K_1\delta - K_2\delta = 0$$

Thus the equivalent spring constant of the side-by-side combination is

$$K_{eq} = \frac{F}{\delta} = K_1 + K_2$$

and the natural frequency is

$$\omega_n = \left|\sqrt{\frac{K_1 + K_2}{M}}\right|$$

For springs in line, the total deflection $\delta$ due to force $F$ is (from Fig. 5.28)

$$\delta = \delta_1 + \delta_2 = \frac{F}{K_1} + \frac{F}{K_2}$$

The equivalent stiffness of the in-line springs is

$$K_{eq} = \frac{F}{\delta} = \left(\frac{1}{K_1} + \frac{1}{K_2}\right)^{-1} = \frac{K_1K_2}{K_1 + K_2}$$

and

$$\omega_n = \left|\sqrt{\frac{K_1K_2}{M(K_1 + K_2)}}\right|$$

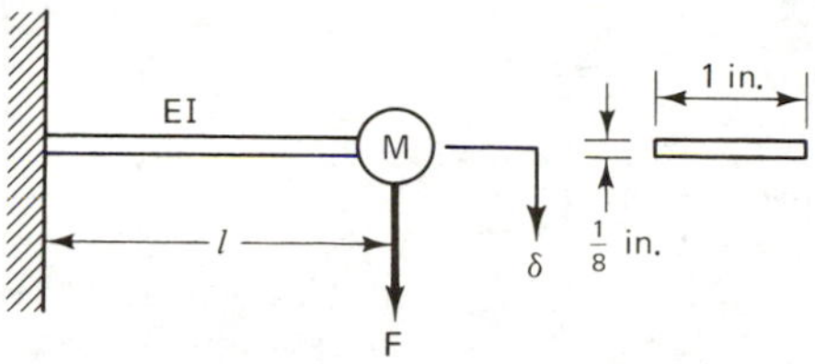


**Figure 5.30** Free vibration of cantilever spring with concentrated mass.

**Example 5.8**

Determine the expression for the natural frequency of the cantilever beam with mass $M$ weighing 9.92 lbf at distance $l$ from the wall (Fig. 5.30). Neglect the mass of the beam.

**Solution** From beam theory we know that the relationship between a force $F$ and the deflection $\delta$ is $\delta = Fl^3/3EI$; thus $K = F/\delta = 3EI/l^3$ and $\omega_n = |\sqrt{3EI/l^3M}|$. If the cross section of the steel beam is 1 in. wide by $\frac{1}{8}$ in. thick and it is 12 in. long, then

$$K = \frac{3(3 \times 10^7 \text{ lbf/in}^2)(0.125)^3 \text{ in}^4}{1728 \text{ in}^3\ (12)} = 8.477 \text{ lbf/in.} = 101.7 \text{ lbf/ft}$$

so that

$$\omega_n = \left|\sqrt{\frac{101.7}{0.310}}\right| = 18.1 \text{ rad/sec} = 2.88 \text{ cycles/sec} = 173 \text{ cycles/min}$$

## 5.8 DECAY OF FREE VIBRATIONS

The accurate analysis of actual physical systems requires that the effect of friction be taken into account. No physical system exhibits free harmonic motion without a decay in the amplitude of its vibration. The simplest frictional force to deal with analytically is one proportional to the velocity. In mechanical systems this is called "viscous damping," since fluid friction obeys this relationship. Then the friction force $F$ is given by $F = C\dot{x}$, in which $C$ is the "damping coefficient," which has the dimensions of force per unit velocity, or force·time/length. The assumption of damping proportional to velocity has been shown experimentally to be a good approximation for many mechanical systems. Damping in a mechanical system is often shown schematically as a dashpot, a device similar to an automobile's oil-filled shock absorber.

We can investigate the decay of free, damped vibrations considering the single-degree-of-freedom system shown in Fig. 5.31a. The system's equation of motion is derived from the free-body diagram of its mass (Fig. 5.31b):

$$M\ddot{x} + C\dot{x} + Kx = 0 \tag{5.71}$$

Since we know from physical considerations that the motion will vanish in time, we must assume an appropriate mathematical expression for the decay. A real expo-

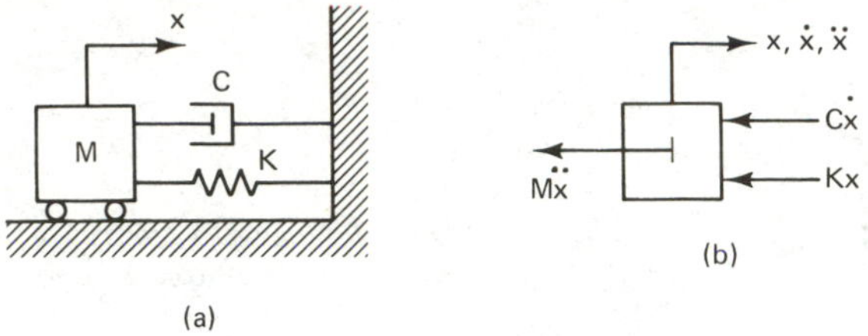


**Figure 5.31** (a) single-degree-of-freedom system with viscous damping; (b) FBD of the mass in the damped system.

nential of the form $x = Ae^{rt}$ will be a solution of Eq. (5.71). Upon substitution we get

$$Mr^2 + Cr + K = 0 \tag{5.72}$$

This quadratic equation is called the "characteristic equation of the system." The solution for $r$ is

$$r = \frac{-C}{2M} \pm \left|\sqrt{\left(\frac{C}{2M}\right)^2 - \frac{K}{M}}\right| = r_1, r_2 \tag{5.73}$$

The "critical damping coefficient" $C_c$ is defined to reduce the radical in Eq. (5.73) to zero:

$$\frac{C_c}{2M} = \sqrt{\frac{K}{M}}, \qquad \text{so that } C_c = 2\sqrt{KM} \tag{5.74}$$

Using the dimensionless damping ratio $\zeta$, defined by

$$\zeta \equiv \frac{C}{C_c} \equiv \frac{C}{2\sqrt{KM}} \tag{5.75}$$

Eq. (5.73) becomes

$$r_{1,2} = \omega_n(-\zeta \pm |\sqrt{\zeta^2 - 1}|) \tag{5.76}$$

We will be concerned with systems having "subcritical damping," for which $\zeta < 1$. Thus

$$\frac{r_{1,2}}{\omega_n} = -\zeta \pm i\,|\sqrt{1 - \zeta^2}| \tag{5.77}$$

The most general solution for $x$ is the superposition of solutions involving $r_1$ and $r_2$:

$$x = Ae^{r_1 t} + Be^{r_2 t} \tag{5.78}$$

for which constants $A$ and $B$ must be found from initial conditions. Substituting (5.77) into (5.78), we have

$$x = e^{-i\zeta\omega_n t}(Ae^{i|\sqrt{1-\zeta^2}|\omega_n t} + Be^{-i|\sqrt{1-\zeta^2}|\omega_n t}) \tag{5.79}$$

Using the identity $e^{i\theta} = \cos\theta + i\sin\theta$, this may be rearranged to give

$$x = De^{-\zeta\omega_n t}\sin(\omega_n\,|\sqrt{1 - \zeta^2}|\,t + \phi) \tag{5.80}$$

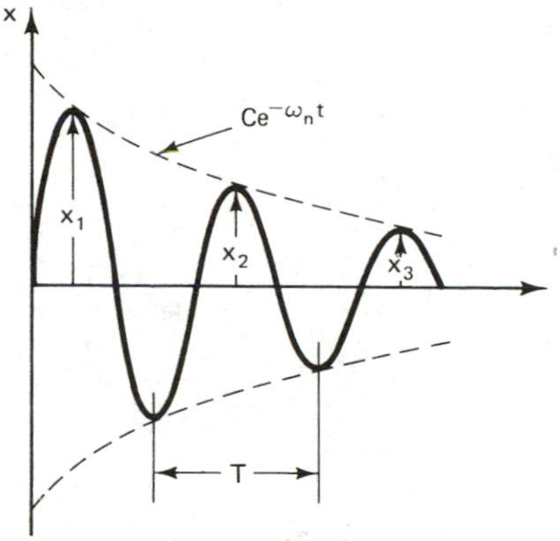


**Figure 5.32** Trace of damped sinusoidal vibration. Note the "damping envelope," $Ce^{-w_n t}$.

where $D$ and $\phi$ are constants to be determined from initial conditions. Equation (5.80) is that of a sine wave with exponentially decreasing amplitude. It is convenient to define the damped natural frequency $\omega_d$ as

$$\omega_d = \omega_n \left|\sqrt{1-\zeta^2}\right| \tag{5.81}$$

Clearly, $\omega_d < \omega_n$, but $\omega_d \simeq \omega_n$ for small $\zeta$. Equation (5.80) is plotted in Fig. 5.32, where $T$ is the period of oscillation in seconds, given by $T = 2\pi/\omega_d$. Note that in this figure $\phi$ is taken equal to zero.

We may find the approximate value for a small damping ratio $\zeta$ experimentally by comparing the change in time of peak amplitudes. Consider two peak amplitudes $x_1$ and $x_2$ separated by $N$ cycles of oscillation. Assume that $x_2 < x_1$; that is,

$$x_2 = x(t_2) < x_1 = x(t_1) \tag{5.82}$$

and

$$t_2 = t_1 + NT \tag{5.83}$$

From Eq. (5.80),

$$\frac{x_2}{x_1} = \frac{e^{-\zeta\omega_n t_2}}{e^{-\zeta\omega_n t_1}} = e^{-\zeta\omega_n NT} = e^{-\zeta 2\pi f_n (t_2 - t_1)} \tag{5.84}$$

where $f_n$ is the natural frequency in revolutions or cycles per second.

For small $\zeta$, $\omega_n \simeq \omega_d$; $T$ will be therefore very nearly the experimentally measured quantity. Taking the logarithm of both sides of Eq. (5.84) gives

$$\ln\left(\frac{x_2}{x_1}\right) = -2\pi f_n (t_2 - t_1)\,\zeta$$

so that

$$\zeta = \frac{\ln(x_1/x_2)}{2\pi f_n (t_2 - t_1)} \tag{5.85}$$

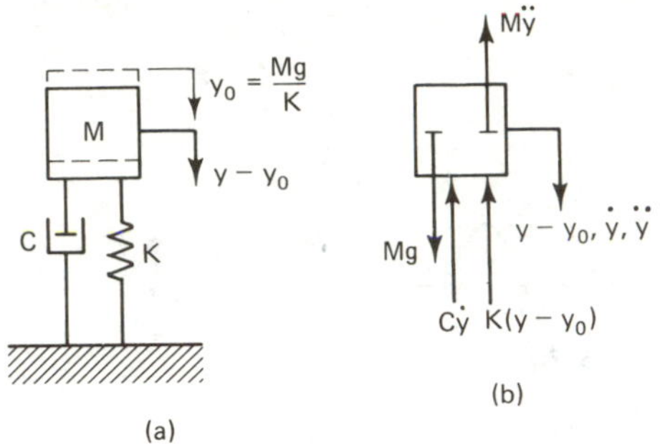


**Figure 5.33** (a) in a vertical system, write the equation of motion for displacement from the static equilibrium position $y_o$; (b) FBD of the single-degree-of-freedom vertically moving system.

These decaying, or "transient" vibrations will not be discussed further here. However, we will find the damping ratio $\zeta$ to be a convenient parameter in discussing forced vibrations.

In a vertically oriented system (Fig. 5.33a) the constant force $W = Mg$ must be included on the free-body diagram (FBD) (Fig. 5.33b). If the downward displacement is measured from the static equilibrium position $y_o$, where

$$Mg = Ky_0, \qquad \text{or} \qquad y_0 = \frac{Mg}{K} \tag{5.86}$$

then $Mg$ is canceled in the equation of motion written for $y$ measured from $y_0$ and

$$M\ddot{y} + C\dot{y} + Ky = 0 \tag{5.87}$$

which is of the same form as Eq. (5.71).

## 5.9 FORCED VIBRATIONS OF SYSTEMS WITH ONE DEGREE OF FREEDOM

In many systems of engineering interest it is important to know the response of a vibratory system to an external periodic force. The amplitude of response to such an external force, often called a "driving force" or a "forcing function," is not only a function of time but of the frequency of vibration itself.

### Undamped Forced Vibrations

We will first consider the simple spring–mass system driven at the mass by a periodic force $F(t)$ given by

$$F = F_0 \sin \omega t = \mathscr{I}(F_0 e^{i\omega t}) \qquad \text{(Fig. 5.34)} \tag{5.88}$$

in which $F_0 > 0$ is the peak amplitude of the varying force. The system is shown in Fig. 5.35.

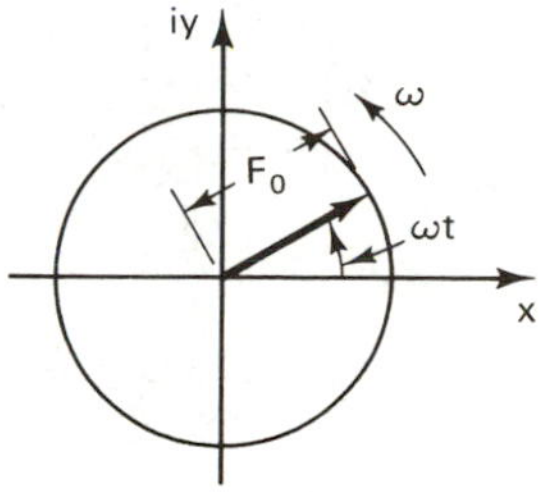


Figure 5.34 Phasor of a sinusoidal vibration.

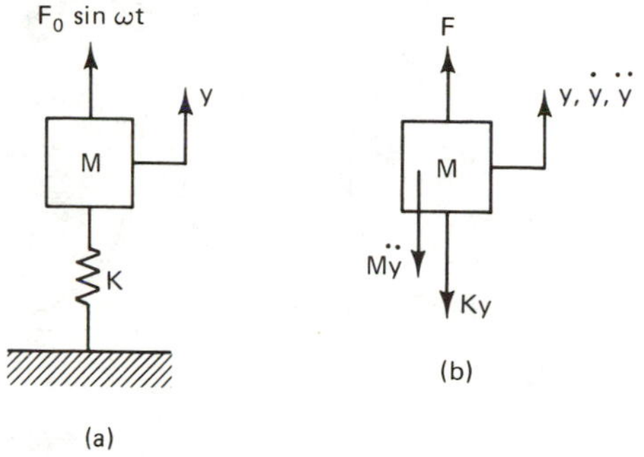


**Figure 5.35** (a) forced system; (b) FBD of mass in forced system. $y$ is measured from static equilibrium position.

The steady-state equation of motion is obtained from the free-body diagram of Fig. 5.35:

$$M\ddot{y} + Ky = F_0 \sin \omega t \tag{5.89}$$

As with the simple harmonic oscillator, we will assume a solution of the form

$$y = A \sin \omega t \tag{5.90}$$

However, we are now interested in the behavior of the amplitude of $A$ as a function of driving frequency $\omega$. Substituting Eq. (5.90) into Eq. (5.89), it can be shown that

$$A = \frac{F_0}{K - M\omega^2} \tag{5.91}$$

It is convenient to define the frequency ratio $\beta$ as

$$\beta = \frac{\omega}{\omega_n} = \omega \left| \sqrt{\frac{M}{K}} \right| \tag{5.92}$$

Dividing Eq. (5.91) by $A_0 = F_0/K$, the static deflection in the spring due to force $F_0$, we obtain the "amplitude ratio" $A/A_0$.

$$\frac{A}{A_0} = \frac{F_0}{KA_0 - MA_0\omega^2}$$

Letting $F_0/K = A_0$,

$$\frac{A}{A_0} = \frac{F_0}{(KF_0/K) - M(F_0/K)\omega^2} = \frac{1}{1 - (M/K)\omega^2}$$

or

$$\frac{A}{A_0} = \frac{1}{1 - \beta^2} \tag{5.93}$$

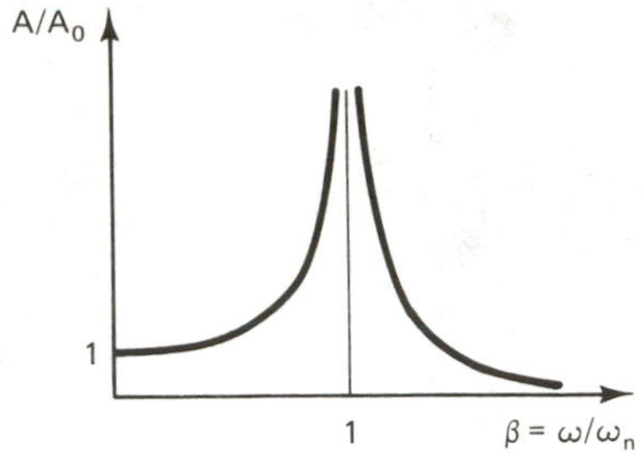


**Figure 5.36** As the forcing frequency $\omega$ approaches the natural frequency $\omega_n$, the amplitude ratio $A/A_o$ approaches infinity in the undamped system.

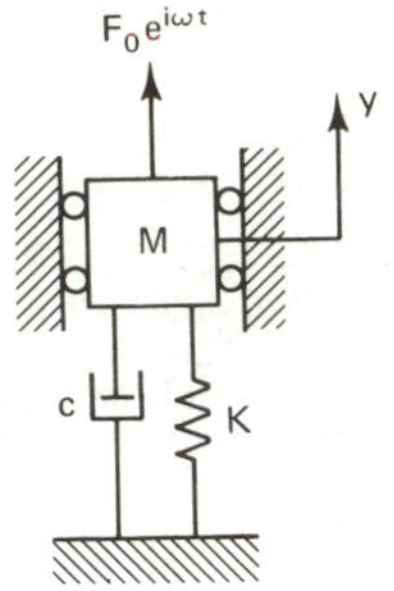


**Figure 5.37** Damped forced system.

Equation (5.93) is plotted in Fig. 5.36. The most important feature of the response is that amplitude $A$ asymptotically approaches infinity as the driving frequency approaches the natural frequency.

## Damped Forced Vibrations

When we include the effects of damping we will find that the maximum response amplitude is finite, as would be expected with real physical systems. Observation of a real system will also show that the resonance frequency is no longer exactly equal to the natural frequency, and that it will be necessary to consider the phase relationship between the force and displacement.

For the damped system of Fig. 5.37, measuring $y$ from the static equilibrium position, the equation of motion is

$$M\ddot{y} + c\dot{y} + ky = F_0 e^{i\omega t} \tag{5.94}$$

where $F_0 e^{i\omega t}$ is a rotating force vector (Fig. 5.34). However, as seen from the friction-free vertical guide in Fig. 5.37, only the vertical component of the force vector is effective in this system. Complex exponential notation has been chosen for mathematical simplicity. The assumed form of $y$ must include a phase angle $\phi$:

$$y = Ae^{i(\omega t - \phi)} \tag{5.95}$$

in which $\phi$ is proportional to the time interval between occurrence of maximum force and maximum $y$. Substituting Eq. (5.95) into (5.94), we have

$$A(Mi^2\omega^2 + ci\omega + K)e^{i(\omega t - \phi)} = A(K - M\omega^2 + ic\omega)\frac{e^{i\omega t}}{e^{i\phi}} = F_0 e^{i\omega t} \tag{5.96}$$

where, from now on, it will be understood that only the imaginary part of the equation is of interest. Dividing by $A(e^{i\omega t}/e^{i\phi})$ and equating the real and imaginary parts, respectively, on each side of this equation we have

$$k - M\omega^2 = \frac{F_0}{A} \cos \phi \tag{5.97}$$

and

$$c\omega = \frac{F_0}{A} \sin \phi \tag{5.98}$$

We can now see why the phase angle is mathematically necessary: The complex equation of motion becomes two nontrivial real equations; hence we must have two unknowns, in this case: $A$ and $\phi$. Employing elementary trigonometric identities Eqs. (5.97) and (5.98) can be readily solved to yield

$$A = \frac{F_0}{|\sqrt{(K - M\omega^2)^2 + (c\omega)^2}|} \tag{5.99}$$

and

$$\phi = \arg (K - M\omega^2 + ic\omega) \tag{5.100}$$

Also, dividing Eq. (5.96) by $e^{i\omega t}/e^{i\phi}$ yields: $A(K - M\omega^2 + ic\omega) = F_0 e^{i\phi}$, from which Eqs. (5.99) and (5.100) follow directly. Substituting the dimensionless parameters of frequency ratio $\beta$ and damping ratio $\zeta$, we can write the solution in the dimensionless form

$$\frac{A}{A_0} = \frac{1}{|\sqrt{(1 - \beta^2)^2 + 4\zeta^2\beta^2}|} \tag{5.101}$$

and

$$\phi = \arg (1 - \beta^2 + i2\zeta\beta) \tag{5.102}$$

These equations are plotted in Fig. 5.38 for various values of $\zeta$. For a given damping ratio, the frequency at which amplitude $A$ is a maximum is defined as the resonant frequency. For small damping ratios (i.e., $\zeta^2 << 1$) the resonant frequency is very close to the natural frequency given by $\beta = 1$.

## Rotary Forcing Function

A rotating imbalance provides a common example of an external driving force. Consider an eccentric mass $m$ at radius $r$ from the center of rotation, as shown in Fig. 5.39. The component of the centrifugal force in the direction of travel is

$$F = mr\omega^2 \sin \omega t \tag{5.103}$$

where $mr\omega^2 = F_0$. Substituting this for $F_0$ in Eq. (5.99) gives

$$A = \frac{mr\omega^2}{|\sqrt{(K - M\omega^2)^2 + (c\omega)^2}|} \tag{5.104}$$

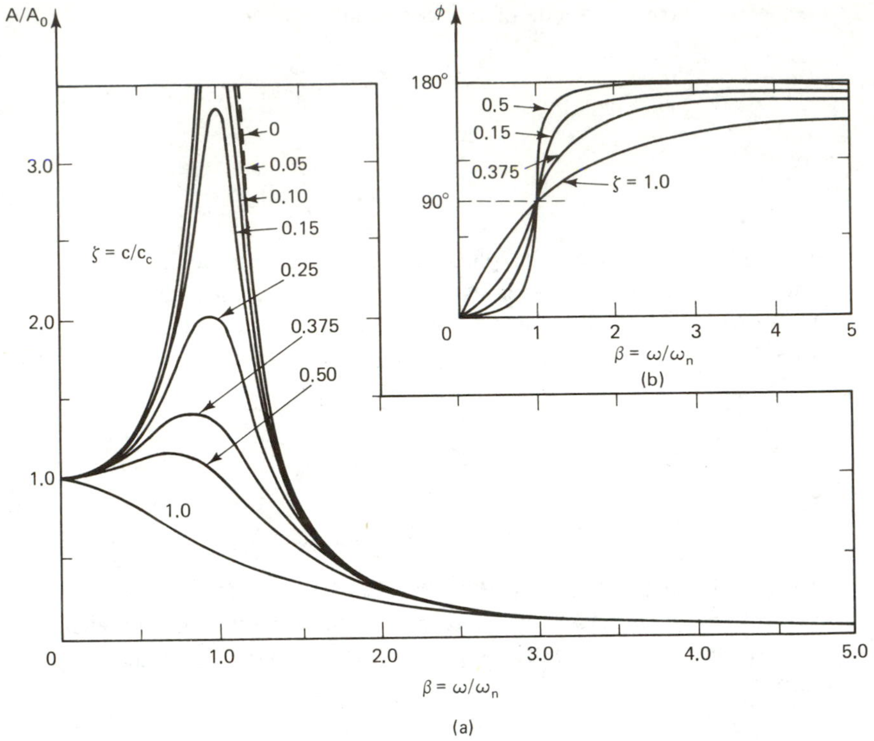


**Figure 5.38** (a) amplitude ratio and (b) phase angle as functions of frequency ratio for various values of damping ratio. [286]

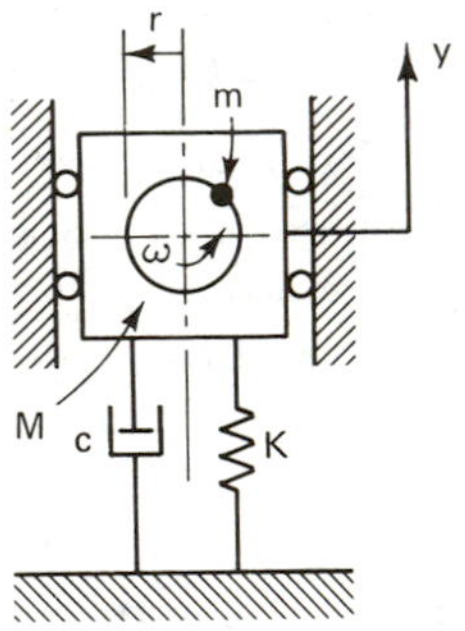


**Figure 5.39** Rotating unbalance as a forcing function.

which may be written in dimensionless form,

$$\frac{MA}{mr} = \frac{\beta^2}{|\sqrt{(1-\beta^2)^2 + 4\zeta^2\beta^2}|} \tag{5.105}$$

This response is plotted in Fig. 5.40, where $\phi = \arg(1 - \beta^2 + i2\zeta\beta)$. In comparing this with Fig. 5.38, note the difference in response for the limits $\beta = 0$ and $\beta \rightarrow \infty$, due to the dependence of the centrifugal force on $\omega^2$.

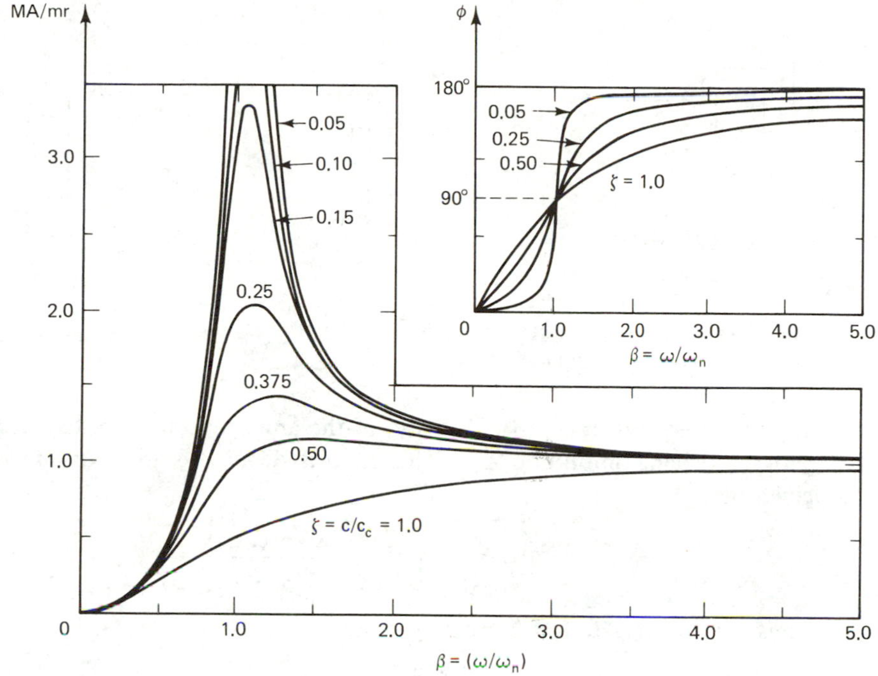


**Figure 5.40** Frequency response of a damped single-degree-of-freedom system with rotating unbalance. [286]

## Base Excitation

The driving force can also come from harmonic motion of the base below the spring and damper, as shown in Fig. 5.41a. The equation of motion of this system is based on the free-body diagram of Fig. 5.41b. Note that $y_1$ and $y_2$ are measured from a static equilibrium position, in which the compression of spring $K$ just balances the weight of $M$. Then $Mg$ is canceled from the equation and we obtain

$$M\ddot{y}_2 + c(\dot{y}_2 - \dot{y}_1) + K(y_2 - y_1) = 0 \tag{5.106}$$

Using complex exponential form for the displacements, that is,

$$y_1 = A_1 e^{i\omega t}, \qquad y_2 = A_2 e^{i(\omega t - \phi)} \tag{5.107}$$

and noting that only the vertical components are effective (Fig. 5.42), we have

$$MA_2 i^2 \omega^2 \frac{e^{i\omega t}}{e^{i\phi}} + c\left(A_2 i\omega \frac{e^{i\omega t}}{e^{i\phi}} - A_1 i\omega e^{i\omega t}\right) + K\left(A_2 \frac{e^{i\omega t}}{e^{i\phi}} - A_1 e^{i\omega t}\right) = 0$$

or

$$A_2(K - M\omega^2 + ic\omega) = A_1(ic\omega + K)e^{i\phi} \tag{5.108}$$

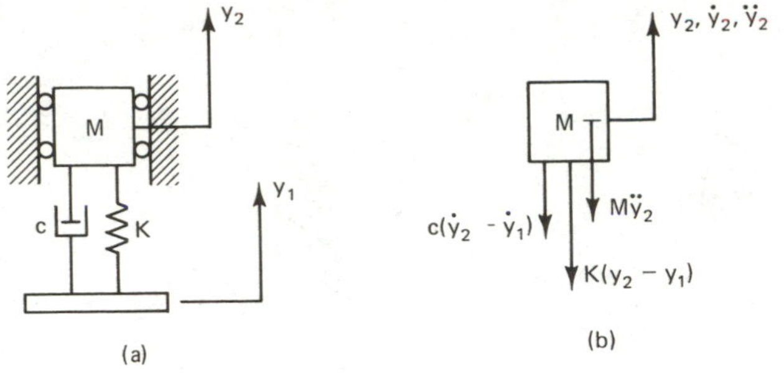


**Figure 5.41** (a) single-degree-of-freedom system with ground excitation and (b) free-body diagram of its mass.

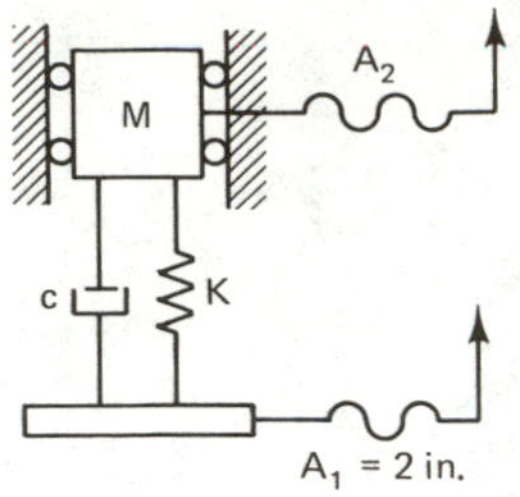


**Figure 5.42** Single-degree-of-freedom system with ground excitation. (See Ex. 5.9)

We are interested in $A_2/A_1$, the ratio of the amplitude transmitted to the mass $A_2$ and the driving amplitude $A_1$. This ratio is defined as the *transmissibility* and is given by

$$T = \frac{A_2}{A_1} = \left| \sqrt{\frac{K^2 + (c\omega)^2}{(K - M\omega^2)^2 + (c\omega)^2}} \right| \tag{5.109}$$

Thus, substituting $\beta$ and $\zeta$, we may write

$$T = \left| \sqrt{\frac{1 + 4\zeta^2\beta^2}{(1-\beta^2)^2 + 4\zeta^2\beta^2}} \right|, \qquad \phi = \arg(1 + \beta^2 + i2\zeta\beta) \tag{5.110}$$

This response is shown in Fig. 5.43. It is interesting to note that

$$T\Big|_{\beta=0} = T\Big|_{\beta=\sqrt{2}} = 1$$

for all values of $\zeta$.

For small damping ratios resonance occurs at $\beta \simeq 1$; thus we may write

$$T_{\max} \simeq \frac{1}{2\zeta}$$

## Force Transmission

In many applications the force $F_T$ transmitted to a rigid base is of interest. Considering the system of Fig. 5.41a, it is easy to show that the ratio of force transmission is given by

$$\frac{F_T}{F_0} = \left| \sqrt{\frac{1 + 4\zeta^2\beta^2}{(1-\beta^2)^2 + 4\zeta^2\beta^2}} \right| = T \tag{5.111}$$

where $F_0$ is the amplitude of a sinusoidal periodic force applied to $M$. Thus the force transmission ratio is equal to the transmissibility $T$ found above for base excitation.

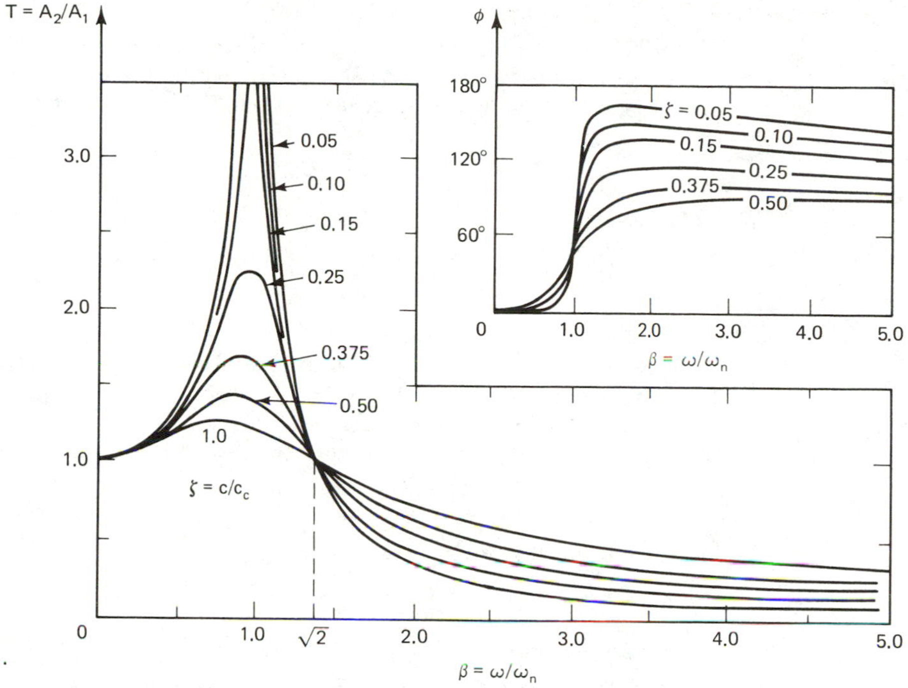


**Figure 5.43** Frequency response of transmissibility and phase angle for the system with ground excitation of Figs. 5.41 and 5.42. [286]

## Applications

The principles of simple vibratory systems presented above have many engineering applications, particularly in machine design. Vibrations occur in many machines and can have such undesirable effects as noise, "fretting" through friction, fatigue failure, tool chatter, damage to sensitive instruments, and so on. Vibration can be controlled by eliminating the source of vibration (i.e., modifying the forcing function). When that is not possible or practical, the vibration can often be reduced to acceptable levels by proper selection of the mass, spring and damping components of the system.

## Vibration Isolation

In the discussion of damped forced vibrations we defined the transmissibility which is a function of spring constant $K$ and damping coefficient $c$, or alternatively, of dimensionless ratios $\beta$ and $\zeta$. In applications where vibrations are present and cannot be eliminated completely, the value of the transmissibility can be so chosen as to minimize the vibrations.

**Example 5.9**

The system of Fig. 5.42 is subject to base excitation with amplitude of $A = 2$ in. (a) For frequency of the base excitation $\omega$ of 8 Hz, find $A_2$; (b) find $\omega$ such that $A_2 =$ 1 in.; (c) suppose that the weight is increased from 9.92 to 12 lbf; find $\omega$ such that $A_2 = 0.5$ in.

**Solution**

(a) If from Eq. (5.75), $\zeta = 0.25$ and from Eq. (5.70), $\omega_n = 9.89$ Hz, then $\beta = 8/9.89 = 0.81$. Substituting into Eq. (5.110), we get

$$T = 2.03$$

Thus $A_2 = TA_1 = 4.06$ in.

(b) We desire to find $\omega$ such that $T = 0.50$. An exact solution requires solving equation 5.110 for $\beta$ given $T = 0.50$ and $\zeta = 0.25$. A satisfactory solution may be obtained by trial and error by first estimating $\beta$ from Fig. 5.43 for $\zeta = 0.25$, $T = 0.5$. We might estimate $\beta \simeq 1.8$. Substitution into Eq. (5.110) gives

$$T\Big|_{\beta=1.8} = 0.557$$

Increasing $\beta$ to 1.9 gives

$$T\Big|_{\beta=1.9} = 0.497$$

If this is satisfactorily accurate, $\omega$ is given by $\omega = 1.9\omega_n = 18.8$ Hz.

(c) As $W$ increases both $\omega_n$ and $\zeta$ change. The procedure is similar to part (b). The solution is left to the reader.

## Dynamic Vibration Absorber

So far we have discussed only systems with a single degree of freedom. Now consider the system of Fig. 5.44, in which the mass $M_1$ is to be protected from the vibration of the base, $y_0$. With the addition of an appropriate mass $M_2$ supported by spring of stiffness $K_2$, we can construct a "dynamic vibration absorber."

The system has two equations of motion, each derived from one free-body diagram for each mass (see Figs. 5.45 and 5.46).

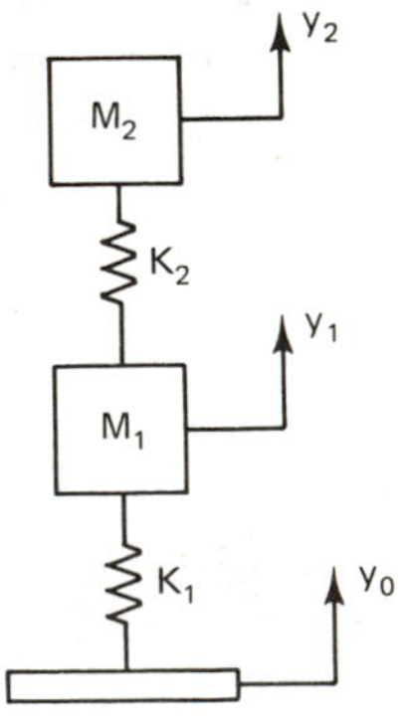


**Figure 5.44** $M_2$, $K_2$ is a "dynamic absorber" if $\frac{K_1}{M_1} = \frac{K_2}{M_2}$.

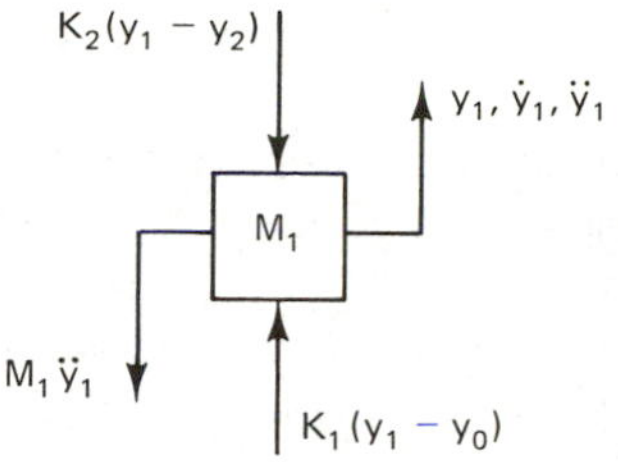


**Figure 5.45** Free-body diagram (FBD) of the main mass of Fig. 5.44.

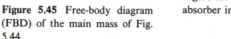

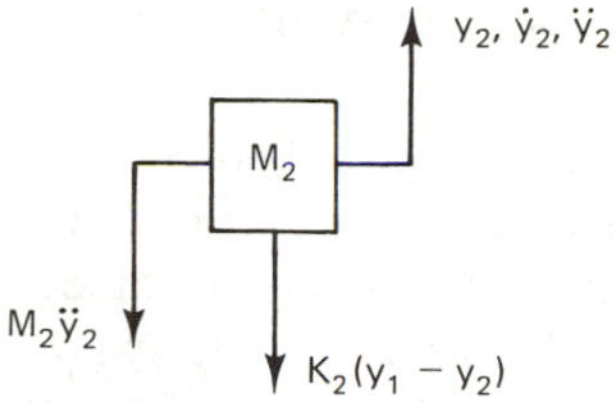


**Figure 5.46** FBD of the dynamic absorber in Fig. 5.44.

$$M_1\ddot{y}_1 + y_1(K_1 + K_2) - y_0K_1 - y_2K_2 = 0 \tag{5.112}$$

$$M_2\ddot{y}_2 - K_2(y_1 - y_2) = 0 \tag{5.113}$$

Assuming displacements of the form $y_j = A_j \sin \omega t$, $j = 0, 1, 2$, Eqs. (5.112) and (5.113) become

$$A_1(K_1 + K_2 - M_1\omega^2) - A_0K_1 - A_2K_2 = 0 \tag{5.114}$$

$$A_2(K_2 - M_2\omega^2) - A_1K_2 = 0 \tag{5.115}$$

If we desire the vibration of mass $M_1$ to vanish, we set $A_1 = 0$. From Eq. (5.115) this implies that

$$\frac{K_2}{M_2} = \omega^2$$

or that the natural frequency of the system ($K_2$, $M_2$) equals the driving frequency of the base. Since it is likely that the greatest vibration problem would occur at the natural frequency of the system ($K_1$, $M_1$), it is generally advisable to choose

$$\frac{K_2}{M_2} = \frac{K_1}{M_1} \tag{5.116}$$

This simple result is really very striking. It means that the resonant vibration of mass–spring system $M_1$, $K_1$ can be eliminated if the natural frequency of the added mass and spring equals the natural frequency of the original system. Moreover, the added mass can be quite small if the spring $K_2$ is soft, as long as relation (5.116) holds. However, from Eq. (5.114), the amplitude of vibration of mass $M_2$ might be quite large:

$$A_2 = -A_0\frac{K_1}{K_2} = -A_0\frac{M_1}{M_2} \tag{5.117}$$

## 5.10 ROTOR BALANCING

The balancing of rigid rotors* is of great importance in the operation of many types of machinery. It is necessary or desirable to balance such components as tires, flywheels, fans, motors, turbines, and so on.

The terms "static" and "dynamic" balancing are commonly used to describe two types of balancing problems, although dynamic balancing also achieves static balance.

### Static Balancing

A "static imbalance" in a rotating system is the effect of an eccentric (or "unbalance") mass which can be detected by quasi-static methods, in which the weight of the unbalance mass reveals its location. The static imbalance in an otherwise symmetric and homogeneous disk can be described as the vector $\mathbf{S}$ directed out from the axis in the direction of the eccentric mass, with its magnitude equal to the eccentric mass times the magnitude of its radius vector, as shown in Fig. 5.47. When several such eccentric masses are present, the net static imbalance $\mathbf{S}_N$ is simply the vector sum:

$$\mathbf{S}_N = \sum_i \mathbf{S}_i = \sum_i m_i \mathbf{r}_i \tag{5.118}$$

in which mass $m_i$ is located at position $\mathbf{r}_i$ from the axis. Static balancing occurs when $\mathbf{S}_N$ vanishes. This may be accomplished by the addition of a single "correction mass" at the appropriate position $\mathbf{r}$. It is elementary to show that static balancing

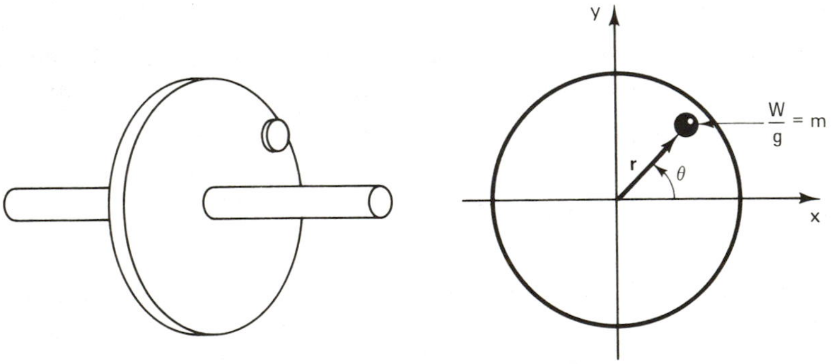


**Figure 5.47** Definition of the *imbalance:* $\mathbf{S} = m\mathbf{r}$

* Flexible rotor balancing, although important in modern machinery, is beyond the scope of this work.

is equivalent to bringing the system's center of gravity to the axis of rotation. In this case the system is the unbalanced disk and the correction mass.

The correction of known static imbalances can be accomplished graphically or numerically. For a graphical solution the various imbalance vectors $\mathbf{S}_i$ are summed to reveal $\mathbf{S}_N$. For a numerical solution, summation of the components of the vectors $\mathbf{S}_i$ yield two scalar equations:

$$S_x = \sum_i m_i x_i \tag{5.119}$$

$$S_y = \sum_i m_i y_i \tag{5.120}$$

where $x_i = r_i \cos \theta_i$, $y_i = r_i \sin \theta_i$. Thus

$$\mathbf{S}_N = S_x + iS_y \qquad |\mathbf{S}_N| = S_x^2 + S_y^2 \tag{5.121}$$

and

$$\theta_n = \arg \mathbf{S}_N \tag{5.122}$$

These principles are exhibited in Ex. 5.10.

**Example 5.10**

A 10.0-in.-diameter disk has three imbalances, as shown in Fig. 5.48. (a) Find the net static imbalance using both numerical and graphical methods; (b) correct the imbalance with a 3-oz weight; (c) correct the imbalance with a weight at $r = 10.0$ in.

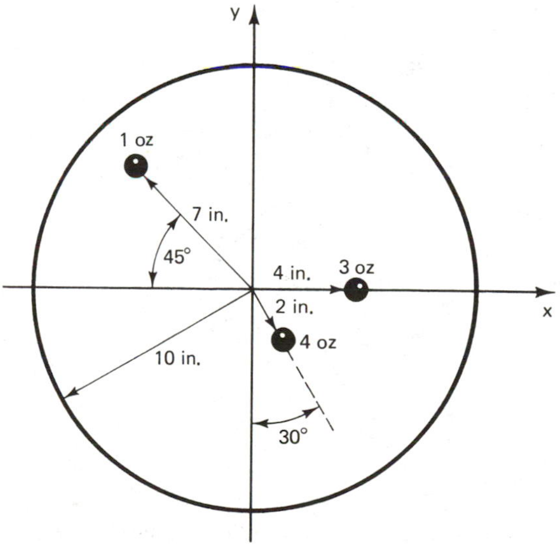


**Figure 5.48**

**Solution**

(a) In order to keep track of all the variables, it is handy to construct the following table, in which all the components are separated:

| Imbalance | $W_i$ (oz) | $r_i$ (in.) | $\theta_i$ (deg) | $x_i$ (in.) | $W_i x_i$ (in·oz) | $y_i$ (in.) | $W_i y_i$ (in·oz) |
|---|---|---|---|---|---|---|---|
| 1 | 3 | 4 | 0 | 4.00 | 12 | 0 | 0 |
| 2 | 1 | 7 | 135 | −4.95 | −4.95 | 4.95 | 4.95 |
| 3 | 4 | 2 | −60 | 1.00 | 4.00 | −1.73 | −6.93 |
| Sum | | | | | 11.05 | | −1.98 |

We find

$$|\mathbf{S}_N| = |\sqrt{(11.05)^2 + (1.98)^2}| = 11.23 \text{ in·oz}$$

$$\theta_N = \arg \mathbf{S}_N = -10.2°$$

$$\mathbf{S}_N = (11.05 - i1.98) \text{ in·oz},$$

Figure 5.49 shows the graphical solution for the total static imbalance.

(b) $r = \dfrac{T_N}{W} = \dfrac{11.23 \text{ in·oz}}{3.0 \text{ oz}} = 3.74 \text{ in.}$

(c) $W = \dfrac{T_N}{r} = \dfrac{11.23 \text{ in·oz}}{10.0 \text{ in.}} = 1.12 \text{ oz}$

$$\theta_N = \arg(-\mathbf{S}_N) = \arg(-11.05 + i1.98) = 169.8°$$

The results of parts (b) and (c) are shown in Figs. 5.50 and 5.51.

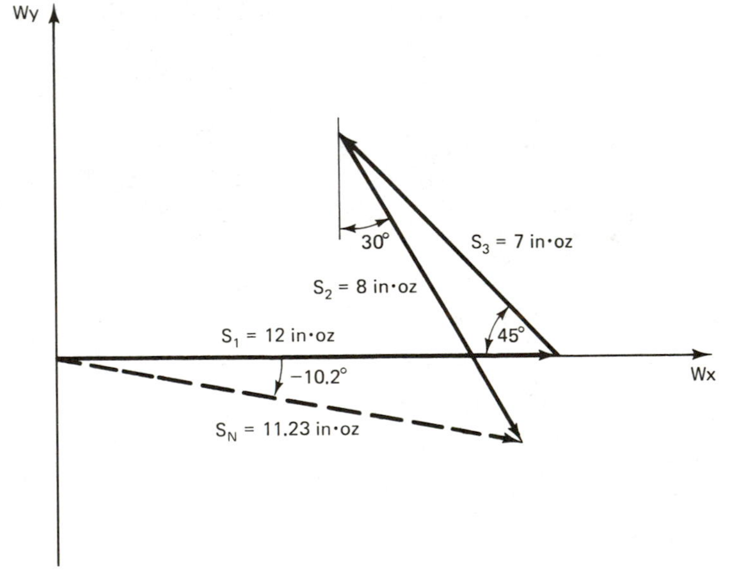


**Figure 5.49** Graphical solution to Ex. 5.10, showing the total imbalance $\mathbf{S}_N$.

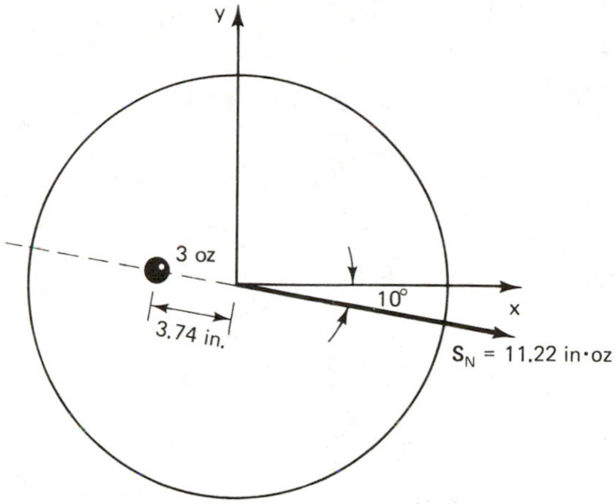


**Figure 5.50** Solution to Ex. 5.10(b), showing counterbalance $-\mathbf{S_N}$ with 3 oz counterweight.

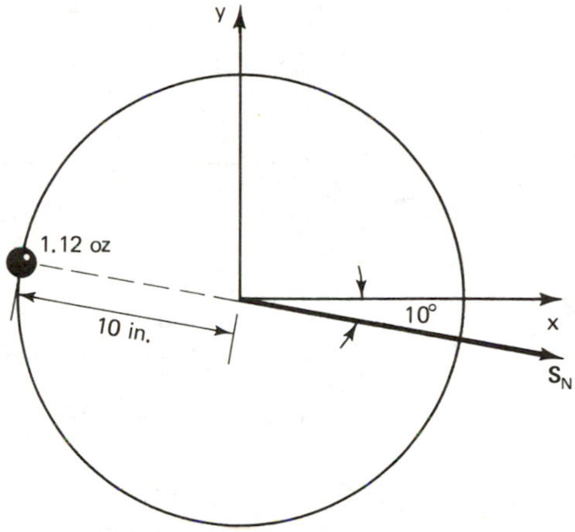


**Figure 5.51** Solution to Ex. 5.10(c), showing counterbalance $-\mathbf{S_N}$ with 10 in. radius.

## Dynamic Imbalance

Even when a rotor is statically balanced, it may exhibit unwanted vibration when rotated about its axis. A "dynamic imbalance" in a rotating object is a torque having an axis perpendicular to the system's axis of rotation. It is due to inertia forces occurring during rotation. The simplest form of a dynamic imbalance is shown in Fig. 5.52. The disk shown is statically balanced, but during rotation at angular velocity $\omega$, the disk is subject to transverse torque $= mr\omega^2L$. When this torque vanishes, the rotor is dynamically balanced. A dynamically balanced rotor is also statically balanced, but, in general, the converse is not true. In practice, a rotor with a length significant compared to its diameter should be dynamically balanced,

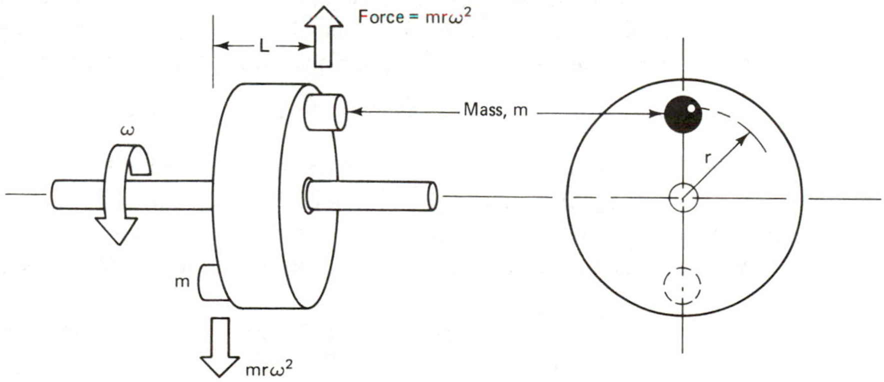


**Figure 5.52** Dynamic imbalance in a statically balanced rotor.

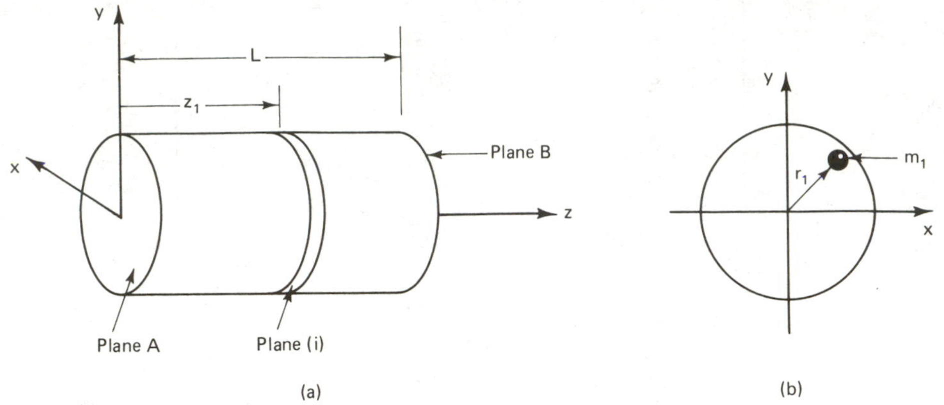


**Figure 5.53** Dynamic imbalance (a) rotor; (b) imbalance in plane ($i$). Dynamic balance is achieved by counterbalances in planes A and B

while static balancing is sufficient for thin disks. However, the nature of the application—speeds, tolerances, strengths of materials, and the like—will dictate the required specifications for balancing.

The dynamic unbalance effect of all eccentric masses in a given rigid rotor may be resolved into simple static unbalances in any pair of planes perpendicular to the axis of rotation. These planes are called "correction planes" since correction weights can be added to the rotor at the proper location in these planes to eliminate the dynamic unbalance (also called "imbalance").

Consider the cylinder of Fig. 5.53 for which the end planes $A$ and $B$ have been chosen as the correction planes. Any plane $i$ within the rotor perpendicular to the axis may contain a static imbalance $\mathbf{S}_i$. When the cylinder is rotating, this imbalance causes a centrifugal force $\mathbf{F}_i$:

$$\mathbf{F}_i = \mathbf{S}_i \omega^2$$

in which $\omega$ is the cylinder's angular velocity. We may now define the dynamic imbalance mass-moment of inertia vector in plane $B$ ($\mathbf{I}_B$) by way of the net torque about the center of plane $A$ due to the various forces $\mathbf{F}_i$. Thus, from Fig. 5.53,

$$\mathbf{I}_B \omega^2 \equiv \sum_i \mathbf{S}_i \omega^2 z_i$$

Since we will discuss rigid rotors only, we can suppress the angular velocity $\omega$, because there are no flexural deflections whose amplitude varies with the $\omega/\omega_n$ (frequency ratio), where $\omega_n$ is the flexural natural frequency of the rotor. Upon defining $\mathbf{I}_A$ by way of the net torque of the forces $\mathbf{F}_i$ acting about the center of plane $B$, we have

$$\mathbf{I}_B = \sum_i \mathbf{S}_i z_i$$

$$\mathbf{I}_A = \sum_i \mathbf{S}_i (L - z_i)$$

It is easy to see that a single static correction in each of planes $A$ and $B$ can be made with dynamic mass-moments of inertia vector equal but opposite to $\mathbf{I}_B$, respectively $\mathbf{I}_A$. In such a configuration the rigid rotor is dynamically balanced.

**Example 5.11**

A 10-in.-long rotor has known imbalances in planes (1) and (2) as shown in Fig. 5.54. The correction planes are $A$ and $B$. (a) Find the dynamic imbalance in the correction planes using both numerical and graphical methods; (b) correct the imbalance with masses placed 6 in. from the axis of rotation.

**Solution**

(a) It is convenient to construct the following table:

| | Plane | $m_i$ (oz) | $r_i$ (in.) | $z_i$ (in.) | $m_i r_i z_i$ (in²·oz) | $\theta_i$ (deg) | $m_i x_i z_i$ (in²·oz) | $m_i y_i z_i$ (in²·oz) |
|---|---|---|---|---|---|---|---|---|
| $\mathbf{I}_B$ | 1 | 5 | 4 | 5 | 100 | 30 | 86.6 | 50 |
| | 2 | 2 | 6 | 8 | 96 | 135 | −67.9 | 67.9 |
| | Sum | | | | | | 18.7 | 117.9 |
| $\mathbf{I}_A$ | 1 | 5 | 4 | 5 | 100 | 30 | 86.6 | 50 |
| | 2 | 2 | 6 | 2 | 24 | 135 | −17.0 | 17.0 |
| | Sum | | | | | | 69.6 | 67.0 |

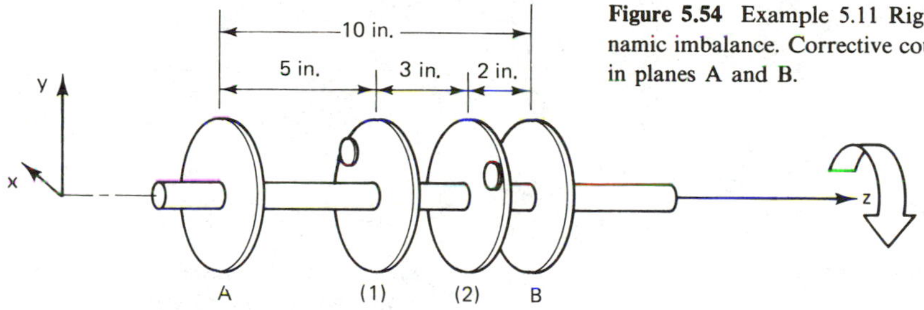


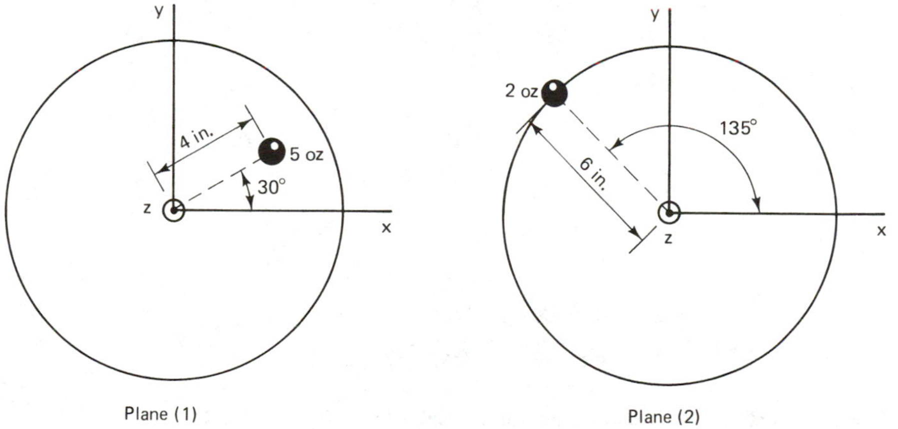


**Figure 5.54** Example 5.11 Rigid rotor with known dynamic imbalance. Corrective counterbalances are desired in planes A and B.

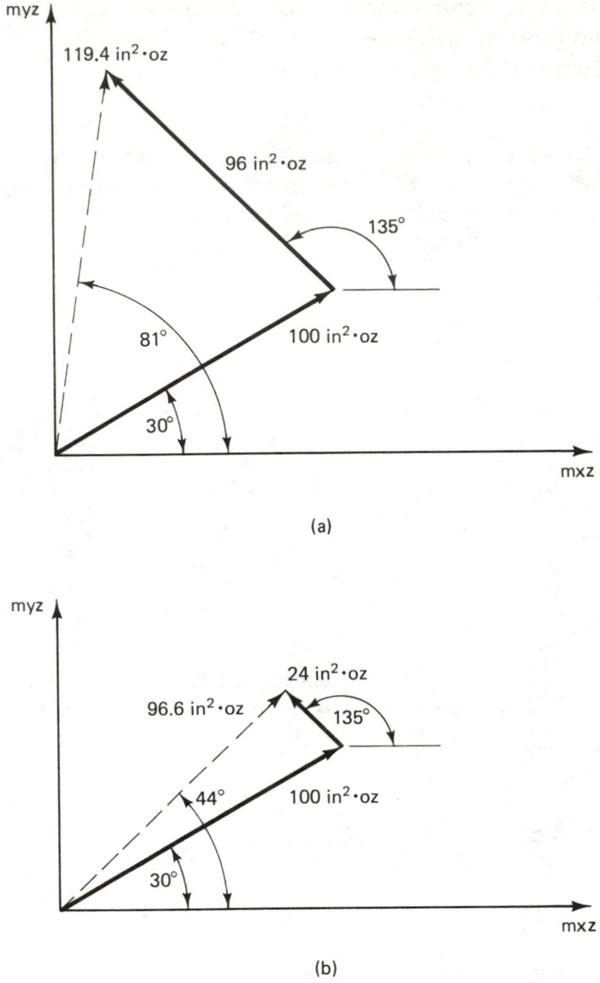


**Figure 5.55** Graphical solution to Ex. 5.11: equivalent imbalances in (a) plane A an (b) plane B.

From the last two columns, we have

$$I_B = |\sqrt{(18.7)^2 + (117.9)^2)}| = 119.4 \text{ in}^2\cdot\text{oz}$$

$$\theta_B = \arg(18.7 + i117.9) = 81.0°$$

and

$$I_A = |\sqrt{(69.6)^2 + (67.0)^2}| = 96.6 \text{ in}^2\cdot\text{oz}$$

$$\theta_A = \arg(69.6 + i67) = 43.9°$$

The graphical solution yielding the same results appears in Fig. 5.55.

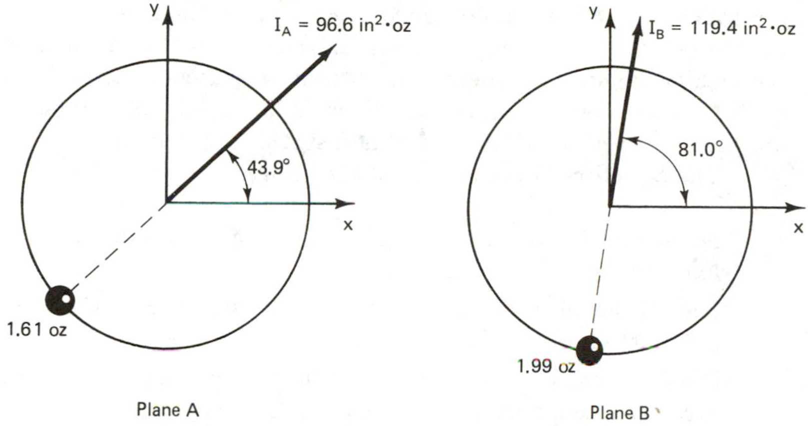


**Figure 5.56** Solution to Ex. 5.11.

(b) Correction masses:

$$m_B = \frac{I_B}{lr} = \frac{119.4 \text{ in}^2\cdot\text{oz}}{10 \text{ in.}\cdot 6 \text{ in.}} = 1.99 \text{ oz}, \qquad \theta_{MB} = \theta_B + 180° = 261°$$

$$m_A = \frac{I_A}{lr} = \frac{96.6 \text{ in}^2\cdot\text{oz}}{10 \text{ in.}\cdot 6 \text{ in.}} = 1.61 \text{ oz}, \qquad \theta_{MA} = \theta_A + 180° = 223.9°$$

These are located opposite the equivalent imbalances in planes $A$ and $B$ (see Fig. 5.56). The reader should verify that $m_A$ can also be found by statically balancing $m_1r_1$, $m_2r_2$, and $m_Br_B$, where $r_B = 6$ in., $\theta_{MB} = 261°$, as found above.

## 5.11 INTRODUCTION TO FORCE AND MOMENT BALANCING OF LINKAGES*

Mechanism design cannot be complete without focusing attention on the interface between that mechanism and its mounting frame. Berkof and Lowen [16–21,172–175] have addressed this problem in depth. Two methods, complementing each other, have been developed, permitting elimination of both shaking forces and shaking moments transmitted to ground.

Force balance is achieved by developing a set of linearly independent time-dependent vectors. These vectors define distributions of mass and locations of the centers of mass such that the center of mass for the entire system remains fixed. Thus the vector sum of the forces transmitted to ground from a force-balanced linkage is zero.

However, force balance does not eliminate shaking moments transmitted to

* The following sections (5.11 to 5.18) on linkage balancing are based on a draft by Dianne Rekow.

the frame. To achieve total balance, moment-of-momentum equations are written for the system. When the vector sum of the moments of momentum becomes zero, the shaking moments are eliminated. This may be accomplished through the addition of inertia counterweights and restrictions on link configurations (when these changes are possible given space constraints of a specific application).

The objectives of Secs. 5.11 to 5.18 are to:

Summarize the work of Berkof and Lowen (and others) for four- and six-bar linkages.

Discuss the effects on torque resulting from increased mass and/or mass redistribution required to achieve balance.

Briefly summarize optimization techniques for cases when design restrictions preclude meeting all the requirements for complete balance.

Describe other methods available to accomplish force and mass balancing.

Discuss the effects of counterweight shapes on torque requirements.

Provide a computer program flow chart for calculating the parameters for force and moment balancing of any four-bar linkage.

## 5.12 FORCE BALANCE

As a linkage moves it transmits forces to its surroundings. Unless it is balanced, these forces result in vibration, noise, wear, and cause fatigue problems. When completely force balanced, the vector sum of the forces acting on the frame is zero. This is accomplished by making the total center of mass for the mechanism stationary. One method to achieve this result is the *method of linearly independent vectors* introduced by Berkof and Lowen, which redistributes link masses so that the time-dependent terms of the equations of motion for the center of mass become zero. This becomes possible if one can obtain a position equation which provides time-dependent vectors that are linearly independent.

**Force balance for four-bar linkages [20].** Figure 5.57 is a four-bar linkage, $O_AABO_B$, containing three moving links of arbitrary mass distribution. Consider an $xoy$ system associated with the linkage system with origin $o$ at $O_A$. Let $S$ be the center of mass for the system of moving links and let $\mathbf{r}_s$ define the position of $S$ with respect to origin $o$. If the total moving mass of the system is given by:

$$M = \sum_{i=2}^{4} m_i \tag{5.123}$$

where $m_i$ is the mass of the $i$th link, then

$$M\mathbf{r}_s = \sum_{i=2}^{4} m_i \mathbf{r}_i \tag{5.124}$$

and $\mathbf{r}_i$ is the vector from the tail of the $i$th link vector $\mathbf{a}_i$ to the center of mass of

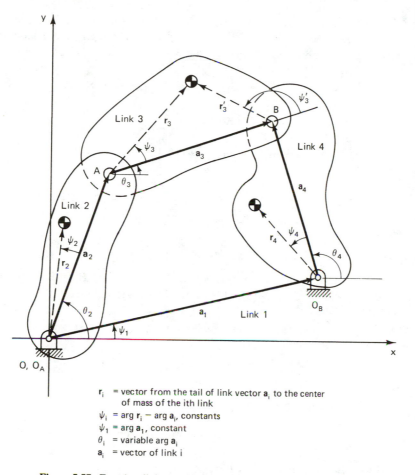

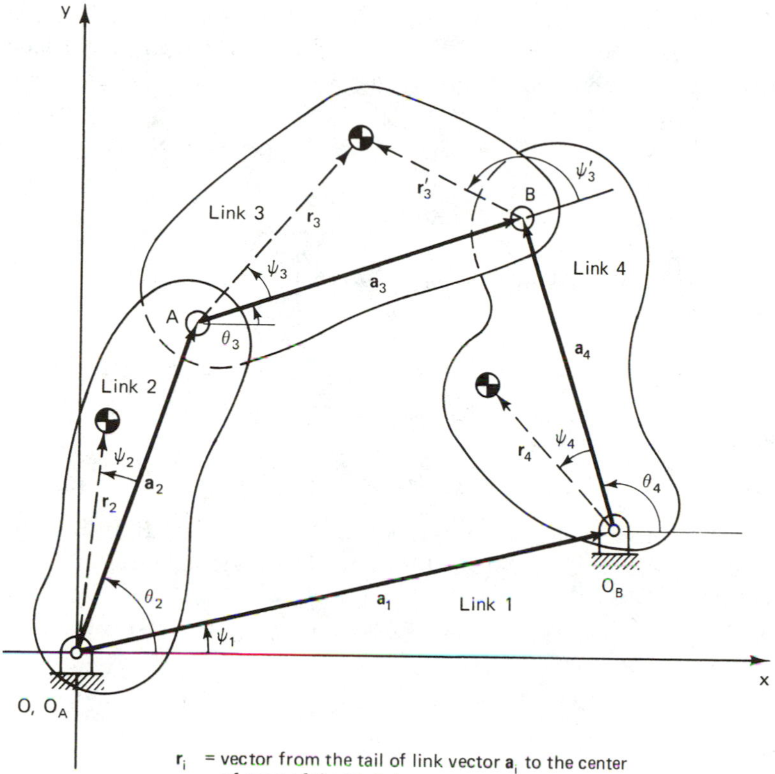


$\mathbf{r}_i$ = vector from the tail of link vector $\mathbf{a}_i$ to the center of mass of the ith link
$\psi_i$ = arg $\mathbf{r}_i$ − arg $\mathbf{a}_i$, constants
$\psi_1$ = arg $\mathbf{a}_1$, constant
$\theta_i$ = variable arg $\mathbf{a}_i$
$\mathbf{a}_i$ = vector of link i

**Figure 5.57** Four-bar linkage with arbitrary distribution of link masses.

the $i$th link. From Fig. 5.57,

$$\mathbf{r}_2 = r_2 e^{i(\theta_2+\psi_2)} \tag{5.125}$$

$$\mathbf{r}_3 = r_3 e^{i(\theta_3+\psi_3)} + a_2 e^{i\theta_2} \tag{5.126}$$

$$\mathbf{r}_4 = r_4 e^{i(\theta_4+\psi_4)} + a_1 e^{i\psi_1} \tag{5.127}$$

Upon substitution of Eqs. (5.125) to (5.127), with rearrangements, Eq. (5.124) becomes

$$M\mathbf{r}_s = m_2 r_2 e^{i(\theta_2+\psi_2)} + m_3(r_3 e^{i(\theta_3+\psi_3)} + a_2 e^{i\theta_2}) + m_4(r_4 e^{i(\theta_4+\psi_4)} + a_1 e^{i\psi_1}) \tag{5.128}$$

We also know from the loop-closure equation that

$$a_2 e^{i\theta_2} + a_3 e^{\theta_3} - a_4 e^{i\theta_4} - a_1 e^{i\psi_1} = 0 \tag{5.129}$$

Equation (5.129) implies that the time-dependent terms of Eq. (5.128) are not linearly independent. However, if Eq. (5.129) is solved for one time-dependent unit vector and substituted into Eq. (5.128), the linearly independent set of time-dependent terms can be obtained. Suppose that we solve Eq. (5.129) for $e^{i\theta_3}$. We obtain

$$e^{i\theta_3} = \frac{1}{a_3}(a_4 e^{i\theta_4} + a_1 e^{i\psi_1} - a_2 e^{i\theta_2}) \tag{5.130}$$

Substitution of Eq. (5.130) into Eq. (5.128) gives

$$\begin{aligned} M\mathbf{r}_s = &\left[ m_2 r_2 e^{i\psi_2} + m_3 a_2 - m_3 r_3 \frac{a_2}{a_3} e^{i\psi_3} \right] e^{i\theta_2} \\ &+ \left[ m_4 r_4 e^{i\psi_4} + \frac{a_4}{a_3} m_3 r_3 e^{i\psi_3} \right] e^{i\theta_4} + \left[ m_4 a_1 + m_3 r_3 \frac{a_1}{a_3} e^{i\psi_3} \right] e^{i\psi_1} \end{aligned} \tag{5.131}$$

which is of the form

$$M\mathbf{r}_s = \mathbf{A} e^{i\theta_2} + \mathbf{B} e^{i\theta_4} + \mathbf{C} \tag{5.132}$$

If the time-dependent terms vanish (i.e., if $\mathbf{A} = \mathbf{B} = \mathbf{0}$), then $M\mathbf{r}_s$ is constant, meeting the criterion required for complete force balance. When $\mathbf{A} = 0$,

$$m_2 r_2 e^{i\psi_2} + m_3 a_2 - m_3 r_3 \frac{a_2}{a_3} e^{i\psi_3} = 0 \tag{5.133}$$

From Fig. 5.57 we see that

$$r_3 e^{i\psi_3} = a_3 + r_3' e^{i\psi_3'} \tag{5.134}$$

Combining Eqs. (5.133) and (5.134) gives

$$m_2 r_2 e^{i\psi_2} = m_3 r_3' \frac{a_2}{a_3} e^{i\psi_3'} \tag{5.135}$$

which can only be satisfied if

$$m_2 r_2 = m_3 r_3' \frac{a_2}{a_3} \qquad \text{and} \qquad \psi_2 = \psi_3' \tag{6.136}$$

When $B = 0$, we obtain

$$m_4 r_4 = m_3 r_3 \frac{a_4}{a_3} \qquad \text{and} \qquad \psi_4 = \psi_3 + \pi \tag{5.137}$$

When these conditions [Eqs. (5.136) and (5.137)] are met, the vector locating the center of gravity for the moving links is given by [see Eq. (5.132)]

$$\mathbf{r}_s = \frac{1}{M}\left( m_4 a_1 + m_3 r_3 \frac{a_1}{a_3} e^{i\psi_3} \right) e^{i\psi_1} = \frac{1}{M}\mathbf{C} \tag{5.138}$$

which is constant.

Since the center of mass of the entire mechanism is kept stationary, full force balance is maintained regardless of variations in input speed. From an applications perspective, Eqs. (5.136) and (5.137) imply that whenever the mass and location of the center of mass of one link of the mechanism is prescribed (link 3 in this derivation), the mass distribution of the remaining links can be arranged to provide full force balance [19]. Further, since the coefficients of the time-dependent variables are zero, this method can be utilized independent of restrictions on mobility or other kinematic constraints for the linkage.

In the case where the links cannot be altered, an equivalent result is obtained by adding two counterweights, attached to any two moving links. Generally, it is most convenient to select the input and output links. There are two constraints on the selection of counterweight position and mass:

Constraint 1:

$$m_i r_i e^{i\psi_i} = m_i^\circ r_i^\circ e^{i\psi_i^\circ} + m_i^* r_i^* e^{i\psi_i^*} \tag{5.139}$$

where $m_i, r_i, \psi_i$ = parameters obtained from Eqs. (5.136) and (5.137)

$m_i^\circ, r_i^\circ, \psi_i^\circ$ = parameters for the unbalanced linkage

$m_i^*, r_i^*, \psi_i^*$ = parameters for the counterweights

Constraint 2:

$$m_i = m_i^\circ + m_i^* \tag{5.140}$$

required only in cases when *both* fixed pivot links (2 and 4) are chosen to receive counterweights. Solving Eq. (5.139) for $m_i^* r_i^*$ yields

$$m_i^* r_i^* = [(m_i r_i)^2 + (m_i^\circ r_i^\circ)^2 - 2 m_i r_i m_i^\circ r_i^\circ \cos(\psi_i - \psi_i^\circ)]^{1/2} \tag{5.141}$$

and

$$\begin{aligned}\psi_i^* = \arg\{&(m_i r_i \cos\psi_i - m_i^\circ r_i^\circ \cos\psi_i^\circ) \\ &+ i(m_i r_i \sin\psi_i - m_i^\circ r_i^\circ \sin\psi_i^\circ)\}\end{aligned} \tag{5.142}$$

**Example 5.12**

Consider an example (Fig. 5.58) of a four-bar linkage. Link dimensions and masses are given in Table 5.9. Assume for this case that the links can be altered and that we select links 2 and 4 for alteration. Further assume that we elect to change only the masses, leaving distance to the center of mass unchanged. From Eqs. (5.136) and (5.137) we see that $\psi_2 = \psi_3'$ and that $\psi_4 = \psi_3 + \pi$. Since we assume that $r_2^\circ$ is to be kept constant, to determine the required balancing mass Eq. (5.136) becomes

$$(m_2^\circ + x_2) r_2^\circ = m_3^\circ r_3'^\circ \left(\frac{a_2}{a_3}\right) \qquad \text{and} \qquad \psi_2 = \psi_3'^\circ = 163.3^\circ$$

which gives $x_2 = 0.178$ lbm. Thus for balance, the center of mass of link 2 must shift to $\psi_2 = 163.3^\circ$ and increase to 0.280 lbm. This is achieved by altering the links rather than by adding counterweights.

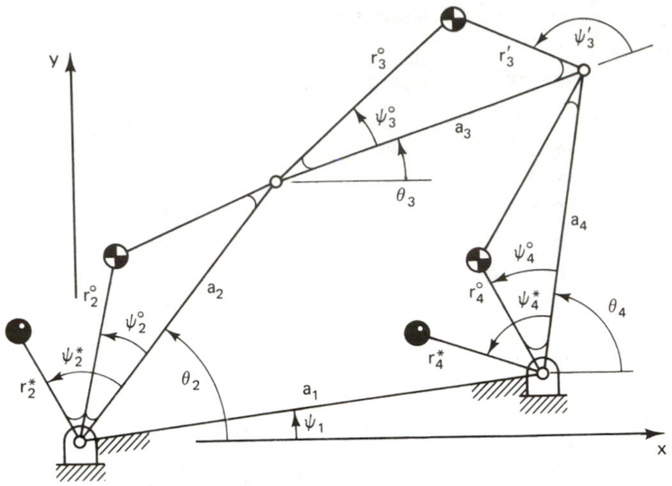


**Figure 5.58** Configuration for unbalanced four-bar mechanism with position of balancing counterweights shown as $r_i^*$ (not to scale) (Ex. 5.12).

**TABLE 5.9** PARAMETERS OF UNBALANCED FOUR-BAR MECHANISM

| Parameter | Link 1 | 2 | 3 | 4 |
|---|---|---|---|---|
| $a$ (in) | 5.500 | 2.000 | 6.000 | 3.000 |
| $r^\circ$ (in.) | — | 1.000 | 3.190 | 1.500 |
| $\psi^\circ$ (deg) | 0 | 0 | 16.0 | 0 |
| $r'^\circ$ (in.) | — | — | 3.063 | — |
| $\psi'^\circ$ (deg) | — | — | 163.3 | — |
| $m^\circ$ (lbm) | — | 0.102 | 0.274 | 0.120 |

Equation (5.137) becomes

$$(m_4^\circ + x_4)r_4^\circ = m_3^\circ r_3^\circ \left(\frac{a_4}{a_3}\right) \qquad \text{and} \qquad \psi_4 = \psi_3 + \pi = 196.0^\circ$$

which gives $x_4 = 0.171$ lbm. Thus we need to shift the center of mass of link 4 to 196° and increase the mass to 0.291 lbm by altering the link (not adding counterweights).

**Example 5.13**

Consider the same four-bar linkage, but suppose now that none of the links can be altered. From Eqs. (5.139) to (5.142), we can determine the parameters for counterweights which can provide the balance. Suppose that we elect to add the counterweights to links 2 and 4. As before, for balance, we must have $\psi_2 = \psi_3'$ and $\psi_4 = \psi_3 + \pi$. Then, from Eq. (5.141), we have

$$m_i^* r_i^* = [(m_i r_i)^2 + (m_i^\circ r_i^\circ)^2 - 2 m_i r_i m_i^\circ r_i^\circ \cos(\psi_i - \psi_i^\circ)]^{1/2}$$

but from Eq. (5.136) we know that

$$m_2 r_2 = m_3^{\circ} r_3'^{\circ}\left(\frac{a_2}{a_3}\right) \qquad \text{and} \qquad \psi_2 = \psi_3'^{\circ} = 163.3^{\circ}$$

Then equation (5.141) becomes

$$m_2^* r_2^* = \left[\left(m_3^{\circ} r_3'^{\circ}\frac{a_2}{a_3}\right)^2 + (m_2^{\circ} r_2^{\circ})^2 - 2\left(m_3^{\circ} r_3'^{\circ}\frac{a_2}{a_3}\right)(m_2^{\circ} r_2^{\circ})\cos(\psi_3'^{\circ} - \psi_2^{\circ})\right]^{1/2} = 0.379$$

From Eq. (5.142),

$$\psi_2^* = \arg\left\{\left(m_3^{\circ} r_3'^{\circ}\frac{a_2}{a_3}\cos\psi_3'^{\circ} - m_2^{\circ} r_2^{\circ}\cos\psi_2^{\circ}\right) + i\left(m_3^{\circ} r_3'^{\circ}\frac{a_2}{a_3}\sin\psi_3'^{\circ} - m_2^{\circ} r_2^{\circ}\sin\psi_2^{\circ}\right)\right\}$$

which yields $\psi_2^* = 154.2^{\circ}$.

Any combination of balancing-counterweight mass ($m_i^*$) and its distance to center of mass ($r_i^*$) can be utilized, as long as their product remains equal to 0.379 lbm·in. and the center of this mass is located at $\psi_2^* = 154.2^{\circ}$. Similar calculations for link 4 utilize Eq. (5.137) to determine $m_4$, $r_4$, and $\psi_4$. If $\psi_4 = \psi_3^{\circ} + \pi$, $\psi_4^* = 168.3^{\circ}$ and $m_4^* r_4^* = 0.595$.

Because the balancing-counterweight masses of links 2 and 4 appear only as their mass–distance products and never as separate terms, the second constraint ($m_i = m_i^* + m_i^{\circ}$) is not required, leaving a very wide range of solutions available to the designer.* Figure 5.59 illustrates schematically (not to scale) one solution to this example.

When counterweights are to be attached to the coupler as well as one of the other links, Eq. (5.137) must first be utilized to obtain $m_3 r_3$ and $\psi_3$. In order to determine

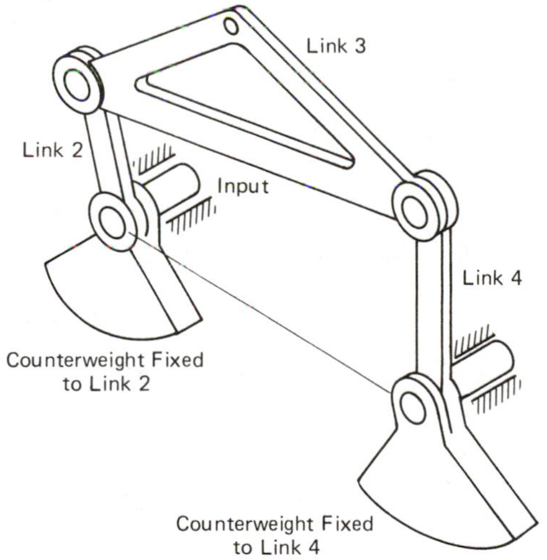


**Figure 5.59** Counterweight balancing of four-bar mechanism.

* L. Dearstyne of Eastman Kodak Company points out that in six-bar examples, the mass constraint equation applies in all cases except when three fixedly pivoted links are chosen to receive counterweights. Thus, two-pivot six-bars are slightly less flexible with regard to choice and placement of counterweights.

$r_3'$ and $\psi_3'$, either $m_3$ or $r_3$ must be arbitrarily selected. Suppose that links 3 and 4 are to be altered. Then, when $m_3$ or $r_3$ is selected, $m_2$, $r_2$, and $\psi_2$ can be established from Eq. (5.136). Then $m_3^* r_3^*$ and $\psi_3^*$ can be established as before. In this case, the constraint on mass is required for link 3 only, since $m_3$ or $r_3$ were selected.

While this method of balancing does ensure that the vector sum of the dynamic forces acting on the mechanism frame is always zero, individual reaction forces at the supports are not zero and are time varying; each support is being subjected to a cyclic shaking force. Indeed, the magnitude of these forces may be greater in the balanced mechanism, because of the added weights, than in the unbalanced case. In general, these reaction forces impart a nonzero, time-varying moment to the frame. This is why we will later take up moment balancing.

**Force balance for six-bar linkages [16,18].** Figure 5.60 is a generalized six-bar planar linkage with arbitrary link-mass distributions. From the figure we see that

$$\begin{aligned}
\mathbf{r}_{s2} &= r_2 e^{i(\theta_2+\psi_2)} \\
\mathbf{r}_{s3} &= r_3 e^{i(\theta_3+\psi_3)} + a_2 e^{i\theta_2} \\
\mathbf{r}_{s4} &= r_4 e^{i(\theta_4+\psi_4)} + a_1 \\
\mathbf{r}_{s5} &= b_2 e^{i(\theta_2+\alpha_2)} + r_5 e^{i(\theta_5+\psi_5)} \\
\mathbf{r}_{s6} &= a_1 + b_4 e^{i\theta_4} + r_6 e^{i(\theta_6+\psi_6)}
\end{aligned} \tag{5.143}$$

where $\alpha_i$ is the angle between two sides on ternary link $i$. We also see that

$$\begin{aligned}
r_3 e^{i\psi_3} &= a_3 + r_3' e^{i\psi_3'} \\
r_5 e^{i\psi_5} &= a_5 + r_5' e^{i\psi_5'} \\
b_4 e^{i\theta_4} &= a_4 e^{i(\theta_4+\alpha_4)} + b_4' e^{i\alpha_4'}
\end{aligned} \tag{5.144}$$

For Fig. 5.60 the loop-closure equations are

$$a_2 e^{i\theta_2} + a_3 e^{i\theta_3} - a_4 e^{i(\theta_4+\alpha_4)} - a_1 = 0 \tag{5.145}$$

$$b_2 e^{i(\theta_2+\alpha_2)} + a_5 e^{i\theta_5} - a_6 e^{i\theta_6} - b_4 e^{i\theta_4} - a_1 = 0 \tag{5.146}$$

Because there are five moving links, two of the $\theta_i$ terms can be eliminated ($\theta_3$ and $\theta_5$) in this derivation. By arguments exactly paralleling those developed for the four-bar, we see that force balance can be achieved only if the center of mass for the system can be made to remain stationary, giving

$$M\mathbf{r}_s = \sum_{i=2}^{6} m_i \mathbf{r}_{si} \tag{5.147}$$

Substituting Eq. (5.143) and (5.144) into the loop-closure equations (5.145) and

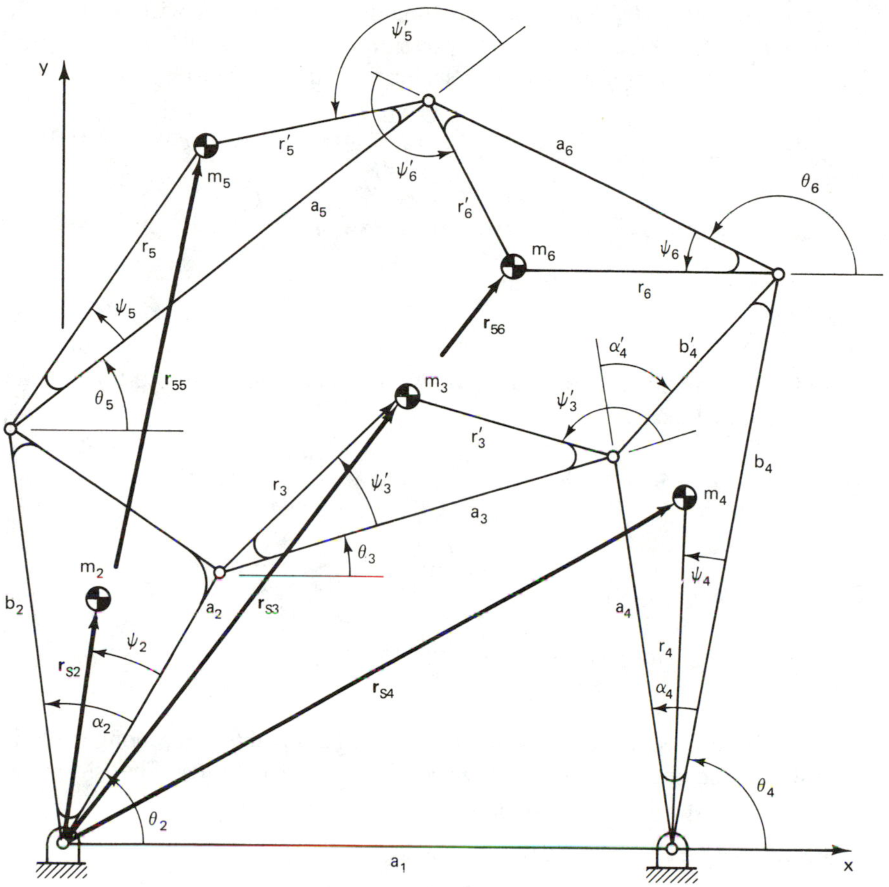


**Figure 5.60** Six-bar linkage with arbitrary link mass distributions.

(5.146), expressing $e^{i\theta_3}$ from Eq. (5.145) and $e^{i\theta_5}$ from Eq. (5.146) and collecting terms, we obtain

$$\begin{aligned} M\mathbf{r}_s = {} & \left[ m_2 r_2 e^{i\psi_2} - m_3 r_3' \frac{a_2}{a_3} e^{i\psi_3'} - m_5 r_5' \frac{b_2}{a_5} e^{i(\psi_5' + \alpha_2)} \right] e^{i\theta_2} \\ & + \left[ m_3 r_3 \frac{a_4}{a_3} e^{i(\psi_3 + \alpha_4)} + m_4 r_4 e^{i\psi_4} + m_5 r_5 \frac{b_4}{a_5} e^{i\psi_5} + m_6 b_4 \right] e^{i\theta_4} \\ & + \left[ m_5 r_5 \frac{a_6}{a_5} e^{i\psi_5} + m_6 r_6 e^{i\psi_6} \right] e^{i\theta_6} \\ & + \left[ m_3 r_3 \frac{a_1}{a_3} e^{i\psi_3} + m_5 r_5 \frac{a_1}{a_5} e^{i\psi_5} + m_4 a_1 + m_6 a_1 \right] \end{aligned} \qquad (5.148)$$

To make $M\mathbf{r}_s$ constant, the time-dependent terms with $e^{i\theta}{}_j$, $=$ 2, 4, 6, must be zero. All three bracketed expressions multiplying $e^{i\theta_j}$ are independent of one another. Therefore, we set each separately equal to zero:

$$m_2 r_2 e^{i\psi_2} - m_3 r'_3 \frac{a_2}{a_3} e^{i\psi'_3} - m_5 r'_5 \frac{b_2}{a_5} e^{i(\psi'_5+\alpha_2)} = 0 \tag{5.149}$$

$$m_3 r_3 \frac{a_4}{a_3} e^{i(\psi_3+\alpha_4)} + m_4 r_4 e^{i\psi_4} + m_5 r_5 \frac{b_4}{a_5} e^{i\psi_5} + m_6 b_4 = 0 \tag{5.150}$$

$$m_5 r_5 \frac{a_6}{a_5} e^{i\psi_5} + m_6 r_6 e^{i\psi_6} = 0 \tag{5.151}$$

Substitution and rearrangement yields the conditions required for balance of the six-bar planar linkage:

$$m_2 \frac{r_2}{a_2} e^{i\psi_2} = m_3 \frac{r'_3}{a_3} e^{i\psi'_3} + m_5 \frac{r'_5}{a_5} \frac{b_2}{a_2} e^{i(\alpha_2+\psi'_5)} \tag{5.152}$$

$$m_4 \frac{r_4}{a_4} e^{i\psi_4} = -m_5 \frac{r_5}{a_5} \frac{b_4}{a_4} e^{i\psi_6} - m_3 \frac{r_3}{a_3} e^{i(\psi_3+\alpha_4)} - m_6 \frac{b_4}{a_4} \tag{5.153}$$

$$\mathrm{m}_5 \frac{r_5}{a_5} e^{i\psi_5} = -m_6 \frac{r_6}{a_6} e^{i\psi_6} \tag{5.154}$$

Again, as in the four-bar, it may not be possible to reconfigure links to achieve balance and one is forced to utilize counterweights instead. That situation is discussed by way of the following example.

**Example 5.14: Balancing of Six-Bar Linkage [18]**

Figures 5.61 and 5.62 give a schematic and an exploded view, respectively, of balancing a six-bar planar linkage. Known parameters for the unbalanced linkage are given in Table 5.10. Achieving balance by reconfiguring links is fairly straightforward in most cases. Only the case of adding counterweights will be considered here.

**TABLE 5.10** PARAMETERS OF UNBALANCED SIX-BAR LINKAGE (Ex. 5.14, Figs. 5.51 and 5.62.)

| Link | 2 | 3 | 4 | 5 | 6 | 1 |
|---|---|---|---|---|---|---|
| $a_i$ (in.) | 2.200 | 4.750 | 0.125 | 5.500 | 1.750 | 4.875 |
| $b_i$ (in.) | 2.300 | — | 0.120 | 6.000 | — | — |
| $\alpha_i$ (deg) | 6.0 | — | 16.0 | 40.4 | — | — |
| $r_i^\circ$ (in.) | 1.125 | 2.480 | 0.122 | 3.290 | 0.776 | — |
| $\psi_i^\circ$ (deg) | 3.0 | 0.0 | 5.0 | 19.0 | 0.0 | 11.0 |
| $r'_i$ (in.) | — | 2.270 | — | 2.618 | — | — |
| $\psi_i^{\circ\prime}$ (deg) | — | 180.0 | — | 155.9 | — | — |
| $m_i^\circ$ (lbf) | 0.134 | 0.182 | 0.167 | 0.382 | 0.087 | — |

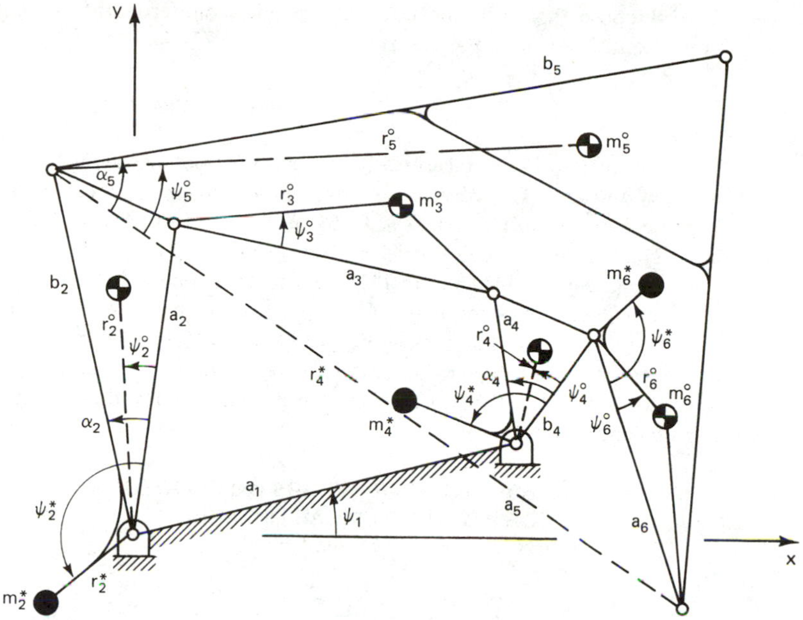


**Figure 5.61** Balancing of six-bar linkage by addition of counterweights to links 2, 4, and 6 (Ex. 5.14).

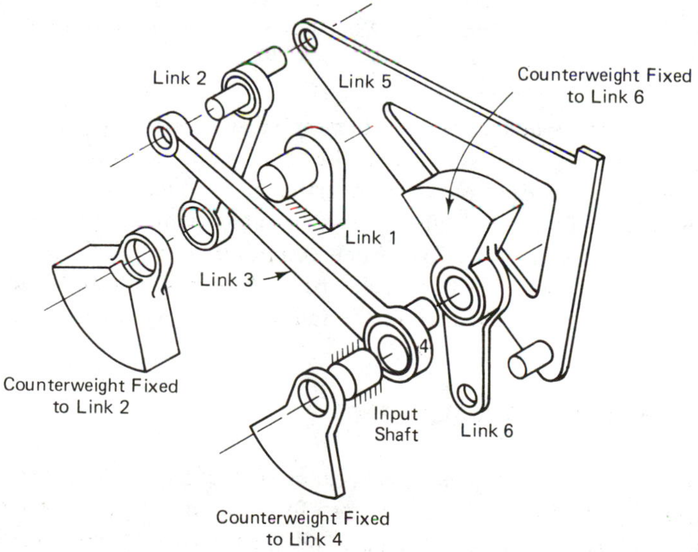


**Figure 5.62** Counterweight balancing of six-bar linkage (Ex. 5.14) (exploded view).

Suppose that it is most feasible to add counterweights to links 2, 4, and 6. From Eq. (5.154) (in scalar form), we have

$$m_6 r_6 = m_5^\circ r_5^\circ \frac{a_6}{a_5} \qquad \text{and} \qquad \psi_6 = \psi_5^\circ + \pi \tag{5.155}$$

From the values in Table 5.10, we can now obtain values for the balanced totals of $m_6 r_6$ and $\psi_6$. In order to continue, either $m_6$ or $r_6$ must be arbitrarily chosen. Suppose that we elect to set $r_6$ at 0.809 in. Then, from Eq. (5.155), we can solve for $m_6$. Then $r_6'$ and $\psi_6'$ are established from geometry [Eq. (5.144)].

From Eqs. (5.141) and (5.142) we can determine $m_6^* r_6^*$ and $\psi_6^*$.

Balanced values for $m_2 r_2$, $\psi_2$, $m_4 r_4$, and $\psi_4$ can now be established from Eqs. (5.152) and (5.153). Again, using Eqs. (5.141) and (5.142), values for the mass, distance to the center of mass, and angles for the counterweights ($m_i^* r_i^*$ and $\psi_i^*$) can be determined for links 2 and 4. Table 5.11 gives the parameters for the linkage when it is fully force balanced.

**TABLE 5.11** PARAMETERS CALCULATED FOR BALANCED SIX-BAR LINKAGE (Ex. 5.14, Figs. 5.61 and 5.62.)

| Link | 2 | 4 | 6 |
|---|---|---|---|
| $m_i r_i$ (lbf·in.) | 0.630 | 0.097 | 0.400[a] |
| $\psi_i$ (deg) | 167.5 | 187.2 | 199.0 |
| $m_i^* r_i^*$ (lbf·in.) | 0.749 | 0.118 | 0.465 |
| $\psi_i^*$ (deg) | 170.6 | 186.8 | 196.3 |
| $m_i^*$ (lbf) | 0.607 | 0.117 | 0.415 |
| $R_i$ (in.) | 2.000 | 1.675 | 1.750 |

[a] $r_6 = 0.809$ by choice leads to $m_6 = 0.494$ lbf, $r_6' = 2.529$ in., $\psi_6' = 186.0°$.

It is to be noted that the foregoing discussions apply only to in-plane balancing. All the links are not moving in the same plane, although they do move in parallel planes. This leads to out-of-plane unbalance, which can be avoided if the mechanism is built with links all of which have the same plane of symmetry (e.g., see connecting rod and slides in Fig. 5.75). Such condition is accomplished by, say, twin couplers of a four-bar rocker-rocker mechanism. Also, the counterbalancing additions should also have their mass center in the same plane of symmetry.

## 5.13 SHAKING MOMENT BALANCING

Force balancing provides a zero vector sum of the inertial forces acting on the frame supports but does not provide zero forces at individual supports. The resultant of these forces will in general be a pure time-varying couple: the shaking moment. If the shaking moment also can be reduced to zero, the mechanism can be completely balanced, avoiding the unpleasant problems of vibration, noise, wear, and fatigue.

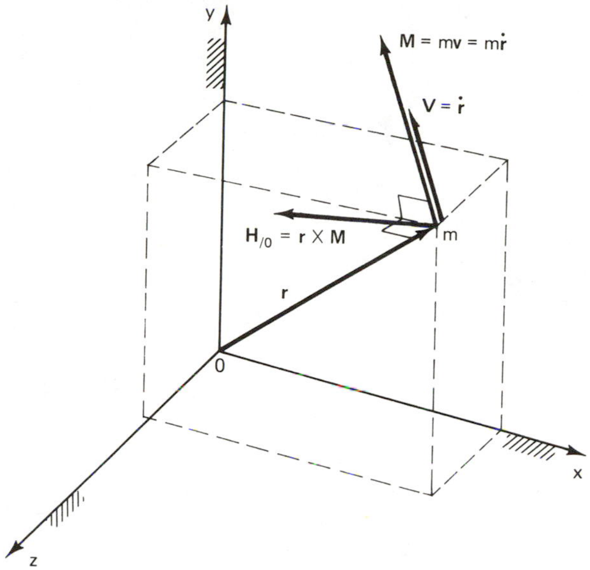


**Figure 5.63** Definition of momentum **M** and moment of momentum (or angular momentum) $H_{/O}$ of the point mass $m$ located at **r** and moving at velocity $\mathbf{v} = \dot{\mathbf{r}}$ with respect to three-dimensional cartesian coordinate system *Oxyz*.

Optimization techniques will permit minimization of shaking moments when complete balance is impossible because of conflicting requirements for force and moment balancing.

**Definition of shaking moment for planar mechanisms [20].** For this study it is necessary to review the concept of angular momentum. From elementary physics, recall the concepts of *momentum* (**M**) and the *moment of momentum* (**H**), which is also called the *angular momentum*. Referring to Fig. 5.63, let the mass point $m$ move with velocity **v** with respect to the three-dimensional coordinate system *Oxyz*. The momentum of $m$ is defined as

$$\mathbf{M} = m\mathbf{v} = m\dot{\mathbf{r}} \tag{5.156}$$

Thus it is seen that the momentum is a vector collinear with the velocity. Its moment about origin $O$ is also a vector, defined by

$$\mathbf{H}_{/O} = \mathbf{r} \times \mathbf{M} = \mathbf{r} \times m\dot{\mathbf{r}} \tag{5.157}$$

which means that $\mathbf{H}_{/O}$, the moment of momentum of $m$ about $O$, is a vector perpendicular to both $\mathbf{r}$ and $\dot{\mathbf{r}}$. According to the rules for vector products (or cross products of vectors) it must obey the right-hand rule, following the sequence $\mathbf{r}$, $\dot{\mathbf{r}}$, $\mathbf{H}_{/O}$ (Fig. 5.63). Note that $\dot{\mathbf{r}}$ is not necessarily perpendicular to $\mathbf{r}$ and that the angular velocity $\dot{\theta}$ is zero because we are dealing with a point mass.

Now consider a finite body of arbitrary shape (having mass $m$ and radius of gyration $k$) in pure rotation about its mass center (see Fig. 5.64) at an angular velocity

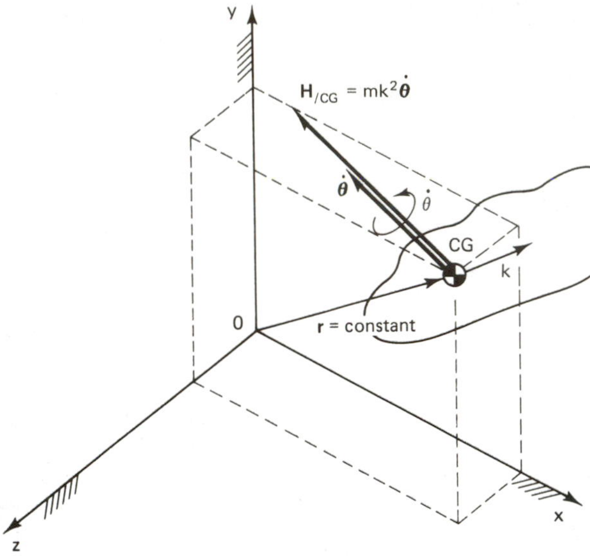


**Figure 5.64** Angular momentum $H_{/CG} = m\,k^2$ of an arbitrary body in three-dimensional space about its center of mass (cg).

of $\dot{\boldsymbol{\theta}}$. Note that here **r** is constant. The angular momentum about the mass center (CG) is

$$\mathbf{H}_{/CG} = mk^2\dot{\boldsymbol{\theta}} \tag{5.158}$$

where $k$ is in the plane of rotation perpendicular to $\dot{\boldsymbol{\theta}}$.

This can be verified by thinking of $m$ as concentrated in a thin annulus at radius $k$ from the center of mass in the plane normal to the vector $\dot{\boldsymbol{\theta}}$. Then every point on this annulus travels at a peripheral speed of $k\dot{\boldsymbol{\theta}}$, which yields the magnitude of the total peripheral linear momentum of the annulus as $mk\dot{\boldsymbol{\theta}}$. The magnitude of the moment of this momentum (or angular momentum) about the center of mass is $mk^2\dot{\boldsymbol{\theta}}$.

Now, if this arbitrary body translates with respect to $O$ as well as rotates about its center of mass, in other words, if **r** is not constant and $\dot{\theta}$ is not zero, the total angular momentum relative to $O$ is

$$\mathbf{H}_{/O} = \mathbf{r} \times m\dot{\mathbf{r}} + \mathbf{H}_{/CG} = \mathbf{r} \times m\dot{\mathbf{r}} + mk^2\dot{\boldsymbol{\theta}} \tag{5.159}$$

In dealing with planar linkages, we observe that the "arbitrary body," link $i$, is in planar motion in the $Oxy$ plane. Therefore, all angular velocity vectors and angular momentum vectors are parallel to the $z$ axis, as shown in Fig. 5.65.

The principle of angular momentum states that the time rate of change of angular momentum equals the sum of the externally applied moments. This is easily shown by recalling Newton's second law. For a point mass, in its original form, it is given by

$$\mathbf{F}\,\Delta t = \Delta(m\mathbf{v}) \tag{5.160}$$

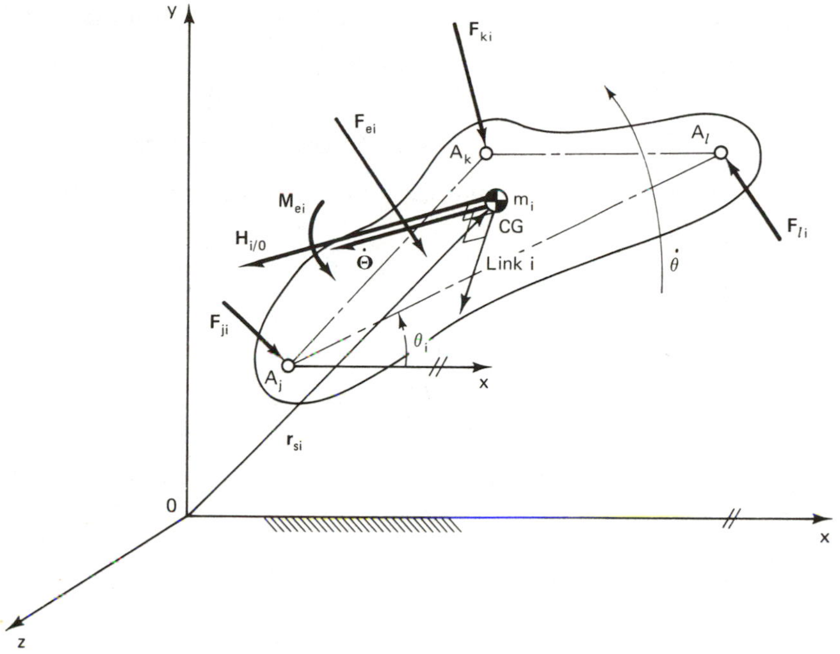


**Figure 5.65** Arbitrary link *i* in rotation about the z axis. *Oxyz:* right-hand spatial cartesian system.

where $\mathbf{F}$ = external force acting on the point mass $m$

$\Delta t$ = time during which $\mathbf{F}$ acts

$\Delta(m\mathbf{v})$ = resulting momentum change of $m$

Now, if $\mathbf{F}$ is a peripheral force acting on $m$ in orbit about $O$ at radius vector $\mathbf{r}$ with peripheral velocity $\mathbf{v}$, the moment of the force $\mathbf{F}$ and the moment-of-momentum change $\Delta(m\mathbf{v})$ is expressed as

$$\mathbf{r} \times \mathbf{F}\,\Delta t = \Delta\mathbf{H}_{/O} \qquad \text{or} \qquad \mathbf{r} \times \mathbf{F} = \frac{\Delta\mathbf{H}_{/O}}{\Delta t} \tag{5.161}$$

which, in the limit, is

$$\mathbf{r} \times \mathbf{F} = \frac{d}{dt}(\mathbf{H}_{/O}) \tag{5.162}$$

Any externally applied moments are given by

$$\mathbf{M}_{i/O} = \mathbf{M}_{ei} + \mathbf{r}_{ei} \times \mathbf{F}_{ei} + \mathbf{M}_{ki} + \mathbf{r}_{ki} \times \mathbf{F}_{ki} + \mathbf{r}_{gi} \times \mathbf{F}_{gi} + \mathbf{M}_{gi} \tag{5.163}$$

where the subscript *ei* implies externally applied

*gi* implies from ground

*ki* implies from another link

and where $r_{ji}$, $j = e, g, k$, are vectors from the origin to a point on the line of action of $\mathbf{F}_{ji}$.

$$\mathbf{F}_{gi} = \mathbf{M}_{gi} = 0 \qquad \text{for links not connected to ground}$$

$$\mathbf{M}_{gi} \neq 0 \qquad \text{for input and output links}$$

Thus, by the principle of angular momentum, the moment acting on the $i$th link ($\mathbf{M}_{i/O}$) is given by

$$\mathbf{M}_{i/O} = \frac{d}{dt}\mathbf{H}_{i/O} \tag{5.164}$$

So for the complete mechanism,

$$\mathbf{M}_O = \frac{d}{dt}\mathbf{H}_O = \frac{d}{dt}\Sigma\,\mathbf{H}_{i/O} = \Sigma\,\mathbf{M}_{i/O} \tag{5.165}$$

Upon substitution of Eqs. (5.159), (5.164), and (5.165) into Eq. (5.163) and summing for the entire mechanism, we have

$$\sum_{i=2}^{n}\{[\mathbf{M}_{ei} + \mathbf{r}_{ei} \times \mathbf{F}_{ei} + \Sigma\,(\mathbf{M}_{ki} + \mathbf{r}_{ki} \times \mathbf{F}_{ki})] + \mathbf{M}_{gi}\} + \sum_{j=a}^{p}(\mathbf{r}_{gj} \times \mathbf{F}_{gj})$$
$$= \frac{d}{dt}\sum_{i=2}^{n}(\mathbf{r}_{si} \times m_i\dot{\mathbf{r}}_{si} + m_i k_i^2 \dot{\boldsymbol{\theta}}_i) = \frac{d}{dt}\mathbf{H}_O \tag{5.166}$$

where $a$ through $p$ are all grounded links and where $i = 2, \ldots, n$ enumerates all moving links.

Since bearing reactions for all adjacent links are equal and opposite ($r_{ki} = -r_{ik}$, $F_{ki} = -F_{ik}$, and $M_{ki} = -M_{ik}$), those terms cancel and Eq. (5.166) reduces to

$$\sum_{i=2}^{n}[(\mathbf{M}_{ei} + \mathbf{r}_{ei} \times \mathbf{F}_{ei}) + \mathbf{M}_{gi}] + \sum_{j=a}^{p}\mathbf{r}_{gj} \times \mathbf{F}_{gj} = \frac{d}{dt}\sum_{i=2}^{n}(\mathbf{r}_{si} \times m_i\dot{\mathbf{r}}_{si} + m_i k_i^2 \dot{\boldsymbol{\theta}}_i) \tag{5.167}$$

For the mechanism, there is a total moment transmitted to ground ($\mathbf{M}_{M/G/O}$) taken with respect to some arbitrary origin $O$. This moment consists of the ground reaction due to the input torque as well as the moment resulting from the ground bearing forces, giving, for the case when there is only one input link and no torque load on the output link:

$$\mathbf{M}_{M/G/O} = \mathbf{M}_{\text{in}\,g} + \sum_{j=a}^{p}(\mathbf{r}_{jg} \times \mathbf{F}_{jg}) = -[\mathbf{M}_{g\ \text{in}} + \sum_{j=a}^{p}(\mathbf{r}_{gj} \times \mathbf{F}_{gj})] \tag{5.168}$$

Combining Eqs. (5.167) and (5.168) gives

$$\sum_{i=2}^{n}(\mathbf{M}_{ei} + \mathbf{r}_{ei} \times \mathbf{F}_{ei}) - \mathbf{M}_{M/G/O} = \frac{d}{dt}\mathbf{H}_O \tag{5.169}$$

If one considers only inertia forces, $\Sigma(\mathbf{M}_{ei} + \mathbf{r}_{ei} \times \mathbf{F}_{ei}) = 0$. The same is true for some linkage interactions in special cases. Then

$$\mathbf{M}_{M/G/O} = -\frac{d}{dt}\mathbf{H}_O \tag{5.170}$$

## Angular Momentum for an Arbitrary Four-Bar Linkage [21]

For a four-bar linkage, we have

$$\mathbf{H}_O = \sum_{i=2}^{4} (\mathbf{r}_{si} \times m_i \dot{\mathbf{r}}_{si} + m_i k_i^2 \dot{\boldsymbol{\theta}}_i) \tag{5.171}$$

which, in scalar form, is

$$H_O = \sum_{i=2}^{4} H_i = \sum_{i=2}^{4} m_i(x_i \dot{y}_i - y_i \dot{x}_i + k_i^2 \dot{\theta}_i) \tag{5.172}$$

where $x_i$ and $y_i$ are the coordinates of the center of mass of the $i$th link with respect to the fixed frame of reference.

If the $x$ axis is in line with the fixed link, $\psi_1 = 0$. Then for link 2 (see Fig. 5.57),

$$x_2 = r_2 \cos(\theta_2 + \psi_2) + x_0 \tag{5.173}$$

$$y_2 = r_2 \sin(\theta_2 + \psi_2) + y_0 \tag{5.174}$$

where $x_0$ and $y_0$ are the coordinates of pin 1 (the ground pivot of link 2). For the coupler (link 3),

$$x_3 = a_2 \cos\theta_2 + r_3 \cos(\theta_3 + \psi_3) + x_0 \tag{5.175}$$

$$y_3 = a_2 \sin\theta_2 + r_3 \sin(\theta_3 + \psi_3) + y_0 \tag{5.176}$$

and for link 4,

$$x_4 = r_4 \cos(\theta_4 + \psi_4) + a_1 + x_0 \tag{5.177}$$

$$y_4 = r_4 \sin(\theta_4 + \psi_4) + y_0 \tag{5.178}$$

Substituting these values [Eqs. (5.173) to (5.178)] into Eq. (5.172) and recombining terms gives

$$\mathrm{H}_O = H_{A0} + x_0 \sum_{i=2}^{4} m_i \dot{y}_i - y_0 \sum_{i=2}^{4} m_i \dot{x}_i \tag{5.179}$$

where

$$\begin{aligned} H_{A0} = {} & m_2(k_2^2 + r_2^2)\dot{\theta}_2 + m_3[a_2^2\dot{\theta}_2 + (k_3^2 + r_3^2)\dot{\theta}_3 \\ & + a_2 r_3 \cos(\theta_2 - \theta_3 - \psi_3)(\dot{\theta}_2 + \dot{\theta}_3)] \\ & + m_4[(k_4^2 + r_4^2) + r_4 a_1 \cos(\theta_4 + \psi_4)]\dot{\theta}_4 \end{aligned} \tag{5.180}$$

Equation (5.180) can be rewritten in terms of $\dot{\theta}_2$, $\dot{\theta}_3$, and $\dot{\theta}_4$, giving

$$\begin{aligned} H_{A0} = {} & [m_2(k_2^2 + r_2^2) + m_3 a_2^2 + m_3 a_2 r_3 \cos(\theta_2 - \theta_3 - \psi_3)]\dot{\theta}_2 \\ & + m_3[(k_3^2 + r_3^2) + a_2 r_3 \cos(\theta_2 - \theta_3 - \psi_3)]\dot{\theta}_3 \\ & + m_4[(k_4^2 + r_4^2) + r_4 a_1 \cos(\theta_4 + \psi_4)]\dot{\theta}_4 \end{aligned} \tag{5.181}$$

This is in the form

$$H_{A0} = \sum_{i=2}^{4} m_i(k_i^2 + r_i^2)\dot{\theta}_i + A\dot{\theta}_2 + B\dot{\theta}_3 + C\dot{\theta}_4 \tag{5.182}$$

where the expressions for $A$, $B$, and $C$ are obvious from comparing Eqs. (5.181) and (5.182).

### Shaking Moment Balancing of a Force-Balanced Four-Bar Mechanism [20,21]

Since for a force-balanced four-bar mechanism $\Sigma_2^4\, m_i x_i$ and $\Sigma_2^4\, m_i y_i$ are constant, $\Sigma_2^4\, m_i \dot{x}_i$ and $\Sigma_2^4\, m_i \dot{y}_i$ vanish, and Eq. (5.179) reduces to $H_O = H_{A0}$ [Eq. (5.182)]. When this angular momentum is constant, the mechanism does not transmit a shaking moment to its frame. Shaking moment balancing is based on this premise. The procedure is as follows.

If, for simplifying the notation, we let

$$\lambda = \frac{a_2}{a_3}, \qquad \mu = \frac{a_4}{a_3}, \qquad \nu = \frac{a_1}{a_3}$$
$$\tau_2 = \sin(\theta_2 - \theta_3) \qquad \text{and} \qquad \tau_3 - \lambda = \cos(\theta_2 - \theta_3) \tag{5.183}$$

and if we note from Fig. 5.57, that $\psi_3$ is related to $\psi_3'$ by

$$a_3 - r_3 \cos\psi_3 = -r_3' \cos\psi_3' \tag{5.184}$$

then, Eq. (5.181) can be rewritten

$$\begin{aligned} H_{A0} = {} & \sum_{i=2}^{4} m_i(k_i^2 + r_i^2)\dot{\theta}_i + m_3 a_2[a_2 + r_3(\tau_3 - \lambda)\cos\psi_3 + r_3\tau_2 \sin\psi_3]\dot{\theta}_2 \\ & + m_3 a_2 r_3[(\tau_3 - \lambda)\cos\psi_3 + \tau_2 \sin\psi_3]\dot{\theta}_3 \\ & + m_4 a_1 r_4[\cos\theta_4 \cos\psi_4 - \sin\theta_4 \sin\psi_4]\dot{\theta}_4 \end{aligned} \tag{5.185}$$

or

$$\begin{aligned} H_{A0} = {} & \sum_{i=2}^{4} m_i(k_i^2 + r_i^2)\dot{\theta}_i + m_3 a_2[a_2 + (\tau_3 - \lambda)(r_3' \cos\psi_3' + a_3) \\ & + r_3\tau_2 \sin\psi_3]\dot{\theta}_2 + m_3 a_2 r_3[(\tau_3 - \lambda)\cos\psi_3 + \tau_2 \sin\psi_3]\dot{\theta}_3 \\ & + m_4 r_4 a_1(\cos\theta_4 \cos\psi_4 - \sin\theta_4 \sin\psi_4)\dot{\theta}_4 \end{aligned} \tag{5.186}$$

This reduces to

$$H_{A0} = \sum_{i=2}^{4} m_i(k_i^2 + r_i^2)\dot{\theta}_i - m_3 a_2 \lambda r_3' \cos \psi_3' \dot{\theta}_2 - m_3 r_3 a_3 \cos \psi_3 \dot{\theta}_3 - m_4 r_4 a_4 \cos \psi_4 \dot{\theta}_4 + V + W \tag{5.187}$$

where

$$V = \left[ m_3 a_2 r_3 \tau_3 \dot{\theta}_2 + m_3 a_2 r_3 \left( \left(\tau_3 - \lambda + \frac{1}{\lambda}\right) \dot{\theta}_3 + m_4 r_4 (a_4 + a_1 \cos \theta_4) \frac{\cos \psi_4}{\cos \psi_3} \dot{\theta}_4 \right] \cos \psi_3 \tag{5.188}$$

and

$$W = \left[ m_3 a_2 r_3 \tau_2 (\dot{\theta}_2 + \dot{\theta}_3) - m_4 r_4 a_1 \sin \theta_4 \frac{\sin \psi_4}{\sin \psi_3} \dot{\theta}_4 \right] \sin \psi_3 \tag{5.189}$$

## Angular Momentum for Force-Balanced In-Line Four-Bar Linkage

For a force-balanced mechanism:

$$\sum m_i x_i = \text{constant} \tag{5.190}$$

and

$$\sum m_i y_i = \text{constant} \tag{5.191}$$

Then

$$\sum m_i \dot{x}_i = \sum m_i \dot{y}_i = 0 \tag{5.192}$$

Further, from Eqs. (5.136) and (5.137), we have

$$m_2 r_2 = m_3 r_3' \frac{a_2}{a_3} \qquad \text{and} \qquad \psi_2 = \psi_3'$$

and

$$m_4 r_4 = m_3 r_3 \frac{a_4}{a_3} \qquad \text{and} \qquad \psi_4 = \psi_3 + \pi$$

Then, from Eq. (5.188), if we assume that $\psi_2 \simeq \psi_4$, for a force-balanced four-bar linkage, we have

$$V = m_3 r_3 \cos \psi_3 \left[ \tau_3 a_2 \dot{\theta}_2 + \frac{\tau_3 \lambda - \lambda^2 - 1}{\lambda} a_2 \dot{\theta}_3 - \mu (a_4 + a_1 \cos \theta_4) \dot{\theta}_4 \right] \tag{5.193}$$

Since

$$\dot{\theta}_3 = \frac{\lambda}{\tau_4} \sin (\theta_1 - \theta_4) \dot{\theta}_1 \tag{5.194}$$

and

$$\dot{\theta}_4 = \frac{\lambda\tau_2}{\mu\tau_4}\dot{\theta}_2 \tag{5.195}$$

where

$$\tau_4 = \lambda \sin(\theta_2 - \theta_4) + \nu \sin\theta_4 \tag{5.196}$$

But Freudenstein's equation [102] gives

$$\lambda^2 + \mu^2 + \nu^2 - 1 = 2\mu\lambda \cos(\theta_2 - \theta_4) + 2\nu(\lambda\cos\theta_2 - \mu\cos\theta_3) \tag{5.197}$$

which, when solved, gives $V = 0$.

From Eq. (5.189), substituting $m_4r_4 = m_3r_3(a_4/a_3)$ gives

$$W = m_3a_3r_3 \sin\psi_3(\lambda\tau_3\dot{\theta}_2 + \lambda\tau_2\dot{\theta}_3 + \mu\nu\sin\theta_3\dot{\theta}_4) \tag{5.198}$$

Substituting $\dot{\theta}_3$ and $\dot{\theta}_4$ from Eqs. (5.194) and (5.195) gives

$$W = 2m_3a_2r_3 \sin\theta_3\lambda\tau_2\dot{\theta}_2 \tag{5.199}$$

Thus, for a force-balanced four-bar linkage, Eq. (5.187) reduces to

$$H_{A0} = \sum_{i=2}^{4} m_i(k_i^2 + r_i^2 - a_ir_i\cos\psi_i)\dot{\theta}_i + 2m_3a_2r_3\sin\psi_3\tau_2\dot{\theta}_2 \tag{5.200}$$

Thus we see that for a force-balanced four-bar linkage, the angular momentum is expressed solely in time-dependent terms, which means that it is *independent* of reference point and of initial position.

### Shaking Moment of Force-Balanced Four-Bar Linkage

From Eq. (5.170) we saw that $\mathbf{M}_{M/G/O} = -(d/dt)\mathbf{H}_O$. But, since $\mathbf{H}_O$ is independent of reference point for a force-balanced mechanism [Eq. (5.200)],

$$\mathbf{M}_{M/G/O} = \mathbf{M}_{M/G} = -\frac{d}{dt}\mathbf{H}_O = -\frac{d}{dt}\mathbf{H} \tag{5.201}$$

Then, from Eq. (5.200),

$$\mathbf{M}_{M/G} = -\sum_{i=2}^{4} m_i(k_i^2 + r_i^2 - a_ir_i\cos\psi_i)\ddot{\theta}_i - 2m_3a_2r_3\sin\psi_3(\tau_2\ddot{\theta}_2 + \dot{\tau}_2\dot{\theta}_2) \tag{5.202}$$

Except for special cases, it is generally not possible to make the shaking moment for a force-balanced linkage vanish. However, individual terms of Eq. (5.202) may become zero. Although that does not necessarily reduce the magnitude of the shaking moment, such reductions do occur when:

1. The input angular velocity ($\dot{\theta}_2$) is constant, yielding $\ddot{\theta}_2 = 0$.
2. The center of mass of the coupler lies on the line connecting pin 2 with pin 3 (gives $\psi_3 = 0$).

3. The mass distribution of any link is a physical pendulum (for a discussion, see Section 5.17), in which the radius of gyration ($k$)* is given by

$$k_i^2 = r_i(a_i \cos \psi_i - r_i) \tag{5.203}$$

4. The mechanism is a parallelogram ($a_2 = a_4$ and $a_3 = a_1$) or rhomboid ($a_1 = a_2 = a_3 = a_4$), yielding $\ddot{\theta}_2 = \ddot{\theta}_4$ and $\ddot{\theta}_3 = 0$.
5. The mechanism is a deltoid linkage ($a_2 = a_1$ and $a_3 = a_4$), yielding $\ddot{\theta}_3 = -\ddot{\theta}_4$, provided that coefficients of these terms are properly adjusted.

Full shaking moment balance can occur in two special cases:

1. A force-balanced parallelogram or rhomboid linkage with an in-line coupler ($\psi_3 = 0$) running at constant velocity
2. A force-balanced deltoid linkage with an in-line coupler running at constant velocity with equal coefficients of $\ddot{\theta}_3$ and $\ddot{\theta}_4$

Full shaking moment balance can also be achieved in a force-balanced four-bar linkage through the use of counterweight inertias, as discussed in the following section.

### Complete Force and Moment Balancing of an In-line Four-Bar Linkage

Utilization of the concepts of inertia counterweights and the physical pendulum permits complete balance of all mass effects (both linear and rotary) but excluding external loads), independent of input angular velocity.

### Concept of Inertia Counterweights

Inertia counterweights permit any unbalanced planar moment, which is proportional to angular acceleration, to be balanced. Since no net inertia forces are introduced by this addition, the shaking force balance [Eq. (5.192)] is unaffected. Unfortunately, the driving torque must increase substantially to drive the now force- and moment-balanced system with the added counterweights.

Let the linkage transmit an unbalanced shaking moment, $I\ddot{\theta}$, to ground. For the balancing counterweights, phase angle and relative inclination of pivots is irrelevant. All that is needed is:

1. A pair of spur gears with gear ratio $n$ such that

$$I_{\text{cwt}} = nI \tag{5.204}$$

where $I_{\text{cwt}}$ = moment of inertia for the counterweight
$I$ = moment of inertia for the unbalanced mechanism referred to the counterweight shaft

* See Eq. (5.33) of Vol. 1.

2. Some additional mass, in any form, as long as it is statically balanced about its axis of rotation. (It must be recognized that backlash and tooth loads may be large and create serious design problems.) However, a number of counterweight schemes are feasible; two are shown in Fig. 5.66.

In Fig. 5.66 the gears of radius $r$ each have a moment of inertia $I_G$. The driving torque is $M_{in}$. The gear bearing reactions are force-balanced and of magnitude $F_B$. The tangential tooth force components, normal to the pivot-to-pivot axis, are also equal to $F_B$:

$$F_B = \frac{I_G + I}{r}\ddot{\theta} \tag{5.205}$$

where $I$ is the moment of inertia of the counterweights about the right-side shaft of Fig. 5.56, equal to that of the linkage referred to this shaft.

The moment of the ground bearing forces is given by

$$M_G = -2rF_B = -2(I_G + I)\ddot{\theta} \tag{5.206}$$

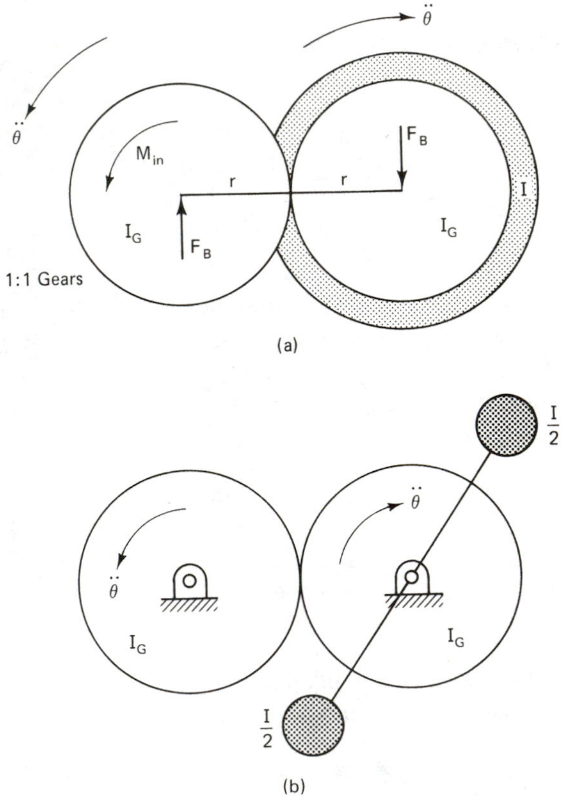


**Figure 5.66** Geared inertia counterweights.

The input torque required originally for the newly added left gear only is

$$M_{\text{in}} = I_G\ddot{\theta} \tag{5.207}$$

So with the addition of the right gear and the counterweights,

$$M_{\text{in cwt}} = I_G\ddot{\theta} + (I_G + I)\ddot{\theta} = (2I_G + I)\ddot{\theta} \tag{5.208}$$

This can be verified by the free-body diagrams shown in Fig. 5.66a. The resulting balancing moment is then given by the sum of the input torque and the moment of the ground bearing forces:

$$M_{M/G} = -(M_{\text{in cwt}} + M_G) = -[(2I_G + I)\ddot{\theta} - 2(I_G + I)\ddot{\theta}] = I\ddot{\theta} \tag{5.209}$$

Thus the resultant shaking moment of $I\ddot{\theta}$ exactly counterbalances a linkage-generated moment of $-I\ddot{\theta}$.

## Concept of Physical Pendulum

The physical pendulum (as defined by Timoshenko and Young [287,288]) is basically an in-line link with the radius of gyration ($k$) related to the position of the center of mass such that (see Fig. 5.67)

$$k^2 = rr' \tag{5.210}$$

Section 5.17 describes two link constructions that can be utilized to make certain links a physical pendulum.

## Application of Inertial Counterweight and Physical Pendulum to the Four-Bar Linkage

From Eq. (5.202), we saw that for a force-balanced, in-line,* four-bar linkage,

$$\mathbf{M}_{M/G} = -\sum_{i=2}^{4} m_i(k_i^2 + r_i^2 - a_i r_i \cos\psi_i)\ddot{\theta}_i - 2m_3 a_2 r_3 \sin\psi_3(\tau_2\ddot{\theta}_2 + \dot{\tau}_2\dot{\theta}_2)$$

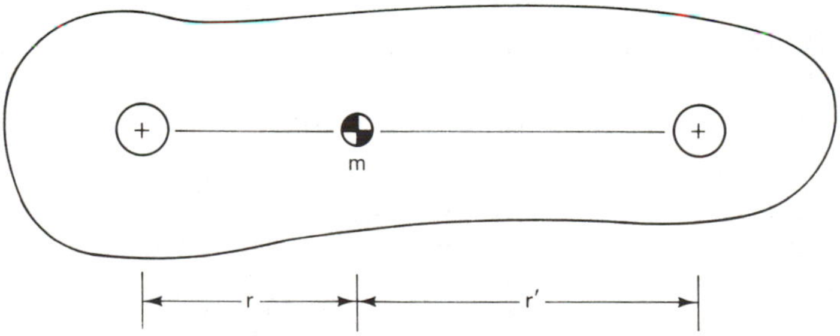


**Figure 5.67** In-line geometry.

* The center of mass of each moving (binary) link must lie on the center line connecting its pivot points. The links need not be symmetric, however. (Ternary links are not considered here.)

which can be simplified to

$$\mathbf{M}_{M/G} = \sum_{i=2}^{4} K_i\ddot{\theta}_i + K_5(\tau_2\ddot{\theta}_2 + \dot{\tau}_2\dot{\theta}_2) \tag{5.211}$$

where

$$K_i = -m_i(k_i^2 + r_i^2 - a_i r_i \cos \psi_i)$$

$$K_5 = -2m_3 a_2 r_3 \sin \psi_3$$

For an in-line coupler which is also a physical pendulum, $\psi_3 = 0$ and $\psi_3' = \pi$, giving $K_5 = 0$.

Adding counterweights $I_2^{**}$ and $I_4^{**}$ geared to the input and output links gives a total shaking moment of

$$\mathbf{M}^{**}_{M/G} = \sum_{i=2}^{4} K_i\ddot{\theta}_i + I_2^{**}\ddot{\theta}_2 + I_4^{**}\ddot{\theta}_4 \tag{5.212}$$

But $M^{**}_{M/G} = 0$ (the shaking moment vanishes) if

$$\begin{aligned} K_2 &= -I_2^{**} \\ K_4 &= -I_4^{**} \\ K_3 &= 0 = -m_3(k_3^2 + r_3^2 - a_3 r_3) \end{aligned} \tag{5.213}$$

$K_3 = 0$ when the coupler is a physical pendulum and $a_3 = 2r_3$.

Summarizing, then, for an in-line four-bar linkage with a physical pendulum coupler and added intertia counterweights, complete moment and force balance can be achieved if the following five conditions can be met:

Requirements for force balance and physical pendulum (link 2):

$$m_2 r_2 = m_3 r_3' \frac{a_2}{a_3} \qquad \text{with} \qquad \psi_2 = \psi_3' = \pi \tag{5.214}$$

Requirements for force balance and physical pendulum (link 4):

$$m_4 r_4 = m_3 r_3 \frac{a_4}{a_3} \qquad \text{with} \qquad \psi_4 = \psi_3 + \pi = \pi \tag{5.215}$$

Requirement for physical pendulum (link 3):

$$k_3^2 = r_3 r_3' \tag{5.216}$$

Requirement for moment balance (counterweight on link 2):

$$I_2^{**} = -K_2 = m_2(k_2^2 + r_2^2 + a_2 r_2) \tag{5.217}$$

Requirement for moment balance (counterweight on link 4):

$$I_4^{**} = -K_4 = m_4(k_4^2 + r_4^2 + a_4 r_4) \tag{5.218}$$

Theoretically, none of the moving links need be in-line originally. However,

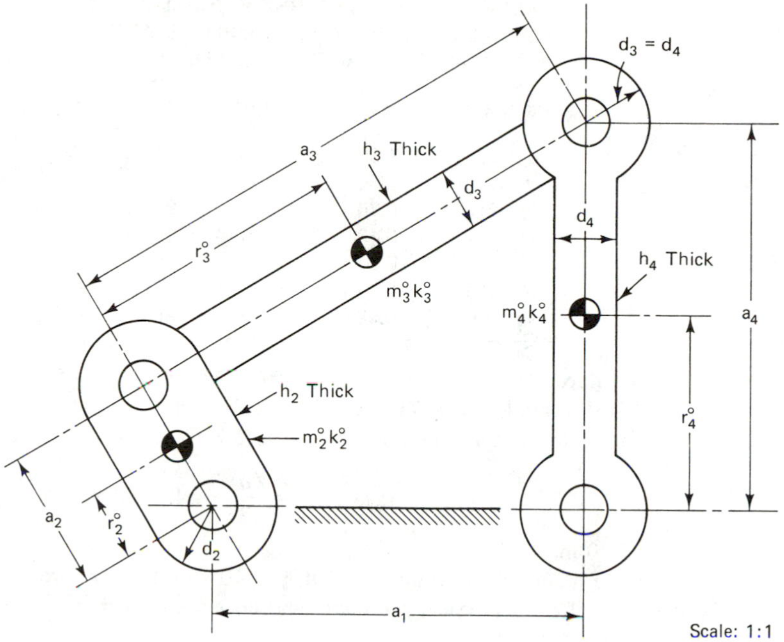


**Figure 5.68** Unbalanced four-bar linkage to be completely force- and moment-balanced.

to achieve complete balance, it must be possible to readjust the mass distribution to obtain the in-line characteristics for all the moving links as well as meeting the criterion of a physical pendulum for the coupler. It is important to note that should the input angular velocity be constant ($\ddot{\theta} = 0$), no counterweight need be added to the input link.

**Example 5.15**

Suppose that we wish to balance completely the unbalanced linkage shown in Fig. 5.68, with parameters given in Table 5.12. The links are steel of density $\gamma = 0.283$ lbf/in³. For convenience, weights rather than masses are used in the calculations since the gravitational constants conveniently cancel in all equations.

**Solution**

*Step 1. Convert the coupler to a physical pendulum.* Section 5.17 outlines the specific requirements for physical pendulums. Since the coupler is an augmented* link, and according to Table 5.12 $a_3/d_3 = 8$, Table 5.16 gives a value of $e_3/d_3 = 1.887$. Thus $e_3 = 0.943$.

The added mass is given by (see Fig. 5.74)

$$m^* = \frac{2\gamma}{g} e_3 d_3 h_3 \tag{5.219}$$

* The reader is advised to review Sec. 5.17 before proceeding.

**TABLE 5.12** PARAMETERS OF UNBALANCED FOUR-BAR THAT IS TO BE COMPLETELY FORCE- AND MOMENT-BALANCED (Fig. 5.68)

| Parameter | Link 2 | 3 | 4 | 1 |
|---|---|---|---|---|
| $a_i$ (in.) | 1.00 | 4.00 | 3.00 | 3.00 |
| $d_i$ (in.) | 0.50 | 0.50 | 0.50 | — |
| $h_i$ (in.) | 0.20 | 0.20 | 0.20 | — |
| $w_i^\circ$ (lbf) | 0.101 | 0.174 | 0.146 | — |
| $r_i^\circ$ (in.) | 0.50 | 2.00 | 1.50 | — |
| $\psi_i^\circ$ (deg) | 0 | 0 | 0 | — |
| $k_i^\circ$ (in.) | 0.593 | 1.577 | 1.261 | — |

Thus the added weight must be $m^*g = w^* = 2\gamma e_3 d_3 h_3 = 0.0534$ lbf, giving the total weight for the coupler of $0.174 + 0.0534 = 0.227$ lbf.

The radius of gyration is given by Eq. (5.216), giving

$$k^2 = rr' = r^2 = \left(\frac{a}{2}\right)^2 = 4.00 \text{ in}^2$$

so $k = 2.00$ in.

If we maintain the condition that $\psi_3 = 0$ and $\psi_3' = \pi$, we satisfy the condition that the coupler is a physical pendulum and does not contribute to the shaking moment [Eq. (5.216)].

*Step 2. Establish force balance.* Assume that the links cannot be altered but that counterweights must be utilized to achieve force balance. Then from Eqs. (5.141) and (5.142), we solve for $m_i^* r_i^*$ and $\psi_i^*$ for links 2 and 4, giving $(m_i^* r_i^*)g = w_i^* r_i^* =$ 0.164 lbf•in. for link 2 and 0.556 lbf•in. for link 4 and $\psi_2^* = \psi_4^* = \pi$.

Suppose that we choose circular counterweights with diameter $2r_i^*$. Then the radius is given by [18]

$$r_i^* = \left(\frac{w_i^* r_i^*}{\gamma \pi h_i^*}\right)^{1/3} \tag{5.220}$$

Suppose further that we select $h^* = 0.50$ in. for both counterweights. Then $r_2^* =$ 0.717 in. and $r_4^* = 1.077$ in. The weight added is given by $w_i^* = (w_i^* r_i^*)/r_i^*$ and total weight of the links ($w_i$) is given by

$$w_i = w_i^\circ + w_i^* \tag{5.221}$$

and the new position of the center of mass is determined by vector addition:

$$w_i \mathbf{r}_i + w_i^\circ \mathbf{r}_i^\circ + w_i^* \mathbf{r}_i^*$$

or

$$\mathbf{r}_i = \frac{w_i^\circ \mathbf{r}_i^\circ + w_i^* \mathbf{r}_i^*}{w_i} \tag{5.222}$$

$$r_i e^{i\psi_i} = \frac{w_i^\circ}{w_i} r_i^\circ e^{i\psi_i^\circ} + \frac{w_i^*}{w_i} r_i^* e^{i\psi_i^*}$$

Now the linkage is completely force balanced [conditions (5.214) and (5.215) are satisfied].

Table 5.13 summarizes the values of the parameters and Fig. 5.69 shows the configuration for the force-balanced linkage.

**TABLE 5.13** PARAMETERS FOR FORCE-BALANCED FOUR-BAR MECHANISM (Fig. 5.69)

| Parameter | Link 2 | 3 | 4 |
|---|---|---|---|
| $w_i^* r_i^*$ (lbf·in.) | 0.164 | — | 0.556 |
| $w_i^*$ (lb) | 0.229 | 0.0534 | 0.516 |
| $r_i^*$ (in.) | 0.717 | — | 1.077 |
| $\psi_i^*$ (deg) | 180.0 | — | 180.0 |
| $w_i$ (lbf) | 0.330 | 0.227 | 0.662 |
| $r_i$ (in.) | 0.650 | 2.000 | 1.171 |
| $\psi_i$ (deg) | 180.0 | 0 | 180.0 |

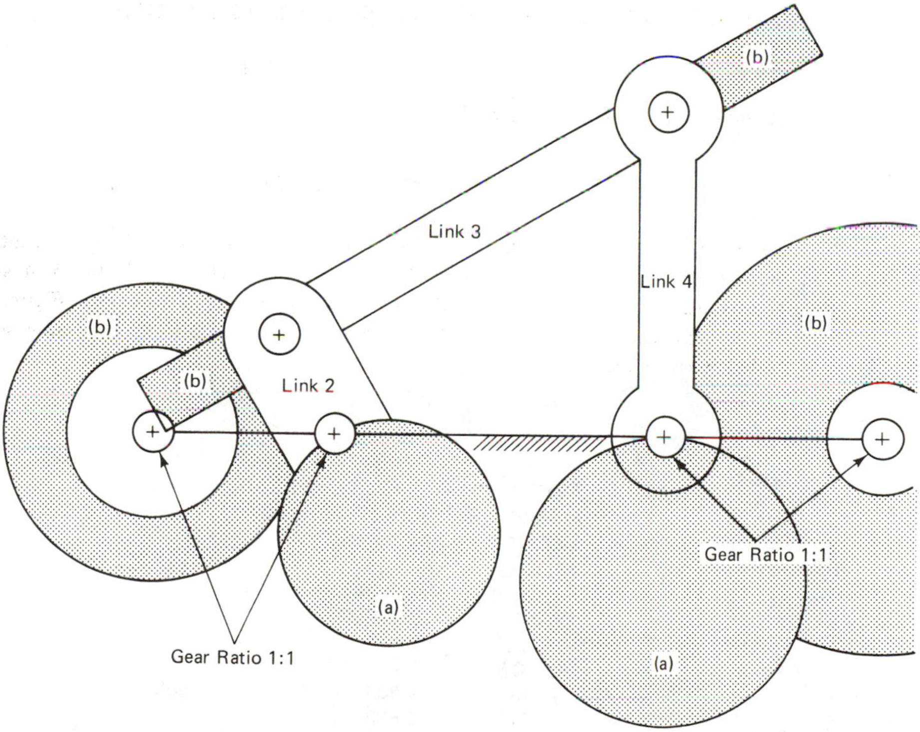


**Figure 5.69** Fully force- and moment-balanced in-line four-bar linkage. Shaded area is material added to achieve (a) force- and (b) moment-balance.

*Step 3. Add inertia counterweights to establish moment balance.* Because we have changed the centers of mass of links 2 and 4, we must calculate the new radius of gyration for those links. Because we have chosen circular counterweights, the contribution from the counterweight is given by

$$k_i^* = \frac{r_i^*}{\sqrt{2}} \tag{5.223}$$

To determine the radius of gyration of a counterbalanced link (Fig. 5.69), we write

$$wk^2 = w^\circ[k^{\circ 2} + |\mathbf{r}^\circ - \mathbf{r}|^2] + w^*[k^{*2} + |\mathbf{r}^* - \mathbf{r}|^2]$$

For an in-link with counterweight (Fig. 5.69), this reduces to

$$wk^2 + w^\circ[k^{\circ 2} + (r^\circ + r)^2] + w^*[k^{*2} + (r^* - r)^2]$$

which leads to Eq. (5.224):

$$k_i^2 = \frac{w_i^\circ}{w_i}[k_i^{\circ 2} + (r_i^\circ + r_i)^2] + \frac{w_i^*}{w_i}[k_i^{*2} + (r_i^* - r_i)^2] \tag{5.224}$$

Equation (5.224) gives $k_2 = 0.960$ in. and $k_4 = 1.199$ in.

The weight moments of inertia of the intertia counterweights using 1:1 gearing as in Fig. 5.66 (conditions 4 and 5) are given by [see Eq. (5.217)]

$$I_i^{**} = w_i(k_i^2 + r_i^2 + a_i r_i) \tag{5.217a}$$

If the counter weight is a circular disk,

$$I_i^{**} = \frac{\pi}{2}(\gamma \rho_i^4 h_i^{**}) \tag{5.225}$$

where $\rho_i$ is the radius and $h_i^{**}$ the thickness of the disk. Suppose that we select $h_2^{**} = h_4^{**} = 1.0$ in. Then $\rho_2 = 1.070$ and $\rho_4 = 1.807$ in. Table 5.14 summarizes the values of $k$, $k^*$, and $I^{**}$ for the fully moment balanced linkage. Figure 5.69 gives the configuration of the final linkage, shown at approximately the same scale as the unbalanced linkage in Fig. 5.68.

**TABLE 5.14** PARAMETERS FOR SHAKING-MOMENT-BALANCED FOUR-BAR MECHANISM—ROTATING COUNTERWEIGHTS (Fig. 5.69)

| Parameter | Link 2 | 3 | 4 |
|---|---|---|---|
| $k_i^*$ (in.) | 0.507 | — | 0.762 |
| $k_i$ (in.) | 0.833 | 2.000 | 1.508 |
| $I_i^{**}$ (lbf·in²) | 0.583 | — | 4.739 |
| $\rho_i$ (in.) | 1.070 | — | 1.807 |
| $w_i^{**}$ (lbf) | 1.090 | — | 5.245 |

### Optimization of Shaking Moments

Unfortunately, one may not be lucky enough to be able to meet all the constraints demanded to totally eliminate shaking moment. Thus optimization techniques must be employed. Berkof (and others) have elected to utilize the method of least squares. Because that analysis quickly becomes complicated in its general form and requires a good deal of curve matching with different combinations of linkage parameters, the reader is referred to Refs. 17, 21, 172, 175, 282, 283.

Analysis of behaviors of a variety of configurations has, however, led to the following observations regarding relative magnitudes of the shaking moment and their minima [19]:

A decrease in distance to the coupler center of mass from the crank pivot tends to decrease the moment.

An increase in ground link length tends to decrease the moment. However, the sensitivity varies greatly with differing configurations.

A decrease in the coupler link length tends to decrease the moment.

No clear trend is evident concerning the effect of output link length.

Shaking moment is proportional to the:
- Fifth power of the input link length $a_2$
- Density of input link 2
- Second power of the input angular velocity

## 5.14 EFFECT OF MOMENT BALANCE ON INPUT TORQUE

As we saw in the example for moment balance, addition of link counterweights and rotating inertial masses to obtain complete balance significantly increased the total weight for the moving parts of the mechanism. This increase in moving weight surely cannot be ignored.

Four different techniques exist for formulating an expression for the input torque of an arbitrary linkage: kinetic energy, classical force analysis, virtual work, and Lagrange's equation. The Lagrange method will now be discussed.

### Kinetic Energy Formulation

For an infinitesimal change in total kinetic energy $T$, an equivalent amount of work is done by input torque $M_{\text{in}}$ moving through angle $\theta$, giving

$$dT = M_{\text{in}}\, d\theta \tag{5.226}$$

For continuous functions, this is equivalent to

$$M_{\text{in}} = \frac{dT}{d\theta} = \frac{dT\, dt}{dt\, d\theta} \tag{5.227}$$

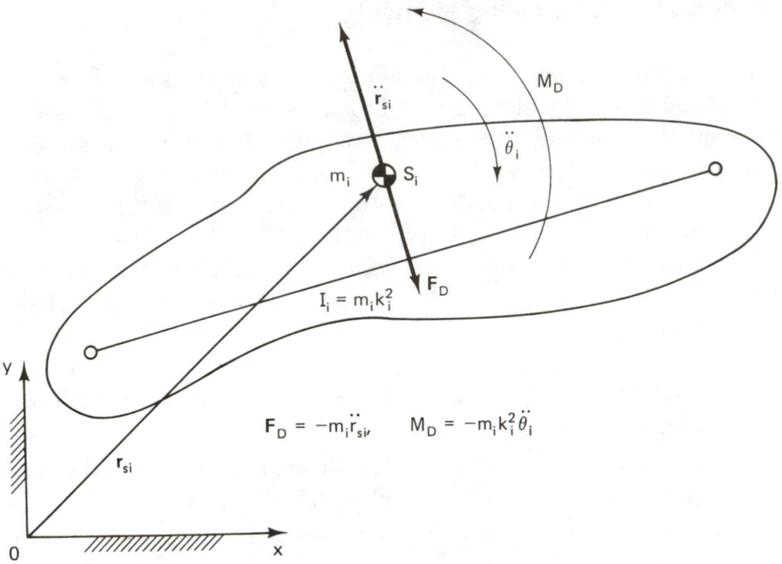


**Figure 5.70** D'Alembert forces and moments for *i*-th link.

giving, for a single-degree-of-freedom force-balanced* mechanisms, where link 2 is the input crank,

$$M_{\text{in}} = \frac{1}{\dot{\theta}_2} \frac{dT}{dt} \tag{5.228}$$

If only D'Alembert forces $F_D$ and moments $M_D$ act on the remaining links, $F_D$ acts through the center of mass $S$ for any link (Fig. 5.70). Then the total kinetic energy $T_i$ for any given link is the sum of the translational and rotational kinetic energies:

$$T_i = \tfrac{1}{2} m_i \dot{r}_{si}^2 + \tfrac{1}{2} m_i k_i^2 \dot{\theta}_i^2 \tag{5.229}$$

where $m_i$ = mass

$\dot{r}_{si}$ = velocity of center of mass $S_i$

$k_i$ = radius of gyration

$\dot{\theta}_i$ = angular velocity, all for the $i$th link

For any mechanism, then, the total kinetic energy is given by

$$T = \sum_{i=2}^{n} T_i = \tfrac{1}{2} \sum_{i=2}^{n} m_i (\dot{r}_{si}^2 + k_i^2 \dot{\theta}_i^2) \tag{5.230}$$

where $i = 2, 3, \ldots, n$ enumerates all moving links.

* Force balancing makes the changes in potential energy of the links cancel out, because the overall center of gravity of all moving links is stationary.

From Eq. (5.228),

$$M_{\text{in}} = \frac{1}{\dot{\theta}_2} \sum_{i=2}^{n} m_i(\dot{r}_{si}\ddot{r}_{si} + k_i^2 \dot{\theta}_i \ddot{\theta}_i) \tag{5.231}$$

where $\ddot{r}_{si}$ = acceleration of the center of mass

$\ddot{\theta}_i$ = angular acceleration of the $i$th link

In terms of Cartesian coordinates,

$$T = \tfrac{1}{2} \sum_{i=2}^{n} m_i(\dot{x}_i^2 + \dot{y}_i^2 + k_i^2 \dot{\theta}_i^2) \tag{5.232}$$

and

$$M_{\text{in}} = \frac{1}{\dot{\theta}_2} \sum_{i=2}^{n} m_i(\dot{x}_i\ddot{x}_i + \dot{y}_i\ddot{y}_i + k_i^2 \dot{\theta}_i \ddot{\theta}_i) \tag{5.233}$$

where $(x_i, y_i)$ are the coordinates of the center of mass of the $i$th link with respect to the fixed frame of reference.

## Instantaneous Power Formulation

The general form of the principle of instantaneous power for driving a mechanism is given by

$$\sum_{i=2}^{n} \mathbf{F}_i \cdot \mathbf{v}_i + \sum_{i=2}^{n} \mathbf{M}_i \cdot \boldsymbol{\omega}_i = 0 \tag{5.234}$$

where $\mathbf{F}_i$ = external force

$\mathbf{M}_i$ = external moment

$\mathbf{v}_i$ = velocity of the point of application of $\mathbf{F}$

$\boldsymbol{\omega}_i$ = angular velocity of the link acted upon by $\mathbf{M}_i$

If friction and external work requirements are disregarded, the D'Alembert force and moments on the $i$th link are given by

$$\mathbf{F}_{Di} = -m_i\ddot{\mathbf{r}}_{si} = -m_i(\ddot{x}\mathbf{i} + \ddot{y}\mathbf{j}) \tag{5.235}$$

where $\mathbf{i}$ and $\mathbf{j}$ are orthonormal unit vectors, or $\mathbf{F}_{Di} = -m_i(\ddot{x}_i + i\ddot{y}_i)$, which is in complex notation for planar motion. Also

$$\mathbf{M}_{Di} = -m_i k_i^2 \ddot{\boldsymbol{\theta}}_i = -m_i k_i^2 \ddot{\theta}_i \mathbf{k} \tag{5.236}$$

in spatial vector notation, or $M_{Di} = -m_i k_i^2 \ddot{\theta}_i$ in the plane. The input torque is

$$\mathbf{M}_{\text{in}} = M_{\text{in}}\mathbf{k} \qquad \text{in space} \tag{5.237}$$

or simply $M_{\text{in}}$ (a real) in planar work. Since neither reaction nor bearing forces produce work, Eq. (5.234) becomes

$$\mathbf{M}_{\text{in}} \cdot \dot{\boldsymbol{\theta}}_2 + \sum_{i=2}^{n} (\mathbf{F}_{Di} \cdot \dot{\mathbf{r}}_s + \mathbf{M}_{Di} \cdot \dot{\boldsymbol{\theta}}_i) = 0 \tag{5.238}$$

where

$$\mathbf{F}_{Di} \cdot \dot{\mathbf{r}}_{si} = -m_i(\dot{x}_i\ddot{x}_i + \dot{y}_i\ddot{y}_i)$$

or, in scalar form for planar motion,

$$M_{\text{in}} = \frac{1}{\dot{\theta}_2} \sum_{i=2}^{n} m_i(\dot{x}_i\ddot{x}_i + \dot{y}_i\ddot{y}_i + k_i^2\dot{\theta}_i\ddot{\theta}_i) \tag{5.239}$$

which is exactly the same form as Eq. (5.233).

### Lagrange's Equation Formulation

For a single-degree-of-freedom force-balanced planar rigid-body linkage, Lagrange's equation is of the form

$$M_{\text{in}} = \frac{d}{dt}\left(\frac{\partial T}{\partial \dot{\theta}_2}\right) - \frac{\partial T}{\partial \theta_2} \tag{5.240}$$

where $M_{\text{in}}$ = generalized force acting on the system

$\theta_2$ = chosen generalized coordinate

$T$ = total kinetic energy

Because $\theta_2$ is the only independent coordinate in a single-degree-of-freedom planar rigid-body mechanism with rotary input, the velocity of the center of mass and angular velocity for the $i$th link is a function only of $\theta_2$ and $\dot{\theta}_2$:

$$\dot{r}_{si} = f_{1i}(\theta_2)\dot{\theta}_2 \tag{5.241}$$

$$\dot{\theta}_i = f_{2i}(\theta_2)\dot{\theta}_2 \tag{5.242}$$

Then total kinetic energy is given by

$$T = f(\theta_2)\dot{\theta}_2^2 \tag{5.243}$$

where $f$ is a function that can be written in terms of the single variable $\theta_2$. Then the terms of Eq. (5.240) can be written in the form

$$\frac{d}{dt}\left(\frac{\partial T}{\partial \dot{\theta}_2}\right) = \frac{d}{dt}(2f\dot{\theta}_2) = 2f\ddot{\theta}_2 + 2\frac{df}{dt}\dot{\theta}_2 \tag{5.244}$$

and

$$\frac{\partial T}{\partial \theta_2} = \dot{\theta}_2^2 \frac{\partial f}{\partial \theta_2} \tag{5.245}$$

Also, since $f = f(\theta_2)$ and $\partial f/\partial\dot{\theta}_2 = 0$,

$$\frac{df}{dt} = \frac{\partial f}{\partial\theta_2}\frac{d\dot{\theta}}{dt}2 + \frac{\partial f}{\partial\theta_2}\frac{d\theta_2}{dt} = \dot{\theta}_2\frac{\partial f}{\partial\theta_2} \tag{5.246}$$

Substitution of Eqs. (5.244), (5.245), and (5.246) into (5.240) yields

$$M_{\text{in}} = 2f\ddot{\theta}_2 + \frac{df}{dt}\dot{\theta}_2 \tag{5.247}$$

But, from Eq. (5.243) $f = T/\dot{\theta}_2^2$, so

$$\frac{df}{dt} = \frac{1}{\dot{\theta}_2^2}\frac{dT}{dt} - 2\frac{\ddot{\theta}_2}{\dot{\theta}_2^3}T \tag{5.248}$$

which, when substituted into Eq. (5.247), gives

$$M_{\text{in}} = \frac{1}{\dot{\theta}_2}\frac{dT}{dt} \tag{5.249}$$

By expressing $f(\theta_2)$ in its complete form, equations for total energy and moment can be obtained:

$$T_{\text{in}} = \tfrac{1}{2}\sum_{i=2}^{n} m_i(\dot{r}_{si}^2 + k_i^2\dot{\theta}_i^2)$$

$$= \tfrac{1}{2}\sum_{i=2}^{n} m_i(\dot{x}_i^2 + \dot{y}_i^2 + k_i^2\dot{\theta}_i^2)$$

and

$$M_{\text{in}} = \frac{1}{\dot{\theta}_2}\sum_{i=2}^{n} m_i(\dot{x}_i\ddot{x}_i + \dot{y}_i\ddot{y}_i + k_i^2\dot{\theta}_i\ddot{\theta}_i)$$

where $(x_i, y_i)$ are the Cartesian coordinates for the center of mass of the $i$th moving link.

### Input Torque Components of In-Line Force-Balanced Four-Bar Linkage

For an in-line four-bar linkage ($\psi_i = 0, 180$),

$$\begin{aligned} \dot{x}_2^2 + \dot{y}_2^2 &= r_2^2\dot{\theta}_2^2 \\ \dot{x}_3^2 + \dot{y}_3^2 &= r_3^2\dot{\theta}_3^2 + a_2^2\dot{\theta}_2^2 + 2a_2r_3\cos(\theta_3 - \theta_2)\dot{\theta}_2\dot{\theta}_3 \\ \dot{x}_4^2 + \dot{y}_4^2 &= r_4^2\dot{\theta}_4^2 \end{aligned} \tag{5.250}$$

where $r_i$, $i = 2, 3, 4$, is the distance to the center of mass from the fixed pivot of links 2 and 4, and from the crankpin of link 3.

Substitution into Eq. (5.230) gives

$$T = \tfrac{1}{2}\{m_2(k_2^2 + r_2^2)\dot{\theta}_2^2 + m_4(k_4^2 + r_4^2)\dot{\theta}_4^2 + m_3[(k_3^2 + r_3^2)\dot{\theta}_3^2 + a_2^2\dot{\theta}_2^2 + 2a_2r_3 \cos(\theta_3 - \theta_2)\dot{\theta}_2\dot{\theta}_3]\} \tag{5.251}$$

which can be written in the form

$$T = -\tfrac{1}{2}\sum_{i=2}^{4} K_i\dot{\theta}_i^2 + Z \tag{5.252}$$

where

$$K_i = -m_i(k_i^2 + r_i^2 + a_ir_i) \tag{5.253}$$

and

$$Z = \tfrac{1}{2}[-m_2r_2a_2\dot{\theta}_2^2 - m_4r_4a_4\dot{\theta}_4^2 - m_3r_3a_3\dot{\theta}_3^2 + m_3(a_2^2\dot{\theta}_2^2 + 2a_2r_3\cos(\theta_3 - \theta_2)\dot{\theta}_3\dot{\theta}_2] \tag{5.254}$$

But from force-balance requirements [Eqs. (5.136) and (5.137)], we have

$$\begin{aligned} m_2r_2 &= m_3(a_3 - r_3)\frac{a_2}{a_3} \\ m_4r_4 &= m_3r_3\frac{a_4}{a_3} \end{aligned} \tag{5.255}$$

Substituting (5.255) into (2.254) gives

$$Z = \frac{m_3r_3}{2a_3}[a_2^2\dot{\theta}_2^2 - a_3^2\dot{\theta}_3^2 - a_4^2\dot{\theta}_4^2 + 2a_2a_3\cos(\theta_3 - \theta_2)\dot{\theta}_2\dot{\theta}_3] \tag{5.256}$$

From loop-closure equations for a four-bar:

$$\begin{aligned} a_2\cos\theta_2 + a_3\cos\theta_3 &= a_4\cos\theta_4 + a_1 \\ a_2\sin\theta_2 + a_3\sin\theta_3 &= a_4\sin\theta_4 \end{aligned} \tag{5.257}$$

By taking time derivatives term by term, squaring both sides and adding the two equations, one obtains

$$a_2^2\dot{\theta}_2^2 + a_3^2\dot{\theta}_3^2 - a_4^2\dot{\theta}_4^2 + 2a_2a_3\cos(\theta_3 - \theta_2)\dot{\theta}_2\dot{\theta}_3 = 0 \tag{5.258}$$

Combining Eqs. (5.256) and (5.258), we obtain: $Z = -m_3r_3a_3\dot{\theta}_3^2$. Substituting this in Eq. (5.252), the total energy for the force-balanced four-bar linkage is given by

$$T = -(\tfrac{1}{2})\sum_{i=2}^{4} K_i\dot{\theta}_i^2 - m_3r_3a_3\dot{\theta}_3^2 \tag{5.259}$$

where

$$K_i = -m_i(k_i^2 + r_i^2 + a_ir_i) \tag{5.260}$$

### Input Torque Components of In-Line Force- and Moment-Balanced Four-Bar Linkage

In moment balance, the effects of the counterweights must be included in the equations in addition to the conditions of force balance. Thus total energy is given by

$$T = -(\tfrac{1}{2}) \sum_{i=2}^{4} K_i \dot{\theta}_i^2 + \tfrac{1}{2}\,[I_2^{**}\dot{\theta}_2^2 + I_4^{**}\dot{\theta}_4^2] \tag{5.261}$$

where $I_j^{**}$ are mass moments of inertia of the rotating moment-balancing masses, and where we assumed a gear ratio of 1:1 for both input and output side sets.

Since, for force and moment balance, the conditions of Eqs. (5.214) to (5.218) are satisfied, if we also assume $r_3 = a_3/2$, substitution in Eq. (5.259) yields

$$K_3 = -m_3\left(\frac{a_3}{4} + \frac{a_3}{4} - \frac{a_3}{2}\right) a_3 = 0$$

Since, from Eqs. (5.217) and (5.218), we also have

$$\begin{aligned} I_2^{**} &= -K_2 \\ I_4^{**} &= -K_4 \end{aligned} \tag{5.262}$$

Equation (5.261) becomes

$$T = -\sum_{i=2}^{4} K_i \dot{\theta}_i^2 \tag{5.263}$$

Comparing Eq. (5.263) with (5.259), we see that the total energy requirement (and hence input torque requirement) doubles as we proceed from force-balanced only to force plus moment balance. A special case occurs when the input has constant velocity. In that case, $K_2$ and $K_3$ are both zero, leaving [from Eq. (5.228)] $M_{in} = -(2/\dot{\theta}_2)[K_4\ddot{\theta}_4]$.

One should note that at complete balance, the single remaining external loading on the linkage is the moment of the ground forces. Since total moment equals zero, these moments must be equal and opposite to the input torque.

In conclusion, then, while complete balance does indeed eliminate shaking forces from being transmitted to the frame, it does so at a price—a significant increase in input torque required, together with increased bearing forces and thus reduced bearing life.

## 5.15 OTHER TECHNIQUES FOR BALANCING LINKAGES

The problem of balancing is worthy of the abundant literature analyzing the problem. Approaches include seeking complete balance, only moment balance, and partial force and/or partial moment balance.

### Complete Balance

The objective of complete force balance is to cause the net shaking force on the frame of a mechanism to vanish. Therefore, the total center of mass of the mechanism must be made stationary. Five techniques for achieving this have been employed:

- Method of linearly independent vectors, as described in this chapter
- Method of static balancing
- Method of principal vectors and extensions
- Cam and duplicate mechanism methods
- Nonlinear programming techniques

**Method of static balancing [173].** The method of static balancing replaces concentrated link masses by statically equivalent systems of masses. By adding counterweights to links, the link centers are progressively brought to end up at several stationary points. The system center of mass is then established by finding the resultant mass center of these stationary masses. The method is generally confined to mechanisms with symmetric mass distributions. Alternatively, the link centers of mass can be brought successively closer to a single ground pivot. This version is not limited to symmetric links.

**Method of principal vectors and extensions [173].** This approach begins by describing the motion of the center of mass analytically and then determining the parameters that influence its resulting trajectory. The method of principal vectors describes the position of the total center of mass of a mechanism by a series of vectors, each directed along one of the links. Using the magnitudes of these principal vectors, a series of binary links can be added, in parallelogram fashion, to the original mechanism. The resulting augmented mechanisms contain a point that coincides with the total center of mass. Adding a counterweighted pantograph device, the augmented mechanism permits the combined center of mass to be brought to a stationary point, thereby balancing the shaking force.

Another method, the double-contour transformation, also enables the center of mass of a given mechanism to be traced out by a point on an attached proportional auxiliary mechanism. By adjusting parameters of the equations relating the auxiliary mechanism to the original mechanism, the total center of mass is made to lie on a point of the proportional mechanism which describes a circle. Thus a redistribution of link masses was required to constrain the center of mass to move within a circle. By adding counterweights, the total center of mass can then be reduced to the required stationary point.

**Cam and duplicate mechanism methods [173].** It is possible to devise cam drives and duplicate mechanisms to ensure that the total center of mass remains stationary. Since each device is necessarily peculiar to its application, the reader is referred to the literature.

**Nonlinear programming techniques [231].** Nonlinear programming techniques permit competing design objectives to be considered through the investigation of trade-offs between those objectives. The technique permits both partial force and partial moment balance to be accomplished simultaneously. In determining the position and size of counterweights to be added, the trade-off between maximum allowed shaking force and maximum allowed shaking moment is examined and the effect of constraining the maximum bearing force on this trade-off is considered.

In nonlinear programming an objective function of the design parameters is determined and its value is minimized by adjusting the independent variables of that function within the constraints established by design conditions or assumed by the designer. The technique is general and allows the use of counterweights, springs, and similar balancing devices. No assumptions are made regarding mass distribution in the unbalanced linkage. Input torque need not be constant and the mechanism need not be planar.

### Shaking Moment Balance [173]

Shaking moment balance has received far less attention in the literature. Beside the technique proposed by Berkof and Lowen, one technique utilizes a cam-actuated oscillating counterweight to provide balance for the shaking moment due to D'Alembert forces in a four-bar linkage. Another method proposes that oscillating counterweights be geared to the crank and follower of the four-bar. Addition of a duplicate mechanism with mirror symmetry to the original mechanism cancels shaking moments due to both D'Alembert forces and D'Alembert moments.

### Partial Balancing Techniques [173]

Partial balancing techniques have been described at length. Nearly all are restricted to planar mechanisms. Many rely on balancing certain harmonics of forces and moments to reduce inertia effects. Others add springs. While the latter technique does not necessarily diminish the vector sum of the shaking forces, it does change the manner by which those forces are transmitted to ground.

Harmonic or *order balance* is often achieved through the addition of supplementary masses. Once counterweight configuration is selected, placement of those additional masses is optimized by elimination of the undesired harmonics. This is accomplished by either *best average* or *best uniform* techniques, utilizing Fourier series and Gaussian least-squares formulations (for best average) or a Chebyshev approach (for best uniform).

To obtain the harmonic form of shaking force components, one first determines the harmonics of the displacement components of the total center of mass of the mechanism. Subsequent differentiations yield velocity, acceleration and hence D'Alembert force and moment variations. A single counterweight attached to the input crank provides partial balance of the first harmonic of the shaking force. The remainder of the unbalance can be minimized with a "Lanchester balancer," a device consisting of two identical counterrotating masses which are symmetrically arranged. When the two counterrotating masses are unequal and placed unsymmetrically, a "generalized Lanchester balancer" results, providing balance for both $x$ and $y$ components of any force harmonic.

Another technique, called a planetary mechanism balancer, capitalizes on the fact that any two-dimensional harmonic resultant describes an elliptic path [283]. Yet another technique adds supplementary masses to existing mechanism links as well as adding an input crank counterweight, permitting complete balance of the first Fourier harmonic. For a spatial linkage, it has been shown that three masses, set on mutually perpendicular bevel gears, can completely balance any three-dimensional shaking-force harmonic.

Any moment harmonic can be balanced by two identical masses, rotating synchronously and 180° out of phase, about two separate pivot points. By shifting the axis of the force counterweight, the first harmonic of the shaking moment due to D'Alembert forces of some force-balanced mechanisms can be eliminated. By adjusting the center distances between somewhat modified rotating masses of the "generalized Lanchester balancer" it is possible to supply the required balancing moments of a given harmonic order in addition to the force balance. Another device utilizes three rotating masses driven by the input shaft. Still another approach utilizes least-squares analysis, resulting in the addition of a single counterweight to the input shaft.

One author, V. A. Kamenskii, [146], addresses the issues concerning criteria by which the quality of balancing may be judged.

Complete force balancing of spatial mechanisms was treated by Kaufman and Sandor with the use of linearly independent spatial vectors [150].

## 5.16 COMPUTER PROGRAM FOR FORCE AND MOMENT BALANCE

Program BALANCE* provides options for either force or force and moment balance of any four-bar linkage. Inputs are the parameters for the unbalanced linkage (link lengths, positions of centers of mass, masses, etc.) as well as the selection of the density of the material to be used in the counterweights and the radius of the counterweights. Currently, it assumes that counterweights will be circular, pivoted on their periphery, and that moment balance can be achieved only for the situation where the coupler (link 3) is the physical pendulum and counterweights are added to links 2 and 4. A flow chart of the program is given in Fig. 5.71 and sample output follows.

### Output of Program BALANCE

Computer output is for the same values as used in Exs. 5.12 and 5.15 of the text.

* For more information, contact Dianne Rekow, University of Minnesota.

```
 ENTER THE MASS OF LINKS 2,3, AND 4 (LBM)
? .102,.274,.120

 ENTER THE LENGTHS OF LINKS 2,3, AND 4
 (PIN-TO-PIN LENGTHS) (INCHES)
? 2,6,3

 ENTER THE DISTANCE FROM THE PIN TO THE
 CENTER OF MASS FOR LINKS 2, 3, AND 4
 (FROM PIN 2 TO CENTER OF MASS FOR LINK 3) (INCHES)
? 1,3.19,1.5

 ENTER THE DISTANCE FROM PIN 3 TO THE
 CENTER OF MASS FOR LINK 3 (INCHES)
? 3.063

 ENTER THE ANGLE BETWEEN THE PIN-TO-PIN
 LENGTH AND THE PIN-TO-CENTER OF MASS LENGTH
 FOR LINKS 2, 3, AND 4 (DEGREES)
? 0,16,0

 ENTER THE ANGLE BETWEEN THE LINE CONNECTIONS
 PINS 2 AND 3 AND THE CENTER OF MASS
 (MEASURING FROM PIN 3) (DEGREES)
? 163.3

 DO YOU WISH TO FORCE BALANCE ONLY OR
 DO YOU WISH TO COMPLETELY BALANCE THE LINKAGE?
 (FORCE ONLY = 1          BOTH=2)
? 1

 WHICH TWO LINKS DO YOU WISH TO CHANGE?
? 2,4

 CAN YOU ALTER THE EXISTING LINK CONFIGURATION?
 (YES=1,NO=0)
? 1

 DO YOU WISH TO ALTER MASS OR DISTANCE TO THE
 CENTER OF MASS? (MASS=1, DISTANCE=2)
? 1
 LINKS TO BE BALANCED VIA ALTERING CURRENT
 DESIGN:              2            4

 PRODUCT OF MASS X DISTANCE FOR LINK 2
 WHEN BALANCED IS         .279754        LBM·INCH

 NEW ANGLE TO CENTER OF MASS OF LINK 2 =       163.300   DEGREES

 PRODUCT OF MASS X DISTANCE FOR LINK 4 WHEN
 BALANCED IS              .437030             LBM·INCH

 NEW ANGLE TO CENTER OF MASS OF LINK 4=         196.000   DEGREES

 DO YOU WISH TO INVESTIGATE THE SAME CASE BUT
 WITHOUT ALTERING THE EXISTING LINKS?
 (YES=1,NO=2)
? 1
 FORCE BALANCE VIA ADDING COUNTERWEIGHTS
```

```
LINKS CHANGED                    2          4

MASS X DISTANCE TO CENTER OF MASS FOR LINK 2
COUNTERWEIGHTS =            .378588          LBM·INCH

PS12=          167.741          DEGREES

MASS X DISTANCE TO CENTER OF MASS FOR LINK 4
COUNTERWEIGHTS =          .612071      LBM·INCH

PS14=          191.351          DEGREES

DO YOU WISH TO INVESTIGATE ANOTHER CASE?
(YES=1,NO=0)
? 1
**************************************************

ENTER THE MASS OF LINKS 2,3, AND 4 (SLUGS)
? .0031,.0054,.0045

ENTER THE LENGTHS OF LINKS 2,3, AND 4
(PIN-TO-PIN LENGTHS) (INCH)
? 1,4,3

ENTER THE DISTANCE FROM THE PIN TO THE
CENTER OF MASS FOR LINKS 2, 3, AND 4
(FROM PIN 2 TO CENTER OF MASS FOR LINK 3) (INCH)
? .5,2,1.5

ENTER THE DISTANCE FROM PIN 3 TO THE
CENTER OF MASS FOR LINK 3 (INCH)
? 2

ENTER THE ANGLE BETWEEN THE PIN-TO-PIN
LENGTH AND THE PIN-TO-CENTER OF MASS LENGTH
FOR LINKS 2, 3, AND 4 (DEGREES)
? 0,0,0

ENTER THE ANGLE BETWEEN THE LINE CONNECTIONS
PINS 2 AND 3 AND THE CENTER OF MASS
(MEASURING FROM PIN 3) (DEGREES)
? 180

DO YOU WISH TO FORCE BALANCE ONLY OR DO
YOU WISH TO COMPLETELY BALANCE THE LINKAGE?
(FORCE ONLY = 1      BOTH=2)
? 2

WHAT IS THE DENSITY OF THE MATERIAL OF THE
COUPLER? (      LBF/CU IN)
? .283

IS THE COUPLER AN IN-LINE RECTANGULAR BAR
WHICH CAN BE RECONFIGURED OR IS THE PIN-TO-PIN
LENGTH SET SO MASS CAN ONLY BE ADDED TO THE ENDS?
(RECONFIGURATION POSSIBLE=1, NOT POSSIBLE=2)
? 2
```

WHAT IS THE WIDTH OF THE LINK (INCHES)?
? .5

WHAT IS THE THICKNESS OF THE LINK? (INCHES)
? .2

ASSUME COUNTERWEIGHTS TO ACHIEVE FORCE BALANCE
ARE CIRCULAR, ARE OF EQUAL THICKNESS, AND ARE
PIVOTED ON THEIR PERIPHERY.
DO YOU WISH TO USE THE SAME MATERIAL AS USED
FOR THE OTHER LINKS?
(YES=1,NO=0)
? 1

WHAT THICKNESS OF THE COUNTERWEIGHTS? (INCHES)
? .5

ASSUME CIRCULAR DISCS PROVIDE THE INTERIA
COUNTERWEIGHTS REQUIRED, GEARED 1:1 TO THE INPUT AND
OUTPUT LINKS 2 AND 4.
IS THE DENSITY OF THESE COUNTERWEIGHTS TO BE
THE SAME AS FOR THE COUPLER, SAME AS THE FORCE
COUNTERWEIGHTS, OR DIFFERENT FROM EITHER OF THESE?
(SAME AS COUPLER=1, AS FORCE COUNTERWEIGHTS=2,
DIFFERENT=3)
? 1

WHAT IS THE DESIRED THICKNESS OF THE COUNTER-
WEIGHTS (ASSUME THEY ARE EQUAL THICKNESS)? (INCHES)
? 1

WHAT IS THE RADIUS OF GYRATION OF THE UNBALANCED
MECHANISM (FOR LINKS 2 AND 4)? (INCHES)
? .593,1.261

PARAMETERS FOR COMPLETELY BALANCED LINKAGE

| LINK | 2 | 3 | 4 |
|---|---|---|---|
| WRSTAR LBF·INCH | .137 | | .478 |
| WSTAR LB | .202 | .053 | .467 |
| RSTAR INCH | .678 | | 1.024 |
| KSTAR INCH | .479 | 2.000 | .724 |
| BALANCED WT LBF | .302 | .227 | .612 |
| BALANCED R IN | .377 | 2.000 | .557 |
| BALANCED K IN | .765 | 2.000 | 1.395 |
| CTWT I LBF·IN**2 | .010 | | .075 |
| RADIUS OF I CWT IN | .391 | | .640 |
| ANGLE TO C OF MASS DEGREES | 180.000 | 0 | 180.001 |

DO YOU WISH TO INVESTIGATE ANOTHER CASE?
(YES=1,NO=0)
? 0

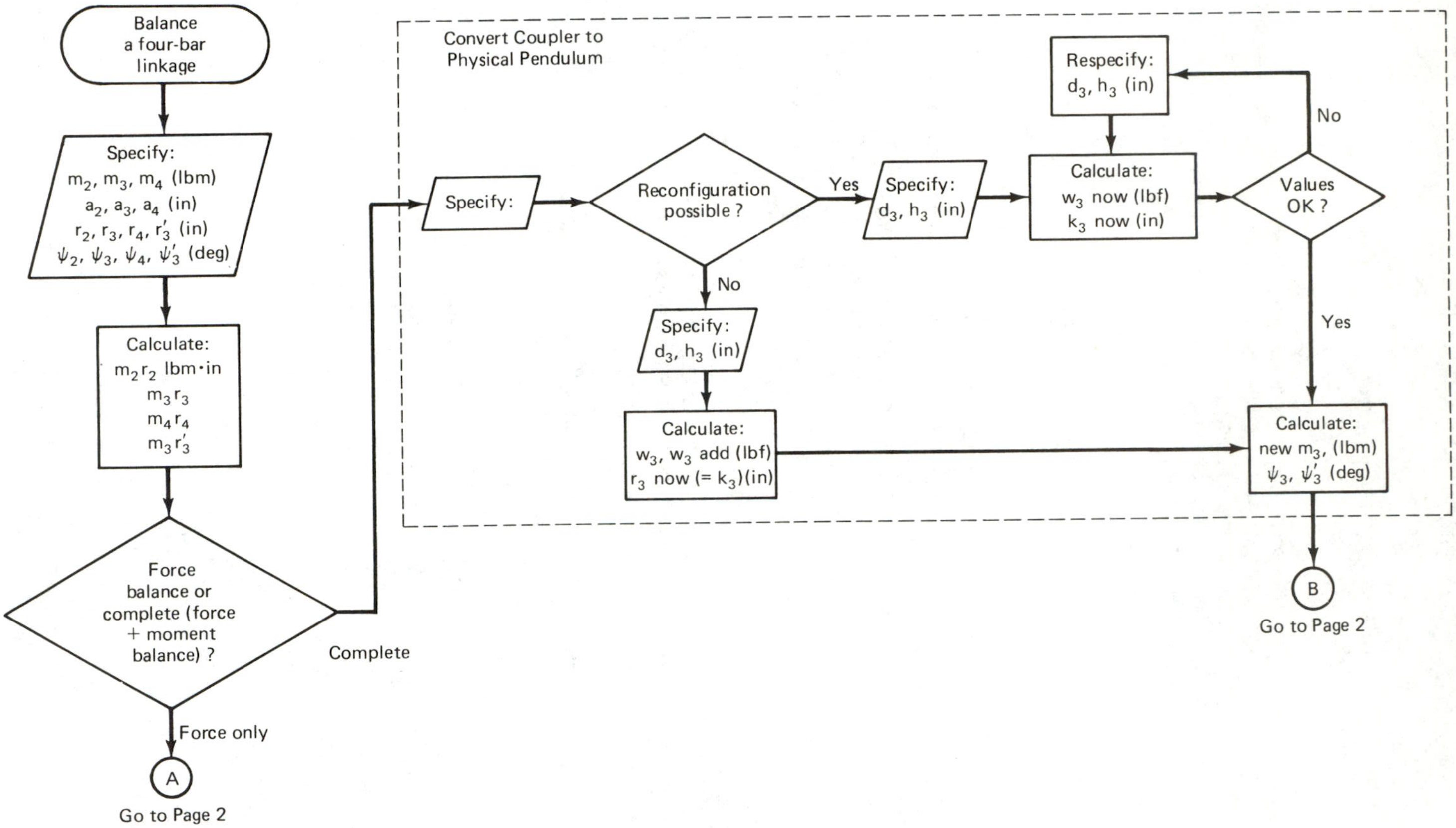


**Figure 5.71** Flowchart of interactive computer program for four-bar linkage balancing.

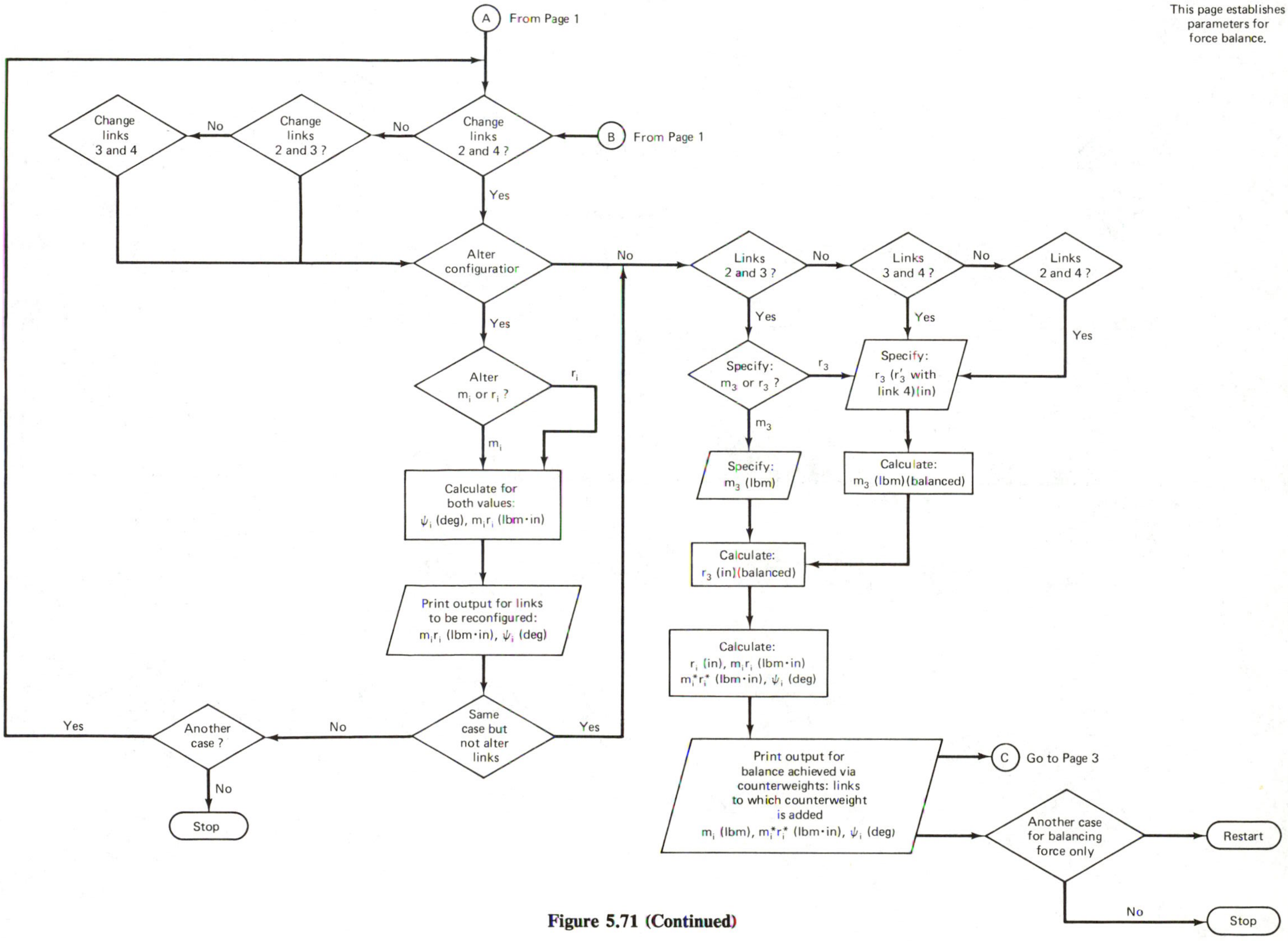


**Figure 5.71 (Continued)**

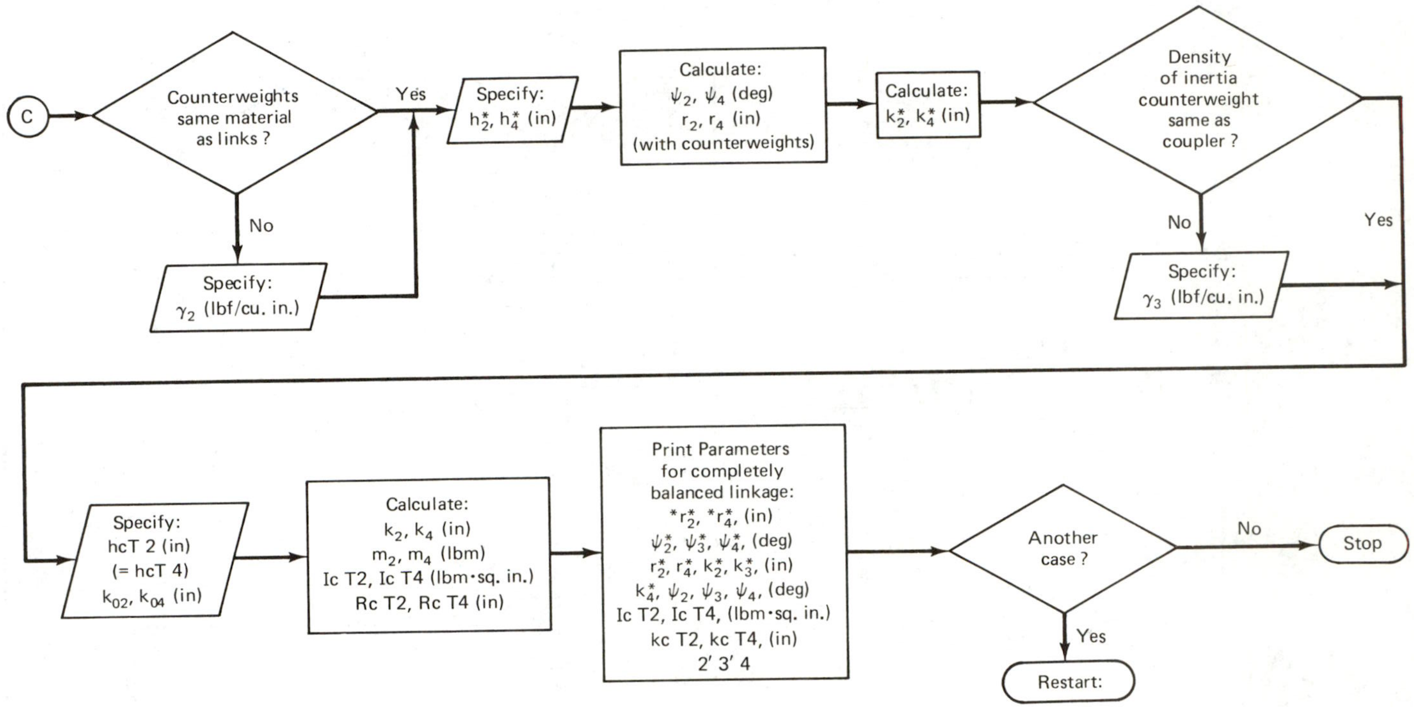


**Figure 5.71 (Continued)**

Nomenclature

| | |
|---|---|
| $m_i$ | = mass of link i (lbm) |
| $a_i$ | = length of link i (in.) |
| $r_i$ | = distance from pin to center of mass (see Fig. 5.58) (in.) |
| $r_i'$ | = distance from pin 3 to center of mass of link 3 (see Fig. 5.58) (in.) |
| $\psi_i$ | = angle between line from pin to pin and line from pin to center of mass of link i (degrees) |
| $\psi_3'$ | = angle between line connecting pins 2 and 3 and center of mass of link 3 (measuring from pin 3) (degrees) |
| $\gamma_3$ | = density of coupler (lbf/in$^3$) |
| $d_3$ | = width of coupler (in.) |
| $h_3$ | = thickness of coupler (in.) |
| $k_i$ | = radius of gyration of ith link (in.) |
| $m_i^* r_i^*$ | = mass X radius for force-balance counterweights ($r_i^*$ measured from same pivot as the center of mass of the ith link) (lbm·in.) |
| $\gamma_2$ | = density of moment-balance counterweight (lbf/in$^3$) |
| $h_i^*$ | = thickness of force-balance counterweight (in.) |
| $k_i^*$ | = radius of gyration of force-balance counterweights (in.) |
| $\gamma_3$ | = density of inertia counterweights (lbf/in$^3$) |
| hcwT2 | = thickness of inertia counterweights (link 2) (in.) |
| hcwT4 | = thickness of inertia counterweights (link 4) (in.) |
| $k_i^\circ$ | = radius of gyration for unbalanced link (in.) |
| IcwT2, IcwT4 | = moment of inertia for inertia counterweights (lbm·in$^2$) |
| kcwT2, kcwT4 | = radius of gyration of counterweight on link 2, link 4, = $\rho/\sqrt{2}$ (in.) |
| $\rho_{2,4}$ | = radius of circular-disk-shaped counterweights on link 2, link 4 = distance of center of mass of counterweight from ground pivot of link 2, link 4 (in.) |

**Figure 5.71 (Continued)**

## 5.17 BALANCING—APPENDIX A: THE PHYSICAL PENDULUM [287,288]

In the present context, a binary link is a "physical pendulum" if, when suspended without friction at either pivot, it has the same period of oscillation as if it were an idealized pendulum with its mass concentrated at its center of gravity on a massless stem of length $k$, where $k$ is the radius of gyration. Thus a physical-pendulum binary link gives the same radius of gyration when suspended by either end. That is, it is symmetric about its midpoint and the center of mass lies on the centerline. When these conditions are met, the radius of gyration ($k$) is given by [18,20]

$$k^2 = rr' \tag{5.264}$$

where $r$ and $r'$ are the distances of the center of gravity from the pivots.

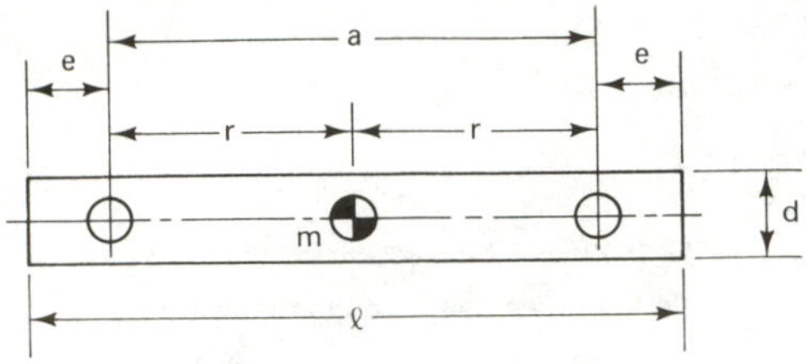


**Figure 5.72** Rectangular bar of uniform thickness. Table 5.15 give proportions to make this a *physical pendulum.*

Two possible configurations meet these general requirements: the rectangular bar and the augmented link. For simplicity, only symmetric links (those with the center of mass midway between the pivots) are considered.

## The Rectangular Bar

The configuration of a rectangular bar is shown in Fig. 5.72. If the link is to be a physical pendulum, it is necessary that

$$\frac{e}{d}=\frac{1}{2}\left[3\left(\frac{a}{d}\right)^2-1\right]^{1/2}-\frac{a}{2d} \tag{5.265}$$

and

$$\frac{l}{d}=\frac{a}{d}+2\left(\frac{e}{d}\right) \tag{5.266}$$

Then the radius of gyration is given by

$$k=rr'=r^2=\left(\frac{a}{2}\right)^2 \tag{5.267}$$

Table 5.15 summarizes some values for these ratios.

**TABLE 5.15** PARAMETERS FOR PHYSICAL PENDULUM—RECTANGULAR BAR (See Fig. 5.72).

| a/d | e/d | ℓ/d |
|---|---|---|
| 2 | 0.658 | 3.317 |
| 3 | 1.050 | 5.099 |
| 4 | 1.428 | 6.856 |
| 5 | 1.801 | 8.602 |
| 6 | 2.072 | 10.344 |
| 7 | 2.542 | 12.083 |
| 8 | 2.910 | 13.820 |
| 9 | 3.278 | 15.556 |
| 10 | 3.646 | 17.292 |

### The Augmented Link

The configuration of an augmented link is shown in Fig. 5.73. To meet the requirements for a physical pendulum, an augmented mass ($m^*$) is added. One configuration of adding $m^*$ is given in Fig. 5.74; many configurations are possible. The requirements for parameters of the link, converting it to a physical pendulum, are that the values of $d$ and $e$ must be selected to give a practical solution to

$$A\left(\frac{e}{d}\right)^3 + B\left(\frac{e}{d}\right)^2 + C\left(\frac{e}{d}\right) + D = 0 \tag{5.268}$$

where $A = 8$

$$B = 12\left(\frac{a}{d}\right) + 24$$

$$C = 24\left(\frac{a}{d}\right) + 26$$

$$D = -2\left(\frac{a}{d}\right)^3 + 13\left(\frac{a}{d}\right) + 12\pi - 10$$

Table 5.16 summarizes selected values of these ratios for links of uniform density and thickness. Note that as density increases or as width $d$ increases, $e$ becomes smaller.

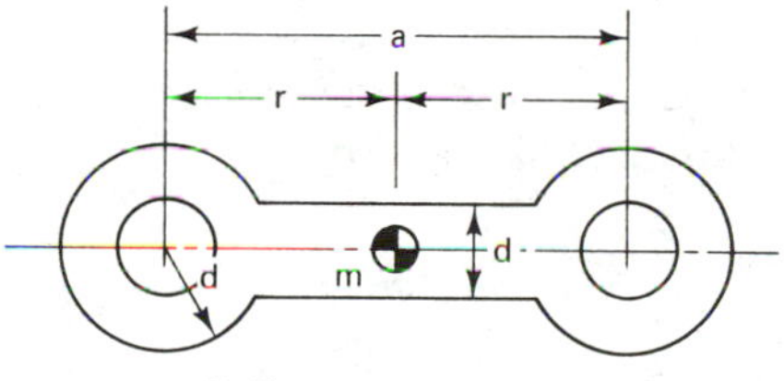


**Figure 5.73** Configuration of augmented link of uniform thickness.

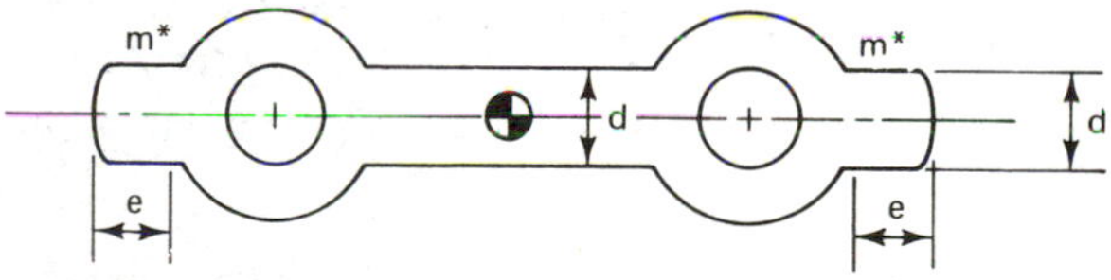


**Figure 5.74** Augmented link of Fig. 5.73 changed to a physical pendulum by the addition of masses m*. See Table 5.16 for proportions.

**TABLE 5.16** PARAMETERS FOR PHYSICAL PENDULUM—AUGMENTED LINK (Fig. 5.74).

| $a/d$ | $e/d$ |
|---|---|
| 4 | 0.330 |
| 5 | 0.740 |
| 6 | 1.130 |
| 7 | 1.511 |
| 8 | 1.887 |
| 9 | 2.260 |
| 10 | 2.631 |

## 5.18 BALANCING—APPENDIX B: THE EFFECT OF COUNTERWEIGHT CONFIGURATION ON BALANCE

As we have seen, the addition of counterweights to achieve force and moment balance has a significant effect on input torque. F. R. Hertrich* [128] has addressed the requirements for minimizing the inertial effects of these added counterweights. He concludes that:

> The material should be as dense as possible.
>
> The contour of the counterweight should be cylindrical with its axis parallel to the center of rotation.
>
> The cylinder should be as long as the particular design will permit.
>
> The cylinder should theoretically be tangent to the center of rotation even when there is a shaft in the center.
>
> If there is free access to the center of rotation, the radius of the counterweight should be determined by

$$R = \left(\frac{A}{\pi \gamma L}\right)^{1/3} \tag{5.269}$$

where $R$ = radius (in.)
$A$ = product of the weight and the distance between the center of mass of the counterweight and its rotational axis (lbf•in)
$\gamma$ = specific weight ($\text{lbf/in}^3$)
$L$ = length of the cylinder (in)

If there is a shaft or hub in the center, the radius of the cylinder should be obtained by

$$R = \frac{r_0}{2 \cos \alpha} \tag{5.270}$$

where $R$ = outside radius of the counterweight (in)
$r_0$ = radius of the shaft or hub in the rotational center (in.)
$\alpha$ = is found from the relationship

$$\frac{A}{L\gamma r_0^3} = \frac{4\alpha - \sin 4\alpha}{16 \cos^3\alpha} \tag{5.271}$$

where $\alpha$ is in radians.

* Reference [265] expands on the minimum inertia counterweight ideas of Hertrich.

## 5.19 ANALYSIS OF HIGH-SPEED ELASTIC MECHANISMS*

### Background

Until now, all considerations in the force analysis of mechanisms, whether static or dynamic, assumed that the links are rigid. The complexity of the mathematical analysis of mechanisms with elastic links has been a deterrent against giving up the rigidity assumption. Vibrations in mechanism links are often disregarded by the designers, machine elements are often overdesigned and quasi-static, rigid-body analysis is preferred for its relative simplicity. Omitting consideration of link deformations under dynamic conditions may contribute to a machine's failure to perform adequately at higher speeds. The area of study pertaining to the motion of mechanisms, with consideration of link elasticity and mass distribution, has been called the kinetoelastodynamics (KED) of mechanisms [74,86].

The effects of mass distribution and elasticity in mechanisms become significant at high speeds. "High speed" may imply different speeds for different mechanisms. One interpretation of this terminology may be speed at which inertial forces become so large that they cannot be ignored. Further, at certain speeds these inertial forces (in addition to the applied forces, for example, the resistive force of a steel plate on the punch of a slider-crank punching machine shown in Fig. 5.75), may excite one or more modes of vibration of the mechanism. The resulting deflections at some critical locations in the mechanism may render the performance of the machine unacceptable. High stress levels together with the large number of stress reversals may cause early failure from fatigue. Other problems associated with high-speed operation are difficulties in balancing, assurance of stability, and avoidance of "knocking" caused by clearances in the bearings of interconnecting members.

Research dealing with the dynamic behavior of mechanisms containing elastic links has been reviewed by Lowen and Jandrasits [174], and by Erdman and Sandor [86]. Some attempts to include elastic effects in the analysis of mechanisms in Refs. [7,8,138–140,154,183,194,285,289,300,327] (the list is by no means complete) have generally been based on the slider-crank mechanism, and in some instances, on the four-bar crank-rocker mechanism. This group of researchers treated the elastic links as continuous systems; the connecting rod, the coupler, or the follower is considered to be the only elastic element in the mechanism. Equations of motion with certain simplifying assumptions [8] are derived for such mechanisms and then solved by analog and/or digital computation to obtain their dynamic response.

A second group of researchers (e.g., [1,65,74,145,153,157,189,226,276,292,314] has treated the elastic links as discrete systems. Here the continuum is divided into a finite number of masses and springs and the resulting finite number of ordinary coupled differential equations are solved. Two of the methods utilizing this approach are the finite-element and finite-difference techniques. The lumped-parameter ap-

* The rest of this chapter is based on a compilation of research related to high-speed mechanisms prepared by Ashok Midha, Purdue University.

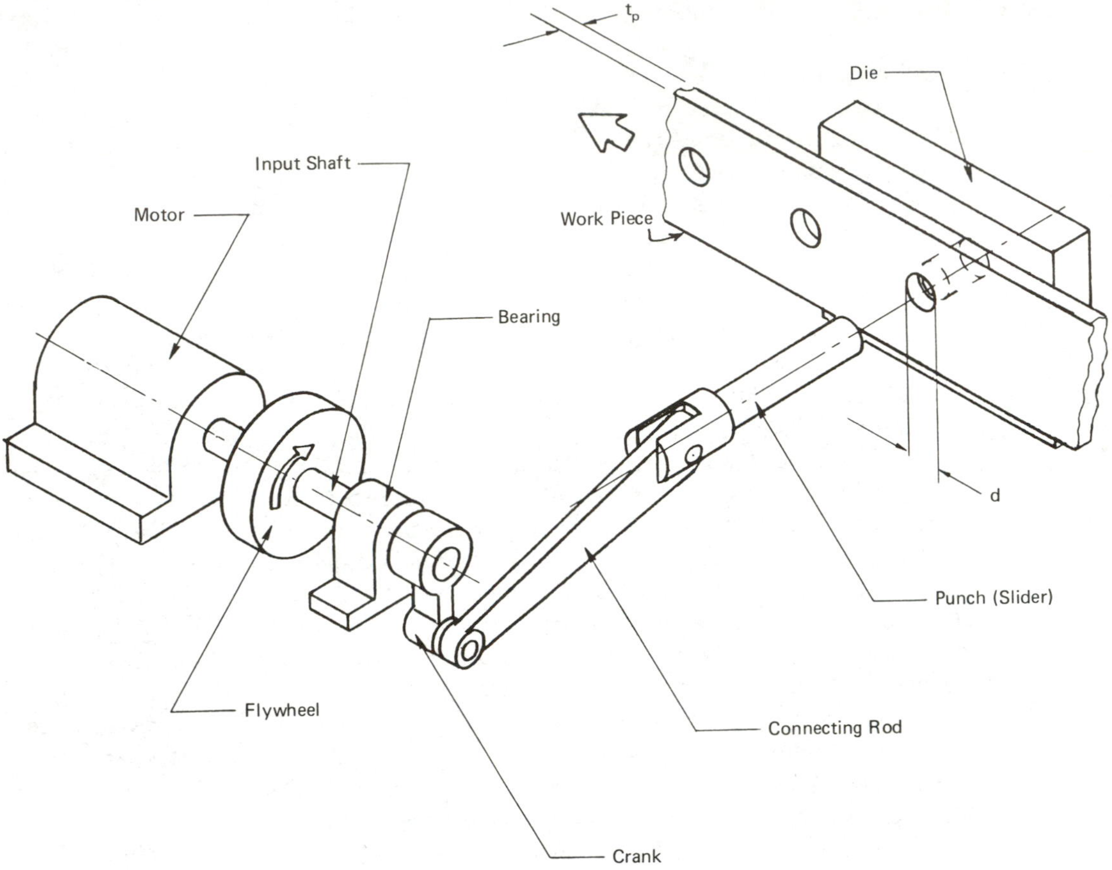


**Figure 5.75** Schematic of a motor-driven slider-crank punching machine.

proach [226,157], utilizing the finite-difference method, has been a useful technique. The finite-element theory of structural analysis (e.g., [48,52,53,130,133,177,182, 206,297,325]) has also been applied to modeling an elastic linkage using both the "force method" [74,90] and the "displacement method" [134,189,292,314]. Using the various analysis methods, researchers have sought to verify their respective techniques by conducting experimental investigations [1,68,112,138,276,292] of high-speed elastic-membered mechanisms. The finite-element method has become well established in engineering and it can be used to model two- and three-dimensional systems. A significant library on finite elements is now available in the literature.

The rigid-body analysis of planar mechanism may be performed graphically, algebraically, or by a combination of the two. Elastic deformation will be accounted for by adding (to the rigid-body degrees of freedom) a number of elastic degrees of freedom to the mechanism. Thus an elastic mechanism has many degrees of freedom, and a corresponding number of coupled differential equations of motion. In general, it is not feasible to analyze elastic mechanism problems without the support of a digital computer. The purpose of this section is to provide the reader with a set of interconnecting building blocks toward the solution of problems involving elastic mechanisms. To this end, some introductory detail of the finite-element method is provided here. The finite element theory texts referred to above should be consulted, especially for modeling complex mechanism systems and usage of more refined element types. References [200,203,326] are also pertinent to finite elements.

## Introduction

A general approach is described in the rest of this chapter for deriving the equations of motion of planar linkages in high-speed machinery. The *displacement finite-element method,* in which nodal elastic displacements are regarded as the unknowns, is applied to develop the mass and stiffness properties of an elastic linkage. To demonstrate the various steps in the analysis, a four-bar linkage is utilized; however, the method is readily extended to other planar multiloop linkages. Starting with a typical elastic planar beam element, the nodal displacement and acceleration expressions are derived, including the terms that couple the elastic and rigid-body motions. The linkage is modeled as consisting of beam elements and its equations of motion are expressed in matrix form. Methods are described for systematic assembly of all elements, resulting in the undamped equations of motion of the total system. Conventional forms of structural damping are reviewed in Sec. 5.27 for inclusion in the equations of motion. Also reviewed are assumptions typically made in order to simplify the analysis as well as facilitate numerical solution. Some numerical methods for obtaining the vibration response are discussed in Sec. 5.30.

In the *force finite-element method,* the internal forces are assumed to be the unknowns and equilibrium conditions are used to derive equations for finding them. Additional equations are based on compatibility conditions.

The *displacement finite-element method* formulation is in general found to be much simpler for a larger number of structural and mechanism problems. For this reason, the latter method is used here. Compatibility conditions in and among the elements are initially satisfied. The equations of motion are expressed in terms of nodal displacements utilizing the equilibrium conditions at each nodal point.

The planar four-bar elastic linkage of Fig. 5.76 is modeled by connecting a series of beam elements. They will be connected in a manner so as to permit the model to account for variations in the linkage geometry. For simplicity, a typical link is regarded as a basic structural (beam) element having uniform cross section across its length. An analytical model of this element will facilitate an adaptation of standard structural dynamics techniques. As shown in Fig. 5.76, small elastic displacements (solid lines) from the rigid-body motion of the linkage (dashed lines) will be modeled.

Each link is modeled by one beam element. This, in conjunction with the assumed displacement shape functions described in a later section, permits at most one inflection point along the beam element. Therefore, the number of modes that

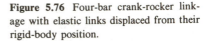

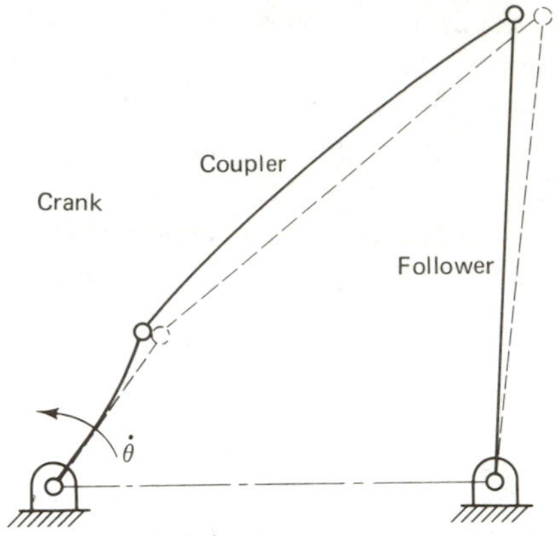


**Figure 5.76** Four-bar crank-rocker linkage with elastic links displaced from their rigid-body position.

can be correctly computed is a function of the number of elements per link. Also, as discussed in Refs. 206 and 292, one-element idealization results in erroneous (higher) modal frequencies. One-element idealization is chosen initially for the sake of simplicity in demonstration. The mass matrix, derived with the help of assumed displacement functions is called the consistent mass matrix.

## 5.20 ELASTIC BEAM ELEMENT IN PLANE MOTION

A general beam element, representing a link of a mechanism, is shown in Fig. 5.77, without the other links. There are two frames of reference: the fixed ($OXY$) and the rotated ($Oxy$) frames, which have a common origin. The $x$ axis of the rotated frame is always parallel to the rigid-body position of the beam element axis throughout its motion. The beam element is shown in its rigid-body position (dashed lines) as well as in its elastically deformed configuration (solid lines). The elastic deformations of the beam element may be completely described by the six generalized nodal displacement coordinates, $u_1$ through $u_6$, illustrated in Fig. 5.77. These displacements, shown in their positive directions with reference to the rigid-body position of the beam element, locate the deformed positions $A'$ and $B'$ of the nodes $A$ and $B$. From Fig. 5.77, the following relationships may be expressed in the fixed ($OXY$) reference frame:

$$\begin{aligned} X_{A'} &= X_A + u_1 \cos\theta - u_2 \sin\theta \\ Y_{A'} &= Y_A + u_1 \sin\theta + u_2 \cos\theta \\ \theta_{A'} &= \theta + u_3 \end{aligned} \tag{5.272}$$

Taking successive derivatives of Eqs. (5.272) with respect to time, expressions for velocity and acceleration of node $A$ may be obtained in the fixed-coordinate system.

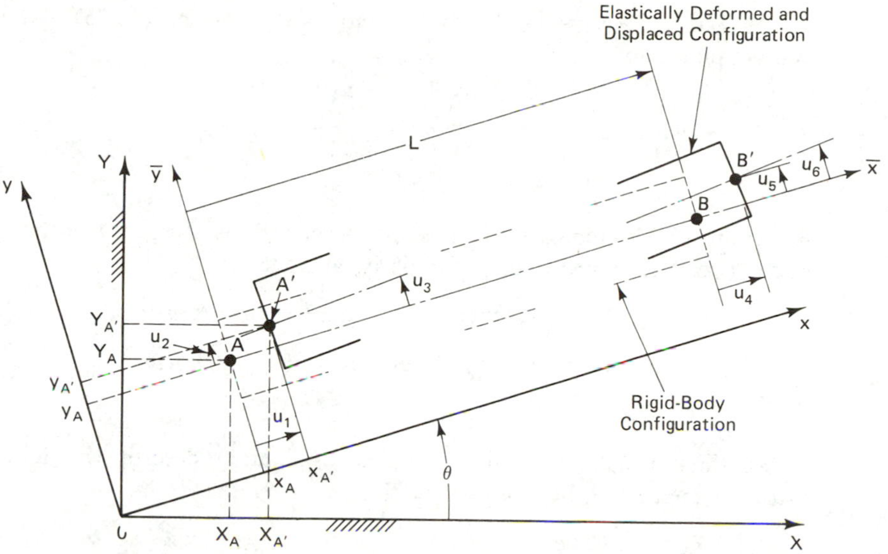


**Figure 5.77** Rigid-body and elastically deformed/displaced configurations of beam element, showing generalized displacements $u_1, u_2, \ldots, u_6$ referred to fixed- and rotated-coordinate systems.

$$\begin{aligned}
\dot{X}_{A'} &= \dot{X}_A + \dot{u}_1 \cos\theta - u_1\dot{\theta}\sin\theta - \dot{u}_2 \sin\theta - u_2\dot{\theta}\cos\theta \\
\dot{Y}_{A'} &= \dot{Y}_A + \dot{u}_1 \sin\theta + u_1\dot{\theta}\cos\theta + \dot{u}_2 \cos\theta - u_2\dot{\theta}\sin\theta \\
\dot{\theta}_{A'} &= \dot{\theta} + \dot{u}_3
\end{aligned} \tag{5.273}$$

and

$$\begin{aligned}
\ddot{X}_{A'} &= \ddot{X}_A + \ddot{u}_1 \cos\theta - 2\dot{u}_1\dot{\theta}\sin\theta - u_1\dot{\theta}^2\cos\theta - u_1\ddot{\theta}\sin\theta - \ddot{u}_2\sin\theta \\
&\quad - 2\dot{u}_2\dot{\theta}\cos\theta + u_2\dot{\theta}^2\sin\theta - u_2\ddot{\theta}\cos\theta \\
\ddot{Y}_{A'} &= \ddot{Y}_A + \ddot{u}_1 \sin\theta + 2\dot{u}_1\dot{\theta}\cos\theta - u_1\dot{\theta}^2\sin\theta + u_1\ddot{\theta}\cos\theta + \ddot{u}_2\cos\theta \\
&\quad - 2\dot{u}_2\dot{\theta}\sin\theta - u_2\dot{\theta}^2\cos\theta - u_2\ddot{\theta}\sin\theta \\
\ddot{\theta}_{A'} &= \ddot{\theta} + \ddot{u}_3
\end{aligned} \tag{5.274}$$

The absolute accelerations in Eqs. (5.274) with respect to the fixed $(OXY)$-coordinate system are expressed in the rotated $(Oxy)$-coordinate system with the help of the following transformation:

$$\begin{aligned}
\ddot{x}_{A'} &= \ddot{X}_{A'} \cos\theta + \ddot{Y}_{A'} \sin\theta \\
\ddot{y}_{A'} &= -\ddot{X}_{A'} \sin\theta + \ddot{Y}_{A'} \cos\theta \\
\ddot{\theta}_{A'} &= \ddot{\theta} + \ddot{u}_3
\end{aligned} \tag{5.275}$$

where we have regarded $\ddot{X}_{A'} = (\ddot{r}_{A'})_X$ and $\ddot{Y}_{A'} = (\ddot{r}_{A'})_Y$. Similarly, if we regard

$\ddot{x}_{A'} = (\ddot{r}_{A'})_x$ and $\ddot{y}_{A'} = (\ddot{r}_{A'})_y$, then equations (5.274) and (5.275) may be combined and simplified to

$$\begin{aligned} \ddot{x}_{A'} &= \ddot{x}_A + \ddot{u}_1 - u_1\dot{\theta}^2 - 2\dot{u}_2\dot{\theta} - u_2\ddot{\theta} \\ \ddot{y}_{A'} &= \ddot{y}_A + \ddot{u}_2 - u_2\dot{\theta}^2 + 2\dot{u}_1\dot{\theta} + u_1\ddot{\theta} \\ \ddot{\theta}_{A'} &= \ddot{\theta} + \ddot{u}_3 \end{aligned} \tag{5.276}$$

Equations (5.276) represent the absolute accelerations of node $A$ of the beam element in the $(Oxy)$ reference frame. Similarly, at node $B$,

$$\begin{aligned} \ddot{x}_{B'} &= \ddot{x}_B + \ddot{u}_4 - u_4\dot{\theta}^2 - 2\dot{u}_5\dot{\theta} - u_5\ddot{\theta} \\ \ddot{y}_{B'} &= \ddot{y}_B + \ddot{u}_5 - u_5\dot{\theta}^2 + 2\dot{u}_4\dot{\theta} + u_4\ddot{\theta} \\ \ddot{\theta}_{B'} &= \ddot{\theta} + \ddot{u}_6 \end{aligned} \tag{5.277}$$

where, the kinematic quantities $\ddot{x}_A$, $\ddot{y}_A$, $\ddot{x}_B$, $\ddot{y}_B$, $\dot{\theta}$, and $\ddot{\theta}$ describe the rigid-body motion of the element, and, in terms of the fixed-coordinate system,

$$\ddot{x}_A = \ddot{X}_A \cos\theta + \ddot{Y}_A \sin\theta \qquad \text{and} \qquad \ddot{y}_A = -\ddot{X}_A \sin\theta + \ddot{Y}_A \cos\theta$$

Let us define the following column vectors:

$$\{\ddot{u}_{ai}\} = \begin{Bmatrix} \ddot{x}_{A'} \\ \ddot{y}_{A'} \\ \ddot{\theta}_{A'} \\ \ddot{x}_{B'} \\ \ddot{y}_{B'} \\ \ddot{\theta}_{B'} \end{Bmatrix} \quad \text{and} \quad \{\ddot{u}_{ri}\} = \begin{Bmatrix} \ddot{x}_A \\ \ddot{y}_A \\ \ddot{\theta} \\ \ddot{x}_B \\ \ddot{y}_{B'} \\ \ddot{\theta} \end{Bmatrix}, \qquad i = 1, 2, \ldots, 6 \tag{5.278}$$

By combining Eqs. (5.276) to (5.278),

$$\begin{Bmatrix} \ddot{u}_{a1} \\ \ddot{u}_{a2} \\ \ddot{u}_{a3} \\ \ddot{u}_{a4} \\ \ddot{u}_{a5} \\ \ddot{u}_{a6} \end{Bmatrix} = \begin{Bmatrix} \ddot{u}_{r1} + \ddot{u}_1 - u_1\dot{\theta}^2 - 2\dot{u}_2\dot{\theta} - u_2\ddot{\theta} \\ \ddot{u}_{r2} + \ddot{u}_2 - u_2\dot{\theta}^2 + 2\dot{u}_1\dot{\theta} + u_1\ddot{\theta} \\ \ddot{u}_{r3} + \ddot{u}_3 + 0 \quad + 0 \quad + 0 \\ \ddot{u}_{r4} + \ddot{u}_4 - u_4\dot{\theta}^2 - 2\dot{u}_5\dot{\theta} - u_5\ddot{\theta} \\ \ddot{u}_{r5} + \ddot{u}_5 - u_5\dot{\theta}^2 + 2\dot{u}_4\dot{\theta} + u_4\ddot{\theta} \\ \ddot{u}_{r6} + \ddot{u}_6 + 0 \quad + 0 \quad + 0 \end{Bmatrix} \tag{5.279}$$

Equations (5.279) may be rewritten as

$$\{\ddot{u}_a\} = \{\ddot{u}_r\} + \{\ddot{u}\} + \{a_n\} + \{a_c\} + \{a_t\} \tag{5.280}$$

where the vectors (from left to right) represent the *absolute, rigid-body,* generalized *relative* (to the rigid-body position of beam element), *normal, Coriolis,* and *tangential* accelerations, respectively. Similar acceleration terms may be identified in Eqs. (5.274) in the fixed $(OXY)$ reference frame. The last three vectors in Eq. (5.280) are the rigid-body and elastic motion coupling terms.

If small generalized displacements are assumed, and if the rigid-body angular velocity and acceleration are also generally small compared to the lowest circular frequency and the frequency squared of the mechanism modal deformation, respectively, then in Eq. (5.280) the product terms in vectors $\{a_n\}$, $\{a_c\}$, and $\{a_t\}$ are considered small compared to corresponding terms in $\{\ddot{u}_r\} + \{\ddot{u}\}$. The normal, Coriolis, and tangential acceleration terms are therefore often neglected. If these terms are not included, the designer should check final displacement, velocity, and acceleration levels to be sure that this assumption is indeed valid. When they are neglected, Eq. (5.280) may be modified as

$$\{\ddot{u}_a\} = \{\ddot{u}_r\} + \{\ddot{u}\} \tag{5.281}$$

Similarly, it may be shown that

$$\{\dot{u}_a\} = \{\dot{u}_r\} + \{\dot{u}\} \tag{5.282}$$

## 5.21 DISPLACEMENT FIELDS FOR BEAM ELEMENT

Each of the six generalized displacement coordinates $\{u_j\}$ in Fig. 5.77 amplifies a displacement field, or shape function, $\phi_j(\bar{x})$ $(j = 1, \ldots, 6)$. $\phi_j(\bar{x})$ are functions of position along the beam only, while their amplitudes $u_j$ are functions of time. The $j$th shape function $\phi_j(\bar{x})$ is defined as the displaced equilibrium configuration of the element when all coordinates are constrained to be zero, and a unit displacement is applied to the $j$th coordinate.

The selection of the shape function [297] must meet the following requirements: (1) possess continuity within the element and across the element boundaries when the element is joined to other elements, (2) include representation of constant (including zero) values of pertinent stresses or strains, and (3) satisfy the displacement compatibility at the boundaries of the adjoining elements.

For the axial deformation of the beam element, a logical choice of the shape function is a linear polynomial in $\bar{x}$. For bending, the displacement field may be described by a cubic polynomial in $\bar{x}$. Reference [133] determines the following:

$$\begin{aligned}
\phi_1(\bar{x}) &= 1 - \frac{\bar{x}}{L} \\
\phi_2(\bar{x}) &= 3\left(\frac{L-\bar{x}}{L}\right)^2 - 2\left(\frac{L-\bar{x}}{L}\right)^3 \\
\phi_3(\bar{x}) &= \bar{x}\left(\frac{L-\bar{x}}{L}\right)^2 \\
\phi_4(\bar{x}) &= \frac{\bar{x}}{L} \\
\phi_5(\bar{x}) &= 3\left(\frac{\bar{x}}{L}\right)^2 - 2\left(\frac{\bar{x}}{L}\right)^3 \\
\phi_6(\bar{x}) &= -(L-\bar{x})\left(\frac{\bar{x}}{L}\right)^2
\end{aligned} \tag{5.283}$$

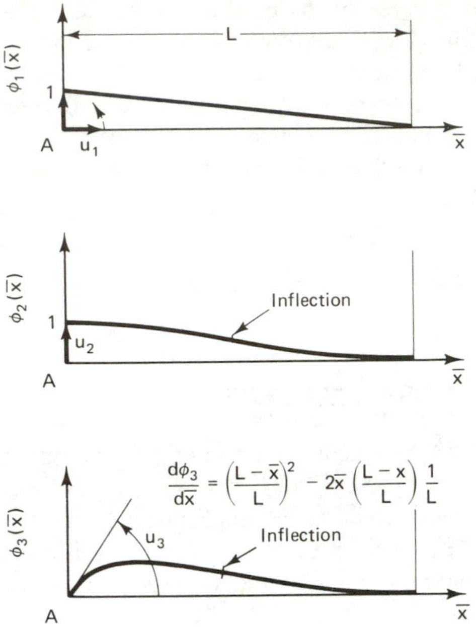


**Figure 5.78** Shape functions of beam element $\phi_1(\bar{x})$, $\phi_2(\bar{x})$ and $\phi_3(\bar{x})$. $\phi_4(\bar{x})$, $\phi_5(\bar{x})$, and $\phi_6(\bar{x})$ are similarly representable as functions of $\bar{x}$.

For graphs of these shape functions, see Fig. 5.78.

For small displacements where the linear theory of elasticity is used, the principle of linear superposition allows the algebraic summing of displacements corresponding to the various shape functions. Thus the transverse displacement, $w(\bar{x}, t)$ (measured along the $\bar{y}$ axis in the $A\bar{x}\bar{y}$ reference frame in Fig. 5.77), may be written as

$$w(\bar{x}, t) = \phi_2(\bar{x})u_2(t) + \phi_3(\bar{x})u_3(t) + \phi_5(\bar{x})u_5(t) + \phi_6(\bar{x})u_6(t) \qquad (5.284)$$

Similarly, the axial displacement,

$$v(\bar{x}, t) = \phi_1(\bar{x})u_1(t) + \phi_4(\bar{x})u_4(t) \qquad (5.285)$$

It must be noted that, from both the physical and mathematical points of view, an important step is the selection of the displacement fields. Elements based on assumed displacement field, as opposed to those based on assumed stress field or a combination of the two, have been found to be most successful and versatile. The importance of assuming displacement fields will become obvious in the following section.

## 5.22 ELEMENT MASS AND STIFFNESS MATRICES*

Selecting $u_i$ ($i = 1, \ldots, 6$) as the generalized coordinates of the problem, the equations of motion of the elastic beam element in Fig. 5.77 may be described by

* A good summary of matrix algebra is Sec. 6.4, *Matrices,* in Ref. 57.

Lagrange's equations (see Sec. 5.6):

$$\frac{d}{dt}\left(\frac{\partial T}{\partial \dot{u}_i}\right) - \frac{\partial T}{\partial u_i} + \frac{\partial U}{\partial u_i} = \overline{Q}_i, \qquad i = 1, 2, \ldots, 6 \tag{5.286}$$

where $\overline{Q}_i$ are the generalized forces acting in the direction of the generalized coordinates, and have no potential; $T$ and $U$ represent the kinetic energy and strain (potential) energy, respectively, and where it is recognized that $U$ is independent of $\dot{u}_i$.

The kinetic energy must be referred to an inertial (fixed reference) frame; for the beam element, it may be expressed as

$$T = \tfrac{1}{2}\int_0^L m(\bar{x})\dot{w}_a(\bar{x}, t)^2\, d\bar{x} + \tfrac{1}{2}\int_0^L m(\bar{x})\dot{v}_a(\bar{x}, t)^2\, d\bar{x} \tag{5.287}$$

where the absolute transverse and axial velocities of the elemental mass $m(\bar{x})\, d\bar{x}$ are

$$\dot{w}_a(\bar{x}, t) = \phi_2(\bar{x})\dot{u}_{a2}(t) + \phi_3(\bar{x})\dot{u}_{a3}(t) + \phi_5(\bar{x})\dot{u}_{a5}(t) + \phi_6(\bar{x})\dot{u}_{a6}(t)$$

$$\dot{v}_a(\bar{x}, t) = \phi_1(\bar{x})\dot{u}_{a1}(t) + \phi_4(\bar{x})\dot{u}_{a4}(t)$$

where $\{\dot{u}_{ai}\}$ is defined by Eq. (5.282), and $m(\bar{x})$ is the mass per unit length.

$$\dot{w}_a(\bar{x}, t)^2 = \sum_i \sum_j \phi_i(\bar{x})\phi_j(\bar{x})\dot{u}_{ai}(t)\dot{u}_{aj}(t), \qquad i, j = 2, 3, 5 \text{ and } 6$$

$$\dot{v}_a(\bar{x}, t)^2 = \sum_k \sum_l \phi_k(\bar{x})\phi_l(\bar{x})\dot{u}_{ak}(t)\, \dot{u}_{al}(t), \qquad k, l = 1 \text{ and } 4 \tag{5.288}$$

The kinetic energy in Eq. (5.287) is then expressed as the sum

$$T = T_w + T_v$$

where

$$T_w = \tfrac{1}{2}\int_0^L \sum_i \sum_j m(\bar{x})\phi_i(\bar{x})\phi_j(\bar{x})\dot{u}_{ai}(t)\dot{u}_{aj}(t)\, d\bar{x}$$

and

$$T_v = \tfrac{1}{2}\int_0^L \sum_k \sum_l m(\bar{x})\phi_k(\bar{x})\phi_l(\bar{x})\dot{u}_{ak}(t)\dot{u}_{al}(t)\, d\bar{x} \tag{5.289}$$

Interchanging the order of integration and summation, we define an element $\bar{m}_{ij}$ of the generalized mass matrix $[\bar{m}]$ as

$$\bar{m}_{ij} = \int_0^L m(\bar{x})\phi_i(\bar{x})\phi_j(\bar{x})\, d\bar{x} \tag{5.290}$$

and similarly

$$\bar{m}_{kl} = \int_0^L m(\bar{x})\phi_k(\bar{x})\phi_l(\bar{x})\, d\bar{x}$$

In matrix form

$$T_w = \tfrac{1}{2}\{\dot{u}_{ai}\}^T[\bar{m}_{ij}]\{\dot{u}_{ai}\} \tag{5.291}$$

and

$$T_v = \tfrac{1}{2}\{\dot{u}_{ak}\}^T[\bar{m}_{kl}]\{\dot{u}_{ak}\}$$

Superimposing $T_w$ and $T_v$, the total kinetic energy becomes

$$T = \tfrac{1}{2}\{\dot{u}_a\}^T[\bar{m}]\{\dot{u}_a\}$$

From Eq. (5.282),

$$T = \tfrac{1}{2}\,[\{\dot{u}_r\} + \{\dot{u}\}]^T\,[\bar{m}][\{\dot{u}_r\} + \{\dot{u}\}] \tag{5.292}$$

Considering the strains associated with the displacement functions, and neglecting those due to temperature variations and any strains initially present, the strain energy of the elastic beam element is

$$U = \tfrac{1}{2}\int_0^L EI(\bar{x})w(\bar{x}, t)^2\,d\bar{x} + \tfrac{1}{2}\int_0^L EA(\bar{x})v(\bar{x}, t)^2\,d\bar{x} \tag{5.293}$$

Taking the derivatives of $w(\bar{x}, t)$ and $v(\bar{x}, t)$, given Eqs. (5.284) and (5.285), with respect to x̄ and defining the elements of the generalized stiffness matrix $[\bar{k}]$ as

$$\bar{k}_{ij} = \int_0^L EI(\bar{x})\phi_i(\bar{x})\phi_j(\bar{x})\,d\bar{x}, \qquad i, j = 2, 3, 5, 6$$

and

$$\bar{k}_{kl} = \int_0^L EA(\bar{x})\phi_k(\bar{x})\phi_l(\bar{x})\,d\bar{x}, \qquad k, l = 1 \text{ and } 4 \tag{5.294}$$

The strain energy (in matrix form) is

$$U = \tfrac{1}{2}\{u\}^T[\bar{k}]\{u\} \tag{5.295}$$

Furthermore, $[\bar{m}]$ and $[\bar{k}]$ can be shown [206] to be

$$[\bar{m}] = \rho AL \begin{bmatrix} \frac{1}{3} & & & \text{Symmetric} & & \\ 0 & \frac{13}{35} & & & & \\ 0 & \frac{11L}{210} & \frac{L^2}{105} & & & \\ \frac{1}{6} & 0 & 0 & \frac{1}{3} & & \\ 0 & \frac{9}{70} & \frac{13L}{420} & 0 & \frac{13}{35} & \\ 0 & \frac{-13L}{420} & \frac{-L^2}{140} & 0 & \frac{-11L}{210} & \frac{L^2}{105} \end{bmatrix} \tag{5.296}$$

where $m(\bar{x}) = m = \rho A$, the uniformly distributed mass per unit length and where rotary inertia is not included, and

$$[\bar{k}] = \begin{bmatrix} \frac{EA}{L} & & & \text{Symmetric} & & \\ 0 & \frac{12EI}{L^3} & & & & \\ 0 & \frac{6EI}{L^2} & \frac{4EI}{L} & & & \\ -\frac{EA}{L} & 0 & 0 & \frac{EA}{L} & & \\ 0 & -\frac{12EI}{L^3} & -\frac{6EI}{L^2} & 0 & \frac{12EI}{L^3} & \\ 0 & \frac{6EI}{L^2} & \frac{2EI}{L} & 0 & -\frac{6EI}{L^2} & \frac{4EI}{L} \end{bmatrix} \tag{5.297}$$

where the elements are derived from engineering beam theory, neglecting shear deformations. It is possible to include the effects of rotatory inertia and shear deformation into the matrices above; Ref. 206 may be consulted for further description of these aspects. With the help of Eq. (5.286), the motion equations for the beam element may be readily derived as

$$[\bar{m}]\{\ddot{u}_a(t)\} + [\bar{k}]\{u(t)\} = \{\bar{Q}\} \tag{5.298}$$

## 5.23 SYSTEM MASS AND STIFFNESS MATRICES

In as much as it is convenient to formulate the mass and stiffness matrices for individual elements in their local coordinates, construction of the total system matrices of the assemblage of arbitrarily oriented elements is best facilitated by defining these element matrices in the global coordinate system. Thus, while we may have several local coordinate systems, we define only one global coordinate system for a given problem.

Consider the general element in Fig. 5.79 with two nodes, 1 and 2. The figure also shows two coordinate systems—the local (element-oriented) and the global (system-oriented) coordinate systems. The latter is defined, with its origin at a given node, parallel to the fixed $(OXY)$ reference frame in Fig. 5.77. In Fig. 5.79, the two coordinate systems $(1\bar{x}\bar{y})$ and $(1\bar{X}\bar{Y})$ are shown at node 1; the same may be defined at node 2. The generalized coordinates are shown labeled in both systems.

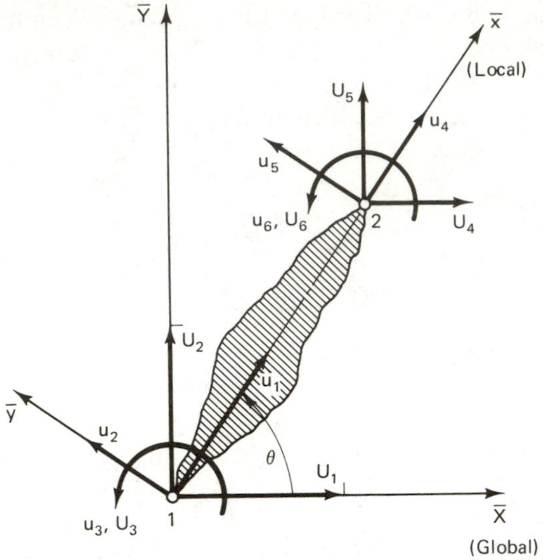


**Figure 5.79** Generalized displacements of beam element in element-oriented (local) and system-oriented (global) coordinates.

It is easy to show that at node 1

$$
\begin{aligned}
u_1 &= U_1 \cos\theta + U_2 \sin\theta \\
u_2 &= - U_1 \sin\theta + U_2 \cos\theta \\
u_3 &= U_3
\end{aligned}
\tag{5.299}
$$

and at node 2

$$
\begin{aligned}
u_4 &= U_4 \cos\theta + U_5 \sin\theta \\
y_5 &= - U_4 \sin\theta + U_5 \cos\theta \\
u_6 &= U_6
\end{aligned}
\tag{5.300}
$$

Equations (5.299) and (5.300) can be expressed in matrix form. To simplify the notation, we define: $\lambda \triangleq \cos\theta$ and $\mu \triangleq \sin\theta$. Then

$$
[R] = \left[\begin{array}{ccc|ccc}
\lambda & \mu & 0 & 0 & 0 & 0 \\
-\mu & \lambda & 0 & 0 & 0 & 0 \\
0 & 0 & 1 & 0 & 0 & 0 \\
\hline
0 & 0 & 0 & \lambda & \mu & 0 \\
0 & 0 & 0 & -\mu & \lambda & 0 \\
0 & 0 & 0 & 0 & 0 & 1
\end{array}\right]
\tag{5.301}
$$

matrix $[R]$ is known as the transformation matrix. With this, the following vector transformations may be expressed:

$$\{u\} = [R]\{U\}$$
$$\{\dot{u}\} = [R]\{\dot{U}\}$$
$$\{\ddot{u}\} = [R]\{\ddot{U}\} \tag{5.302a}$$

and also,

$$\{\dot{u}_a\} = [R]\{\dot{U}_a\}$$
$$\{\ddot{u}_a\} = [R]\{\ddot{U}_a\}$$

where, referring to Fig. 5.77

$$\{\dot{U}_{ai}\} = \begin{Bmatrix} \dot{X}_{A'} \\ \dot{Y}_{A'} \\ \dot{\theta}_{A'} \\ \dot{X}_{B'} \\ \dot{Y}_{B'} \\ \dot{\theta}_{B'} \end{Bmatrix}, \qquad \{\ddot{U}_{ai}\} = \begin{Bmatrix} \ddot{X}_{A'} \\ \ddot{Y}_{A'} \\ \ddot{\theta}_{A'} \\ \ddot{X}_{B'} \\ \ddot{Y}_{B'} \\ \ddot{\theta}_{B'} \end{Bmatrix}, \qquad i = 1, \ldots, 6 \tag{5.302b}$$

The kinetic energy in Eq. (5.292) is rewritten with the help of Eq. (5.302b)

$$T = \tfrac{1}{2}\{\dot{U}_a\}^T[m]\{\dot{U}_a\}$$

where $[m]$, now referred to the global coordinates, is:

$$[m] = [R]^T[\bar{m}][R] \tag{5.303}$$

Similarly, with the help of Eq. (5.302a), the strain energy in Eq. (5.295) may be rewritten as

$$U = \tfrac{1}{2}\{U\}^T[k]\{U\},$$

where $k$ is now defined in the global coordinates as

$$[k] = [R]^T[\bar{k}][R]. \tag{5.295a}$$

The need for this sort of transformation of coordinates, which allows the motion of all moving links of a mechanism to be referred to one global coordinate system, will become more evident in sections to follow.

## 5.24 ELASTIC LINKAGE MODEL

For the finite-element analysis of a linkage, the following sequence of steps is used:

1. An idealization of the linkage structure is needed. This will require selection of the type and size of the finite elements to form the system mesh.
2. The system-oriented element mass and stiffness matrices, referred to a global coordinate system, are generated for each element.

3. These element mass and stiffness matrices are superposed systematically to develop the mass and stiffness matrices of the total structural system of the linkage.
4. Determination of the unknown nodal displacements on the linkage is made by solving a system of coupled ordinary differential equations of motion. These equations are obtained by using the equilibrium conditions at the nodes.
5. All desired results, such as stresses and strains, associated with the problem are computed.

As an example, consider the four-bar linkage of Fig. 5.80, in which each link is modeled by one finite-beam element. The input shaft is assumed to be connected to a flywheel with high inertia, ensuring that no undue fluctuations occur in the input angular velocity. Then the input crank is reduced to a rotating cantilever beam allowing us to treat the moving mechanism as an "instantaneous structure" in each position. It should be pointed out, however, that this assumption is made only to simplify the computational rigor described here.

In Fig. 5.81, system-oriented generalized displacements are labeled to describe the structural deformation of the linkage as well as to maintain compatibility between the elements at the nodes. For example, at node 2, $U_1$ and $U_2$ are required to describe the nodal translations from the rigid-body position. Two more independent displacements, $U_3$ and $U_4$, are necessary at node 2 to describe the rotational deformations in each of the two elements, 1 and 2 with respect to their respective rigid-body orientations. The displacements above include two rotations, one for each adjacent link, and thus simulate a pin joint. A rigid connection between two elements will be simulated by only one rotational displacement.

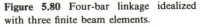

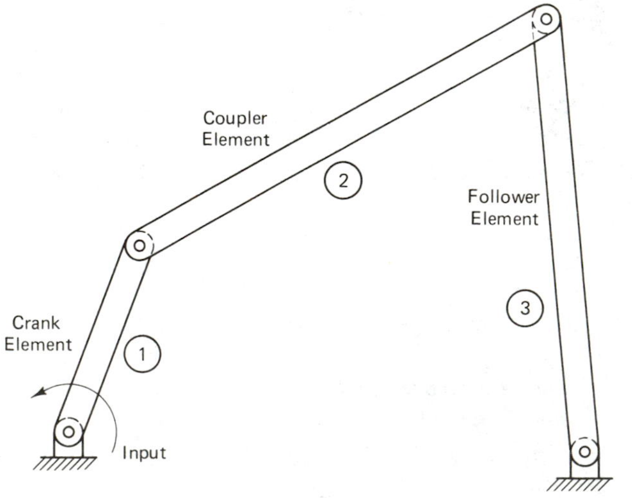


**Figure 5.80** Four-bar linkage idealized with three finite beam elements.

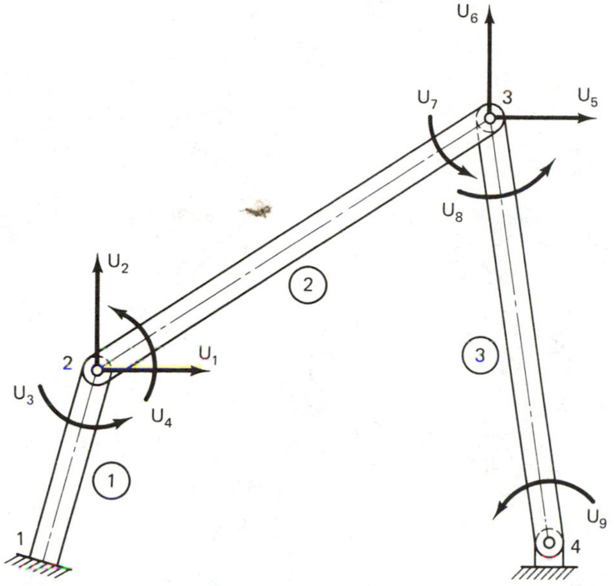


**Figure 5.81** Generalized displacements in system-oriented coordinates with nodal compatibility.

## 5.25 CONSTRUCTION OF TOTAL SYSTEM MATRICES

It may be shown that $[R]^{-1} = [R]^T$. Then, from Eq. (5.302a), the following relationships are given:

$$\{\ddot{U}\} = [R]^T\{\ddot{u}\} \tag{5.304}$$

and

$$\{\ddot{U}_a\} = [R]^T\{\ddot{u}_a\}$$

With the help of Eq. (5.281),

$$\{\ddot{U}_a\} = [R]^T\{\ddot{u}_r\} + [R]^T\{\ddot{u}\}$$

This equation may be rewritten

$$\{\ddot{U}_a\} = \{\ddot{U}_r\} + \{\ddot{U}\} \tag{5.305}$$

This equation may also be directly derived, upon careful observation and simplification, from Eqs. (5.274) and (5.302b).

Using Eqs. (5.302a), and premultiplying by $[R]^T$, equations of motion (5.298) of the beam element are rewritten in the system-oriented coordinates as

$$[m]\{\ddot{U}_a(t)\} + [k]\{U(t)\} = \{Q\} \tag{5.306}$$

where $[m] = [R]^T[\bar{m}][R]$, $[k] = [R]^T[\bar{k}][R]$ and

$$\{Q\} = [R]^T\{\bar{Q}\}$$

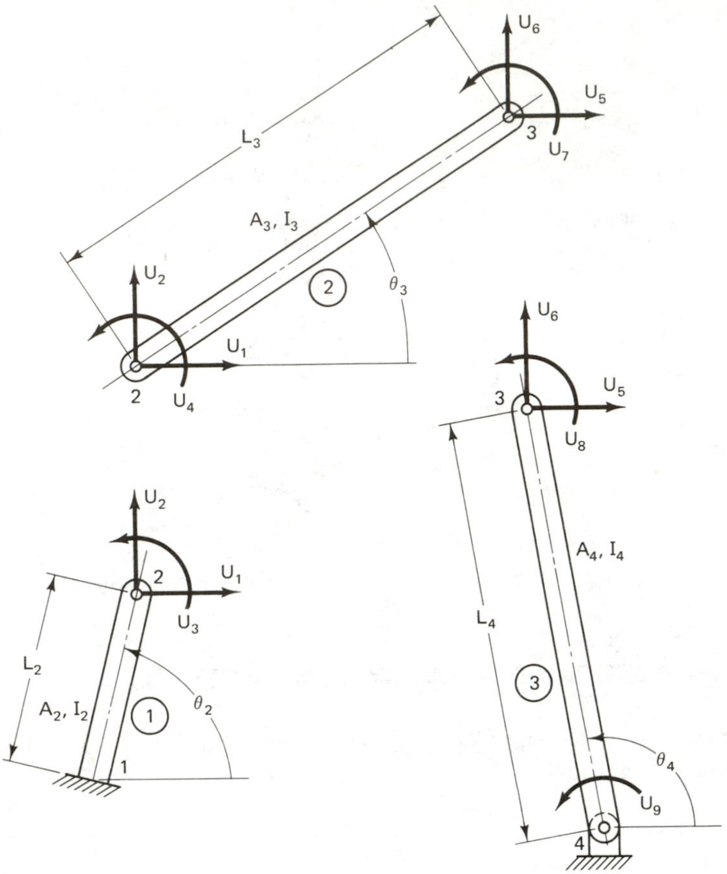


**Figure 5.82** Generalized element displacements in system-oriented (global) coordinates with nodal compatibility.

In Fig. 5.82, the finite elements of Fig. 5.81 are shown separated. Appropriate displacements are labeled on each while retaining compatibility at the nodes. The global system oriented mass and stiffness matrices of the $i$th element are

$$[m]_i = [R]_i^T[\bar{m}]_i[R]_i$$

and (5.307)

$$[k]_i = [R]_i^T[\bar{k}]_i[R]_i, \qquad i = 1, 2, 3$$

Equation (5.307) is valid for an element represented by six generalized coordinates. The kinetic energy for the crank element (numbered 1 in Fig. 5.82) may be expressed, using Eq. (5.303), as

$$T_1 = \tfrac{1}{2}\{\dot{U}_a\}_1^T[m]_1\{\dot{U}_a\}_1 \tag{5.308}$$

Further, $\{\dot{U}_a\}_1$ is defined as

$$\{\dot{U}_a\}_1 = \begin{Bmatrix} 0 \\ 0 \\ 0 \\ --- \\ \dot{U}_{a1} \\ \dot{U}_{a2} \\ \dot{U}_{a3} \end{Bmatrix} = \begin{Bmatrix} \dot{U}_z \\ --- \\ \dot{U}_{nz} \end{Bmatrix}_1 \tag{5.309}$$

Subscripts $z$ and $nz$ represent the zero and nonzero components of the velocity vector, respectively. Note that the kinetic energy associated with the rigid-body crank rotation, due to mass moment of inertia of the crank shaft, has been left out of Eq. (5.308). For the problem at hand, it is inconsequential in deriving the equations of motion in the following section. This is due to the fact that no generalized coordinate is assigned corresponding to this rotation. This is consistent with the assumption of no fluctuations in the prescribed input shaft speed.

Equation (5.308) may be rewritten as

$$T_1 = \frac{1}{2} \begin{Bmatrix} \dot{U}_z \\ -- \\ \dot{U}_{nz} \end{Bmatrix}_1^T \left[ \begin{array}{c|c} m_{zz} & m_{z-nz} \\ \hline m_{nz-z} & m_{nz-nz} \end{array} \right]_1 \begin{Bmatrix} \dot{U}_z \\ \dot{U}_{nz} \end{Bmatrix}_1$$

Since $\{\dot{U}_z\} = \bar{0}$,

$$T_1 = \tfrac{1}{2}\{\dot{U}_{nz}\}_1^T [m_{nz-nz}]_1 \{\dot{U}_{nz}\}_1$$

or, expanded,

$$T_1 = \frac{1}{2} \begin{Bmatrix} \dot{U}_{a1} \\ \dot{U}_{a2} \\ \dot{U}_{a3} \end{Bmatrix}^T \begin{bmatrix} m_{11}^1 & m_{12}^1 & m_{13}^1 \\ m_{21}^1 & m_{22}^1 & m_{23}^1 \\ m_{31}^1 & m_{32}^1 & m_{33}^1 \end{bmatrix} \begin{Bmatrix} \dot{U}_{a1} \\ \dot{U}_{a2} \\ \dot{U}_{a3} \end{Bmatrix} \tag{5.310}$$

It must be remembered that $[m]_1$ is a sixth-order square matrix. The matrix $[m_{nz-nz}]_1$ represents this mass matrix after having eliminated the rows and columns that correspond to zero generalized coordinates of the element. Only nonzero generalized coordinates have been labeled in Fig. 5.82. Clearly, then, the mass matrix in Eq. (5.310) corresponds to the three coordinates shown at node 2 of element 1; that is, the elements of this matrix correspond to the fourth, fifth, and sixth rows and columns of the system mass matrix $[m]_1$. However, these elements have been subscripted to correspond to the nonzero coordinates, strictly as a matter of convenience. The superscript 1 has been added in Eq. (5.310), to represent mass matrix elements of the element numbered 1. Kinetic energies $T_2$ and $T_3$ for the elements 2 and 3, respectively, can be similarly derived, and finally

$$T = T_1 + T_2 + T_3.$$

In matrix form,

$$T = \tfrac{1}{2}\{\dot{U}_{ai}\}^T[M]\{\dot{U}_{ai}\}, \qquad i = 1, 2, \ldots, 9 \tag{5.311}$$

Equation (5.311) is shown again in its expanded form.

$$T=\begin{Bmatrix}\dot{U}_{a1}\\ \dot{U}_{a2}\\ \dot{U}_{a3}\\ \dot{U}_{a4}\\ \dot{U}_{a5}\\ \dot{U}_{a6}\\ \dot{U}_{a7}\\ \dot{U}_{a8}\\ \dot{U}_{a9}\end{Bmatrix}^T \begin{bmatrix} m^1_{11}+m^2_{11} & m^1_{12}+m^2_{12} & m^1_{13} & m^2_{14} & m^2_{15} & m^2_{16} & m^2_{17} & 0 & 0\\ m^1_{21}+m^2_{21} & m^1_{22}+m^2_{22} & m^1_{23} & m^2_{24} & m^2_{25} & m^2_{26} & m^2_{27} & 0 & 0\\ m^1_{31} & m^1_{32} & m^1_{33} & 0 & 0 & 0 & 0 & 0 & 0\\ m^2_{41} & m^2_{42} & 0 & m^2_{44} & m^2_{45} & m^2_{46} & m^2_{47} & 0 & 0\\ m^2_{51} & m^2_{52} & 0 & m^2_{54} & m^2_{55}+m^3_{55} & m^2_{56}+m^3_{56} & m^2_{57} & m^3_{58} & m^3_{59}\\ m^2_{61} & m^2_{62} & 0 & m^2_{64} & m^2_{65}+m^3_{65} & m^2_{66}+m^3_{66} & m^2_{67} & m^3_{68} & m^3_{69}\\ m^2_{71} & m^2_{72} & 0 & m^2_{74} & m^2_{75} & m^2_{76} & m^2_{77} & 0 & 0\\ 0 & 0 & 0 & 0 & m^3_{85} & m^3_{86} & 0 & m^3_{88} & m^3_{89}\\ 0 & 0 & 0 & 0 & m^3_{95} & m^3_{96} & 0 & m^3_{98} & m^3_{99} \end{bmatrix} \begin{Bmatrix}\dot{U}_{a1}\\ \dot{U}_{a2}\\ \dot{U}_{a3}\\ \dot{U}_{a4}\\ \dot{U}_{a5}\\ \dot{U}_{a6}\\ \dot{U}_{a7}\\ \dot{U}_{a8}\\ \dot{U}_{a9}\end{Bmatrix} \tag{5.312}$$

As indicated before, the superscript denotes the particular beam element contributing to the mass element.

Matrix $[M]$ is the total system mass matrix. Similarly, from strain energy considerations, the total system stiffness matrix $[K]$ may be derived by superposing the strain energies of the individual elements.

$$U=\tfrac{1}{2}\{U_i\}^T[K]\{U_i\}, \qquad i=1, 2, \ldots, 9 \tag{5.313}$$

Differential equations for beam displacements, used in engineering beam theory, have been utilized in determining the stiffness properties of the elements. Rotatory inertia and shear deformation of the elements have been neglected for convenience. In the displacement finite-elements method, these effects are easily included [206] by appropriate modifications of the system matrices.

Having defined, in principle, the construction of the total system mass and stiffness matrices, a systematic computational programming procedure must be sought. It must allow convenient formulation of the system-oriented matrices for each of the linkage elements, elimination of the appropriate rows and columns corresponding to the zero generalized coordinates, and finally, superposing them to form the system matrices. This procedure must duly regard the boundary conditions dictated at the nodes.

One such method that may be utilized is the extension of an approach called the *permutation vector method* used by Oakberg [328] in the analysis of frames, and extended by Imam [134] and by Imam, Sandor, and Kramer [137] to the analysis of elastic linkages. Owing to space limitations, the method is not described here. Some of its major advantages are: (1) the determination of the conventional force-transfer and connectivity matrices is not required; (2) the total system matrices obtained by this method are reduced to correspond only to the nonzero generalized coordinates; and (3) the approach is sufficiently general to accommodate both single and multiloop planar linkages.

## 5.26 EQUATIONS OF MOTION

If the system-oriented coordinates $\{U\}$ are utilized as the generalized coordinates to describe the structural deformation of the linkage from its rigid-body position, as in Fig. 5.81, its equations of motion can be derived from Lagrange equations

$$\frac{d}{dt}\left(\frac{\partial T}{\partial \dot{U}_i}\right) - \frac{\partial T}{\partial U_i} + \frac{\partial U}{\partial U_i} = Q_i, \qquad i = 1, 2, \ldots, 9 \tag{5.314}$$

The first, second, and the third terms on the left-hand side reduce to

$$\sum_{j=1}^{9} M_{ij}\ddot{U}_{aj}, \qquad 0, \qquad \text{and} \qquad \sum_{j=1}^{9} K_{ij}U_j$$

respectively. In matrix form, the equations of motion are

$$[M]\{\ddot{U}_a\} + [K]\{U\} = \{Q\} \tag{5.315}$$

In the absence of any frictional or driving (external loading) forces at the nodes, $Q_i$ ($i = 1, \ldots, 9$) may be taken as zero. Note that although a driving input torque at the crank shaft is anticipated, it is inconsequential here since the corresponding generalized coordinate has been suppressed. Also, the reactive forces at the nodes of interconnecting members do not appear in $\{Q\}$ since the net change of potential due to these remains zero at all times.

If the structural damping matrix for the linkage is represented by $[C]$, then by including the damping forces, and utilizing Eq. (5.308), the equations of motion become

$$[M]\{\ddot{U}\} + [C]\{\dot{U}\} + [K]\{U\} = -[M]\{\ddot{U}_r\} \tag{5.316}$$

The coefficient matrices $[M]$, $[C]$, and $[K]$ are functions of the linkage geometry and vary as the input (crank) angle is varied. These values are repeated after each motion cycle of the linkage. Thus a periodicity is imposed on the coefficient matrices of the differential equation (5.316) with respect to crank angle. For constant or periodic input velocity, these would be time-periodic. The same is true of the rigid-body acceleration vector $\{\ddot{U}_r\}$.

The constancy of the mass matrix $[M]$, in arriving at Eq. (5.315) from Eq. (5.314), has been implicitly assumed. This is consistent with the commonly made assumption in numerical methods that the time-dependent system parameters are held constant over each discretized time interval.

The foregoing method used to derive the equations of motion is quite general, and it can be readily extended to model multiloop planar linkages as well as those possessing links with complex geometries. The latter may be facilitated by the more complex element types available in the library of finite elements. Generation of the system mesh should be regarded critically, since increasing the number of nodes significantly increases computation time and costs.

## 5.27 DAMPING IN LINKAGES

The integration procedures utilized to arrive at the mass and stiffness matrices may be extended to compute the damping matrix, provided that some fundamental damping properties of the structural material are known. Generally, these are uncertain. Customarily, damping ratios are measured experimentally in several modes of vibration of a structure that is similar to the one under investigation. These ratios are then assumed as applicable in the modal analysis of the new structure.

For most structures, the exact form of the damping matrix is unknown since the sources of energy loss are complicated. Also, in most cases, the effect of damping on the vibration mode shapes of the structure is small. Therefore, an assumption as to the form of this matrix is justifiable. The damping can often be approximated by a suitable viscous damping ratio for several modes. An exception to this is when nonlinear response or modal coupling are involved. For most structural analysis purposes, it is assumed that the orthogonality of the damping forces is maintained. This assumption is made to facilitate the uncoupling of the homogeneous equations of motion; this, in turn, expedites the solving of these equations. The forms of damping matrix generally employed in structural dynamic analyses are discussed in this section. Since the linkage during its motion is regarded as a structure in numerous positions, the adaptation of the conventional structural damping matrices to linkage problems is deemed appropriate. In uncoupling Eq. (5.316), the first step involves determining the modal frequencies and the corresponding mode shapes of the structure for "undamped free vibrations." This involves the solution of the following equation, also known as the eigenvalue problem, for its characteristic values:

$$[K]\{\Phi_n\} = \omega_n^2 [M]\{\Phi_n\} \tag{5.317}$$

where $\omega_n$ is the frequency (eigenvalue) of the $n$th mode and $\{\Phi_n\}$ the $n$th-mode shape (eigenvector).* The mode shapes satisfy the orthogonality conditions

$$\{\Phi_n\}^T[M]\{\Phi_m\} = 0$$

and

$$\{\Phi_n\}^T[K]\{\Phi_m\} = 0, \qquad m \neq n \tag{5.318}$$

The mode shapes may be normalized such that

$$\{\Phi_n\}^T[M]\{\Phi_n\} = 1 \tag{5.319}$$

The following are then true:

$$[\Phi]^T[M][\Phi] = [I]$$

and

$$[\Phi]^T[K][\Phi] = [\omega^2] \tag{5.320}$$

where $[\Phi]$ is the modal matrix, whose columns are the natural modes of the system,

* For a brief refresher on eigenvalues and eigenvectors see footnote following Eq. (5.65).

$[I]$ is a unit matrix, and $[\omega^2]$ is a diagonal matrix of the natural frequency squared.

For the uncoupling procedure, the following coordinate transformation is made:

$$\{U\} = [\Phi]\{\eta\} \tag{5.321}$$

where $\{\eta\}$ represents the modal amplitude vector, also referred to as the normal coordinates vector. Also,

$$\{\dot{U}\} = [\Phi]\{\dot{\eta}\}$$

and

$$\{\ddot{U}\} = [\Phi]\{\ddot{\eta}\} \tag{5.322}$$

Substituting Eqs. (5.321), and (5.322) into Eq. (5.316) and premultiplying by $[\Phi]^T$, the following equations result:

$$[\Phi]^T[M][\Phi]\{\ddot{\eta}\} + [\Phi]^T[C][\Phi]\{\dot{\eta}\} + [\Phi]^T[K][\Phi]\{\eta\} = -[\Phi]^T[M]\{\ddot{U}_r\} \tag{5.323}$$

In order that the orthogonality of the damping forces be secured, the following must be true:

$$\begin{aligned} \{\Phi_n\}^T[C]\{\Phi_m\} &= 0, \qquad m \neq n \\ \{\Phi_n\}^T[C]\{\Phi_m\} &= \alpha_n, \qquad m = n \end{aligned} \tag{5.324}$$

where $\alpha_n$ is not necessarily zero. Equation (5.323) may now be written as

$$[M^\eta]\{\ddot{\eta}\} + [C^\eta]\{\dot{\eta}\} + [K^\eta]\{\eta\} = \{N\} \tag{5.325}$$

where

$$\begin{aligned} [M^\eta] &= [\Phi]^T[M][\Phi] \\ [C^\eta] &= [\Phi]^T[C][\Phi] \\ [K^\eta] &= [\Phi]^T[K][\Phi] \\ \{N\} &= -[\Phi]^T[M]\{\ddot{U}_r\} \end{aligned} \tag{5.326}$$

Equation (5.325) may also be expressed as

$$M^\eta_{ii}\ddot{\eta}_i + C^\eta_{ii}\dot{\eta}_i + K^\eta_{ii}\eta_i = N_i, \qquad i = 1, 2, \ldots n. \tag{5.327}$$

Here $n$ represents the number of equations (9 for the four-bar linkage problem).

With the discussion above at hand, some damping forms are listed here:

1. A simplified damping form is often used. This is obtained by defining $\alpha_r$ in Eq. (5.324) as

$$\alpha_r = 2\xi_r\omega_r = C^\eta_{rr} \qquad \text{and} \qquad [\alpha] = [C^\eta] \tag{5.328}$$

where $\xi$ is the damping ratio for the $r$th mode and $\omega_r$ its frequency. $[C]$ may be found from Eq. (5.326).

2. The Wilson method computes a damping matrix, $[C_r]$, for each mode of vibration, $r$

$$[C_r] = 2\xi_r \omega_r \{\theta_r\}\{\theta_r\}^T, \qquad \text{where} \qquad \{\theta_r\} = [M]\{\Phi_r\} \tag{5.329}$$

   The total damping matrix is obtained by superposition of the desired number ($n$) of modal contributions. Thus

$$[C] = \sum_{r=1}^{n} [C_r] \tag{5.330}$$

   Modes not included here remain undamped.

3. Wilson and Penzien have described a form of damping matrix [329] satisfying Eqs. (5.324). It is defined as a series of matrices, each of which is proportional to one of the matrices of the Caughey series [330] and satisfies the orthogonal condition; thus

$$[C] = [M] \sum_{b=0}^{n-1} a_b [[M]^{-1}[K]]^b \tag{5.331}$$

   Coefficients $a_b$ are selected to provide the desired damping ratio in a corresponding number of modes. Their values are computed from a set of $n$ simultaneous equations of the following form:

$$\xi_r = \frac{1}{2}\left[\frac{a_0}{\omega_r} + a_1\omega_r + \cdots + a_{n-1}\omega_r^{2n-3}\right] \tag{5.332}$$

   where $r = 1, 2, \ldots, n$.

4. If in Eq. (5.331) only the first two terms of the series are retained, the resulting damping is known as Rayleigh damping. Here, the damping matrix is a linear combination of the mass and stiffness matrices. It may also be represented by

$$[C] = \alpha_1[K] + \alpha_2[M] \tag{5.333}$$

   where $\alpha_1$ and $\alpha_2$ are weighting coefficients.

5. Damping proportional to stiffness is known as "structural" damping. The damping force vector is proportional to the displacement vector—however, in phase with the velocity. Damping ratios remain constant for all modes since the constant of proportionality includes the damping ratio.

6. Yet another form of damping is defined proportional to the mass matrix and is encountered in Ref. 74. Thus

$$[C] = [M][C'] \tag{5.334}$$

where $C'_{ij} = 2\xi(\text{signum } K_{ij})(|K_{ij}/M_{ij}|)^{1/2}$

signum $K_{ij} = (K_{ij}/|K_{ij}|)$.

Nonproportional damping as well as other variations of the forms listed above have also been used in the dynamic analysis of structures. A discussion on the damping forms may be found in Ref. 133.

## 5.28 RIGID-BODY ACCELERATION

A computer-programmable method is essential for obtaining the generalized rigid-body acceleration vector in Equation (5.305). The various kinematic quantities, namely the angular velocities and accelerations [in Eqs. (3.84), (3.85), (4.21), and (4.22) of Vol. 1], as computed for links of a four-bar linkage (Fig. 5.82) are listed below:

$$\omega_3 = -\frac{L_2}{L_3}\omega_2 \frac{\sin(\theta_4 - \theta_2)}{\sin(\theta_4 - \theta_3)}$$

$$\omega_4 = -\frac{L_2}{L_4}\omega_2 \frac{\sin(\theta_3 - \theta_2)}{\sin(\theta_4 - \theta_3)}$$

$$\alpha_3 = \frac{[-L_2\alpha_2 \sin(\theta_4 - \theta_2) + L_2\omega_2^2 \cos(\theta_4 - \theta_2) + L_3\omega_3^2 \cos(\theta_4 - \theta_3) - L_4\omega_4^2]}{L_3 \sin(\theta_4 - \theta_3)} \tag{5.335}$$

$$\alpha_4 = \frac{[-L_2\alpha_2 \sin(\theta_3 - \theta_2) + L_2\omega_2^2 \cos(\theta_3 - \theta_2) - L_4\omega_4^2 \cos(\theta_3 - \theta_4) + L_3\omega_3^2]}{L_4 \sin(\theta_4 - \theta_3)}$$

where $\omega_i$ and $\alpha_i$ denote the angular velocity and acceleration of the $i$th link, respectively; $\omega_2$ and $\alpha_2$ are the prescribed input values; and $L_j$, $j = 2, 3, 4$, are link lengths.

With the aid of the quantities listed above, the generalized rigid-body acceleration vector $\{\ddot{U}_r\}$ may be determined. Thus $\ddot{U}_{r1}$ is derived corresponding to the generalized coordinate $U_1$ (Fig. 5.81):

$$\ddot{U}_{r1} = -L_2\omega_2^2 \cos\theta_2 - L_2\alpha_2 \sin\theta_2$$

where $L_2$ is the length and $\theta_2$ the angle measured from the global $X$ axis of the input crank. Similarly,

$$\ddot{U}_{r2} = -L_2\omega_2^2 \sin\theta_2 + L_2\alpha_2 \cos\theta_2 \qquad \text{and} \qquad \ddot{U}_{r3} = \alpha_2 \tag{5.336}$$

Others may be found similarly.

## 5.29 STRESS COMPUTATION

From Eqs. (5.283) and (5.285), the axial displacement is (Fig. 5.79)

$$v(\bar{x}, t) = \left(1 - \frac{\bar{x}}{L}\right) u_1(t) + \frac{\bar{x}}{L} u_4(t)$$

The axial strain at the neutral axis is

$$\epsilon_{\bar{x}} = \frac{\partial v(\bar{x}, t)}{\partial x} = -\frac{1}{L} u_1(t) + \frac{1}{L} u_4(t)$$

or in matrix form, the axial stress is

$$\sigma_{\bar{x}} = E\begin{bmatrix} -\frac{1}{L} & \frac{1}{L} \end{bmatrix}\begin{Bmatrix} u_1(t) \\ u_4(t) \end{Bmatrix} \tag{5.337}$$

From beam theory, the bending moment is

$$M = -EI\frac{\partial^2 w(\bar{x}, t)}{\partial \bar{x}^2}$$

where $w(\bar{x}, t)$ has been defined in Eq. (5.284), and if $\bar{y}$ be the distance of a point from the neutral axis, the bending stress at this point is

$$\sigma_{\bar{x}} = -E\bar{y}\frac{\partial^2 w(\bar{x}, t)}{\partial \bar{x}^2}$$

Or, in matrix form,

$$\sigma_{\bar{x}} = -E\bar{y}\begin{bmatrix} \frac{\partial^2 \phi_2}{\partial \bar{x}^2} & \frac{\partial^2 \phi_3}{\partial \bar{x}^2} & \frac{\partial^2 \phi_5}{\partial \bar{x}^2} & \frac{\partial^2 \phi_6}{\partial \bar{x}^2} \end{bmatrix}\begin{Bmatrix} u_2(t) \\ u_3(t) \\ u_5(t) \\ u_6(t) \end{Bmatrix} \tag{5.338}$$

where $\phi_i$ are defined in Eqs. (5.283),

$$\begin{aligned} \frac{\partial^2 \phi_2}{\partial \bar{x}^2} &= -\frac{6}{L^3}(L - 2\bar{x}) \\ \frac{\partial^2 \phi_3}{\partial \bar{x}^2} &= \frac{2}{L^2}(2L - 3\bar{x}) \\ \frac{\partial^2 \phi_5}{\partial \bar{x}^2} &= \frac{6}{L^3}(L - 2\bar{x}) \\ \frac{\partial^2 \phi_6}{\partial \bar{x}^2} &= -\frac{2}{L^2}(L - 3\bar{x}) \end{aligned} \tag{5.338a}$$

For the total normal stress, the results of Eqs. (5.337) and (5.338) must be superposed.

It should be noted that even though the transverse displacement of the beam element may be predicted fairly accurately by Eq. (5.284), the required second derivatives, e.g., in Eqs. (5.293) and (5.338), may not be accurately represented. Also, the assumed displacement fields need not be polynomials. Sine series are conveniently applied under certain conditions. Further discussion on these may be found in Ref. 133.

## 5.30 METHOD OF SOLUTION

Ideally, an infinite number of rigid-body positions are associated with the simulation of one continuous motion cycle of a linkage. The linkage, since it traverses the same rigid-body positions during its next motion cycle, is position-periodic. A distinct set of mass, damping, stiffness, and inertia properties is associated with each such position. These parameters are position-periodic; if the input to the linkage is a constant angular velocity, these parameters are also time-periodic. In numerical methods, this continuous motion is replaced by a number of discretized steps. The concept is analogous to finite-element theory, where the elastic medium itself is discretized. During each time step, the system parameters (mass, damping, and stiffness) are assumed to remain constant in solving the equations of motion. While this produces only an approximate solution, the true solution is approached as the step size tends to zero.

Thus the discretization of the continuous motion cycle serves a twofold purpose. First, it substantially reduces the number of costly computations over the cycle, and second, it facilitates solving the differential equations of motion with time-dependent coefficients as differential equations possessing constant coefficients over each discretized interval.

Briefly described below is a numerical method [186] for computing the periodic solutions of vibrating systems governed by equations similar to that described by Eq. (5.316). We now apply this method to treat a single-degree-of-freedom, mass–dashpot–spring system whose governing differential equation of motion is a linear, second-order equation with time-dependent and periodic coefficients. The system is excited by a periodic forcing function. The method is appropriate since the method of modal (response) analysis of structures lends itself to this application by uncoupling the differential equations of motion [Eq. (5.316)]. The uncoupled form of the equations is shown in Eqs. (5.325) and (5.327). The modal analysis method is also referred to as the "mode superposition method."

The classical single-degree-of-freedom vibration model (with constant parameters) is governed by the equation of motion

$$m\ddot{x} + c\dot{x} + kx = F(t) \tag{5.339}$$

where $F(t)$ is a periodic function (only when a periodic solution is sought) integrable over its time period. The complete response [299] at time $t$ is given as

$$\begin{aligned} x(t) = {} & Ae^{-\xi\omega_n t} \sin \omega_d t + Be^{-\xi\omega_n t} \cos \omega_d t \\ & + \frac{1}{m\omega_d} \int_0^t e^{-\xi\omega_n(t-\tau)} \sin [\omega_d(t-\tau)] F(\tau)\, d\tau \end{aligned} \tag{5.340}$$

The velocity $\dot{x}(t)$ is

$$\begin{aligned} \dot{x}(t) = {} & A[-\xi\omega_n e^{-\xi\omega_n t} \sin \omega_d t + \omega_d e^{-\xi\omega_n t} \cos \omega_d t] \\ & + B[-\xi\omega_n e^{-\xi\omega_n t} \cos \omega_d t - \omega_d e^{-\xi\omega_n t} \sin \omega_d t] \\ & + \frac{1}{m\omega_d} \int_0^t [-\xi\omega_n e^{-\xi\omega_n(t-\tau)} \sin [\omega_d(t-\tau)] \\ & + \omega_d e^{-\xi\omega_n(t-\tau)} \cos [\omega_d(t-\tau)] F(\tau)\, d\tau \end{aligned} \tag{5.341}$$

where, the natural frequency $\omega_n = (k/m)^{1/2}$, damped natural frequency $\omega_d = \omega_n(1 - \xi^2)^{1/2}$, and damping ratio, $\xi = c/2m\omega_n$.

Assuming damping to be small (less than critical), the general form of the homogeneous solution is represented by the first two terms in Eq. (5.340), while the last term represents the particular solution. The latter is obtained by using the Duhamel integral. $A$ and $B$ are arbitrary constants determined by initial conditions. The periodic solution is now sought for the differential equation of motion,

$$m(t)\ddot{x} + c(t)\dot{x} + k(t)x = F(t) \tag{5.342}$$

such that

$$m(t) = m(t + T_f)$$

$$c(t) = c(t + T_f)$$

$$k(t) = k(t + T_f)$$

$$\omega_n(t) = \left[\frac{k(t)}{m(t)}\right]^{1/2} = \omega_n(t + T_f)$$

$$F(t) = F(t + T_f)$$

where $T_f$ is the fundamental time period of linkage motion.

Numerically, the problem of periodic solution may be tackled by discretizing the continua of the system parameters. For convenience, let the time period $T_f$ be divided into $n$ "equal" intervals, such that

$$\Delta t = \frac{T_f}{n}$$

The process of discretization of the system parameters, and their labeling in each interval, is illustrated in Figs. 5.83 and 5.84. Having assumed these parameters to

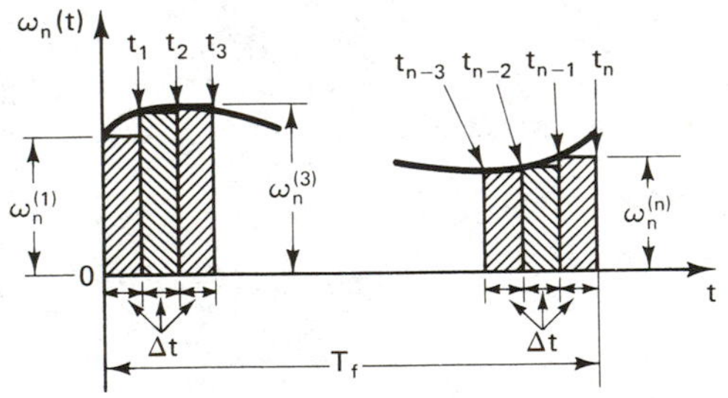


**Figure 5.83** Discretization of time-dependent natural frequency.

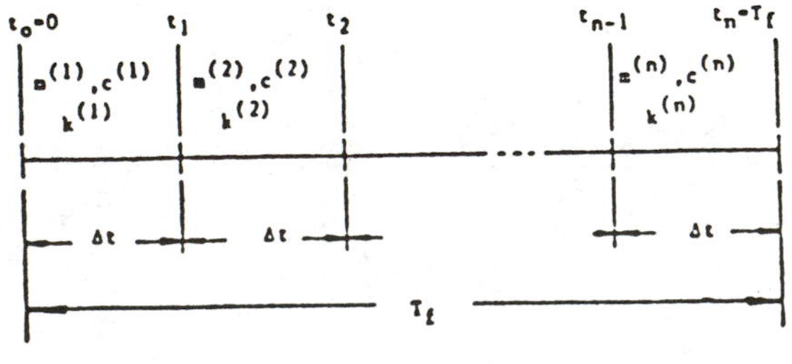


**Figure 5.84** Discretization of time-dependent system parameters.

be constant over the duration of each interval, the solution form reduces to that of the linear second-order differential equation of motion with "constant" coefficients.

The general form of displacement and velocity in the $m$th time interval is expressed using Eqs. (5.340) and (5.341) as

$$\begin{aligned} x(t) = {} & A^{(m)} e^{-\xi\omega_n^{(m)}(t-t_{m-1})} \sin[\omega_d^{(m)}(t-t_{m-1})] \\ & + B^{(m)} e^{-\xi\omega_n^{(m)}(t-t_{m-1})} \cos[\omega_d^{(m)}(t-t_{m-1})] \\ & + \frac{1}{m^{(m)}\omega_d^{(m)}} \int_{t_{m-1}}^{t} e^{-\xi\omega_n^{(m)}(t-\tau)} \sin[\omega_d^{(m)}(t-\tau)] F(\tau)\, d\tau \end{aligned} \tag{5.343}$$

and

$$\begin{aligned} \dot{x}(t) = {} & A^{(m)}[-\xi\omega_n^{(m)} e^{-\xi\omega_n^{(m)}(t-t_{m-1})} \sin[\omega_d^{(m)}(t-t_{m-1})] \\ & + \omega_d^{(m)} e^{-\xi\omega_n^{(m)}(t-t_{m-1})} \cos[\omega_d^{(m)}(t-t_{m-1})]] \\ & + B^{(m)}[-\xi\omega_n^{(m)} e^{-\xi\omega_n^{(m)}(t-t_{m-1})} \cos[\omega_d^{(m)}(t-t_{m-1})] \\ & - \omega_d^{(m)} e^{-\xi\omega_n^{(m)}(t-t_{m-1})} \sin[\omega_d^{(m)}(t-t_{m-1})]] \\ & + \frac{1}{m^{(m)}\omega_d^{(m)}} \int_{t_{m-1}}^{t} [-\xi\omega_n^{(m)} e^{-\xi\omega_n^{(m)}(t-\tau)} \sin[\omega_d^{(m)}(t-\tau)] \\ & + \omega_d^{(m)} e^{-\xi\omega_n^{(m)}(t-\tau)} \cos[\omega_d^{(m)}(t-\tau)]] F(\tau)\, d\tau \end{aligned} \tag{5.343a}$$

for

$$(t_{m-1} \le t \le t_m)$$

and

$$m = 1, \ldots\ldots, n$$

$A^{(m)}$ and $B^{(m)}$ describe the $2n$ system unknowns and are determined by the initial conditions at time $t = t_{m-1}$.

Each of the equations (5.343) and (5.343a) may be utilized in evaluating the displacement and velocity at both ends of each of the $n$ time intervals. In order to maintain the continuity of the solution through the fundamental time period $T_f$, the conditions of "compatibility" must be imposed at all the intermediate time points. For example, the displacement and velocity at time $t_1$ may be evaluated with the help of Eqs. (5.343) and (5.343a) for both $m = 1$ and 2, since this time point is common to both intervals 1 and 2. The two displacements and velocities must be equal, respectively. This compatibility condition at $t_1$ gives two equations with four unknowns, $A^{(1)}$, $B^{(1)}$, $A^{(2)}$, and $B^{(2)}$. When this condition is imposed at all the intermediate node points [i.e., $(t_1, t_2, \ldots, t_{n-1})$], $2n$ unknowns result in

$(2n-2)$ equations. Furthermore, from "periodicity" of the solution, two additional equations are derived by equating the displacements and velocities, respectively, at node points $t_0$ and $t_n$. The $2n$ conditions now form a set of simultaneous linear equations containing $2n$ unknowns. For details of these equations, the reader is referred to Ref. 186. In matrix form, these equations are written as

$$\begin{bmatrix} a_1^{(1)} & b_1^{(1)} & c_1^{(2)} & d_1^{(2)} & 0 & 0 & \cdots & 0 & 0 & 0 & 0 \\ a_2^{(1)} & b_2^{(1)} & c_2^{(2)} & d_2^{(2)} & 0 & 0 & & 0 & 0 & 0 & 0 \\ 0 & 0 & a_1^{(2)} & b_1^{(2)} & c_1^{(3)} & d_1^{(3)} & & 0 & 0 & 0 & 0 \\ 0 & 0 & a_2^{(2)} & b_2^{(2)} & c_2^{(3)} & d_2^{(3)} & \cdots & 0 & 0 & 0 & 0 \\ \vdots & & & & & \vdots & & \vdots & & & \vdots \\ 0 & 0 & 0 & 0 & 0 & 0 & \cdots & a_1^{(n-1)} & b_1^{(n-1)} & c_1^{(n)} & d_1^{(n)} \\ 0 & 0 & 0 & 0 & 0 & 0 & & a_2^{(n-1)} & b_2^{(n-1)} & c_2^{(n)} & d_2^{(n)} \\ c_1^{(1)} & d_1^{(1)} & 0 & 0 & 0 & 0 & & 0 & 0 & a_1^{(n)} & b_1^{(n)} \\ c_2^{(1)} & d_2^{(1)} & 0 & 0 & 0 & 0 & \cdots & 0 & 0 & a_2^{(n)} & b_2^{(n)} \end{bmatrix} \begin{Bmatrix} A^{(1)} \\ B^{(1)} \\ A^{(2)} \\ B^{(2)} \\ \vdots \\ A^{(n-1)} \\ B^{(n-1)} \\ A^{(n)} \\ B^{(n)} \end{Bmatrix} = \begin{Bmatrix} e_1^{(1)} \\ e_2^{(1)} \\ e_1^{(2)} \\ e_2^{(2)} \\ \vdots \\ e_1^{(n-1)} \\ e_2^{(n-1)} \\ e_1^{(n)} \\ e_2^{(n)} \end{Bmatrix} \tag{5.344}$$

where

$$a_1^{(i)} = -e^{-\xi\omega_n^{(i)}\Delta t} \sin \omega_d^{(i)}\, \Delta t$$

$$b_1^{(i)} = -e^{-\xi\omega_n^{(i)}\Delta t} \cos \omega_d^{(i)}\, \Delta t$$

$$c_1^{(i)} = \frac{1}{m^{(i)}\omega_d^{(i)}} \int_{t_{i-1}}^{t_i} e^{-\xi\omega_n^{(i)}(t_i-\tau)} \sin \omega_d^{(i)}(t_i-\tau) F(\tau)\, d\tau$$

$$a_2^{(i)} = -[-\xi\omega_n^{(i)} e^{-\xi\omega_n^{(i)}\Delta t} \sin \omega_d^{(i)}\, \Delta t + \omega_d^{(i)} e^{-\xi\omega_n^{(i)}\Delta t} \cos \omega_d^{(i)} \Delta t]$$

$$b_2^{(i)} = -[-\xi\omega_n^{(i)} e^{-\xi\omega_n^{(i)}\Delta t} \cos \omega_d^{(i)}\, \Delta t - \omega_d^{(i)} e^{-\xi\omega_n^{(i)}\Delta t} \sin \omega_d^{(i)}\, \Delta t]$$

$$e_2^{(i)} = \frac{1}{m^{(i)}\omega_d^{(i)}} \int_{t_{i-1}}^{t_i} [-\xi\omega_n^{(i)} e^{-\xi\omega_n^{(i)}(t_i-\tau)} \sin \omega_d^{(i)}(t_i-\tau)$$
$$+ \omega_d^{(i)} e^{-\xi\omega_n^{(i)}(t_i-\tau)} \cos \omega_d^{(i)}(t_i-\tau)] F(\tau)\, d\tau$$

$$c_1^{(i)} = 0$$

$$d_1^{(i)} = 1$$

$$c_2^{(i)} = \omega_d^{(i)}$$

$$d_2^{(i)} = -\xi\omega_n^{(i)} \qquad i = 1, 2, \ldots, n$$

The matrix equation (5.344) can be solved for the unknowns $A^{(m)}$ and $B^{(m)}$ uniquely by any one of the numerical techniques described in Ref. 33. Equations (5.343) and (5.343a) may now be utilized to determine the periodic solution at any time in the $n$ intervals of the fundamental time period.

A transient solution is also possible for Eq. (5.316), via its uncoupled form in Eq. (5.327), with the help of Eqs. (5.343) and (5.343a). $A^{(m)}$ and $B^{(m)}$ are now determined by way of the initial conditions for the $m$th time interval (i.e., the displacement and velocity at time $t = t_{m-1}$). More specifically,

$$B^{(m)} = x(t_{m-1}) \tag{5.345}$$

and

$$A^{(m)} = \frac{\dot{x}(t_{m-1}) + \xi\omega_n^{(m)}x(t_{m-1})}{\omega_d^{(m)}}$$

Let the Wilson method be used here to compute a damping matrix $[C_r]$ for each mode of vibration $r$:

$$[C_r] = 2\xi_r\omega_r\{\theta_r\}\{\theta_r\}^T \tag{5.329}$$

where

$$\{\theta_r\} = [M]\{\Phi_r\}$$

or

$$[C_r] = 2\xi_r\omega_r[M]\{\Phi_r\}\{\Phi_r\}^T[M]^T \tag{5.346}$$

Equation (5.346) is appropriately mutliplied by $\{\Phi_r\}$:

$$\{\Phi_r\}^T[C_r]\{\Phi_r\} = 2\xi_r\omega_r\{\Phi_r\}^T[M]\{\Phi_r\}\{\Phi_r\}^T[M]^T\{\Phi_r\}$$

Using Eq. (5.319), the equation may be written as

$$C^{\eta}_{rr} = 2\xi_r\omega_r \tag{5.347}$$

The uncoupled equations of motion are

$$M^{\eta}_{ii}(t)\ddot{\eta}_i(t) + C^{\eta}_{ii}(t)\dot{\eta}_i(t) + K^{\eta}_{ii}(t)\eta_i(t) = N_i(t), \qquad 1 \le i \le n \tag{5.327}$$

If constant or periodic input angular velocity to the linkage is assumed, the periodicity conditions on the system parameters are

$$M^{\eta}_{ii}(t) = M^{\eta}_{ii}(t + T_f)$$
$$C^{\eta}_{ii}(t) = C^{\eta}_{ii}(t + T_f)$$
$$K^{\eta}_{ii}(t) = K^{\eta}_{ii}(t + T_f)$$
$$N_i(t) = N_i(t + T_f)$$

where $T_f$ is the time period of the linkage motion cycle.

Equations (5.327) represent $n$ one-degree-of-freedom problems, and the differential equation of each is a linear second-order equation with time-periodic coefficients. For the periodic solution, each of these differential equations may be solved in the manner of Eq. (5.344) resulting in solution $\{\eta_i\}$ in the (decoupled) $\eta$-coordinate system.

Equations (5.327) may also be solved for the transient solution by utilizing Eqs. (5.343) and (5.345). Generally, the initial conditions are taken in the physical coordinates (i.e., $\{U(0)\}$ and $\{\dot{U}(0)\}$). In order to solve Eqs. (5.327), which are expressed in the normal coordinates, these initial conditions must be transformed into the normal coordinates thus:

$$\eta_i(0) = \frac{1}{M^{\eta}_{ii}} \{\Phi_i\}^T[M]\{U(0)\}$$
$$\dot{\eta}_i(0) = \frac{1}{M^{\eta}_{ii}} \{\Phi_i\}^T[M]\{\dot{U}(0)\} \tag{5.348}$$

Having solved for $\eta_i$, the solution $\{U\}$ in physical coordinates may be obtained by the coordinate transformation,

$$\{U\} = [\Phi]\{\eta\} \tag{5.321}$$

In the case of a periodic solution, both $[\Phi]$ and $\{\eta\}$ are periodic, resulting in a periodic $\{U\}$.

In order that Eqs. (5.327) be solved by way of Eqs. (5.343) and (5.344), it is desirable that the forcing term, $N_i(t)$, be available as an analytical function of time. This is easily done by obtaining a Fourier series for the $n$ discrete (periodic) values $N_i$, computed at the time nodes of the fundamental period. The $N_i$ values, for various modes, are determined from Eqs. (5.326). The integral terms in Eq. (5.344) may be alternatively approximated by the trapezoidal method of integration. In this case, only discrete values $N_i$ are required.

A somewhat less complicated solution of Eqs. (5.327) is possible if $N_i$ is also assumed to remain constant during each time interval, its value (as well as the system parameters) being modified in transition from one interval to another. The response to such step-function input is given [134,314] for the $i$th mode as

$$\eta(t) = \frac{N_i}{K^{\eta}_{ii}} - \frac{e^{-\xi_i \omega_i t}}{(1-\xi_i^2)^{1/2}} \left[ \frac{N_i}{K^{\eta}_{ii}} \{\xi_i \sin \omega_{di} t + (1-\xi_i^2)^{1/2} \cos \omega_{di} t\} \right.$$
$$\left. + \eta_i(0)\{\xi_i \sin \omega_{di} t - (1-\xi_i^2)^{1/2} \cos \omega_{di} t\} - \dot{\eta}_i(0) \left\{ \frac{1}{\omega_i} \sin \omega_{di} t \right\} \right] \tag{5.349}$$

where

$$\omega_i = \left( \frac{K^{\eta}_{ii}}{M^{\eta}_{ii}} \right)^{1/2} \quad \text{and} \quad \omega_{di} = \omega_i (1-\xi_i^2)^{1/2} \tag{5.349}$$

All other steps in the total system response analysis remain unchanged. Equation (5.349) implicitly assumes that the forcing as well as system parameters will take on "constant" values, valid for the interval corresponding to time $t$. Also, the final solutions for the preceding time interval serve as the initial conditions for the current time interval.

A second approach is to integrate Eq. (5.316) numerically by proceeding in a series of time steps $\Delta t$ while computing the solution at each step. Modal analysis (mode superposition) is preferable when a limited number of lower-frequency modes sufficiently describe the response. When the linkage structure is subjected to short-duration loads that would tend to excite many modes, or when nonlinearity in structure is involved, the numerical integration methods are more desirable. More detailed discussions on the advantages and disadvantages of these methods of solution may be found in Refs. 47 and 312. The integration methods may be used in conjunction with the modal analysis method; they may be used to evaluate the Duhamel integral [57,186,299] and its time derivative, as in Eqs. (5.340) and (5.341), to solve the uncoupled differential Eqs. (5.327). These uncoupled equations may also be directly solved by numerical integration.

Step-by-step integration methods are prone to bring about cumulative errors as well as numerical instability. This can be minimized by the use of small $\Delta t$, which in turn results in long computational times. The time step $\Delta t$ is often estimated as a fraction of the shortest time period of vibration of the structure. Numerical analysts, unlike structural analysts, prefer the fourth-order Runge–Kutta method. Formulas for the Runge–Kutta methods may be found in Refs. 33 and 34, as in numerous other references (see Sec. 5.31). Some numerical integration schemes are enumerated and compared in Refs. 12, 113 and 197. Several schemes have also been described clearly, for linear as well as nonlinear problems, by Wilson and coworkers [311,312]. The Wilson-$\theta$ method [12] is somewhat complex but can be shown to be unconditionally stable. The Newmark method [196] is simple because a linearly varying acceleration is assumed. To exemplify the step-by-step integration methods, one form of the Newmark method is discussed briefly below.

The Newmark method in its simplest form assumes that the acceleration varies linearly during a fixed time interval $\Delta t$; the length of this interval may be estimated by

$$\Delta t \simeq \frac{1}{4}\frac{1}{2\pi\omega_{\max}} \tag{5.350}$$

where $\omega_{\max}$ is the highest natural frequency of the system. The algorithm is started, using initial conditions, at $t = 0$. The following vectors and mass matrix are defined:

$$\begin{aligned} \{\alpha\} &= \{\dot{U}(0)\} + \frac{\Delta t}{2}\{\ddot{U}(0)\} \\ \{\beta\} &= \{U(0)\} + \Delta t\,\{\dot{U}(0)\} + \frac{\Delta t^2}{3}\{\ddot{U}(0)\} \end{aligned} \tag{5.351}$$

$$[\bar{M}] = [M] + \frac{\Delta t}{2}[C] + \frac{\Delta t^2}{6}[K] \tag{5.352}$$

For a structure with constant geometry, $[\bar{M}]$ is constant and need be computed only once. Next, for the first time step,

$$\{\bar{F}\} = \{F\}_{t=\Delta t} - [C]\{\alpha\} - [K]\{\beta\} \tag{5.353}$$

where $\{F\}$ is the forcing vector; in Eq. (5.316)

$$\{F\} = -[M]\{\ddot{U}_r\}$$

The motion of the structure at ($t = \Delta t$) is determined as

$$\{\ddot{U}\}_{t=\Delta t} = [\bar{M}]^{-1}\{\bar{F}\}_{t=\Delta t}$$

$$\{\dot{U}\}_{t=\Delta t} = \{\alpha\} + \frac{\Delta t}{2}\{\ddot{U}\}_{t=\Delta t} \tag{5.354}$$

$$\{U\}_{t=\Delta t} = \{\beta\} + \frac{\Delta t^2}{6}\{\ddot{U}\}_{t=\Delta t}$$

This completes one iteration of the Newmark algorithm. One would next return to Eqs. (5.351) with the new values for $\{U\}$, $\{\dot{U}\}$, and $\{\ddot{U}\}$, and $\{\bar{F}\}$ will be calculated using $\{F\}_{t=2\Delta t}$ in Eq. (5.353). For the linkage structure, $[\bar{M}]$ will change with time and will need to be calculated for each time step.

**Example 5.16**

In Fig. 5.85, an in-line slider-crank mechanism is shown. The crank-shaft is rotating in the counterclockwise direction at a constant angular velocity $\omega_2$. Consider the crank to be rigid and the connecting rod elastic and of lengths $L_2$ and $L_3$, their reference angles being $\theta_2$ and $\theta_3$, respectively. The connecting rod is assumed to have uniform cross-sectional area $A_3$, and area moment of inertia $I_3$. The slider, of mass $M$, moves in its path without friction. Its center of gravity coincides with point $C$. No externally applied forces act on the slider. The masses of bearing assembly at points $B$ and $C$ are assumed to be $m_B$ and $m_C$, respectively.

Discretize the connecting rod into two beam elements and develop the total system mass and stiffness matrices, in the global coordinates, expressing them in terms of the reference angles shown in Fig. 5.85. Neglecting the damping forces in the mechanism, write the equations of motion and derive the terms appearing in the forcing vector.

**Solution** In Fig. 5.86, the generalized displacement coordinates $U_1$ through $U_6$ are required to describe the mechanism deformation relative to the rigid-body configuration. Note that the prescribed boundary conditions are accounted for. Since the crank is rigid, no generalized coordinates are required at point $B$ at the crank end. However, one rotation $U_1$ is needed to describe the angular deformation of the connecting rod at point $B$.

It must be noted that the transformation matrix $[R]$ for either beam element segment of the connecting rod is the same. Since the two beam elements are identical,

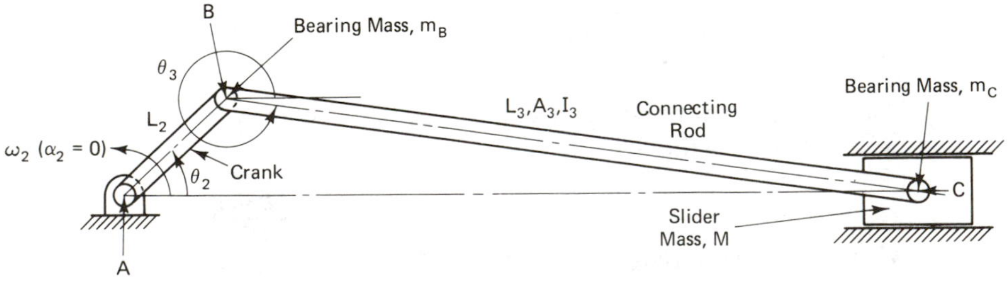


**Figure 5.85** Slider-crank mechanism with elastic coupler analyzed in Ex. 5.16.

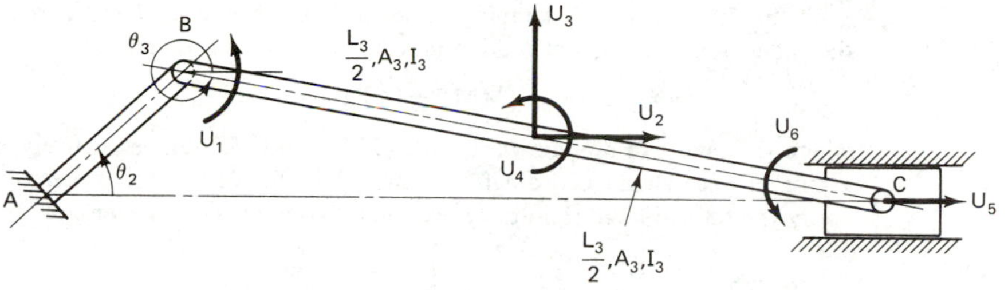


**Figure 5.86** Finite element discretization and definition of elastic displacements of the coupler. (See Ex. 5.16.)

their system-oriented mass and stiffness matrices in the global coordinates are identical as well. Now, if one follows steps similar to those leading up to the formulation of the total mass matrix in Eq. (5.312), and adding lumped masses along the diagonal wherever appropriate, one obtains

$$[M] = \rho A_3 L \begin{bmatrix} \frac{L^2}{105} & & & & & \text{Symmetric} \\ -\frac{13L}{420}\mu & 1+\frac{8}{105}\mu^2 & & & & \\ \frac{13L}{420}\lambda & -\frac{8}{105}\lambda\mu & 1+\frac{8}{105}\lambda^2 & & & \\ -\frac{L^2}{140} & 0 & 0 & \frac{2L^2}{105} & & \\ 0 & 1+\frac{4}{105}\lambda^2 & \frac{4}{105}\lambda\mu & -\frac{13L}{140}\mu & 1+\frac{4}{105}\mu^2+(M+m_c) & \\ 0 & \frac{13L}{420}\mu & -\frac{13L}{420}\lambda & -\frac{L^2}{140} & \frac{11L}{210}\mu & \frac{L^2}{105} \end{bmatrix} \tag{5.355}$$

where $\rho$ is mass density, $\lambda = \cos\theta_3$ and $\mu = \sin\theta_3$, and $L = L_3/2$. The total stiffness matrix is obtained similarly:

$$[K] = \frac{E}{L} \begin{bmatrix} 4I_3 & & & & & \text{Symmetric} \\ \frac{6I_3}{L}\mu & 2A_3\lambda^2+\frac{24I_3}{L^2}\mu^2 & & & & \\ -\frac{6I_3}{L}\lambda & \left(2A_3-\frac{24I_3}{L^2}\right)\lambda\mu & 2A_3\mu^2+\frac{24I_3}{L^2}\lambda^2 & & & \\ 2I_3 & 0 & 0 & 8I_3 & & \\ 0 & -A_3\lambda^2-\frac{12I_3}{L^2}\mu^2 & \left(-A_3+\frac{12I_3}{L^2}\right)\lambda\mu & \frac{6I_3}{L}\mu & A_3\lambda^2+\frac{12I_3}{L^2}\mu^2 & \\ 0 & -\frac{6I_3}{L}\mu & \frac{6I_3}{L}\lambda & 2I_3 & \frac{6I_3}{L}\mu & 4I_3 \end{bmatrix} \tag{5.356}$$

In the absence of damping forces in the mechanism and of external forces on the slider, the equations of motion may be expressed in matrix form as

$$[M]\{\ddot{U}\} + [K]\{U\} = -[M]\{\ddot{U}_r\} \tag{5.357}$$

where $[M]$ and $[K]$ are shown in Eqs. (5.355) and (5.356), respectively, and $\{U\}$ is the vector of generalized coordinates illustrated in Fig. 5.86. It is left to the reader to verify the terms in the rigid-body acceleration vector, $\{\ddot{U}_r\}$, which are

$$\{\ddot{U}_r\} = \begin{Bmatrix} \ddot{U}_{r1} \\ \ddot{U}_{r2} \\ \ddot{U}_{r3} \\ \ddot{U}_{r4} \\ \ddot{U}_{r5} \\ \ddot{U}_{r6} \end{Bmatrix} = \begin{Bmatrix} \alpha_3 \\ -L_2\omega_2^2 \cos\theta_2 - \dfrac{L_3}{2}(\omega_3^2 \cos\theta_3 + \alpha_3 \sin\theta_3) \\ -L_2\omega_2^2 \sin\theta_2 + \dfrac{L_3}{2}(\alpha_3 \cos\theta_3 - \omega_3^2 \sin\theta_3) \\ \alpha_3 \\ -L_2\omega_2^2 \cos\theta_2 - L_3(\alpha_3 \sin\theta_3 + \omega_3^2 \cos\theta_3) \\ \alpha_3 \end{Bmatrix} \tag{5.358}$$

**Example 5.17**

In this example, by Turcic [292], the dimensions, section, and material properties of the members of an elastic four-bar (Fig. 5.81) crank-rocker linkage are found in Table 5.17. Investigate the natural frequencies of this mechanism as well as the midpoint strains for the coupler and output links.

**TABLE 5.17** LINKAGE PARAMETERS FOR EXS. 5.17 AND 5.18

| | Crank (2) | Coupler (3) | Follower (4) |
|---|---|---|---|
| Length ($L_i$) | 4.25 in (10.80 cm) | 11.00 in (27.94 cm) | 10.65 in (27.05 cm) |
| Area ($A_i$) | 0.167 in² (1.077 cm²) | 0.063 in² (0.406 cm²) | 0.063 in² (0.406 cm²) |
| Area moment of inertia ($I_i$) for bending | $3.881 \times 10^{-4}$ in⁴ ($1.616 \times 10^{-2}$ cm⁴) | $2.084 \times 10^{-5}$ in⁴ ($8.674 \times 10^{-4}$ cm⁴) | $2.084 \times 10^{-5}$ in⁴ ($8.674 \times 10^{-4}$ cm⁴) |
| Distance between ground pivots ($L_1$): | 10.00 in (25.40 cm) | | |
| Lumped weight of bearing assembly ($W_2 = W_3$): | 0.0925 lbf (0.42 N) | | |
| Modulus of elasticity ($E$) (material = aluminum): | $10.3 \times 10^6$ psi ($7.10 \times 10^7$ kPa) | | |
| Weight density ($\rho$): | 0.098 lbf/in³ ($2.66 \times 10^{-2}$ N/Cm³) | | |

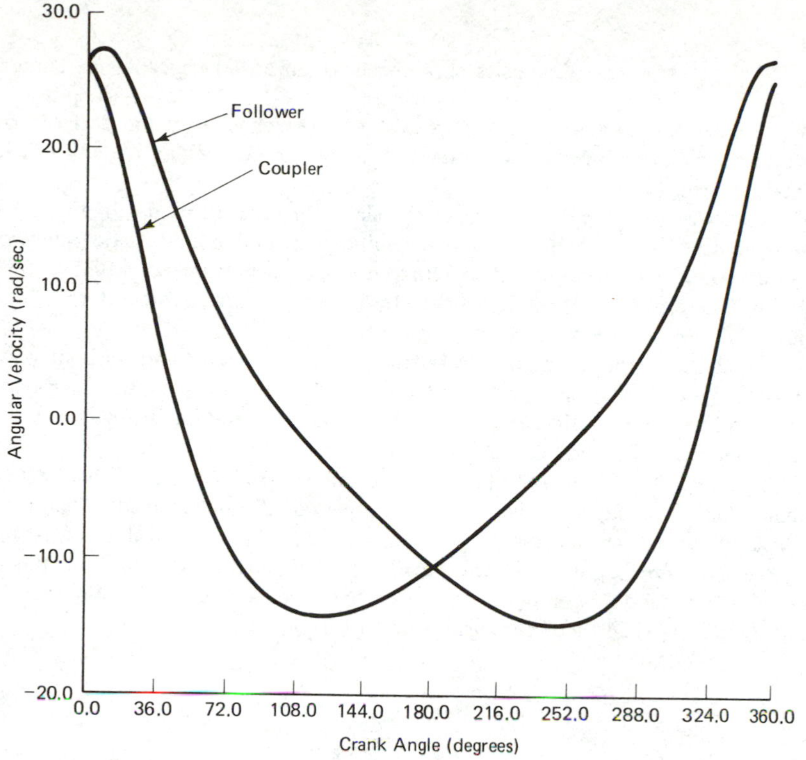


**Figure 5.87** Coupler and follower rigid-body angular velocity versus crank angle for Ex. 5.17.

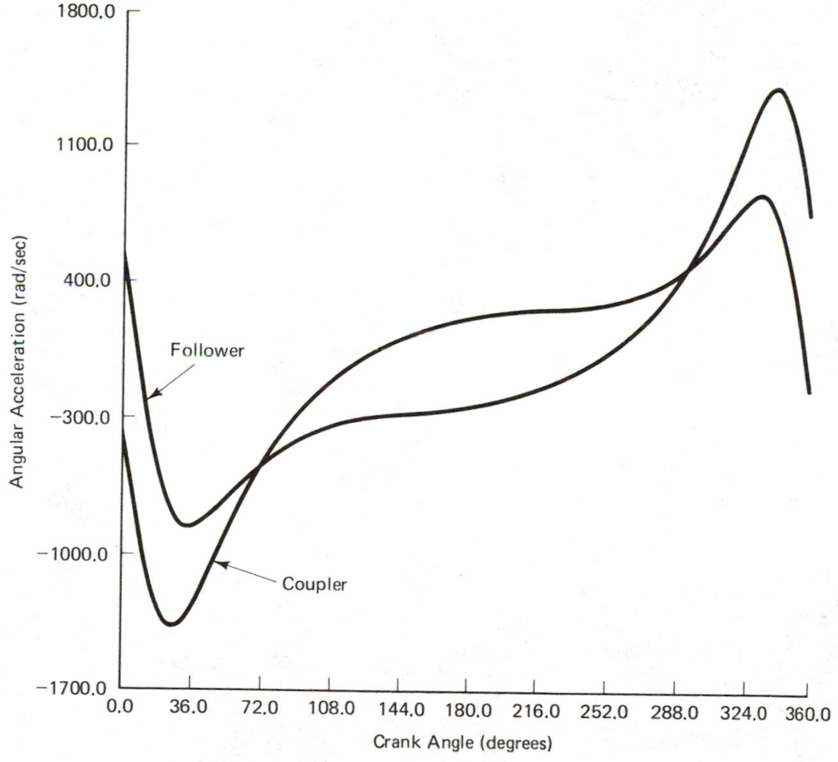


**Figure 5.88** Coupler and follower rigid-body angular acceleration versus crank angle for Ex. 5.17.

**Solution** Figures 5.87 and 5.88 show the rigid-body angular velocity and acceleration of the coupler and the follower for a constant clockwise crank angular velocity of 340 rpm (35.60 rad/sec).

Three increasingly complex models of the elastic linkage are used. In Fig. 5.89, each link is idealized by a single beam element, resulting in a total of nine elastic (nonzero) degrees-of-freedom. The circled numbers represent the element numbers. Figures 5.90 and 5.91 illustrate examples of multielement idealization per link, representing a total of 18 and 27 degrees-of-freedom, respectively.

Among the assumptions made are bearings without friction and without play. Only planar motion is assumed, and small elastic deformations from the rigid-body equilibrium positions are considered. The gravitational acceleration is assumed small compared to the rigid-body and elastic accelerations.

In the results that follow, only the first three modes are included. Ninety equal-time subintervals of the fundamental forcing time period, $T_f$, are deemed sufficient to yield reasonably accurate steady-state response (strain) of the linkage. Material damping of the system is estimated by equivalent small modal damping ratio for each mode; this is the Wilson method for computing the damping matrix, as suggested in Eq. (5.332). The damping ratio in each mode is considered to be 0.03.

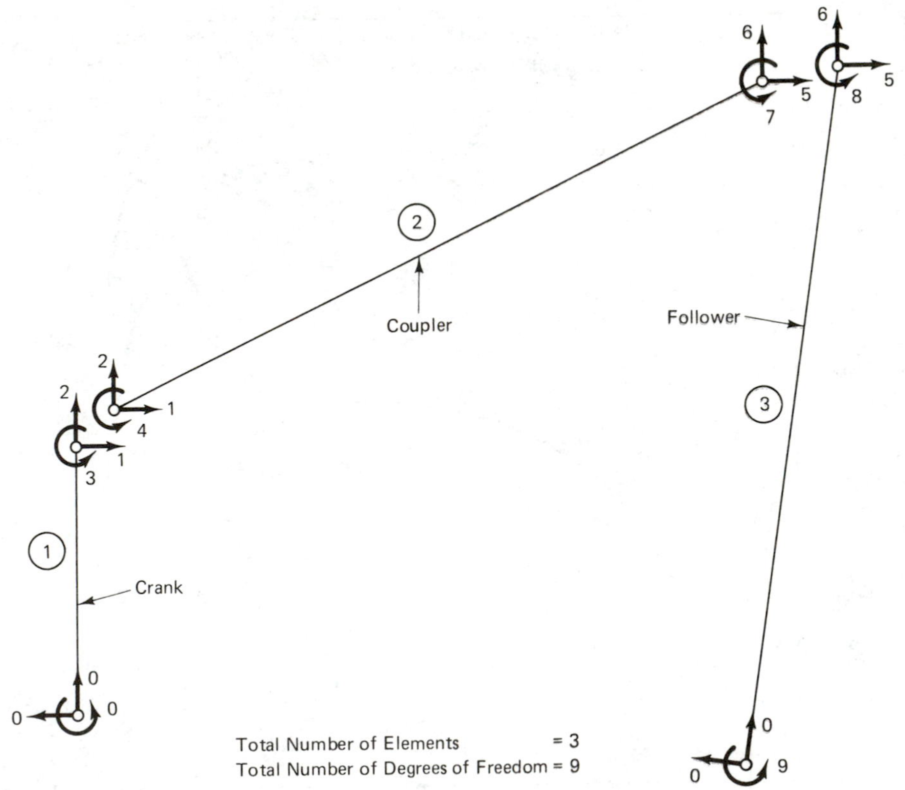


**Figure 5.89** Elastic mechanism model number one for Ex. 5.17.

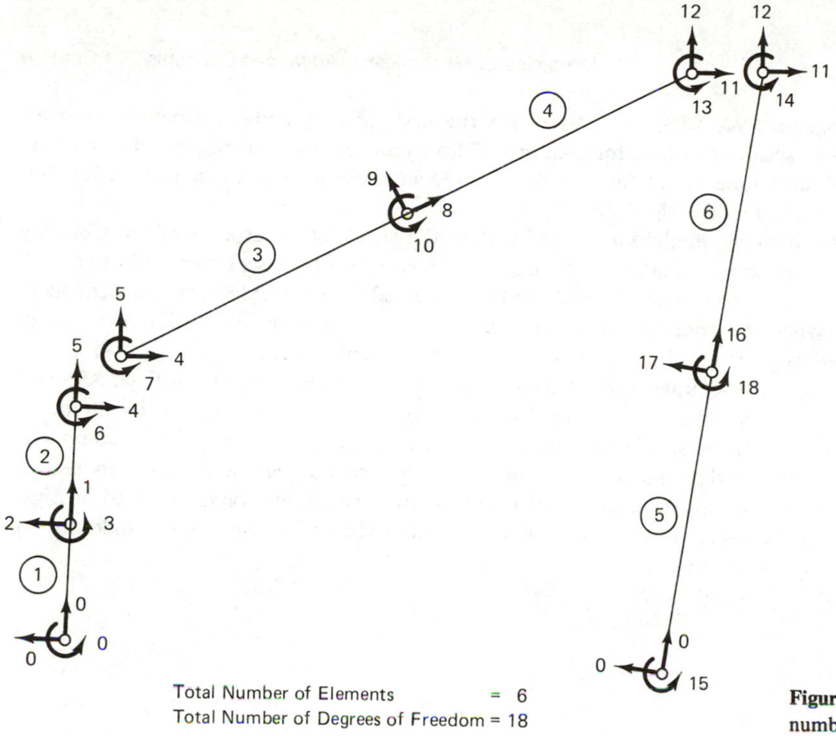


**Figure 5.90** Elastic mechanism model number two for Ex. 5.17.

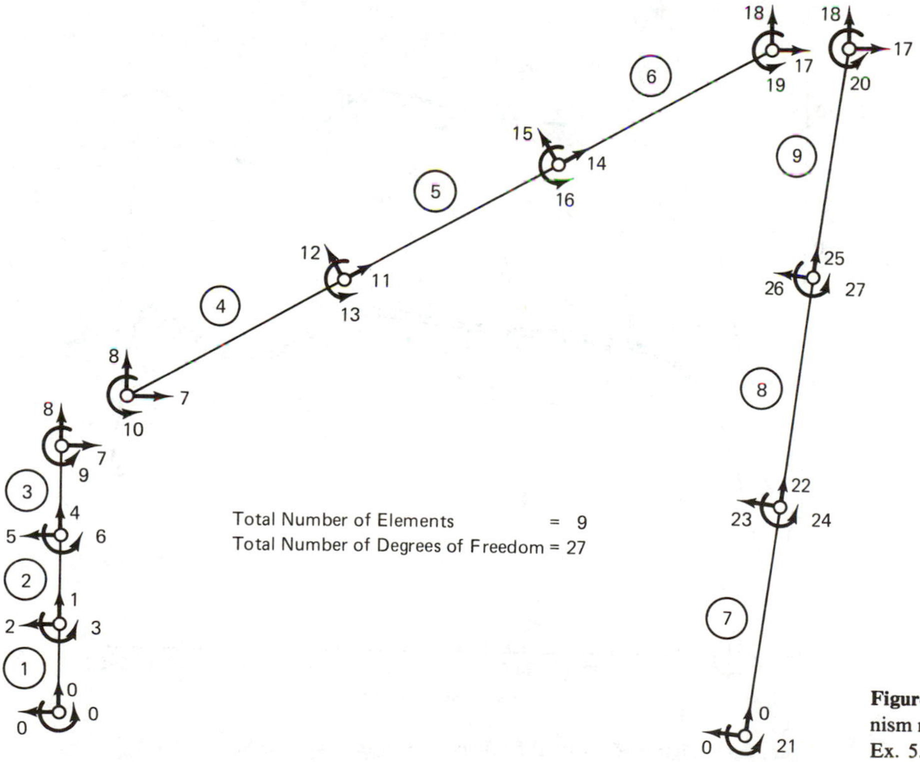


**Figure 5.91** Elastic mechanism model number three for Ex. 5.17.

Figures 5.92, 5.93, and 5.94 depict the first, second, and third natural (modal) frequency variations in time for each of the three models described earlier. The convergence of such time variations can be observed in these figures for increasing number of elements in linkage idealization.

Based on the method developed earlier, the steady-state response of the vibrating linkage is computed. Figures 5.95 through 5.98 compare the coupler and follower midpoint strains for the three different models discussed. The coupler strain is defined as positive when the midpoint of the coupler is deflected upward and the follower strain is defined as positive when the follower midpoint is deflected inward.

Clearly, the coupler and follower midpoint (steady-state) strains in Figs. 5.95 and 5.96, respectively, manifest considerable discrepancy in response between the three-element (model one) and six-element (model two) idealization of the linkage. The computed strains, however, do tend toward convergence as the number of elements increases. This may be observed from the coupler and follower midpoint strains recorded in Figs. 5.97 and 5.98, respectively, for six-element (model two) and nine-element (model three) idealizations of the linkage.

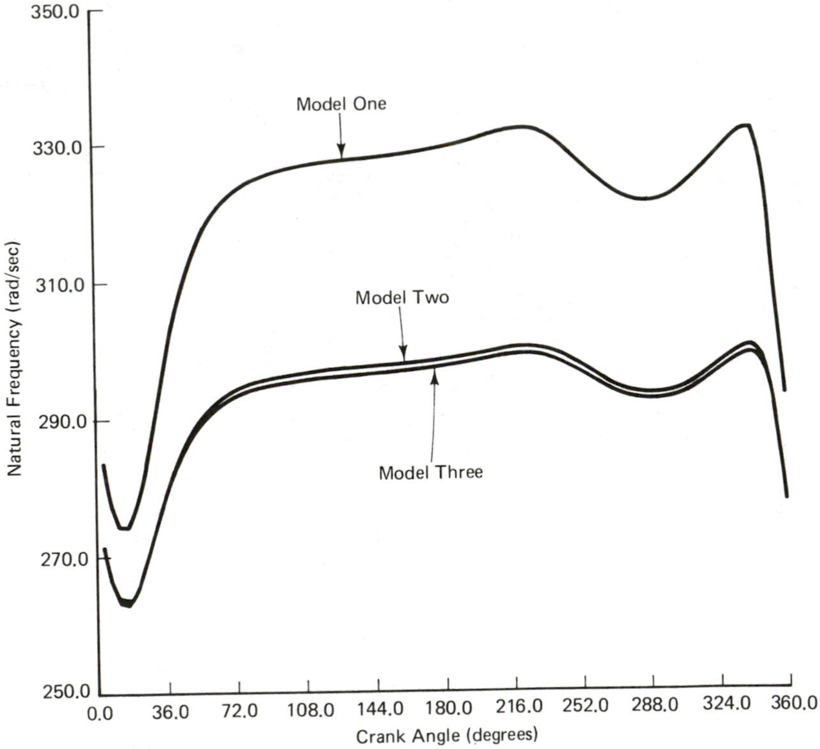


**Figure 5.92** First mode natural frequency versus crank angle for Ex. 5.17.

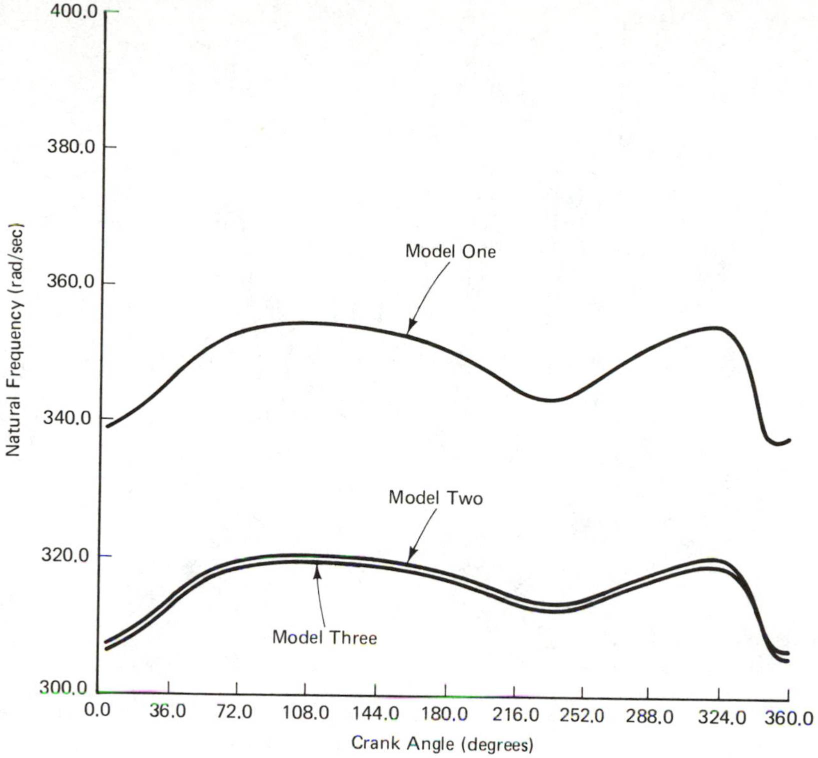


**Figure 5.93** Second mode natural frequency versus crank angle for Ex. 5.17.

**Figure 5.94** Third mode natural frequency versus crank angle for Ex. 5.17.

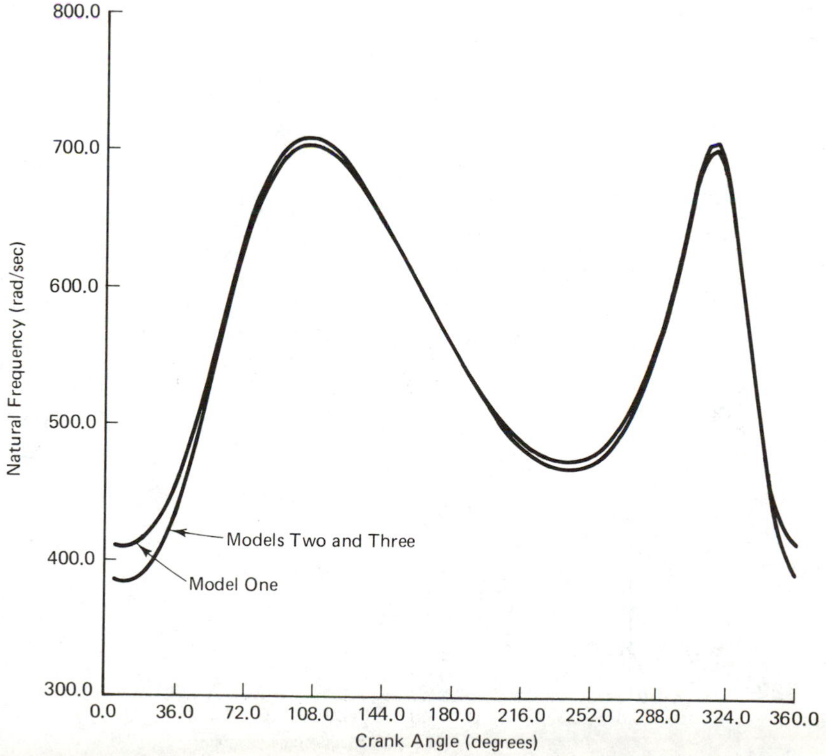

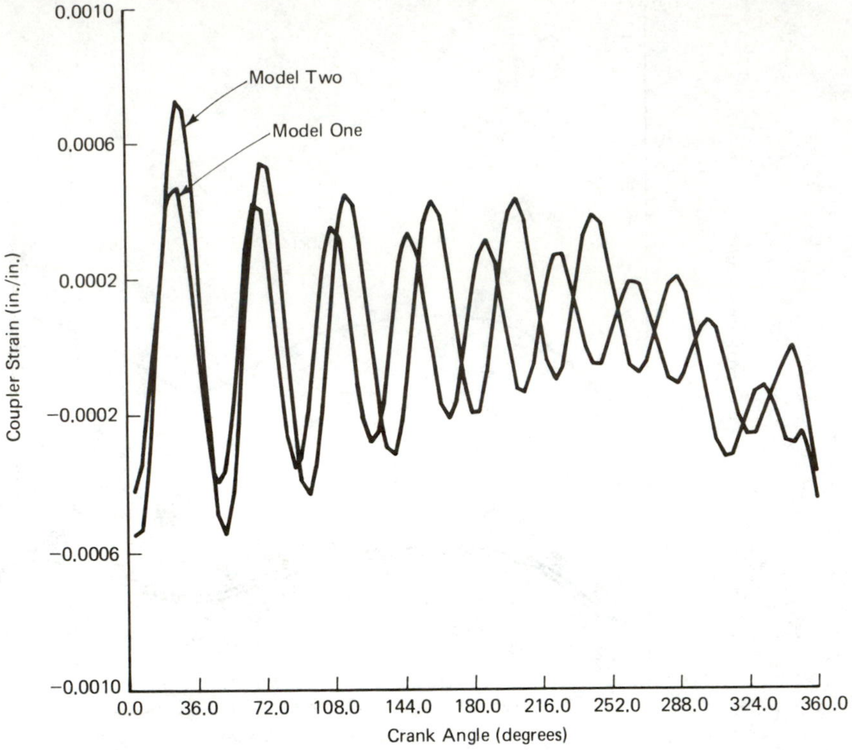


**Figure 5.95** Comparison of the coupler midpoint strains for models one and two (Ex. 5.17).

**Figure 5.96** Comparison of the follower midpoint strains for models one and two (Ex. 5.17).

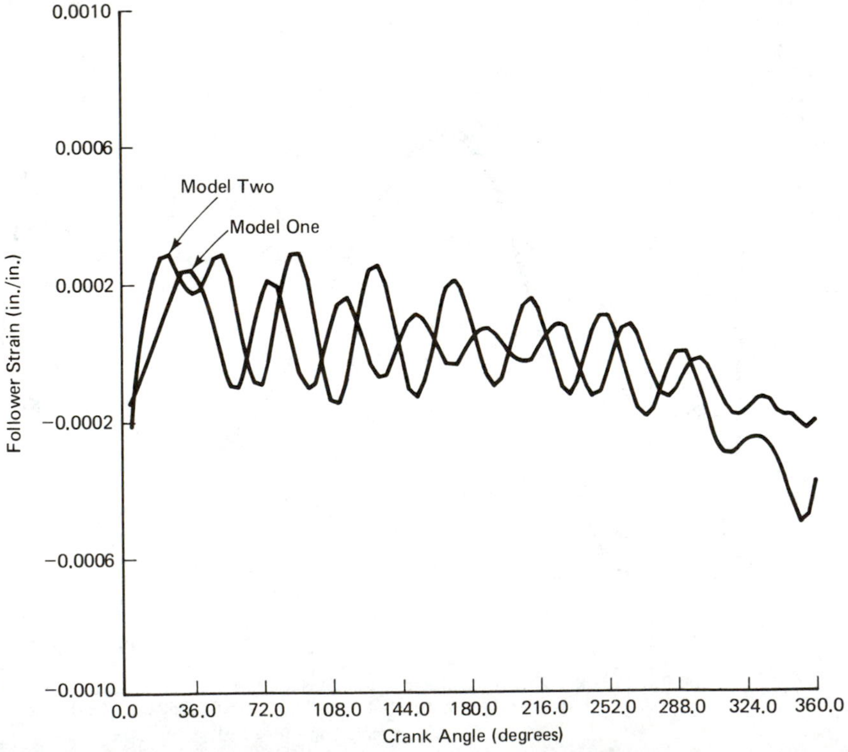

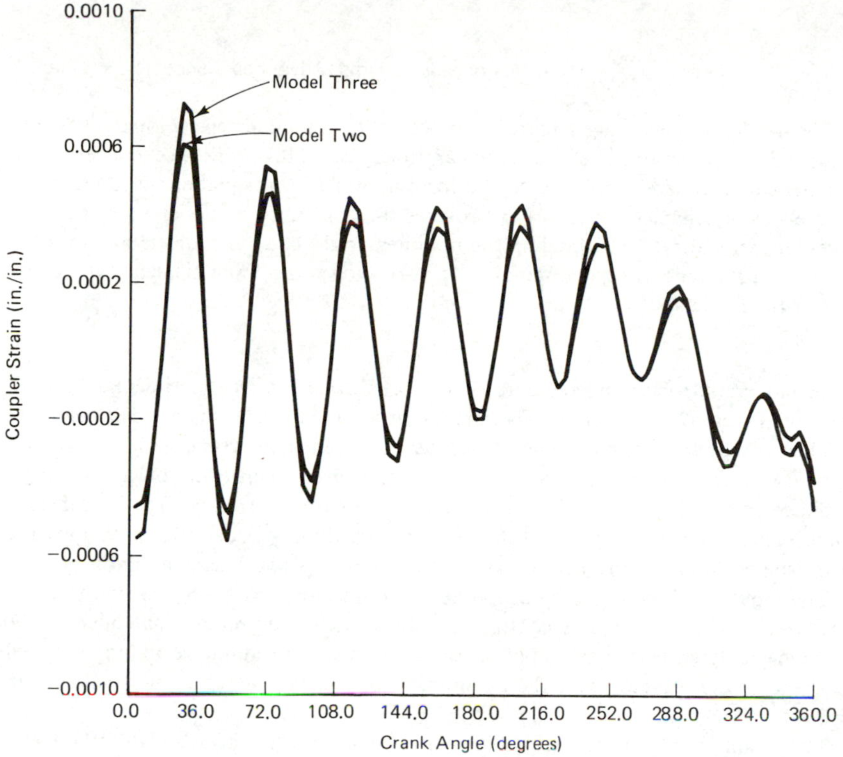


**Figure 5.97** Comparison of the coupler midpoint strains for models two and three (Ex. 5.17).

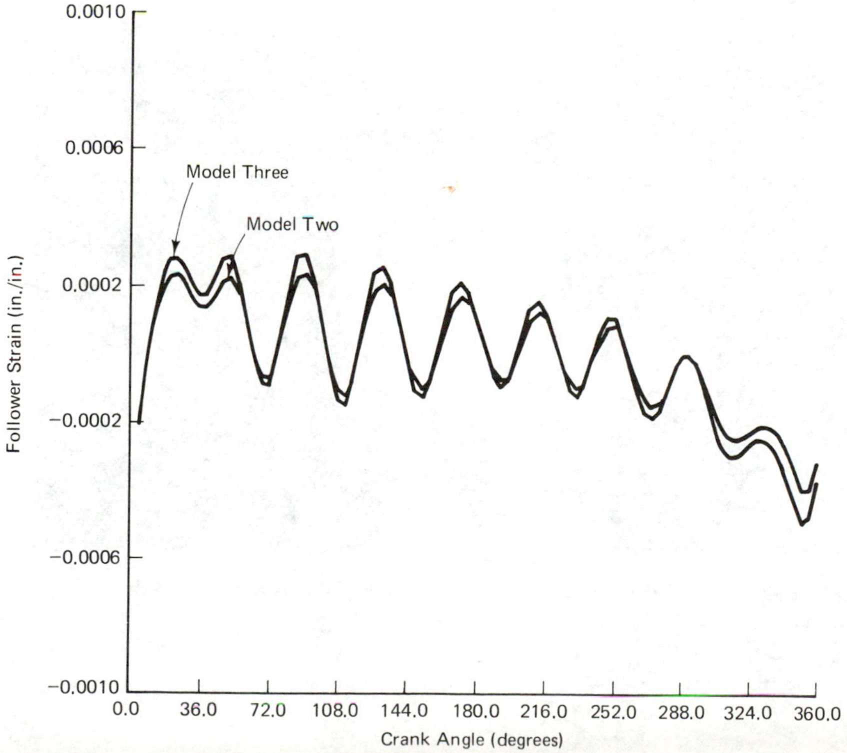


**Figure 5.98** Comparison of the follower midpoint strains for models two and three (Ex. 5.17).

The reader is encouraged to refer to Ref. 292 for an extensive analytical and experimental investigation of elastic four-bar linkages. This work also concerns the development of a general approach to the formulation of the equations of motion for complex elastic mechanism systems. It provides the capability to model a general two- or three-dimensional system, including the nonlinear rigid-body and elastic motion coupling terms in its general representation. It also allows any finite element type to be utilized in the model.

**Example 5.18**

The four-bar crank-rocker linkage data of Ex. 5.17 describes an experimental linkage, reported by Turcic [292], wherein experimental and analytical results are compared.

This linkage (Fig. 5.99) is constructed using a split input crank with the coupler passing between the two crank links. The input shafts are supported in journal bearings. The linkage is attached to a 3.5-in. thick oak foundation. This is then bolted to a steel and concrete support which is anchored to the floor. Thus, all precautions are taken to isolate the mechanism from extraneous vibrations and resonant conditions.

The input crank is driven by a Bodine $\frac{1}{4}$ HP dc motor through two timing belts, one for each shaft; the motor is controlled with a Bodine dc motor controller. Two 11-in. diameter flywheels are mounted on the input shafts to minimize the input speed fluctuations for consistency, with the assumption of constant input angular velocity in Ex. 5.17.

The strains are measured with the help of a four active arm bridge circuit and Micro Measurement EA-13-250 BG-120 strain gages. The coupler and follower midpoint strains are measured using a Nicolet Model 206 Digital oscilloscope with a floppy disk storage system. The crank "zero" position is recorded using a GE H21A2 photon-coupled Interrupter Module with an eight microsecond on time. The photocell is trig-

**Figure 5.99** Photo of the experimental crank-rocker linkage. (Courtesy of Ashok Midha.)

gered each time the crank passes through its zero position identified by a rotating disk with a single small hole.

In order that additional analytical strains from Ref. 292 be presented, the following discussion is necessary. In preceding developments in this section, the beam element transverse displacements are assumed independent of the axial displacements. In reality, however, a compressive axial deformation will tend to increase the transverse deformation of the beam, thus effectively decreasing its transverse stiffness. Similarly, a tensile axial deformation will have the opposite effect. An approximate method to correct this deficiency, in the stiffness matrix $[\bar{k}]$ in Eq. (5.297), is described in Ref. 206. To include the above coupling effect, a "geometric stiffness matrix," $[\bar{k}_G]$, is simply added to the matrix $[\bar{k}]$, above. The axial forces in the beam elements are approximated by a quasi-static force analysis of the linkage and are subsequently used in deriving $[\bar{k}_G]$.

The geometric stiffness matrix $[\bar{k}_G]$ in Ref. 206 is given for a beam element as:

$$[\bar{k}_G] = \frac{F}{L}\begin{bmatrix} 0 & & & & & \\ 0 & 6/5 & & & & \\ 0 & L/10 & 2L^2/15 & & \text{Symmetric} & \\ 0 & 0 & 0 & 0 & 6/5 & \\ 0 & -6/5 & -L/10 & 0 & & \\ 0 & L/10 & -L^2/30 & 0 & -L/10 & 2L^2/15 \end{bmatrix} \tag{5.359}$$

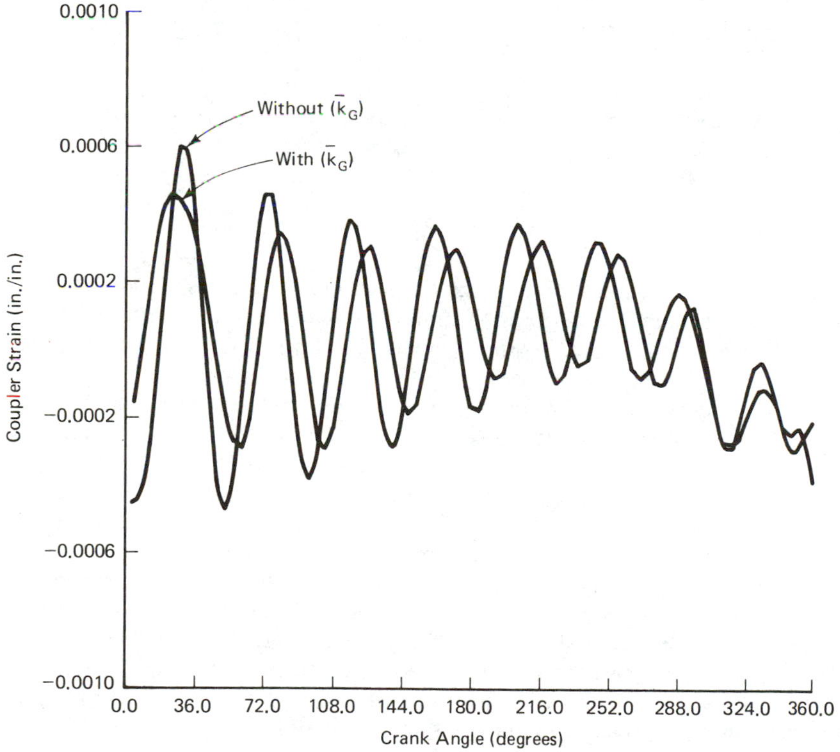


**Figure 5.100** Comparison of the model three coupler midpoint strains with and without the inclusion of the geometrical stiffness matrix $(\bar{k}_G)$.

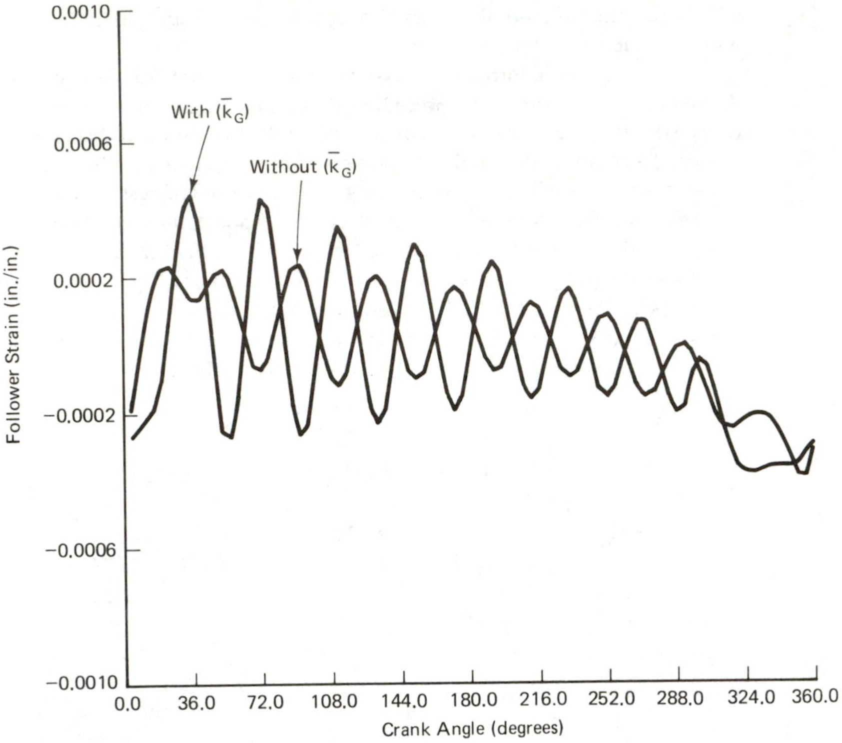


**Figure 5.101** Comparison of the model three follower midpoint strains with and without the inclusion of the geometrical stiffness matrix ($\bar{k}_G$).

where $F$ is equal to the axial force in the beam element. Figures 5.100 and 5.101 manifest the significant differences in midpoint (steady-state) strains for both the coupler and follower, respectively, with and without inclusion of $[\bar{k}_G]$ effects. Model three (Fig. 5.91) is used for these demonstrations.

The following are comparisons of experimental and analytical strains at the input shaft speed of 308.48 rpm. Figures 5.102a and b depict the experimental and analytical coupler midpoint strains, without and with $[\bar{k}_G]$ effects, respectively. Figure 5.103 presents their comparison in amplitude versus normalized frequency rather than the time response comparisons, as in Fig. 5.102. The normalized frequency is defined as the ratio of signal frequency content to the fundamental frequency (input speed). Figure 5.104 represents corresponding midpoint steady-state strains for the follower, and Fig. 5.105 shows its amplitude versus normalized frequency comparisons. For extensive results and comparisons, at varying speeds, the reader is referred to Ref. 292.

The above comparisons show very good agreement between the experimental and analytical results at all operating speeds considered. Clearly, the inclusion of the $[\bar{k}_G]$ terms in the analysis considerably improves these comparisons.

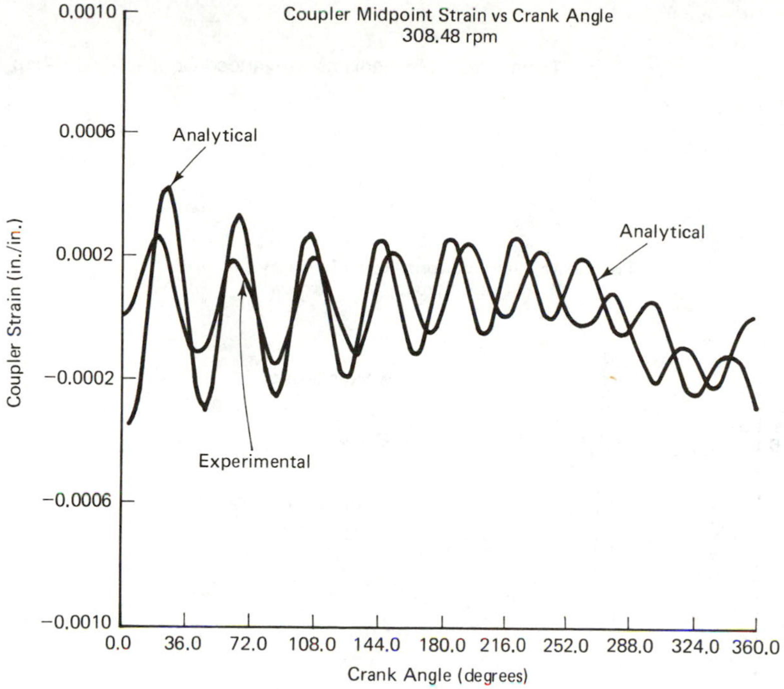


(a)

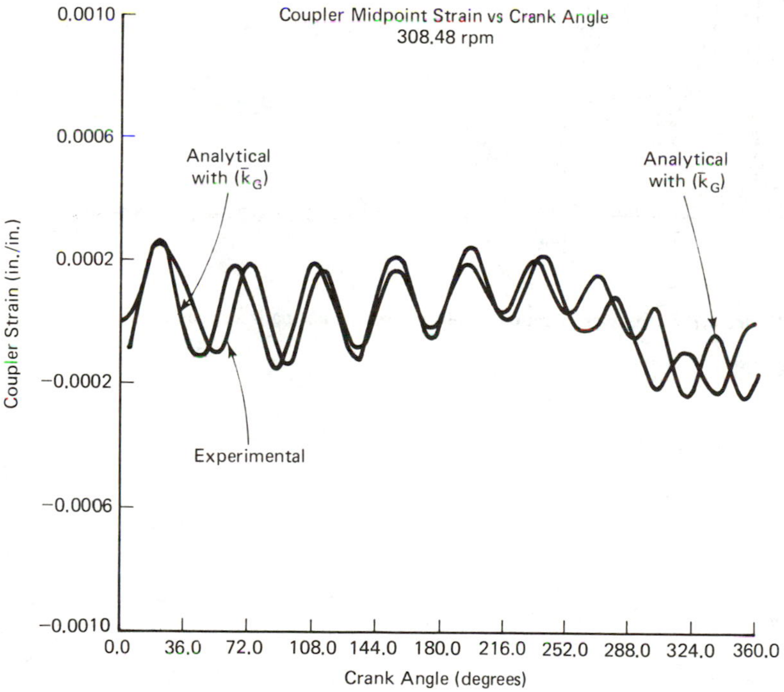


(b)

**Figure 5.102** Comparison of the experimental and analytical coupler midpoint strains for an input speed of 308.48 RPM.

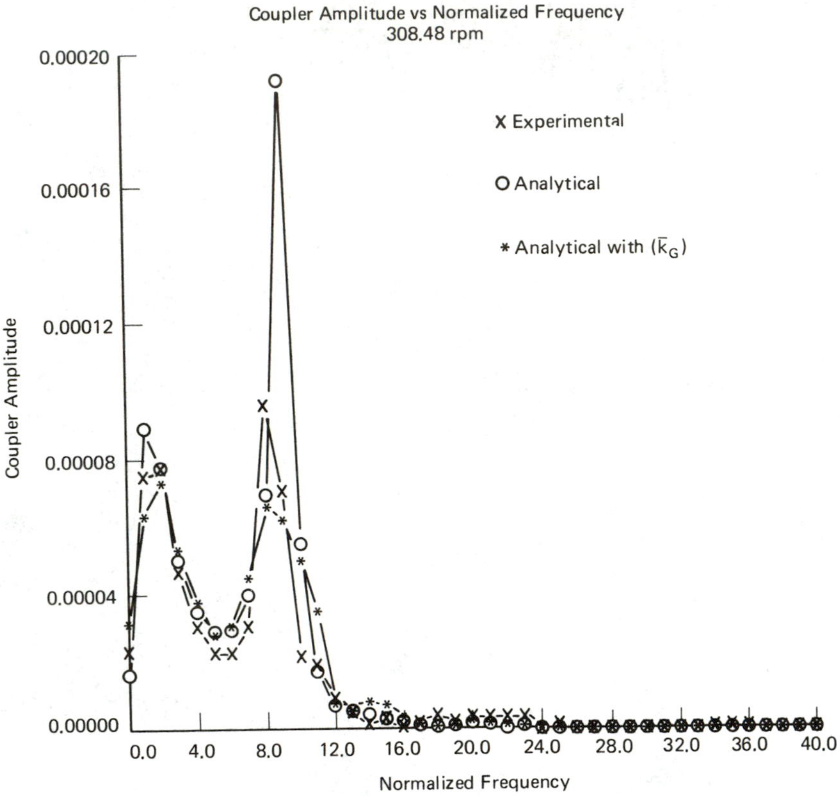


**Figure 5.103** Coupler vibration amplitude versus normalized frequency.

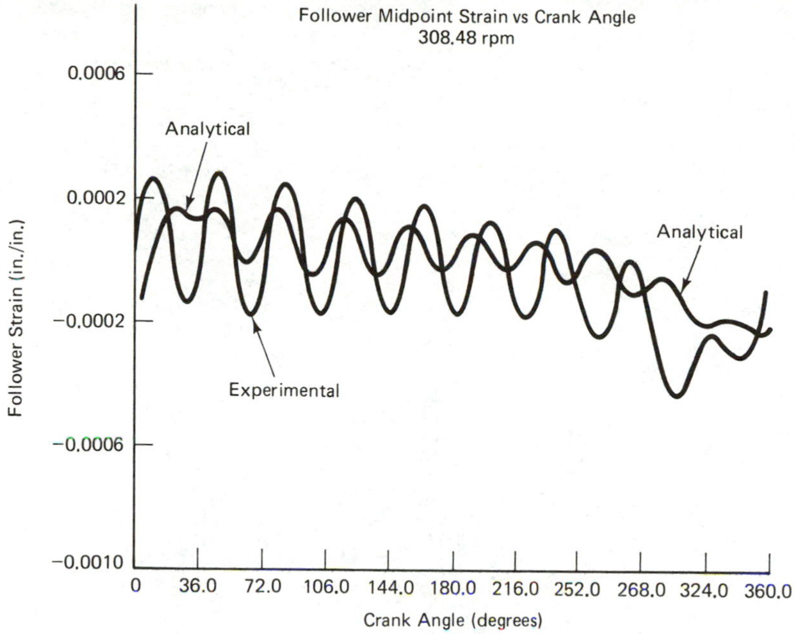


(a)

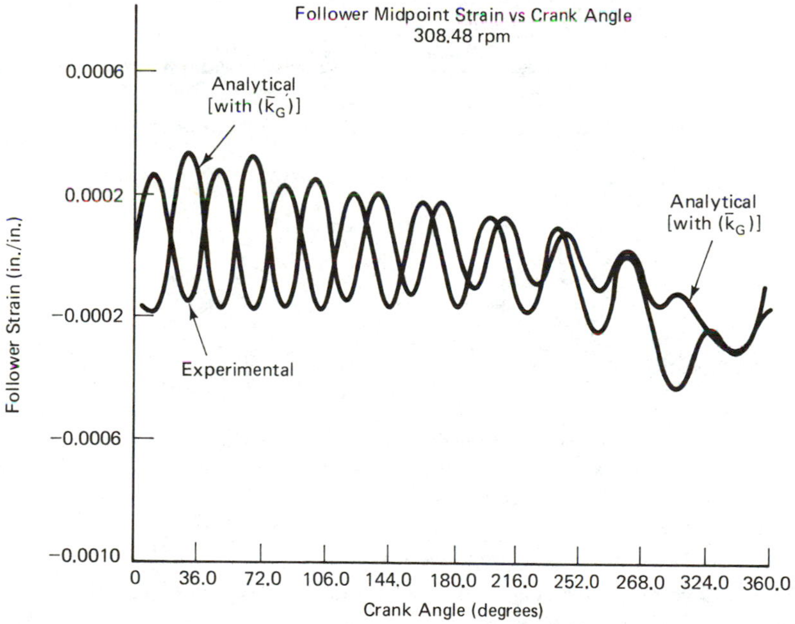


(b)

**Figure 5.104** Comparison of the experimental and analytical follower midpoint strains for an input speed of 308.48 RPM.

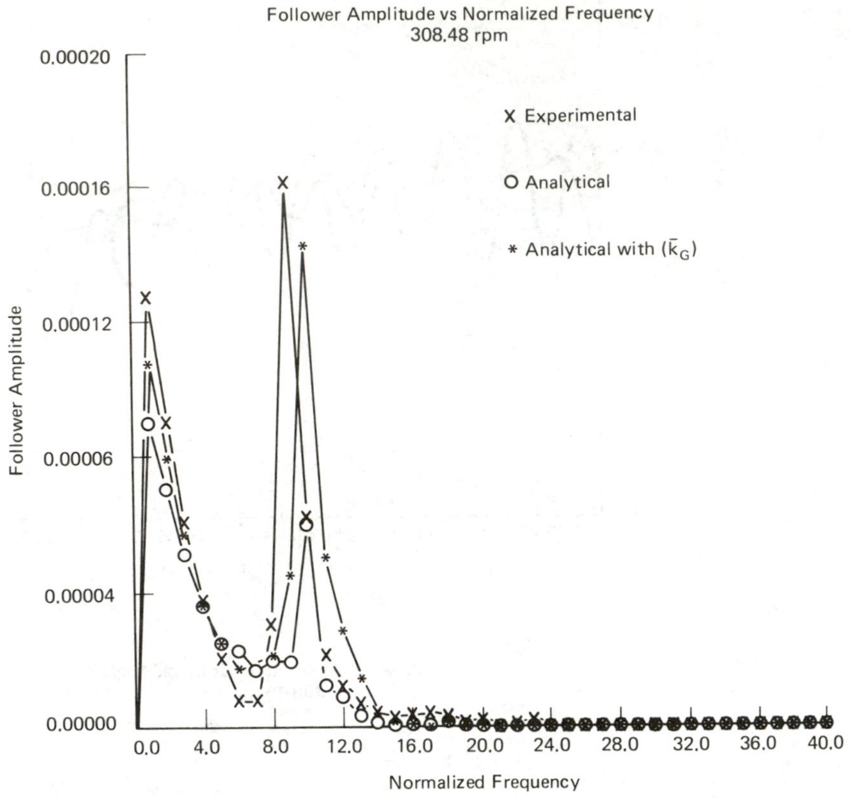


**Figure 5.105** Follower vibration amplitude versus normalized frequency.

## Problems

**5.1.** In Fig. P5.1 known external force $S$ and inertia force $P$ are applied to links 3 and 4, respectively. Links 2 and 3 are assumed to have very small inertia as compared to forces $P$ and $S$. What torque must be applied to link 2 for equilibrium?

**(a)** Use free-body diagrams and solve graphically.

**(b)** Use the virtual work method.

**5.2.** The four-bar linkage in Fig. P5.2 has two forces acting on it, an inertia force $P = 100$ lbf acting at point $B$ and an external torque $T_4 = -600$ in·lbf acting on link 4. Find the required input torque $T_2$ and the pin forces at $O_2$, $A$, $B$, and $O_4$.

**(a)** Use the graphical method based on free-body diagrams.

**(b)** Use the virtual work formulation.

**(c)** Use the complex-number method.

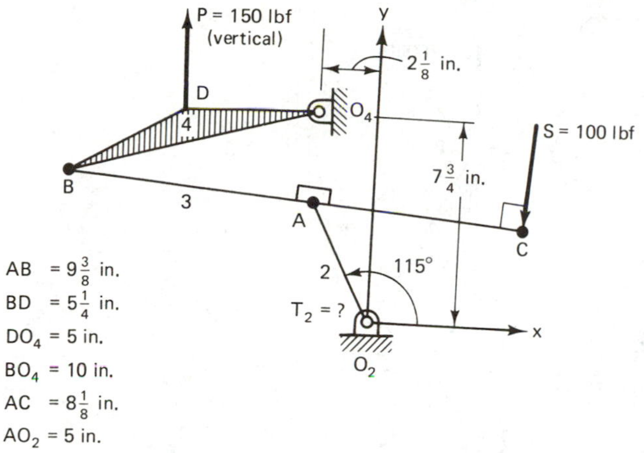

$$AB = 9\tfrac{3}{8} \text{ in.}$$
$$BD = 5\tfrac{1}{4} \text{ in.}$$
$$DO_4 = 5 \text{ in.}$$
$$BO_4 = 10 \text{ in.}$$
$$AC = 8\tfrac{1}{8} \text{ in.}$$
$$AO_2 = 5 \text{ in.}$$

**Figure P5.1**

$$O_2 A = 6 \text{ in.}$$
$$AB = 9 \text{ in.}$$
$$AC = 15 \text{ in.}$$
$$O_4 C = 8.25 \text{ in.}$$
$$O_2 O_4 = 17.13 \text{ in.}$$

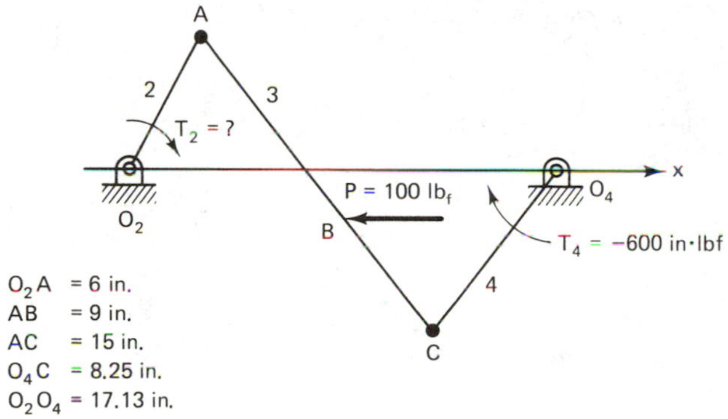

**Figure P5.2**

**5.3.** Figure P5.3 shows a double-piston mechanism which is assumed to be in equilibrium under the action of external forces $P$ and $S$ and an input torque $T_2$. Find the input torque $T_2$ and the pin forces at $A$, $C$, and $O_2$ as well as the contact force between the pistons and their guides.

    **(a)** Use the graphical method based on free-body diagrams.

    **(b)** Use the virtual work formulation.

    **(c)** Use the complex-number method.

**5.4.** Find the input torque $T_2$ applied to link 2 to maintain equilibrium of the linkage shown in Fig. P5.4 with external force $P$ acting on it.

    **(a)** Use the graphical method based on free-body diagrams.

    **(b)** Use the virtual work formulation.

    **(c)** Use the complex-number method.

**5.5.** The four-bar mechanism of Fig. P5.5 has one external force $P = 200$ lbf and one inertia force $S = 150$ lbf acting on it. The system is in dynamic equilibrium as a result of torque $T_2$ applied to link 2. Find $T_2$ and the pin forces.

    **(a)** Use the graphical method based on free-body diagrams.

    **(b)** Use the virtual work formulation.

    **(c)** Use the complex-number method.

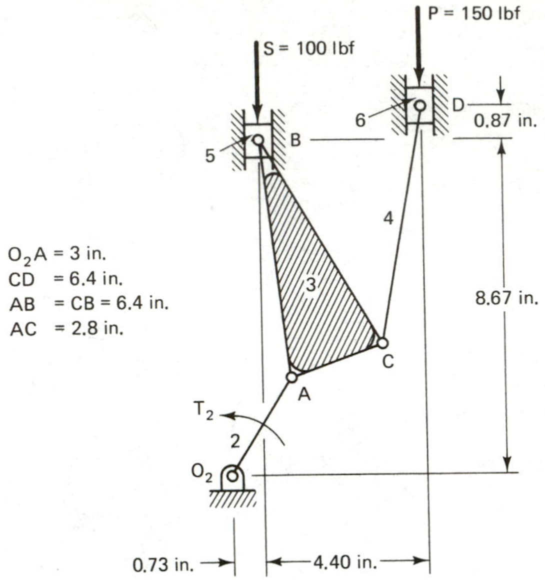


**Figure P5.3**

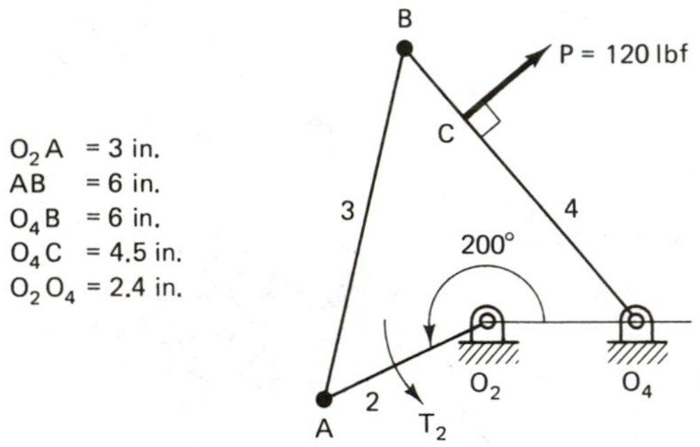


**Figure P5.4**

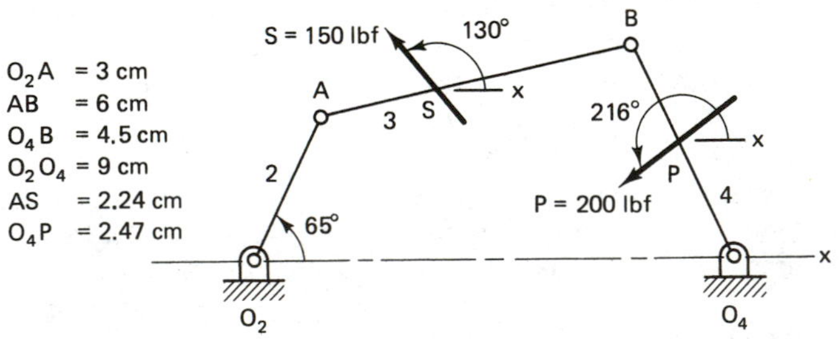


**Figure P5.5**

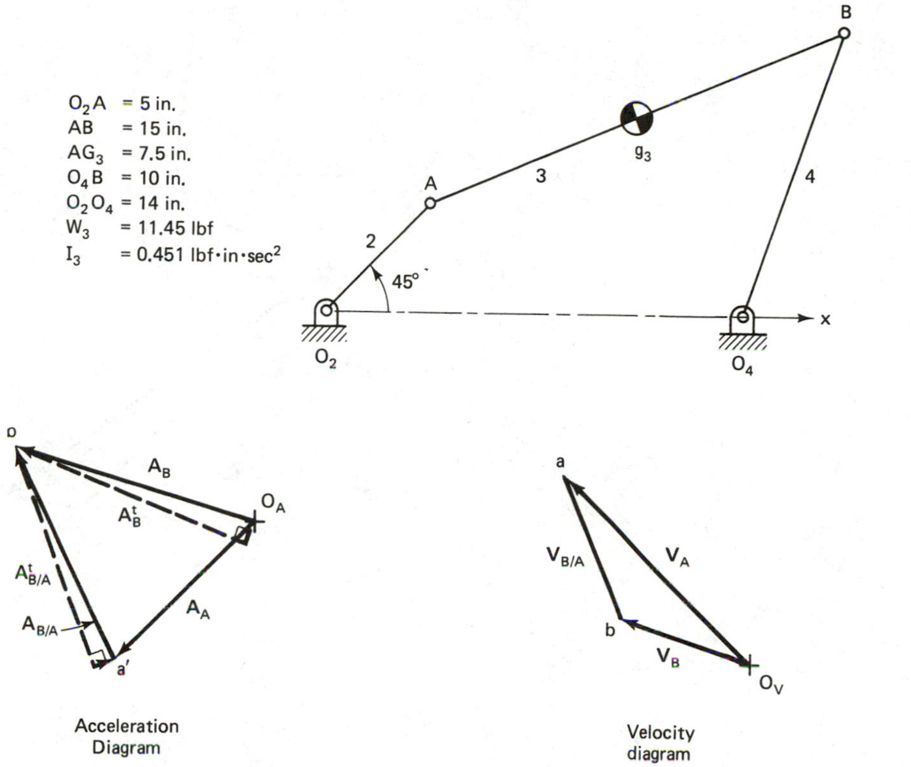


**Figure P5.6**

**5.6.** The input crank of the four-bar mechanism shown in Fig. P5.6 rotates at a constant speed $\omega_2 = 300$ rpm ccw. Only the coupler link is considered to have significant inertia. The velocity and acceleration diagrams are given in the figure.

**(a)** Determine the linear acceleration of the center of gravity of link 3 and the angular acceleration $\alpha_3$.

**(b)** Find the inertia force $\mathbf{F}_{O3}$ of the coupler link.

**(c)** Using the virtual work method find the directions and magnitudes of the pin forces at $A$ and $B$.

**(d)** Determine the required input torque to drive this mechanism in this position and under the conditions described above.

**(e)** Find the instantaneous value of the shaking forces and shaking moment transmitted to ground.

**5.7.** The input crank of the four-bar linkage of Fig. P5.7 rotates at a constant speed of $\omega_2 = 500$ rad/sec cw. Each link has significant inertia. The velocity and acceleration diagrams are provided in the figure.

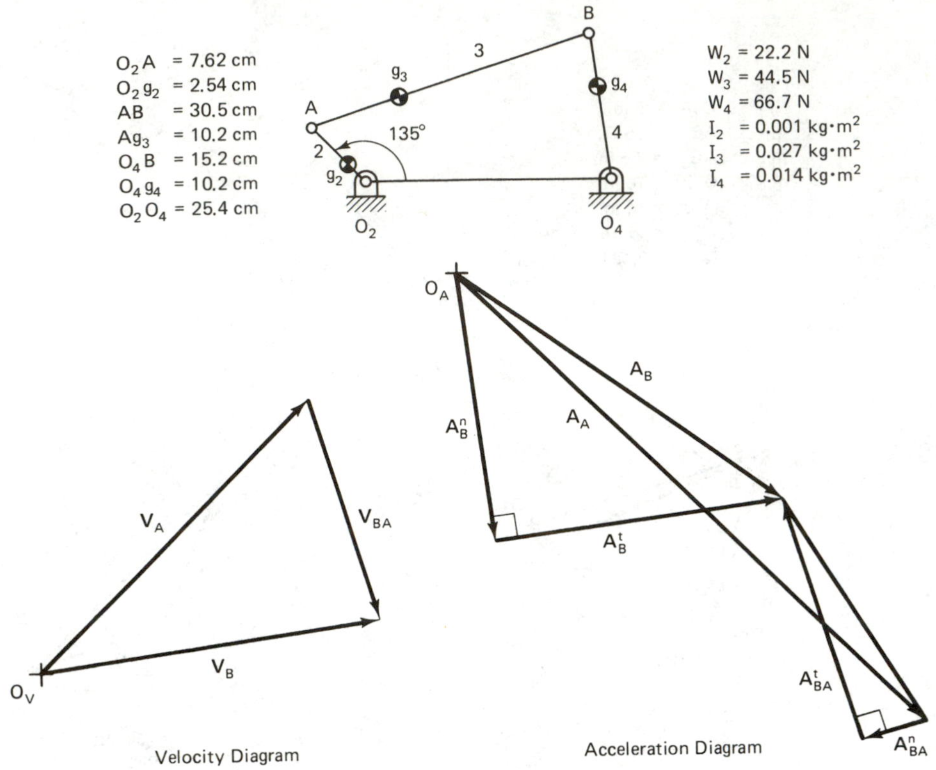


**Figure P5.7**

**(a)** Find the inertia forces $\mathbf{F}_{O2}$, $\mathbf{F}_{O3}$, and $\mathbf{F}_{O4}$.

**(b)** Using the virtual work method, find the directions and magnitudes of the pin forces at $A$ and $B$.

**(c)** Determine the required input torque to drive this mechanism in this position and under the conditions described above.

**(d)** Find the instantaneous value of the shaking forces and shaking moment transmitted to ground.

**5.8.** The slider-crank mechanism of Fig. P5.8 is to be analyzed to determine the effect of the inertia of the connecting rod (link 3). The velocity and acceleration diagrams are shown in the figure and the magnitude of $\mathbf{V}_A$ is given.

**(a)** Find the inertia force $\mathbf{F}_{O3}$ of the coupler link.

**(b)** Find the directions and magnitudes of the pin forces at $A$ and $B$.

**(c)** Using the virtual work method, determine the required input torque to drive this mechanism in this position and under the conditions described above.

**(d)** Find the instantaneous value of the shaking forces and shaking moment transmitted to ground.

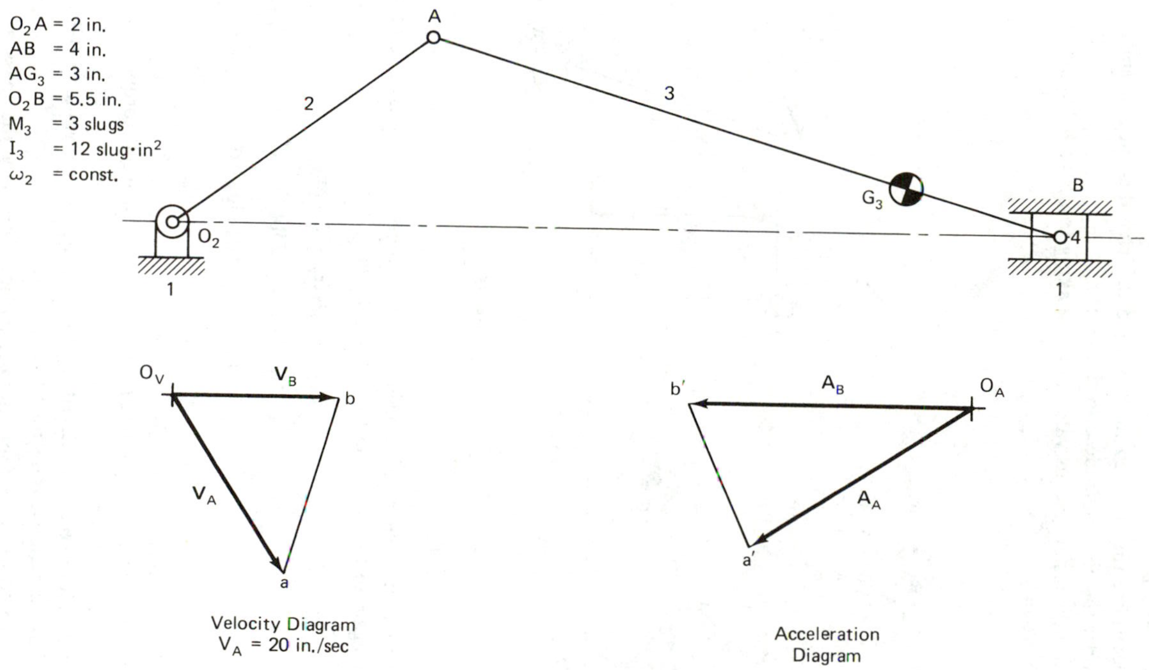

O2A = 2 in.
AB = 4 in.
AG3 = 3 in.
O2B = 5.5 in.
M3 = 3 slugs
I3 = 12 slug·in2
ω2 = const.
A
B
2
3
4
G3
O2
1
1
Ov
VB
b
VA
a
Velocity Diagram
VA = 20 in./sec
b′
AB
OA
AA
a′
Acceleration
Diagram


**Figure P5.8**

**5.9.** Determine the effect of the inertia of the coupler link of the four-bar linkage in Fig. P5.9. The pertinent data and the acceleration diagram are shown in the figure.

**(a)** Find the inertia force $\mathbf{F}_{O3}$ of the coupler link.

**(b)** Using the virtual work method, find the directions and magnitudes of the pin forces at $A$ and $B$.

**(c)** Determine the required input torque to drive this mechanism in this position and under the conditions described above.

**(d)** Find the instantaneous value of the shaking forces and shaking moment transmitted to ground.

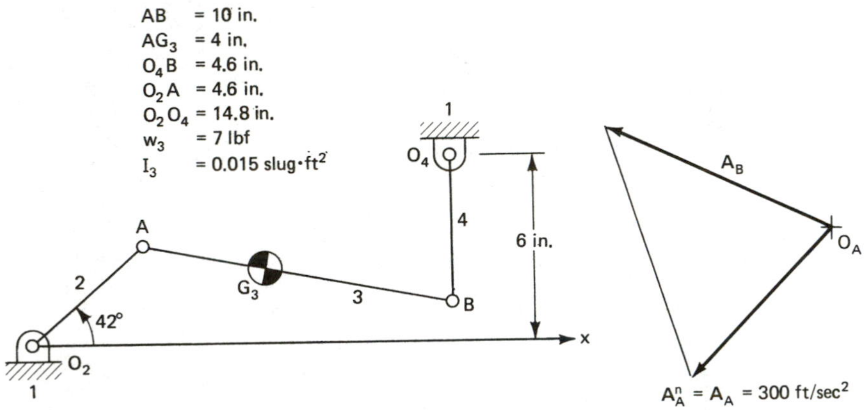


**Figure P5.9**

**5.10.** Determine the effect of considering the inertia of links 2 and 4 in Prob. 5.9. Refer again to Fig. P5.9, and add the following data:

$$O_2G_2 = O_4G_4 = 2.5 \text{ in.}$$

$$W_2 = W_4 = 5 \text{ lbf}$$

$$I_2 = I_4 = 0.01 \text{ slug}\cdot\text{ft}^2$$

**(a)** Find the inertia forces $\mathbf{F}_{O2}$ and $\mathbf{F}_{O4}$.

**(b)** With all inertias considered, find the required input torque using the virtual work method.

**(c)** With all inertias considered, find the instantaneous value of the shaking forces and shaking moment transmitted to ground.

**5.11.** Figure P5.10 shows a four-bar mechanism with only the mass and inertia of link 3 considered pertinent. For the given data, determine $\theta_3$ and $\theta_4$ and then:

**(a)** Find the inertia force $\mathbf{F}_{O3}$ of the coupler link.

**(b)** Using the virtual work method, find the directions and magnitudes of the pin forces at $A$ and $B$.

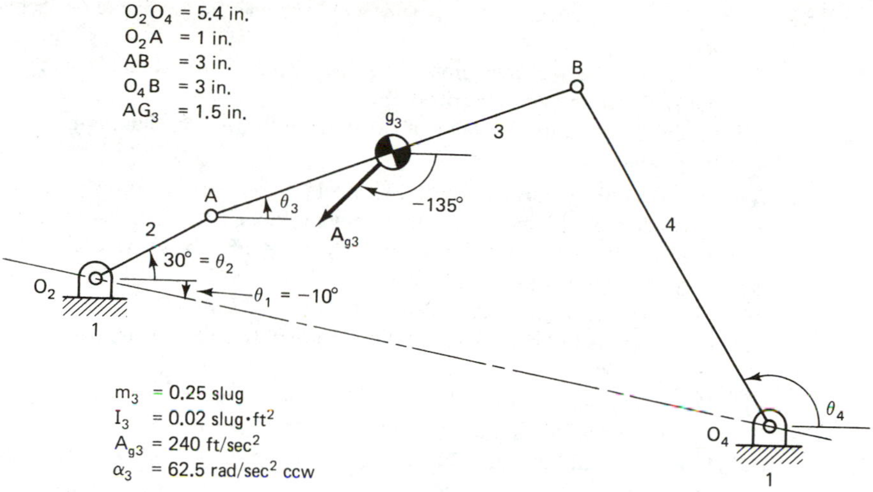


**Figure P5.10**

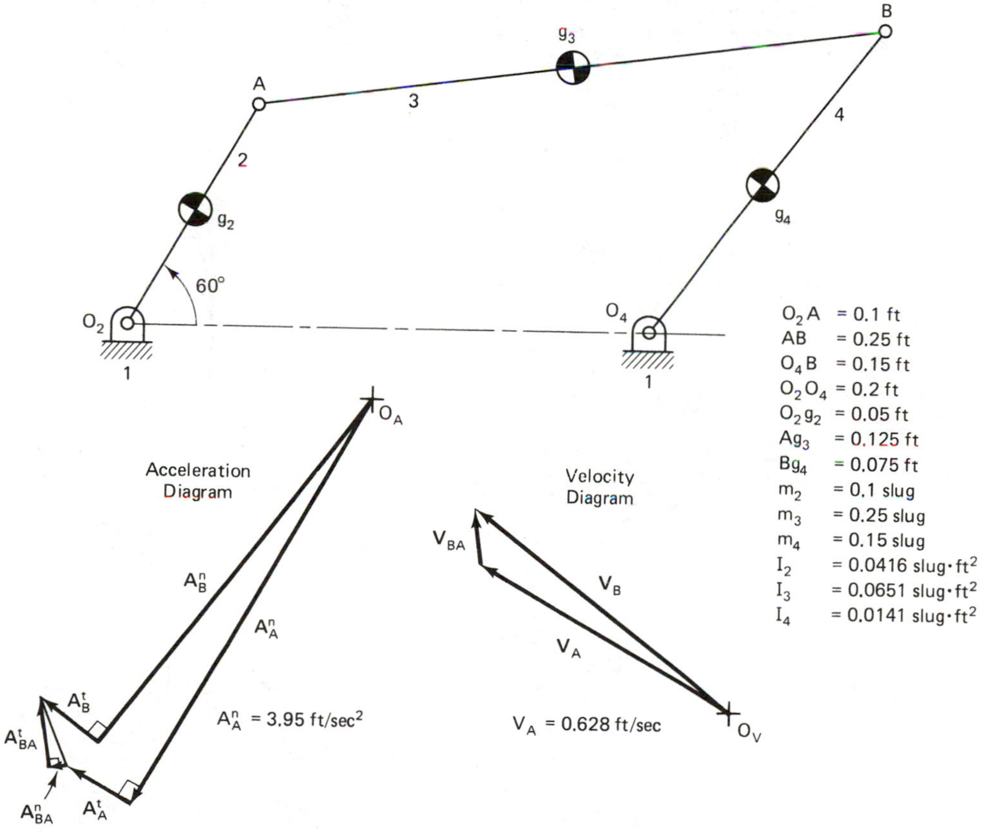


**Figure P5.11**

(c) Determine the required input torque to drive this mechanism in this position and under the conditions described above.

(d) Find the instantaneous value of the shaking forces and shaking moment transmitted to ground.

**5.12.** The four-bar linkage mechanism in Fig. P5.11 is to be analyzed instantaneously to determine the effect of the inertia of three moving links. If the input angular velocity and acceleration are $\omega_2 = 2\pi$ rad/sec ccw and $\alpha_2 = 2\pi$ rad/sec ccw and the velocity and acceleration diagrams are given, then:

(a) Find the inertia force $\mathbf{F}_{O2}$, $\mathbf{F}_{O3}$, and $\mathbf{F}_{O4}$.

(b) Using the virtual work method, find the directions and magnitudes of the pin forces at $A$ and $B$.

(c) Determine the required input torque to drive this mechanism in this position and under the conditions described above.

(d) Find the instantaneous value of the shaking forces and shaking moment transmitted to ground.

**5.13.** Figure P5.12 shows a four-bar mechanism that has a counterweight on link 2 so that the center of mass of link 2 is at $O_2$. The radius of gyration of link 3 about $g_3 = 4.5$ in. while the radius of gyration of link 4 about $O_4$ is 3.9 in. The input angular velocity is constant at $\omega_2 = 40$ rad/sec ccw.

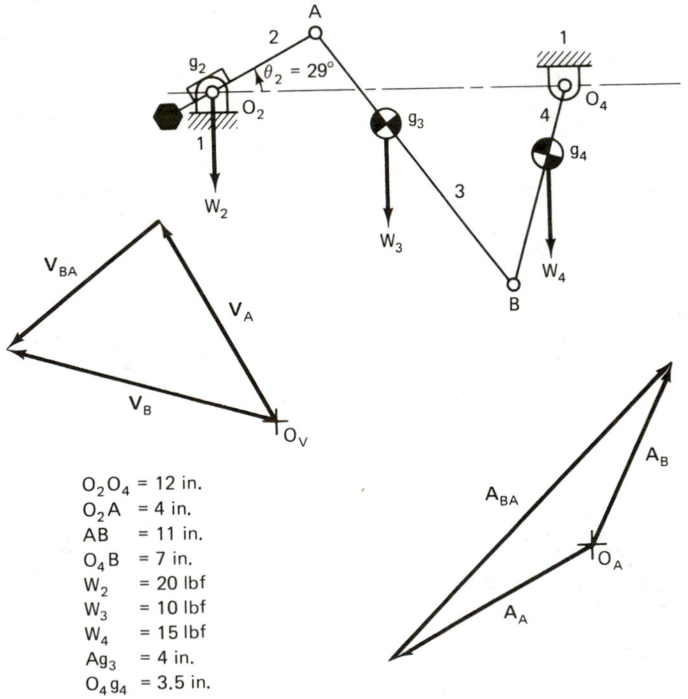


**Figure P5.12**

(a) Find the inertia forces $\mathbf{F}_{O2}$, $\mathbf{F}_{O3}$, and $\mathbf{F}_{O4}$.

(b) Using the virtual work method, find the directions and magnitudes of the pin forces at $A$ and $B$.

(c) Determine the required input torque to drive this mechanism in this position and under the conditions described above.

(d) Find the instantaneous value of the shaking forces and shaking moment transmitted to ground.

**5.14.** A slider-crank mechanism shown in Fig. P5.13 is the main drive of a compressor. The input crank is partially counterbalanced with a weight so that $g_2$ is $\frac{1}{2}$ in. from $O_2$ toward the balancing weight. The radius of gyration of link 3 about $g_3$ is 4.1 in. The only external force is the given gas force $\mathbf{F}_4$ acting on the face of the piston. All four inertia forces are also given. (The mechanism is moving in the horizontal plane without friction so the link weights have no effect.)

(a) Using virtual work, find the directions and magnitudes of the pin forces at $A$ and $B$.

(b) Determine the required input torque to drive this mechanism in this position and under the conditions described above.

(c) Find the instantaneous value of the shaking forces and shaking moment transmitted to ground.

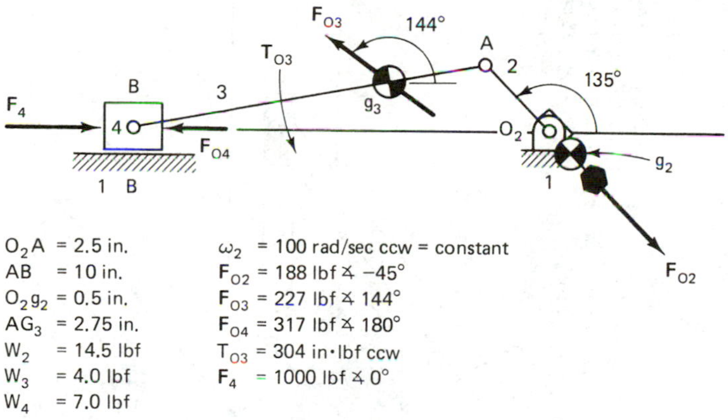


**Figure P5.13**

**5.15.** Same question as Prob. 5.14 but add friction between links 1 and 4 with $\mu = 0.1$. (*Hint:* Sometimes the principle of superposition does not yield correct results when friction is present. Recall that superposition is applicable only when dealing with linear systems. Therefore, use virtual work to determine the increase in pin forces and in $T_2$.) Find $\mathbf{V}_B$ by the instant center method. Note that arg $\mathbf{V}_B = 180°$.

**5.16.** A scoop mechanism is shown in Fig. P5.14. We assume that the inertia of link 3 has little effect in the force analysis. Link 2 has mass of 5 kg and mass moment of inertia about $g_2$ of 24 kg·cm². Both the direction of the input force $\mathbf{F}_{in}$ and its point of application are known.

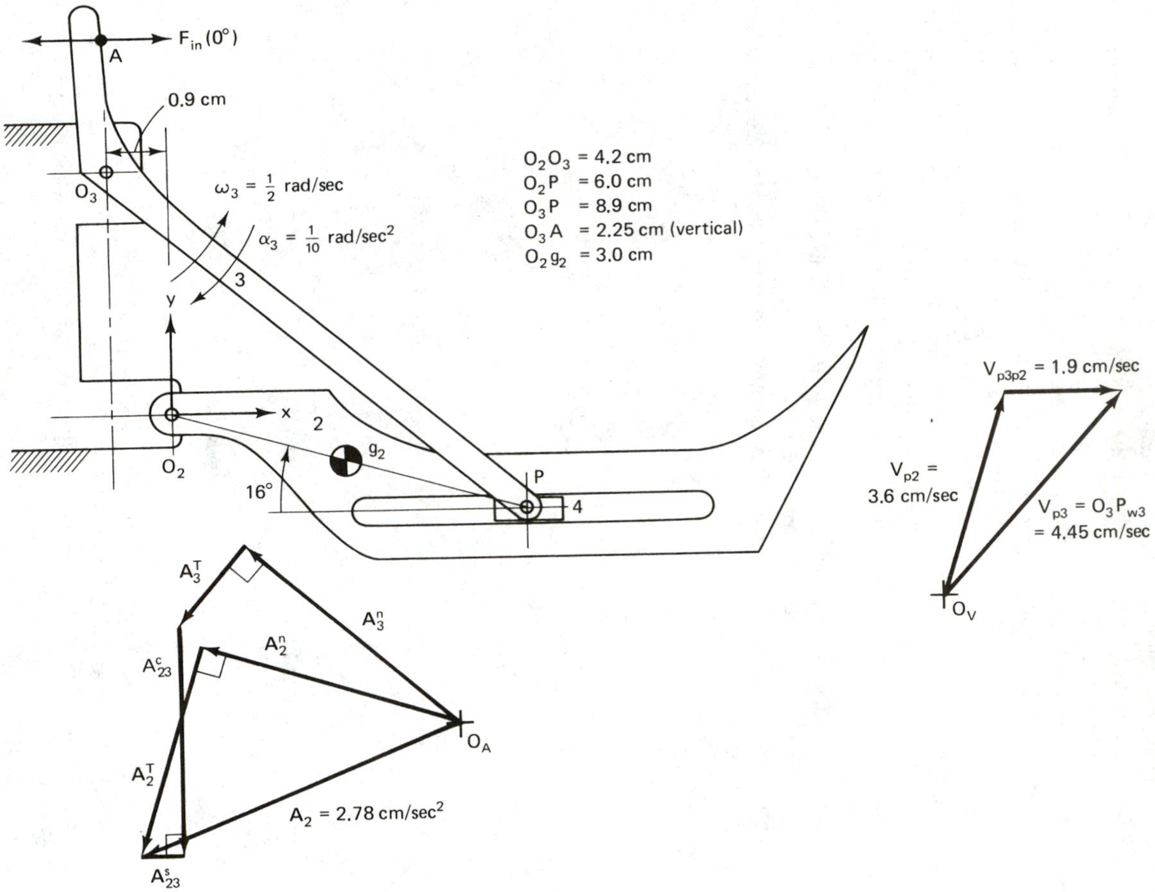

$F_{in}$ (0°)
A
0.9 cm
$O_3$
$\omega_3 = \frac{1}{2}$ rad/sec
$\alpha_3 = \frac{1}{10}$ rad/sec$^2$
3
y
x
2
$O_2$
$g_2$
16°
P
4
$O_2O_3$ = 4.2 cm
$O_2P$ = 6.0 cm
$O_3P$ = 8.9 cm
$O_3A$ = 2.25 cm (vertical)
$O_2g_2$ = 3.0 cm
$V_{p3p2}$ = 1.9 cm/sec
$V_{p2}$ = 3.6 cm/sec
$V_{p3} = O_3P_{w3}$ = 4.45 cm/sec
$O_V$
$A_3^T$
$A_3^n$
$A_2^n$
$A_{23}^c$
$A_2^T$
$A_{23}^s$
$O_A$
$A_2$ = 2.78 cm/sec$^2$


Figure P5.14

**(a)** Find the inertia force $\mathbf{F}_{O2}$ of the output link.

**(b)** Using the virtual work method, find the directions and magnitudes of the pin forces at $A$ and $B$.

**(c)** Determine the required input torque to drive this mechanism in this position and under the conditions described above.

**(d)** Find the instantaneous value of the shaking forces and shaking moment transmitted to ground.

**5.17.** Same questions as Prob. 5.16 but in this case a load of dirt is in the scoop with center of mass 6 cm from $O_2$ on a line 0° to the right of $O_2$. The mass of the load is 10 kg and the mass moment of the load inertia about the center of gravity of the load is 20 kg·cm². Also include in the analysis the mass and inertia of link 3: $O_3g_3 = 4$ cm on the centerline $O_3P$, $m_3 = 2$ kg, and $I_3 = 15$ kg·cm². How do the resulting pin forces and input force compare with Prob. 5.16?

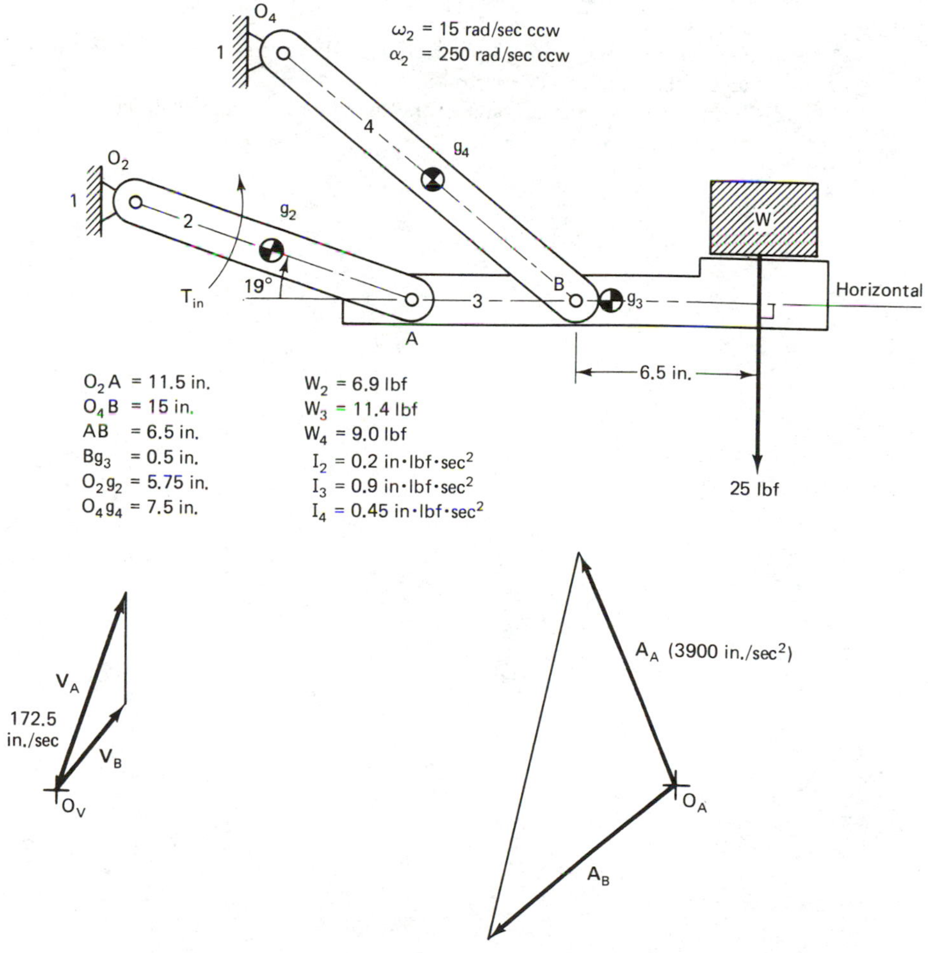


**Figure P5.15**

**5.18.** A lift mechanism is shown in Fig. P5.15. For this problem assume that link 3 only has significant mass and inertia. (Disregard the mass and inertia of links 2 and 4 and the load $W$.)

**(a)** Find the inertia force $\mathbf{F}_{O3}$.

**(b)** Using the virtual work method, find the directions and magnitudes of the pin forces at $A$ and $B$.

**(c)** Determine the required input torque to drive this mechanism in this position and under the conditions described above.

**(d)** Find the instantaneous value of the shaking forces and shaking moment transmitted to ground.

**5.19.** Referring to Prob. 5.18, include the effects of the mass and inertia of links 2 and 4 as well as the load $W$.

**(a)** Find the inertia force $\mathbf{F}_{O2}$ and $\mathbf{F}_{O4}$.

**(b)** Using the virtual work method, find the directions and magnitudes of the pin forces at $A$ and $B$.

**(c)** Determine the required input torque to drive this mechanism in this position and under the conditions described above.

**(d)** Find the instantaneous value of the shaking forces and shaking moment transmitted to ground.

# 6 Spatial Mechanisms—With an Introduction to Robotics

## 6.1 INTRODUCTION

The major part of this textbook has been devoted to linkages and other mechanisms that move in two dimensions. Although most mechanical linkages have such planar motion, there are many cases where three-dimensional (or spatial) movement is required. Possible link connection types which allow movement in three-dimensional space can now be expanded from those introduced in Table 1.2 (see Table 6.1 for the commonly used joint pairs). The *revolute* and *prismatic* (*slider*) joints are quite familiar from planar linkages. They both allow a single degree of freedom of motion between the links they connect. Keep in mind, however, in spatial mechanisms, that the axes of these joints need not be parallel or perpendicular to the axes of other joints. Thus general spatial motions may be obtained with these joints.

Another single-degree-of-freedom joint, the *helix* (*or screw*) *pair,* is introduced. Notice that there is a linear relationship between the axial translation and the angle of rotation of the screw relative to the nut. The *cylindrical pair* has no coupling between the sliding and rotational positions so that this pair permits two degrees of freedom of relative motion. The *spherical* and *plane joints* allow three degrees of freedom of relative motion—three rotations for the spherical and two translations and one rotation for the planar joint.

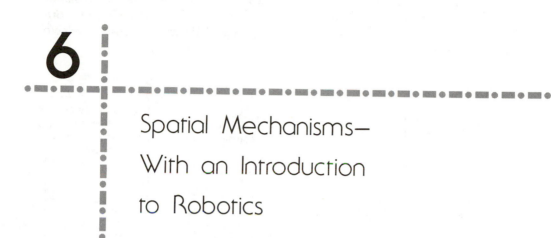

These joint pairs and three-dimensional links may be combined in countless combinations to yield spatial mechanisms. Some practical applications of space mechanisms are shown in Figs. 6.1 to 6.15. Figures 6.1 to 6.7 and 6.13 to 6.15 are examples of three-dimensional mechanisms used for a variety of purposes. A sample of three-dimensional mechanisms for coupling together misaligned shafts are shown in Figs. 6.8 to 6.12. (Reference 132 covers a general theory for constant velocity shaft couplings.)

**TABLE 6.1**

| Name of pair | Geometric form | Schematic representations | Relative degrees of freedom between elements of pair |
|---|---|---|---|
| Revolute ($R$) | 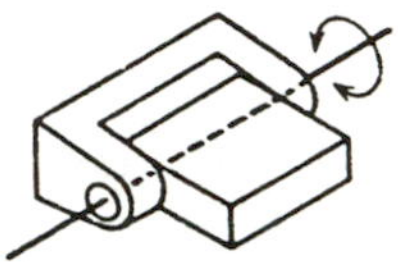 | 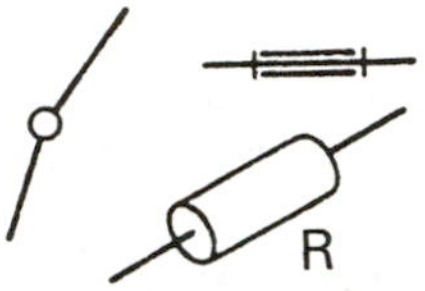  | 1 |
| Prismatic ($P$) (slide) | 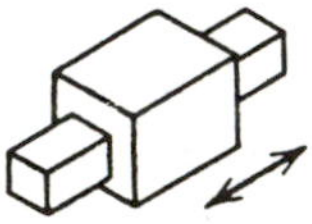 | 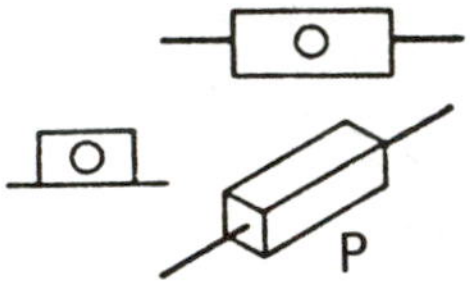  | 1 |
| Helical ($H$) (screw) | 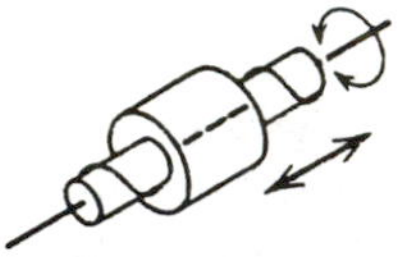 | 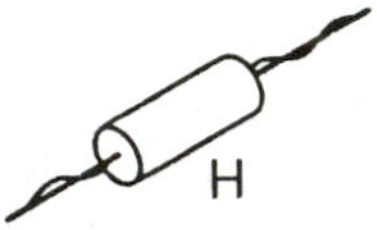  | 1 |
| Cylinder ($C$) | 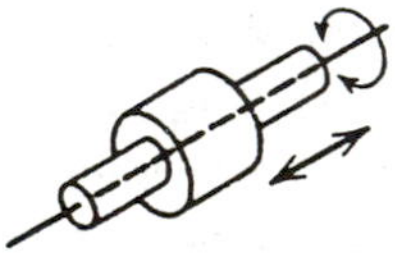 | 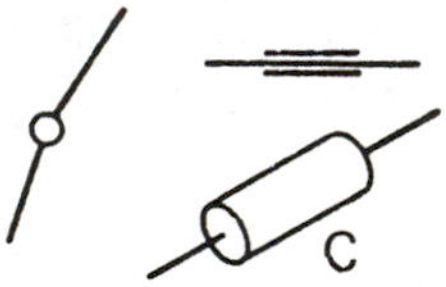  | 2 |
| Sphere ($S$) |  | 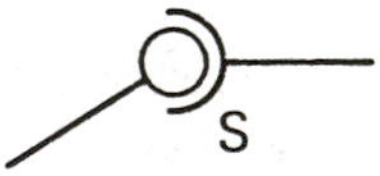  | 3 |
| Plane ($P_L$) | 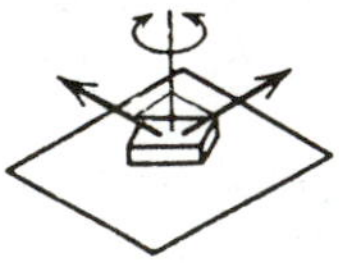 | 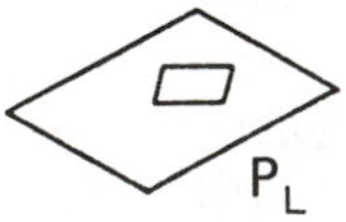  | 3 |

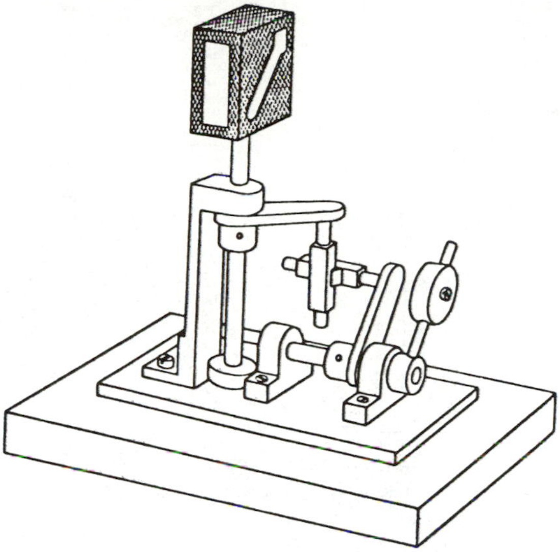

**Figure 6.1** Railway signal mechanism. (From ref. 62.)

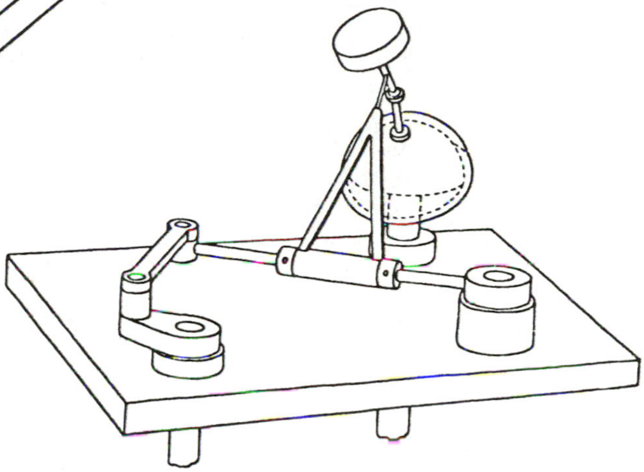

**Figure 6.2** Lens polishing machine. (From ref. 62.)

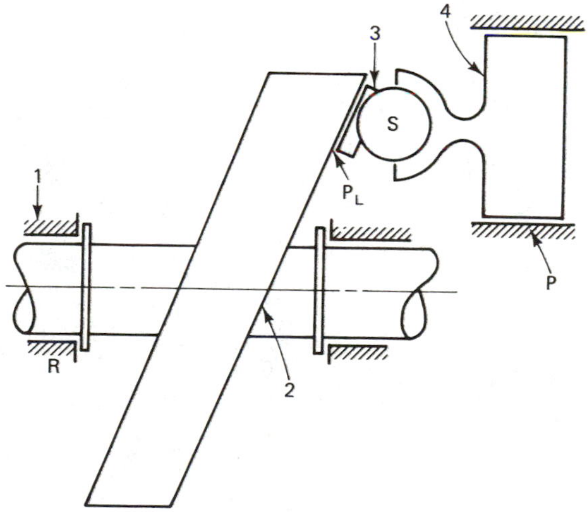


**Figure 6.3** Swash plate drive (part of a hydraulic pump and/or motor. (From ref. 105.)

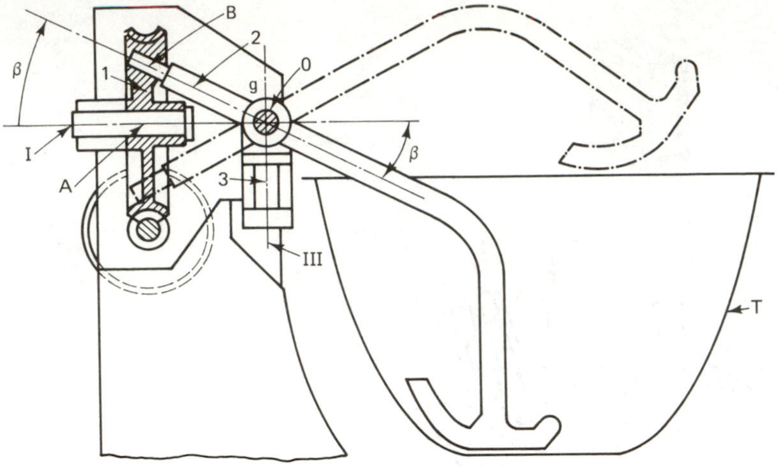


**Figure 6.4** Four-link dough kneading mechanism based on Hook's joint. The wobbling motion of link 2 kneads dough in the tank. The worm drive is not considered part of the four-link mechanism. The four links are: 1. the crank (the wormgear); 2. the kneader; 3. the follower link; 4. the frame. (From ref. 46.)

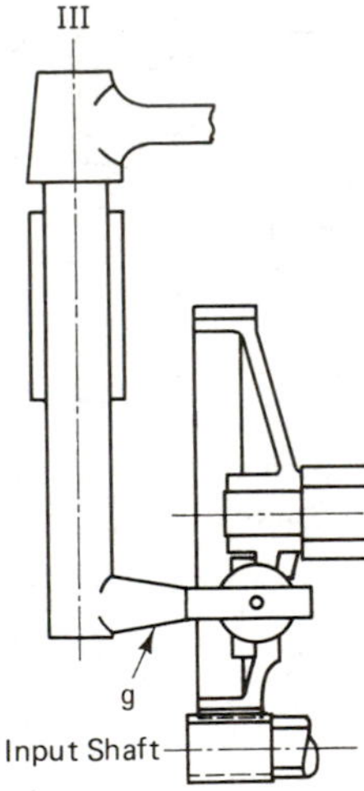


**Figure 6.5** Agitator mechanism. A practical variation of the three-dimensional crank-slider is the agitator mechanism. As input gear I rotates, link g swivels (and also lifts) shaft III. Hence the link g has both an oscillating rotary motion and a sinusoidal harmonic translation in the direction of its axis of rotation. Link g performs what is essentially a nonuniform screw motion with continuously changing lead-to-rotation ratio in each cycle. (From ref. 46.)

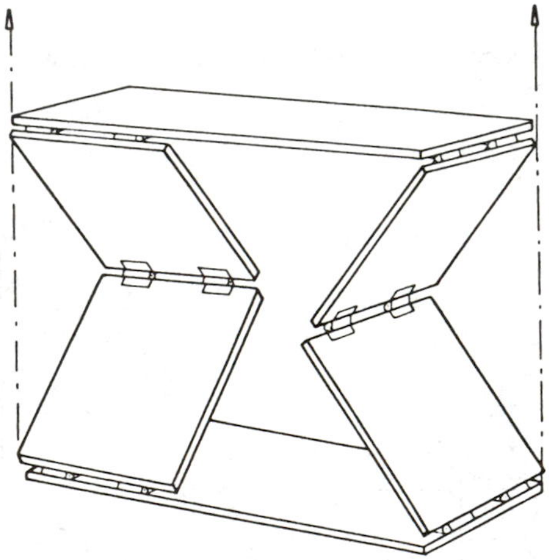

**Figure 6.6** Spatial parallel-guidance linkage of Sarrut. (From ref. 62.)

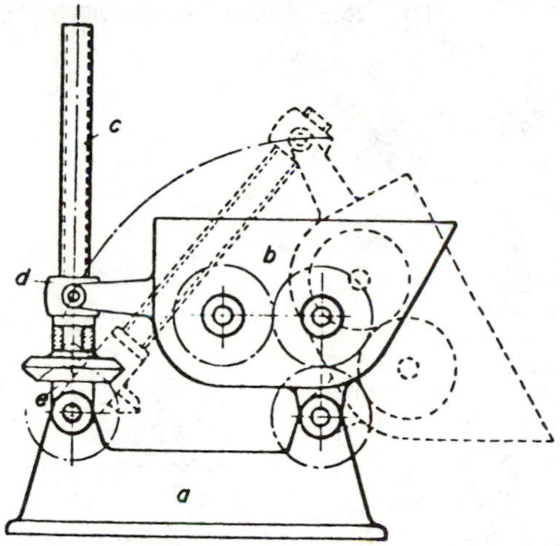


**Figure 6.7** Screw mechanism as used for unloading a mixing machine. (From ref. 62.)

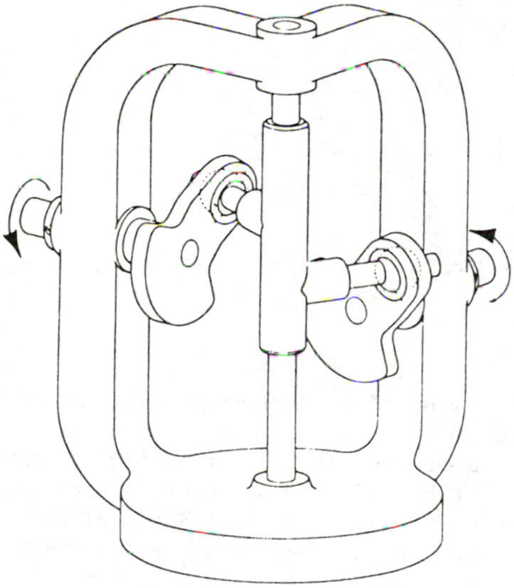

**Figure 6.8** Almond coupling. (From ref. 62.)

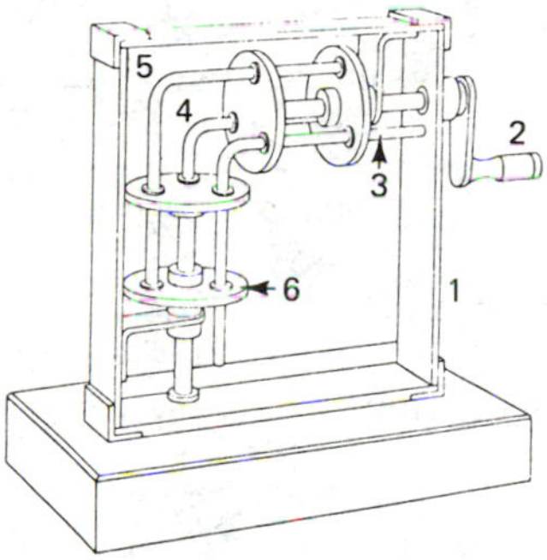


**Figure 6.9** Bent pin coupling. (From ref. 62.)

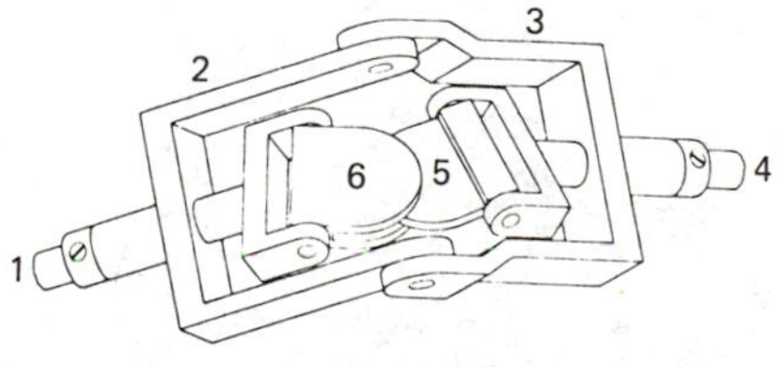


**Figure 6.10** Plate coupling. (From ref. 62.)

Fig. 6.11a. $RRP_LRR$ (Tracta)

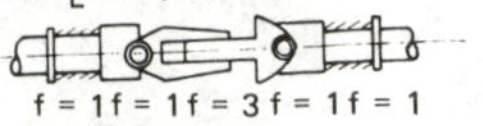


Fig. 6.11b. RRSRR (Clemens)

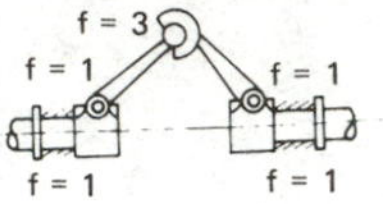


Fig. 6.11c. RPSPR (Altmann)

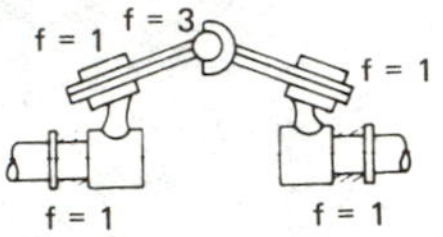


Fig. 6.11d. RCRCR (Myard)

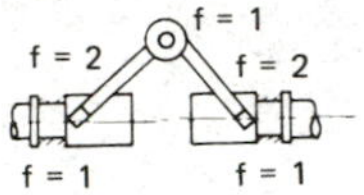


Fig. 6.11e. RCPCR

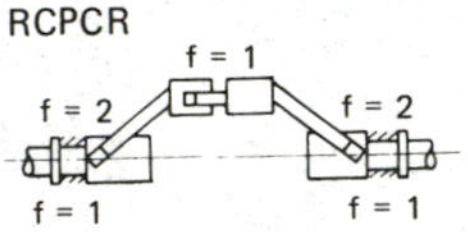


Fig. 6.11f. RRRRRRR (Myard, Voss, Wachter & Rieger)

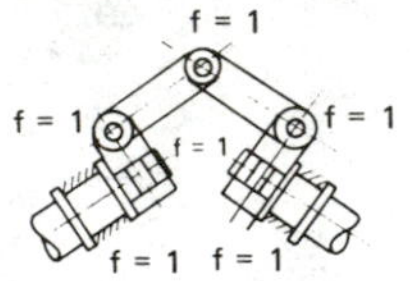


Fig. 6.11g. RRPRPRR (Derby, S.W. Industries)

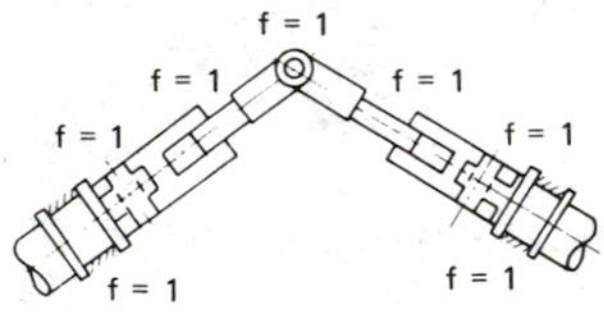


**Figure 6.11** Constant-velocity shaft couplings from ref. [105], also called *universal joints,* because they permit misalignment of shafts. (a) $RRP_LPR$ (Tracta) (see also [307]); (b) RRSRR (Clemens); (c) RPSPR (Altmann); (d) RCRCR (Myard); (e) RCPCR; (f) RRRRRRR (Myard, Voss, Wachter, & Rieger); (g) RRPRPRR (Derby, S. W. Industries).

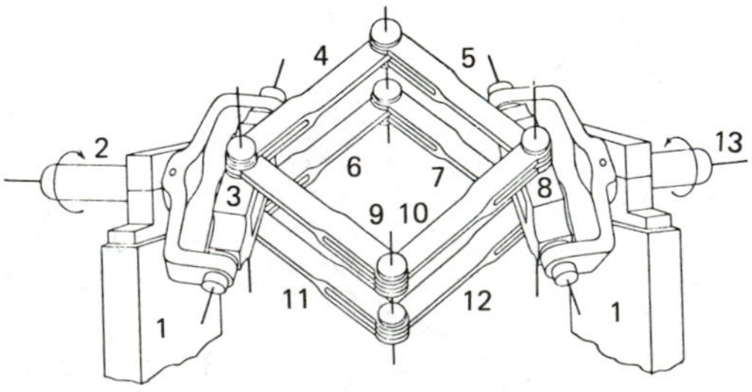


**Figure 6.12** Cross coupling of Krupp. (From ref. 62.)

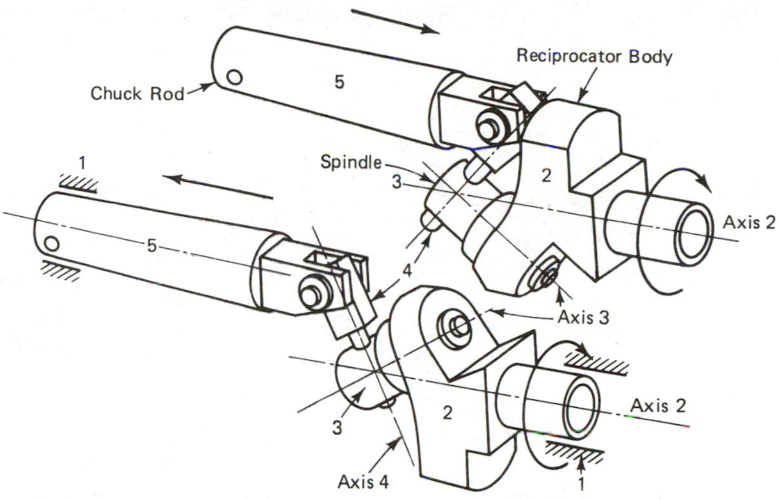


**Figure 6.13** Angled link converts rotary motion to reciprocation. Reciprocator body 2 contains a spindle 3 that is mounted on an angle to the axis of body 2. As body 2 rotates, spindle 3 rotates about its own axis, but link 5 always remains in the vertical plane. Thus chuck rod 5 is caused to reciprocate. (Used in a pneumatic saw by Thomas C. Wilson, Inc., Long Island City, N.Y.) (*From Design and Development/Scanning the Field for Ideas, September 24, 1964, p. 158.*)

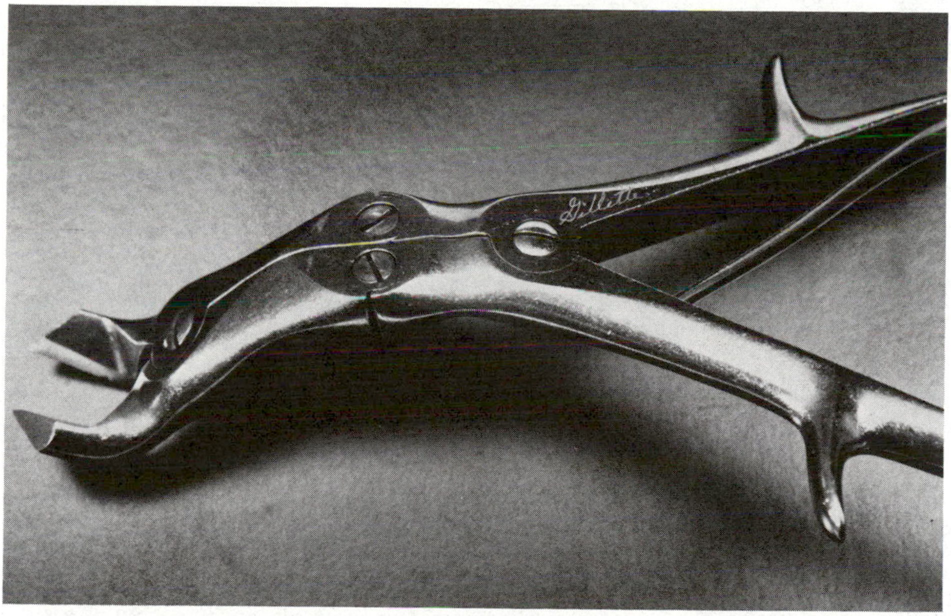

**Figure 6.14** Surgeon's bone-cutting rougeur: a spherical four-bar mechanism. All four pin axes are intersect at a single point.

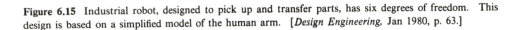

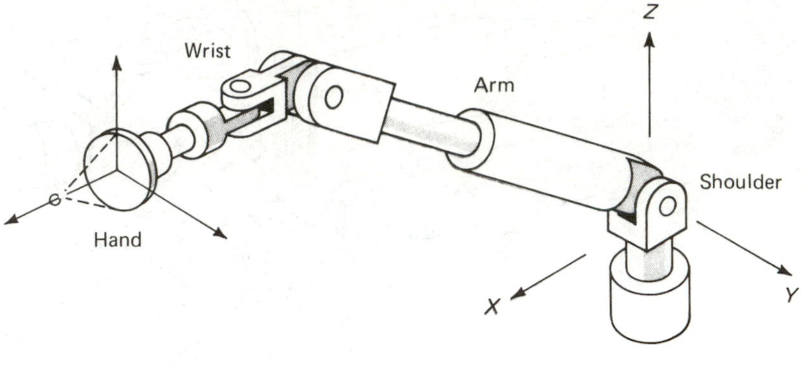


**Figure 6.15** Industrial robot, designed to pick up and transfer parts, has six degrees of freedom. This design is based on a simplified model of the human arm. [*Design Engineering*, Jan 1980, p. 63.]

Figure 6.15 shows one of many "open-loop" spatial mechanisms used as a robot or manipulator. The increasing need for higher productivity in industry as well as the complex demands on moving products and components in dangerous environments (high temperature, noise, pollution, radiation, etc.), point toward the need for a great variety of mechanical robots. However, for certain highly repetitive tasks of limited complexity, open loop robots with multiple actuators are greatly "over-qualified," and hence less economical than closed-loop single-input mechanisms. This chapter provides some of the fundamentals involved in the mechanical design of open-loop, multi-degree-of-freedom, multiple-input industrial robots, as well as closed-loop, single-input spatial mechanisms.

## Degrees of Freedom

In Chap. 1 we covered Gruebler's equation for degrees of freedom of planar mechanisms. For movement in three dimensions (including the joint pairs of Table 6.1) a different equation* is required for determining degrees of freedom [105]. Let

$F =$ degree of freedom of the mechanism

$\ell =$ number of links in the mechanism (including the fixed link; all links are considered as rigid bodies having at least two joints: if several machine parts are assembled as a rigid part, the assembly is considered as a single link)

* See also the appendix to Chap. 2.

$j$ = number of joints in the mechanism; each joint is assumed as binary (i.e., connecting two links); joints connecting more than two links will be treated as was done with Gruebler's equation (i.e., a separate joint between each two links); thus a joint connecting 3 links would count as two joints

$f_i$ = degrees of freedom of the $i$th joint, as shown in Table 6.1; this is the number of degrees of freedom of the relative motion between the connected links

$\lambda$ = degrees of freedom of the space within which the mechanism operates; for plane motion and for motion on a curved surface $\lambda = 3$ and for spatial motions $\lambda = 6$ (exceptional cases are discussed in Ref. 105)

$L_{\mathrm{IND}}$ = number of independent circuits or closed loops in the mechanism

The following degree-of-freedom equations then apply to a large class of mechanisms:

$$F = \lambda(\ell - j - 1) + \sum_{i=1}^{j} f_i \tag{6.1}$$

$$L_{\mathrm{IND}} = j - \ell + 1 \tag{6.2}$$

Combining (6.1) and (6.2), we have

$$\sum f_i = F + \lambda L_{\mathrm{IND}} \tag{6.3}$$

For example, for the plane slider-crank mechanism shown in Fig. 1.3,

$$\ell = j = 4$$

$$\sum f_i = 1 + 1 + 1 + 1 = 4, \qquad \lambda = 3$$

Hence $F = 3(4 - 4 - 1) + 4 = 1$.

In the application of these equations certain rules need to be kept in mind:

1. Joints connecting $n$ links, where $n > 2$, are called multiple joints and are counted as $(n - 1)$ binary joints.
2. Certain linkages commonly thought of as having one degree of freedom may have an $F$ value greater than 1. This can occur in spatial linkages, for example, having links with two joints of types spherical-spherical (S-S), spherical-cylindrical (S-C), and spherical-planar (S-$P_L$). Such links can have a "redundant" or "passive" freedom of rotation about the axis connecting the joints, which is independent of the motion of the mechanism as a whole (e.g., see Ex. 6.1).
3. Some highly significant mechanisms do not obey the general degree-of-freedom equations given above. These are mechanisms that depend on special dimensions or proportions for their mobility. Mechanisms with mixed plane/spatial portions (variable $\lambda$) are usually exceptional. In spatial linkages, special cases are often associated with parallel, intersecting, or perpendicular joint axes. There are no simple rules that will predict whether a mechanism obeys Eq. (6.1). Here experience is very helpful, and the presence of one or more of the above-listed special characteristics is a signal to be watchful.

**Example 6.1**

Determine the degrees of freedom of the swash plate drive of Fig. 6.3.

$$\ell = j = 4$$

$$\sum f_i = 1 + 3 + 3 + 1$$

Since $\lambda = 6$, then from Eq. (6.3), $F = 2$. One would anticipate that this mechanism has a single degree of freedom (say the rotation of link 2 if the swash plate mechanism is used as a hydraulic motor). The second degree of freedom is the rotation of link 3 about an axis through the center of the sphere and normal to the face of the swash plate. This passive degree of freedom does not interfere with the desired input–output kinematic relationship of the drive (although it certainly plays a role in the lubrication and wear of the mating surfaces).

## 6.2 TRANSFORMATIONS DESCRIBING PLANAR FINITE DISPLACEMENTS*

As a forerunner to modeling spatial displacements, various mathematical models are described for representing finite planar displacements. These models are called "transformations," because they transform the starting position coordinates of a body in planar motion into the final-position coordinates. Once these are clearly understood, their extension to three-dimensional motion readily follows, and is covered in later sections.

## 6.3 PLANAR FINITE TRANSFORMATIONS

To begin our discussion of transformations, we start with the planar finite rotation operator used in Chap. 2 for synthesis by complex numbers. Consider the two complex vectors of Fig. 6.16. They represent two finitely separated positions ($\mathbf{A}_1$ and $\mathbf{A}_2$) of a rigid link that rotates about the origin. We seek a transformation that rotates $\mathbf{A}_1$ by $\theta_{12}$ to a new position at $\mathbf{A}_2$. From Chap. 2 we know that the operator $e^{i\theta_{12}}$ will do this, but let us rederive it to find out more about this operator.

Let us assume that there is an unknown operator $\mathbf{R}_{12}$, whose product with $\mathbf{A}_1$ yields $\mathbf{A}_2$:

$$\mathbf{A}_2 = \mathbf{R}_{12}\mathbf{A}_1 \tag{6.4}$$

This may be rearranged by complex division to

$$\mathbf{R}_{12} = \frac{\mathbf{A}_2}{\mathbf{A}_1}$$

* The authors wish to acknowledge the contribution of William Dahlof in the development of parts of Secs. 6.2–6.10.

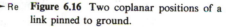

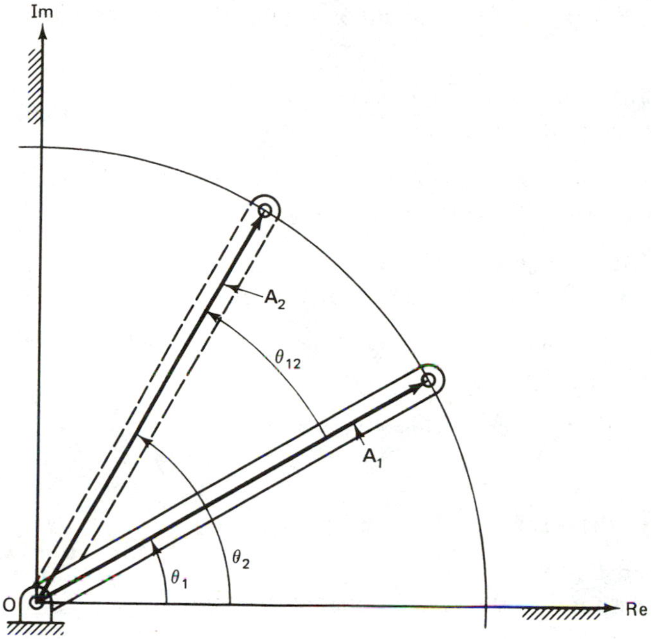


**Figure 6.16** Two coplanar positions of a link pinned to ground.

The complex vectors $\mathbf{A}_1$ and $\mathbf{A}_2$ may be written as

$$\mathbf{A}_1 = |\mathbf{A}_1| (\cos \theta_1 + i \sin \theta_1)$$

$$\mathbf{A}_2 = |\mathbf{A}_2| (\cos \theta_2 + i \sin \theta_2)$$

Substitution leads to

$$\mathbf{R}_{12} = \frac{|\mathbf{A}_2| (\cos \theta_2 + i \sin \theta_2)}{|\mathbf{A}_1| (\cos \theta_1 + i \sin \theta_1)} \tag{6.5}$$

This form of the transformation is not very useful. Normally, the initial position of a link ($\mathbf{A}_1$) is known and we wish to rotate it by a known angle $\theta_{12}$. Therefore, a form of Eq. (6.5) that is only a function of $\theta_{12}$ is desired. In Fig. 6.16 notice that only rotation is occurring and that the origins of the vectors remain at the origin $O$ of the coordinate system, while their tips describe a circular arc centered at $O$. Therefore, the length of the rigid link must remain constant, so

$$|\mathbf{A}_1| = |\mathbf{A}_2|$$

and these two terms cancel in Eq. (6.5). Next, the denominator must be rationalized by multiplying it and the numerator by the complex conjugate of the denominator:

$$\mathbf{R}_{12} = \frac{(\cos \theta_2 + i \sin \theta_2)(\cos \theta_1 - i \sin \theta_1)}{(\cos \theta_1 + i \sin \theta_1)(\cos \theta_1 - i \sin \theta_1)}$$

Adopting the shorthand notation ($c\alpha = \cos\alpha$, $s\alpha = \sin\alpha$), expanding, and regrouping terms yields

$$\mathbf{R}_{12} = \frac{(c\theta_2\, c\theta_1 + s\theta_2\, s\theta_1) + i(c\theta_1\, s\theta_2 - c\theta_2\, s\theta_1)}{(c^2\theta_1 + s^2\theta_1) + i(c\theta_1\, s\theta_1 - c\theta_1\, s\theta_1)} \tag{6.6}$$

By inspection you can see that the denominator simplifies to unity. To further simplify Eq. (6.6), we must first complicate it by using $\theta_2 = \theta_1 + \theta_{12}$ to eliminate $\theta_2$, yielding

$$\begin{aligned}\mathbf{R}_{12} = {} & [c(\theta_1 + \theta_{12})\, c\theta_1 + s(\theta_1 + \theta_{12})\, s\theta_1] \\ & + [s(\theta_1 + \theta_{12})\, c\theta_1 - c(\theta_1 + \theta_{12})\, s\theta_1]i\end{aligned}$$

which is a function containing only $\theta_1$ and $\theta_{12}$. Finally, the $\theta_1$ terms may be eliminated by employing the trigonometric identities:

$$\cos(\alpha - \beta) = \cos\alpha \cos\beta + \sin\alpha \sin\beta$$

and

$$\sin(\alpha - \beta) = \sin\alpha \cos\beta - \cos\alpha \sin\beta$$

and letting $\alpha = (\theta_1 + \theta_{12})$ and $\beta = \theta_1$.
This reduces to the well-known form of the planar rotation operator:

$$\mathbf{R}_{12} = \cos\theta_{12} + i \sin\theta_{12} \tag{6.7}$$

Using Euler's formula, we have

$$e^{i\alpha} = \cos\alpha + i\sin\alpha$$

Equation (6.7) becomes

$$\mathbf{R}_{12} = e^{i\theta_{12}} \tag{6.8}$$

which is the form of the finite rotation operator used in Chap. 2. So, using the fact that the link (represented by $\mathbf{A}_1$ and $\mathbf{A}_2$) is a rigid link rotating about the origin (i.e., $|A_1| = |A_2|$) and that the angle of the link in its second position is just the angle of the first position plus the rotation angle ($\theta_2 = \theta_1 + \theta_{12}$), we have shown that the rotation operator is strictly a function of the angle of rotation ($\theta_{12}$). That is, the transformation describes the displacement and is independent of the actual positions of the link. It is important in our understanding of mechanism kinematics that we think of this and other transformations presented later as mathematical means of describing motion.* This is one role of these operators. Another role we will see is that they will transform the location vector of a point expressed in one coordinate system into the location vector of the same point expressed in another coordinate system. This second role is important when we know the position vector of a tracer point $P$ in the coordinate system attached to the moving tracer link and wish to find the position vector of $P$ expressed in the coordinate system of the fixed frame of reference.

* This applies when all positions and vectors are referred to one and the same coordinate system.

## 6.4 IDENTITY TRANSFORMATION

Let us now switch to matrix notation and take a look at several more planar operators. Consider the 2 × 2 identity matrix. When used as a transformation, its effect is to leave the vector unchanged.

$$\begin{bmatrix} 1 & 0 \\ 0 & 1 \end{bmatrix} \begin{Bmatrix} x \\ y \end{Bmatrix} = \begin{Bmatrix} 1x + 0y \\ 0x + 1y \end{Bmatrix} = \begin{Bmatrix} x \\ y \end{Bmatrix} \tag{6.9}$$

Although this does not appear very useful at this stage, it is worth a brief examination. Note that the second entry in the first row of the transformation represents the contribution of $y$ to the new value of $x$. Similarly, the first entry in the second row represents the contribution of $x$ to the new value of $y$. Thus with a 2 × 2 matrix operator, both final values are functions of each initial value. The usefulness of the identity transformation will become apparent later when basic transformations are used as building blocks to form more complex ones.

## 6.5 PLANAR MATRIX OPERATOR FOR FINITE ROTATION

Most of the mathematics used in this chapter is based on matrix techniques, so we seek now a planar matrix rotation operator. After deriving the complex rotation operator, we found that a finite rotation operator does not depend on either the initial or final position of the vector, only on the angle of rotation. But the coordinates* of the final vector are each a function of both the real and imaginary coordinates of the initial vector. In matrix notation, it follows logically that there must be a similar arrangement. We saw with the 2 × 2 identity matrix that the 2 × 2 matrix form of a transformation indeed satisfies this arrangement. Thus, we begin by assuming a 2 × 2 matrix form.

Let $R(\theta_{12})$ be a 2 × 2 matrix, whose product with $\mathbf{A}_1$ yields $\mathbf{A}_2$:

$$\begin{Bmatrix} x_2 \\ y_2 \end{Bmatrix} = \begin{bmatrix} r_{11} & r_{12} \\ r_{21} & r_{22} \end{bmatrix} \begin{Bmatrix} x_1 \\ y_1 \end{Bmatrix} \tag{6.10}$$

where $x_1$ and $y_1$ are the coordinates of $\mathbf{A}_1$ and $x_2$ and $y_2$ are those of $\mathbf{A}_2$, both expressed in the same coordinate system. Without going through a rigorous derivation of this transformation, we may determine the four components of $R$ by direct comparison with the complex rotation operator. Expanding Eq. (6.10) by way of matrix multiplication yields

$$\begin{aligned} x_2 &= r_{11}x_1 + r_{12}y_1 \\ y_2 &= r_{21}x_1 + r_{22}y_1 \end{aligned} \tag{6.11}$$

* Coordinates of a vector are the scalar values of its components referred to the axes of a coordinate system. For example (Fig. 6.17), the coordinates of $\mathbf{z}$ are $z_x = -2$ and $z_y = 2.5$.

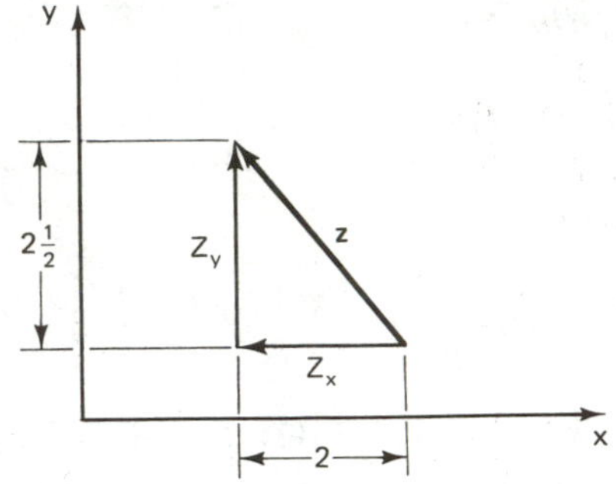


**Figure 6.17** Vector **z** in the x-y plane.

Expanding similar vectors in complex form and using the operator of Eq. (6.7) yields

$$\text{Real}_2 + i\ \text{Imaginary}_2 = \mathcal{R}e_2 + i\ \mathcal{I}m_2 = (\cos\theta_{12} + i\sin\theta_{12})(\mathcal{R}e_1 + i\ \mathcal{I}m_1)$$

which when expanded yields

$$\begin{aligned}\mathcal{R}e_2 &= \cos\theta_{12}\ \mathcal{R}e_1 + i^2 \sin\theta_{12}\ \mathcal{I}m_1 \\ \mathcal{I}m_2 &= \sin\theta_{12}\ \mathcal{R}e_1 + \cos\theta_{12}\ \mathcal{I}m_1\end{aligned} \tag{6.12}$$

Comparison of Eqs. (6.11) and (6.12) yields

$$[R(\theta_{12})] = \begin{bmatrix}\cos\theta_{12} & -\sin\theta_{12} \\ \sin\theta_{12} & \cos\theta_{12}\end{bmatrix} \tag{6.13}$$

We can verify this and look at the contribution of each entry of $R(\theta_{12})$ by example.

**Example 6.2**

Derive the 2 × 2 matrix rotation operator for the planar rotation of the link in Fig. 6.16, letting $\theta_1 = 30°$ and $\theta_2 = 60°$.

**Solution** The initial position of the link is represented by vector $\mathbf{A}_1 = 0.866 + 0.5i$ while the final position is $\mathbf{A}_2 = 0.5 + 0.866i$. Since the angle of rotation is $\theta_{12} = 30°$, the 2 × 2 matrix is

$$[R(\theta_{12})] = \begin{bmatrix}\cos 30 & -\sin 30 \\ \sin 30 & \cos 30\end{bmatrix} = \begin{bmatrix}0.866 & -0.5 \\ 0.5 & 0.866\end{bmatrix}$$

The reader is asked to verify Eq. (6.10) for this example.

## 6.6 HOMOGENEOUS COORDINATES AND FINITE PLANAR TRANSLATION

If all finite displacements are referred to one and the same coordinate system, we need to distinguish five kinematically possible displacement sequences that a rigid body may go through. There is the trivial case of no motion. There is pure rotation and pure translation. There is a rotation followed by a translation, and finally, a translation followed by a rotation. We have found mathematical operators that de-

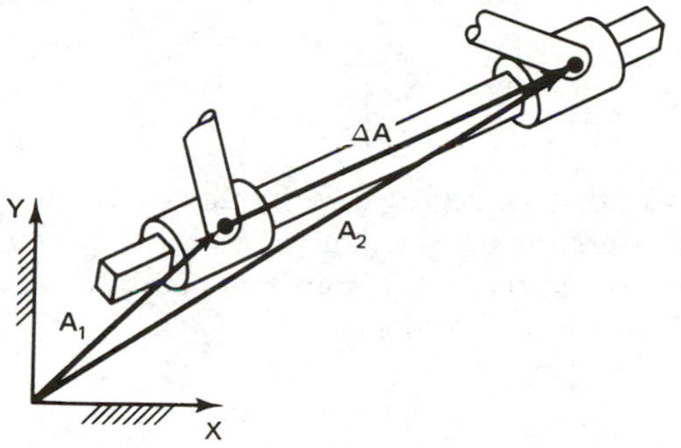


**Figure 6.18** Pure translation.

scribe the first two motions (the 2 × 2 identity matrix and the finite rotation matrix). In this subsection we seek a transformation that describes pure translation. These transformations (also called operators) will be combined in the next section to form a single operator for each of the remaining two displacement sequences.

When a particle undergoes pure translation, the vector coordinates representing its location in a coordinate system will change. Consider the slider pivot of Fig. 6.18. It has undergone a pure translation (represented by $\Delta\mathbf{A}$) from its initial position at $\mathbf{A}_1$ to a final postion at $\mathbf{A}_2$. As before, we seek a transformation $T$ whose product with $\mathbf{A}_1$ yields $\mathbf{A}_2$:

$$\mathbf{A}_2 = [T]\mathbf{A}_1 \tag{6.14}$$

Using vector addition, it should be obvious that

$$\mathbf{A}_2 = \mathbf{A}_1 + \Delta\mathbf{A}$$

or

$$x_2 = x_1 + \Delta x$$

and

$$y_2 = y_1 + \Delta y$$

One way that this translation can be represented is by using the 2 × 2 matrix with an additional column tacked on (representing $\Delta\mathbf{A}$) and converting $\mathbf{A}_1$ from a 2 × 1 column vector to a 3 × 1 homogeneous column vector,* which we will designate by $\{A_1\}$:

$$\begin{Bmatrix} x_2 \\ y_2 \end{Bmatrix} = \begin{bmatrix} 1 & 0 & \Delta x \\ 0 & 1 & \Delta y \end{bmatrix} \begin{Bmatrix} x_1 \\ y_1 \\ 1 \end{Bmatrix} \qquad \text{or} \qquad \mathbf{A}_2 = [T]\{A_1\}$$

* In this case, the 3 × 1 homogeneous representation of the 2 × 1 matrix notation may be thought of as a means of including translation (or vector addition) conveniently in a transformation. In general, an $n$-dimensional space is represented by an $n + 1$ set of homogeneous coordinates. Other computer applications of homogeneous coordinates such as scaling may be found in Refs. 195 and 220.

Multiplying this out yields

$$\begin{Bmatrix} x_2 \\ y_2 \end{Bmatrix} = \begin{Bmatrix} x_1 + 0y_1 + \Delta x \\ 0x_1 + y_1 + \Delta y \end{Bmatrix}$$

which is what we wanted. But note that $\mathbf{A}_2$ represented by its coordinates $x_2$ and $y_2$ only is a $2 \times 1$ column matrix rather than a $3 \times 1$ homogeneous column matrix like $\{A_1\}$. To remedy this, we may expand the $2 \times 3$ translation operator to a $3 \times 3$ matrix, keeping everything in homogeneous coordinates:

$$[T] = \begin{bmatrix} 1 & 0 & \Delta x \\ 0 & 1 & \Delta y \\ 0 & 0 & 1 \end{bmatrix} \tag{6.15}$$

Most of the methods presented in this chapter use this homogeneous coordinate notation, including a 1 as a third coordinate of a planar vector, or as a fourth coordinate of a spatial vector.

## 6.7 CONCATENATION OF FINITE DISPLACEMENTS

Suppose that a pivot or a particle goes through a "rotation" describing a circular path around the origin of a coordinate system and then a straight-line translation, both displacements being expressed in one and the same coordinate system. Can a single operator describe this motion? Consider the links of Fig. 6.19. The slider 2 on link 1, initially located by $\mathbf{A}_1$ with respect to a fixed coordinate system attached to the center of rotation of link 1, rotates and translates to its final position at $\mathbf{A}_2$. Since positional kinematics is concerned only with the initial and final positions,

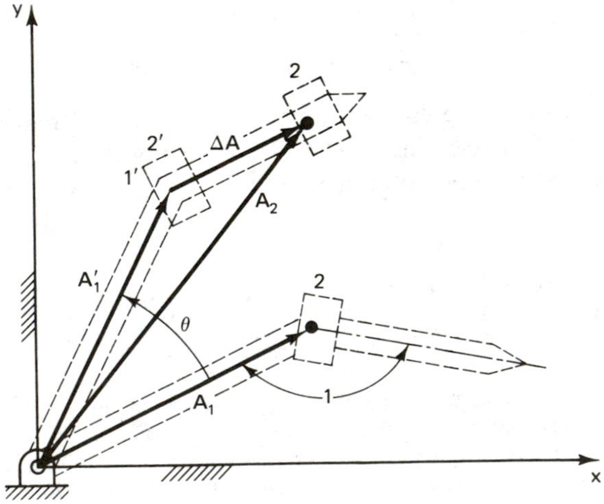


**Figure 6.19** Rotation followed by translation.

this motion may be considered as a rotation to $\mathbf{A}'_1$, followed by a translation to $\mathbf{A}_2$ by $\Delta\mathbf{A}$. Mathematically, we could represent this in two steps by

$$\{A'_1\} = [R]\{A_1\}$$

$$\{A_2\} = [T]\{A'_1\}$$

Or we could combine these two equations, yielding

$$\{A_2\} = [T][R]\{A_1\}$$

If both the translation and the rotation operators are expressed as $3 \times 3$ matrices, an operation called *concatenation* may be performed. Concatenation is basically a matrix operation in which two or more transformations are combined by premultiplying the first operator by the second, the product of these two by the third, and so on, to reduce the number of operators to a single operator. If we expand the $2 \times 2$ rotation operator of Eq. (6.13) to a $3 \times 3$ matrix by adding four zeros and a 1 (the need for the 1 will be apparent soon), a planar general displacement operator $[D]$ may be found by matrix multiplication:

$$\begin{aligned}[D] &= \begin{bmatrix} 1 & 0 & \Delta x \\ 0 & 1 & \Delta y \\ 0 & 0 & 1 \end{bmatrix} \begin{bmatrix} \cos\theta & -\sin\theta & 0 \\ \sin\theta & \cos\theta & 0 \\ 0 & 0 & 1 \end{bmatrix} \\ &= \begin{bmatrix} \cos\theta & -\sin\theta & \Delta x \\ \sin\theta & \cos\theta & \Delta y \\ 0 & 0 & 1 \end{bmatrix}\end{aligned} \tag{6.16}$$

Thus, for the case of planar motion consisting of a rotation followed by a translation, both expressed in the same coordinate system, the final position of a vector representing a particle or a pivot may be found using a single $3 \times 3$ matrix operator when both the initial and final positions are expressed in homogeneous coordinates.

Before verifying this operator with an example, two points are worth noting. First, note that if $\theta = 0$ (no rotation), $\cos\theta = 1$, $\pm = \sin\theta = 0$, $\{D\}$ reduces to the translation operator. Second, if $\Delta x = 0$ and $\Delta y = 0$ (no translation), $\{D\}$ reduces to the rotation operator. If there is no rotation ($\theta = 0$) and no translation ($\Delta x = \Delta y = 0$), then $\{D\}$ reduces to the identity matrix. Thus $\{D\}$ is an operator representing four possible finite planar displacements of a point: (1) no motion, (2) pure rotation about the origin, (3) pure translation, and (4) rotation about the origin followed by translation, all defined in the same coordinate system.

For the fifth case, when translation occurs first and is followed by rotation, the concatenation is (using c = cos and s = sin)

$$\begin{bmatrix} c\theta & -s\theta & 0 \\ s\theta & c\theta & 0 \\ 0 & 0 & 1 \end{bmatrix} \begin{bmatrix} 1 & 0 & \Delta x \\ 0 & 1 & \Delta y \\ 0 & 0 & 1 \end{bmatrix} = \begin{bmatrix} c\theta & -s\theta & c\theta\,\Delta x - s\theta\,\Delta y \\ s\theta & c\theta & s\theta\,\Delta x + c\theta\,\Delta y \\ 0 & 0 & 1 \end{bmatrix}$$

This is different from $[D]$ for case 4, Eq. (6.16). Let us demonstrate these by an example similar to the displacement shown in Fig. 6.19. Let $\mathbf{A}_1 = 2 + i$, $\theta =$

36.87°, and $\Delta\mathbf{A} = 1 + 0.5i$. Thus, if the rotation is followed by the translation,

$$[D] = \begin{bmatrix} \cos 36.87^\circ & -\sin 36.87^\circ & 1 \\ \sin 36.87^\circ & \cos 36.87^\circ & 0.5 \\ 0 & 0 & 1 \end{bmatrix}$$

$$\{A_2\} = \begin{Bmatrix} x_2 \\ y_2 \\ 1 \end{Bmatrix} = \begin{bmatrix} 0.8 & -0.6 & 1 \\ 0.6 & 0.8 & 0.5 \\ 0 & 0 & 1 \end{bmatrix} \begin{Bmatrix} 2 \\ 1 \\ 1 \end{Bmatrix} = \begin{Bmatrix} 1.6 - 0.6 + 1 \\ 1.2 + 0.8 + 0.5 \\ 1 \end{Bmatrix}$$

$$\{A_2\} = \begin{Bmatrix} 2 \\ 2.5 \\ 1 \end{Bmatrix} \qquad \text{and} \qquad \mathbf{A}_2 = 2 + 2.5i$$

This result can be checked by scaling Figs. 6.19 and 6.20.

On the other hand, if the translation by the same $\Delta\mathbf{A} = 1 + 0.5i$ occurs first, followed by the rotation by $\theta = 36.87°$, we have

$$\{A_3\} = \begin{Bmatrix} x_3 \\ y_3 \\ 1 \end{Bmatrix} = \begin{bmatrix} 0.8 & -0.6 & 0.8 - 0.3 \\ 0.6 & 0.8 & 0.6 + 0.4 \\ 0 & 0 & 1 \end{bmatrix} \begin{Bmatrix} 2 \\ 1 \\ 1 \end{Bmatrix} = \begin{Bmatrix} 1.6 - 0.6 + 0.5 \\ 1.2 + 0.8 + 1 \\ 1 \end{Bmatrix}$$

Thus $\mathbf{A}_3 = 1.5 + 3i$. This operation is shown in Fig. 6.20 in dashed lines.

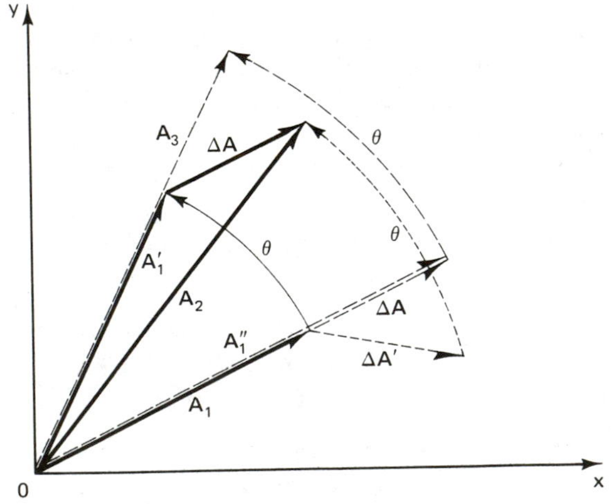


**Figure 6.20** Starting from $\mathbf{A}_1$, rotation by $\theta$ followed by translation by $\Delta\mathbf{A}$ results in final position $\mathbf{A}_2$. When translation by $\Delta\mathbf{A}$ precedes notation by $\theta$, the result is $\mathbf{A}_3$.

If we want to model a case 5 *combined displacement,* model the motion link 2 from its starting position to its final position shown in Fig. 6.19, that is, we wish to have the translation take place first followed by the rotation, then we must define the translation along the first position of link 1. $\Delta\mathbf{A}' = e^{-i36.87°}\Delta\mathbf{A} = 1.1 - 0.2i.$ With this:

$$\{A_4\} = \begin{Bmatrix} x_4 \\ y_4 \\ 1 \end{Bmatrix} = \begin{bmatrix} 0.8 & -0.6 & 0.8(1.1) - 0.6(-0.2) \\ 0.6 & 0.8 & 0.6(1.1) + 0.8(-0.2) \\ 0 & 0 & 1 \end{bmatrix} \begin{Bmatrix} 2 \\ 1 \\ 1 \end{Bmatrix}$$

$$= \begin{Bmatrix} 1.6 - 0.6 + 0.88 + 0.12 \\ 1.2 + 0.8 + 0.66 - 0.16 \\ 1 \end{Bmatrix}$$

$$\mathbf{A}_4 = 2 + 2.5i = \mathbf{A}_2$$

## 6.8 ROTATION ABOUT AN AXIS NOT THROUGH THE ORIGIN

Up to this point, the rotation operators discussed have been for rotation of a vector pivoted about the origin. This can also be interpreted as the rotation of a free vector with respect to a fixedly oriented coordinate system with the origin attached to the tail of the free vector. For instance, in the synthesis chapter, loop-closure equations were applied to the vectors representing a linkage. This involves head-to-tail addition of vectors, each vector being a free vector with its own origin and own coordinate system (see Fig. 6.21). Each of these systems has axes parallel with the corresponding axes of all other such systems, including that of the fixed frame of reference, and each system translates with the tail of its respective vector.

Let point $A_1$ be located in the fixed $x$, $y$ system of Fig. 6.22a by the vector $\mathbf{A}_1$. Now let point $A_1$ describe the arc $\theta$ about point $Q$ at $\mathbf{Q}$. We seek a single

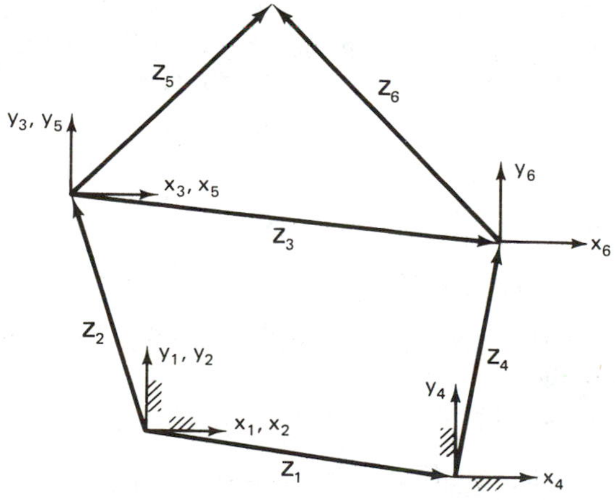


**Figure 6.21** Coordinate systems for each vector representing a planar four-bar linkage.

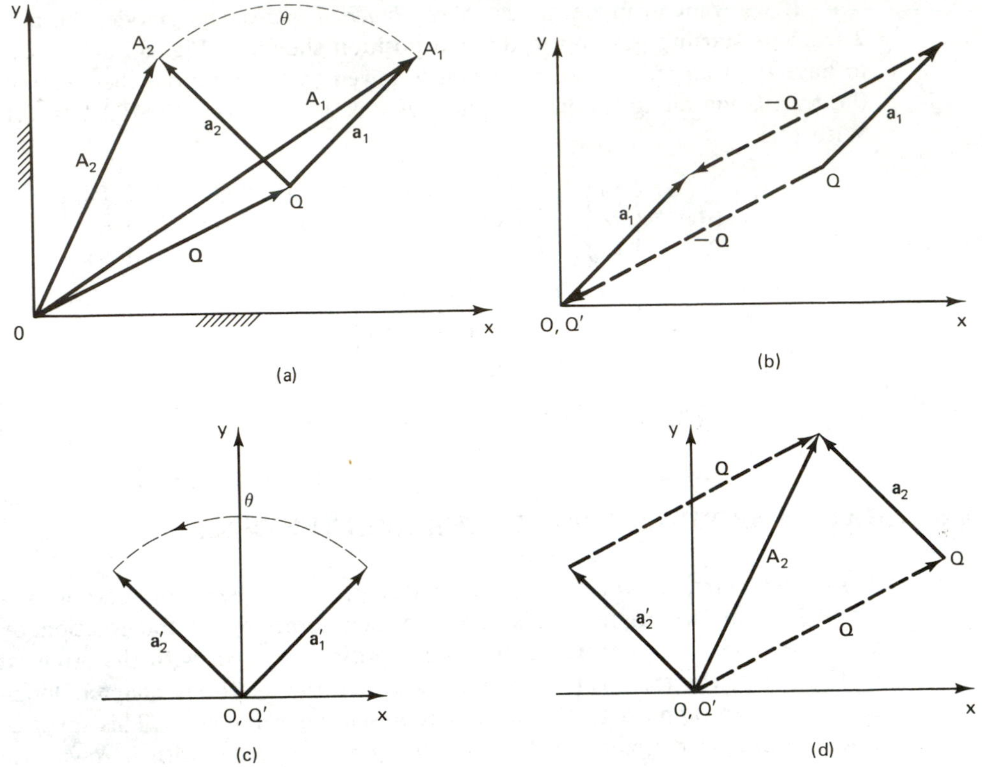


**Figure 6.22** Illustration of transformation steps representing the movement of point A from position 1 to 2.

operator that will describe the absolute motion of point $A$. We may construct a free vector $\mathbf{a}_1$ emanating from point $Q$ to the particle initially at $\mathbf{A}_1$. If point $Q$ (and the tail of $\mathbf{a}_1$) is translated back to the origin, $\mathbf{a}_1$ becomes a vector that is free to rotate about the origin. This step, shown in Fig. 6.22b, is represented mathematically by

$$\{a_1'\} = [T(-\mathbf{Q})]\{a_1\}$$

The rotation by angle $\theta$ (Fig. 6.22c) is represented by

$$\{a_2'\} = [R(\theta)]\{a_1'\}$$

Finally, point $Q'$ may be translated back to its original position (with $\mathbf{a}_2'$ attached to it) using the translation operator

$$\{A_2\} = [T(+\mathbf{Q})]\{a_2'\}$$

as shown in Fig. 6.22d. These three steps may be concatenated to form a single

$3 \times 3$ matrix operator:

$$\{\mathbf{A}_2\} = [D(\theta, Q_x, Q_y)]\{\mathbf{A}_1\}$$

where

$$[D] = [T(+\mathbf{Q})][R(\theta)][T(-\mathbf{Q})]$$

$$[D] = \begin{bmatrix} 1 & 0 & Q_x \\ 0 & 1 & Q_y \\ 0 & 0 & 1 \end{bmatrix} \begin{bmatrix} \cos\theta & -\sin\theta & 0 \\ \sin\theta & \cos\theta & 0 \\ 0 & 0 & 1 \end{bmatrix} \begin{bmatrix} 1 & 0 & -Q_x \\ 0 & 1 & -Q_y \\ 0 & 0 & 1 \end{bmatrix}$$

$$= \begin{bmatrix} 1 & 0 & Q_x \\ 0 & 1 & Q_y \\ 0 & 0 & 1 \end{bmatrix} \begin{bmatrix} \cos\theta & -\sin\theta & -Q_x\cos\theta + Q_y\sin\theta \\ \sin\theta & \cos\theta & -Q_x\sin\theta - Q_y\cos\theta \\ 0 & 0 & 1 \end{bmatrix} \tag{6.17}$$

$$[D] = \begin{bmatrix} \cos\theta & -\sin\theta & Q_x - (Q_x\cos\theta - Q_y\sin\theta) \\ \sin\theta & \cos\theta & Q_y - (Q_x\sin\theta + Q_y\cos\theta) \\ 0 & 0 & 1 \end{bmatrix}$$

The verification of this by a numerical example is left to the student as an exercise (see Exer. 6.1).

## 6.9 RIGID-BODY TRANSFORMATIONS

Thus far we have discussed various transformations to describe the displacements of pivots, particles, or points. How can the operators already found be applied to the displacements of a system of particles such as a rigid body? We used a $3 \times 1$ homogeneous column matrix to describe a vector representing a single point, pivot, or particle. One good feature of the planar $3 \times 3$ translation, rotation, and general displacement matrix operators is that they may easily be programmed on a computer to manipulate a $3 \times n$ matrix of $n$ column vectors representing $n$ particles of a planar rigid body. Since the distance of each particle of a rigid body from every other particle of that rigid body is constant, the vectors locating each particle of a rigid body must undergo the same transformation when the rigid body moves and the proper axis, angle, and/or translation is specified to represent its motion. For example, the general planar motion of three points of a planar rigid body, from their initial positions (**A**, **B**, **C**) to their final positions (**A**′, **B**′, **C**′), can be represented by

$$\begin{bmatrix} A'_x & B'_x & C'_x \\ A'_y & B'_y & C'_y \\ 1 & 1 & 1 \end{bmatrix} = \begin{bmatrix} d_{11} & d_{12} & d_{13} \\ d_{21} & d_{22} & d_{23} \\ d_{31} & d_{32} & d_{33} \end{bmatrix} \begin{bmatrix} A_x & B_x & C_x \\ A_y & B_y & C_y \\ 1 & 1 & 1 \end{bmatrix} \tag{6.18}$$

## 6.10 SPATIAL TRANSFORMATIONS

### Rotation

All points on a rigid body undergoing pure rotation describe arcs in planes perpendicular to a fixed line, the axis of rotation. In a planar two-dimensional $x,y$ system, the axis is always perpendicular to the $x,y$ plane and thus parallel to the $z$ axis. Therefore, the axis may be completely defined by specifying its intersection with the $x,y$ plane (e.g., the origin $O$ or point $Q$ used in the preceding section) and by agreeing that $x,y$ and the axis are parallel to the axes of an $xyz$ right-hand system. Thus the axis always points toward the observer, and right-hand (positive) rotations appear counterclockwise. In the general spatial case, motion is not constrained to the $xy$ plane and the axis of rotation may be oriented in any direction. Therefore, the location and direction of the axis must be incorporated into the spatial rotation operator.

One way an axis in space may be specified is by defining the location of a point on the axis and a unit vector in the positive direction along the axis. If, in deriving a spatial rotation operator, we specify that the axis must pass through the origin and if the three coordinates of the unit vector are used explicitly in the operator, the axis will be fully incorporated into the operator.

Also, using a unit vector to define the orientation of the axis allows us to establish a sign convention for the rotation. In a planar system, if the thumb of the right-hand points in the positive $z$ direction (i.e., toward the observer), the fingers curl in a counterclockwise direction. Thus a planar angle is usually defined as positive for a counterclockwise rotation. We adopt the same right-hand rule here: When the right thumb points in the direction of the unit vector of the axis, a positive rotation about the axis is in the direction of the curled fingers.

Before introducing a spatial $3 \times 3$ matrix rotation operator, let us consider this unit vector in more detail. We use the symbol $\hat{\mathbf{u}}$ to represent it. The "hat" above it indicates that it is a unit vector (its magnitude is 1). Referring to Fig. 6.23, the three coordinates are the *direction cosines* of the line:

$$\begin{aligned} u_x &= \cos \alpha \\ u_y &= \cos \beta \\ u_z &= \cos \gamma \end{aligned} \tag{6.19}$$

where $\alpha$, $\beta$, and $\gamma$ are the angles of the vector measured from the positive branch of the $x$, $y$, and $z$ axes, respectively.

Figure 6.23a is an axonometric sketch of $\hat{\mathbf{u}}$ and its coordinates. Figure 6.23*b*, *c*, and *d* are orthographic projection views taken by facing the arrow heads of the coordinate axes. For example, view A-A shows the positive $z$ axis pointing toward us, which is indicated by the encircled dot representing the apex of its arrowhead. (If the positive axis were pointing away from us, a cross would indicate that we are looking at the tail of the arrow.) Views A-A and B-B also show how the true

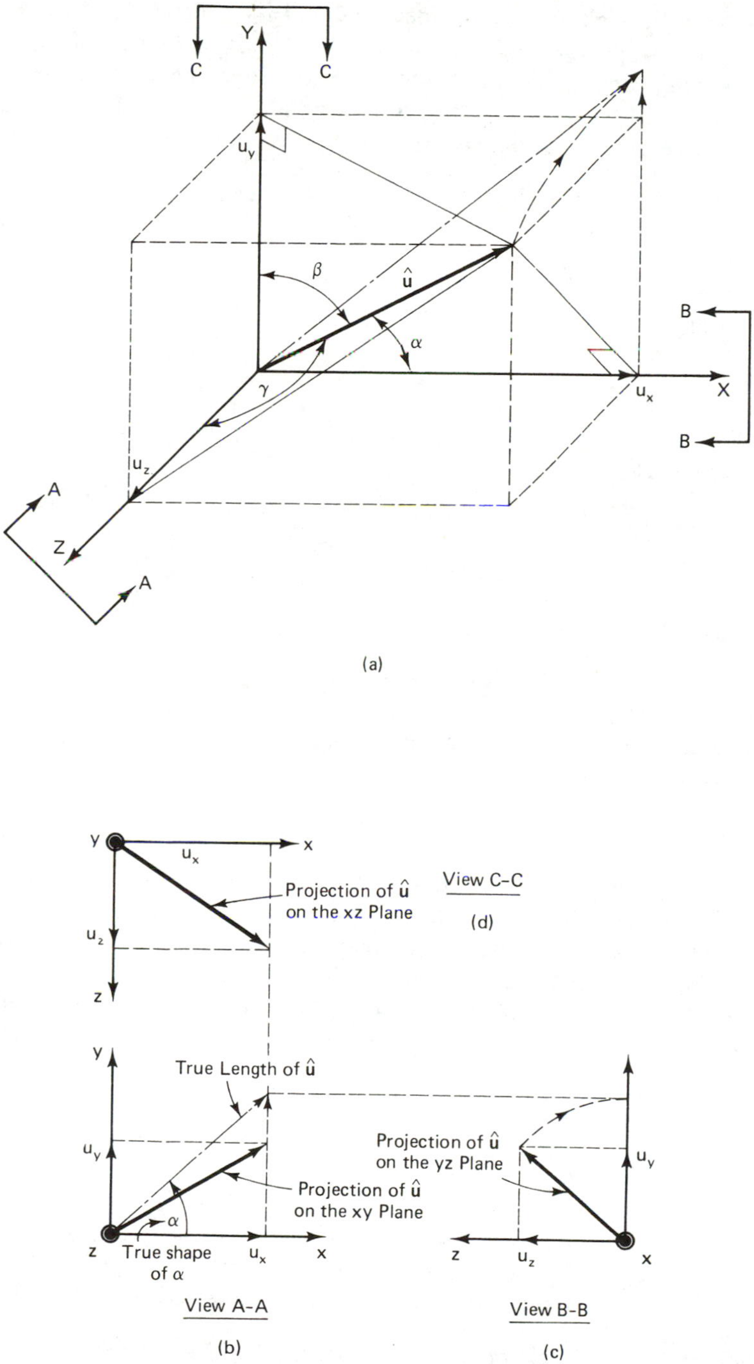


**Figure 6.23** Unit vector **u** and its direction cosines.

shape of the right triangle formed by $\hat{\mathbf{u}}$ and $u_x$, is determined by rotating their plane into the $xy$ picture plane. This is a well-known method of descriptive geometry, displaying the true length of $\hat{\mathbf{u}}$ and the true shape of the angle $\alpha$.

A unit vector $\hat{\mathbf{u}}$ pointing from a point $A$ on an axis in space toward point $B$ anywhere on that same axis may be found by normalizing the vector **AB:**

$$\mathbf{u} = \frac{\mathbf{B} - \mathbf{A}}{|\overrightarrow{AB}|}$$

$$u_x = \frac{B_x - A_x}{|\overrightarrow{AB}|}$$

$$u_y = \frac{B_y - A_y}{|\overrightarrow{AB}|}$$

$$u_z = \frac{B_z - A_z}{|\overrightarrow{AB}|}$$

where

$$|\overrightarrow{AB}| = [(B_x - A_x)^2 + (B_y - A_y)^2 + (B_z - A_z)^2]^{1/2}$$

The fact that it is a unit vector will be used as a constraint later in the mechanism synthesis Sec. 6.19:

$$|\hat{\mathbf{u}}| = [u_x^2 + u_y^2 + u_z^2]^{1/2} = 1 \tag{6.21}$$

At this stage we introduce the spatial $3 \times 3$ matrix rotation operator. The interested student is referred to a later subsection for its derivation. Keep in mind that this, like the planar $2 \times 2$ rotation operator, is only for rotation about an axis passing through the origin. Also, it does not change the magnitude of a vector and it is a function of both the angle of rotation and of the coordinates of the unit vector of the axis of rotation. It may be completely symbolized by $[R(\theta, \hat{\mathbf{u}})]$, whose elements are

$$\begin{aligned}[R(\theta, \hat{\mathbf{u}})] &= \begin{bmatrix} r_{11} & r_{12} & r_{13} \\ r_{21} & r_{22} & r_{23} \\ r_{31} & r_{32} & r_{33} \end{bmatrix} \\ &= \begin{bmatrix} u_x^2\, v\theta + c\theta & u_x u_y\, v\theta - u_z\, s\theta & u_x u_z\, v\theta + u_y\, s\theta \\ u_x u_y\, v\theta + u_z\, s\theta & u_y{}^2\, v\theta + c\theta & u_y u_z\, v\theta - u_x\, s\theta \\ u_x u_z\, v\theta - u_y\, s\theta & u_y u_z\, v\theta + u_x\, s\theta & u_z^2\, v\theta + c\theta \end{bmatrix}\end{aligned} \tag{6.22}$$

where $\theta$ is the angle of rotation about $\hat{\mathbf{u}}$ (right-hand rule with $\hat{\mathbf{u}}$), $c\theta = \cos\theta$, $s\theta = \sin\theta$, $v\theta = \text{versine}(\theta) = 1 - \cos\theta$, and $\hat{\mathbf{u}} = [u_x, u_y, u_z]^T$.

**Example 6.3**

Use Eq. (6.22) to find the operators representing rotation about the $x$, $y$, and $z$ axes.

**Solution** *Rotation about the x axis.* Here $u_x = 1$, $u_y = u_z = 0$. A positive rotation $\theta$ is ccw when looking toward the origin from a point on the positive $x$ axis.

$$[R(\theta, u_x)] = \begin{bmatrix} 1^2(1-c\theta)+c\theta & 1(0)v\theta - 0s\theta & 1(0)v\theta + 0s\theta \\ 1(0)v\theta + 0s\theta & 0^2\,v\theta + c\theta & 0(0)v\theta - 1s\theta \\ 1(0)v\theta - 0s\theta & 0(0)v\theta + 1s\theta & 0^2\,v\theta + c\theta \end{bmatrix}$$
$$= \begin{bmatrix} 1 & 0 & 0 \\ 0 & \cos\theta & -\sin\theta \\ 0 & \sin\theta & \cos\theta \end{bmatrix} \quad (6.23)$$

Recall that in rotation about the $x$ axis, all points move in planes perpendicular to the $x$ axis. Therefore, the $x$ coordinate of a vector does not change. Note that the top row of the operator is the same as the top row of the $3 \times 3$ identity matrix; thus the $x$ coordinate remains unchanged. You can easily verify this by multiplying out an example of your own choice. Figure 6.24 shows such an example both in matrix form and by three-view orthographic projections. The reader is encouraged to assign appropriate values to $a_x$, $a_y$, $a_z$, and $\theta$, carry out the operation both graphically and analytically, and compare results (see Exer. 6.2). Also note the similarity of the four lower-right entries (the lower right $2 \times 2$ minor) in $R[(\theta, u_x)]$ with the planar $2 \times 2$ rotation operator [Eq. (6.13)].

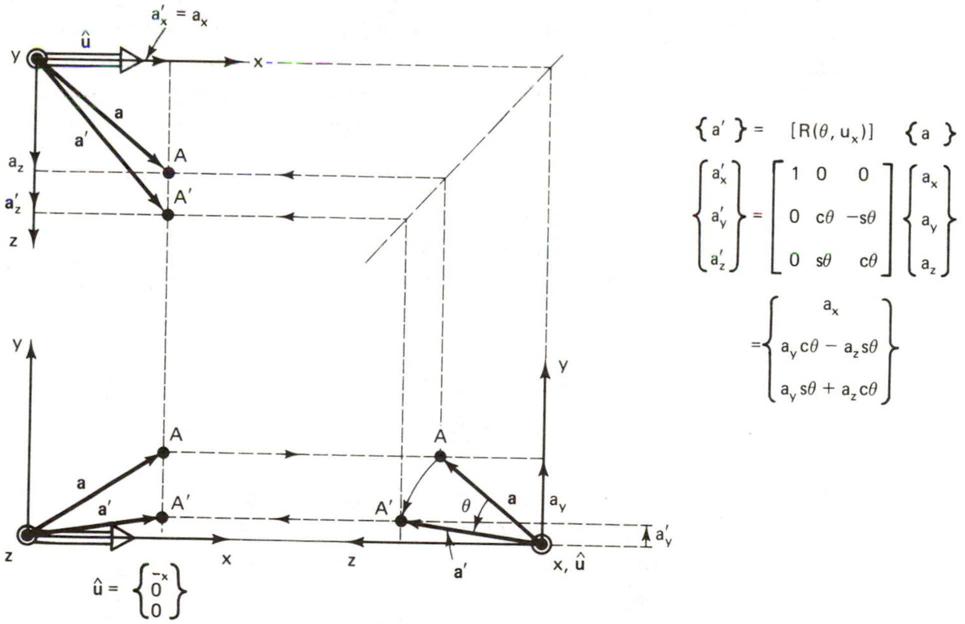


**Figure 6.24** Three orthographic views of vector **a** rotated about the x-axis to **a′**.

*Rotation about the y axis.* Here $u_x = 0$, $u_y = 1$, $u_z = 0$. A positive rotation $\theta$ is ccw when looking toward the origin from a point on the positive $y$ axis. Substitution as before yields

$$[R(\theta, u_y)] = \begin{bmatrix} \cos\theta & 0 & \sin\theta \\ 0 & 1 & 0 \\ -\sin\theta & 0 & \cos\theta \end{bmatrix} \tag{6.24}$$

Note that this is similar to the planar operator except that the sine terms are interchanged.

*Rotation about the z axis.* Here $u_x = u_y = 0$, $u_z = 1$. You can easily verify that

$$[R(\theta, u_z)] = \begin{bmatrix} \cos\theta & -\sin\theta & 0 \\ \sin\theta & \cos\theta & 0 \\ 0 & 0 & 1 \end{bmatrix} \tag{6.25}$$

This last operator can be seen as the $xy$ planar case extended to three-dimensional space, and it is identical with the planar rotation operator in homogeneous $3 \times 3$ matrix form. This yields an interesting interpretation of the homogeneous $3 \times 1$ column matrix form of planar vectors. In three-dimensional space, consider a plane parallel with the $xy$ plane which intersects the $z$ axis at $z = 1$. Then the three-dimensional radius vector **A** locating any point $A$ in this plane will be

$$\{A\} = \begin{bmatrix} A_x \\ A_y \\ 1 \end{bmatrix} \tag{6.26}$$

which is the same as the homogeneous $3 \times 1$ column matrix form of the planar vector $A_x = iA_y$ (i.e., the projection of the spatial vector **A** onto the $xy$ plane). Thus, if we operate on $\{A\}$ of Eq. (6.26) with the rotation operator $[R(\theta, u_z)]$ of Eq. (6.25), the result of this operation will be to rotate the projection of **A** on the $xy$ plane by $\theta$. This is identical with the homogeneous form of the planar case described earlier.

**Example 6.4**

Determine the combined rotation matrix for rotating a free vector in the fixed $Oxyz$ system by the following rotations in this order: $\alpha$ about the $z$ axis, $\beta$ about the $y$ axis, and $\gamma$ about the $x$ axis:

**Solution**

$$R(\alpha, u_z) = \begin{bmatrix} c\alpha & -s\alpha & 0 \\ s\alpha & c\alpha & 0 \\ 0 & 0 & 1 \end{bmatrix}$$

$$R(\beta, u_y) = \begin{bmatrix} c\beta & 0 & s\beta \\ 0 & 1 & 0 \\ -s\beta & 0 & c\beta \end{bmatrix}$$

$$R(\gamma, u_x) = \begin{bmatrix} 1 & 0 & 0 \\ 0 & c\gamma & -s\gamma \\ 0 & s\gamma & c\gamma \end{bmatrix}$$

Concatenating these three matrices in the given order results in

$$R(\theta, \hat{u}) = [R(\gamma, u_x)][R(\beta, u_y)][R(\alpha, u_z)]$$

$$= \begin{bmatrix} c\beta\, c\alpha & -c\beta\, s\alpha & s\beta \\ c\alpha\, s\beta\, s\gamma + c\gamma\, s\alpha & -s\alpha\, s\beta\, s\gamma + c\gamma\, c\alpha & -s\gamma\, c\beta \\ -c\alpha\, s\beta\, c\gamma + s\gamma\, s\alpha & s\alpha\, s\beta\, c\gamma + s\gamma\, c\alpha & c\gamma\, c\beta \end{bmatrix}$$

**Example 6.5**

Using Eq. (6.22) and the results of the previous examples, determine $\hat{\mathbf{u}}$ and $\theta$ for the specific case where $\alpha = \beta = \gamma = 30°$.

**Solution** Substituting $\alpha = \beta = \gamma = 30°$ into the resulting rotation matrix in Example 6.4, we have

$$R(\theta, \hat{\mathbf{u}}) = \begin{bmatrix} 0.750 & -0.433 & 0.500 \\ 0.650 & 0.625 & -0.433 \\ -0.125 & 0.650 & 0.750 \end{bmatrix}$$

For the following derivations, we will refer to the elements of the rotation matrix by $r_{ij}$.

$$R(\theta, \hat{\mathbf{u}}) = \begin{bmatrix} r_{11} & r_{12} & r_{13} \\ r_{21} & r_{22} & r_{23} \\ r_{31} & r_{32} & r_{33} \end{bmatrix}$$

$$= \begin{bmatrix} u_x^2\, v\theta + c\theta & u_x u_y\, v\theta - u_z\, s\theta & u_x u_z\, v\theta + u_y\, s\theta \\ u_x u_y\, v\theta + u_z\, s\theta & u_y^2\, v\theta + c\theta & u_y u_z\, v\theta - u_x\, s\theta \\ u_x u_z\, v\theta - u_y\, s\theta & u_y u_z\, v\theta + u_x\, s\theta & u_z^2\, v\theta + c\theta \end{bmatrix}$$

Adding elements (1, 1), (2, 2), and (3, 3) on both sides of this equation and simplifying results in an expression from $\theta$ in terms of the diagonal elements:

$$\theta = \cos^{-1} \tfrac{1}{2}(r_{11} + r_{22} + r_{33} - 1)$$

Substituing from above, we have

$$\theta = \cos^{-1} \tfrac{1}{2}(0.750 + 0.625 + 0.750 - 1) = 55.77°$$

Subtracting element $r_{23}$ from $r_{32}$ yields

$$r_{32} - r_{23} = u_y u_z\, v\theta + u_x\, s\theta - (u_y u_z\, v\theta - u_x\, s\theta) = 2u_x \sin\theta$$

$$u_x = \frac{r_{32} - r_{23}}{2 \sin\theta} = 0.655$$

Similarly,

$$u_y = \frac{r_{13} - r_{31}}{2 \sin\theta} = 0.378$$

$$u_z = \frac{r_{21} - r_{12}}{2 \sin\theta} = 0.655$$

Note that $u_x$, $u_y$, and $u_z$ could also be obtained from the diagonal terms of the rotation matrix.

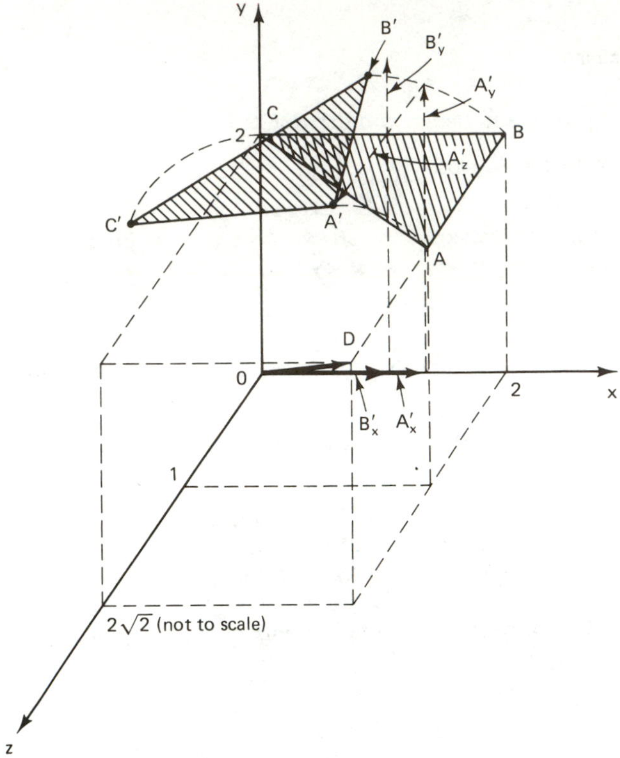


**Figure 6.25** Axonmetric drawing of triangle ABC that rotates about axis *OD* to the primed position. Note that axis *OD* goes through the origin.

### Example 6.6

Points *A*, *B*, and *C* on a rigid body (in the axonmetric drawing shown in Fig. 6.25), rotate by 30° about an axis that passes through the origin pointing toward point *D*. Find the new coordinates (*A′*, *B′*, and *C′*) if the coordinates of point *D* are (2, 2, $2\sqrt{2}$). Check the results graphically (see Exer. 6.3).

**Solution**

$$\mathbf{A}=[2,2,1]^T,\qquad \mathbf{B}=[2,2,0]^T,\qquad \mathbf{C}=[0,2,0]^T$$

$$\hat{\mathbf{u}}=\frac{[2,2,2\sqrt{2}]^T}{[2^2+2^2+2^2(2)]^{1/2}}=\frac{[2,2,2\sqrt{2}]^T}{\sqrt{16}}=\left[\frac{1}{2},\frac{1}{2},\frac{\sqrt{2}}{2}\right]^T$$

$$[R(30°,\hat{\mathbf{u}})]=\begin{bmatrix}\frac{1}{4}(0.134)+0.866 & \frac{1}{4}(0.134)-\frac{\sqrt{2}}{2}(0.5) & \frac{\sqrt{2}}{4}(0.134)+\frac{1}{2}(0.5)\\ \frac{1}{4}(0.134)+\frac{\sqrt{2}}{2}(0.5) & \frac{1}{4}(0.134)+0.866 & \frac{\sqrt{2}}{4}(0.134)-\frac{1}{2}(0.5)\\ \frac{\sqrt{2}}{4}(0.134)-\frac{1}{2}(0.5) & \frac{\sqrt{2}}{4}(0.134)+\frac{1}{2}(0.5) & \frac{1}{2}(0.134)+0.866\end{bmatrix}$$

$$[\mathbf{A}'\quad \mathbf{B}'\quad \mathbf{C}']=[R(30°),\hat{\mathbf{u}}][\mathbf{A}\quad \mathbf{B}\quad \mathbf{C}]=\begin{bmatrix}0.900 & -0.320 & 0.297\\ 0.387 & 0.900 & -0.203\\ -0.203 & 0.297 & 0.933\end{bmatrix}\begin{bmatrix}2 & 2 & 0\\ 2 & 2 & 2\\ 1 & 0 & 0\end{bmatrix}$$

$$=\begin{bmatrix}1.457 & 1.180 & -0.640\\ 2.371 & 2.574 & 1.800\\ 1.121 & 0.188 & 0.594\end{bmatrix}$$

## Translation

General motion in a plane includes two translational degrees of freedom. Recall that in Sec. 6.7 we desired a matrix operator which, when operating on a vector, would add these two translations ($\Delta x$ and $\Delta y$) to the coordinates of that vector. This was accomplished by combining $\Delta x$ and $\Delta y$ with a $3 \times 3$ identity matrix and using homogeneous notation to represent the vector. A spatial translation operator may be formed, using the same logic, by including the three translational degrees of freedom defined by

$$[D] = \begin{Bmatrix} \Delta x \\ \Delta y \\ \Delta z \end{Bmatrix} \tag{6.27}$$

in a $4 \times 4$ identity matrix and using the $4 \times 1$ homogeneous notation to represent the vector **D**. Thus the spatial translation operator is

$$[T] = \begin{bmatrix} 1 & 0 & 0 & \Delta x \\ 0 & 1 & 0 & \Delta y \\ 0 & 0 & 1 & \Delta z \\ 0 & 0 & 0 & 1 \end{bmatrix} \tag{6.28}$$

The homogeneous representation of a single vector locating point $A$ in three-dimensional space with respect to an $xyz$ coordinate system is

$$\mathbf{r}_A = \{r_A\} = \begin{Bmatrix} x_A \\ y_A \\ z_A \\ 1 \end{Bmatrix} \tag{6.29}$$

**Example 6.7**

Point $P$, shown axonometrically in Fig. 6.26, is to be translated by $\Delta x = -1$, $\Delta y = -2$, and $\Delta z = +2$. Using Eq. (6.29), the initial position of point $P$ is

$$\mathbf{r}_1 = \{r_1\} = \begin{Bmatrix} 1 \\ 4 \\ 2 \\ 1 \end{Bmatrix}$$

Using the spatial translation operator of Eq. (6.28), we have

$$\mathbf{r}_2 = \{r_2\} = \begin{bmatrix} 1 & 0 & 0 & -1 \\ 0 & 1 & 0 & -2 \\ 0 & 0 & 1 & 2 \\ 0 & 0 & 0 & 1 \end{bmatrix} \begin{Bmatrix} 1 \\ 4 \\ 2 \\ 1 \end{Bmatrix}$$

so that

$$\{r_2\} = \begin{Bmatrix} 0 \\ 2 \\ 4 \\ 1 \end{Bmatrix}$$

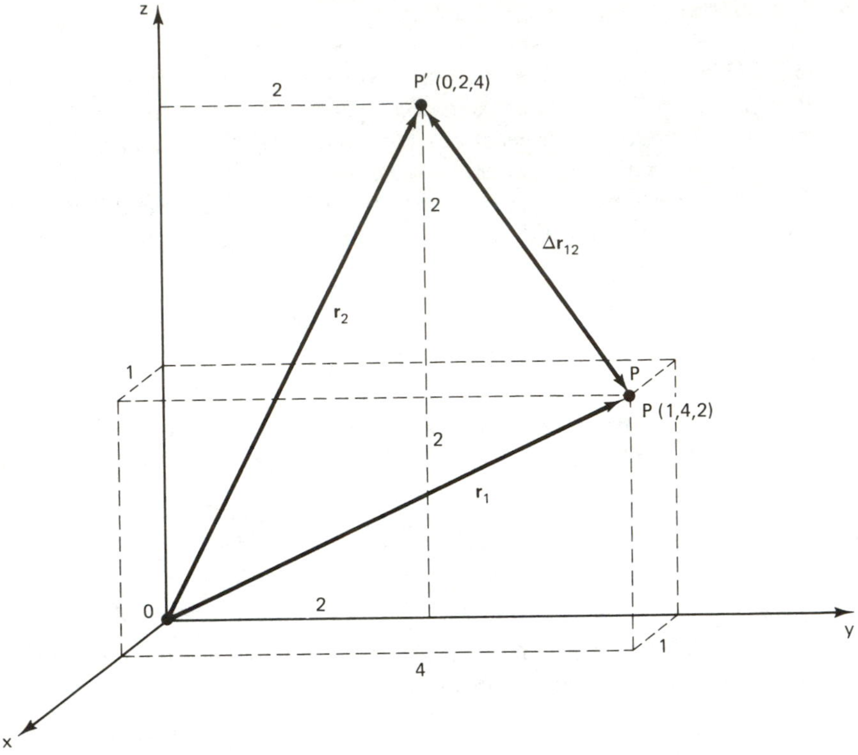


**Figure 6.26** Pure translation of point P.

## The 4 × 4 Translation and Rotation Matrix for Axis Through the Origin

The next logical step is to combine the 3 × 3 rotation matrix having its axis through the origin [Eq. (6.22)] with the 4 × 4 translation matrix of Eq. (6.28), written for translation along the axis of rotation. Expanding the 3 × 3 rotation matrix $[r_{ij}]$ to a 4 × 4 as was done in the planar case [see Eq. (6.16)], and premultiplying it by the 4 × 4 translation matrix, the *screw operator* for an axis through the origin is found for describing a rotation followed by axial translation.

$$[S_{12}] = [T][R(\theta, \hat{\mathbf{u}})] = \begin{bmatrix} 1 & 0 & 0 & \Delta x \\ 0 & 1 & 0 & \Delta y \\ 0 & 0 & 1 & \Delta z \\ 0 & 0 & 0 & 1 \end{bmatrix} \begin{bmatrix} r_{11} & r_{12} & r_{13} & 0 \\ r_{21} & r_{22} & r_{23} & 0 \\ r_{31} & r_{32} & r_{33} & 0 \\ 0 & 0 & 0 & 1 \end{bmatrix}$$

$$= \begin{bmatrix} r_{11} & r_{12} & r_{13} & \Delta x \\ r_{21} & r_{22} & r_{23} & \Delta y \\ r_{31} & r_{32} & r_{33} & \Delta z \\ 0 & 0 & 0 & 1 \end{bmatrix} \tag{6.30}$$

The resulting *screw matrix* in Eq. (6.30) can be expressed in symbolic form by partioning the matrix

$$[S_{12}] = \left[\begin{array}{c|c} R & D \\ \hline 0 & I \end{array}\right] \tag{6.31}$$

where $[R]$ is defined by Eq. (6.22), $[D]$ by Eq. (6.27), where $[I] = 1$, and where $[D] = D\hat{\mathbf{u}}$, or $\Delta x = |D| u_x$, $\Delta y = |D| u_y$, and $\Delta z = |D| u_z$.

The screw operator is so named because it acts as a nut and bolt: a rotation about the axis of the bolt together with a translation along this axis. The direction of the screw axis is along $\hat{\mathbf{u}}$, which emanates from the origin. Recall that $\hat{\mathbf{u}}$ was defined in Eq. (6.21) and Fig. 6.23a.

The general rigid-body motion of a body in space from one position to another can be described uniquely by a single screw displacement whose axis does not necessarily pass through the origin [11,13,60,223,267].

Three examples will illustrate the foregoing discussion.

**Example 6.8**

A particle $P$ describes a 45° arc in space from an initial position at (0, 1, 3) about an axis through the origin with unit vector

$$\hat{\mathbf{u}} = [0, \cos 45°, \cos 45°]^T$$

as shown in Fig. 6.27. The particle is also translated parallel to this axis by

$$[D] = \begin{Bmatrix} 0 \\ 1 \\ 1 \end{Bmatrix} = \sqrt{2}\,\hat{\mathbf{u}}$$

Find the rotation matrix $R$ and the final position of the particle by use of Eq. (6.30).

**Solution** Notice that

$$u_x = 0, \qquad \cos\theta_{12} = \frac{1}{\sqrt{2}}$$

$$u_y = \frac{1}{\sqrt{2}}, \qquad \sin\theta_{12} = \frac{1}{\sqrt{2}}$$

$$u_z = \frac{1}{\sqrt{2}}, \qquad \text{vers}\,\theta_{12} = 1 - \frac{1}{\sqrt{2}}$$

The rotation operator [Eq. (6.22)] becomes

$$[R_{12}] = \begin{bmatrix} 0 + \frac{1}{\sqrt{2}} & 0 - \left(\frac{1}{\sqrt{2}}\right)^2 & 0 + \left(\frac{1}{\sqrt{2}}\right)^2 \\ 0 + \left(\frac{1}{\sqrt{2}}\right)^2 & \left(\frac{1}{\sqrt{2}}\right)^2\left(1 - \frac{1}{\sqrt{2}}\right) + \frac{1}{\sqrt{2}} & \left(\frac{1}{\sqrt{2}}\right)^2\left(1 - \frac{1}{\sqrt{2}}\right) - 0 \\ 0 - \left(\frac{1}{\sqrt{2}}\right)^2 & \left(\frac{1}{\sqrt{2}}\right)^2\left(1 - \frac{1}{\sqrt{2}}\right) + 0 & \left(\frac{1}{\sqrt{2}}\right)^2\left(1 - \frac{1}{\sqrt{2}}\right) + \frac{1}{\sqrt{2}} \end{bmatrix}$$

With this,

$$\mathbf{r}_1' = \begin{Bmatrix} r_{1x}' \\ r_{1y}' \\ r_{1z}' \end{Bmatrix} = \begin{bmatrix} 0.707 & -0.500 & 0.500 \\ 0.500 & 0.854 & 0.146 \\ -0.500 & 0.146 & 0.854 \end{bmatrix} \begin{Bmatrix} 0 \\ 1 \\ 3 \end{Bmatrix}$$

$$= \begin{Bmatrix} 0 - 0.500 + 1.500 \\ 0 + 0.854 + 0.439 \\ 0 + 0.146 + 2.562 \end{Bmatrix} = \begin{Bmatrix} 1.00 \\ 1.293 \\ 2.708 \end{Bmatrix}$$

which is the position of point $P'$ in the figure. This can be checked by realizing that, since there is only rotation,

$$|\mathbf{r}_1| = |\mathbf{r}_1'|$$

Solving each side yields

$$|\mathbf{r}_1| = [0^2 + 1^2 + 3^2]^{1/2} = \sqrt{10}$$

$$|\mathbf{r}_1'| = [1.000^2 + 1.292^2 + 2.708^2]^{1/2} = \sqrt{10.005}$$

so the magnitudes are the same except for minor rounding error.

Using Eq. (6.30), the final position of the particle is found:

$$\mathbf{r}_2 = \{r_2\} = [S_{12}]\{r_1\} = \begin{bmatrix} 0.707 & -0.500 & 0.500 & 0 \\ 0.500 & 0.854 & 0.146 & 1 \\ -0.500 & 0.146 & 0.854 & 1 \\ 0 & 0 & 0 & 1 \end{bmatrix} \begin{Bmatrix} 0 \\ 1 \\ 3 \\ 1 \end{Bmatrix}$$

$$= \begin{Bmatrix} 1.00 \\ 2.29 \\ 3.70 \\ 1.00 \end{Bmatrix}$$

Recall that $[S_{12}]$ was derived for rotation about an axis through the origin followed by coaxial translation. Does axial translation followed by coaxial rotation (axis through the origin) lead to a different operator? (See Fig. 6.27.)

**Example 6.9***

Determine the unique screw displacement matrix given the coordinates of each of four points of a rigid body in its initial and final positions. The homogeneous coordinates for the four points in their initial positions are chosen as

$$[\mathbf{A}_1 \quad \mathbf{B}_1 \quad \mathbf{C}_1 \quad \mathbf{D}_1] = \begin{bmatrix} 0 & 2 & 0 & -2 \\ 3 & 7 & 5 & 5 \\ 7 & 10 & 10 & 7 \\ 1 & 1 & 1 & 1 \end{bmatrix}$$

* This and the next example contributed by Dan Olson.

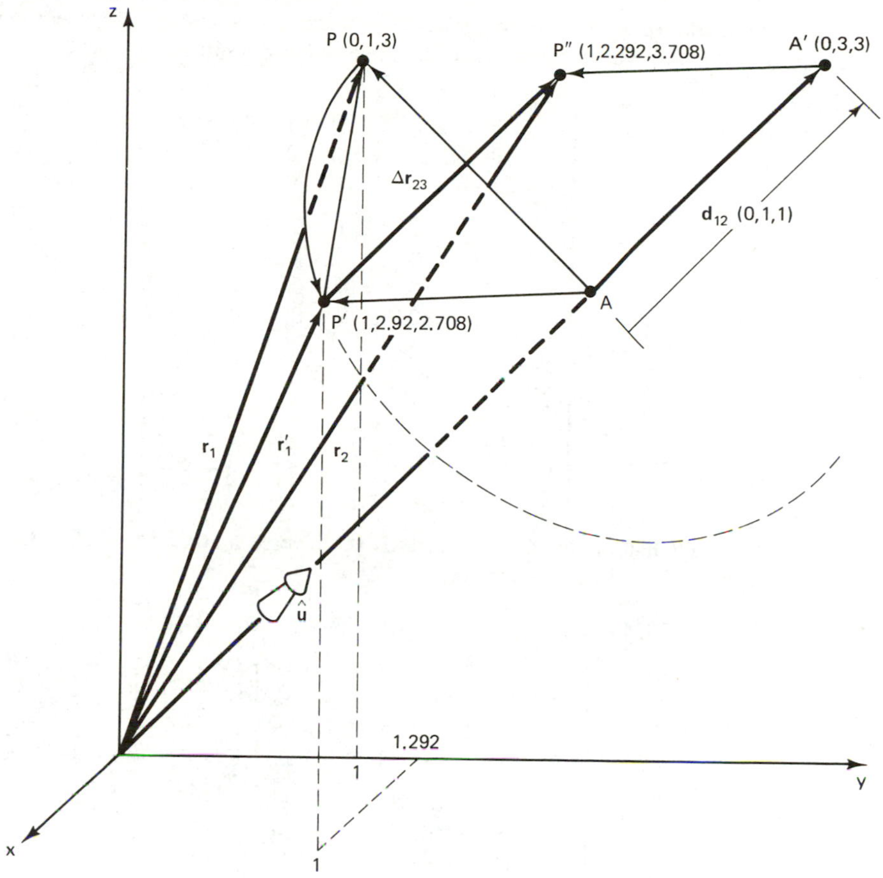


**Figure 6.27** Screw displacement of point P about axis which goes through the origin.

and their final positions are

$$[\mathbf{A}_2 \quad \mathbf{B}_2 \quad \mathbf{C}_2 \quad \mathbf{D}_2] = \begin{bmatrix} 1.90 & 3.29 & 4.26 & 3.92 \\ 11.23 & 14.44 & 13.41 & 9.83 \\ 7.19 & 11.29 & 8.84 & 8.59 \\ 1 & 1 & 1 & 1 \end{bmatrix}$$

**Solution** We seek to determine the screw matrix $S_{12}$ such that $[\mathbf{A}_2 \quad \mathbf{B}_2 \quad \mathbf{C}_2 \quad \mathbf{D}_2] = [S_{12}][\mathbf{A}_1 \quad \mathbf{B}_1 \quad \mathbf{C}_1 \quad \mathbf{D}_1]$. Postmultiplying both sides by $[\mathbf{A}_1 \quad \mathbf{B}_1 \quad \mathbf{C}_1 \quad \mathbf{D}_1]^{-1}$ results in

$$[S_{12}] = [\mathbf{A}_2 \quad \mathbf{B}_2 \quad \mathbf{C}_2 \quad \mathbf{D}_2][\mathbf{A}_1 \quad \mathbf{B}_1 \quad \mathbf{C}_1 \quad \mathbf{D}_1]^{-1}$$

The inverse matrix $[\mathbf{A}_1 \quad \mathbf{B}_1 \quad \mathbf{C}_1 \quad \mathbf{D}_1]^{-1}$ can be determined (if $[\mathbf{A}_1 \quad \mathbf{B}_1 \quad \mathbf{C}_1 \quad \mathbf{D}_1]$ is nonsingular) by forming a $4 \times 8$ matrix by post-adjoining the $4 \times 4$ identity matrix, and then reducing it to *row-echelon* form using elementary row operations until the $4 \times 4$ identity matrix appears in the pre-adjoining position:

$$\begin{bmatrix} 0 & 2 & 0 & -2 & 1 & 0 & 0 & 0 \\ 3 & 7 & 5 & 5 & 0 & 1 & 0 & 0 \\ 7 & 10 & 10 & 7 & 0 & 0 & 1 & 0 \\ 1 & 1 & 1 & 1 & 0 & 0 & 0 & 1 \end{bmatrix}$$

$$= \begin{bmatrix} 1 & 1 & 1 & 1 & 0 & 0 & 0 & 1 \\ 0 & 1 & 0 & -1 & \frac{1}{2} & 0 & 0 & 0 \\ 0 & -19 & -5 & -14 & 0 & -7 & 3 & 0 \\ 7 & 10 & 10 & 7 & 0 & 0 & 1 & 0 \end{bmatrix}$$

$$\vdots$$

$$= \begin{bmatrix} 1 & 0 & 0 & 0 & \frac{1}{4} & -\frac{1}{4} & -\frac{1}{6} & \frac{35}{12} \\ 0 & 1 & 0 & 0 & \frac{1}{4} & \frac{1}{4} & -\frac{1}{6} & \frac{5}{12} \\ 0 & 0 & 1 & 0 & -\frac{1}{4} & -\frac{1}{4} & \frac{1}{2} & -\frac{33}{12} \\ 0 & 0 & 0 & 1 & -\frac{1}{4} & \frac{1}{4} & -\frac{1}{6} & \frac{5}{12} \end{bmatrix}$$

The right half of this $4 \times 8$ matrix is the inverse matrix, $[\mathbf{A}_1 \quad \mathbf{B}_1 \quad \mathbf{C}_1 \quad \mathbf{D}_1]^{-1}$ (see Exer. 6.15). The screw matrix

$$[S_{12}] = [\mathbf{A}_2 \quad \mathbf{B}_2 \quad \mathbf{C}_2 \quad \mathbf{D}_2][\mathbf{A}_1 \quad \mathbf{B}_1 \quad \mathbf{C}_1 \quad \mathbf{D}_1]^{-1}$$

$$= \begin{bmatrix} 1.90 & 3.29 & 4.26 & 3.92 \\ 11.23 & 14.44 & 13.41 & 9.83 \\ 7.19 & 11.29 & 8.84 & 8.59 \\ 1 & 1 & 1 & 1 \end{bmatrix} \begin{bmatrix} \frac{1}{4} & -\frac{1}{4} & -\frac{1}{6} & \frac{35}{12} \\ \frac{1}{4} & \frac{1}{4} & -\frac{1}{6} & \frac{5}{12} \\ -\frac{1}{4} & -\frac{1}{4} & \frac{1}{2} & -\frac{33}{12} \\ -\frac{1}{4} & \frac{1}{4} & -\frac{1}{6} & \frac{5}{12} \end{bmatrix}$$

$$= \begin{bmatrix} -0.748 & 0.263 & 0.612 & -3.17 \\ 0.608 & -0.093 & 0.788 & 5.99 \\ 0.263 & 0.963 & -0.092 & 4.94 \\ 0 & 0 & 0 & 1 \end{bmatrix}$$

To verify this result, see Exer. 6.16.
(*Note:* The coordinates of $\mathbf{A}_2$, $\mathbf{B}_2$, $\mathbf{C}_2$, and $\mathbf{D}_2$ are rounded off to two decimal places.)

**Example 6.10**

For Example 6.9 determine $\hat{\mathbf{u}}$, $\theta$, $\mathbf{D}$, and the point at which the screw axis intersects the $yz$ plane (recall that the unique screw axis does not necessarily pass through the origin), using Eq. (6.22) and several translation operators.

**Solution** Let $P(x, y, z)$ be a point on the screw axis. Let $Q_1$ be a point to undergo a screw displacement to $Q_2$ (see Fig. 6.28). The rotation portion of the screw matrix [Eq. (6.31)] is referred to an axis through the origin, so we first translate $P$ to the origin and obtain $Q_1'$.

$$\{Q_1'\} = \begin{bmatrix} 1 & 0 & 0 & -x \\ 0 & 1 & 0 & -y \\ 0 & 0 & 1 & -z \\ 0 & 0 & 0 & 1 \end{bmatrix} \{Q_1\}$$

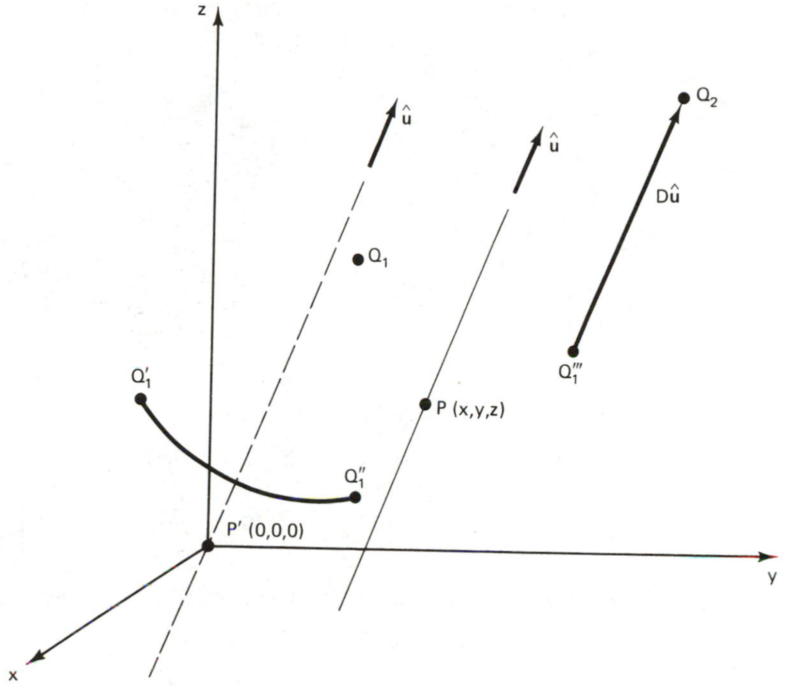


**Figure 6.28** Example 6.10.

We then obtain $Q_1''$ from $Q_1'$ by applying the rotation operator:

$$\{Q_1''\} = \begin{bmatrix} r_{11} & r_{12} & r_{13} & 0 \\ r_{21} & r_{22} & r_{23} & 0 \\ r_{31} & r_{32} & r_{33} & 0 \\ 0 & 0 & 0 & 1 \end{bmatrix} \{Q_1'\}$$

Next, we translate $P$ back to its original position and obtain $Q_1'''$.

$$\{Q_1'''\} = \begin{bmatrix} 1 & 0 & 0 & x \\ 0 & 1 & 0 & y \\ 0 & 0 & 1 & z \\ 0 & 0 & 0 & 1 \end{bmatrix} \{Q_1''\}$$

Finally, we translate parallel to the screw axis to obtain $Q_2$.

$$\{Q_2\} = \begin{bmatrix} 1 & 0 & 0 & Du_x \\ 0 & 1 & 0 & Du_y \\ 0 & 0 & 1 & Du_z \\ 0 & 0 & 0 & 1 \end{bmatrix} \{Q_1'''\}$$

Concatenating these four matrices in their proper order results in the general screw displacement matrix with its axis passing through the point $(x, y, z)$.

$$[S_{12}] = \begin{bmatrix} 1 & 0 & 0 & Du_x \\ 0 & 1 & 0 & Du_y \\ 0 & 0 & 1 & Du_z \\ 0 & 0 & 0 & 1 \end{bmatrix} \begin{bmatrix} 1 & 0 & 0 & x \\ 0 & 1 & 0 & y \\ 0 & 0 & 1 & z \\ 0 & 0 & 0 & 1 \end{bmatrix}$$

$$\begin{bmatrix} r_{11} & r_{12} & r_{13} & 0 \\ r_{21} & r_{22} & r_{23} & 0 \\ r_{31} & r_{32} & r_{33} & 0 \\ 0 & 0 & 0 & 1 \end{bmatrix} \begin{bmatrix} 1 & 0 & 0 & -x \\ 0 & 1 & 0 & -y \\ 0 & 0 & 1 & -z \\ 0 & 0 & 0 & 1 \end{bmatrix}$$

$$= \begin{bmatrix} r_{11} & r_{12} & r_{13} & Du_x + x - (r_{11}x + r_{12}y + r_{13}z) \\ r_{21} & r_{22} & r_{23} & Du_y + y - (r_{21}x + r_{22}y + r_{23}z) \\ r_{31} & r_{32} & r_{33} & Du_z + z - (r_{31}x + r_{32}y + r_{33}z) \\ 0 & 0 & 0 & 1 \end{bmatrix}$$

In this particular case, $S_{12}$ was derived in Ex. 6.9

$$[S_{12}] = \begin{bmatrix} -0.748 & 0.263 & 0.612 & -3.17 \\ 0.608 & -0.093 & 0.788 & 5.99 \\ 0.263 & 0.963 & -0.092 & 4.94 \\ 0 & 0 & 0 & 1 \end{bmatrix}$$

$\theta$, $u_x$, $u_y$, and $u_z$ can be determined by the formulas derived in Ex. 6.5:

$$\theta = \cos^{-1} [\tfrac{1}{2}(r_{11} + r_{22} + r_{33} - 1)]$$
$$= \cos^{-1} [\tfrac{1}{2}(-0.748 - 0.093 - 0.092 - 1)]$$
$$= 165°$$

$$u_x = \frac{r_{32} - r_{23}}{2 \sin \theta} = 0.333$$

$$u_y = \frac{r_{13} - r_{31}}{2 \sin \theta} = 0.667$$

$$u_z = \frac{r_{21} - r_{12}}{2 \sin \theta} = 0.667$$

With $u_x$, $u_y$, and $u_z$ determined (see Fig. 6.29), we can equate elements $r_{14}$, $r_{24}$, and $r_{34}$ of our two expressions for the screw matrix, providing us with three equations in four unknowns—$D$, $x$, $y$, and $z$. One of these unknowns, say $x$, can be assumed arbitrarily.

To determine the point at which the screw axis intersects the $yz$ plane, we set $x$ equal to zero, and solve the three equations simultaneously for $D$, $y$, and $z$.

$$D(0.333) - (0.263)y - (0.612)z = -3.17$$
$$D(0.667) + (1.093)y - (0.788)z = 5.99$$
$$D(0.667) - (0.963)y + (1.092)z = 4.96$$

The screw axis intersects the $yz$ plane at (0, 6, 6).

$$D = 6.25$$

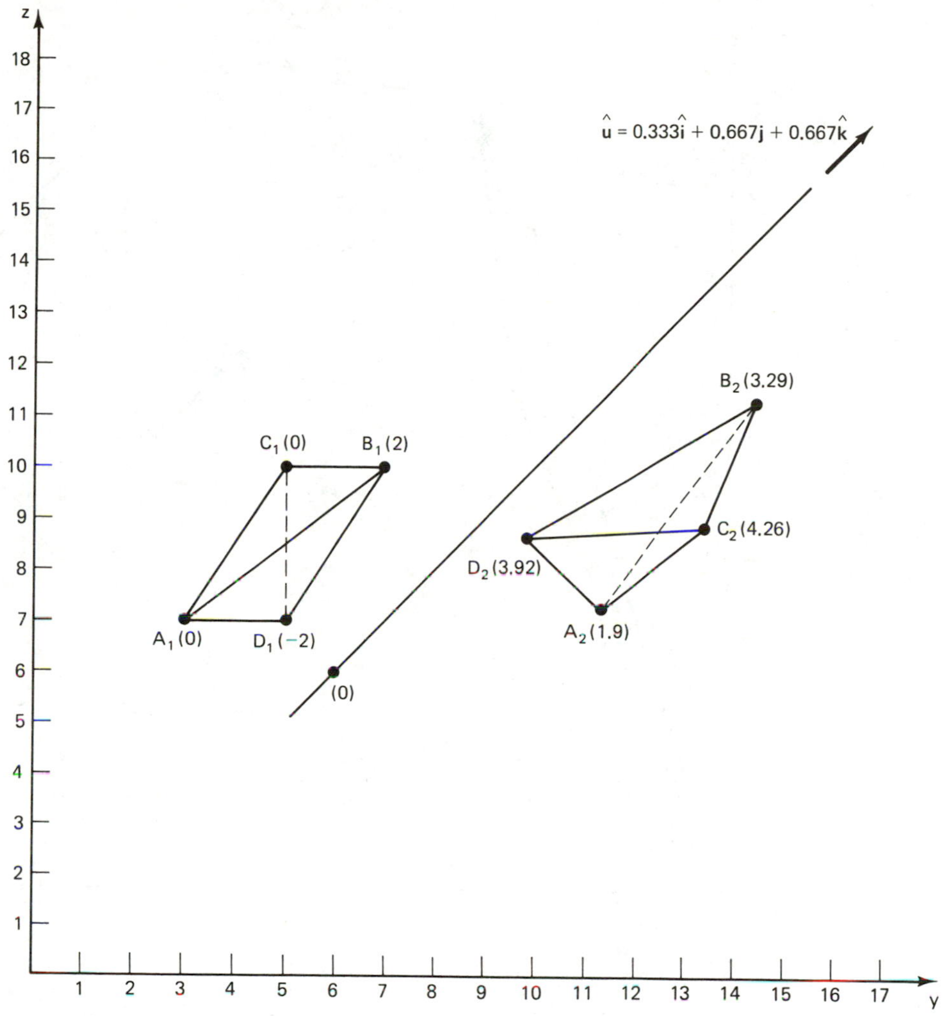


**Figure 6.29** Example 6.10: projection onto the *YZ* plane (*x* coordinate in parentheses).

## Derivation of Rotation Matrix with Axis through Origin*

In order to derive the rotation matrix with its axis through the origin [Eq. (6.22)], consider the movement of particle *P* in Fig. 6.30. Notice that *P* rotates about a point *A* on an axis through the origin. The axis is described by a unit vector **û**,

* This subsection may be skipped without loss in flow of continuity of the text.

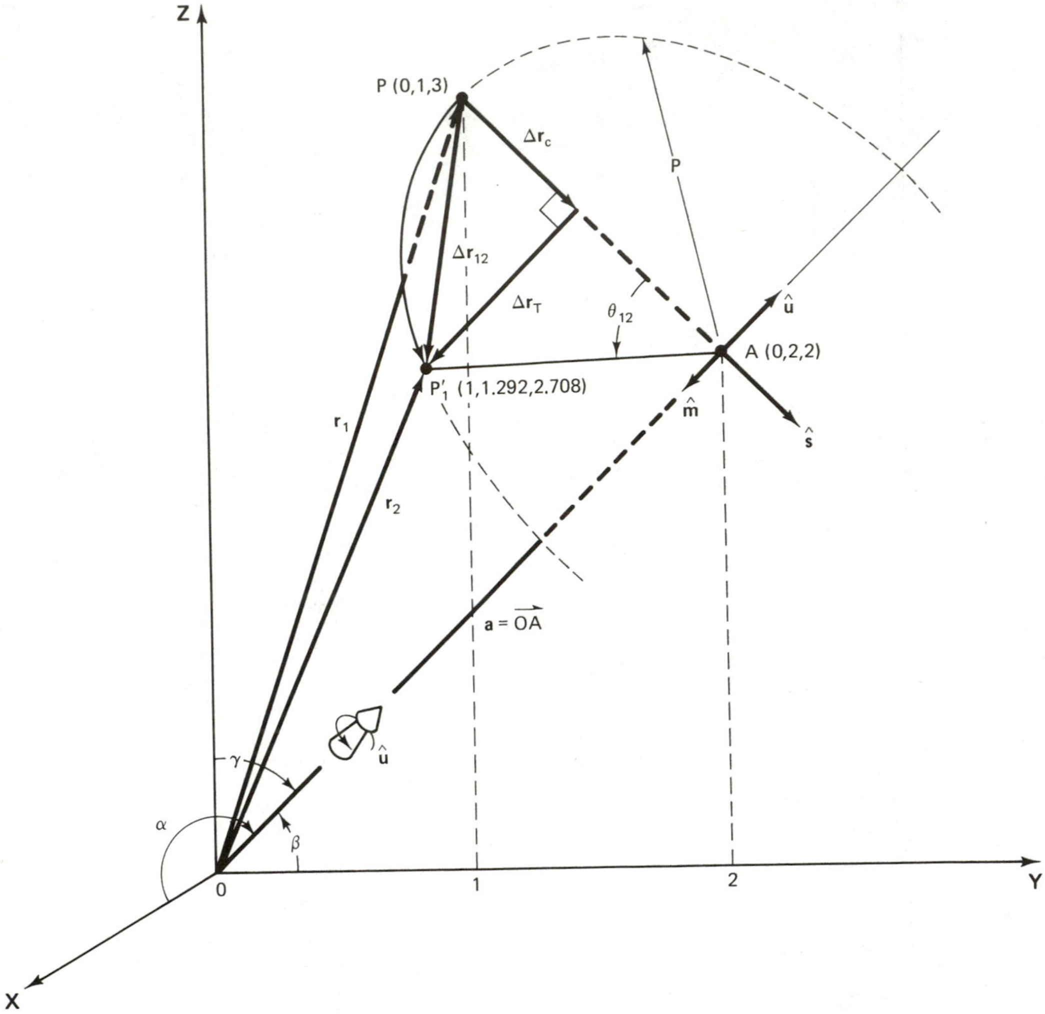


**Figure 6.30** Rotation of point P about an axis **u** which goes through the origin.

which is equivalent to describing the slope of this line in space. The coordinates of $\hat{\mathbf{u}}$ are the direction cosines of the axis:

$$\begin{aligned} u_x &= \cos\alpha \\ u_y &= \cos\beta \\ u_z &= \cos\gamma \end{aligned} \tag{6.32}$$

where $\alpha$, $\beta$, and $\gamma$ are measured from the positive $x$, $y$, and $z$ coordinate axes respectively to the vector $\hat{\mathbf{u}}$ erected at the origin. Note that

$$u_x^2 + u_y^2 + u_z^2 = 1 \qquad \text{or} \qquad \cos^2\alpha + \cos^2\beta + \cos^2\gamma = 1 \tag{6.33}$$

Let us derive a general expression for $P'$ in terms of $P$, $\hat{\mathbf{u}}$, and $\theta$, the angle of rotation. In this case,

$$\mathbf{P} = [0, 1, 3]^T \qquad \text{and} \qquad \mathbf{A} = [0, 2, 2]^T$$

so the axis is in the *yz* plane and is described by

$$\hat{\mathbf{u}} = [0, \cos 45°, \cos 45°]^T$$

Particle $P$ rotates around the axis by an angle

$$\theta_{12} = 45°$$

(measured positive using the right-hand rule with the thumb pointing in the same direction as $\hat{\mathbf{u}}$) at a radius:

$$P = |\mathbf{P} - \mathbf{A}| = \sqrt{2}$$

These choices are made for convenience of illustration, but the derivation is general. What we seek is a spatial rotation operator $R_{12}$ such that

$$\mathbf{r}_2 = R_{12}\mathbf{r}_1$$

where $\mathbf{r}_1$ for this case is $\mathbf{P}$.

To begin, several vectors must be defined. A vector from the origin to point $A$ is defined as $\mathbf{a}$, where $\mathbf{a}$ is the projection of $\mathbf{r}_1$ onto the axis $\hat{\mathbf{u}}$. To find $\mathbf{a}$, imagine a right triangle, *POA*, where $\mathbf{a}$ is one of the sides. The magnitude of $\mathbf{a}$ is $|\mathbf{r}_1| \cos (\measuredangle POA)$, the dot product of $\mathbf{r}_1$ and $\hat{\mathbf{u}}$. Since $\mathbf{a}$ is in the same direction and sense as $\hat{\mathbf{u}}$ and since $|\hat{\mathbf{u}}| = 1$, $\mathbf{a}$ is found to be

$$\mathbf{a} = (\mathbf{r}_1 \cdot \hat{\mathbf{u}})\hat{\mathbf{u}} \tag{6.34}$$

Also needed are two unit vectors defining the directions of the components of $\Delta\mathbf{r}_{12}$ in the plane of rotation. These are the centripetal ($\Delta\mathbf{r}_c$) and the tangential ($\Delta\mathbf{r}_t$) components. For this purpose, an orthonormal coordinate system may be established at $A$, with the mutually perpendicular unit vectors $\hat{\mathbf{m}}$, $\hat{\mathbf{s}}$, and $\hat{\mathbf{u}}$. The unit vector $\hat{\mathbf{s}}$ may be found by normalizing the vector $\overrightarrow{PA}$, which is $\mathbf{a} - \mathbf{r}_1$. To normalize this vector, divide it by its magnitude, the radius ($p$). Therefore, $\hat{\mathbf{s}}$ is defined as

$$\hat{\mathbf{s}} = \frac{\mathbf{a} - \mathbf{r}_1}{p} \tag{6.35}$$

Since $\hat{\mathbf{m}}$, $\hat{\mathbf{s}}$, and $\hat{\mathbf{u}}$ are orthonormal, $\hat{\mathbf{m}}$ may be defined as

$$\hat{\mathbf{m}} = \hat{\mathbf{s}} \times \hat{\mathbf{u}} = -\hat{\mathbf{u}} \times \hat{\mathbf{s}} \tag{6.36}$$

The magnitudes of $\Delta\mathbf{r}_T$ and $\Delta\mathbf{r}_c$ may be determined by referring to Fig. 6.31, which shows two triangles in the plane of rotation. It is obvious from the figure that

$$|\Delta\mathbf{r}_T| = p \sin \theta_{12} \tag{6.37}$$

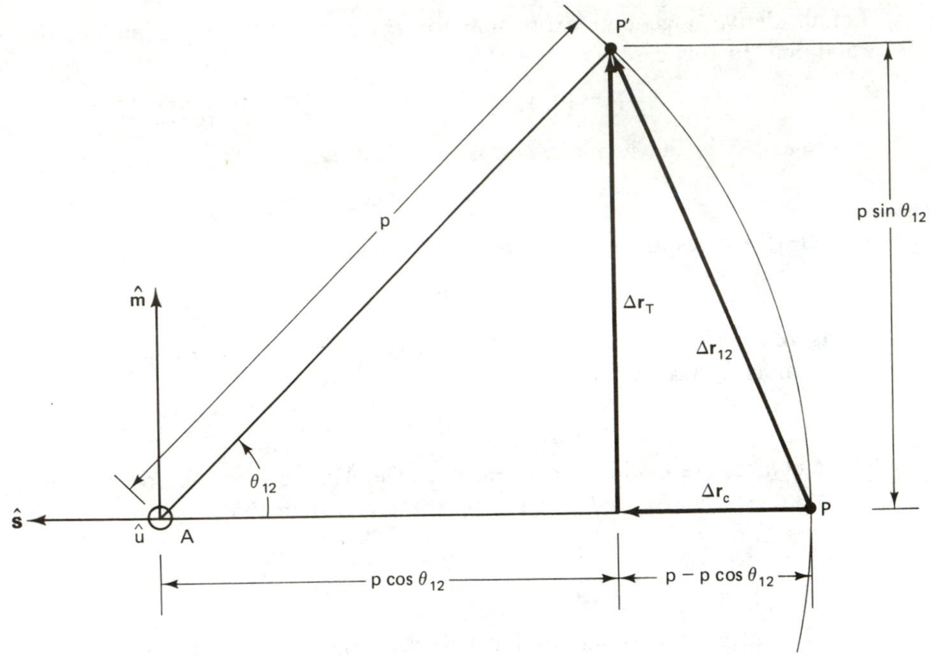


**Figure 6.31** View in the $\hat{\mathbf{s}}$,$\hat{\mathbf{m}}$ plane of Fig. 6.30.

Also note from the figure that

$$|\Delta \mathbf{r}_c| = p - p\cos\theta_{12} = p(1-\cos\theta_{12}) \tag{6.38}$$

Since $\hat{\mathbf{m}}$ and $\hat{\mathbf{s}}$ are unit vectors, the two components of $\Delta\mathbf{r}_{12}$ are found to be

$$\Delta\mathbf{r}_T = p\sin\theta_{12}\,\hat{\mathbf{m}} \qquad \text{and} \qquad \Delta\mathbf{r}_c = p \text{ vers } \theta_{12}\,\hat{\mathbf{s}} \tag{6.39}$$

where

$$\text{vers } \theta_{12} \triangleq 1-\cos\theta_{12}$$

and where vers is short for versine. Substituting for $\hat{\mathbf{m}}$ and $\hat{\mathbf{s}}$ yields

$$\Delta\mathbf{r}_t = p\sin\theta_{12}\,(\hat{\mathbf{s}}\times\hat{\mathbf{u}}) = p\sin\theta_{12}\,\frac{(\mathbf{a}-\mathbf{r}_1)\times\hat{\mathbf{u}}}{p} \tag{6.40}$$

and

$$\Delta\mathbf{r}_c = p \text{ vers } \theta_{12}\,\frac{\mathbf{a}-\mathbf{r}_1}{p}$$

which simplifies to

$$\Delta \mathbf{r}_t = \sin\theta_{12}[(\mathbf{a}\times\hat{\mathbf{u}}) - (\mathbf{r}_1\times\hat{\mathbf{u}})]$$
$$\Delta \mathbf{r}_c = \text{vers}\,\theta_{12}(\mathbf{a}-\mathbf{r}_1) \tag{6.41}$$

Since **a** and $\hat{\mathbf{u}}$ are in the same direction,

$$\mathbf{a}\times\hat{\mathbf{u}} = 0 \tag{6.42}$$

and because

$$-\mathbf{r}_1\times\hat{\mathbf{u}} = \hat{\mathbf{u}}\times\mathbf{r}_1 \tag{6.43}$$

$\Delta \mathbf{r}_t$ can be further simplified to

$$\Delta \mathbf{r}_t = \sin\theta_{12}(\hat{\mathbf{u}}\times\mathbf{r}_1) \tag{6.44}$$

Substituting for **a** yields

$$\Delta \mathbf{r}_c = \text{vers}\,\theta_{12}[(\mathbf{r}_1\cdot\hat{\mathbf{u}})\hat{\mathbf{u}} - \mathbf{r}_1] \tag{6.45}$$

The rotation of $\mathbf{r}_1$ to $\mathbf{r}_2$ can be described by the displacement vector $\Delta\mathbf{r}_{12}$, where

$$\Delta\mathbf{r}_{12} = \mathbf{r}_2 - \mathbf{r}_1$$

or

$$\mathbf{r}_2 = \mathbf{r}_1 + \Delta\mathbf{r}_{12} \tag{6.46}$$

In this case, $\Delta\mathbf{r}_{12}$ is known in terms of the angle of rotation $\theta_{12}$, the axis $\hat{\mathbf{u}}$, and $\mathbf{r}_1$, so we can derive the rotation operator $R_{12}$ as a function of these. Equation (6.46) can now be written in matrix form:

$$\begin{Bmatrix} r_{2x} \\ r_{2y} \\ r_{2z} \end{Bmatrix} = \begin{bmatrix} 1 & 0 & 0 \\ 0 & 1 & 0 \\ 0 & 0 & 1 \end{bmatrix} \begin{Bmatrix} r_{1x} \\ r_{1y} \\ r_{1z} \end{Bmatrix} + \sin\theta_{12} \begin{vmatrix} \mathbf{i} & \mathbf{j} & \mathbf{k} \\ u_x & u_y & u_z \\ r_{1x} & r_{1y} & r_{1z} \end{vmatrix}$$
$$+ \text{vers}\,\theta_{12} \begin{bmatrix} u_x(r_{1x}u_x + r_{1y}u_y + r_{1z}u_z) - r_{1x} \\ u_y(r_{1x}u_x + r_{1y}u_y + r_{1z}u_z) - r_{1y} \\ u_z(r_{1x}u_x + r_{1y}u_y + r_{1z}u_z) - r_{1z} \end{bmatrix} \tag{6.47}$$

After solving the determinant and factoring out $\mathbf{r}_1$ in the second two terms, Eq. (6.47) becomes

$$\begin{Bmatrix} r_{2x} \\ r_{2y} \\ r_{2z} \end{Bmatrix} = \left( \begin{bmatrix} 1 & 0 & 0 \\ 0 & 1 & 0 \\ 0 & 0 & 1 \end{bmatrix} + \begin{bmatrix} 0 & -u_z & u_y \\ u_z & 0 & -u_x \\ -u_y & u_x & 0 \end{bmatrix} \sin\theta_{12} \right.$$
$$\left. + \begin{bmatrix} u_x^2 - 1 & u_x u_y & u_x u_z \\ u_x y_y & u_y^2 - 1 & u_y u_z \\ u_x u_z & u_y u_z & u_z^2 - 1 \end{bmatrix} \text{vers}\,\theta_{12} \right) \begin{Bmatrix} r_{1x} \\ r_{1y} \\ r_{1z} \end{Bmatrix} \tag{6.40}$$

The rotation operator is within the parentheses. After simplifying, $[R_{12}]$ becomes:

$$[R_{12}] = \begin{bmatrix} u_x^2 \text{ vers } \theta_{12} + \cos\theta_{12} & u_x u_y \text{ vers } \theta_{12} - u_z \sin\theta_{12} & u_x u_z \text{ vers } \theta_{12} + u_y \sin\theta_{12} \\ u_x u_y \text{ vers } \theta_{12} + u_z \sin\theta_{12} & u_y^2 \text{ vers } \theta_{12} + \cos\theta_{12} & u_y u_z \text{ vers } \theta_{12} - u_x \sin\theta_{12} \\ u_x u_z \text{ vers } \theta_{12} - u_y \sin\theta_{12} & u_y u_z \text{ vers } \theta_{12} + u_x \sin\theta_{12} & u_z^2 \text{ vers } \theta_{12} + \cos\theta_{12} \end{bmatrix} \quad (6.22)$$

## 6.11 ANALYSIS OF SPATIAL MECHANISMS*

In this and succeeding sections the 4 × 4 transformation matrices introduced in previous sections are adopted to model spatial mechanisms for the purposes of analysis. The modeling procedure discussed below begins by establishing a local Cartesian coordinate system for each link. The location and orientation of the local coordinate system on each link is arbitrary; however, it is usually chosen at a joint center. The modeling task then becomes one of describing how the coordinate systems of connected links move *relative* to one another. Consequently, transformation matrices must be defined to describe these relative motions. It will be shown that they are similar to the 4 × 4 matrices introduced earlier. The methods introduced in this section are especially useful when programmed for a digital computer. Most of the operations are matrix and vector manipulations.

For example, a typical task of displacement analysis is to determine the position of a point on a moving link in the fixed coordinate system attached to the frame or ground link. To illustrate, let point $P$ be attached to link 2 (see Fig. 6.32) and its location be defined in the *local coordinate system* $x_2y_2z_2$ embedded in the moving

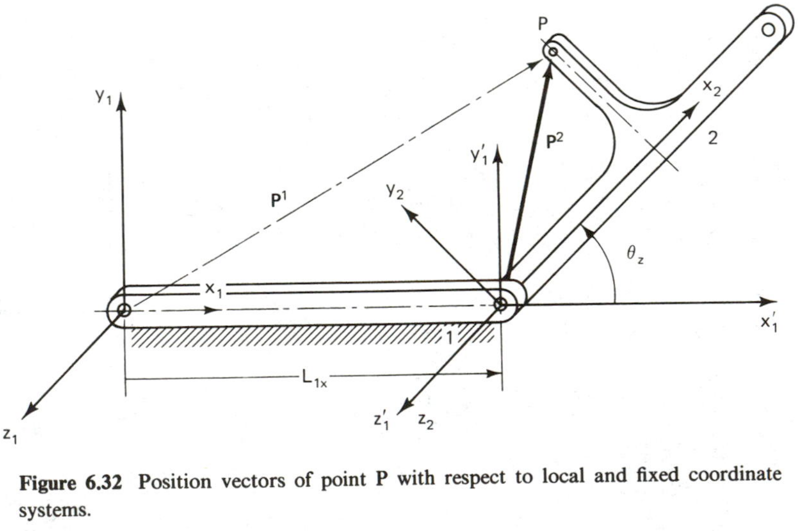


**Figure 6.32** Position vectors of point P with respect to local and fixed coordinate systems.

* The authors wish to acknowledge the contribution of Robert Williams, employed at Control Data Corporation, Minneapolis, Minnesota, to the development of parts of Secs. 6.11 to 6.17.

link 2, by the vector $\mathbf{P}^2$. To find the location of $P$ with respect to the *global coordinate system* $x_1y_1z_1$ embedded in the fixed link 1, namely $\mathbf{P}^1$, after link 2 has been rotated by the angle $\theta_z$, we must account for the fact that system $x_2y_2z_2$ has been obtained from $x_1y_1z_1$ by combined rotation and translation. In the following sections we derive the necessary transformation matrices for various spatial link shapes and joint motions to accomplish the transformation as follows:

$$\mathbf{P}^1 = T_{12}\mathbf{P}^2 \tag{6.49}$$

It is important to observe how the matrix operator $T_{12}$ is constructed to obtain $\mathbf{P}^1$ from $\mathbf{P}^2$. Observe that $\mathbf{P}^2$ is defined in the $x_2y_2$ coordinate system. Also note that $x_2y_2$ is obtained from $x_1'y_1'$ by a rotation through the angle $\theta_z$ about $z_1'$. Therefore, the argument of $\mathbf{P}^2$ expressed in the $x_1'y_1'$ system is increased by $\theta_z$ from its argument in the $x_2y_2$ system. Hence $(\mathbf{P}^2)' = [R(\theta_z)]\mathbf{P}^2$. Next, observe that $x_1'y_1'$ came from $x_1y_1$ by a translation $L_{1x}$. Thus the $x$ coordinate of any vector in the $x_1'y_1'$ system is increased by $L_{1x}$ when the vector is expressed in the $x_1y_1$ system. As a result, $\mathbf{P}^1 = [T(L_{1x})](\mathbf{P}^2)' = [T(L_{1x})]\,[R(\theta_z)]\mathbf{P}^2$. So we see that the operator $T_{12}$ is obtained by the concatenation of a rotation and a translation matrix in that order:

$$T_{12} = [T(L_{1x})]\,[R(\theta_z)] \tag{6.50}$$

### Elementary 4 × 4 Transformation Matrices

There are seven elementary forms of 4 × 4 transformation matrices which, when used in suitable combination, describe most joints found in mechanisms (spatial or planar). They can also be used to describe complex link shapes between the joints. The seven matrices are shown schematically in Fig. 6.33. The first three are combined rotation–translation matrices [refer to Eqs. (6.23, 6.24, 6.25 and 6.28)]. They are intended for transforming a position vector $\mathbf{P}^j$ of point $P$ embedded in the $O_jx_jy_jz_j$ coordinate system to the vector $\mathbf{P}^i$, locating the same point $P$ in the $O_ix_iy_iz_i$ coordinate system.

Since, in general, the $j$th system is obtained from the $i$th system by a translation $(a, b, c)$ to $O_j$ and is then rotated about $O_j$ to its $j$th system orientation, the transformation from $\mathbf{P}^j$ to $\mathbf{P}^i$ is obtained similarly to the previously described planar example, namely, by concatenating the rotation and translation matrices in that order. For example, the translation [from Eq. (6.28)] is described as

$$T(\mathbf{D}) = \begin{bmatrix} 1 & 0 & 0 & a \\ 0 & 1 & 0 & b \\ 0 & 0 & 1 & c \\ 0 & 0 & 0 & 1 \end{bmatrix} \tag{6.51}$$

and the rotation about the $x$ axis as

$$R(\theta_x) = \begin{bmatrix} 1 & 0 & 0 & 0 \\ 0 & c\theta_x & -s\theta_x & 0 \\ 0 & s\theta_x & c\theta_x & 0 \\ 0 & 0 & 0 & 1 \end{bmatrix} \tag{6.52}$$

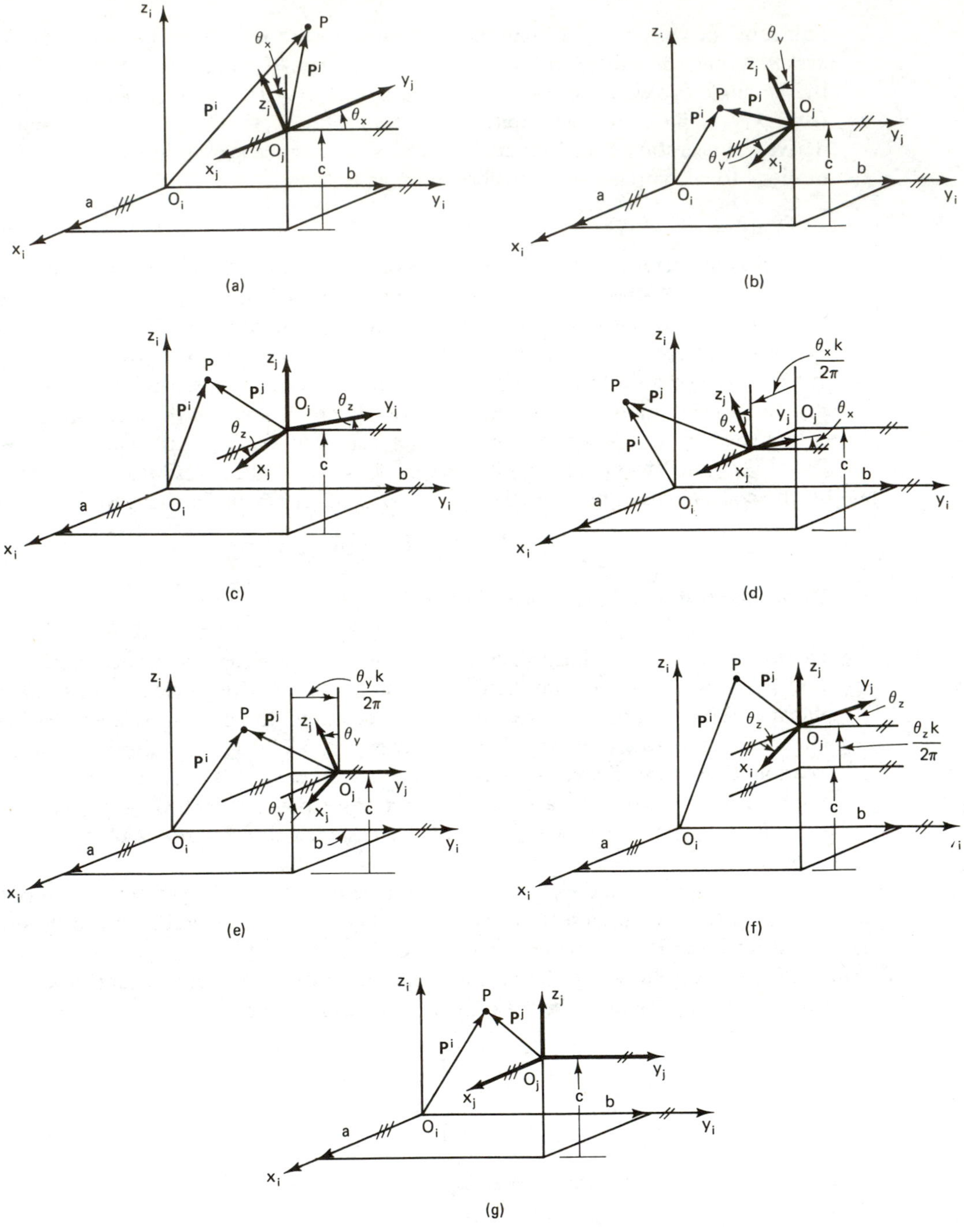


**Figure 6.33** Three-dimensional representations for the seven elementary transformation matrices.

Concatenating these for rotation followed by translation will yield the matrix shown in Fig. 6.33a:

$$T^1_{ij}\,(a, b, c, \theta_x) = \begin{bmatrix} 1 & 0 & 0 & a \\ 0 & 1 & 0 & b \\ 0 & 0 & 1 & c \\ 0 & 0 & 0 & 1 \end{bmatrix} \begin{bmatrix} 1 & 0 & 0 & 0 \\ 0 & c\theta_x & -s\theta_x & 0 \\ 0 & s\theta_x & c\theta_x & 0 \\ 0 & 0 & 0 & 1 \end{bmatrix} = \begin{bmatrix} 1 & 0 & 0 & a \\ 0 & c\theta_x & -s\theta_x & b \\ 0 & s\theta_x & c\theta_x & c \\ 0 & 0 & 0 & 1 \end{bmatrix} \tag{6.53}$$

The same translation followed by rotations about the $y$ and $z$ axes are found in a similar manner and are shown in Fig. 6.33b and c, respectively. These are followed by the three matrices representing the same translation following a screw motion about the $x$, $y$, and $z$ axes, respectively, in Fig. 6.33d, e, and f [refer to Eq. (6.30)]. Finally, the last matrix, the seventh, is the pure translation [see Eq. (6.28)], shown in Fig. 6.33g.

1. Case 1: System $x_jy_jz_j$ is obtained from $x_iy_iz_i$ by translation $\mathbf{D} = [a, b, c,]^T$ followed by rotation about its translated origin at $(a, b, c)$ by the angle $\theta_x$ around $\hat{\mathbf{u}} = [u_x, 0, 0]^T$. Thus for this case, the transformation to obtain $\mathbf{P}^i$ from $\mathbf{P}^j$ is: $\mathbf{P}^i = T^1_{ij}\mathbf{P}^j$ where, from Eq. (6.53).

$$T^1_{ij}(a, b, c, \theta_x) = \begin{bmatrix} 1 & 0 & 0 & a \\ 0 & c\theta_x & -s\theta_x & b \\ 0 & s\theta_x & c\theta_x & c \\ 0 & 0 & 0 & 1 \end{bmatrix}$$

2. Case 2: Similar to case 1, except the rotation is $\theta_y$. Here $\mathbf{P}^i = T^2_{ij}\mathbf{P}^j$, where

$$T^2_{ij}(a, b, c, \theta_y) = \begin{bmatrix} c\theta_y & 0 & s\theta_y & a \\ 0 & 1 & 0 & b \\ -s\theta_y & 0 & c\theta_y & c \\ 0 & 0 & 0 & 1 \end{bmatrix}$$

3. Case 3 is also similar to case 1 except for rotation about the $z$ axis. Again, $\mathbf{P}^i = T^3_{ij}\mathbf{P}^j$, where

$$T^3_{ij}(a, b, c, \theta_z) = \begin{bmatrix} c\theta_z & -s\theta_z & 0 & a \\ s\theta_z & c\theta_z & 0 & b \\ 0 & 0 & 1 & c \\ 0 & 0 & 0 & 1 \end{bmatrix}$$

4. Case 4: The system $x_jy_jz_j$ is obtained from $x_iy_iz_i$ by translation $\mathbf{D} = [a, b, c,]^T$ followed by a screw motion in the $x$ direction. Now $\mathbf{P}^i = T^4_{ij}\mathbf{P}^j$, where

$$T^4_{ij}(a, b, c, \theta_x, k) = \begin{bmatrix} 1 & 0 & 0 & a + \dfrac{\theta_x k}{2\pi} \\ 0 & c\theta_x & -s\theta_x & b \\ 0 & s\theta_x & c\theta_x & c \\ 0 & 0 & 0 & 1 \end{bmatrix}$$

and where $k$ is the lead of the screw.

5. Case 5 is similar to case 4, except that the screw motion of the coordinate system takes place in the $y$ direction. Here $\mathbf{P}^i = T_{ij}^5\mathbf{P}^j$, where

$$T_{ij}^5(a, b, c, \theta_y, k) = \begin{bmatrix} c\theta_y & 0 & s\theta_y & a \\ 0 & 1 & 0 & b + \dfrac{\theta_y k}{2\pi} \\ -s\theta_y & 0 & c\theta_y & c \\ 0 & 0 & 0 & 1 \end{bmatrix}$$

6. Case 6 applies when the screw motion in the $z$ direction of the coordinate system follows its $a$, $b$, $c$ translation. Thus $\mathbf{P}^i = T_{ij}^6\mathbf{P}^j$, where

$$T_{ij}^6(a, b, c, \theta_z, k) = \begin{bmatrix} c\theta_z & -s\theta_z & 0 & a \\ s\theta_z & c\theta_z & 0 & b \\ 0 & 0 & 1 & c + \dfrac{\theta_z k}{2\pi} \\ 0 & 0 & 0 & 1 \end{bmatrix}$$

7. Case 7 applies when the system $x_j\, y_j\, z_j$ is obtained from $x_i y_i z_i$ by translation $\mathbf{D} = [a, b, c]^T$ only. For this case $\mathbf{P}^i = T_{ij}^7\mathbf{P}^j$, where

$$T_{ij}^7(a, b, c) = \begin{bmatrix} 1 & 0 & 0 & a \\ 0 & 1 & 0 & b \\ 0 & 0 & 1 & c \\ 0 & 0 & 0 & 1 \end{bmatrix}$$

It is important to note that the translations and rotations in these matrices describe *relative* motion of a coordinate system moving from coincidence with the $i$th system to coincidence with the $j$th system. This differs from earlier interpretations of these transformations which were defined as absolute translations and rotations since they were "relative" to the fixed $(x_0, y_0, z_0)$ reference coordinate system. Even though the interpretation of the quantities is different here, the form of the matrices does not change and will thus be used without further derivation.

It may be instructive to investigate one of these translation-plus-rotational matrices further to show in more detail the motion described. Consider the third rotational matrix—translation of the coordinate system plus its rotation about the translated $z$ axis—for the case when $c = 0$. This reduces the motion to two dimensional planar motion of the type (in the $xy$ plane) used earlier in the text. Figure 6.34 shows this planar simplification. As an example, let us find the location in the $i$th coordinate system of point $\mathbf{P}_{1j}$ defined in the $j$th coordinate system. Recall that the $j$th coordinate system was obtained from the $i$th coordinate system by way of a translation of $(a, b)$ and rotation $\theta_z$ of the $i$th system. The location of point $P_1$, defined by $\mathbf{P}_{1j}$ in the $j$th system, is identified as $\mathbf{P}_{1i}$ referred to the $i$th system.

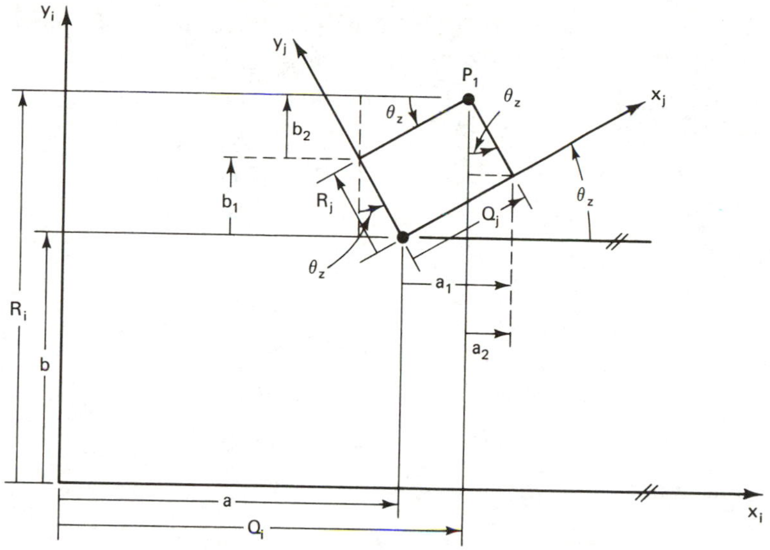


**Figure 6.34** Components of point $P_1$ in two coordinate systems.

Expressed as a matrix product,

$$\{P_{1i}\} = [S_{ij}]\{P_{1j}\} \tag{6.54}$$

where

$$\{P_{1j}\} = \begin{bmatrix} Q_j \\ R_j \\ 0 \\ 1 \end{bmatrix} \quad \text{and} \quad \{P_{1i}\} = \begin{bmatrix} Q_i \\ R_i \\ 0 \\ 1 \end{bmatrix} \tag{6.55}$$

Note that four homogeneous coordinates are used as before. Let us assume that $S_{ij}$ is the third elementary matrix shown in Fig. 6.33c. Equation (6.54) becomes

$$\{P_{1i}\} = T^3_{ij}(a, b, 0, \theta_z)\{P_{1j}\} \tag{6.56}$$

After substitution

$$\begin{bmatrix} Q_i \\ R_i \\ 0 \\ 1 \end{bmatrix} = \begin{bmatrix} c\theta_z & -s\theta_z & 0 & a \\ s\theta_z & c\theta_z & 0 & b \\ 0 & 0 & 1 & 0 \\ 0 & 0 & 0 & 1 \end{bmatrix} \begin{bmatrix} Q_j \\ R_j \\ 0 \\ 1 \end{bmatrix} \tag{6.57}$$

Expanding this matrix product, $\{P_{1i}\}$ is defined as

$$\{P_{1i}\} = \begin{bmatrix} Q_i \\ R_i \\ 0 \\ 1 \end{bmatrix} = \begin{bmatrix} Q_j\, c\theta_z - R_j\, s\theta_z + a \\ Q_j\, s\theta_z + R_j\, c\theta_z + b \\ 0 \\ 1 \end{bmatrix} \tag{6.58}$$

Let us now look closer at Fig. 6.34, where components of $\mathbf{P}_1$ have been defined in both the $i$th and the $j$th coordinate systems.

From Fig. 6.34,

$$Q_i = a + a_1 - a_2$$
$$R_i = b + b_1 + b_2 \tag{6.59}$$

where

$$a_1 = Q_j \, c\theta_z$$
$$a_2 = R_j \, s\theta_z$$
$$b_1 = R_j \, c\theta_z \tag{6.60}$$
$$b_2 = Q_j \, s\theta_z$$

Substituting into (6.59), we obtain

$$Q_i = a + Q_j \, c\theta_z - R_j \, s\theta_z$$
$$R_i = b + R_j \, c\theta_z + Q_j \, s\theta_z \tag{6.61}$$

which is the same as shown in Eq. (6.58).

Continuing with our discussion of the remaining elementary transformation matrices, the $T_{ij}^4$, $T_{ij}^5$, and $T_{ij}^6$ matrices in Fig. 6.33 describe a screw motion of the coordinate system along a translated coordinate axis. These screw motions consist of a second translation of the system along an already translated axis and a simultaneous, proportional rotation about the same axis. Such screw motion is described by the lead of the screw ($k$) expressed as units of length per rotation of the screw. The screw motion of the coordinate system follows a translation of that system. It is left to the reader to investigate these matrices in more detail as was done for the translation-plus-rotational matrices.

The final matrix $T_{ij}^7$ shown in Fig. 6.33g is a "translation only" matrix. It describes a pure translation of a coordinate system form the $i$th to the $j$th position.

## 6.12 LINK AND JOINT MODELING WITH ELEMENTARY MATRICES

Some illustrative examples demonstrating the use of the elementary matrices for modeling links and joints of mechanisms are shown in this section. Perhaps the most common joint is the revolute. An example is shown in Fig. 6.35, connecting two links numbered $i$ and $j$. Before the joint can be modeled, local coordinate systems on the two links must be defined. In the figure these coordinate systems are denoted as $(x_\ell, y_\ell, z_\ell)$ and $(x_m, y_m, z_m)$. The identification of the link (e.g., $i$) will in general not necessarily correspond to the identification of the coordinate system (e.g., $\ell$) as some joints and links require intermediate coordinate systems in this definition (see the next example). Link $i$ and the revolute joint $ij$ in Fig. 6.35 is described using a type 3 elementary matrix, since the rotation is about a $z$ axis. Prior to the rotation

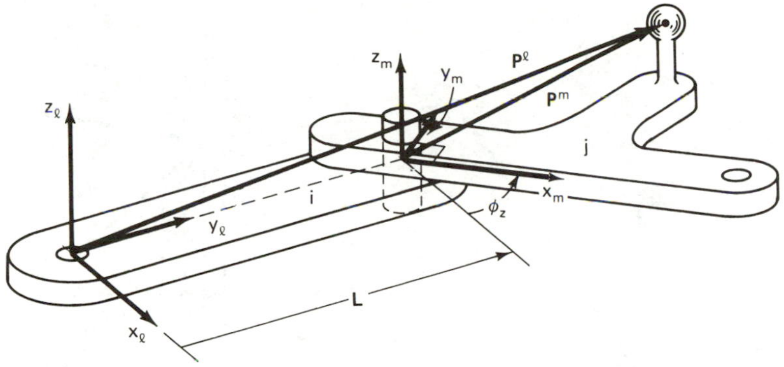


**Figure 6.35** Revolute joint on a binary link.

of this joint, there is a translation in the $y_\ell$ direction of length $L$. Therefore, the description of the $i$th link with the $ij$ joint is

$$S_{\ell m}(\phi_z) = T^3_{\ell m}(O, L, O, \phi_z) = \begin{bmatrix} c\phi_z & -s\phi_z & 0 & 0 \\ s\phi_z & c\phi_z & 0 & L \\ 0 & 0 & 1 & 0 \\ 0 & 0 & 1 & 1 \end{bmatrix} \tag{6.62}$$

The symbol $S_{\ell m}(\phi_z)$ is used to represent a general $4 \times 4$ transformation matrix. It is used throughout the remainder of this chapter to represent a general matrix. Furthermore, a quantity or quantities in parentheses following the symbol $S$ denote the relative motion variables in the matrix. Since the link length $L$ is a constant, although it appears in the matrix of Eq. (6.62), it is omitted from the symbol of the operator $S_{\ell m}(\phi_z)$.

Here is how this operator works. Suppose that we know the location of point $P$ of link $j$ with respect to the coordinate system $x_m y_m z_m$ embedded in link $j$. Let this position vector be denoted as $\mathbf{P}^m$. Now, if we want to know the position of $P$ referred to the $x_\ell y_\ell z_\ell$ system of link $i$, namely $\mathbf{P}^\ell$, we obtain it by

$$\mathbf{P}^\ell = S_{\ell m}(\phi_z)\mathbf{P}^m \tag{6.63}$$

This notion indicates that, while $\mathbf{P}^m$ is a constant (because it is embedded in the rigid link $j$), $\mathbf{P}^\ell$ is a function of $\phi_z$ and therefore changes as link $j$ rotates relative to link $i$.

A cylindrical joint is shown in Fig. 6.36 between links $m$ and $n$. To describe link $m$ and this joint, two elementary matrices are required. One matrix describes the translation along link $m$ and the translation along the axis of the joint. A second matrix covers the rotation of the joint. Thus the link $m$ and the cylindrical joint between links $m$ and $n$ shown in Fig. 6.36 are defined as follows:

$$S_{j\ell}(B, \psi_y) = T^7_{jk}(A, B, 0)\; T^2_{k\ell}(0, 0, 0, \psi_y) \tag{6.64}$$

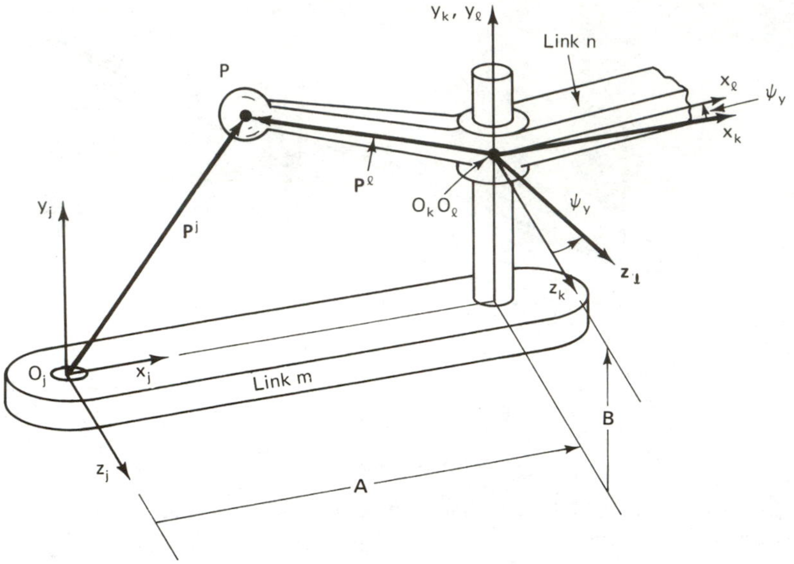


**Figure 6.36** Cylindrical joint with a binary link.

Or, after multiplying the matrices,

$$S_{j\ell}(B, \psi_y) = \begin{bmatrix} c\psi_y & 0 & s\psi_y & A \\ 0 & 1 & 0 & B \\ -s\psi_y & 0 & c\psi_y & 0 \\ 0 & 0 & 0 & 1 \end{bmatrix} \tag{6.65}$$

With this, in Fig. 6.36,

$$\mathbf{P}^j = S_{j\ell}(B, \psi_y)\mathbf{P}^\ell \tag{6.66}$$

Note, that for this case, $B$ and $\psi_y$ represent the degrees of relative freedom in this joint.

A *screw joint* is shown in Fig. 6.37. The screw axis is in the $z_j$ direction, so a type 6 matrix from Fig. 6.33 describes this joint. This matrix is written as

$$S_{ij}(\theta_z) = T^6_{ij}(A, B, C, \theta_z, k) = \begin{bmatrix} c\theta_z & -s\theta_z & 0 & A \\ s\theta_z & c\theta_z & 0 & B \\ 0 & 0 & 1 & C + \dfrac{k\theta_z}{2\pi} \\ 0 & 0 & 0 & 1 \end{bmatrix} \tag{6.67}$$

and in Fig. 6.37,

$$\mathbf{P}^i = S_{ij}(\theta_z)\mathbf{P}^j \tag{6.68}$$

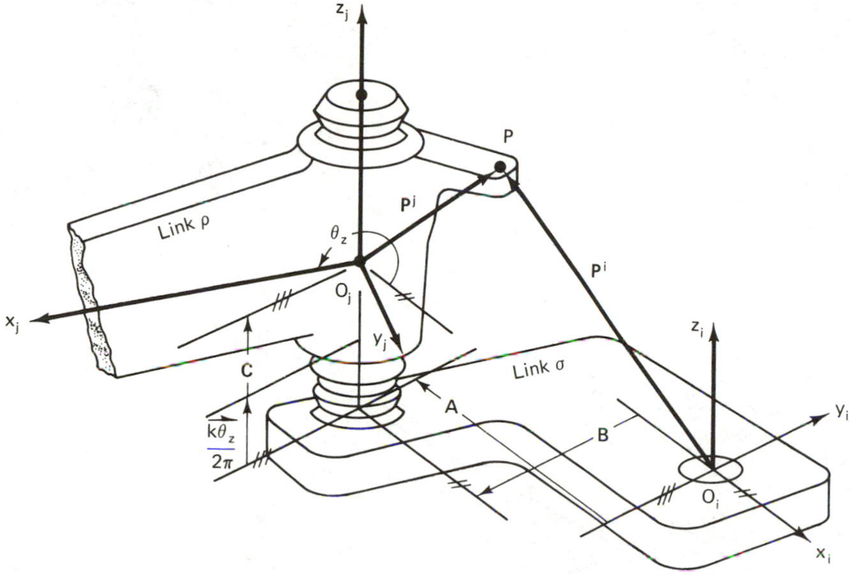


**Figure 6.37** Screw joint with a binary link.

Note that for this joint two of the constant translations are in the negative directions of $x_i$ and $y_i$; consequently, $A$ and $B$ would have negative values. Also, the screw joint has rotated by a little less than 270°. This leaves the $x_j$ axis pointing in nearly the negative direction of the $y_i$ axis and the rotation has caused a translation in the direction of the screw of $k\theta_z/2\pi$. Thus, if this joint were reverse-rotated from its current position by the amount $\theta_z$, the $x_jy_jz_j$ coordinate system would be parallel to $x_iy_iz_i$ and its origin would be a distant $C$ above the $x_iy_i$ plane. The single degree of relative freedom in this joint is $\theta$, because $k$, the lead of the screw, is constant. For a right-hand screw, $k$ is positive; for a left-hand screw, $k$ is negative.

## Spherical Joint

A common joint appearing in many spatial mechanisms is the spherical joint. An example is shown in Fig. 6.38. This joint allows freedom of rotation around all three axes of one link relative to another. Typically, these rotations are described using *Euler angles.* The elementary rotational matrices shown in Fig. 6.33 are used to describe the spherical joint in a manner similar to that resulting from an Eulerian rotation. For purposes of discussion, the spherical joint is represented by its kinematic equivalent of three revolute joints whose axes of rotation intersect at the center of the spherical joint. This equivalent representation is shown in Fig. 6.39 together with two added coordinate systems used to aid in the description of the joint. The transformation matrices defining this joint are

$$S_{i\ell} = T^3_{ij}(0, L_i, 0, \theta_{zj})T^1_{jk}(0, 0, 0, \psi_{xj})T^3_{k\ell}(0, 0, 0, \phi_{zk}) \tag{6.69}$$

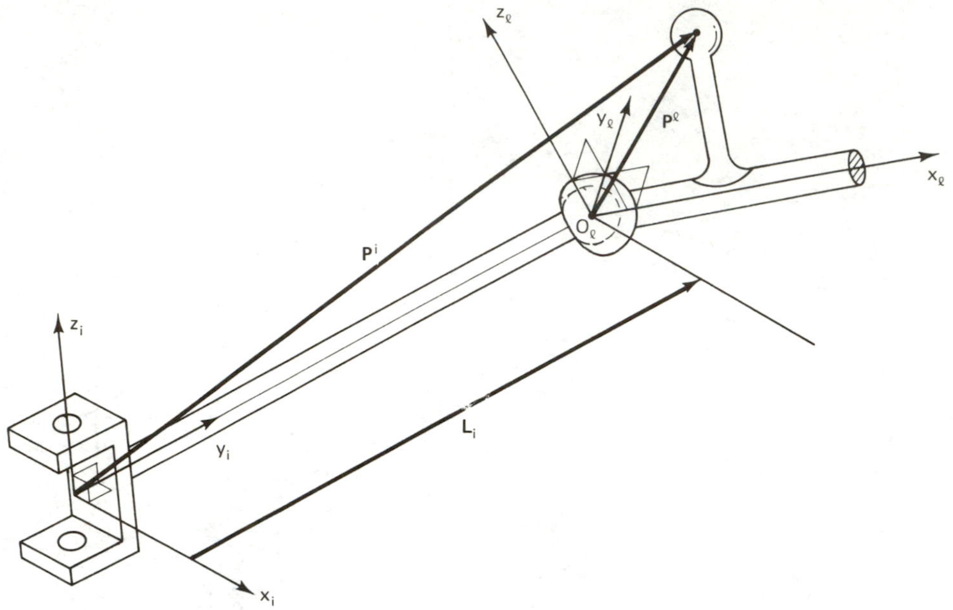


**Figure 6.38** Spherical joint.

The first matrix defines the translation of the coordinate system along the $y_i$ axis by distance $L_i$ followed by a rotation $\theta_{zj}$ about an axis parallel to the $z_i$ axis. This defines the $x_j y_j z_j$ coordinate system. The second matrix defines a rotation of the $j$th system about the $x_j$ axis by the angle $\psi_{xj}$, which locates the $x_k y_k z_k$ coordinate system. The final matrix represents a rotation of the $k$th system about the $z_k$ axis by the angle $\phi_{zk}$, which defines the final $x_\ell y_\ell z_\ell$ coordinate system. Upon concatenating the matrice defined in Eq. (6.69), the following operator is found (see Exer. 6.33):

$$S_{i\ell}(0, L_i, 0, \theta_{zj}, \psi_{xj}, \phi_{zk}) =$$
$$\begin{bmatrix} c\theta_{zj}c\phi_{zk} - s\theta_{zj}c\psi_{xj}s\phi_{zk} & -c\theta_{zj}s\phi_{zk} - s\theta_{zj}c\psi_{xj}c\phi_{zk} & s\theta_{zj}s\psi_{xj} & 0 \\ s\theta_{zj}c\phi_{zk} + c\theta_{zj}c\psi_{xj}s\phi_{zk} & -s\theta_{zj}s\phi_{zk} + c\theta_{zj}c\psi_{xj}c\phi_{zk} & -c\theta_{zj}s\psi_{xj} & L_i \\ s\psi_{xj}s\phi_{zk} & s\psi_{xj}c\phi_{zk} & c\psi_{xj} & 0 \\ 0 & 0 & 0 & 1 \end{bmatrix} \tag{6.70}$$

With this, in Figs. 6.38 and 6.39,

$$\mathbf{P}^i = S_{i\ell}(\theta_{zj}, \psi_{xj}, \phi_{zk})\mathbf{P}^\ell \tag{6.71}$$

The $3 \times 3$ rotational matrix in (6.70) is the *Eulerian rotation transformation.*

The three elementary matrices used above to describe the spherical joint comprise only one combination of many that can be used. For example, if the coordinate axes shown in Fig. 6.38 were renamed, but still retained their relative orientation, the successive rotations described in (6.69) would be rotations about different axes. This would cause the operator $S_{i\ell}$ to appear different; however, it would still define the same overall relative translation plus rotation.

Visualization of Euler angles can be made easy by the use of equivalent revolute joints shown axonometrically in Fig. 6.39 together with drawn-to-scale orthographic projections of the successive translation and rotations of the $xyz$ coordinate system originally embedded in link $i$. Let us follow through the construction of these orthographic projections (o.p.'s) in detail.

Refer first to Fig. 6.39 and observe that the first o.p. is to be the view A-A, looking against the positive $x_i$ axis. Thus in Fig. 6.40 view A-A shows $x_i$ as an encircled dot, and it shows the $y_i$ and $z_i$ axes in true length. It also shows that the true length of link $i$ as the vector $\mathbf{L}_i$.

Now we return to Fig. 6.39 and see that view B-B looks against the positive $z_i$. Thus in Fig. 6.40, view B-B shows $z_i$ as an encircled dot and $x_i$ and $y_i$ in true length. This view also shows the translation of the origin of the $xyz$ coordinate system from $O_i$ to $O_j$ by $L_i = (0, 50.8, 0)$ mm, and the rotation $\theta_{zj} = 30°$, both in true shape. These take the system to $O_jx_jy_jz_j$, with $z_j$ appearing as an encircled dot and $x_j$ and $y_j$ shown in true length.

Back again to the axonometric drawing in Fig. 6.39, we note that view C-C is to look against the positive $x_j$ axis. Thus in the orthographic view C-C of Fig. 6.40, $x_j$ appears as an encircled dot and $y_j$ and $z_j$ are shown in true length, as well as the rotation $\psi_{xj} = 30°$, which takes the system to $x_ky_kz_k$, where $x_k \equiv x_j$. Observe that $O_k$ remains at $O_j$.

Returning once more to the orthographic picture, we see that view D-D will look against the positive $z_k$ axis. Therefore, in the o.p. view D-D, $z_k$ shows up as an encircled dot, $x_k$ and $y_k$ are true length, and so is the third Euler angle $\phi_{zk} = 30°$. With this last rotation the system has reached the position $O_\ell x_\ell y_\ell z_\ell$.

Observe now that in the o.p.'s constructed so far, all three Euler angles and all coordinate axes—original, intermediate, and final—are shown in true shape, without distortion. However, to show $z_\ell$ in true length also, we need to construct one more o.p. This is view E-E in Fig. 6.40.

With the five o.p.'s of Fig. 6.40, we have developed a clearly visualizable geometric method to deal with Euler angle transformation. Let us now use this new tool to verify our analytical method represented by Eqs. (6.70) and (6.71).

**Example 6.11**

Let the spatial dyad of Fig. 6.38 consisting of links $i$ and $\ell$ be connected by a spherical joint, modeled mathematically by the transformation $S_{i\ell}(\theta_{zj}, \psi_{xj}, \phi_{zk})$, where $L_i = 50.8$ mm, and all three Euler angles, $\theta_{zj}$, $\psi_{xj}$, and $\phi_{zk}$, are variable, representing the three rotational degrees of freedom of the spherical joint. The kinematically equivalent four-link open chain, with three revolutes having concurrent axes, shown in Fig. 6.39, separates these three Euler rotations. Furthermore, these rotations appear in true shape in the orthographic projections of Fig. 6.40.

To verify Eq. (6.71), we start by embedding the point $P$ in the final, $x_\ell y_\ell z_\ell$ system

**Figure 6.39** Three-revolute equivalent of a spherical joint.

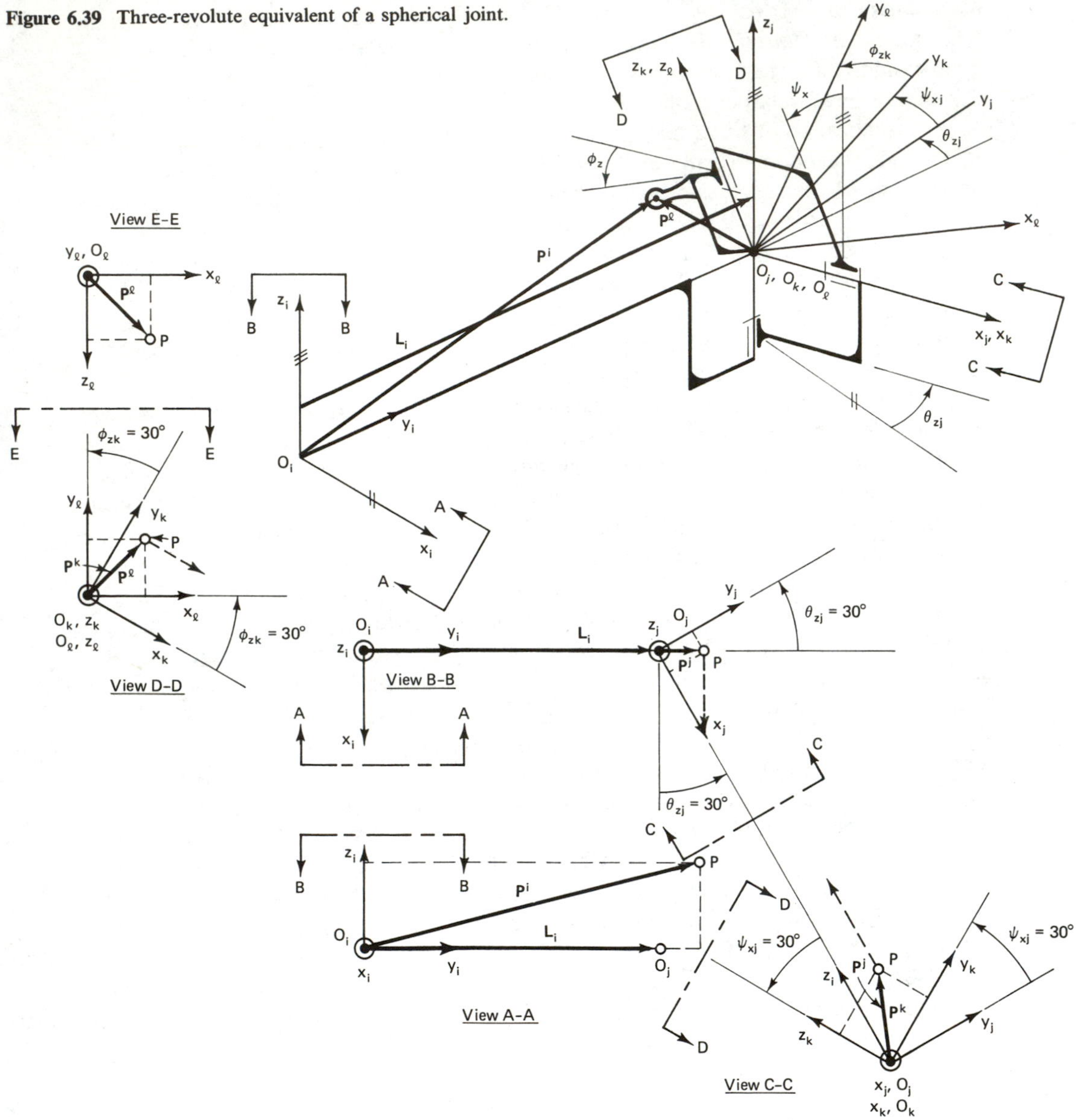


**Figure 6.40** Orthographic views from fig. 6.39.

at $\mathbf{P}^{\ell} = (10, 10, 10)$ (all Cartesian coordinates are in millimeters), shown in views D-D and E-E. In the $k$th system,

$$\mathbf{P}^{k} = T^{3}_{k\ell}(0, 0, 0, \phi_{zk})\mathbf{P}^{\ell} = \begin{bmatrix} c\phi_{zk} & -s\phi_{zk} & 0 & 0 \\ s\phi_{zk} & c\phi_{zk} & 0 & 0 \\ 0 & 0 & 1 & 0 \\ 0 & 0 & 0 & 1 \end{bmatrix} \begin{bmatrix} 10 \\ 10 \\ 10 \\ 1 \end{bmatrix} = \begin{bmatrix} 3.660 \\ 13.660 \\ 10.000 \\ 1 \end{bmatrix}$$

where $\quad c\phi_{zk} = \cos 30° = 0.866$

$$s\phi_{zk} = \sin 30° = 0.500$$

The coordinates of $\mathbf{P}^{k}$ in the $k$th system can be verified by scaling in views D-D and C-C. Note that the $P^{k}_{y}$ coordinate was projected back to view C-C from view D-D and the $P^{k}_{z}$ coordinate, which is equal to $P^{\ell}_{z} = 10$ mm, was laid out in view C-C parallel with the $z_k$ axis.

Next, we compute

$$\mathbf{P}^{j} = T^{1}_{jk}(0, 0, 0, \psi_{xj})\mathbf{P}^{k} = \begin{bmatrix} 1 & 0 & 0 & 0 \\ 0 & c\psi_{xj} & -s\psi_{xj} & 0 \\ 0 & s\psi_{xj} & c\psi_{xj} & 0 \\ 0 & 0 & 0 & 1 \end{bmatrix} \begin{bmatrix} 3.660 \\ 13.660 \\ 10.000 \\ 1 \end{bmatrix} = \begin{bmatrix} 3.660 \\ 6.830 \\ 15.490 \\ 1 \end{bmatrix}$$

where we used $c\psi_{xj} = 0.866$ and $s\psi_{xj} = 0.500$. The coordinates of $\mathbf{P}^{j}$ can be verified by scaling in views C-C and B-B. Note that the coordinate $P^{j}_{y}$ was projected up from view C-C to view B-B and that in view B-B the coordinate $P^{j}_{x}$ was laid off equal to $P^{k}_{x} = 3.660$ mm.

Finally, we transform $\mathbf{P}^{j}$ to $\mathbf{P}^{i}$:

$$\mathbf{P}^{i} = T^{3}_{ij}(0, L_i, 0, \theta_{zj})\mathbf{P}^{j} = \begin{bmatrix} c\theta_{zj} & -s\theta_{zj} & 0 & 0 \\ s\theta_{zj} & c\theta_{zj} & 0 & L_i \\ 0 & 0 & 1 & 0 \\ 0 & 0 & 0 & 1 \end{bmatrix} \begin{bmatrix} 3.660 \\ 6.830 \\ 15.490 \\ 1 \end{bmatrix} = \begin{bmatrix} -0.245 \\ 56.715 \\ 15.490 \\ 1 \end{bmatrix}$$

where we used $L_i = 50.8$, $\theta_{zj} = 30°$. The coordinates of $\mathbf{P}^{i}$ appear true length in views A-A and B-B. Here $P^{i}_{y}$ was projected down from view B-B to view A-A and $P^{i}_{z}$ was laid off in view A-A equal to $P^{k}_{z} = 15.490$.

Now to verify the overall transformation operator of Eq. (6.70), where we have concatenated the foregoing three separate transformations, we substitute in Eq. (6.71) as follows:

$$\mathbf{P}^{i} = \begin{bmatrix} (0.866)(0.866) - (0.5)(0.866)(0.5) & -(0.866)(0.5) - (0.5)(0.866)(0.866) & (0.5)(0.5) & 0 \\ (0.5)(0.866) + (0.5)(0.5)(0.866) & -(0.5)(0.5) + (0.866)^{3} & -(0.866)(0.5) & 50.8 \\ (0.5)(0.5) & (0.5)(0.866) & 0.866 & 0 \\ 0 & 0 & 0 & 1 \end{bmatrix} \begin{bmatrix} P^{\ell}_{x} \\ P^{\ell}_{y} \\ P^{\ell}_{z} \\ 1 \end{bmatrix}$$

$$= \begin{bmatrix} 0.533 & -0.808 & 0.25 & 0 \\ 0.650 & 0.399 & -0.433 & 50.8 \\ 0.250 & 0.433 & 0.866 & 0 \\ 0 & 0 & 0 & 1 \end{bmatrix} \begin{bmatrix} 10 \\ 10 \\ 10 \\ 1 \end{bmatrix} = \begin{bmatrix} -0.250 \\ 56.960 \\ 15.490 \\ 1 \end{bmatrix}$$

which checks with our previous result within rounding error.

In the screw joint example (Fig. 6.37), it was shown how the elementary matrices describe the shape of a link. The displacements $A_i$ and $B_i$, defined in system $i$, describe the location of the $z_j$ axis relative to the $z_i$ axis. Also, the displacement $C_i$ says something about the shape of the "nut" side of the screw joint. The next example shows even more vividly the usefulness of the elementary matrices in describing link shapes.

A complex link shape is shown in Fig. 6.41. It is shown in much the way it may appear on a detailed engineering drawing, but only the dimensions required to define its kinematic shape are shown. The link is a binary link labeled $m$ and it is defined below together with the revolute joint between links $m$ and $n$. Three matrices are required to define link $m$ with revolute joint $mn$:

$$S_{i\ell}(\gamma_{zk}) = T^2_{ij}(0, 0, 0, \alpha_{yi})T^1_{jk}(0, A_j, B_j, \beta_{xk})T^3_{k\ell}(0, 0, 0, \gamma_{zk}) \tag{6.72}$$

Concatenating the matrices (see Exer. 6.6), (6.72) becomes

$$S_{i\ell}(\gamma_{zk}) = \begin{bmatrix} c\alpha_{yi}\, c\gamma_{zk} + s\alpha_{yi}\, s\beta_{xk}\, s\gamma_{zk} & -c\alpha_{yi}\, s\gamma_{zk} + s\alpha_{yi}\, s\beta_{xk}\, c\gamma_{zk} & s\alpha_{yi}\, c\beta_{xk} & +B\, s\alpha_{yi} \\ c\beta_{xk}\, s\gamma_{zk} & c\beta_{xk}\, c\gamma_{zk} & -s\beta_{xk} & A \\ -s\alpha_{yi}\, c\gamma_{zk} + c\alpha_{yi}\, s\beta_{xk}\, s\gamma_{zk} & s\alpha_{yi}\, s\gamma_{zk} + c\alpha_{yi}\, s\beta_{xk}\, c\gamma_{zk} & c\alpha_{yi}\, c\beta_{xk} & +B\, c\alpha_{yi} \\ 0 & 0 & 0 & 1 \end{bmatrix} \tag{6.73}$$

The first two matrices define the twist, bend, length and offset of link $m$, while the third matrix contains the degree of freedom $\gamma_{zk}$ of the revolute joint. The complexity of the rotational part of the matrix shown in (6.73) exemplifies the advantage of using the elementary matrices. With these, a point $P$ embedded in link $n$ at $\mathbf{P}^\ell$ can be located with respect to link $m$ by vector $\mathbf{P}^i = S_{i\ell}(\gamma_{zk})\mathbf{P}^\ell$ (see Fig. 6.41 and Exer. 6.7).

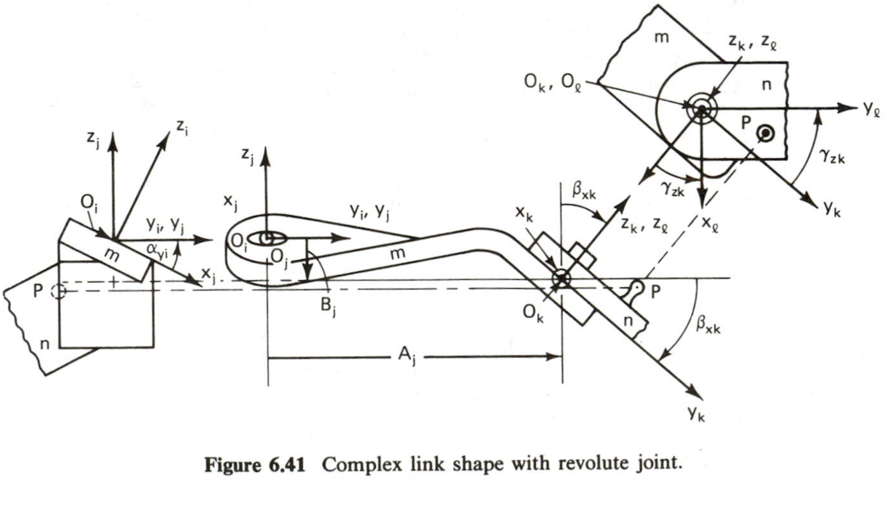


**Figure 6.41** Complex link shape with revolute joint.

### Modeling Spatial Mechanisms

Analytical techniques are used in Chaps. 2 and 3 to model planar mechanisms. The techniques consisted of defining closed vector loops for the mechanism. These are complex vector equations of loop closure, expressing the fact that the vectors around the loop add up to zero. This is sufficient for planar mechanisms since all rotations take place in parallel planes. When modeling spatial mechanisms, additional constraints are required since rotations may occur about axes or in planes of different orientations. Therefore, in modeling spatial mechanisms, we need an approach that reflects rotations about any axis and translations in any of three mutually perpendicular directions.

The $4 \times 4$ transformation matrix is well suited to define such constraints. As noted earlier [Eq. (6.31)], the $4 \times 4$ matrix can be thought of as containing four parts. Two of these are considered here. The $3 \times 3$ rotational matrix $[R]$ defines the orientation of a second coordinate system at a second joint of a link referred to the coordinate system at the first joint of that link, and the translational vector $(D)$ defines the relative locations of their origins, defined in the first system. If a kinematic loop describing a mechanism is closed, the coordinate system at the end of the loop must be parallel to and located at the same point as the coordinate system at the beginning. This implies that the concatenated product of the rotational matrices around the loop must equal the $3 \times 3$ indentity matrix, and the translational vectors around the loop must have zero resultant. Hence the combined result is the $4 \times 4$ indentity matrix.

Before considering a typical spatial mechanism, the planar mechanism shown in Fig. 6.42 will be modeled to demonstrate the techniques involved. Also, it is easy to compare with results determined earlier.

When modeling a mechanism using the elementary matrices, local coordinate systems on each link should be established first. This has been done as shown in Fig. 6.42. $x_0y_0z_0$ is the coordinate system for link 1 (the fixed link). Consequently, this coordinate system is a reference coordinate system. Links 2, 3, and 4 have coordinate systems $x_1y_1z_1$, $x_2y_2z_2$, and $x_3y_3z_3$, respectively, fixed to them. For convenience in describing translations from joint to joint, the $x$ axis is defined along the length of each link. A type 3 matrix from Fig. 6.33c is used to model each link-and-joint combination of this mechanism. Consequently, the $x_1y_1z_1$ coordinate system is defined relative to the $x_0y_0z_0$ coordinate system by

$$S_{01}(\theta_1) = T^3_{01}(0, 0, 0, \theta_1) \tag{6.74}$$

where $\theta_1$ is defined in the system 0. Furthermore, the $x_2y_2z_2$ coordinate system is defined relative to the $x_1y_1z_1$ system by

$$S_{12}(\theta_2) = T^3_{12}\,(L_2, 0, 0, \theta_2) \tag{6.75}$$

where $L_2$ and $\theta_2$ are defined in system 1, and in similar fashion,

$$S_{23}(\theta_3) = T^3_{23}\,(L_3, 0, 0, \theta_3) \tag{6.76}$$

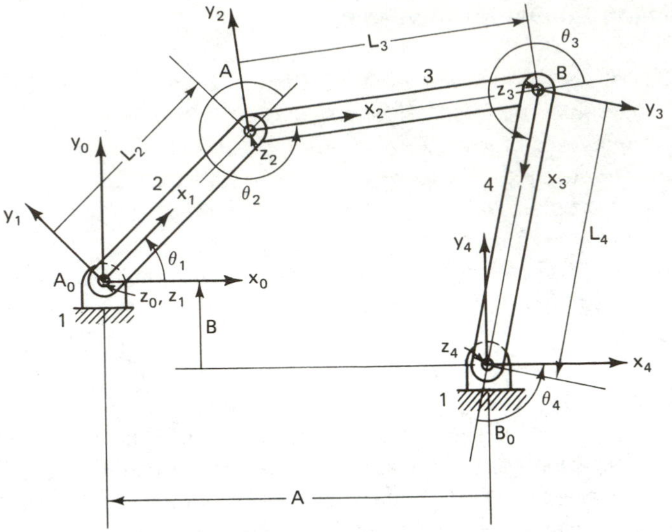


**Figure 6.42** A special-case of spatial mechanisms—the planar four-bar.

where $L_3$ and $\theta_3$ are defined in system 2, and

$$S_{34}(\theta_4) = T^3_{34}\,(L_4, 0, 0, \theta_4) \tag{6.77}$$

where $L_4$ and $\theta_4$ are defined in system 3.

How are these independent link-joint descriptions put together to define the mechanism? Look first at the $x_2y_2z_2$ coordinate system. How is it related to $x_0y_0z_0$? We know the relative relationships ($S_{01}$, $S_{12}$), and by concatenating these (see Exer. 6.8) we get

$$S_{02} = S_{01}S_{12} = \begin{bmatrix} c(\theta_1+\theta_2) & -s(\theta_1+\theta_2) & 0 & L_2c\theta_1 \\ s(\theta_1+\theta_2) & c(\theta_1+\theta_2) & 0 & L_2s\theta_1 \\ 0 & 0 & 1 & 0 \\ 0 & 0 & 0 & 1 \end{bmatrix} \tag{6.78}$$

Note that the translational vector of this matrix product defines the joint center of $x_2y_2z_2$ and the sum of relative angular rotations ($\theta_1 + \theta_2$) defines its orientation. If the matrix product of the joint transformation is continued, the location and orientation of $x_4y_4z_4$ can be obtained (see Exer. 6.9).

$$S_{04} = S_{01}S_{12}S_{23}S_{34}$$

$$= \begin{bmatrix} c(\theta_1+\theta_2+\theta_3+\theta_4) & -s(\theta_1+\theta_2+\theta_3+\theta_4) & 0 & [L_2\,c\theta_1 + L_3\,c(\theta_1+\theta_2) + L_4\,c(\theta_1+\theta_2+\theta_3)] \\ s(\theta_1+\theta_2+\theta_3+\theta_4) & c(\theta_1+\theta_2+\theta_3+\theta_4) & 0 & [L_2\,s\theta_1 + L_3\,s(\theta_1+\theta_2) + L_4\,s(\theta_1+\theta_2+\theta_3)] \\ 0 & 0 & 1 & 0 \\ 0 & 0 & 0 & 1 \end{bmatrix} \tag{6.79}$$

However, this does not close the loop for this mechanism. One additional transformation matrix is required to get from $x_4y_4z_4$ to $x_0y_0z_0$. This is a constant translation-only matrix:

$$S_{40} = T^7_{40}(A, B, 0) \tag{6.80}$$

where $A$ and $B$ are defined in the $(x_4y_4z_4)$ coordinate system as shown in Fig. 6.42. This will return us to the $x_0y_0z_0$ coordinate system. Since the coordinate system defined by stepping around the loop must be the same as the one at the beginning in terms of location and orientation, the product of the transformation matrices must be the indentity matrix (see Exer. 6.10):

$$S_{01}S_{12}S_{23}S_{34}S_{40} = I \tag{6.81}$$

or

$$\begin{bmatrix} c(\theta_1+\theta_2+\theta_3+\theta_4) & -s(\theta_1+\theta_2+\theta_3+\theta_4) & 0 & L_2\,c(\theta_1)+L_3\,c(\theta_1+\theta_2)+L_4\,c(\theta_1+\theta_2+\theta_3)-A\,c(\theta_1+\theta_2+\theta_3+\theta_4)-B\,s(\theta_1+\theta_2+\theta_3+\theta_4) \\ s(\theta_1+\theta_2+\theta_3+\theta_4) & c(\theta_1+\theta_2+\theta_3+\theta_4) & 0 & L_2\,s(\theta_1)+L_3\,s(\theta_1+\theta_2)+L_4\,s(\theta_1+\theta_2+\theta_3)+B\,c(\theta_1+\theta_2+\theta_3+\theta_4)-A\,s(\theta_1+\theta_2+\theta_3+\theta_4) \\ 0 & 0 & 1 & 0 \\ 0 & 0 & 0 & 1 \end{bmatrix}$$

$$= \begin{bmatrix} 1 & 0 & 0 & 0 \\ 0 & 1 & 0 & 0 \\ 0 & 0 & 1 & 0 \\ 0 & 0 & 0 & 1 \end{bmatrix} \tag{6.82}$$

The first observation that can be made from Fig. 6.42 is that the sum of the *relative* angular rotations $(\theta_1+\theta_4+\theta_3+\theta_4)$ is 0°, 360°, or some multiple of 360°. Therefore the sine of the sum is 0 and the cosine is 1. Second, the $x_0$ and $y_0$ components of the resultant translational vector must be zero. The latter results in two equations (see Exer. 6.11):

$$\begin{aligned} L_2\,c\theta_1 + L_3\,c(\theta_1+\theta_2) + L_4\,c(\theta_1+\theta_2+\theta_3) - A &= 0 \\ L_2\,s\theta_1 + L_3\,s(\theta_1+\theta_2) + L_4\,s(\theta_1+\theta_2+\theta_3) + B &= 0 \end{aligned} \tag{6.83}$$

Two other equations are known from the rotational part of the matrix:

$$c(\theta_1+\theta_2+\theta_3+\theta_4) = 1 \qquad \text{and} \qquad s(\theta_1+\theta_2+\theta_3+\theta_4) = 0$$

It is left to the reader to show that these are the same equations that are obtained using complex numbers. (Note that normally the angle $\theta$ contained in the complex rotational operator $e^{i\theta}$ is referred to the fixed axis while the angles in Eq. (6.83) are relative angles between successive coordinate systems).

Next let us focus our attention on the spatial mechanism shown in Fig. 6.43.

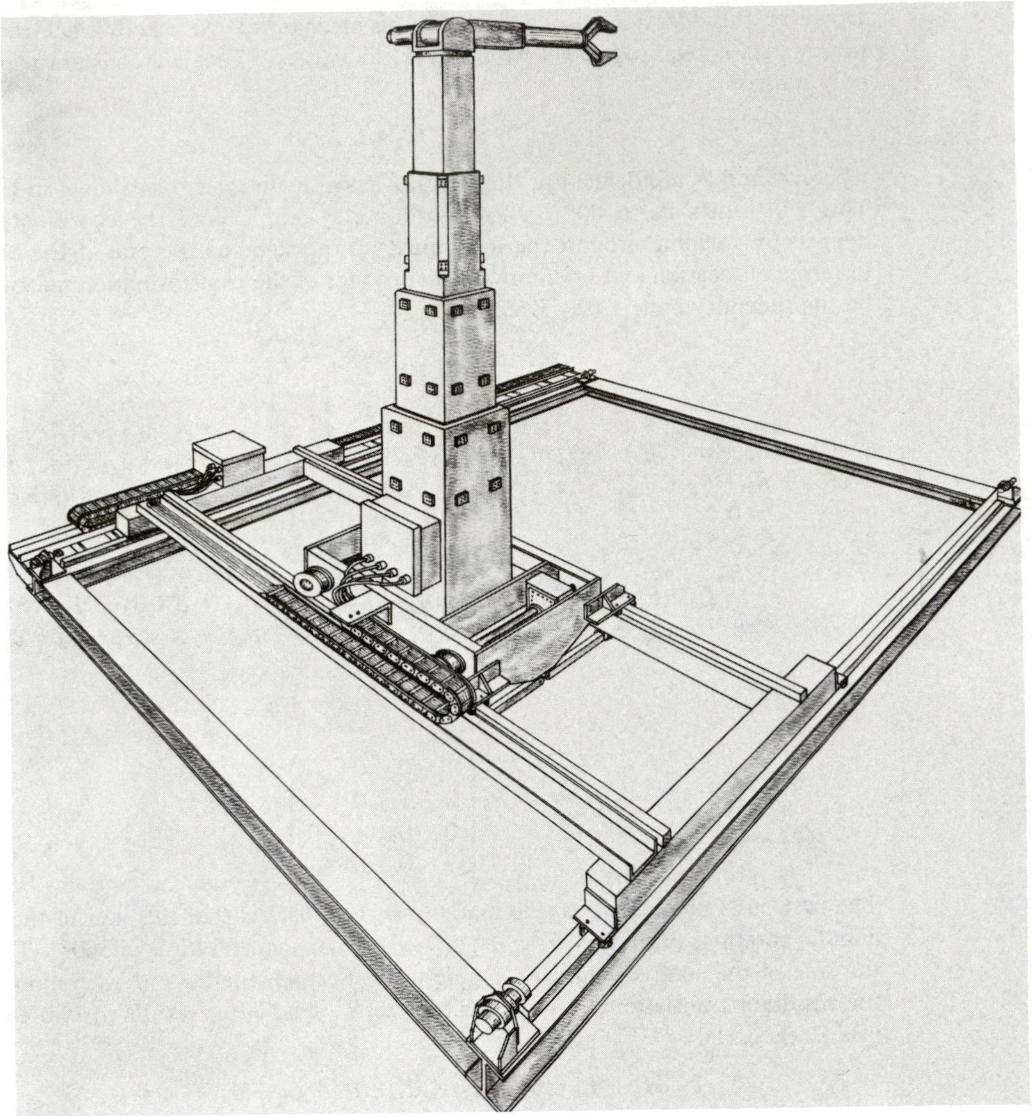

**Figure 6.43** Spatial Mechanism—an industrial robot (Courtesy of *GCA/PaR* Systems.)

## 6.13 KINEMATIC ANALYSIS OF AN INDUSTRIAL ROBOT*

Figure 6.43 shows an industrial robot with six degrees of freedom plus a gripper. An artist's conception depicting its use in a factory setting is illustrated in Fig. 6.44. The first three degrees of freedom are translations followed by three rotations as shown in Fig. 6.45. We wish to determine the six independent joint movements ($a$, $c$, $b$, $\theta_1$, $\theta_2$, $\theta_3$) given the final position of the hand with respect to a "home position" (these variables are zero in the home position).

Before modeling this robot using elementary matrices, local coordinate systems on each link were established as shown in Fig. 6.45. Notice that this spatial mechanism is *open loop*. The last, 7th coordinate system, that of the hand opening, need not be coincident with the first. Each coordinate system $x_i y_i z_i$ is defined relative to the $x_{i-1} y_{i-1} z_{i-1}$ coordinate system by

$$\begin{aligned} S_{01} &= T^7_{01}(a, 0, 0) \\ S_{12} &= T^7_{12}(0, 0, c) \\ S_{23} &= T^7_{23}(0, L_1+b, 0) \\ S_{34}(\theta_1) &= T^2_{34}(0, 0, 0, \theta_1) \end{aligned} \tag{6.84}$$

where $\theta_1$ is a rotation about the $y_3$ axis,

$$S_{45}(\theta_2) = T^3_{45}(0, 0, 0, \theta_2)$$

where $\theta_2$ is a rotation about the $z_4$ axis,

$$S_{56}(\theta_3) = T^2_{56}(0, L_2, 0, \theta_3)$$

where $\theta_3$ is the hand rotation about the $y_5$ axis,

$$S_{67} = T^7_{67}\,(0, 0, d)$$

where $L_1$ and $L_2$ are constants, and $a$, $c$, $b$, $d$, $\theta_1$, $\theta_2$, and $\theta_3$ are variables ($d$ is the gripper opening).

To model this robot, the position vector $\mathbf{P}^i$ of a point $P$ expressed in the $i$th coordinate system needs to be expressed in terms of coordinate system $x_0 y_0 z_0$ by

$$\mathbf{P}^{\circ} = S_{0i}\mathbf{P}^i \tag{6.85}$$

* This section was prepared with the help of Sern Hong Wang. More details can be found in [271].

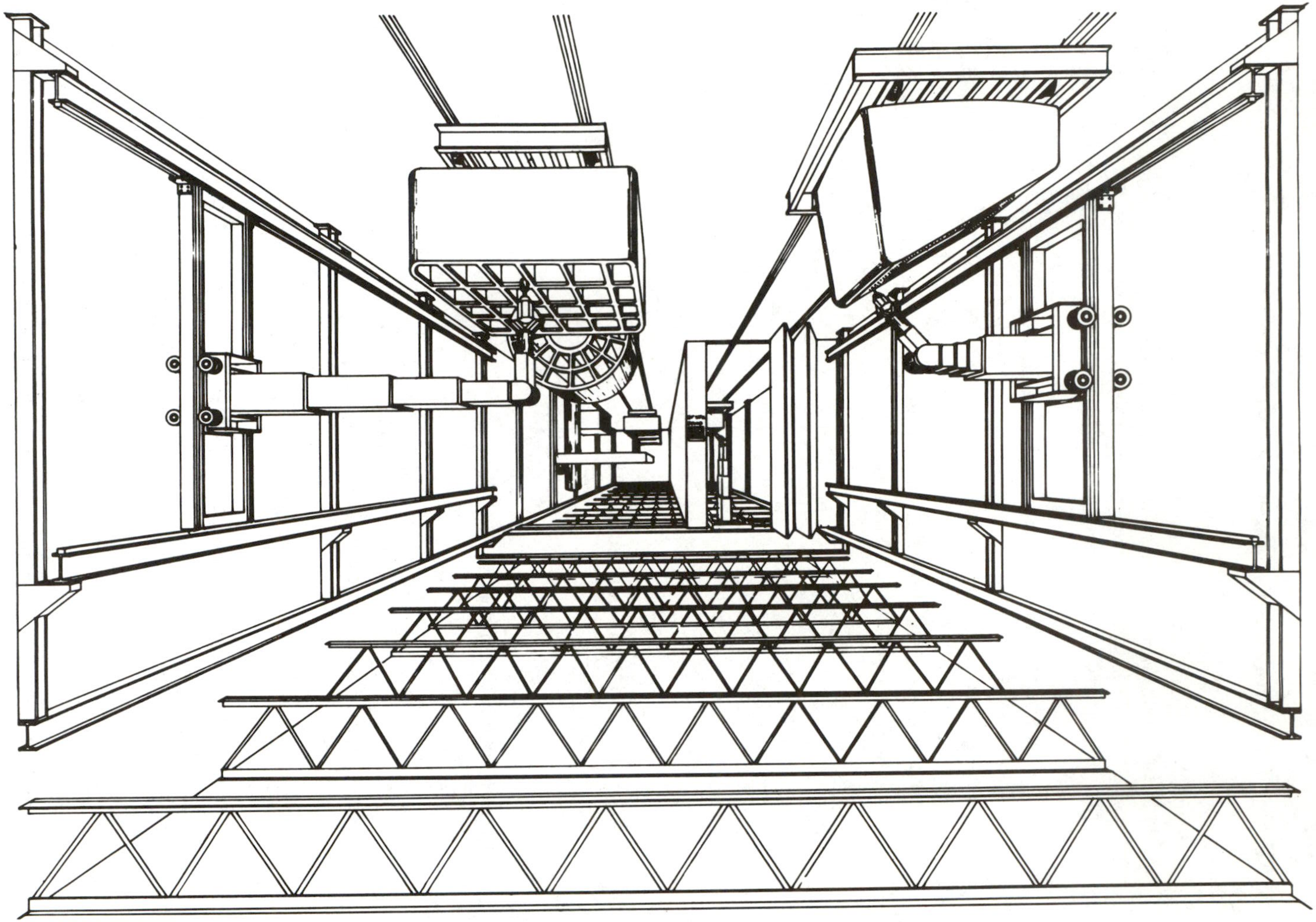

**Figure 6.44** Artist's conception of factory setting for the industrial robot of Fig. 6.43. (Courtesy of *GCA/PaR* Systems.)

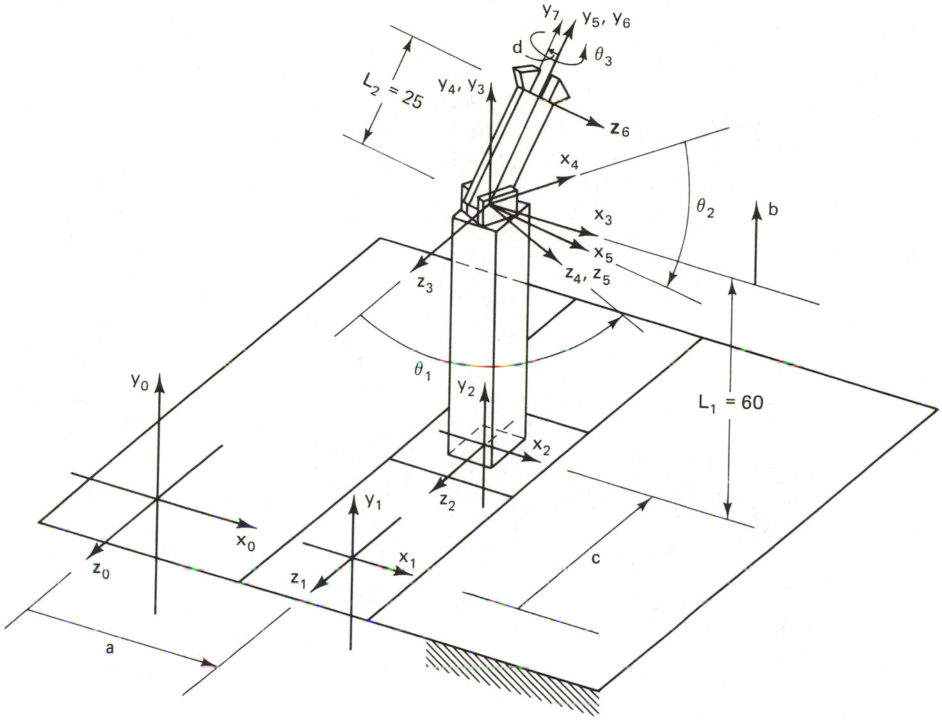


**Figure 6.45** Definition of coordinate systems for the robot of fig. 6.43. Note that translation *c*, defined with respect to the $z_1$ axis, is a negative quantity. Similarly, $\theta_2$, a rotation about the $z_4$ axis, is a negative angle.

The concatenated matrix, $S_{0i}$, for each link $i$ is

$$S_{01} = \begin{bmatrix} 1 & 0 & 0 & a \\ 0 & 1 & 0 & 0 \\ 0 & 0 & 1 & 0 \\ 0 & 0 & 0 & 1 \end{bmatrix} \tag{6.86}$$

$$S_{02} = S_{01}S_{12} = \begin{bmatrix} 1 & 0 & 0 & a \\ 0 & 1 & 0 & 0 \\ 0 & 0 & 1 & c \\ 0 & 0 & 0 & 1 \end{bmatrix} \tag{6.87}$$

$$S_{03} = \{S_{01}S_{12}\}S_{23}$$

$$= S_{02}S_{23}$$

$$= \begin{bmatrix} 1 & 0 & 0 & a \\ 0 & 1 & 0 & L_1 + b \\ 0 & 0 & 1 & c \\ 0 & 0 & 0 & 1 \end{bmatrix} \tag{6.88}$$

$$S_{04} = S_{03}S_{34}$$

$$= S_{03}\begin{bmatrix} \cos\theta_1 & 0 & \sin\theta_1 & 0 \\ 0 & 1 & 0 & 0 \\ -\sin\theta_1 & 0 & \cos\theta_1 & 0 \\ 0 & 0 & 0 & 1 \end{bmatrix}$$

$$= \begin{bmatrix} \cos\theta_1 & 0 & \sin\theta_1 & a \\ 0 & 1 & 0 & b + L_1 \\ -\sin\theta_1 & 0 & \cos\theta_1 & c \\ 0 & 0 & 0 & 1 \end{bmatrix} \tag{6.89}$$

$$S_{05} = S_{04}S_{45}$$

$$= S_{04}\begin{bmatrix} \cos\theta_2 & -\sin\theta_2 & 0 & 0 \\ \sin\theta_2 & \cos\theta_2 & 0 & 0 \\ 0 & 0 & 1 & 0 \\ 0 & 0 & 0 & 1 \end{bmatrix}$$

$$= \begin{bmatrix} \cos\theta_1\cos\theta_2 & -\cos\theta_1\sin\theta_2 & \sin\theta_1 & a \\ \sin\theta_2 & \cos\theta_2 & 0 & b + L_1 \\ -\sin\theta_1\cos\theta_2 & \sin\theta_1\sin\theta_2 & \cos\theta_1 & c \\ 0 & 0 & 0 & 1 \end{bmatrix} \tag{6.91*}$$

$$S_{06} = S_{05}S_{56}$$

$$= S_{05}\begin{bmatrix} \cos\theta_3 & 0 & \sin\theta_3 & 0 \\ 0 & 1 & 0 & L_2 \\ -\sin\theta_3 & 0 & \cos\theta_3 & 0 \\ 0 & 0 & 0 & 1 \end{bmatrix}$$

$$= \begin{bmatrix} (\cos\theta_1\cos\theta_2\cos\theta_3 - \sin\theta_1\sin\theta_3) & -\cos\theta_1\sin\theta_2 & (\cos\theta_1\cos\theta_2\sin\theta_3 + \sin\theta_1\cos\theta_3) & (-L_2\cos\theta_1\sin\theta_2 + a) \\ \sin\theta_2\cos\theta_3 & \cos\theta_2 & \sin\theta_2\sin\theta_3 & (L_2\cos\theta_2 + b + L_1) \\ (-\sin\theta_1\cos\theta_2\cos\theta_3 - \cos\theta_1\sin\theta_3) & \sin\theta_1\sin\theta_2 & (-\sin\theta_1\cos\theta_2\sin\theta_3 + \cos\theta_1\cos\theta_3) & (L_2\sin\theta_1\sin\theta_2 + c) \\ 0 & 0 & 0 & 1 \end{bmatrix} \tag{6.92}$$

* Eq. 6.90 omitted.

$$S_{07} = S_{06}S_{67}$$

$$= S_{06}\begin{bmatrix} 1 & 0 & 0 & 0 \\ 0 & 1 & 0 & 0 \\ 0 & 0 & 1 & d \\ 0 & 0 & 0 & 1 \end{bmatrix}$$

$$= \begin{bmatrix} (\cos\theta_1\cos\theta_2\cos\theta_3 & -\cos\theta_1\sin\theta_2 & (\cos\theta_1\cos\theta_2\sin\theta_3 & \{-L_2\cos\theta_1\sin\theta_2 + a \\ -\sin\theta_1\sin\theta_3) & & +\sin\theta_1\cos\theta_3) & +d(\cos\theta_1\cos\theta_2\sin\theta_3 \\ & & & +\sin\theta_1\cos\theta_3)\} \\ \sin\theta_2\cos\theta_3 & \cos\theta_2 & \sin\theta_2\sin\theta_3 & \{L_2\cos\theta_2 + b + L_1 \\ & & & +d(\sin\theta_2\sin\theta_3)\} \\ (-\sin\theta_1\cos\theta_2\cos\theta_3 & \sin\theta_1\sin\theta_2 & (-\sin\theta_1\cos\theta_2\sin\theta_3 & \{(L_2\sin\theta_1\sin\theta_2 + c) \\ -\cos\theta_1\sin\theta_3) & & +\cos\theta_1\cos\theta_3) & +d(-s3\sin\theta_1\cos\theta_2\sin\theta_3 \\ & & & +\cos\theta_1\cos\theta_3)\} \\ 0 & 0 & 0 & 1 \end{bmatrix} \quad (6.93)$$

When specific values are given for the seven degrees of freedom: $a$, $c$, $b$, $d$, $\theta_1$, $\theta_2$, and $\theta_3$, they can be substituted into $T^7_{01}$, $T^7_{12}$, $T^7_{23}$, $T^2_{34}$, . . . [Eq. (6.84)]. Then the $S_{0i}$'s can be computed in a sequence. For instance,

$$\begin{bmatrix} S_{05} = S_{01}\,S_{12}\,S_{23}\,S_{34}\,S_{45} \\ \quad = T^7_{01}\,T^7_{12}\,T^7_{23}\,T^2_{34}\,T^3_{45} \end{bmatrix} \quad (6.94)$$

We obtain a position vector $\mathbf{P}^5$ of a point expressed in system 5 in terms of the fixed-coordinate system $x_0y_0z_0$ as follows:

$$\begin{bmatrix} x_0 \\ y_0 \\ z_0 \\ 1 \end{bmatrix} = S_{05}\begin{bmatrix} x_5 \\ y_5 \\ z_5 \\ 1 \end{bmatrix} \quad (6.95)$$

Figure 6.46* shows the home position of the robot and two pedestals. An object is to be picked up from the first pedestal in the three-dimensional position shown and placed at a predetermined orientation on the second pedestal. Figures 6.47* and 6.48* show the robot near the pickup position and returning from the final position of the object, respectively.

* These figures were originally generated on a Terak microcomputer.

$$
\begin{aligned}
a &= 0.0 \\
b &= 0.0 \\
c &= 0.0 \\
d &= 4.0 \\
\theta_1 &= 0° \\
\theta_2 &= 0° \\
\theta_3 &= 0°
\end{aligned}
$$

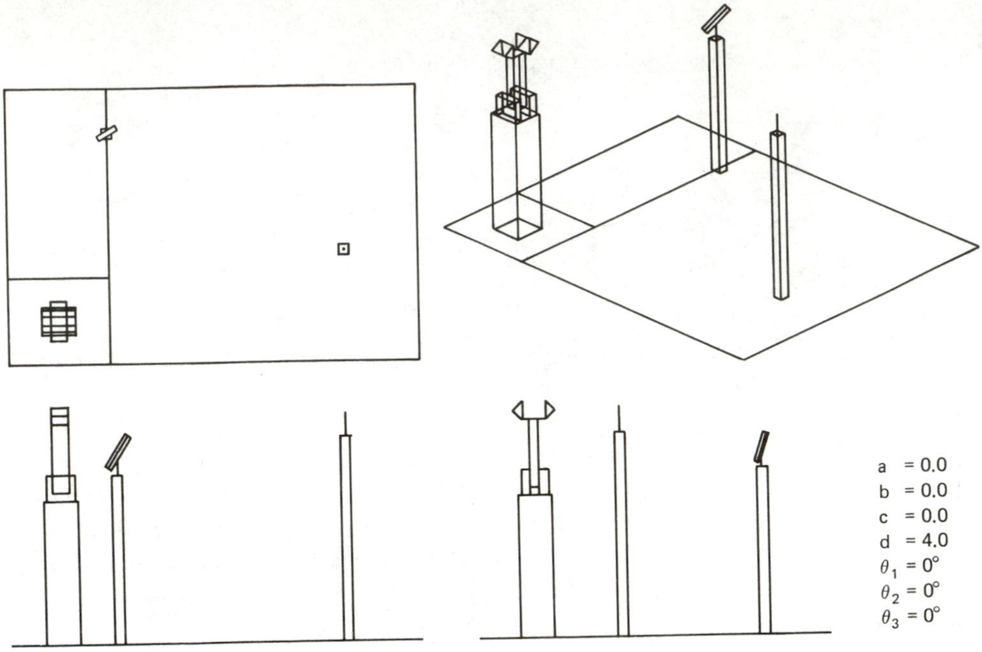

**Figure 6.46** Micro-computer-generated display of the robot of Fig. 6.43 in the home position with the hand open by four units.

$$
\begin{aligned}
a &= 5.71 \\
b &= -14.58 \\
c &= -71.75 \\
d &= 4.0 \\
\theta_1 &= 30.0° \\
\theta_2 &= -15.0° \\
\theta_3 &= 0°
\end{aligned}
$$

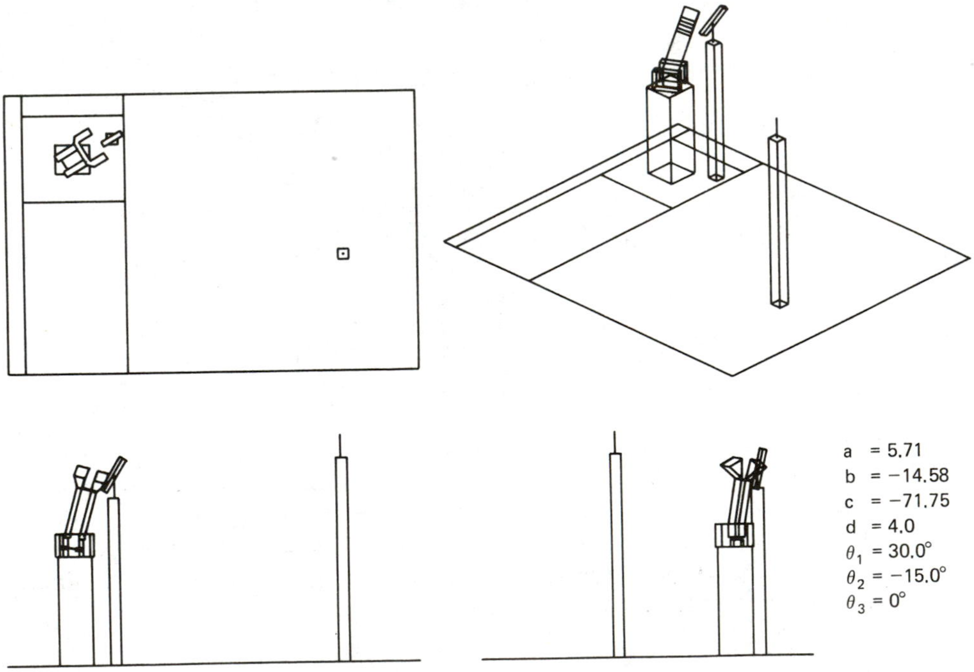

**Figure 6.47** Micro-computer-generated display of the robot of Fig. 6.43 near the first pedestal.

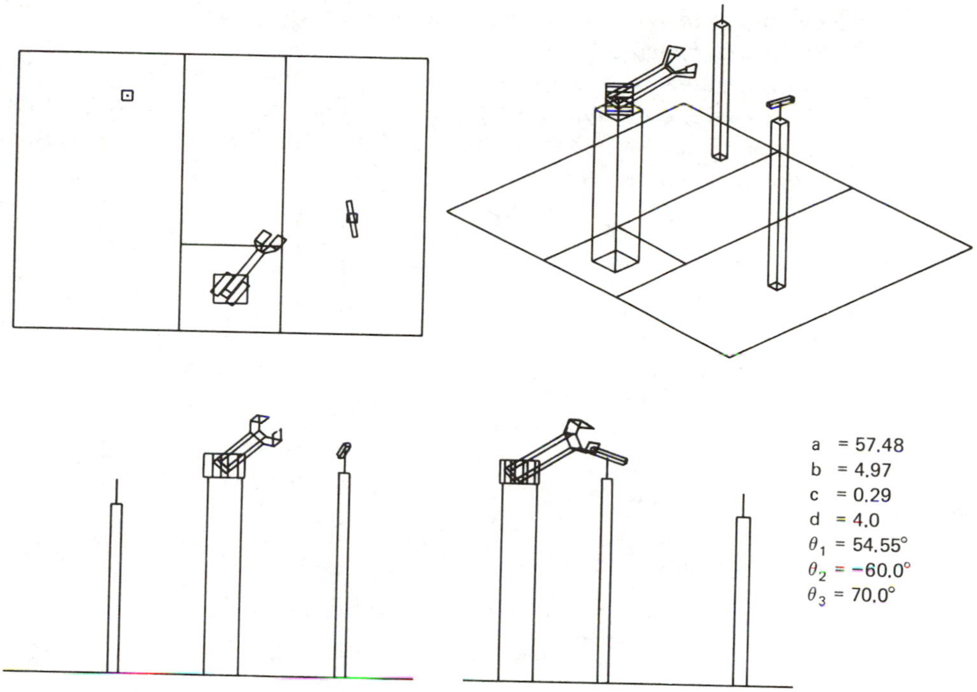


**Figure 6.48** Robot returning from drop-off position.

## The Inverse Kinematics Problem

The preceding section illustrated how an industrial robot could be analyzed using elementary matrices. Note that the relative translations and rotations of each degree of freedom ($a$, $c$, $b$, $d$, $\theta_1$, $\theta_2$, and $\theta_3$) were given with respect to successively translated and rotated coordinate systems, and by concatenation, with respect to a home position. In many cases, the reverse is required. When programming a robot, the displacements at the individual degrees of freedom are to be determined, given a required robot gripper (*end effector*) position translated and rotated away from the home position. This is sometimes referred to as the "inverse kinematics problem." In many cases, this task results in cumbersome equations with multiple solutions [71,199,322]. For this example, however, since there are three translations followed by three rotations, the inverse kinematics problem is more easily solved.

The prescribed position of the gripper in the fixed coordinate system, $x_0y_0z_0$ of Fig. 6.45, is given by $x_g$, $y_g$, $z_g$, $\theta_{1g}$, $\theta_{2g}$, $\theta_{3g}$. Since the three rotational degrees of freedom are the same as the rotations of the gripper, we only solve for the three translations: $a$, $b$, and $c$.*

In order to allow the gripper to pick up the object, the origin of link 6 needs to have the same coordinate (in the fixed coordinate system) and orientation as the object does. The coordinates of the origin of system 6 need to be expressed in the fixed-coordinate system by

$$\begin{bmatrix} x_{06} \\ y_{06} \\ z_{06} \\ 1 \end{bmatrix} = S_{06} \begin{bmatrix} 0 \\ 0 \\ 0 \\ 1 \end{bmatrix} \tag{6.96}$$

where $(S_{06})$ was defined by Eq. (6.92). Thus

$$\begin{bmatrix} x_{06} \\ y_{06} \\ z_{06} \\ 1 \end{bmatrix} = \begin{bmatrix} -L_2 \cos\theta_1 \sin\theta_2 + a \\ L_2 \cos\theta_2 + b + L_1 \\ L_2 \sin\theta_1 \sin\theta_2 + c \\ 1 \end{bmatrix} \tag{6.97}$$

So we can obtain $a$, $b$, and $c$ simply by

$$\begin{aligned} a &= x_{06} + (L_2 \cos\theta_1 \sin\theta_2) \\ b &= y_{06} - (L_2 \cos\theta_2 + L_1) \\ c &= z_{06} - (L_2 \sin\theta_1 \sin\theta_2) \end{aligned} \tag{6.98}$$

Here $x_{06}$, $y_{06}$, $z_{06}$, $\theta_1$, and $\theta_2$ are knowns, and

$$x_{06} = x_g \tag{6.99}$$

$$y_{06} = y_g \tag{6.100}$$

$$z_{06} = z_g \tag{6.101}$$

$$\theta_1 = \theta_{1g} \tag{6.102}$$

$$\theta_2 = \theta_{2g} \tag{6.103}$$

For the object pickup position (Fig. 6.47 shows the robot nearly at this position)

$$x_g = 20, \qquad y_g = 80, \qquad z_g = -80$$

$$\theta_{1g} = 30°, \qquad \theta_{2g} = -30°, \qquad \theta_{3g} = 40°$$

* It has been shown (e.g., Duffy [71]) that the rotations in a spatial mechanism can be treated independently of translations, as the rotations of an equivalent spherical mechanism. This is accomplished by translating all nonintersecting axes parallel with themselves into concurrence at one point and analyzing the rotations of the resulting spherical mechanism by way of spherical trigonometry. The determination of translations in this section utilizes this decoupling of translations from rotations.

$a$, $b$, and $c$ can be obtained by

$$a = 20 + (25)(\cos 30°)[\sin(-30°)] = 9.17$$

$$b = 80 - (25)[\cos(-30°)] - 60 = -1.65$$

$$c = -80 - (25)(\sin 30°)[\sin(-30°)] = -73.75$$

For the object in the second pedestal position

$$x_g = 100, \quad y_g = 95, \quad z_g = -30$$

$$\theta_{1g} = 100°, \quad \theta_{2g} = -110°, \quad \theta_{3g} = 70°$$

$$a = 104.08$$

$$b = 43.55$$

$$c = -6.86$$

### Spatial Slider-Crank Mechanism Example

The planar example (Fig. 6.42) illustrated the basic principles involved in modeling closed-loop mechanisms using the elementary matrices shown in Fig. 6.33. The spatial slider-crank mechanism shown in Fig. 6.49 is considered next. This mechanism contains a revolute joint, a slider joint, and two spherical joints. To describe this mechanism, a series of matrices that relate the relative motion of adjacent links is required. Then, these *link-joint* descriptions are concatenated in a closed loop such that the first and last coordinate systems defined are parallel and their origins are located at the same point; in other words, they coincide.

First, we establish a local coordinate system for each link. It is advantageous to orient coordinate systems to coincide with the geometry and physical constraints of the joints of the mechanism. This has been done as shown in Fig. 6.49. Not all coordinate systems required to define the spherical joint motions have been shown for the sake of clarity. Only those attached to the links are shown.

The revolute joint between links 1 and 2 is defined with respect to $x_0y_0z_0$ by the transformation

$$S_{01} = T_{01}^{2}(0,\ L_1,\ 0,\ \theta_1) \tag{6.104}$$

$S_{01}$ represents a rotation about the $y_0$ axis following a translation of $L_1$ in the $y_0$ direction. Consequently, a type 2 elementary matrix is required.

Definition of the spherical joint between links 2 and 3 starts with a translation $L_2 = (L_2,\ 0,\ 0)$ defined in the $x_1y_1z_1$ system. This translates $x_1y_1z_1$, parallel with itself, to the center of joint (2, 3), becoming $x_1'y_1'z_1'$ (not shown). This system is now rotated by the first Euler angle $\theta_2$ about its own $z_1'$ axis to $x_2y_2z_2$ (not shown). Next $x_2y_2z_2$ is rotated by the second Euler angle $\theta_3$ about its own $x_2$ axis to $x_3y_3z_3$ (not shown), followed by the third Euler rotation $\theta_4$ about $z_3$ and thus becoming

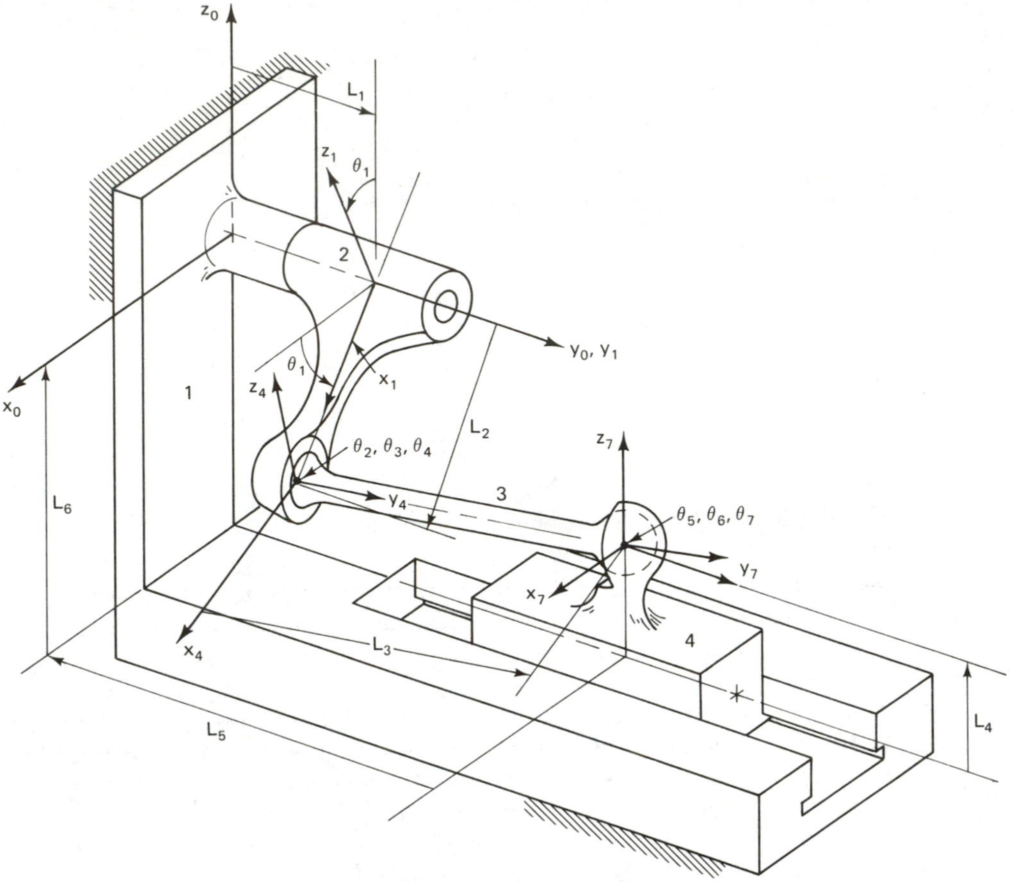


**Figure 6.49** Spatial slider-crank mechanism.

$x_4y_4z_4$, which is shown. This sequence of coordinate displacements is modeled by

$$S_{14} = T^3_{12}(L_2,\ 0,\ 0,\ \theta_2)T^1_{23}(0,\ 0,\ 0,\ \theta_3)T^3_{34}(0,\ 0,\ 0,\ \theta_4) \tag{6.105}$$

The spherical joint connecting links 3 and 4 is defined in a similar manner using the Euler angles $\theta_5$ about $z'_4$, $\theta_6$ about $x_5$, and $\theta_7$ about $z_6$:

$$S_{47} = T^3_{45}(0,\ L_3,\ 0,\ \theta_5)T^1_{56}(0,\ 0,\ 0,\ \theta_6)T^3_{67}(0,\ 0,\ 0,\ \theta_7) \tag{6.106}$$

Here the initial translation $L_3$ is in the $y_4$ direction. Again the intermediate coordinate systems $x'_4y'_4z'_4$, $x_5y_5z_5$, and $x_6y_6z_6$ are not shown but are defined similarly to the intermediate systems at joint 23.

Because of the orientation of the slider in this mechanism, coordinate system $x_7y_7z_7$ is defined parallel to the $x_0y_0z_0$ coordinate system. Therefore, the final joint, between links 4 and 1, is a translation-only type:

$$S_{70} = T^7_{70}(0,\ L_5,\ [L_6 - L_4]) \tag{6.87}$$

The displacement in the $y_7$ ($y_0$) direction is the slider displacement. Consequently, the distance noted as $L_5$ (in the negative $y_7$ direction) will change as the mechanism assumes different configurations.

The closing of the loop defined by these link-joint transformations is expressed as

$$S_{00} = S_{01}S_{12}S_{23}S_{34}S_{45}S_{56}S_{67}S_{70} = I \tag{6.88}$$

This equation states that, as the loop is traversed, and the dimensions of each link and the rotations and translations of each joint are considered, the coordinate system at the end is congruent with the one at the beginning. That is, the product of all the matrices is equal to the identity matrix. This equation is called the loop-closure equation. It is a nonlinear matrix equation that must be solved during the position analysis of a mechanism. For a general mechanism, the dependent variables, which define the configuration of the mechanism, are expressed as nonlinear functions of the input variables and constants of the mechanism. Consequently, this equation is difficult to solve for the general case.

A general position analysis procedure is outlined later in this chapter. It is based on a Newton–Raphson iterative procedure commonly used for the solution of nonlinear equations. For this procedure, the *Jacobian matrix* for the nonlinear equations is required. This is a matrix of the first partial derivatives of the equations with respect to the dependent variables of the mechanism. It is shown in the next section how these derivatives can be expressed as matrix products.

Before discussing taking the partial derivatives of the loop-closure equation, one further comment should be made about Eq. (6.88). It is used here to describe the mechanism and in this form it can be solved, as will be shown, to provide the position solution of the mechanism. Taking the derivative of this equation with respect to time provides a velocity-loop-closure equation. This equation can be used in the velocity analysis of the mechanism. The time derivative is constructed using the chain rule and involves taking partial derivatives similar to those taken during the position analysis. Going one step further, and taking derivatives of the velocity-loop-closure equation with respect to time provides an acceleration-loop-closure equation. This is the basis for acceleration analysis. Again, taking the derivative is achieved by application of the chain rule and results in similar partial derivatives, as discussed above. Consequently, the next section, on taking derivatives of matrices, is basic to the entire kinematic analysis of a general spatial mechanism.

## Partial Derivatives of the Elementary Matrices

Taking the partial derivatives of the elementary matrices shown in Fig. 6.33 with respect to the variable quantity contained in the matrix is defined in this section. The variable in the rotational matrices and the screw matrices is the angle of rotation contained in the matrix, provided that the matrix does not define a constant rotation as shown in the complex link shape example discussed earlier. Similarly, the translation along any one axis in a translation only matrix may be a variable. It is mathematically convenient to describe the operation of taking the partial derivative as a matrix product. Let us see how this is done.

Define the partial derivative of a general elementary matrix $S_{ij}$ as follows:

$$\frac{\partial S_{ij}(q_m)}{\partial q_m} = S_{ij}(q_m)Q_l \tag{6.109}$$

where $q_m$ represents the variable quantity contained within the matrix, and the matrix $Q_l$ is a derivative operator associated with the type of elementary matrix defined as $S_{ij}$. To derive the derivative operator, we form the partial of the matrix by replacing each element in the matrix by its partial derivative with respect to the one variable of interest. For a type 1 matrix (Eq. 6.53), with $\theta$ as the variable of interest, this results in

$$\frac{\partial T^1_{ij}(a, b, c, \theta)}{\partial \theta} = \begin{bmatrix} 0 & 0 & 0 & 0 \\ 0 & -s\theta & -c\theta & 0 \\ 0 & c\theta & -s\theta & 0 \\ 0 & 0 & 0 & 0 \end{bmatrix} \tag{6.110}$$

because $a$, $b$, and $c$ are constants describing rigid link dimensions. However, even if they are variable, their partial derivative with respect to $\theta$ is zero. Designating the unknown derivative operator for the type 1 matrix as $(Q_1)$, Eq. (6.110) becomes

$$\frac{\partial T^1_{ij}}{\partial \theta} = T^1_{ij}Q_1 = \begin{bmatrix} 0 & 0 & 0 & 0 \\ 0 & -s\theta & -c\theta & 0 \\ 0 & c\theta & -s\theta & 0 \\ 0 & 0 & 0 & 0 \end{bmatrix} \tag{6.111}$$

or

$$\frac{\partial T^1_{ij}}{\partial \theta} = \begin{bmatrix} 1 & 0 & 0 & a \\ 0 & c\theta & -s\theta & b \\ 0 & s\theta & c\theta & c \\ 0 & 0 & 0 & 1 \end{bmatrix} Q_1 = \begin{bmatrix} 0 & 0 & 0 & 0 \\ 0 & -s\theta & -c\theta & 0 \\ 0 & c\theta & -s\theta & 0 \\ 0 & 0 & 0 & 0 \end{bmatrix} \tag{6.112}$$

The solution for $Q_1$ can be obtained in a number of ways, the easiest is perhaps by inspection. Observe that columns 1 and 4 of the final matrix are zero vectors. This can be achieved by setting the first and fourth columns of $Q_1$ to zero. This does not affect columns 2 and 3. Column 2 of the final matrix is equal to column 3 of the original elementary matrix. Consequently, if the second column of $Q_1$ contains a 1 in the third row (all other elements equal zero), the required result will be realized. Column 3 of the final matrix is equal to $-1$ times column 2 of the original matrix. This leads to the conclusion that column 3 of $Q_1$ should be zero except for $-1$ in row 2. Putting all this together, we get

$$Q_1 = \begin{bmatrix} 0 & 0 & 0 & 0 \\ 0 & 0 & -1 & 0 \\ 0 & +1 & 0 & 0 \\ 0 & 0 & 0 & 0 \end{bmatrix} \tag{6.113}$$

The derivative operator matrices for elementary matrices type 2 and 3 can be found in a similar manner (see Exer. 6.12):

$$Q_2 = \begin{bmatrix} 0 & 0 & +1 & 0 \\ 0 & 0 & 0 & 0 \\ -1 & 0 & 0 & 0 \\ 0 & 0 & 0 & 0 \end{bmatrix}, \qquad Q_3 = \begin{bmatrix} 0 & -1 & 0 & 0 \\ +1 & 0 & 0 & 0 \\ 0 & 0 & 0 & 0 \\ 0 & 0 & 0 & 0 \end{bmatrix} \tag{6.114}$$

The operator matrices for the elementary screw matrices (types 4, 5, and 6) will be slightly different since the variable $\theta$ of the matrix appears in the translation vector as well as in the rotational portion of the matrix. Consider a type 4 matrix. Taking the derivative of each entry with respect to $\theta$ yields

$$\frac{\partial T^4_{ij}(a, b, c, \theta, k)}{\partial \theta} = \begin{bmatrix} 0 & 0 & 0 & \dfrac{k}{2\pi} \\ 0 & -s\theta & -c\theta & 0 \\ 0 & c\theta & -s\theta & 0 \\ 0 & 0 & 0 & 0 \end{bmatrix} \tag{6.115}$$

Expressed as a matrix product, this becomes

$$\begin{bmatrix} 1 & 0 & 0 & a + \dfrac{k\theta}{2\pi} \\ 0 & c\theta & -s\theta & b \\ 0 & s\theta & c\theta & c \\ 0 & 0 & 0 & 1 \end{bmatrix} Q_4 = \begin{bmatrix} 0 & 0 & 0 & \dfrac{k}{2\pi} \\ 0 & -s\theta & -c\theta & 0 \\ 0 & c\theta & -s\theta & 0 \\ 0 & 0 & 0 & 0 \end{bmatrix} \tag{6.116}$$

The first three columns of $Q_4$ are found in the same way the first three columns of $Q_1$ were found. Upon further inspection the fourth column is defined as follows:

$$Q_4 = \begin{bmatrix} 0 & 0 & 0 & \dfrac{k}{2\pi} \\ 0 & 0 & -1 & 0 \\ 0 & +1 & 0 & 0 \\ 0 & 0 & 0 & 0 \end{bmatrix} \tag{6.117}$$

The term $k/2\pi$ is placed in the row of the fourth column corresponding to the direction of the screw translation. The derivative operator matrices for elementary screw matrices types 5 and 6 can be shown to be (see Exer. 6.13)

$$Q_5 = \begin{bmatrix} 0 & 0 & +1 & 0 \\ 0 & 0 & 0 & \dfrac{k}{2\pi} \\ -1 & 0 & 0 & 0 \\ 0 & 0 & 0 & 0 \end{bmatrix}, \qquad Q_6 = \begin{bmatrix} 0 & -1 & 0 & 0 \\ +1 & 0 & 0 & 0 \\ 0 & 0 & 0 & \dfrac{k}{2\pi} \\ 0 & 0 & 0 & 0 \end{bmatrix} \tag{6.118}$$

The translation-only matrix requires three derivative operators since this matrix can be used to describe a variable translation along any one direction. If the variable translation is in the $x$ direction, taking the partial derivative of the translation-only matrix with respect to the $x$ directional translation will result in

$$\frac{\partial T^7_{ij}(a, b, c)}{\partial a} = \begin{bmatrix} 0 & 0 & 0 & 1 \\ 0 & 0 & 0 & 0 \\ 0 & 0 & 0 & 0 \\ 0 & 0 & 0 & 0 \end{bmatrix} \tag{6.119}$$

The derivative operator for this case is obtained from the following equation:

$$\begin{bmatrix} 1 & 0 & 0 & a \\ 0 & 1 & 0 & b \\ 0 & 0 & 1 & c \\ 0 & 0 & 0 & 1 \end{bmatrix} Q_7 = \begin{bmatrix} 0 & 0 & 0 & 1 \\ 0 & 0 & 0 & 0 \\ 0 & 0 & 0 & 0 \\ 0 & 0 & 0 & 0 \end{bmatrix} \tag{6.120}$$

By inspection,

$$Q_7 = \begin{bmatrix} 0 & 0 & 0 & 1 \\ 0 & 0 & 0 & 0 \\ 0 & 0 & 0 & 0 \\ 0 & 0 & 0 & 0 \end{bmatrix} \tag{6.121}$$

If the variable translation is in the $y$ direction ($Q_8$) or the $z$ direction ($Q_9$) the derivative operator matrices are

$$Q_8 = \begin{bmatrix} 0 & 0 & 0 & 0 \\ 0 & 0 & 0 & 1 \\ 0 & 0 & 0 & 0 \\ 0 & 0 & 0 & 0 \end{bmatrix}, \qquad Q_9 = \begin{bmatrix} 0 & 0 & 0 & 0 \\ 0 & 0 & 0 & 0 \\ 0 & 0 & 0 & 1 \\ 0 & 0 & 0 & 0 \end{bmatrix} \tag{6.122}$$

Now that taking derivatives of the elementary matrices has been defined as a matrix product, we are ready to take the derivative of the loop-closure equation.

### Taking the Derivative of the Loop-Closure Equation

A general loop-closure equation is written as

$$S_{00} = S_{01}S_{12} \cdots S_{(k-1)k}(q_m)S_{k(k+1)} \cdots S_{(n-1)n}S_{n0} \tag{6.123}$$

Any one variable in a mechanism described by a loop-closure equation is wholly contained in only one of the matrices making up the product. In the example shown, the variable denoted by $q_m$ is contained in matrix $S_{(k-1)k}$ only. Upon taking the derivative of Eq. (6.123) with respect to $q_m$:

$$\frac{\partial S_{00}}{\partial q_m} = S_{01}S_{12} \cdots \frac{\partial S_{(k-1)k}(q_m)}{\partial q_m} S_{k(k+1)} \cdots S_{(n-1)n}S_{n0} \tag{6.124}$$

Replacing the derivative (6.124) by its equivalent matrix product:

$$\frac{\partial S_{00}}{\partial q_m} = S_{01}S_{12} \cdots S_{(k-1)k}(q_m)Q_i S_{k(k+1)} \cdots S_{(n-1)n}S_{n0} \tag{6.125}$$

where $Q_i$ is the proper type of derivative operator matrix for the $S_{(k-1)k}$ matrix. This reduces the taking of derivative on the right side of the equation to a convenient matrix product. However, the same derivative can be obtained by operating on the left side of the equation. Here is how.

Let us insert a specially constructed identity matrix following the derivative operator in (6.125). This is equivalent to inserting unity in a product of scalars. The identity matrix to be inserted is defined as

$$I = (S_{01}S_{12} \cdots S_{(k-1)k})^{-1}(S_{01}S_{12} \cdots S_{(k-1)k}) \tag{6.126}$$

which is a common definition for the identity matrix: a matrix times its inverse. Equation (6.125) then becomes

$$\frac{\partial S_{00}}{\partial q_m} = S_{01}S_{12} \cdots S_{(k-1)k}(q_m)Q_i\,(S_{01}S_{12} \cdots S_{(k-1)k})^{-1}(S_{01}S_{12} \cdots S_{(k-1)k})S_{k(k+1)} \cdots S_{(n-1)n}S_{n0} \tag{6.127}$$

Observe that the matrix product at the end of (6.127) is $S_{00}$ itself. Consequently,

$$\frac{\partial S_{00}}{\partial q_m} = D_m S_{00} \tag{6.128}$$

where

$$D_m = (S_{01}S_{12} \cdots S_{(k-1)k})Q_i(S_{01}S_{12} \cdots S_{(k-1)k})^{-1} \tag{6.129}$$

$D_m$ is defined as the derivative operator matrix with respect to the $m$th variable in the loop. However, a formidable calculation needs to be performed in Eq. (6.129) before realizing the mathematical benefits shown in Eq. (6.128). This can be avoided by the analytical determination of Eq. (6.129). To this end, we first require an analytical definition of the inverse matrix shown. A general $4 \times 4$ matrix made up of a product of $4 \times 4$ elementary transformation matrices [e.g., see Eq. (6.129)] is expressed as

$$S_{0k} = \begin{bmatrix} r_{11} & r_{12} & r_{13} & \Delta_x \\ r_{21} & r_{22} & r_{23} & \Delta_y \\ r_{31} & r_{32} & r_{33} & \Delta_z \\ 0 & 0 & 0 & 1 \end{bmatrix} \tag{6.130}$$

as noted in Eq. (6.30). The general $4 \times 4$ matrix used here has a special form in that the $3 \times 3$ rotational part $[R]$ is an orthogonal matrix. This implies that the inverse of this part is equal to its transpose ($[R]^{-1} = [R]^T$). This is well known from the matrix theory of linear systems. It can be found, for instance, in Ref.

259. It is shown there that the inverse of Eq. (6.130) is (see Exer. 6.14)

$$S_{ok}^{-1} = \begin{bmatrix} r_{11} & r_{21} & r_{31} & [-(\Delta_x r_{11} + \Delta_y r_{21} + \Delta_z r_{31})] \\ r_{12} & r_{22} & r_{32} & [-(\Delta x r_{12} + \Delta_y r_{22} + \Delta_z r_{32})] \\ r_{13} & r_{23} & r_{33} & [-(\Delta x r_{13} + \Delta_y r_{23} + \Delta_z r_{33})] \\ 0 & 0 & 0 & 1 \end{bmatrix} \tag{6.131}$$

With this matrix defined, the derivative operator matrices ($D_m$) corresponding to each type of differential operator can be expressed in analytical form by expressing the product shown in (6.129) for each $Q_i$ defined earlier. When this is done the following results are obtained. The superscript in each of the terms shown below corresponds to the number of the derivative operator used. Note that these expressions lend themselves to being efficiently and easily programmed for a digital computer.

The derivative operators for the rotational matrices $T_{jk}^1$, $T_{jk}^2$, and $T_{jk}^3$ are

$$D_m^i = S_{ok} Q_i S_{ok}^{-1} = \begin{bmatrix} 0 & -r_{3i} & r_{2i} & \Delta_y r_{3i} - \Delta_z r_{2i} \\ r_{3i} & 0 & -r_{1i} & \Delta_z r_{1i} - \Delta_x r_{3i} \\ -r_{2i} & r_{1i} & 0 & \Delta_x r_{2i} - \Delta_y r_{1i} \\ 0 & 0 & 0 & 0 \end{bmatrix}_{i=1,2,3} \tag{6.132}$$

For the screw matrices $T_{jk}^4$, $T_{jk}^5$, and $T_{jk}^6$ the derivative operators are

$$D_m^i = S_{ok} Q_i S_{ok}^{-1}$$

$$= \begin{bmatrix} 0 & -r_{3(i-3)} & r_{2(i-3)} & \frac{k}{2\pi} r_{1(i-3)} + (\Delta_y r_{3(i-3)} - \Delta_z r_{2(i-3)}) \\ r_{3(i-3)} & 0 & -r_{1(i-3)} & \frac{k}{2\pi} r_{2(i-3)} + (\Delta_z r_{1(i-3)} - \Delta_x r_{3(i-3)}) \\ -r_{2(i-3)} & r_{1(i-3)} & 0 & \frac{k}{2\pi} r_{3(i-3)} + (\Delta_x r_{2(i-3)} - \Delta_y r_{1(i-3)}) \\ 0 & 0 & 0 & 0 \end{bmatrix} \tag{6.133}$$

where $i = 4, 5, 6$.

Three derivative operators are defined for the translation-only matrix $T_{jk}^7$, one each for a variable translation in the $x(i = 7)$, $y(i = 8)$, and $z(i = 9)$ directions.

$$D_m^i = S_{ok} Q_i S_{ok}^{-1} = \begin{bmatrix} 0 & 0 & 0 & r_{1(i-6)} \\ 0 & 0 & 0 & r_{2(i-6)} \\ 0 & 0 & 0 & r_{3(i-6)} \\ 0 & 0 & 0 & 0 \end{bmatrix}, \qquad i = 7, 8, 9 \tag{6.134}$$

In summary, we have derived a procedure for taking the derivative of a matrix loop-closure equation with respect to any variable contained within it. Furthermore, this is done by premultiplying the equation by a derivative operator matrix as shown in Eq. (6.128). To form this matrix product, the derivative operator matrix must be formulated as stated in Eq. (6.132). In order to form these expressions, the location of the variable must be known and the type of elementary matrix must be known to assure that the proper expression in Eq. (6.132) is used. This completes the discussions on taking the derivatives of the elementary matrices and of the loop-closure equation. We see next how this technique is used in the position analysis of a general spatial mechanism.

## 6.14 POSITION ANALYSIS

The relative joint motion variables (or "joint variables") used in the description of a mechanism may be considered as belonging to one of two groups. The first consists of the vector of independent motion variables ($\bar{q}_I$) of the mechanism. These are often called the degrees of freedom of the mechanism or the inputs for the mechanism. The second group of joint variables forms the vector of dependent variables ($\bar{q}_D$). The values of the dependent variables are functions of the input (independent) variables and the geometric constraints of the mechanism. These constraints are expressed mathematically by the matrix loop-closure equation, which is a nonlinear matrix equation in terms of the joint variables. Consequently, a procedure for obtaining the values of the dependent variables is one that solves the loop-closure equation for these values. This procedure is called the *position analysis* of a mechanism. A general scheme used to analyze a mechanism is developed in this section. The loop-closure equation is solved using standard numerical analysis techniques. In particular, a Newton–Raphson iterative procedure is applied to the position analysis of a general spatial mechaism.

The matrix loop-closure equation was expressed earlier as

$$S_{00} = S_{01}S_{12}S_{23} \cdots S_{(n-1)n}S_{n0} = I \tag{6.135}$$

It equals the identity matrix $I$ *only* when the dependent variables have their correct values. Expressed in terms of the variables as defined above, the nonlinear equation used for position analysis is

$$S_{00}(\bar{q}_D, \bar{q}_I) - I = 0 \tag{6.136}$$

Consequently, we must find the "roots" or "zeros" of this matrix equation. A Newton–Raphson iterative procedure [33,50] is well suited for solving this type of problem. Beginning with an initial estimate of the dependent variables ($\hat{\bar{q}}_D$), a small correction ($\Delta\bar{q}_D$) can be found such that when it is added to the estimate, a value closer to the correct value results. This iterative procedure is derived below. It begins by performing a first-order linear expansion of Eq. (6.136) about ($\hat{\bar{q}}_D$). This yields

$$S_{00}(\hat{\bar{q}}_D, \bar{q}_I) - I + \sum_{i=1}^{m} \frac{\partial S_{00}(\hat{\bar{q}}_D, \bar{q}_I)}{\partial q_i} \Delta q_i = 0 \tag{6.137}$$

where $m$ is the total number of all independent plus all dependent variables. Note that for the time being all variables $q_i$ and $\Delta q_i$ are being considered regardless of whether or not they are independent or dependent variables. It becomes evident why we wish to do this later when we will divide the variables into their respective groups. With the mathematical tools developed in earlier sections, Eq. (6.137) can be expressed in a more convenient form.

Using the derivative operator matrix $D_i$ (Eq. (6.132) for differentiating the loop-closure equation, Eq. (6.137) becomes

$$S_{00} + \sum_{i=1}^{m} (D_i S_{00}\, \Delta q_i) - I = 0 \tag{6.138}$$

Postmultiplying by $S_{00}^{-1}$ yields

$$I + \sum_{i=1}^{m} (D_i\, \Delta q_i) - S_{00}^{-1} = 0 \tag{6.139}$$

or

$$\sum_{i=1}^{m} D_i\, \Delta q_i = S_{00}^{-1} - I \tag{6.140}$$

This is a 4 × 4 matrix equation. From it, 16 scalar equations can be formed by equating the two sides, element by element. However, all 16 equations are not meaningful and independent. Recall the general form of the derivative operator matrix for the $\ell$th variable in the loop from Eq. (6.132):

$$D_\ell = \begin{bmatrix} 0 & D(1,2)_\ell & D(1,3)_\ell & D(1,4)_\ell \\ D(2,1)_\ell & 0 & D(2,3)_\ell & D(2,4)_\ell \\ D(3,1)_\ell & D(3,2)_\ell & 0 & D(3,4)_\ell \\ 0 & 0 & 0 & 0 \end{bmatrix} \tag{6.141}$$

The bottom row and the diagonal elements are all zero and therefore cannot be used. Furthermore, the 3 × 3 rotational part of the matrix is antisymmetric: the elements above the diagonal equal the negative of those below the diagonal [see Eq. (6.132)]. Consequently, only three of the six nonzero elements of the rotational part can be used to form independent scalar equations. The three translational elements can also be used, so there are a total of six elements from $D_i$ which can be used to formulate independent equations. They are: $D(1,3)_\ell$, $D(2,1)_\ell$, $D(3,2)_\ell$, $D(1,4)_\ell$, $D(2,4)_\ell$, and $D(3,4)_\ell$.

Next consider the term on the right side of (6.140). It can be expressed as

$$S_{00}^{-1} - I = \begin{bmatrix} E(1,1)-1 & E(1,2) & \underline{E(1,3)} & \underline{E(1,4)} \\ \underline{E(2,1)} & E(2,2)-1 & E(2,3) & \underline{E(2,4)} \\ E(3,1) & \underline{E(3,2)} & E(3,3)-1 & \underline{E(3,4)} \\ 0 & 0 & 0 & 0 \end{bmatrix} \tag{6.142}$$

The underlined elements are those used to form the six scalar equations for the iteration procedure. At convergence, these elements will equal zero. This leaves six elements of Eq. (6.142) which are not included in the solution procedure. Because the rotational matrix of $S_{00}$ (and thus $S_{00}^{-1}$) is orthogonal, the following elements also equal zero when $E(2,1)$, $E(1,3)$, and $E(3,2)$ are zero:

$$E(1,2), \quad E(3,1), \quad \text{and} \quad E(2,3) \tag{6.143}$$

This is determined by taking the cross product of two columns and setting it equal to the third column of the 3 × 3 rotational matrix. The diagonal terms are not constrained, however. In fact, the following is true:

$$\begin{aligned} E(1,1)E(2,2) &= E(3,3) \\ E(2,2)E(3,3) &= E(1,1) \\ E(3,3)E(1,1) &= E(2,2) \end{aligned} \tag{6.144}$$

and

$$\begin{aligned} E(1,1)^2 &= 1 \\ E(2,2)^2 &= 1 \\ E(3,3)^2 &= 1 \end{aligned} \tag{6.145}$$

For these relationships to hold, the values of any two diagonal terms may be $-1$ instead of the $+1$ value required in (6.142). Consequently, these terms must be considered explicitly in the iteration procedure. This is done as is shown in (6.146) using the form suggested by Uicker [294–296].

Summarizing, the scalar equations for the iteration procedure can be written in matrix form as

$$\begin{bmatrix} D(1,3)_1 & D(1,3)_2 & \cdots & D(1,3)_m \\ D(2,1)_1 & D(2,1)_2 & \cdots & D(2,1)_m \\ D(3,2)_1 & D(3,2)_2 & \cdots & D(3,2)_m \\ D(1,4)_1 & D(1,4)_2 & \cdots & D(1,4)_m \\ D(2,4)_1 & D(2,4)_2 & \cdots & D(2,4)_m \\ D(3,4)_1 & D(3,4)_2 & \cdots & D(3,4)_m \end{bmatrix} \begin{bmatrix} \Delta q_1 \\ \Delta q_2 \\ \cdot \\ \cdot \\ \cdot \\ \Delta q_m \end{bmatrix} = \begin{bmatrix} E(1,3) + E(1,1) + E(3,3) - 2 \\ E(2,1) + E(1,1) + E(2,2) - 2 \\ E(3,2) + E(2,2) + E(3,3) - 2 \\ E(1,4) \\ E(2,4) \\ E(3,4) \end{bmatrix} \tag{6.146}$$

or

$$[S](\overline{\Delta q}) = (\bar{R}) \tag{6.147}$$

Notice that the diagonal terms of Eq. (6.142) are included in the first three equations by taking them two at a time. The matrix $[S]$ is called the system geometry matrix. It contains significant information about a mechanism, as will be shown.

The procedure as outlined thus far is for mechanisms described with one loop-closure equation. For more complex mechanisms requiring multiple loops, a set of six scalar equations per loop can be formulated as above. Consequently, the technique discussed here is applicable to multiple-loop mechanisms.

A distinction between the independent and dependent joint variables must now be made in order that changes in the dependent variables can be determined. Also the question of how many joint variables can be independent must be addressed. This is equivalent to asking how many degrees of freedom a mechanism has. For example, a planar four-bar linkage (a one-degree-of-freedom mechanism) requires four joint variables in its description. Three must be dependent if one is independent.

As mentioned earlier in this chapter, Gruebler's equation may not provide correct results when special geometry is used for a mechanism. The matrix $[S]$ in (6.147) is dependent on the geometry of the mechanism, and contains the required information of configuration and dimensions. The size of $[S]$ is $6 \times m$, where the number of columns, $m$, is the number of joint variables. The number that are dependent is determined by the rank $r$ of matrix $[S]$. It is this quantity that indicates the number of independent equations in Eq. (6.146). Note that the maximum allowed for a single loop mechanism is 6. In summary, there are $(m - r)$ independent joint variables or $(m - r)$ degrees of freedom for the mechanism.

When analyzing a mechanism, exactly $(m - r)$ joint variables must be specified as inputs. If more than $(m - r)$ variables are specified, the mechanism cannot be analyzed since it is overconstrained. If fewer than $(m - r)$ variables are specified, the mechanism can be analyzed; however, the solution is not unique, as the mechanism is underconstrained. Consequently, the joint variables are divided into *three* groups. The first contains the $r$-dependent variables $(\bar{q}_D)$. The second contains the specified independent variables $(\bar{q}_I)$. Assuming that the number in this group is $s$, a third group consists of $(m - r - s)$ unspecified independent variables $(\bar{q}_U)$. The number of variables in this third group is known only after the rank of $[S]$ is determined. However, in determining the rank of $[S]$, it is definitely known that the columns corresponding to $(\bar{q}_I)$ are not to be considered. This can be done by partitioning the $6 \times m$ matrix $[S]$ and by rearranging the columns as follows:

$$[S_{D,U} \mid S_I] \begin{bmatrix} \bar{q}_{D,U} \\ \bar{q}_I \end{bmatrix} = (\bar{R}) \tag{6.148}$$

The rank of $[S_{D,U}]$ is determined next. This leads to a further partitioning of the $[S]$ matrix as follows:

$$\left[\begin{array}{c|c|c} S_D & S_U & S_I \\ (r \times r) & r \times (m - r - s) & r \times s \\ \hline \text{A} & \multicolumn{2}{c}{\text{B}} \\ (6 - r) \times r & \multicolumn{2}{c}{(6 - r) \times (m - r)} \end{array}\right] \begin{bmatrix} \Delta\bar{q}_D \\ \Delta\bar{q}_U \\ \Delta\bar{q}_I \end{bmatrix} = \begin{bmatrix} \bar{R}_1 \\ \bar{R}_2 \end{bmatrix} \tag{6.149}$$

Rewriting the $r$-independent equations, we have

$$[S_D](\Delta\bar{q}_D) = (\bar{R}_1) - [S_U](\Delta\bar{q}_U) - [S_I](\Delta\bar{q}_I) \tag{6.150}$$

Since $(\Delta\bar{q}_I)$ is zero (independent variables are not changed) and arbitrarily setting $(\Delta\bar{q}_U)$ equal to zero, (6.150) can be solved for $(\Delta\bar{q}_D)$:

$$(\Delta\bar{q}_D) = [S_D]^{-1}(\bar{R}_1) \tag{6.151}$$

These changes in the dependent variables are added to the current estimate:

$$\hat{\bar{q}}_D \leftarrow \hat{\bar{q}}_D + \Delta\bar{q}_D \tag{6.152}$$

thus obtaining a new estimate which is closer to the correct value of the dependent variables. This iteration procedure continues until the changes $(\Delta\bar{q}_D)$ are small and $S_{00}$ approaches the identity matrix.

A further note on the significance of $(\bar{q}_U)$. Much of this text has dealt with the analysis of fully constrained mechanisms. For these cases, there are no unspecified independent variables. When such variables do occur, the mechanism is usually handled as a dynamic time-response problem, as discussed in Chap. 5. In these cases the mechanism usually has forces and torques acting on it to produce an unknown motion that is determined by a time-integration scheme.

## 6.15 VELOCITY ANALYSIS

Once the position analysis of a mechanism is completed, it can be analyzed further to determine velocities, i.e., time rates of change of the joint variables. This consists of finding the velocity of the dependent joint variables ($\dot{\bar{q}}_D$), given the known velocities of the independent joint variables ($\dot{\bar{q}}_I$). If there were unspecified independent joint variables ($\bar{q}_U$) found during the position analysis, their velocities ($\dot{\bar{q}}_U$) must be specified to perform a velocity analysis. It will be assumed here that this is the case.

Taking the derivative of the loop-closure equation with respect to time provides a system of equations that will be used for the velocity analysis of a mechanism. Using the chain rule we get

$$\frac{dS_{00}}{dt} = \sum_{i=1}^{m} \frac{\partial S_{00}}{\partial q_i}\frac{dq_i}{dt} = 0 \tag{6.153}$$

Replacing the partial derivative of the loop-closure equation with its matrix product equivalent, and recognizing that $S_{00} = I$ after the position analysis is completed, Eq. (6.153) becomes

$$\sum_{i=1}^{m} D_i S_{00} \dot{q}_i = \sum_{i=1}^{m} D_i \dot{q}_i = 0 \tag{6.154}$$

This again is a $4 \times 4$ matrix equation from which scalar equations can be formulated. Using the same elements of $D_i$ as were used earlier for the position analysis, we arrive at six equations for the velocity analysis:

$$\begin{bmatrix} D(1,3)_1 & D(1,3)_2 & \cdots & D(1,3)_m \\ D(2,1)_1 & D(2,1)_2 & \cdots & D(2,1)_m \\ D(3,2)_1 & D(3,2)_2 & \cdots & D(3,2)_m \\ D(1,4)_1 & D(1,4)_2 & \cdots & D(1,4)_m \\ D(2,4)_1 & D(2,4)_2 & \cdots & D(2,4)_m \\ D(3,4)_1 & D(3,4)_2 & \cdots & D(3,4)_m \end{bmatrix} (\dot{\bar{q}}) = 0 \tag{6.155}$$

or

$$[S](\dot{\bar{q}}) = 0 \tag{6.156}$$

The $[S]$ matrix is the same system geometry matrix found earlier. Discussion concerning the rank of this matrix and partitioning it to arrive at a set of linear independent equations applies here as well. The results will be a matrix partitioned as follows:

$$\left[\begin{array}{c|c|c} S_D & S_U & S_I \\ \hline A & \multicolumn{2}{c}{B} \end{array}\right] \begin{bmatrix} \dot{\bar{q}}_D \\ \dot{\bar{q}}_U \\ \dot{\bar{q}}_I \end{bmatrix} = 0 \tag{6.157}$$

or

$$[S_D]\,\dot{\vec{q}}_D = -([S_U]\,\dot{\vec{q}}_U + [S_I]\,\dot{\vec{q}}_I) \tag{6.158}$$

Solving for the velocity of the dependent variables results in

$$\dot{\vec{q}}_D = -[S_D]^{-1}\,([S_U]\,\dot{\vec{q}}_U + [S_I]\,\dot{\vec{q}}_I) \tag{6.159}$$

Note that similar terms appear in the equations for position and velocity analyses. Consequently, this technique is well suited for programming on a digital computer.

## 6.16 ACCELERATION ANALYSIS

Once the velocity analysis has been completed, an acceleration analysis of a mechanism can be performed. This consists of determining the acceleration of the dependent joint variables ($\ddot{\vec{q}}_D$) given known accelerations of the independent joint variables ($\ddot{\vec{q}}_I$) and the unspecified independent joint variables ($\ddot{\vec{q}}_U$). It is again assumed that the question of unspecified joint variables has been resolved and they have been given known accelerations, thus making the mechanism fully constrained. In a dynamic time-response problem, these accelerations ($\ddot{\vec{q}}_U$) are determined from the effects of the forces acting on the mechanism.

Taking the derivative of the velocity-loop-closure equation (6.153) with respect to time yields a system of equations containing the accelerations of the joint variables in a mechanism. This can be written as

$$\frac{d^2S_{00}}{dt^2} = \sum_{i=1}^{m} \frac{\partial S_{00}}{\partial q_i}\,\ddot{q}_i + \sum_{i=1}^{m}\sum_{j=1}^{m} \frac{\partial^2 S_{00}}{\partial q_i\,\partial q_j}\,\dot{q}_i\dot{q}_j = 0 \tag{6.160}$$

The first summation term in this equation is similar to the term appearing in Eq. (6.153). Replacing the derivative by its equivalent matrix product and recognizing that $S_{00} = I$, we have

$$\sum_{i=1}^{m} \frac{\partial S_{00}}{\partial q_i}\,\ddot{q}_i = \sum_{i=1}^{m} D_i\ddot{q}_i \tag{6.161}$$

The second derivative of $S_{00}$ in (6.160) must be defined further since it has not come up before. The loop-closure equation can be expanded and differential operators for the $i$th and $j$th joint variable inserted at the proper locations.

$$\frac{\partial^2 S_{00}}{\partial q_i\,\partial q_j} = S_{01}S_{12}\cdots S_{(k-1)k}(q_i)Q_iS_{k(k+1)}\cdots S_{(l-1)l}(q_j)Q_jS_{l(l+1)}\cdots S_{(n-1)n}S_{n0} \tag{6.162}$$

Inserting two forms of the identity matrix in this equation yields

$$\frac{\partial^2 S_{00}}{\partial q_i \, \partial q_j} = S_{01}S_{12} \cdots$$

$$S_{(k-1)k}(q_i)Q_i \underbrace{(S_{01} \cdots S_{(k-1)k})^{-1}(S_{01} \cdots S_{(k-1)k})}_{I} S_{k(k+1)} \cdots$$

$$S_{(l-1)l}(q_j)Q_j \underbrace{(S_{01} \cdots S_{(l-1)l})^{-1}(S_{01} \cdots S_{(l-1)l})}_{I} S_{l(l+1)} \cdots S_{(n-1)n}S_{n0} \tag{6.163}$$

This equation can be rewritten in terms of derivative matrix operators,

$$\frac{\partial^2 S_{00}}{\partial q_i \, \partial q_j} = D_i D_j S_{00}, \qquad i < j \tag{6.164}$$

if $q_i$ appears in the loop-closure equation before $q_j$. Writing Eq. (6.164) in a more general way, we have

$$\frac{\partial^2 S_{00}}{\partial q_i \, \partial q_j} = \overline{D_i D_j} = \begin{cases} D_i D_j, & i \leq j \\ D_j D_i, & j < i \end{cases} \tag{6.165}$$

where it is again recognized that $S_{00} = I$. Summarizing these steps, the matrix $G_m$ is defined as follows:

$$[G_m] = \sum_{i=1}^{m} \sum_{j=1}^{m} \frac{\partial^2 S_{00}}{\partial q_i \, \partial q_j} \dot{q}_i \dot{q}_j = \sum_{i=1}^{m} \sum_{j=1}^{m} \overline{D_i D_j} \dot{q}_i \dot{q}_j \tag{6.166}$$

Note that this term is known after the velocity analysis is complete. It contains the "normal" acceleration terms and the "Coriolis" acceleration terms. Computing the double summation in Eq. (6.166) can be very costly on a computer for large values of $m$. However, the following recursive scheme can be used.

Define

$$W_{k-1} = \sum_{i=1}^{k-1} D_i \dot{q}_i \tag{6.167}$$

then

$$W_k = \sum_{i=1}^{k} D_i \dot{q}_i = W_{k-1} + D_k \dot{q}_k$$

Furthermore, define

$$G_{k-1} = \sum_{i=1}^{k-1} \sum_{j=1}^{k-1} \overline{D_i D_j} \dot{q}_i \dot{q}_j$$

Then

$$
\begin{aligned}
G_k &= \sum_{i=1}^{k}\sum_{j=1}^{k} \overline{D_i D_j \dot{q}_i \dot{q}_j} \\
&= D_1 D_1 \dot{q}_1^2 + 2D_1 D_2 \dot{q}_1 \dot{q}_2 + \cdots + 2D_1 D_{k-1}\dot{q}_1\dot{q}_{k-1} + 2D_1 D_k \dot{q}_1 \dot{q}_k \\
&\quad + D_2 D_2 \dot{q}_2^2 + 2D_2 D_3 \dot{q}_2 \dot{q}_3 + \cdots + 2D_2 D_{k-1}\dot{q}_2\dot{q}_{k-1} + 2D_2 D_k \dot{q}_2 \dot{q}_k \\
&\quad + \cdots + D_{k-1} D_{k-1}\dot{q}_{k-1}^2 + \cdots + 2D_{k-1} D_k \dot{q}_{k-1}\dot{q}_k + D_k D_k \dot{q}_k^2 \\
&= G_{k-1} + 2W_{k-1} D_k \dot{q}_k + D_k D_k \dot{q}_k^2
\end{aligned}
\tag{6.168}
$$

or

$$G_k = G_{k-1} + (2W_{k-1} + D_k \dot{q}_k) D_k \dot{q}_k$$

where

$$W_0 = G_0 = 0$$

Returning to our acceleration analysis, and combining Eqs. (6.160), (6.161) and (6.166), we get

$$\sum_{i=1}^{m} D_i \ddot{q}_i + [G_m] = 0 \tag{6.169}$$

Here again we have a $4 \times 4$ matrix equation from which at most six independent linear equations can be formulated. In matrix form they are

$$
\begin{bmatrix}
D(1,3)_1 & D(1,3)_2 & \cdots & D(1,3)_m \\
D(2,1)_1 & D(2,1)_2 & \cdots & D(2,1)_m \\
D(3,2)_1 & D(3,2)_2 & \cdots & D(3,2)_m \\
D(1,4)_1 & D(1,4)_2 & \cdots & D(1,4)_m \\
D(2,4)_1 & D(2,4)_2 & \cdots & D(2,4)_m \\
D(3,4)_1 & D(3,4)_2 & \cdots & D(3,4)_m
\end{bmatrix}
(\ddot{q}) = -
\begin{bmatrix}
G_m(1,3) \\ G_m(2,1) \\ G_m(3,2) \\ G_m(1,4) \\ G_m(2,4) \\ G_m(3,4)
\end{bmatrix}
\tag{6.170}
$$

or

$$[S](\ddot{q}) = -(\overline{G}) \tag{6.171}$$

Again we have the system geometry matrix $[S]$. Partitioning it as before yields

$$
\left[\begin{array}{c|c|c} S_D & S_U & S_I \\ \hline A & \multicolumn{2}{c}{B} \end{array}\right]
\begin{bmatrix} \ddot{q}_D \\ \ddot{q}_U \\ \ddot{q}_I \end{bmatrix}
=
\begin{bmatrix} -\overline{G}_{m1} \\ -\overline{G}_{m2} \end{bmatrix}
\tag{6.172}
$$

where $\overline{G}_{m1}$ is the vector associated with the dependent variables and $\overline{G}_{m2}$ with the rest of the variables. The independent equations can be written as

$$[S_D](\ddot{q}_D) = -(\overline{G}_{m1} + [S_U]\ddot{q}_U + [S_I]\ddot{q}_I) \tag{6.173}$$

and the accelerations of dependent joint variables can be found as

$$\ddot{q}_D = -[S_D]^{-1}(\overline{G}_{m1} + [S_U]\ddot{q}_U + [S_I]\ddot{q}_I) \tag{6.174}$$

Now that the position, velocity and acceleration of all joint variables have been determined, the kinematics of particular points on a mechanism can be found. Following the next section, the foregoing methods will be illustrated by an example, in which a spatial mechanism will be analyzed.

## 6.17 POINT KINEMATICS IN THREE-DIMENSIONAL SPACE

Up to this point, the kinematics of an entire mechanism has been considered. The analysis techniques developed provide information about the joint variables but not about particular points on the links in the mechanism. In this section the kinematics of points defined on links is considered. The position, velocity, and acceleration of these points are determined in terms of the joint variables obtained in the kinematic analysis of the entire mechanism.

The position of a point may be described in two ways: (1) by a set of *xyz* coordinates defined in the local coordinate system of a link, and (2) in terms of global coordinates or those measured with respect to a fixed reference frame.

Typically, the location of a point is known and is fixed in the local coordinate system of a link, and the motion of this point in a global reference coordinate system is desired as the link moves through the constrained motion of a mechanism. Accordingly, a point will be defined in terms of four homogeneous coordinates to make them compatible with the $4 \times 4$ matrix notation. Thus the local definition of a point $P$ on link $i$ is

$$\mathbf{r}_{ip} = \begin{bmatrix} x_i \\ y_i \\ z_i \\ 1 \end{bmatrix} \tag{6.175}$$

See Fig. 6.50 for the definition of point $P$ on link 3. In this figure, the coordinate systems shown $(1, 2, 3 \ldots, n)$ are the local coordinate systems of the links. All of the intermediate coordinate systems defined by the transformation matrices describing the links and the joints between the links are not shown. Consequently, notation used to define the local coordinate systems is required. The transformation matrix describing the location and orientation of a coordinate system fixed to link $i$ is defined as

$$A_i = S_{01}S_{12} \cdots S_{k-1,k} \tag{6.176}$$

Equation (6.176) is the product of $k$ elementary matrices. With this, the location of the point in the fixed reference frame is defined as

$$\mathbf{R}_p = \begin{bmatrix} x_p \\ y_p \\ z_p \\ 1 \end{bmatrix} = A_i\mathbf{r}_{ip} \tag{6.177}$$

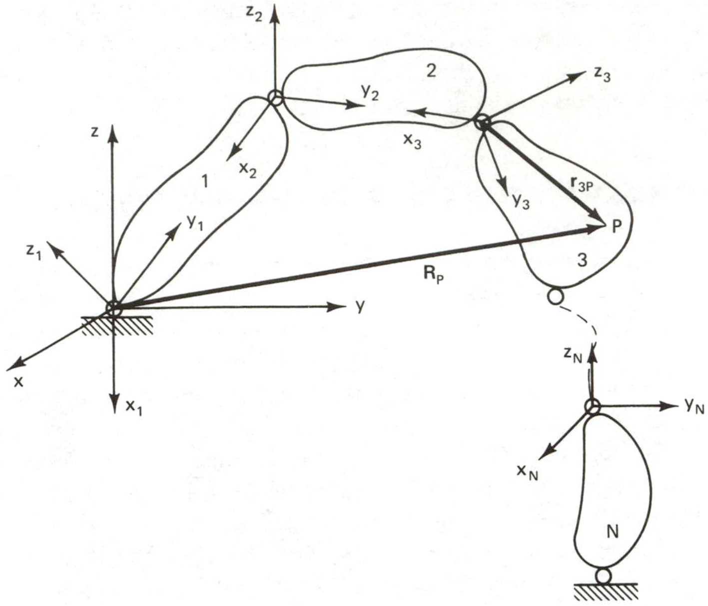


**Figure 6.50** Point $P$ defined by a local vector $\mathbf{r}_{3P}$ and a global vector $\mathbf{R}_P$.

The velocity of this point is determined by taking the time derivative of Eq. (6.177). Using the chain rule on the right-hand side results in

$$\dot{\mathbf{R}}_p = \begin{bmatrix} \dot{x}_p \\ \dot{y}_p \\ \dot{z}_p \\ 0 \end{bmatrix} = \sum_{\text{l}=1}^{m} \frac{\partial A_i}{\partial q_{\text{l}}} \dot{q}_{\text{l}} \mathbf{r}_{ip} \tag{6.178}$$

where it is recognized that there are $m$ variable matrices in $A_i$ and that $\mathbf{r}_{ip}$, defined in the local coordinate system, is constant (rigid link assumption). Taking the time derivative of $A_i$ is achieved with the use of the derivative operator matrix; the correct type $Q_n$ is inserted at the appropriate location:

$$\frac{\partial A_i}{\partial q_{\text{l}}} = S_{01} S_{12} \cdots S_{n-1,n} Q_n S_{n,n+1} \cdots S_k \tag{6.179}$$

Using the definition of $D_{\text{l}}$ expressed in Eq. (6.129), this becomes

$$\frac{\partial A_i}{\partial q_{\text{l}}} = D_{\text{l}} A_i \tag{6.180}$$

Consequently, (6.178) becomes

$$\dot{\mathbf{R}}_p = \left( \sum_{\text{l}=1}^{m} D_{\text{l}} \dot{q}_{\text{l}} \right) A_i \mathbf{r}_{ip} = \left( \sum_{\text{l}=1}^{m} D_{\text{l}} \dot{q}_{\text{l}} \right) \mathbf{R}_p \tag{6.181}$$

where it is recognized that $A_i$ is constant in the summation and where Eq. (6.177) has been utilized. However, the summation on the right has been calculated earlier during the acceleration analysis; see Eqs. (6.167) and (6.168) for the definition of $W_k$. If these intermediate matrices are saved during the acceleration analysis, no further calculations are necessary to determine the velocity of point $P$ except for the forming of the matrix-vector product:

$$\dot{\mathbf{R}}_p = W_m \mathbf{R}_p \tag{6.182}$$

The acceleration of a point is determined by taking the first time derivative of Eq. (6.178):

$$\ddot{\mathbf{R}}_p = \begin{bmatrix} \ddot{x} \\ \ddot{y} \\ \ddot{z} \\ 0 \end{bmatrix} = \sum_{l=1}^{m} \frac{\partial A_i}{\partial q_l} \ddot{q}_l \mathbf{r}_{ip} + \sum_{j=1}^{m} \sum_{l=1}^{m} \frac{\partial^2 A_i}{\partial q_l \, \partial q_j} \dot{q}_l \dot{q}_j \mathbf{r}_{ip} \tag{6.183}$$

Considering the first term on the right and utilizing Eq. (6.180), we obtain

$$\sum_{l=1}^{m} \frac{\partial A_i}{\partial q_l} \ddot{q}_l \mathbf{r}_{ip} = \sum_{l=1}^{m} D_l \, \ddot{q}_l \mathbf{R}_p \tag{6.184}$$

Let us now define $\alpha_m$ as follows:

$$\alpha_m \triangleq \sum_{l=1}^{m} D_l \ddot{q}_l \tag{6.185}$$

Next, consider the second term on the right side of (6.183). Again a comparison between the second derivative shown here and the second derivative of a loop-matrix equation is made [see Eqs. (6.162) to (6.166)]. Applying the same procedure will result in

$$\sum_{j=1}^{m} \sum_{l=1}^{m} \frac{\partial^2 A}{\partial q_l \; q_j} \dot{q}_l \dot{q}_j \mathbf{r}_{ip} = \sum_{j=1}^{m} \sum_{l=1}^{m} \overline{D_j D_l} \dot{q}_l \dot{q}_j \mathbf{R}_p \tag{6.186}$$

However, the double-summation term was defined earlier during the acceleration analysis [see Eqs. (6.167) and (6.168)]:

$$G_m = \sum_{j=1}^{m} \sum_{l=1}^{m} \overline{D_j D_l} \dot{q}_l \dot{q}_i \tag{6.187}$$

Substituting Eqs. (6.185) and (6.187) into (6.183) yields the following expression for the acceleration of point $P$:

$$\ddot{\mathbf{R}}_p = (\alpha_m + G_m)\mathbf{R}_p \tag{6.188}$$

The calculation of $\alpha_m$ is required for this expression. However, for a particular position of the mechanism it is the same for any point on a rigid link, as is $W_m$ and $G_m$. So for the calculation of the velocity and acceleration of many points on one link, these matrices need be calculated only once. Consequently, these procedures are very efficient when properly programmed for digital computation.

## 6.18 EXAMPLE: KINEMATIC ANALYSIS OF A THREE-DIMENSIONAL MECHANISM

The mechanism shown in Fig. 6.49 will be used as an example to demonstrate the analysis procedures outlined in this chapter. The mechanism is repeated in Fig. 6.51 showing the dimensions used in the calculation. The figure is not drawn to scale. The results for this example were obtained using a spatial mechanism analysis program developed using the procedures outlined here. The joints are defined as discussed earlier, with two exceptions. The two spherical joints have been redefined to eliminate the passive degree-of-freedom rotation of link 3 about its long axis (see Sec. 1.6). The spherical joint between links 2 and 3 is described as a *U*-joint with

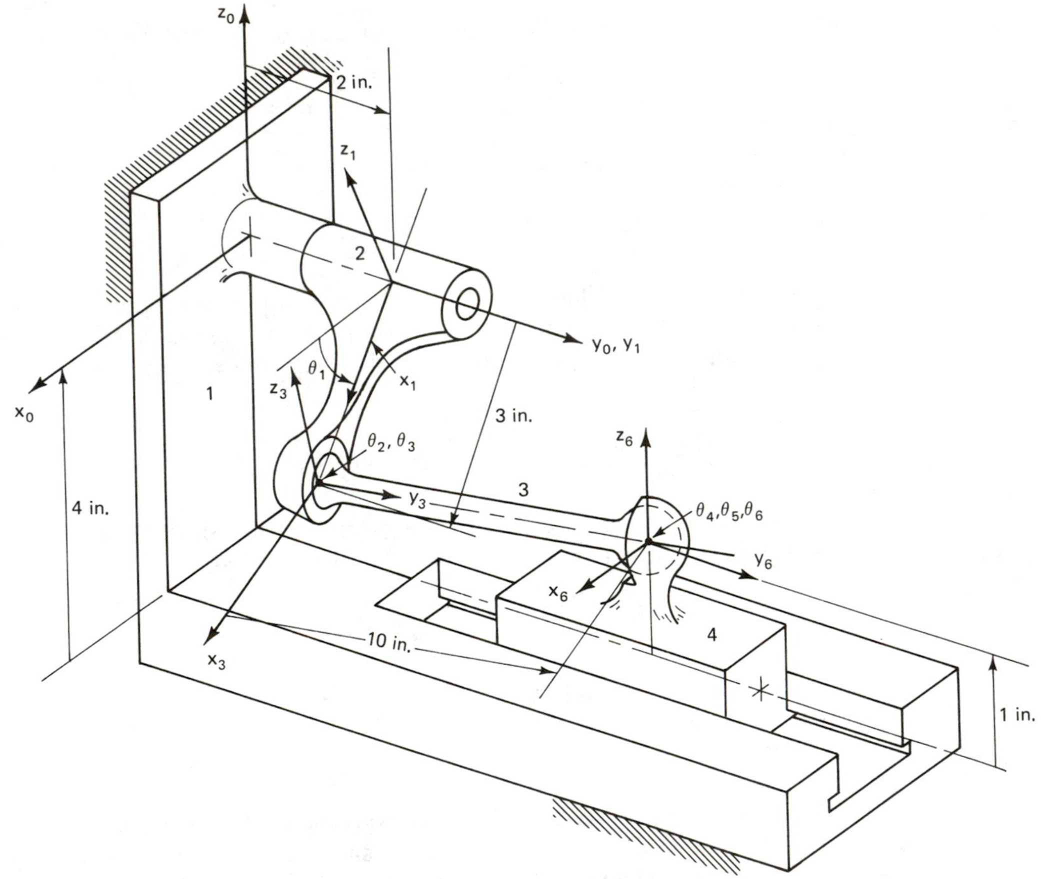


**Figure 6.51** Spatial slider-crank mechanism with key dimensions shown (Drawing is not to scale).

two rotational degrees of freedom. This is achieved by eliminating the third rotation discussed in the earlier example or a rotation about a $z$ axis. Because of this change, the spherical joint between links 3 and 4 is also changed in order to more easily define the orientation of the $x_6y_6z_6$ coordinate system to be parallel to the $x_0y_0z_0$ system. The second spherical joint is defined with a series of three rotations, the first being a rotation about the $y_3$ axis; the second, a rotation about the $z_4$ axis (not shown); and the third, a rotation about the $x_5$ axis (not shown).

The computer program mentioned above is called SAM (*S*patial *A*nalysis of *M*echanisms).* It has a special input command language to allow for ease of problem definition. Only two commands will be discussed here to show the definition of the example mechanism. The mechanism is described by a series of LINK and TRAN commands which define the kinematic topology of the mechanism. The LINK command indicates which links are connected to one another and the TRAN command defines the transformation matrices describing the links and joints. For the spatial slider-crank mechanism the input commands are:

| COMMAND | EXPLANATION |
|---|---|
| LINK/L2,L1/ | Link 2 (L2) is connected to link 1 (L1). |
| TRAN/T1,RY,V,0,2,0,90/ | A variable rotation about the $y$ axis (RY). following a translation of $(0, 2, 0)^T$. Starting value: $\theta_1 = 90°$ |
| LINK/L3,L2/ | Link 3 (L3) is connected to Link 2 |
| TRAN/T2,RZ,V,3/ | A variable rotation about the $z$ axis following a translation of $(3, 0, 0)^T$. |
| TRAN/T3,RX,V/ | A variable rotation about the $x$ axis. |
| LINK/L4,L3/ | Link 4 (L4) is connected to link 3. |
| TRAN/T4,RY,V,0,10,0,−90/ | A variable rotation about the $y$ axis following a translation of $(0, 10, 0)^T$. Starting value: $\theta_4 = -90°$ |
| TRAN/T5,RZ,V/ | A variable rotation about the $z$ axis. |
| TRAN/T6,RX,V/ | A variable rotation about the $x$ axis. |
| LINK/L1,L4/ | Close the loop by defining link 1 with respect to link 4. |
| TRAN/T7,TY,V,0,−12,3/ | A variable translation in the $Y_7$ direction (TY) and a constant translation in the $Z_7$ direction (3). Starting value: $Y_7 = -12$ |

The program is set up to recognize that a kinematic loop has been closed whenever a previously defined link is used as the first parameter of a LINK command. The program automatically sets up the control logic required to define the loop matrix equation and all other analysis requirements. Although a simple single-loop mechanism is shown, the program has the capability to analyze multiloop spatial mechanisms, as well as to perform dynamic analysis of open- and closed-loop mechanisms. A flowchart of the kinematic analysis portion of the program is shown in Fig. 6.52.

The example mechanism† was analyzed with link 2 having a constant angular velocity of 3 rad/sec. The initial angle of this link is 90°, and it was rotated for a little more than one complete revolution (430°). The position, velocity, and acceleration of the slider is plotted with respect to the angle of link 2. These plots are

* For more information, contact Robert Williams through second author.

† This example was prepared with the help of Sern Hong Wang.

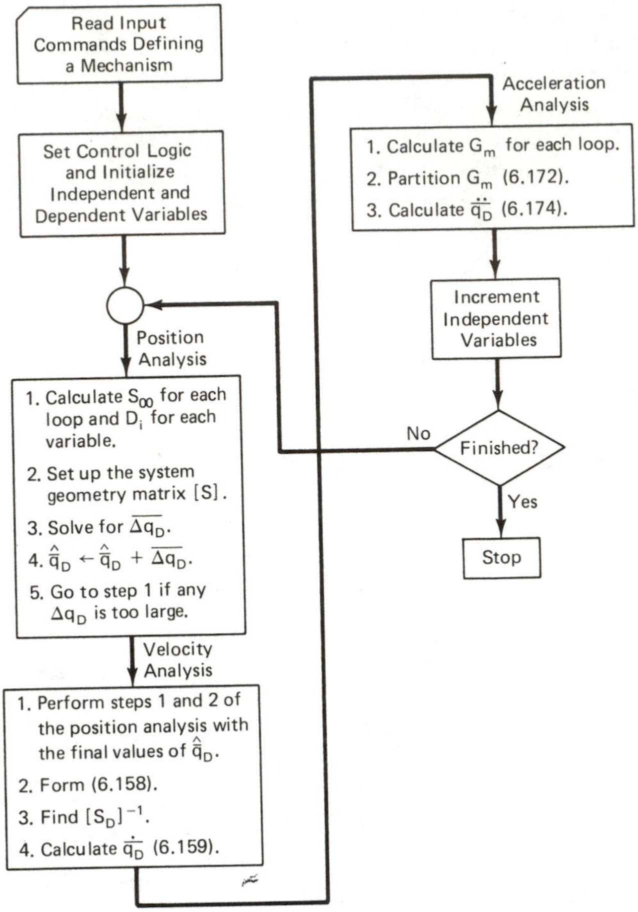


**Figure 6.52** Flowchart of the *Spatial Analysis of Mechanisms Program* (SAM) (kinematic portion only).

shown in Fig. 6.53. The SAM program also has a graphic display capability which allows "stick figure" representations to be generated for each link in a mechanism and then the entire mechanism displayed. Examples of these outputs are shown in Fig. 6.54, where six positions of the mechanism are shown. These hard copies off the graphics terminal are visual checks of the position analysis of the mechanism, which can be invaluable for complex spatial mechanism analysis.

There are other commercially available computer codes that perform kinematic and dynamic analysis of spatial mechanisms. The ADAMS (*A*utomatic *D*ynamic *A*nalysis of *M*echanism *S*ystems) [37,309] and IMP (*I*ntegrated *M*echanisms *P*rogram) [255,256,294–296] are two such programs.

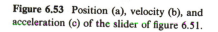

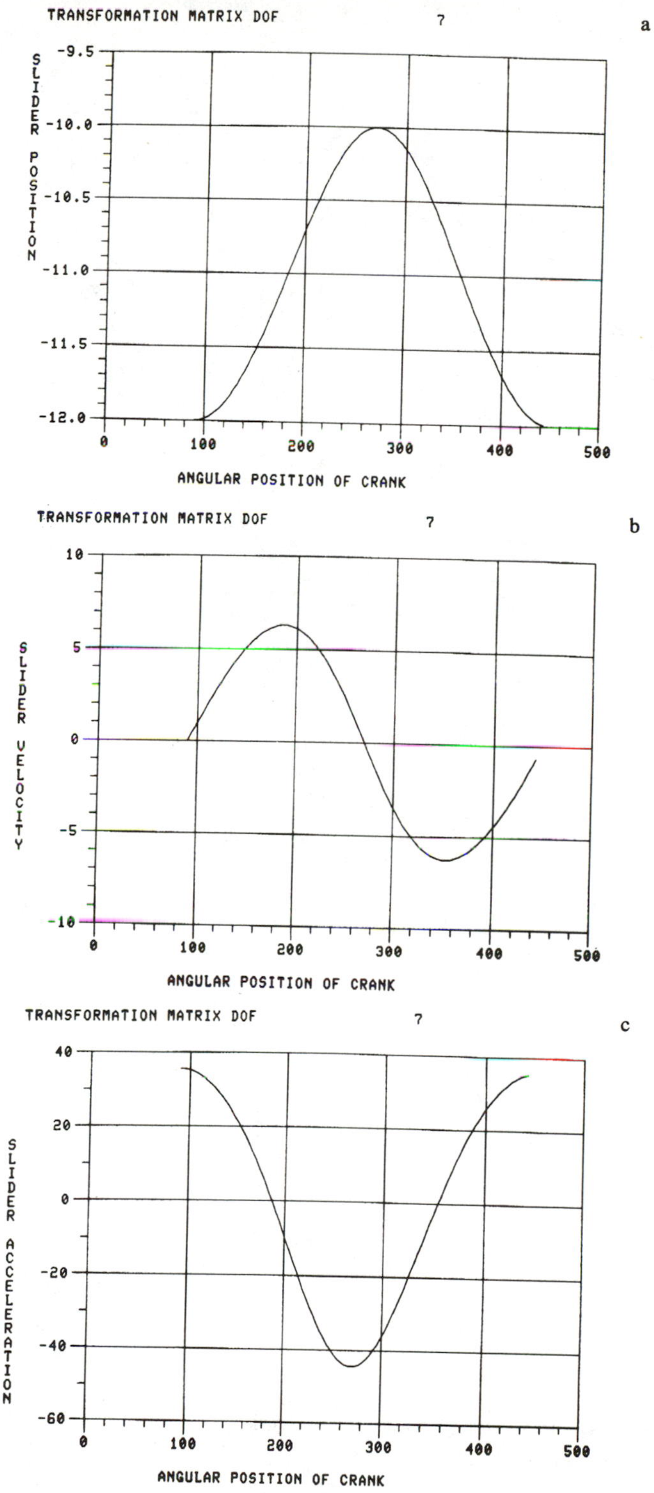


**Figure 6.53** Position (a), velocity (b), and acceleration (c) of the slider of figure 6.51.

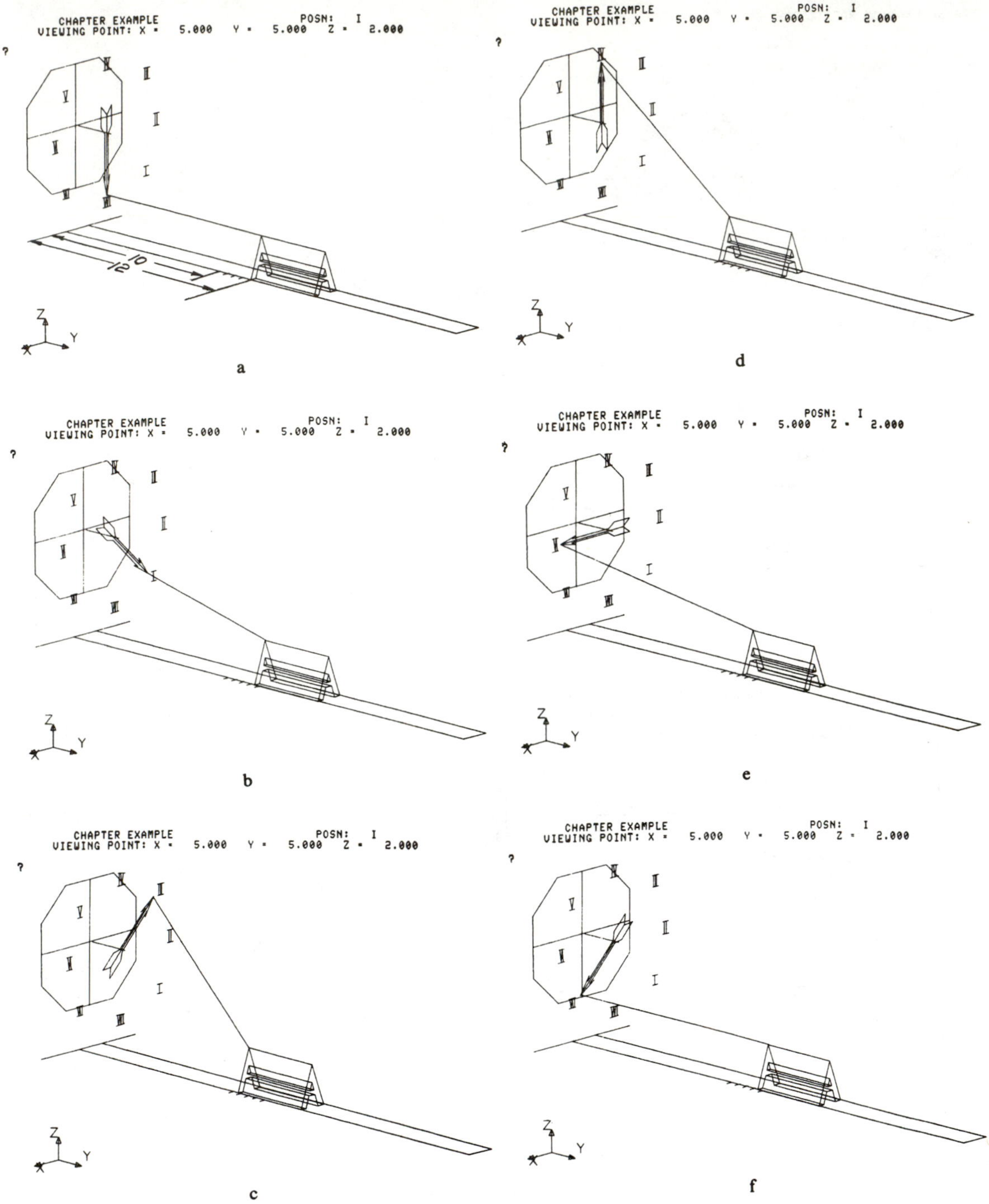


**Figure 6.54** Positions of the slider-crank mechanism of figure 6.51 for crank positions at 90°(a), 135°(b), 225°(c), 270°(d), 0°(e) and 45°(f).

## 6.19 SYNTHESIS OF SPATIAL MECHANISMS

In Chap. 2 standard-form synthesis equations were developed to design planar mechanisms analytically. The standard-form equation was based on the loop closure of the complex vectors representing the mechanism. By specifying various parameters, this general tool could be used to solve for the remaining unknowns, thus generating candidate dyads to form linkages for all forms of planar path-, motion-, or function-generator mechanisms. Because of the wide variety of spatial mechanisms and spatial joint types which yield nonlinear constraint equations, no such standard-form method has yet been found for spatial synthesis. Two methods are introduced, however, which attack the spatial synthesis problem by way of spatial dyads. In this section the tools for a procedure for spatial synthesis, based in part on the work of C. H. Suh and C. W. Radcliffe [272–275], is presented. This method was chosen because it provides the beginning student with a flexible tool that is fairly easy to comprehend and implement on a computer. Section 6.20 presents a method based on a vector approach. The interested student who wishes to investigate spatial synthesis further is referred to the bibliography for other works such as ref's [2–4,13,41,54,60,62, 124,125,127,141,151,159,167,215,223,238,239,249,269,291,319].

### How to Build Synthesis Equations

Various constraint equations may be used like building blocks to construct mathematical models for the synthesis of spatial linkages. The result is a system of nonlinear equations that describe joint constraints. Table 6.2 lists five of the most useful "building blocks" that can form spatial dyads. These may be combined to form many types of useful linkages, just as planar dyads were combined to form planar linkages. Table 6.2 lists explicitly the constraint equations readily programmed for computation and solved for such dyadic synthesis. In the next several paragraphs, each type of constraint is discussed.

### The Building Blocks

The reader will recall that, in the plane, the basic dyadic "standard form" was developed by asking this question: Given the desired coplanar motion of a rigid plane, can we find coplanar dyads to guide the moving plane through several prescribed positions? Adapting this philosophy to space, we ask: Given the desired spatial motion of a rigid body, can we find spatial dyads to guide the moving body through several prescribed spatial positions? To answer this question, we need mathematical expressions of the spatial constraints of various types of spatial dyads. These are illustrated and described in Table 6.3.

Suppose that we wish to guide a point $A$ through $n$ prescribed positions in space. We would therefore specify the position vectors

$$\mathbf{A}_j = (x_j\mathbf{i} + y_j\mathbf{j} + z_j\mathbf{k}), \qquad j = 1, 2, \ldots, n$$

**TABLE 6.2** LINK CONSTRAINTS—BUILDING BLOCKS FOR SYNTHESIS OF SPATIAL MECHANISMS (see Table 6.3 for constraint equations).

| Joint types of dyad | Constraint equations | Number of scalar equations for maximum number of prescribed positions | Maximum number of positions/scalar equations | Scalar unknowns | Number of scalar unknowns the designer is free to specify |
|---|---|---|---|---|---|
| S-S: 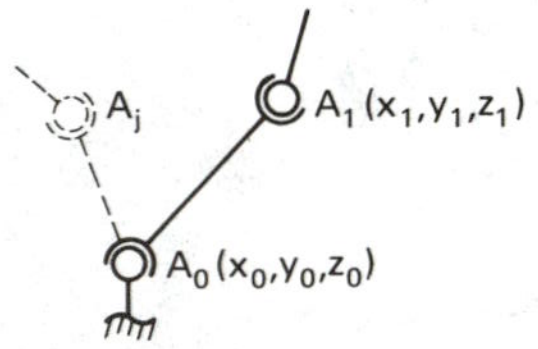  | Constant length | 6 | 7/6 | $x_0\ y_0\ z_0$<br>$x_1\ y_1\ z_1$ | None |
| R-S: 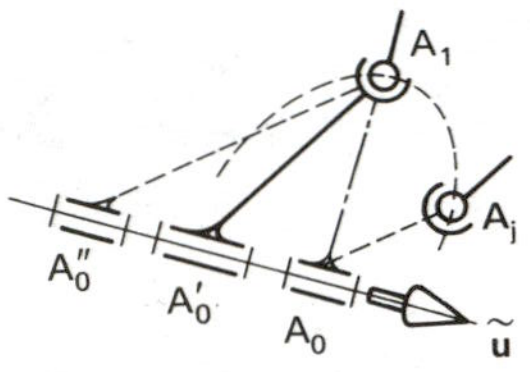  | Constant length<br>Planar motion<br>Direction cosines | 3<br>4<br>1 | 4/8 | $x_0\ y_0\ z_0$<br>$x_1\ y_1\ z_1$<br>$u_x\ u_y\ u_z$ | One |
| R-R: 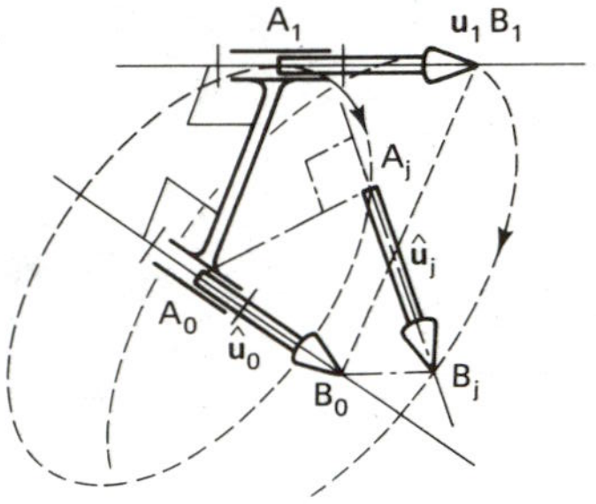  | Constant length $(\overrightarrow{A_0A_j})$<br>Planar motion $(\overrightarrow{A_0A_j} \perp \hat{u}_0, \hat{u}_j)$<br>Direction cosines<br>Constant length (of auxiliary link) $(\overrightarrow{B_0B_j})$ | 2<br>6<br>2<br>2 | 3/12 | $x_0\ y_0\ z_0$<br>$x_1\ y_1\ z_1$<br>$u_{x0}\ u_{y0}\ u_{z0}$<br>$u_{x1}\ u_{y1}\ u_{z1}$ | None |

| | | | | | |
|---|---|---|---|---|---|
| R-C: 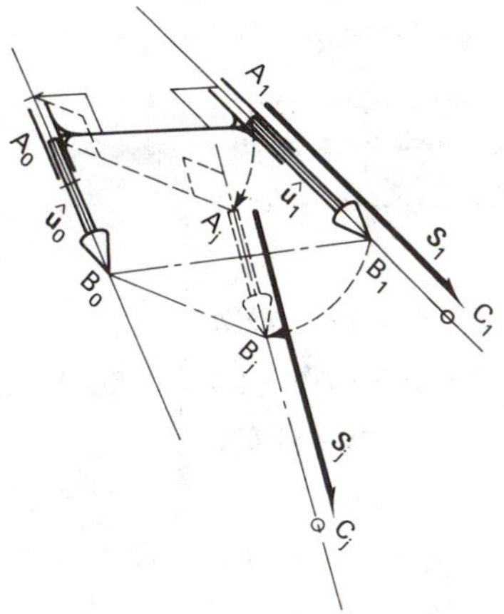 | Constant length $(\overrightarrow{A_0A_j})$<br>Planar motion $(\overrightarrow{A_0A_j} \perp \hat{\mathbf{u}}_o)$<br>Direction cosines<br>Constant length (of auxiliary link)<br>Translational motion | 2<br>3<br>2<br>2 | 3/12 | $x_0\ y_0\ z_0$<br>$x_1\ y_1\ z_1$<br>$u_{x0}\ u_{y0}\ u_{z0}$<br>$u_{x1}\ u_{y1}\ u_{z1}$<br>$S_2 - S_1,\ S_3 - S_1$ | Two |
| C-C: 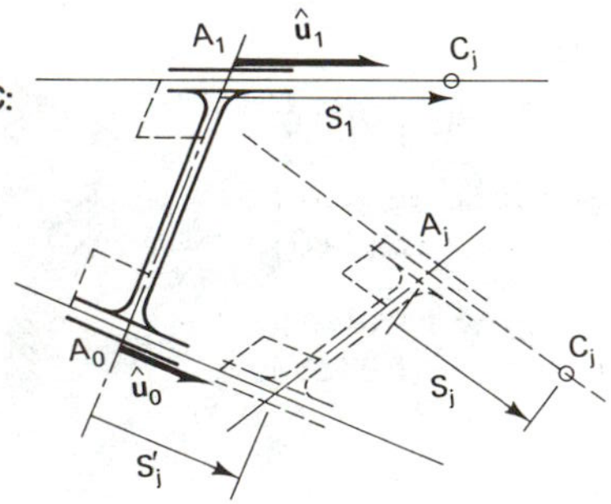 | Constant length<br>$\overrightarrow{A_0A_j} \perp \hat{\mathbf{u}}_0$<br>Direction cosines<br>Constant length (of auxiliary link)<br>Two translational motions | 4<br>5<br>2<br>4<br>5 | 5/20 | $x_0\ y_0\ z_0$<br>$x_1\ y_1\ z_1$<br>$u_{x0}\ u_{y0}\ u_{z0}$<br>$u_{x1}\ u_{y1}\ u_{z1}$<br>$S_2\ S_3\ S_4\ S_5$<br>$S_2 - S_1,\ S_3 - S_1,$<br>$S_4 - S_1,\ S_5 - S_1$ | None |

**TABLE 6.3** CONSTRAINT EQUATIONS FOR SPATIAL SYNTHESIS (see Table 6.2 for illustrations of *S-S*, *R-S* and *R-R* links).

| Dyad type (number of positions)<br>Constraint equations | Kind of equation | Number of equations |
|---|---|---|
| *S-S* ($n \le 7$): | | |
| $(x_1 - x_0)^2 + (y_1 - y_0)^2 + (z_1 - z_0)^2$<br>$= (x_j - x_0)^2 + (y_j - y_0)^2 + (z_j - z_0)^2 \quad (j = 2, 3, \ldots, n)$ | Constant length | 6 |
| *R-S* ($n \le 4$): | | |
| $(x_1 - x_0)^2 + (y_1 - y_0)^2 + (z_1 - z_0)^2 = (x_j - x_0)^2 + (y_j - y_0)^2 + (z_j - z_0)^2 \quad (j = 2, 3, \ldots, n)$ | Constant length | 3 |
| $u_x(x_j - x_0) + u_y(y_j - y_0) + u_z(z_j - z_0) = 0 \quad (j = 1, 2, \ldots, n)$ | Planar motion | 4 |
| $u_x^2 + u_y^2 + u_z^2 = 1$ | Direction cosines | 1 |
| *R-R* ($n \le 3$): | | |
| $(x_1 - x_0)^2 + (y_1 - y_0)^2 + (z_1 - z_0)^2 = (x_j - x_0)^2 + (y_j - y_0)^2 + (z_j - z_0)^2 \quad (j = 2, 3)$ | Constant length | 2 |
| $u_{x0}(x_j - x_0) + u_{y0}(x_j - x_0) + u_{z0}(z_j - z_0) = 0 \quad (j = 1, 2, 3)$ | Planar motion of $A_1$–$A_j$ | 3 |
| $u_{x0}(x_j + u_{xj} - x_0 - u_{x0}) + u_{y0}(y_j + u_{yj} - y_0 - u_{y0})$<br>$+ u_{z0}(x_j + u_{zj} - z_0 - u_{z0}) = u_{x0}(x_1 + u_{x1} - x_0 - u_{x0}) \quad (j = 1, 2, 3)$<br>$+ u_{y0}(y_1 + u_{y1} - y_0 - u_{y0}) + u_{z0}(z_1 + u_{z1} - z_0 - u_{z0})$ | Planar motion of $B_1 - B_j$ | 3 |
| $u_{x0}^2 + u_{y0}^2 + u_{z0}^2 = 1, \quad u_{x1}^2 + u_{y1}^2 + u_{z1}^2 = 1$ | Directional cosines | 2 |
| $[(x_1 + u_{x1}) - (x_0 + u_{x0})]^2 + [(y_1 + u_{y1}) - (y_0 + u_{y0})]^2 + [(z_1 + u_{z1}) - (z_0 + u_{z0})]^2$<br>$= [(x_j + u_{xj}) - (x_0 + u_{z0})]^2 + [(y_j + u_{yj}) - (y_0 + u_{y0})]^2$<br>$+ [(z_j + u_{zj}) - (z_0 + u_{z0})]^2 \quad (j = 2, 3)$ | Constant length | 2 |

If we take a coordinate system centered at $A_1$, we need only to specify the $n - 1$ vectors

$$\mathbf{a}_j = \mathbf{A}_j - \mathbf{A}_1$$

With these, we can let $\mathbf{A}_1$ be an unknown and prescribe the relative position of $\mathbf{A}_j$, $j = 2, 3, \ldots n$, with respect to $\mathbf{A}_1$ by specifying $\mathbf{a}_j$, $j = 2, 3, \ldots n$.

To guide the point $A$ through positions $\mathbf{A}_j$, $j = 1, 2, \ldots n$, consider the sphere-sphere binary link of Table 6.2. When neither end is fixed, it has five degrees of freedom plus a sixth passive rotational freedom about the axis of the link between the centers of the joints. The only constraint inherent in this link is that the moving joint in all its prescribed positions ($\mathbf{A}_1$, $\mathbf{A}_j$) must remain on the surface of a sphere, centered at some unknown ground joint $A_0$. Thus the link length $|\overrightarrow{A_0A_j}|$ must be constant for all positions:

$$(x_1 - x_0)^2 + (y_1 - y_0)^2 + (z_1 - z_0)^2$$
$$= (x_j - x_0)^2 + (y_j - y_0)^2 + (z_j - z_0)^2, \qquad j = 2, 3, \ldots n \tag{6.189}$$

where $x_j = x_1 + a_{xj}$, $y_j = y_1 + a_{yj}$, $z_j = z_1 + a_{zj}$, and where $a_{xj}$, $a_{yj}$, $a_{zj}$ are the coordinates of the displacement vector $\mathbf{a}_j$. Here $n$ is the number of prescribed positions; that is, the coordinates of $A_j(x_j, y_j, z_j)$, $j = 1, 2, \ldots n$, are given. If $\mathbf{A}_j$, the position vector of $A_j$, is expressed in terms of $\mathbf{A}_1$ as $\mathbf{A}_j = \mathbf{A}_1 + \mathbf{a}_j$, where $\mathbf{a}_j$ is a given displacement vector with $j = 2, 3, \ldots n$, Eqs. (6.189) become a set of nonlinear design equations that may be solved by computation for the unknown initial position of the binary link ($x_0$, $y_0$, $z_0$, $x_1$, $y_1$, $z_1$). If the ground pivot is specified by giving $\mathbf{A}_0(x_0, y_0, z_0)$, then the three unknowns ($x_1$, $y_1$, $z_1$) (coordinates of the unknown $\mathbf{A}_1$) may be found for four prescribed positions ($n = 4$) or three prescribed displacements $\mathbf{a}_j$, $j = 2, 3, 4$. A maximum of seven positions (six displacements) may be used if the values of all six coordinates of $\mathbf{A}_1$ and $\mathbf{A}_0$ are to be determined by computation (see Table 6.3).

For the same point guidance in space, consider now the revolute-sphere (R-S) links of Table 6.2. It should be intuitively obvious from the illustration in the table that the location of the pivot along the revolute axis is irrelevant, and that all three links shown in the table are kinematically equivalent: For all three ground-pivot locations ($A_0$, $A_0'$, and $A_0''$), the spherical joint is constrained to travel on the same plane perpendicular to the axis of the revolute joint. Actually, any grounded revolute joint on this axis may replace any joint that is a synthesis solution, without changing the mechanism. If we choose the unique joint at $A_0$ (the intersection of the axis and the plane of sphere-joint motion) a simple constraint equation results. Referring now to Table 6.2, a vector $\overrightarrow{A_0A_j}$ must always be perpendicular to a unit vector $\hat{\mathbf{u}}$, the axis of the grounded Revolute. Recall that the dot product of two perpendicular vectors is zero. Then the right-angle constraint equation is $\hat{\mathbf{u}} \cdot \overrightarrow{A_0A_j} = 0$, or

$$u_x(x_j - x_0) + u_y(y_j - y_0) + u_z(z_j - z_0) = 0, \qquad j = 1, 2, \ldots, n \tag{6.190}$$

where again $n$ is the number of prescribed positions. The fact that $\hat{\mathbf{u}}$ is a unit vector may also be used to form a constraint equation. The components of $\hat{\mathbf{u}}$ are the direction cosines of the revolute axis; thus this constraint is called the direction-cosine equation:

$$u_x^2 + u_y^2 + u_z^2 = 1 \tag{6.191}$$

You may have noticed that the R-S link is similar to the S-S link, but with two additional degrees of freedom removed, since the revolute ground pivot permits only one rotation. The distance from the unknown $A_0$ to all the given $A_j$ is again fixed. Therefore, the constant-length equations [Eq. (6.189)] must also be applied to this link.

To completely specify an R-S link in its initial position, nine unknown scalar components must be determined: $x_0$, $y_0$, $z_0$, $x_1$, $y_1$, $z_1$, $u_x$, $u_y$, and $u_z$. As with the S-S link, if the $\mathbf{A}_j$'s are specified in terms of $\mathbf{A}_1$, then Eqs. (6.189) and (6.190) are functions containing only these unknowns. The direction-cosine equation (6.191) need only be applied once. So for the maximum of four positions, specified as $\mathbf{A}_j(x_j, y_j, z_j)$, $j = 1, 2, 3, 4$, there are three constant-length equations, four right-angle equations (planar motion), and one direction-cosine equation. For four prescribed positions, $\mathbf{A}_j$, $j = 1, 2, 3, 4$, there will be eight scalar equations with nine scalar unknowns. These constraint equations are shown in Table 6.3. One of the unknowns may be specified arbitrarily. For instance, the $x$ component of the revolute joint may be specified to ensure adequate distance from an adjacent joint.

The *revolute-revolute* link is similar to the R-S link, except a second axis of rotation must be determined. Using the same logic as with the R-S link, the unique link perpendicular to both axes is chosen to simplify the equations. The actual position of the fixed revolute joint may be set elsewhere along its axis at the discretion of the designer. The only two cases where this unique link does not exist is the trivial one where the two axes are coincident and the planar case where the axes are parallel. The basic R-R link is shown in two positions in Table 6.2. Neglecting the planar case, each rotation of the link about the fixed joint changes the direction of the moving axis ($\hat{\mathbf{u}}_j$) as well as the coordinates of the moving joint location ($A_j$). Assuming that $\hat{\mathbf{u}}_j$ and $\mathbf{A}_j$ can be described by functions of $\hat{\mathbf{u}}_1$ and $\mathbf{A}_1$, we are left with 12 scalar quantities that are necessary to describe the link in its initial position ($x_0$, $y_0$, $z_0$, $x_1$, $y_1$, $z_1$, $u_{x0}$, $u_{y0}$, $u_{z0}$, $u_{x1}$, $u_{y1}$, $u_{z1}$).

Just as with the S-S and R-S links, the rigidity constraint between the two joints requires that $A_0$ and $A_1$ maintain a constant relative distance. So Eq. (6.189) (constant-length equation) provides the first subset of our constraint equations. Also, vector $\overrightarrow{A_0A_j}$ must remain in a plane perpendicular to $\hat{\mathbf{u}}_0$, and therefore the right-angle constraint equations [Eq. (6.190)] apply here. When we defined the link as the unique shortest link between the axes, the second axis ($\hat{\mathbf{u}}_j$) also became normal to the link. This is a third constraint that may be applied using the right-angle constraint equation. The direction-cosine equation [Eq. (6.191)] may be applied once to each axis resulting in two more design equations.

Finally, in addition to these constraints, a less obvious constraint must be applied. The equations mentioned above do not preclude the possibility that the moving joint at $A_j$ may rotate 180° about the axis of the link $\overrightarrow{A_0A_j}$. A positive rotation (right-hand rule) of the link about $\mathbf{u}_j$ of the revolute joint is the same as a negative rotation about $-\hat{\mathbf{u}}_j$. Therefore, to keep the results consistent, an auxiliary link is defined between the two joints. To keep from introducing any more unknowns, the end points of this link ($\overrightarrow{B_0B_j}$) are defined as

$$\mathbf{B}_0 = \mathbf{A}_0 + \hat{\mathbf{u}}_0$$

$$\mathbf{B}_j = \mathbf{A}_j + \hat{\mathbf{u}}_j$$

as shown in Table 6.2. Since $\overrightarrow{A_0A_j}$ is a rigid link and $\hat{\mathbf{u}}_0$ and $\hat{\mathbf{u}}_j$ are constant-length vectors rigidly attached to $A_0$ and $A_j$, a constant-length constraint may be applied to the auxiliary link $\overrightarrow{B_0B_j}$. This assures that $\hat{\mathbf{u}}_0$ and $\hat{\mathbf{u}}_j$ maintain a constant relative orientation, thus completing the mathematical model of the R-R link constraint. The system of synthesis equations for the maximum of three positions is a set of 12 nonlinear equations as follows:

| | |
|---|---|
| 2 constant-length equations | ($\overrightarrow{A_0A_j}$) |
| 3 right-angle equations | ($\overrightarrow{A_0A_j} \cdot \hat{\mathbf{u}}_0$) |
| 3 right-angle equations | ($\overrightarrow{A_0A_j} \cdot \hat{\mathbf{u}}_1$) |
| 2 direction-cosine equations | ($\hat{\mathbf{u}}_0$ and $\hat{\mathbf{u}}_1$) |
| 2 constant-length equations | ($\overrightarrow{B_0B_j}$) |

Constraint equations are shown in Table 6.3, for S-S, R-S, and R-R links. R-C and C-C links are treated similarly in developing the equations that model these joints, and establishing the data for entries in Tables 6.2 and 6.3.

Spatial mechanisms made up the link building blocks of Table 6.2 may be synthesized by using the constraint equations of Table 6.3. These equations are nonlinear in the unknowns that specify the initial position of the mechanism. Numerical techniques are required for solution of these equations. The next section describes a similar approach using vectors for solving constraint equations.

## 6.20 VECTOR SYNTHESIS OF SPATIAL MECHANISMS*

### Description of the Mechanism in Vector Notation

Figure 6.55 shows a schematic of the mechanism to be synthesized. It consists of the two grounded links with fixed-axis revolutes at $A_0$ and $B_0$ and moving spheric joints $A$ and $B$, (binary R-S links 2 and 3); the one grounded link with fixed-axis cylindric joint $C_0$ and spheric joint $C$ (binary C-S link 4); the coupler with three spheric joints $A$, $B$, and $C$ (ternary S-S-S link 5) and the fixed frame (ternary R-R-C link 1). The fixed orthonormal coordinate system of reference $OXYZ$ is

* G. N. Sandor, D. Kohli, C. F. Reinholtz, and A. Ghosal, "Closed-Form Analytic Synthesis of a Five-Link Spatial Motion Generator," *Proceedings*, 7-th Conference on Apld. Mechanisms, Kansas City, Sept. 1982, pp. xxvii–7; *J. Mechanism and Machine Theory*, 1983 or 84.

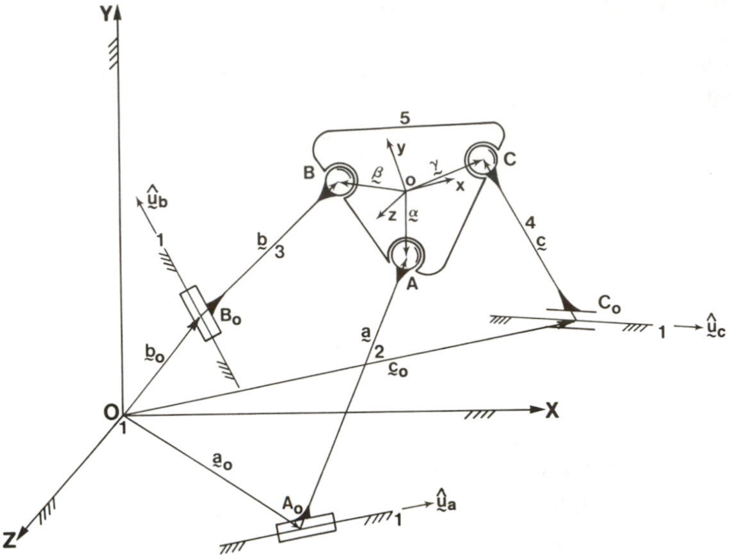


**Figure 6.55** The RSSR-SC mechanism and associated vectors.

embedded in link 1, and the moving orthonormal coordinate system o*xyz* is embedded in the coupler (link 5). The use of only these two coordinate systems differs from many conventional space-mechanism theories in which each and every link requires at least one separate local coordinate system.

This is a single-degree-of-freedom mechanism with one link fixed. Therefore, the rotational position of, say, link 2 determines the position of all other moving links.

The desired body motion is prescribed by specifying successive positions of the moving origin, o, and corresponding orientations of the moving o*xyz* coordinate system.

In Fig. 6.55 the fixed vectors $\mathbf{a}_o$, $\mathbf{b}_o$, and $\mathbf{c}_o$ locate the two grounded revolutes at $A_o$ and $B_o$ and the grounded cylindric joint at $C_o$. The directions of the axes of these joints are denoted by unit vectors $\hat{\mathbf{u}}_a$, $\hat{\mathbf{u}}_b$, and $\hat{\mathbf{u}}_c$. Vectors **a**, **b**, and **c** represent the grounded links 2, 3, and 4, expressed in O*XYZ*. Observe that vectors **a**, **b**, and **c** are perpendicular to unit vectors $\hat{\mathbf{u}}_a$, $\hat{\mathbf{u}}_b$, and $\hat{\mathbf{u}}_c$, respectively. Owing to the *R*-joints at $A_o$ and $B_o$, link vectors **a** and **b** can only rotate about their respective revolute axes $\hat{\mathbf{u}}_a$ and $\hat{\mathbf{u}}_b$. Link vector **c** can not only rotate about $\hat{\mathbf{u}}_c$, but can also translate along $\hat{\mathbf{u}}_c$, albeit always remaining attached and perpendicular to it. Vectors, $\boldsymbol{\alpha}$, $\boldsymbol{\beta}$ and $\boldsymbol{\gamma}$, embedded in the coupler, locate spheric joints *A*, *B*, and *C* with respect to the similarly embedded moving coordinate system o*xyz*. These spheric joints must remain connected to the three grounded dyads. Therefore, "vector dyads" $(\mathbf{a},\boldsymbol{\alpha})$ $(\mathbf{b},\boldsymbol{\beta})$, and $(\mathbf{c},\boldsymbol{\gamma})$ must remain connected at their respective tips throughout the motion.

## Specifying the Prescribed Motion

In Fig. 6.56 the first and the $j$th prescribed positions of the moving coordinate system are shown as $o_1x_1y_1z_1$ and $o_jx_jy_jz_j$. Their origins $o_1$ and $o_j$ are located by the given position vectors $\mathbf{o}_1$ and $\mathbf{o}_j$, while their orientations are specified by given $3 \times 3$ rotation matrices $[R_1]$ and $[R_j]$ in terms of Euler angles with respect to the fixed $OXYZ$ frame of reference.* In order to express these conditions mathematically, let $\mathbf{V}^1$ be a vector expressed in the moving system in the first position and $\mathbf{V}^0$ be the same vector expressed in the fixed global coordinate system. The relation between $\mathbf{V}^1$ and $\mathbf{V}^0$ is given by

$$\mathbf{V}^0 = [R_1]\mathbf{V}^1. \tag{6.192}$$

Or, in general when $\mathbf{V}^j$ is the vector expressed in the moving system in the $j$th position,

$$\mathbf{V}^0 = [R_j]\mathbf{V}^j. \tag{6.193}$$

Thus, by specifying $\mathbf{o}_1$, $\mathbf{o}_j$, $[R_1]$, and $[R_j]$ for $j = 2, 3, \ldots, n$, we have prescribed $n$ discrete positions of the coupler (link 5 in Fig. 6.55).

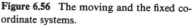

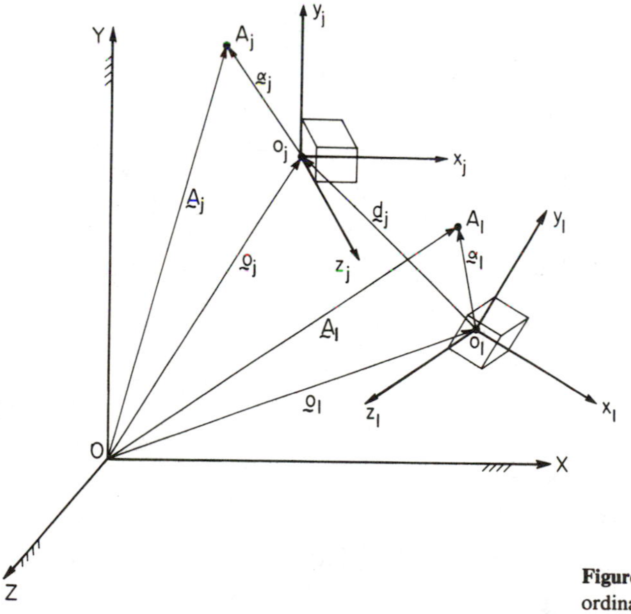


**Figure 6.56** The moving and the fixed coordinate systems.

* See, for example, the 3 × 3 rotation submatrix of Eq. (6.70).

## The Dyadic Displacement Polygon

Figure 6.57 shows three of the five links of the mechanism of Fig. 6.55. They are:

Link 1: the fixed link with coordinate system $OXYZ$ and with vectors $\mathbf{a}_o$ locating the point $A_o$ on the fixed revolute axis $\hat{\mathbf{u}}_a$

Link 2: the grounded R-S link with vector $\mathbf{a}_1$ connecting its moving spheric joint $A$ to the fixed revolute in position 1 and

Link 5: the coupler, with embedded coordinate system $oxyz$ and embedded vector $\boldsymbol{\alpha}$ connecting moving origin o to the spheric joint $A$, also in position 1, marked $o_1x_1y_1z_1$ and $\boldsymbol{\alpha}_1$

In addition to these first or starting positions, Fig. 6.57 also shows the $j$th displaced positions of links 2 and 5, as follows:

Link 5: with its origin o translated by the vector $\mathbf{o}_j - \mathbf{o}_1$ from position 1 to position $j$ and its embedded coordinate system $o_jx_jy_jz_j$, including the embedded vector $\boldsymbol{\alpha}$, rotated from orientation 1 to orientation $j$ (from $\boldsymbol{\alpha}_1$ to $\boldsymbol{\alpha}_j$) and

Link 2: with vector $\mathbf{a}$ rotated about $\hat{\mathbf{u}}_a$ from position $\mathbf{a}_1$ to position $\mathbf{a}_j$ by the angle $\phi_j$.

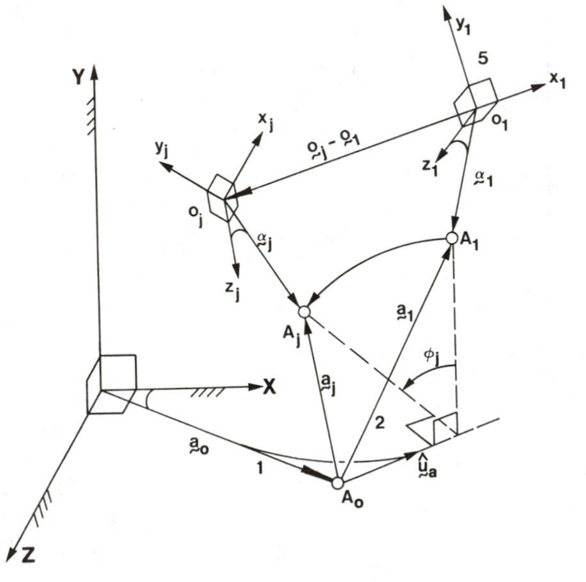


**Figure 6.57** Schematic of the first and the $j$th positions of links 2 and 5.

Note that in Figure 6.57 we have the following known features: Prescribed are $\mathbf{o}_1$, $\mathbf{o}_j$, $o_1x_1y_1z_1$ [see Eq. 6.192] $o_jx_jy_jz_j$ [see Eq. 6.193]. Unknown are $\boldsymbol{\alpha}_j$, $\mathbf{a}_1$, $\mathbf{a}_j$, $\mathbf{a}_o$. In addition, as will be shown later, for 3 prescribed positions $\boldsymbol{\alpha}_1$ may be assumed arbitrarily.

We now define the *dyadic displacement polygon* as the closed vector polygon with vertices $A_o$, $A_1$, $o_1$, $o_j$, and $A_j$. Closure of this polygon can be written by expressing all its vector sides in the *OXYZ* system and equating their sum with zero. Going counter-clockwise in Fig. 6.57, we have

$$\mathbf{a}_1 - [R_1]\,\boldsymbol{\alpha}_1 + \mathbf{o}_j - \mathbf{o}_1 + [R_j]\,\boldsymbol{\alpha}_1 - \mathbf{a}_j = 0 \qquad (6.194)$$

In Eq. (6.194) $\mathbf{o}_j$, $\mathbf{o}_1$, $[R_1]$, and $[R_j]$ are given, and all other quantities are unknown. However, as will be shown, for $j = 2, 3$ (i.e., 3 prescribed positions), $\boldsymbol{\alpha}_1$ is a free choice.

## Finding a Suitable Revolute Axis for Link 2

The preceding discussion shows that positions of a body in space are specified by giving the location and orientation of the moving coordinate system embedded in the body at each position. Thus, we have given:

$$\mathbf{o}_j \text{ and } [R_j], \qquad j = 1, 2, 3. \qquad (6.195)$$

Now we turn our attention to the *R-S* dyad shown in vector form in Fig. 6.57. By assuming a value for $\boldsymbol{\alpha}$ we have located the spherical joint relative to the fixed frame of reference, namely $[R_1]$ and in the $j$th position, $\boldsymbol{\alpha}_j$ will be given by:

$$\boldsymbol{\alpha}_j = [R_j]\,\boldsymbol{\alpha}_1, \qquad j = 1, 2, 3 \qquad (6.196)$$

which is known, because $[R_1]$ and $[R_j]$ were prescribed.

Referring to Figure 6.56, the vector $\mathbf{A}_j$ in the *OXYZ* system can be expressed as the sum of $\mathbf{o}_j$ and $[R_j]\,\boldsymbol{\alpha}_1$, or

$$\mathbf{A}_j = \mathbf{o}_j + [R_j]\,\boldsymbol{\alpha}_1, \qquad j = 1, 2, 3. \qquad (6.197)$$

For an R-S pair with three prescribed positions, the vectors $(\mathbf{A}_j - \mathbf{A}_1)$, $j = 2, 3$, must lie in a plane perpendicular to the axis of the revolute joint. Thus, the unit vector along the axis of the revolute joint, $\hat{\mathbf{u}}_a$, is given by the following relation:

$$\hat{\mathbf{u}}_a = \frac{(\mathbf{A}_2 - \mathbf{A}_1) \times (\mathbf{A}_3 - \mathbf{A}_1)}{|(\mathbf{A}_2 - \mathbf{A}_1) \times (\mathbf{A}_3 - \mathbf{A}_1)|} \qquad (6.198)$$

## Finding the Location of the Fixed Revolute $A_o$

Since the grounded link $\mathbf{a}$ rotates about the revolute axis $\hat{\mathbf{u}}_a$, no kinematic generality is lost by making it perpendicular to $\hat{\mathbf{u}}_a$. Therefore, we can say that the vectors $\mathbf{a}_j$, $j = 1, 2, 3$, also lie in the plane defined by $(\mathbf{A}_j - \mathbf{A}_1)$ for $j = 2, 3$. In addition, the axis $\hat{\mathbf{u}}_a$, which is normal to this plane, intersects it at $A_o$, located by the unknown vector $\mathbf{a}_o$ in the *OXYZ* system. This is the unknown position of the revolute joint

$A_o$ in the global coordinate system. The point $A_o$ can be found by locating the center of the circle which passes through the three points $A_1$, $A_2$, and $A_3$. This can be done in the following way.

Referring to Fig. 6.58, first find the unit vectors $\hat{\mathbf{p}}_2$ and $\hat{\mathbf{p}}_3$ which are perpendicular to $(\mathbf{A}_2 - \mathbf{A}_1)$ and $(\mathbf{A}_3 - \mathbf{A}_2)$, respectively, and also to $\hat{\mathbf{u}}_a$.

$$\hat{\mathbf{p}}_2 = \hat{\mathbf{u}}_a \times (\mathbf{A}_2 - \mathbf{A}_1) \,/\, |\mathbf{A}_2 - \mathbf{A}_1| \tag{6.199}$$

$$\hat{\mathbf{p}}_3 = \hat{\mathbf{u}}_a \times (\mathbf{A}_3 - \mathbf{A}_2) \,/\, |\mathbf{A}_3 - \mathbf{A}_2| \tag{6.200}$$

Now let the lines of action of these two vectors line up along the perpendicular bisectors of $(\mathbf{A}_2 - \mathbf{A}_1)$ and $(\mathbf{A}_3 - \mathbf{A}_2)$ respectively. Then these lines will intersect at $A_o$. By denoting the distance from $(\mathbf{A}_2 - \mathbf{A}_1)$ to $A_o$ along $\hat{\mathbf{p}}_2$ as $\lambda_2$ and similarly the distance from $(\mathbf{A}_3 - \mathbf{A}_2)$ to $A_o$ along $\hat{\mathbf{p}}_3$ as $\lambda_3$, we express $\mathbf{a}_o$ in two ways:

$$\begin{aligned}\mathbf{a}_o &= \lambda_2\hat{\mathbf{p}}_2 + \mathbf{A}_1 + \frac{\mathbf{A}_2 - \mathbf{A}_1}{2} \\ &= \lambda_3\hat{\mathbf{p}}_3 + \mathbf{A}_2 + \frac{\mathbf{A}_3 - \mathbf{A}_2}{2}\end{aligned} \tag{6.201}$$

To solve for $\lambda_3$, first precrossmultiply the second and third expressions in Eq. (6.201) with $\hat{\mathbf{p}}_2$ to give

$$\lambda_2\hat{\mathbf{p}}_2 \times \hat{\mathbf{p}}_2 + (\hat{\mathbf{p}}_2 \times \mathbf{A}_1) + \mathbf{p}_2 \times \frac{\mathbf{A}_2 - \mathbf{A}_1}{2} = \lambda_3\hat{\mathbf{p}}_2 \times \hat{\mathbf{p}}_3 + (\hat{\mathbf{p}}_2 \times \mathbf{A}_2) + \mathbf{p}_2 \times \frac{\mathbf{A}_3 - \mathbf{A}_2}{2} \tag{6.202}$$

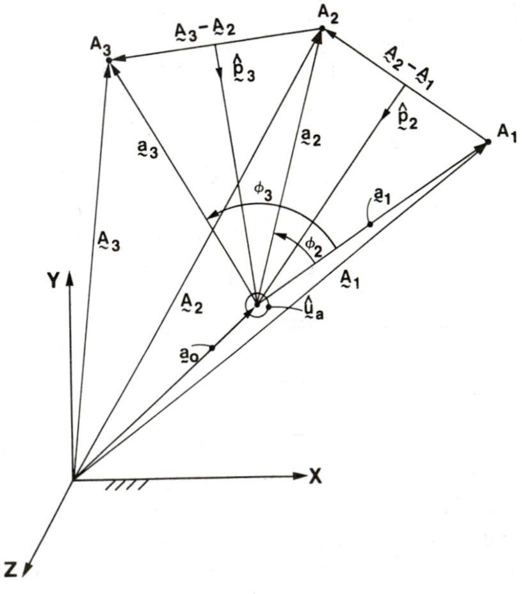


**Figure 6.58** The three positions of the *R-S* link and the associated vectors.

Observe that in Eq. (6.202) the first term on the left side is zero, which eliminates $\lambda_2$.

Next take the dot product of both sides of Eq. (6.202) with $\hat{\mathbf{p}}_2 \times \hat{\mathbf{p}}_3$ and divide by $\hat{\mathbf{p}}_2 \times \hat{\mathbf{p}}_3$ dot $\hat{\mathbf{p}}_2 \times \hat{\mathbf{p}}_3$, making the coefficient of $\lambda_3$ equal to one. Solving explicitly for $\lambda_3$ gives:

$$\lambda_3 = \frac{\hat{\mathbf{p}}_2 \times \hat{\mathbf{p}}_3 \cdot \left[\hat{\mathbf{p}}_2 \times \mathbf{A}_1 + \left(\mathbf{p}_2 \times \frac{\mathbf{A}_2 - \mathbf{A}_1}{2}\right) - (\hat{\mathbf{p}}_2 \times \mathbf{A}_2) - \hat{\mathbf{p}}_2 \times \frac{\mathbf{A}_3 - \mathbf{A}_2}{2}\right]}{\hat{\mathbf{p}}_2 \times \hat{\mathbf{p}}_3 \cdot \hat{\mathbf{p}}_2 \times \hat{\mathbf{p}}_3} \tag{6.203}$$

The value of $\lambda_3$ thus obtained can be substituted into Eq. (6.201) to find $\mathbf{a}_o$:

$$\mathbf{a}_o = \lambda_3 \mathbf{p}_3 + \mathbf{A}_2 + \frac{\mathbf{A}_3 - \mathbf{A}_2}{2} \tag{6.204}$$

Knowing $\mathbf{a}_o$, $\mathbf{a}_1$ can be found by

$$\mathbf{a}_1 = \mathbf{A}_1 - \mathbf{a}_o \tag{6.205}$$

This result gives the location of the fixed *R* joint and the moving *S* joint of link 2 in the starting position. This completely determines the first *R-S* dyad consisting of $\mathbf{a}_1$ and $\boldsymbol{\alpha}_1$.

### Finding the Direction and Location of the Fixed Revolute of Link 3

The second *R-S* dyad is found in an identical manner to the first *R-S* dyad, except that $\boldsymbol{\beta}$, $\mathbf{B}$, and $\mathbf{b}$ are substituted for $\boldsymbol{\alpha}$, $\mathbf{A}$, and $\mathbf{a}$ with all subscripts in Eqs. (6.194) through (6.205). Note that, like $\boldsymbol{\alpha}_1$, $\boldsymbol{\beta}_1$ can also be assumed arbitrarily at the designer's discretion.

### Finding the Axis and the Initial Location of the Cylindric Joint of Link 4

Figure 6.59 shows the *C-S* dyad in vector form. As was the case for the *R-S* dyad, we are given the position and orientation of the moving coordinate system at each of the three prescribed positions; this is given by Eq. (6.195).

Again we assume the location of the spherical joint, in this case joint *C*, in the moving *oxyz* system located by vector $\boldsymbol{\gamma}_1$ and embedded in the coupler link. We can now write an equation like Eq. (6.196) to give $\boldsymbol{\gamma}_j$ in the fixed frame of reference:

$$\boldsymbol{\gamma}_j = [R_j]\,\boldsymbol{\gamma}_1, \qquad j = 1, 2, 3 \tag{6.206}$$

Note that the $[R_j]$-*s* represent the prescribed orientation of the moving body, and

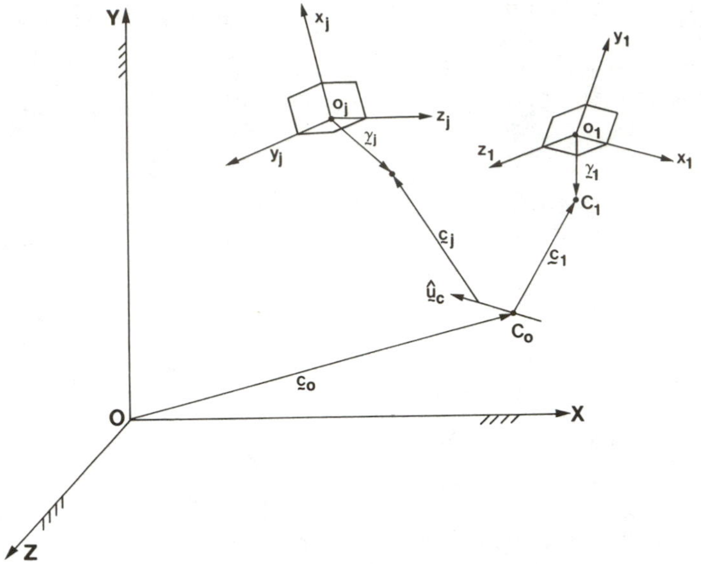


**Figure 6.59** The first and the *j*th position of the *C-S* pair.

are the same as for the *R-S* link. Thus, everything on the right-hand side of Eq. (6.206) is known.

At this point we must assume two additional scalar quantities in order to synthesize the *C-S* pair. This can be done by either assuming the two displacements of the cylindric joint axis, namely $S_2$ and $S_3$, or by assuming the direction of the cylindric joint axis, $\hat{\mathbf{u}}_c$. Note that $\hat{\mathbf{u}}_c$, being a unit vector, is specified by giving only two of its three scalar components. The remaining component can then be found from

$$u_{cx}^2 + u_{cy}^2 + u_{cz}^2 = 1 \tag{6.207}$$

The relationship between the vector $\hat{\mathbf{u}}_c$ and the scalar displacements $S_2$ and $S_3$ can be determined from Fig. 6.60 to be:

$$S_2 = \hat{\mathbf{u}}_c \cdot (\mathbf{C}_2 - \mathbf{C}_1) \tag{6.208}$$

$$S_3 = \hat{\mathbf{u}}_c \cdot (\mathbf{C}_3 - \mathbf{C}_1) \tag{6.209}$$

If we assume values of $S_2$ and $S_3$, Eqs. (6.207), (6.208), and (6.209) form a set of three equations in the three unknown components of $\hat{\mathbf{u}}_c$. On the other hand, if we assume the direction of $\hat{\mathbf{u}}_c$ by assuming two of its components, we can directly solve Eq. (6.207) for the third component of $\hat{\mathbf{u}}_c$. With the vector $\hat{\mathbf{u}}_c$ determined, Eq. (6.208) and (6.209) can be directly solved for the displacements $S_2$ and $S_3$. In either case the problem is simple to solve. Thirdly, we may assume one component of $\hat{\mathbf{u}}_c$ and one axial displacement.

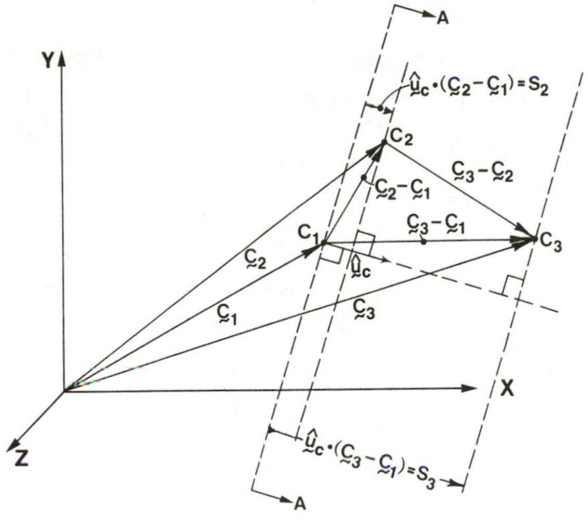


**Figure 6.60** The three positions of the *C-S* link and the associated vectors.

Figure 6.61 shows section *A-A* taken from Fig. 6.60. This is the plane defined by the unit normal $\hat{\mathbf{u}}_c$ and passing through the point $C_1$. The points $C_2'$ and $C_3'$ are the projections of points $C_2$ and $C_3$ onto this plane and the vectors $\mathbf{C}_2'$ and $\mathbf{C}_3'$ from the origin are given as follows:

$$\mathbf{C}_2' = \mathbf{C}_2 - \mathbf{S}_2\hat{\mathbf{u}}_c \tag{6.210}$$

$$\mathbf{C}_3' = \mathbf{C}_3 - \mathbf{S}_3\hat{\mathbf{u}}_c \tag{6.211}$$

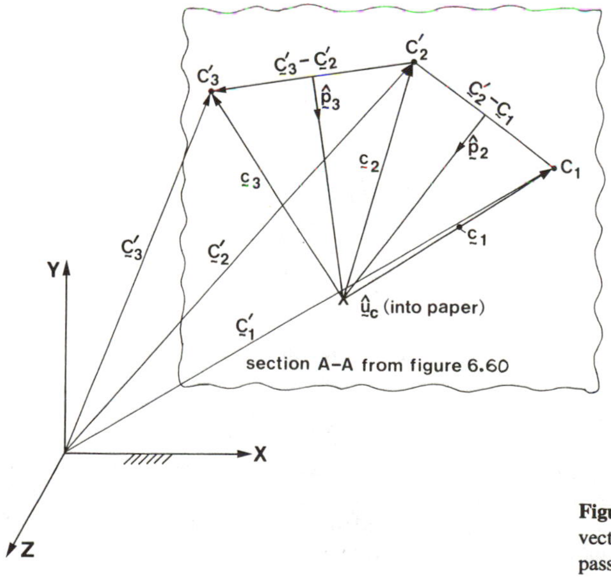


**Figure 6.61** Projections of the *C-S* link vectors onto the plane normal to $\hat{\mathbf{u}}_c$ and passing through point $C_1$.

Notice that all values on the right-hand side of these equations are known, also, note $C_1' = C_1$.

In the *C-S* dyad, the cylindric joint constrains the spheric joint to move on the surface of a cylinder whose axis is defined by $\hat{\mathbf{u}}_c$. The projections of points on the cylinder onto a plane normal to $\hat{\mathbf{u}}_c$ will, therefore, lie on a circle. Thus, points $C_1$, $C_2'$, and $C_3'$ define a circle with its center on axis $\hat{\mathbf{u}}_c$.

The problem of finding the initial location of the cylindric joint, $\mathbf{C}_o$, has now been reduced to the problem of finding the location $C_o$ of the revolute axis $\hat{\mathbf{u}}_c$ for an *R-S* dyad.

For the *C-S* dyad, the synthesis Eqs. (6.203), (6.204), and (6.205) become:

$$\lambda_3 = \frac{\hat{\mathbf{p}}_2 \times \hat{\mathbf{p}}_3 \cdot \left[\hat{\mathbf{p}}_2 \times \mathbf{C}_1 + \left(\hat{\mathbf{p}}_2 \times \dfrac{\mathbf{C}_2' - \mathbf{C}_1}{2}\right) - (\hat{\mathbf{p}}_2 \times \mathbf{C}_2') - \hat{\mathbf{p}}_2 \times \dfrac{\mathbf{C}_3' - \mathbf{C}_2'}{2}\right]}{\hat{\mathbf{p}}_2 \times \hat{\mathbf{p}}_3 \cdot \hat{\mathbf{p}}_2 \times \hat{\mathbf{p}}_3} \tag{6.212}$$

$$\mathbf{C}_o = \lambda_3 \hat{\mathbf{p}}_3 + \mathbf{C}_2' + [(\mathbf{C}_3' - \mathbf{C}_2')/2] \tag{6.213}$$

$$\mathbf{c}_1 = \mathbf{C}_1 - \mathbf{c}_o. \tag{6.214}$$

Thus, the dimensions of the *C-S* dyad are fully determined.

## Numerical Example of Closed-Form Synthesis of a Spatial Mechanism by Vector Notation

**Example 6.12**

Design an *RSSR-SC* mechanism to guide a rigid body through three finitely-separated positions in space. ($o_{jz} \equiv 0$ does not affect generality, because 3 points define a plane.)
Input Data:
Moving origin:

$$o_j = \begin{bmatrix} 1.50 & 1.50 & 0 \\ 0.75 & 1.25 & 0 \\ 0.35 & 0.80 & 0 \end{bmatrix} = \begin{Bmatrix} \mathbf{o}_1^T \\ \mathbf{o}_2^T \\ \mathbf{o}_3^T \end{Bmatrix}$$

Euler angles of the moving plane (in degrees):

$$\begin{bmatrix} 18.5 & 21 & 30 \\ 10 & 20 & 30 \\ 0 & 5 & 10 \end{bmatrix} = \begin{bmatrix} \lambda_{1z} & \lambda_{2z} & \lambda_{3z} \\ \mu_{1x'} & \mu_{2x'} & \mu_{3x'} \\ \zeta_{1z''} & \zeta_{2z''} & \zeta_{3z''} \end{bmatrix}$$

Here the first subscript denotes the position and the second subscript denotes the axis of rotation. For example $\lambda_{1z}$ is the rotation about the $z$ axis in the initial (first) position, $\mu_{2x'}$ is the rotation about the new $x'$ (rotated $x$) axis in the second position.

The positions of the moving body are illustrated in Fig. 6.62 by means of a set of orthographic projections of the body at each of the three prescribed positions. A cube with edges 1 unit long is used to describe the position and orientation of the moving body in the three prescribed positions. The front and the top view of the cube in the three prescribed positions are shown in the figure.

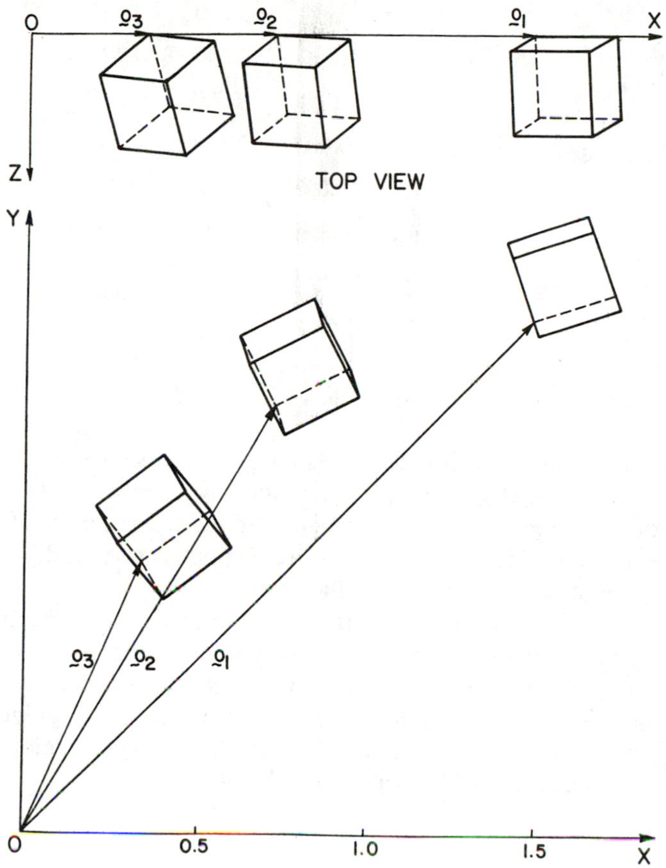


**Figure 6.62** The three prescribed positions used in Ex. 6.12.

Location of spheric joints *A, B* and *C* in the moving frame of reference

$$\begin{Bmatrix} \boldsymbol{\alpha}^T \\ \boldsymbol{\beta}^T \\ \boldsymbol{\gamma}^T \end{Bmatrix} = \begin{bmatrix} -1.5 & 0 & 0 \\ 1.5 & 0 & 0 \\ 0 & 0 & -1.5 \end{bmatrix}$$

Direction of the cylindrical joint axis, $\hat{\mathbf{u}}_c$

$$\hat{\mathbf{u}}_c = .632\hat{\mathbf{i}} - .632\hat{\mathbf{j}} + .447\hat{\mathbf{k}}$$

Dimensions of the synthesized mechanism are as follows (all vectors are transposed and are expressed in the fixed frame of reference):

| | *X* | *Y* | *Z* |
|---|---|---|---|
| $\mathbf{a}_o =$ | 0.287 | −0.055 | −0.130 |
| $\mathbf{a}_1 =$ | −0.21 | 1.080 | 0.130 |

| | | | |
|---|---|---|---|
| $\mathbf{b}_o =$ | 2.89 | −1.230 | 1.530 |
| $\mathbf{b}_1 =$ | 0.037 | 3.200 | −1.53 |
| $\mathbf{c}_o =$ | 0.958 | −0.751 | −4.360 |
| $\mathbf{c}_1 =$ | 0.460 | 2.500 | 2.880 |
| $\hat{\mathbf{u}}_a =$ | 0.008 | −0.118 | 0.993 |
| $\hat{\mathbf{u}}_b =$ | 0.009 | 0.431 | 0.902 |

$S_1 = -0.497$
$S_2 = -0.644$ scalar quantities

The synthesized mechanism is shown in Fig. 6.63.

This section presented the development of procedures for synthesis of an *RSSR-SC* five link spatial mechanism for motion *generation*. The developed procedure of synthesis is simple and may be extended to include the design for multiply separated spatial positions of the rigid body. The technique is also readily adaptable for interactive computation with visual graphic display.

Since the equations of synthesis are linear, little difficulty is encountered in computation. However crank rotatability and branching problems are encountered in the computation. Since there are several arbitrary choices available to the designer (for example $\boldsymbol{\alpha}$, $\boldsymbol{\beta}$, $\boldsymbol{\gamma}$, $S_2$, and $S_3$) a mechanism which guides the body through three prescribed positions and also satisfies additional constraints of crank rotatability and avoidance of branching can be obtained.

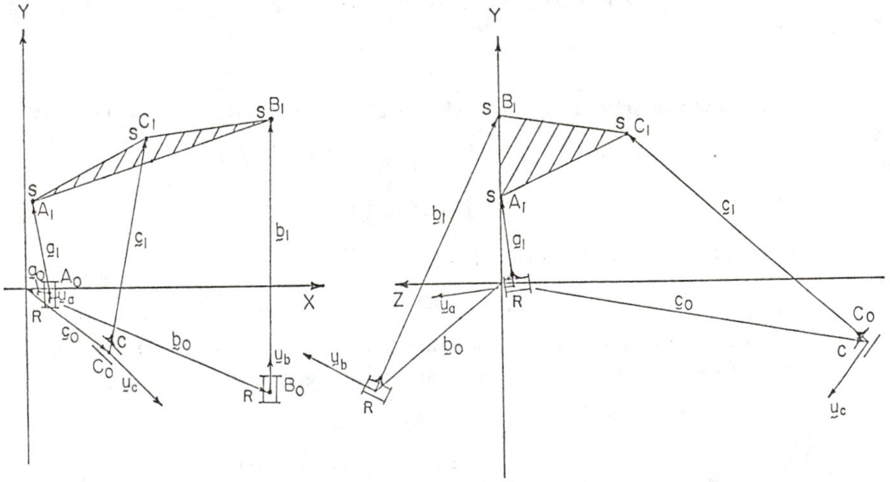


**Figure 6.63** Two orthographic views of the mechanism synthesized in Example 6.12.

## PROBLEMS

**6.1.** Figure P6.1 shows the MacPherson strut suspension system, which is used on many automobiles [49].

(a) Using the joint pairs in Table 6.1, draw a three-dimensional kinematic diagram of this suspension.

(b) Using Eqs. (6.1) to (6.3), verify the degrees of freedom of this mechanism.

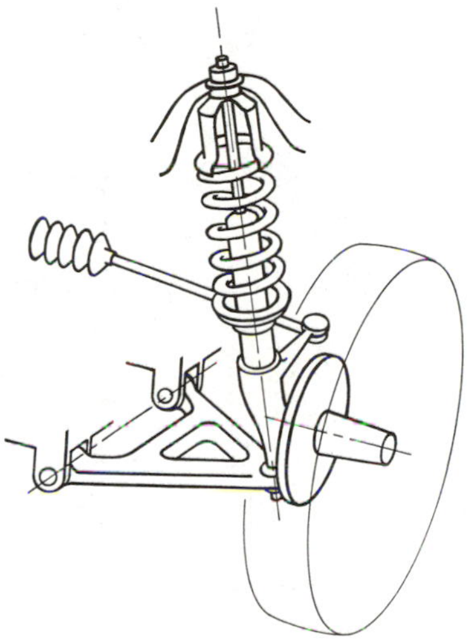

**Figure P6.1** MacPherson strut suspension.

**6.2.** For selected figures (6.1 to 6.15):

(a) Describe the function of the mechanism.

(b) Draw a three-dimensional kinematic diagram of the mechanism using the joint pairs of Table 6.1.

(c) Using Eqs. (6.1) to (6.3), verify the degrees of freedom of this mechanism.

(d) Does the mechanism have redundant (or passive) degrees of freedom?

(e) Does mobility depend on special proportions? Which are they?

(f) Could any links be removed without affecting the functioning of the mechanism? Which, if any? Why are these used at all?

**6.3.*** Determine the degrees of freedom for the spatial parallel-guidance linkage of Fig. 6.6.

**6.4.** If in Fig. P6.2, $\mathbf{A}_1 = 0.707 + 0.707i$:

(a) Find $\mathbf{A}_2$ if $\theta_{12} = 165°$. Use the matrix rotation operator method.

(b) What is $\theta_1$?

(c) What is $\theta_2$?

**6.5.** Given three points on a planar rigid body ($\mathbf{A} = 0.866 + 5i$, $\mathbf{B} = 1 + i$, $\mathbf{C} = 1.5 + i$), and its motion given by $\theta = 90°$, $\Delta X = 1$, and $\Delta Y = 0.5$. Find the final positions ($\mathbf{A}'$, $\mathbf{B}'$, $\mathbf{C}'$) by using Eq. (6.18). Make a step-by-step scale drawing of the transformation. (*Hint:* Show the rotation first, then the translations.)

* This and the following problems were prepared with the help of Chris Paulson, IBM, San José, California.

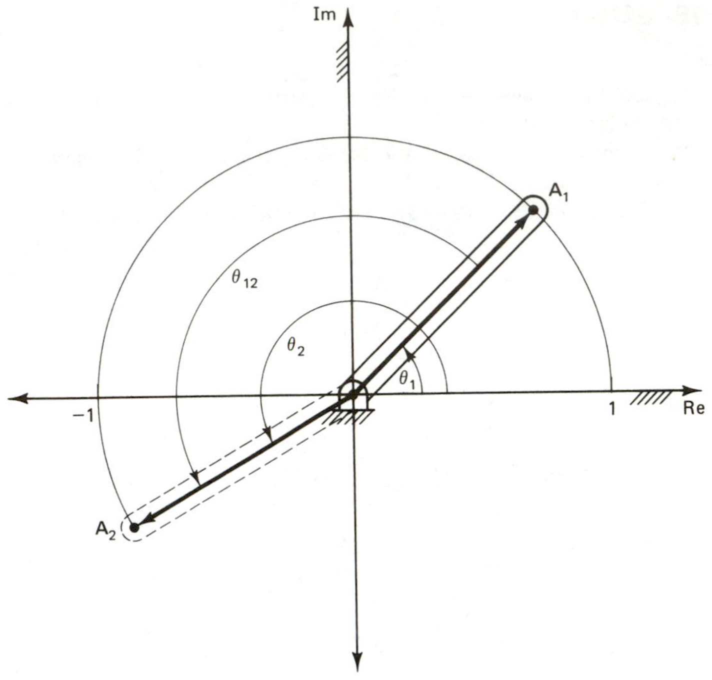


**Figure P6.2**

**6.6.** Use the rotation matrix from Ex. 6.5 (in the chapter) to transform the coordinates of a tetrahedron defined by four points, $A$, $B$, $C$, and $D$, to its final position. The four points in their first position are

$$\{\mathbf{A}_1 \quad \mathbf{B}_1 \quad \mathbf{C}_1 \quad \mathbf{D}_1\} = \begin{bmatrix} 2 & 2 & -2 & 0 \\ 10 & 6 & 8 & 8 \\ 8 & 8 & 6 & 10 \end{bmatrix}$$

**6.7.** The spatial mechanism in Fig. P6.3 consists of a Cardano* joint constrained to rotate about the $x$ and $y$ axes. If $\theta_x$ is rotated first, followed by a rotation of $\psi_y$: (a) determine the transformation operator for this mechanism by combining elementary transformation matrices. Then let the point $P$ be at the following starting position:

$$\{P\} = \begin{bmatrix} 18 \\ 32 \\ 25 \\ 1 \end{bmatrix}$$

Let $\theta_x = 33°$ and $\psi_y = 44°$, and (b) find the rotated position of $P$ for $\theta_x$ followed by $\psi_y$ and then (c) starting from the same original position, but rotate by $\psi_y$ first and then by $\theta_x$. (d) Are the two final positions the same? In other words, are the $\theta_x$ and $\psi_y$ rotations interchangeable?

* Two revolutes with mutually perpendicular concurrent axes.

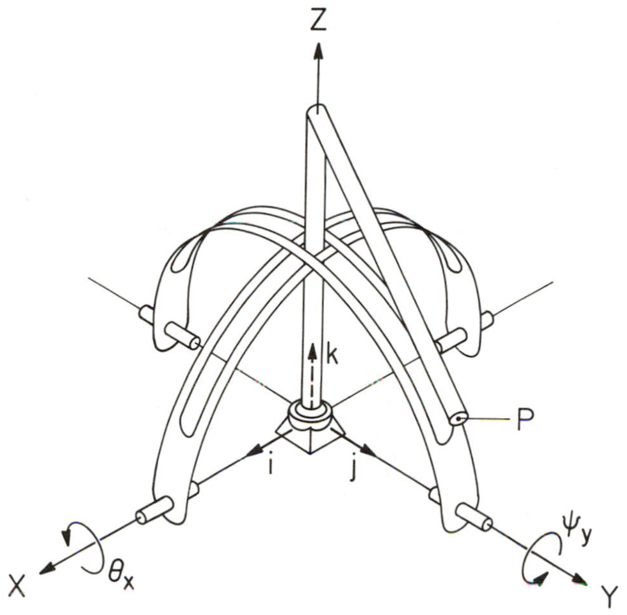


**Figure P6.3**

**6.8.** By combining elementary transformation matrices, determine the transformation operator for the spatial open loop mechanism in Fig. P6.4 from the local moving coordinate system of link $k$ to the fixed coordinate system of link $i$. Note that $\theta_{yj}$ and $\psi_{zk}$ are motion variables, because link j is an R-R link.

**6.9.*** Find $\mathbf{P}^i$ expressed in the fixed $i$th coordinate system for the spatial mechanism in Fig. P6.4 given that, expressed in the $k$th local (moving) system, $\mathbf{P}^k = (9, 0, 0)^T$, $\theta_{yi} = 30°$, $C = 36$, and $\psi_{zj} = 180°$. Verify your solution by drawing the final position in the $x_i y_i z_i$ system.

**6.10.** Determine the transformation operator for the screw joint mechanism in Fig. P6.5, which will transform a vector $\mathbf{P}^j$ expressed in the $j$th moving system to the vector $\mathbf{P}^i$ expressed in the $i$th "fixed" system.

**6.11.*** Find $\mathbf{P}^i$ for the screw joint mechanism of Fig. P6.5 if $\mathbf{P}^j = (0, 2, 0)^T$, $\theta_{zj} = 180°$, $A = 2$, $B = -3$, $C = 2$, and the lead $k = 0.125$. Verify your solution by a drawing.

**6.12.** Determine the operator that will transform a vector $\mathbf{P}^l$ into the vector $\mathbf{P}^i$ for the planar open loop mechanism in Fig. P6.6 by combining elementary transformation matrices.

**6.13.*** Find $\mathbf{P}^i$ for the planar open loop mechanism in Fig. P6.6 given that $\mathbf{P}^l = (2, 0, 0)^T$, $\theta_{yi} = -90°$, $\psi_{yk} = 180°$, $\phi_{y1} = 180°$, $A = 4$, and $B = 4$. Verify your results by a scaled drawing.

**6.14.** Determine the operator that will transform a vector $\mathbf{P}^l$ into the vector $\mathbf{P}^i$ for the spatial open loop mechanism in Fig. P6.7 by combining elementary transformation matrices.

* Use the operator derived in the preceding problem.

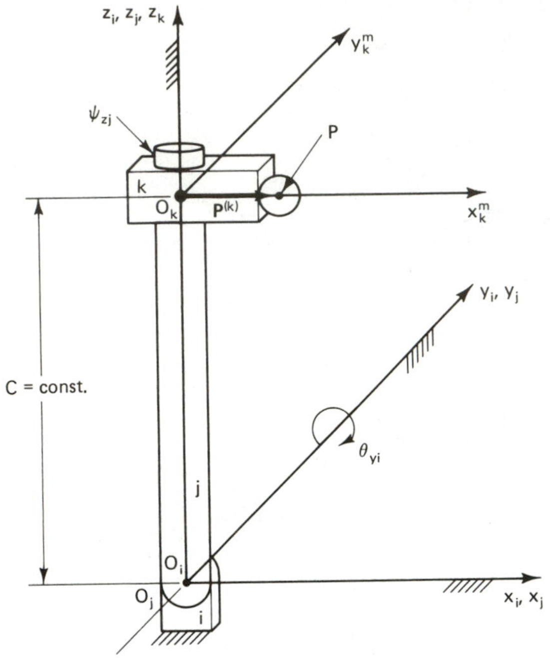


Figure P6.4

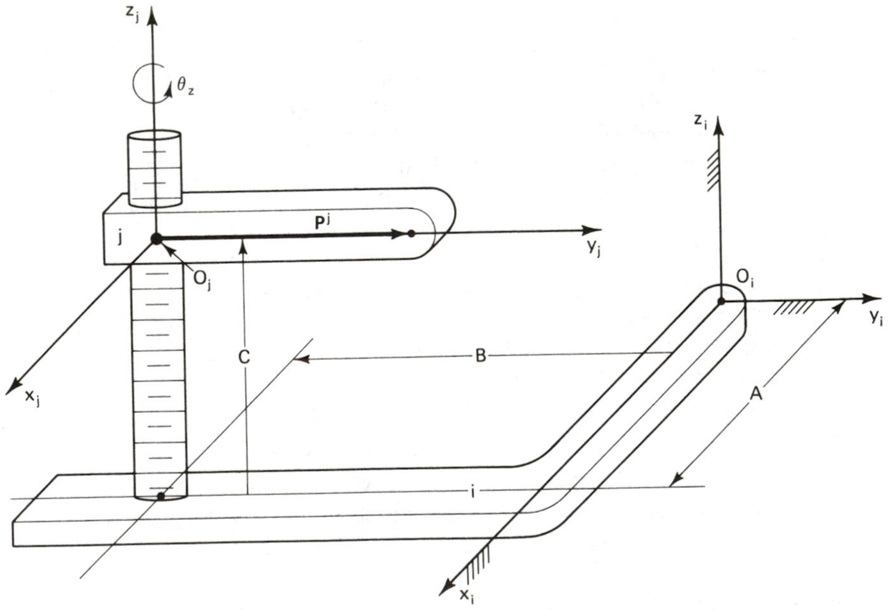


Figure P6.5

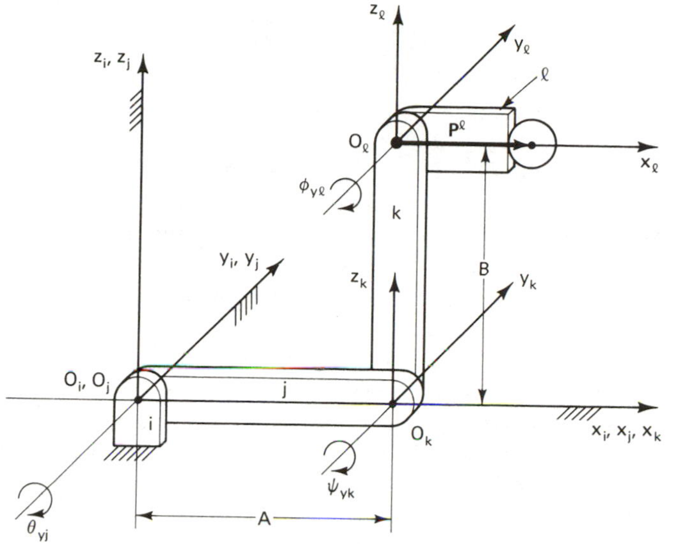


**Figure P6.6**

**6.15.*** Find $\mathbf{P}^i$ for the spatial open loop mechanism in Fig. P6.7 given that $\mathbf{P}^\ell = (0, 0, -0.5)^T$ $\theta_z = 180°$, $\psi_{yk} = -90°$, $\phi_{x1} = -90°$, $A = 0.5$, and $B = 1.0$. Verify your results with a scaled drawing.

**6.16.** Find $\mathbf{P}^i$ expressed in the $i$th coordinate system for the spatial open loop mechanism in Fig. P6.4 given that, expressed in the $k$th local (moving) system, $\mathbf{P}^k = (1.5, 0, 0)^T$, $C = 6$, $\theta_{yj} = -45°$, and $\psi_{zk} = 90°$ by using the transformation operator obtained in Prob. 6.8. Verify your solution by drawing the final position in the $x_i y_i z_i$ system.

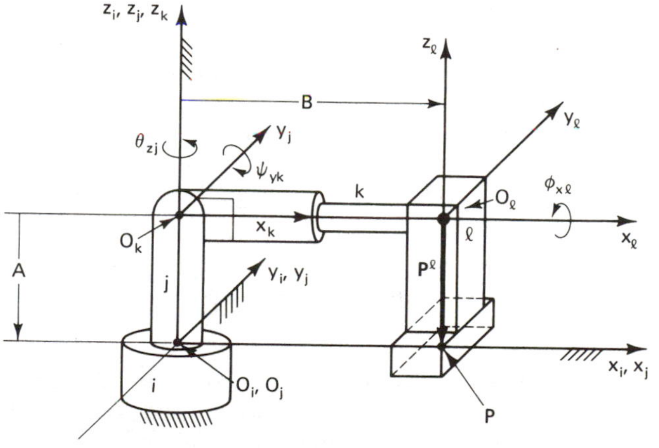


**Figure P6.7**

* Use the operator derived in the preceding problem.

**6.17.** Find $\mathbf{P}^i$ for the screw joint mechanism of Fig. P6.5 if $\mathbf{P}^j = (0, 12, 0)^T$, $\theta_{zj} = 14\pi$ rad, $A = 12$, $B = -18$, $C = 12$, and the lead $k = \frac{1}{12}$, by using the transformation operator obtained in Prob. 6.10. Verify your solution by a scaled drawing.

**6.18.** Find $\mathbf{P}^i$ for the planar open loop mechanism in Fig. P6.6 given that $\mathbf{p}^{\ell} = (1, 0, 0)^T$, $\theta_{yj} = -135°$, $\psi_{yk} = 180°$, $\phi_{y\ell} = 45°$, $A = 2$, and $B = 2$, by using the transformation operator obtained in Problem 6.12. Verify the solution by a scaled drawing.

**6.19.** Find $\mathbf{P}^i$ for the spatial open loop mechanism in Fig. P6.7, given that $\mathbf{P}^{(\ell)} = (0, 0, -2)^T$, $\theta_{zj} = 180°$, $\psi_{yk} = -45°$, $\phi_{x\ell} = 0°$, $A = 2$, and $B = 4$, by using the transformation operator obtained in Prob. 6.14. Verify the solution by a drawing.

**6.20.** Find $\mathbf{p}^i$ expressed in the $i$th coordinate system for the spatial mechanism in Fig. P6.4 given that, expressed in the $k$th local (moving) system, $\mathbf{p}^k = (1, 0, 0)^T$, $C = 4$, $\theta_{yj} = -60°$, and $\psi_{zk} = 135°$ by using the transformation operator obtained in Prob. 6.8. Verify the solution by a scaled drawing.

**6.21.** Find $\mathbf{p}^i$ expressed in the $i$th coordinate system for the spatial mechanism in Fig. P6.4 given that, expressed in the $k$th local (moving) system, $\mathbf{p}^k = (16, 0, 0)^T$, $C = 64$, $\theta_{yj} = 30°$, and $\psi_{zk} = -45°$ by using the transformation operator obtained in Prob. 6.8. Verify the solution by a scaled drawing.

**6.22.** Find $\mathbf{p}^i$ for the screw joint mechanism of Fig. P6.5 if $\mathbf{p}^j = (0, 4.5, 0)^T$, $\theta_{zj} = -3.5\pi$ rad., $A = 4.5$, $B = -6.75$, $C = 5$, and the lead $k = \frac{1}{16}$, by using the transformation operator obtained in Prob. 6.10. Verify the solution by a scaled drawing.

**6.23.** Find $\mathbf{p}^i$ for the screw joint mechanism of Fig. P6.5 if $\mathbf{p}^j = (0, 15, 0)^T$, $\theta_{zj} = -6.25\pi$ rad., $A = 15$, $B = -22.5$, $C = 16$, and the lead $k = \frac{1}{8}$, by using the transformation operator obtained in Prob. 6.10. Verify the solution by a scaled drawing.

**6.24.** Find $\mathbf{p}^i$ for the planar mechanism in Fig. P6.6 given that $\mathbf{p}^{\ell} = (2.75, 0, 0)^T$, $\theta_{yj} = -60°$, $\psi_{yk} = 120°$, $\phi_{y\ell} = -45°$, $A = 5.5$, and $B = 5.5$, by using the transformation operator obtained in Prob. 6.12. Verify the solution by a scaled drawing.

**6.25.** Find $\mathbf{p}^i$ for the planar mechanism in Fig. P6.6 given that $\mathbf{p}^{\ell} = (18, 0, 0)^T$, $\theta_{jy} = -30°$, $\psi_{yk} = 45°$, $\phi_{y\ell} = -90°$, $A = 40$, and $B = 40$ by using the transformation operator obtained in Prob. 6.12. Verify the solution by a scaled drawing.

**6.26.** Find $\mathbf{p}^i$ for the spatial mechanism in Fig. P6.7, given that $\mathbf{p}^{\ell} = (0, 0, -4.5)^T$, $\theta_{zj} = 45°$, $\psi_{yk} = -60°$, $\phi_{x\ell} = 30°$, $A = 4.5$, and $B = 9$ by using the transformation operator obtained in Prob. 6.14. Verify the solution by a scaled drawing.

**6.27.** Find $\mathbf{p}^i$ for the spatial mechanism in Fig. P6.7 given that $\mathbf{p}^{(\ell)} = (0, 0, -7.5)^T$, $\theta_{zj} = -150°$, $\psi_{yk} = -45°$, $\phi_{x\ell} = -30°$, $A = 7.5$ and $B = 14.5$, by using the transformation operator obtained in Prob. 6.14. Verify the solution by a scaled drawing.

**6.28.** Often one is required to interpret a patent or blueprint that contains details that mask the structure of a mechanism. Reference 31 cites such an example including the following wording: "Mechanism for converting rotary motion into reciprocating motion [Boxall 1939]. Figure P6.8 (Boxall's Fig. 4) illustrates the basic concept. The patent describes this figure as follows:

> A plunger 10 of a high speed reciprocating pump is slidable in a cylinder . . . supported by a frame member 12. It is required to reciprocate the plunger 10 by rotating a shaft 13 which is borne in a bearing . . . carried by a frame member 15. The frame members . . . are supported in a housing 17 of any convenient construction.

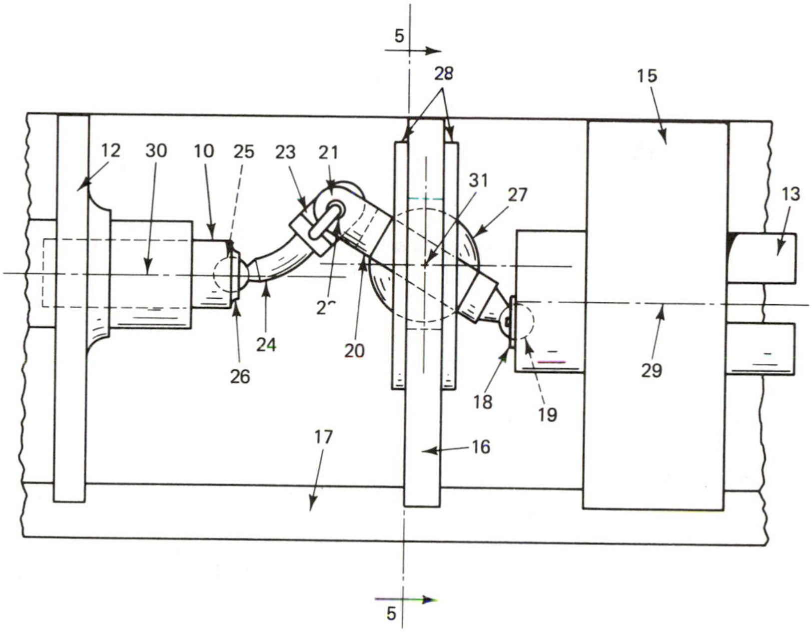


**Figure P6.8** Boxall Patent Fig. 4, ref. 31,105.

The end of the shaft 13 is formed with an eccentric recess containing a cup 18, which receives the ball-shaped end 19 of a cylindrical rod 20, the other end of which is forked at 21 to engage a tubular gudgeon-pin 22 upon which the tongue-end 23 of a bent coupling rod 24 is pivoted. The other end of the rod is formed with a ball-end 25 which engages a cup 26 in the end of the plunger 10.

The cylindrical rod 20 is slidable in a diametral bore in a ball 27, which is supported for tilting movement in part-spherical recesses in plates 28 secured to a frame-member 16.

**(a)** Determine the basic kinematic structure of this mechanism, and draw a sketch of its kinematic diagram.

**(b)** Determine the degrees of freedom of this device.

**(c)** Does the answer to part (b) suggest an alternative design?

**6.29.** Determine the degrees of freedom of the 3D mechanism of Fig. P6.9.

**6.30.** Determine the degrees of freedom of the 3D mechanism of Fig. P6.10.

**6.31.** Determine the degrees of freedom of the 3D mechanism of Fig. P6.11.

**6.32.** Figure P6.12 shows a five-axis industrial robot. Using the elementary matrices of Secs. 6.11 and 6.12, determine a mathematical model for the position of the hand of this robot with respect to a fixed coordinate system.

**6.33.** Figure P6.13 shows a three-axis (RPP) industrial robot. Using the elementary matrices of Secs. 6.11 and 6.12, determine a mathematical model for the position of the hand of this robot with respect to a fixed coordinate system.

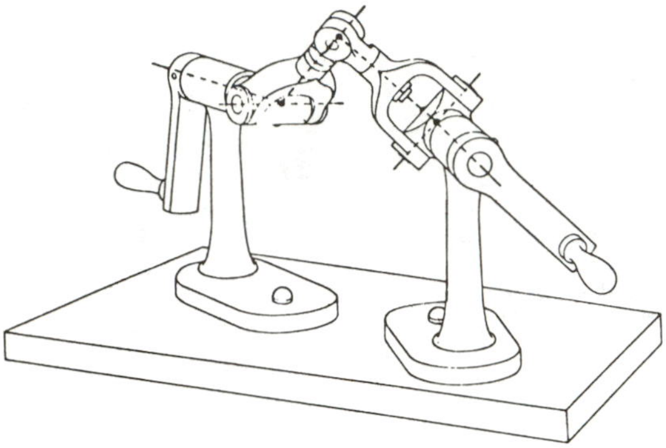

**Figure P6.9** Clemens coupling from ref. 62.

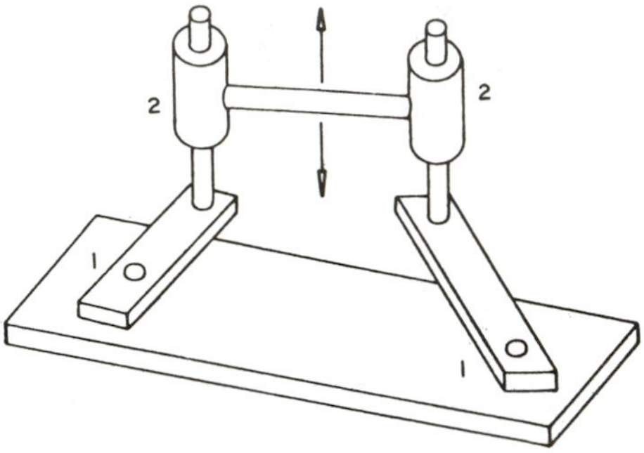


**Figure P6.10** Mechanism from ref. 62.

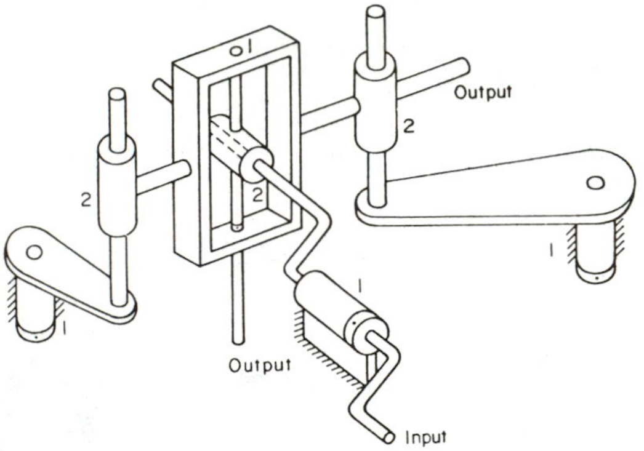


**Figure P6.11** Mechanism from ref. 62.

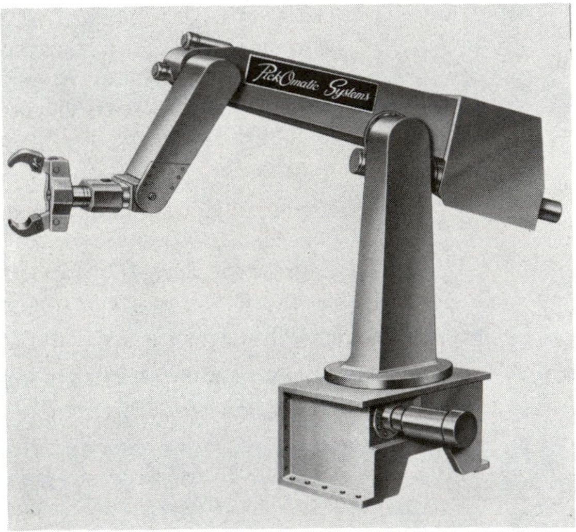


**Figure P6.12** Five-axis industrial robot. (*Courtesy of Pickomatic Systems—Fraser Automation.*)

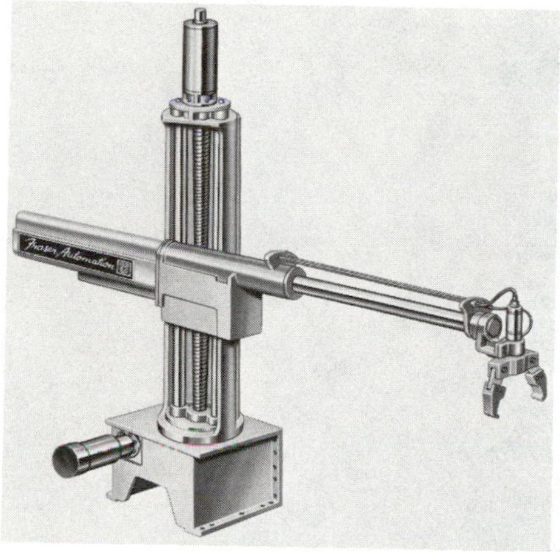


**Figure P6.13** Three-axis industrial robot. (*Courtesy of Pickomatic Systems—Fraser Automation.*)

**6.34.** Figure P6.14 shows a five-axis industrial robot. Using the elementary matrices of Secs. 6.11 and 6.12, determine a mathematical model for the position of the hand of this robot with respect to a fixed coordinate system.

**6.35.** Figure P6.15 shows a six-axis industrial robot (three mutually perpendicular revolute joints are located at the end of the arm). Using the elementary matrices of Secs. 6.11 and 6.12, determine a mathematical model for the position of the hand of this robot with respect to a fixed coordinate system.

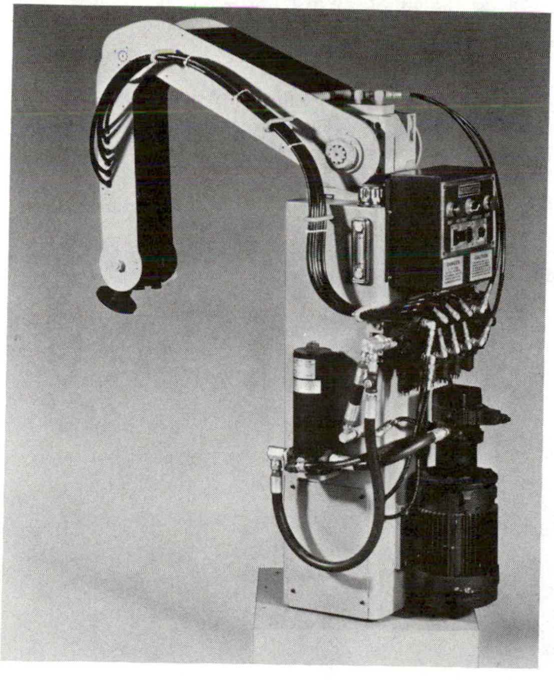

**Figure P6.14** Five-axis industrial robot. (*Courtesy of Thermwood Corp., Inc.*)

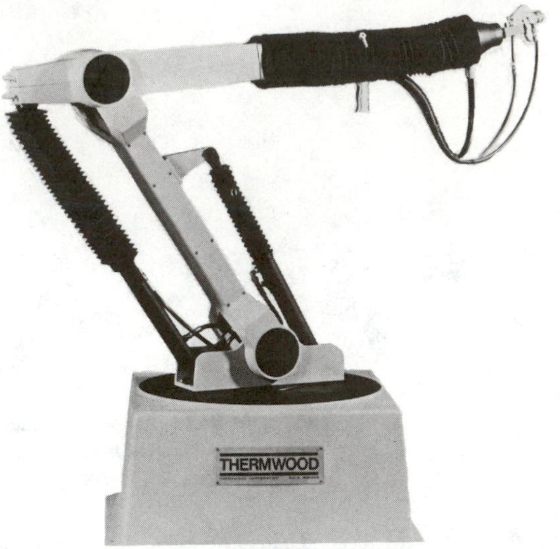


**Figure P6.15** Six-axis industrial robot. (*Courtesy of Thermwood Corp., Inc.*)

**6.36.** Figure P6.16 shows a five-axis industrial robot (RRPRR). Using the elementary matrices of Secs. 6.11 and 6.12, determine a mathematical model for the position of the hand of this robot with respect to a fixed coordinate system.

**6.37.** Figure P6.17 shows a five-axis industrial robot that contains a closed-loop chain. What chain is it? Is the analysis of this robot made simpler based on its type? How would you analyze this robot? Set up the equations for the position of the grip with respect to a fixed coordinate system.

**6.38.** Figure P6.12 and Fig. P6.14 are both five-axis robots. Compare their type. How would the work space of these robots compare with the robot in Fig. P6.16?

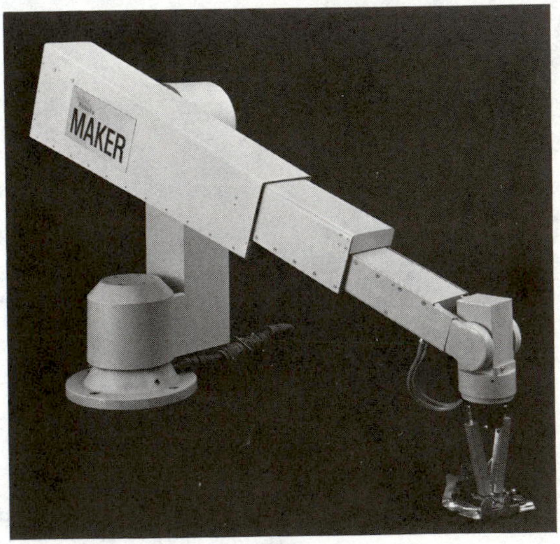


**Figure P6.16** Industrial robot. (*Courtesy of United States Robots.*)

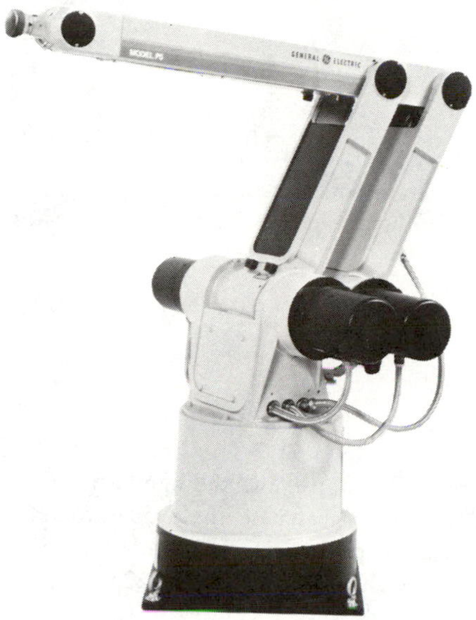

**Figure P6.17** Five-axis industrial robot. *(Courtesy of General Electric Inc.)*

**6.39.** Determine the position, velocity, and acceleration characteristics of the slider of the spatial crank slider mechanism in Fig. 6.49. (One could use the SAM computer program.) The link lengths in Fig. 6.49 are defined as follows:

$$L_1 = 3''$$

$$L_2 = 4''$$

$$L_3 = 9''$$

$$L_4 = 1''$$

$$L_5 = 12''$$

$$L_6 = 4''$$

$$\theta_1 = 90°$$

$$\theta_5 = -90°$$

The given input constant angular velocity is 6.3 rad/sec for link 2. The initial angle of link 2 is 90°.

**6.40.** Same as Prob. 6.39 except $L_3 = 8''$. Is this change in length by 1 in. a problem?

**6.41.** Same as Prob. 6.39 except:

$$L_1 = 1''$$

$$L_2 = 4''$$

$$L_3 = 30''$$

$$L_4 = 1''$$

$$L_5 = 31''$$

$$L_6 = 5''$$

$$\theta_1 = 0°$$

$$\theta_5 = 0°$$

**6.42.** Figure 6.11 shows seven different constant-velocity couplings. As a designer who has a need for such a mechanism, list the positive and negative features of each type.

## EXERCISES

**6.1.** In Fig. 6.22a, let $\mathbf{A}_1 = 78 + 52i$, $\mathbf{Q} = 52 + 26i$, and $\theta = 90°$. Express $\mathbf{A}_1$ and $\mathbf{Q}$ in homogeneous $3 \times 1$ column vector form, find the homogeneous $3 \times 3$ matrix operators $[T(-Q)]$, $[R(\theta)]$, and $[T(+Q)]$, concatenate these operators to find $[D(\theta, Q_x, Q_y)]$, and then apply this concatenated operator to $\{A_1\}$ to find $\{A_2\}$. Reconvert $\{A_2\}$ to complex vector form and verify your results by scaling Fig. 6.22a.

**6.2.** Refer to Fig. 6.24 and:

**(a)** Perform the exercise described in connection with this figure in the text, placing $A$ in the first octant of space (i.e., $a_x > 0$, $a_y > 0$, $a_z > 0$).

**(b)** Same as part (a), but place $A$ in the second octant of space (i.e., $a_x < 0$, $a_y > 0$, $a_z > 0$).

**(c)** Same as part (a), but place $A$ in the third octant (i.e., $a_x < 0$, $a_y < 0$, $a_z > 0$).

**(d)** Same as part (a), but place $A$ in the fourth octant (i.e., $a_x > 0$, $a_y < 0$, $a_z > 0$).

**(e)** Same as part (a), but place $A$ in the fifth octant (i.e., $a_x > 0$, $a_y > 0$, $a_z < 0$).

**(f)** Same as part (a), but place $A$ in the sixth octant (i.e., $a_x < 0$, $a_y > 0$, $a_z < 0$).

**(g)** Same as part (a), but place $A$ in the seventh octant (i.e., $a_x < 0$, $a_y < 0$, $a_z < 0$).

**(h)** Same as part (a), but place $A$ in the eighth octant (i.e., $a_x > 0$, $a_y < 0$, $a_z < 0$).

**6.3.** Check the solution of Ex. 6.6 using the method of orthographic projection and descriptive geometry demonstrated in Figs. 6.23b, c, d, and 6.24.

**6.4.** By concatenation, derive the operator for axial translation (axis through the origin) followed by coaxial rotation. Does this differ from the screw operator of Eq. (6.30)? Illustrate your result with an example using data from Ex. 6.8.

**6.5.** Write Eq. (6.69) in full array and verify Eq. (6.70) by matrix multiplication.

**6.6.** Write Eq. (6.72) in full array and verify Eq. (6.73) by matrix multiplication.

**6.7.** In Fig. 6.41, let the point $P$ of link $n$ be located by $\mathbf{P}^j = (4.\ 12,\ 9)^T$ (all coordinates are in millimeters). If in Fig. 6.41, $A_j = 58$, $B_j = -9$, $\alpha_{yi} = -30°$, $\beta_{xk} = -45°$, and $\gamma_{zk} = 45°$, determine $\mathbf{P}^i$ according to Eq. (6.73) and verify your results graphically by way of drawn-to-scale orthographic projections similar to those of Fig. 6.40. (*Hint:* Review Ex. 6.11.)

**6.8.** Prove Eq. (6.78) by concatenation (i.e., by premultiplying $S_{12}$ by $S_{01}$).

**6.9.** Prove Eq. (6.79) by concatenation.

**6.10.** Prove by concatenation that the left sides of Eqs. (6.81) and (6.82) are equal.

**6.11.** Verify Eq. (6.83).

**6.12.** Verify Eq. (6.114) by taking derivatives of the respective elementary matrices.

**6.13.** Verify Eq. (6.118) by taking derivatives of the respective elementary matrices.

**6.14.** By using well-known rules for inverting a nonsingular square matrix, prove Eq. (6.131).

**6.15.** Prove that the inverse $[A]^{-1}$ of an $nxn$ nonsingular matrix $[A]$ can be found by forming the matrix $[A \,\vdots\, I]$ of $n$ rows and $2n$ columns and then performing elementary row operations until obtaining $[I \,\vdots\, B]$, where $[B]\,[A] = [A]\,[B] = [I]$, and thus $[B] = [A]^{-1}$.

**6.16.** Verify the value of $S_{12}$ obtained in Exer. 6.8 by performing the transformation $[S_{02}]\,[\mathbf{A}_1\ \mathbf{B}_1\ \mathbf{C}_1\ \mathbf{D}_1] = [\mathbf{A}_2\ \mathbf{B}_2\ \mathbf{C}_2\ \mathbf{D}_2]$. Assume appropriate values for $\theta_1$, $\theta_2$, and $L_2$ and point-position vectors. Check your result by a scaled drawing.

# References


1. Alexander, R.M., and K.L. Lawrence, "An Experimental Investigation of the Dynamic Response of an Elastic Mechanism," *Journal of Engineering for Industry,* 96, no. 1 (February 1974), pp. 268–74.
2. Alizade, R.I., I.G. Novruzebekov, F. Freudenstein, and G.N. Sandor, "Optimum Synthesis of Path Generator Spatial Slider-Crank Mechanism," *Zhurnal Machinovedenie,* no. 1 (1980) (in Russian).
3. ______, A.V.M. Rao, and G.N. Sandor, "Optimum Synthesis of Two-Degree-of-Freedom Planar and Spatial Function Generating Mechanisms Using the Penalty Function Approach," *Journal of Engineering for Industry,* 97, no. 2 (May 1975), pp. 629–34.
4. ______, G.N. Sandor, and A.V.M. Rao, "Optimum Synthesis of Four-Bar and Offset Slider-Crank Planar and Spatial Mechanisms Using the Penalty Function Approach with Inequality and Equality Constraints," *Journal of Engineering for Industry,* 97, no. 3 (August 1975), pp. 785–90.
5. Alt, H., "The Geometry of Linkage Dwell Mechanisms," *Ing.-Arch.,* 3 (1932), pp. 394–411 (in German).
6. ______, "Linkages of Dwell Mechanisms," *Zeitschrift des VDI,* 76 (1932), pp. 456–62, 533–37 (in German).
7. Badlani, M., and W. Kleinhenz, "Dynamic Stability of Elastic Mechanisms," *Journal of Mechanical Design,* 101, no. 1 (January 1979), pp. 149–53.
8. ______, and A. Midha, "A Hierarchy of Equations of Motion of Planar Mechanisms with Elastic Link," *Proceedings of the Sixth Applied Mechanisms Conference (Denver, Colorado, October 1979),* pp. XXII-1–4. Stillwater, Oklahoma: Oklahoma State University, 1979.
9. Bagci, C., "Dynamic Force and Torque Analysis of Mechanisms Using Dual Vectors and 3 × 3 Screw Matrix," *Journal of Engineering for Industry,* 94 (1972), pp. 738–45.

10. ———, "Force and Torque Analyses of Plane Mechanisms by Matrix Displacement—Direct Element Method," *International Symposium on Linkages and Computer Design Methods* (*Bucharest, Romania, June 1973*), pp. 77–97.

11. Ball, R.S., *A Treatise on the Theory of Screws.* Cambridge: Cambridge University Press, 1900.

12. Bathe, K.J., and E.K. Wilson, "Stability and Accuracy Analysis of Direct Integration Methods," *International Journal of Earthquake Engineering and Structural Dynamics,* 1, no. 3 (1973), pp. 283–91.

13. Beggs, J.S., *Mechanism.* New York: McGraw-Hill Book Company, 1955.

14. Benedict, C.E., and D. Tesar, "Analysis of a Mechanical System Using Kinematic Influence Coefficients," *Proceedings of the First Applied Mechanisms Conference* (*Stillwater, Oklahoma, July 1969*), Paper No. 37. Stillwater, Oklahoma: Oklahoma State University, 1969.

15. ———, and D. Tesar, "Optimal Torque Balancing for a Complex Stamping and Indexing Machine," ASME Paper No. 70-Mech-82, 1970.

16. Berkof, R.S., "Force Balancing of a Six-Bar Linkage," *Proceedings of the Fifth World Congress on Theory of Machines and Mechanisms,* pp. 1082–85. New York: American Society of Mechanical Engineers, 1979.

17. ———, "The Input Torque in Linkages," *Journal of Mechanism and Machine Theory,* 14 (1979), pp. 61–73.

18. ———, "On the Optimization of Mass Distribution in Mechanisms," Ph.D. thesis, City University of New York, 1969.

19. ———, "Complete Force and Moment Balancing of Inline Four-Bar Linkages," *Journal of Mechanism and Machine Theory,* 8, no. 3 (1973), pp. 397–410.

20. ———, and G.G. Lowen, "A New Method for Completely Force Balancing Simple Linkages," *Journal of Engineering for Industry,* no. 1 (1969), pp. 21–6.

21. ———, and G.G. Lowen, "Theory of Shaking Moment Optimization of Force-Balanced Four-Bar Linkages," *Journal of Engineering for Industry,* 93B, no. 1 (1971), pp. 53–60.

22. ———, J. Boomgaarden, A.G. Erdman, et al., "Mechanism Case Studies VI," ASME Paper No. 82-DET-47, 1982.

23. ———, A.G. Erdman, D. Hewitt, T. Bjorklund, D. Harvey, et al., "Mechanism Case Studies IV," ASME Paper No. 78-DET-19, 1978.

24. Beyer, R., *Kinematic Synthesis of Mechanisms.* Translated by Kuenzel, Chapman, and A.S. Hall, Jr., 1962. New York: McGraw-Hill Book Company, 1963.

25. Bickford, J.H., *Mechanisms for Intermittent Motion.* New York: Industrial Press, Inc., 1972.

26. Bisshopp, K.E., "Rodigues' Formula and the Screw Matrix," *Journal of Engineering for Industry,* 91, no. 1 (February 1969), pp. 178–84.

27. Bjorklund, T., and A.G. Erdman, "Case Study: The Thing," *Proceedings of the Fifth Applied Mechanisms Conference* (*Oklahoma City, Oklahoma, November 1977*). Stillwater, Oklahoma: Oklahoma State University, 1977.

28. Black, Ted, "Designing Geared Five-Bar Mechanisms," *Design Engineering,* (May 1980), pp. 51–4.

29. Bobillier, E., *Cours de géométrie,* 12th ed., p. 232, 1870.

30. Bonhom, W.E., "Calculating the Response of a Four-Bar Linkage," ASME Paper No. 70-Mech-69, 1970.

31. Boxall, F.C., *Mechanism for Converting Rotary Motion into Reciprocating Motion,* U.S. Patent No. 2,173,247, October 31, 1939.

32. Broniarek, C.A., and G.N. Sandor, "Dynamic Stability of an Elastic Parallelogram Linkage," *Journal of Nonlinear Vibration Problems,* 12, pp. 315–25. Warsaw: Polish Academy of Science/Institute of Basic Technical Problems, 1971.

33. Carnahan, B., H.A. Luther, and J.O. Wilkes, *Applied Numerical Methods.* New York: John Wiley & Sons, Inc., 1969.

34. Carpenter, W.C., "Viscoelastic Stress Analysis," *International Journal of Numerical Methods in Engineering,* 4, no. 3 (1972), pp. 357–66.

35. Carson, W.L., and J.M. Trummel, "Time Response of Lower Pair Spatial Mechanisms Subjected to General Forces," ASME Paper No. 68-Mech-57, 1968.

36. Caughey, T.K., "Classical Normal Modes in Damped Linear Systems," *Journal of Applied Mechanics,* 27 (1960), pp. 269–71.

37. Cayley, A., "On Three-bar Motion," *Proceedings of the London Mathematical Society,* Vol. 7, 1876.

38. Chace, M.A., "Using DRAM and ADAMS Programs to Simulate Machinery, Vehicles," *Agricultural Engineering,* (November 1978), pp. 16–18.

39. ______, "Vector Analysis of Linkages," *Journal of Engineering for Industry,* 84, no. 2 (May 1963), pp. 289–96.

40. ______, and J.C. Angell, "Interactive Simulation of Machinery with Friction and Impact Using DRAM," SAE Paper No. 770050, 1977.

41. ______, and J.C. Angell, "User's Guide to DRAM (Dynamic Response of Articulated Machinery)," Design Engineering Computer Aids Lab, University of Michigan, 1972.

42. Chakraborty, J., and S.G. Dhande, *Kinematics and Geometry of Planar and Spatial Cam Mechanisms.* New Delhi, India: Wiley Eastern Limited, 1972.

43. Chase, T., A.G. Erdman, and D. Riley, "Synthesis of Six Bar Linkages Using an Interactive Package," *Proceedings of the Seventh Applied Mechanisms Conference (Kansas City, Missouri, December 1981),* Paper No. LI, 5 pgs. Stillwater, Oklahoma: Oklahoma State University, 1981.

44. Chebyshev, P.L., "Théorie des mécanismes connus sous le nom de parallélogrammes (1953)," in *Oeuvres de P.L. Tchebychef,* Vol. 1. St. Petersburg: Markoff et Sonin, 1899; *Modern Mathematical Classics: Analysis,* S730. Ed. Richard Bellom. New York: Dover Publications, Inc., 1961.

45. Chen, P., and B. Roth, "A Unified Theory for the Finitely and Infinitesimally Separated Position Problems of Kinematic Synthesis," *Journal of Engineering for Industry,* 91 (1969).

46. Chironis, N.P., *Machine Devices and Instrumentation.* New York: McGraw-Hill Book Company, 1966.

47. ______, *Mechanisms, Linkages, and Mechanical Controls.* New York: McGraw-Hill Book Company, 1965.

48. Clough, R.W., "Analysis of Structural Vibrations and Dynamic Response," *U.S.–Japan Seminar on Recent Advances in Matrix Models of Structural Analyses and Design (Tokyo, Japan, 1969),* pp. 441–86. R.H. Gallagher, Y. Yamada, and J.T. Oden, eds. Huntsville, Alabama: University of Alabama Press, 1971.

49. Cook, R.D., *Concepts and Applications of Finite Element Analysis.* New York: John Wiley & Sons, Inc., 1974.

50. Cronin, D.L., "MacPherson Strut Kinematics," *Journal of Mechanism and Machine Theory,* 16, no. 6 (1981), pp. 631–44.

51. Dahlquist, G., and A. Bjorck, *Numerical Methods.* Englewood Cliffs, New Jersey: Prentice-Hall, Inc., 1974.

52. Denavit, J., "Displacement Analysis of Mechanisms Based on (2 × 2) Matrices of Dual Numbers," *VDI Berichte,* 29 (1958).

53. Desai, C.S., *Elementary Finite Element Method.* Englewood Cliffs, New Jersey: Prentice-Hall, Inc., 1974.

54. ———, and J.F. Abel, *Introduction to the Finite Element Method.* New York: Van Nostrand Reinhold Company, 1979.

55. Dhande, S.H., B.S. Bhadoria, and J. Chakraborty, "A Unified Approach to the Analytical Design of Three-Dimensional Cam Mechanisms," ASME Paper No. 74-DET-8, 1974.

56. Dijksman, E.A., *Motion Geometry of Mechanisms.* Cambridge: Cambridge University Press, 1976.

57. ———, "Six-Bar Cognates of Watt's Form," *Journal of Engineering for Industry,* 93, no. 1 (1971), pp. 183–90.

58. Dimarogonas, A.D., *Vibrating Engineering,* p. 568. St. Paul, Minnesota: West Publishing Company, 1976.

59. ———, G.N. Sandor, and A.G. Erdman, "Synthesis of a Geared *N*-Bar Linkage," *Journal of Engineering for Industry,* 93, no. 1 (February 1971), pp. 157–64.

60. Dimentberg, F.M., "A General Method for the Investigation of Finite Displacements of Spatial Mechanisms and Certain Cases of Passive Joints," *Akademiya Nauk SSSR,* Trudy Sem. Teor. Mash. Mekh., 5, no. 17 (1948), pp. 5–39.

61. ———, *The Screw Calculus and Its Applications in Mechanics.* Moscow: Izdatel'stvo "Nauka," Glavnaya Redaktsiya Fiziko-Matematicheskoy Literatury, 1965.

62. Dixon, M.W., and C.O. Huey, Jr., "Fundamentals of Kinematic Synthesis," *1973 Textile Engineering Conference (Charlotte, North Carolina, April 1973),* pp. 1–31.

63. Diziuglu, B., "Theory and Practice of Spatial Mechanisms with Special Positions of the Axes," *Journal of Mechanism and Machine Theory,* 13 (1978), pp. 139–53.

64. Dobrovolskii, V.V., "General Investigation of Motion of the Links in a Seven-Link Spatial Mechanism by the Method of Spherical Images," *Akademiya Nauk SSSR,* Trudy Sem. Teor. Mash. Mekh., 12, no. 47 (1952), pp. 52–62.

65. Dubowsky, S., "On Predicting the Dynamic Effects of Clearance in Planar Mechanisms," *Journal of Engineering for Industry,* 96, no. 1 (February 1974), pp. 317–23.

66. ———, and F. Freudenstein, "Dynamic Analysis of Mechanical Systems with Clearances, Part I: Formation of Dynamic Model; Part II: Dynamic Response," *Journal of Engineering for Industry,* 93, no. 1 (February 1971), pp. 305–16.

67. ———, and T.N. Gardner, "Dynamic Interactions of Link Elasticity and Clearance Connection," *Journal of Engineering for Industry,* 97, no. 2 (May 1975), pp. 652–61.

68. ———, and T.N. Gardner, "Design and Analysis of Multi-Link Flexible Mechanisms with Multiple Clearance Connections," *Journal of Engineering for Industry,* 99, no. 1 (February 1977), pp. 88–96.

69. ———, and M.F. Moening, "An Experimental and Analytical Study of Impact Forces in Elastic Mechanical Systems with Clearances," *Journal of Mechanism and Machine Theory,* 13 (1978), pp. 451–65.

70. ______, and S.C. Young, "An Experimental and Analytical Study of Connection Forces in High-Speed Mechanisms," *Journal of Engineering for Industry,* 97, no. 4 (1975), pp. 1166–74.

71. Duffy, J., "An Analysis of Five, Six and Seven-Link Spatial Mechanisms," *Proceedings of the Third World Congress for the Theory of Machines and Mechanisms* (*Kupari, Yugoslavia, 1971*), Vol. C, pp. 83–98.

72. ______, *Analysis of Mechanisms and Robot Manipulators.* New York: John Wiley & Sons, Inc., 1981.

73. ______, "A Derivation of Dual Displacement Equations for Spatial Mechanisms Using Spherical Trigonometry," *Revue Roumaine des Sciences Techniques, Série de Mecanique Appliquée,* Part 1, 6 (1971); Part 2, 1 (1972); Part 3, 3 (1972).

74. Erdman, A.G., "Dynamic Synthesis of a Variable Speed Drive," *Proceedings of the Third Applied Mechanisms Conference* (*Stillwater, Oklahoma, November 1973*), Paper No. 38. Stillwater, Oklahoma: Oklahoma State University, 1973.

75. ______, "A General Method for Kineto-elastodynamic Analysis and Synthesis of Mechanisms," Doctoral dissertation, Rensselaer Polytechnic Institute, 1972.

76. ______, "A Guide to Mechanism Dyanmics," *Proceedings of the Third Applied Mechanisms Conference* (*Stillwater, Oklahoma, November 1973*), Paper No. 6. Stillwater, Oklahoma: Oklahoma State University, 1973.

77. ______, "Three and Four Precision Point Kinematic Synthesis of Planar Linkages," *Journal of Mechanism and Machine Theory,* 16, (1981), pp. 227–45.

78. ______, and W.L. Carson, "Teaching Unit on Complex Numbers as Applied to Linkage Modeling," in *Monograph on Mechanism Design,* Paper No. 12, 7 pgs. New York: McGraw-Hill Book Company; NSF Report No. GK-36624, 1977.

79. ______, and J.E. Gustafson, "LINCAGES: *L*inkage *IN*teractive *C*omputer *A*nalysis and *G*raphically *E*nhanced *S*ynthesis Package," ASME Paper No. 77-DTC-5, 1977.

80. ______, and Dana Lonn, "Synthesis of Planar Six-Bar Linkages for Five Precision Conditions by Complex Numbers," in *Monograph on Mechanism Design,* Paper No. 59, 7 pgs. New York: McGraw-Hill Book Company; NSF Report No. GK-36624, 1977.

81. ______, and Dana Lonn, "A Unified Synthesis of Planar Six-Bar Mechanisms Using Burmester Theory," *Proceedings of the Fourth World Congress on the Theory of Machines and Mechanisms* (*Newcastle-upon-Tyne, England, September 1975*).

82. ______, and A. Midha, "Man-Made Mechanism Models Multiply Mental Motivation," *Proceedings of the Third Applied Mechanics Conference* (*Stillwater, Oklahoma, November 1973*), Paper No. 3. Stillwater, Oklahoma: Oklahoma State University, 1973.

83. ______, and J.A. Peterson, "Operator for a Casement-Type Window," U.S. Patent No. 4,266,371, May 12, 1981.

84. ______, and D. Riley, "Computer Aided Linkage Design Using the LINCAGES Package," ASME Paper No. 81-DET-121, 1981.

85. ______, and D. Riley, "Computer Graphics and Computer Aided Design in Mechanical Engineering at the University of Minnesota," *Computers in Education,* 5 (Summer 1981), pp. 229–43.

86. ______, and G.N. Sandor, "Kinematic Synthesis of a Geared-Five-Bar Function Generator," *Journal of Engineering for Industry,* 93B, no. 1 (February 1971), pp. 157–64.

87. ———, and G.N. Sandor, "Kineto-elastodynamics—A Review of the State of the Art and Trends," *Mechanism and Machine Theory,* 7 (1972), pp. 19–33.

88. ———, and G.N. Sandor, "Applied Kineto-elastodynamics," *Proceedings of the Second Applied Mechanisms Conference,* pp. 21–1 to 21–17. Stillwater, Oklahoma: Oklahoma State University, 1971.

89. ———, and P. Starr, "Towards Technology Transfer: Kinematic and Dynamic Analysis of Linkages for the Design Engineer," *Proceedings of the First ASME Design Technology Conference in Machine Design (October 1974),* pp. 335–46. New York: American Society of Mechanical Engineers, 1974.

90. ———, E. Nelson, J. Peterson, and J. Bowen, "Type and Dimensional Synthesis of Casement Window Mechanisms," ASME Paper No. 80-DET-78, 1980; *Mechanical Engineering,* (December 1981), pp. 46–55.

91. ———, G.N. Sandor, and R.G. Oakberg, "A General Method for Kineto-elastodynamic Analysis and Synthesis of Mechanisms," *Journal of Engineering for Industry,* 94, no. 4 (February 1971), pp. 11–16.

92. ———, C. Tekse, and D. Ferguson, "Application of Burmester Theory—An Eight-Bar Inserting Head Mechanism," *Proceedings of the Fourth Applied Mechanisms Conference (Chicago, November 1975),* pp. 11–1 to 11–7. Stillwater, Oklahoma: Oklahoma State University, 1975.

93. ———, R.K. Westby, G.R. Fichtinger, and F.R. Tepper, "Mechanisms Case Studies II: A New Derailleur Mechanism," ASME Paper No. 74-DET-56, 1974; *Mechanical Engineering,* 97 (July 1975), pp. 36–7.

94. Ferguson, D.E., A.G. Erdman, and D.A. Frohrib, "Design Analysis of Feedbowls," *Proceedings of the Fifth World Congress on the Theory of Machines and Mechanisms (Montreal, July 1979),* pp. 61–4.

95. Fiacco, A.V., and G.P. McCormick, *Nonlinear Programming—Sequential Unconstrained Minimization Technique.* New York: John Wiley & Sons, Inc., 1968.

96. Fichtinger, G., R. Westby, and A. Erdman, "Combination of Design Disciplines Offers Series of Novel Mechanisms," *Product Engineering,* 45, no. 12 (December 1974), pp. 35–6.

97. Filemon, E., "In Addition to the Burmester Theory," *Proceedings of the Third World Congress for the Theory of Machines and Mechanisms (Kupari, Yugoslavia, 1971),* Vol. D, pp. 63–78.

98. ———, "Useful Ranges of Centerpoint Curves for Design of Crank-and-Rocker Linkages," *Journal of Mechanism and Machine Theory,* 7 (1972), pp. 47–53.

99. Fletcher, R., and M.J.D. Powell, "A Rapidly Convergent Descent Method for Minimization," *Computer Journal,* 6, no. 2 (1963), pp. 163–68.

100. Fox, R.L., *Optimization Methods for Engineering Design,* 2nd ed. Reading, Massachusetts: Addison-Wesley Publishing Company, Inc., 1973.

101. ———, and K.C. Gupta, "Optimization Technology as Applied to Mechanism Design," *Journal of Engineering for Industry,* 95 (1973), pp. 657–63.

102. ———, and K.D. Willmert, "Optimum Design of Curve-Generating Linkage with Inequality Constraints," *Journal of Engineering for Industry,* 89, no. 1 (February 1967), pp. 144–52.

103. Freudenstein, F., "An Analytical Approach to the Design of Four-Link Mechanisms," *Transactions of the ASME,* 76 (1954), pp. 483–92.

104. ______, "Harmonic Analysis of Crank-and-Rocker Mechanisms with Applications," *Journal of Applied Mechanics,* 81E (1959), pp. 673–75.

105. ______, "Structural Error Analysis in Plane Kinematic Synthesis," *Journal of Engineering for Industry,* 81, no. 1 (January 1959), pp. 15–22.

106. ______, and E.R. Maki, "The Creation of Mechanisms According to Kinematic Structure and Function," General Motors Research Publications, GMR-3073, September 1979; *International Journal for the Science of Architecture and Design* (1980).

107. ______, and G.N. Sandor, "Kinematics of Mechanisms," in *Mechanical Design and System Handbook,* pp. 4.1–4.68. New York: McGraw-Hill Book Company, 1964.

108. ______, and G.N. Sandor, "On the Burmester Points of a Plane," *Journal of Applied Mechanics,* 28, no. 1 (March 1961), pp. 41–9; Discussion and Authors' Closure, *Journal of Applied Mechanics,* 28, no. 3 (September 1961), pp. 473–75.

109. ______, and G.N. Sandor, "Synthesis of Path-Generating Mechanisms by Means of a Programmed Digital Computer," *Journal of Engineering for Industry,* 81B, no. 2 (May 1959), pp. 159–68.

110. Garrett, R.E., and A.S. Hall, Jr., "Effect of Tolerance and Clearance in Linkage Design," *Journal of Engineering for Industry,* 91B (1969), pp. 198–202.

111. Giese, R. C., "Three Position Dyad Linkage Synthesis Summary," Master's Paper, University of Minnesota, August 1982.

112. Goldstein, H., *Classical Mechanics.* Reading, Massachusetts: Addison-Wesley Publishing Company, Inc., 1957.

113. Golebiewski, E.P., and J.P. Sadler, "Analytical and Experimental Investigation of Elastic Slider-Crank Mechanisms," *Journal of Engineering for Industry,* 98, no. 4 (November 1976), pp. 1266–71.

114. Goudreau, G.L., and R.L. Taylor, "Evaluation of Numerical Integration Methods in Elastodynamics," *Computer Methods in Applied Mechanisms and Engineering,* 2, no. 1 (1973), pp. 69–97.

115. Gupta, K.C., "Design of Four-Bar Function Generators with Mini-Max Transmission Angles,"*Journal of Engineering for Industry,* 99, no. 2 (1977), pp. 360–66.

116. ______, "A General Theory for Synthesizing Crank-Type Four Bar Function Generators with Transmission Angle Control," *Journal of Applied Mechanics,* 45, no. 2 (June 1968).

117. ______, "Synthesis of Position, Path and Function Generating 4-Bar Mechanisms with Completely Rotatable Driving Links," *Journal of Mechanism and Machine Theory,* 15 (1980), pp. 93–101.

118. Gupta, V.K., "Kinematic Analysis of Plane and Spatial Mechanisms," *Transactions of the ASME, Series B,* 95, no. 2 (1973), pp. 481–86.

119. Hagen, D., A.G. Erdman, D. Harvey, and J. Tacheny, "Rapid Algorithms for Kinematic and Dynamic Analysis of Planar Rigid Linkages with Revolute Joints," ASME Paper No. 78-DET-64, 1978.

120. Hain, K., *Applied Kinematics,* 2nd ed. New York: McGraw-Hill Book Company, 1967.

121. Hall, A.S. Jr., *Kinematics and Linkage Design.* West Lafayette, Indiana: Balt Publishers, 1961.

122. Halter, J.M., "Force Synthesis to Produce a Desired Time Response of Mechanisms," Ph.D. dissertation, University of Missouri—Columbia, May 1975.

123. Han, C.-Y., "Balancing of High Speed Machinery," *Journal of Engineering for Industry*, 89, no. 1 (1967), pp. 111–18.

124. Harris, C.M., and C.E. Crede, "Numerical Methods of Analysis," *Shock and Vibration Handbook*, 1st ed., Vol. 2, pp. 28–32 to 28–36. New York: McGraw-Hill Book Company, 1961.

125. Harrisberger, L., "A Number Synthesis Survey of Three-Dimensional Mechanisms," *Journal of Engineering for Industry*, 87, no. 2 (May 1965), pp. 213–20.

126. Hartenberg, R.S., and J. Denavit, *Kinematic Synthesis of Linkages*. New York: McGraw-Hill Book Company, 1964.

127. Hartmann, W., "A New Method for Seeking the Circle of Curvature," *Zeitschrift des VDI*, (1983), p. 95 (in German).

128. Hernandez, M., G.N. Sandor, and D. Kohli, "Closed-Form Analytic Synthesis of R-Sp-R Three-Link Function Generator for Multiply Separated Positions," ASME Paper No. 82-DET-75, 1982.

129. Hertrich, F.R., "How to Balance High-Speed Mechanisms with Minimum-Inertia Counterweights," *Machine Design*, 35, no. 6 (March 1963), pp. 160–64.

130. Hinkle, Rolland T., *Kinematics of Machines*, 2nd ed. Englewood Cliffs, New Jersey: Prentice-Hall, Inc., 1960.

131. Holand, I., and K. Bell, eds., *Finite Element Methods in Stress Analysis*. Trondheim, Norway: Tapir Press, 1969.

132. Hrones, J.A., and G.L. Nelson, *Analysis of the Four-Bar Linkage*. New York: The Technology Press of MIT and John Wiley & Sons, Inc., 1951.

133. Hunt, K.H., "Constant-Velocity Shaft Couplings—A General Theory," *Journal of Engineering for Industry*, (1973), pp. 455–64.

134. Hurty, W.C., and M.F. Rubinstein, *Dynamics of Structures*. Englewood Cliffs, New Jersey: Prentice-Hall, Inc. 1964.

135. Imam, I., "A General Method for Kineto-elastodynamic Analysis and Design of High-Speed Mechanisms," Doctoral dissertation, Rensselaer Polytechnic Institute, 1973.

136. ———, "High-Speed Mechanism Design—A General Analytical Approach," *Journal of Engineering for Industry*, 97, no. 2 (May 1975), pp. 609–28.

137. ———, and G.N. Sandor, "A General Method of Kineto-elastodynamic Design of High Speed Mechanisms," *Journal of Mechanism and Machine Theory*, 8, no. 4 (Winter 1973), pp. 497–516.

138. ———, G.N. Sandor, and S.N. Kramer, "Deflection and Stress Analysis in High-Speed Planar Mechanisms with Elastic Links," *Journal of Engineering for Industry*, 95, no. 2 (May 1973), pp. 541–48.

139. Jandrasits, W.G., and G.G. Lowen, "The Elastic–Dynamic Behavior of a Counterweighted Rocker Link with an Overhanging Endmass in a Four-Bar Linkage, Part I: Theory; Part II: Application and Experiment," *Journal of Mechanical Design*, 101, no. 1 (January 1979), pp. 77–98.

140. Jasinski, P.W., H.C. Lee, and G.N. Sandor, "Stability and Steady-State Vibrations in a High-Speed Slider-Crank Mechanism," *Journal of Applied Mechanics*, 37, no. 4 (December 1970), pp. 1069–76.

141. ______, H.C. Lee, and G.N. Sandor, "Vibration of Elastic Connecting Rod of a High-Speed Slider-Crank Mechanism," *Journal of Engineering for Industry,* 93, no. 2 (May 1971), pp. 636–44.

142. Jenkins, E.M., F.R.E. Crossley, and K.H. Hunt, "Gross Motion Attributes of Certain Spatial Mechanisms," *Journal of Engineering for Industry,* 90, no. 1 (February 1969), pp. 83–90.

143. Johnson, R.C., "Impact Forces in Mechanisms," *Machine Design,* 30 (1958), pp. 138–46.

144. ______, and K. Towfigh, "Application of Number Synthesis to Practical Problems in Creative Design," ASME Paper No. 65-WA/MD-9, 1965; in *Mechanical Design Synthesis with Optimization Applications.* Ed. R.C. Johnson. New York: Van Nostrand Reinhold Company, 1971.

145. Jones, F.D., H.L. Horton, and J.A. Newell, eds. *Ingenious Mechanisms for Designers and Inventors,* 4 vols. New York: The Industrial Press, 1968.

146. Kalaycioglu, S., and C. Bagci, "Determination of the Critical Operating Speeds of Planar Mechanisms by the Finite Element Using Planar Actual Line Elements and Lumped Mass Systems," *Journal of Mechanical Design,* 101, no. 2 (April 1979), pp. 210–23.

147. Kamenskii, V.A., "On the Question of the Balancing of Plane Linkages," *Journal of Mechanisms,* 3 (1968), pp. 303–22.

148. Kaufman, R.E., "Mechanism Design by Computer," *Machine Design,* (October 1978), pp. 94–100.

149. ______, "Singular Solution in Burmester Theory," ASME Paper No. 72-Mech-23, 1972.

150. ______, and G.N. Sandor, "Bicycloidal Crank: A New Four Link Mechanism," *Journal of Engineering for Industry,* 91, no. 1 (February 1969), pp. 91–6.

151. ______, and G.N. Sandor, "Complete Force Balancing of Spatial Linkages," *Journal of Engineering for Industry,* 93, no. 2 (1971), pp. 620–26.

152. ______, and G.N. Sandor, "Operators for Kinematic Synthesis of Mechanisms by Stretch Rotation Techniques," ASME Paper No. 70-Mech-79, 1970.

153. Keler, M.L., "Kinematics and Statics Including Friction in Single Loop Mechanisms by Screw Calculus and Dual Vectors," *Transactions of the ASME, Series B,* 95, no. 2 (1973), pp. 471–80.

154. Khan, M.R., and K.D. Willmert, "Vibrational Analysis of Mechanisms Using Constant Length Finite Elements," ASME Paper No. 76-WA/DE-21, 1976.

155. Khotin, B.M., "A Kinematic Analysis of Mechanisms with Consideration of Linkage Elasticity," *Sbornik Trudy,* Leningradshii Institut Inzhenerov Zheleznodorozhnogo Transporta, No. 218 (1964), pp. 214–19 (in Russian).

156. Kobrinsky, A.E., "Die Dynamik der Maschinen und Systeme mit Vibroschlagwirkung," *VDI Berichte,* 127 (1969).

157. Kohli, D., and G.N. Sandor, "Elastodynamics of Planar Linkages Including Torsional Vibrations of Input and Output Shafts and Elastic Deflections at Supports," *Proceedings of the Fourth World Congress on the Theory of Machines and Mechanisms (Newcastle-upon-Tyne, England, September 1975),* Vol. 2, pp. 247–52.

158. ______, and G.N. Sandor, "Lumped Parameter Approach for Kineto-elastodynamic Analysis of Elastic Spatial Mechanisms," *Proceedings of the Fourth World Congress on the Theory of Machines and Mechanisms (Newcastle-upon-Tyne, England, September 1975),* pp. 253–58.

159. ———, and G.N. Sandor, "Nonlinear Vibration Analysis of Elastic Spatial Mechanisms Using a Lumped Parameter Approach," in *Topics in Contemporary Mechanics,* Courses and Lectures No. 210, pp. A223–27. Udine, Italy: International Centre for Mechanical Sciences.

160. ———, and A.H. Soni, "Synthesis of Spatial Mechanisms via Successive Screw Displacements and Pair Geometry Constraints," *Proceedings of the Fourth World Congress on the Theory of Machines and Mechanisms (Newcastle-upon-Tyne, England, 1975),* Vol. 4, no. 132, pp. 711–16.

161. ———, D. Hunter, and G.N. Sandor, "Elastodynamic Analysis of a Completely Elastic System," *Journal of Engineering for Industry,* 99, no. 3 (1977), pp. 604–9.

162. ———, A.E. Thompson, and G.N. Sandor, "Design of Four-Bar Linkages with Specified Motion Characteristics of All Moving Links," in *Monograph on Mechanical Design,* Paper No. 50. New York: McGraw-Hill Book Company, 1977; NSF Report No. GK36624.

163. Kotelnikoff, A.P., "Screw Calculus and Some Applications of the Same to Geometry and Mechanics," *Annals of the Imperial University of Kazan,* No. 9 (1895), pp. 79–152; No. 10 (1895), pp. 1–90; No. 11 (1895), pp. 139–58; No. 1 (1896), pp. 110–44.

164. Kramer, S.N., and G.N. Sandor, "Finite Kinematic Synthesis of a Cycloidal-Crank Mechanism for Function Generation," *Journal of Engineering for Industry,* 92, no. 3 (August 1970), pp. 531–36.

165. ———, and G.N. Sandor, "Kinematic Synthesis of Watt's Mechanism," ASME Paper No. 70-Mech-50, 1970.

166. ———, and G.N. Sandor, "Selective Precision Synthesis—A General Method of Optimization for Planar Mechanisms," *Journal of Engineering for Industry,* 97, no. 2 (May 1975), pp. 689–701.

167. Lee, T.W., "On the Kinetmatics and Dynamic Synthesis of a Variable-Speed Drive," ASME Paper No. 77-DET-124, 1977.

168. Levitskii, N.I., and S.H. Shakvasian, "Synthesis of Spatial Four Link Mechanisms with Lower Pairs," *Akademiya Nauk SSSR,* Trudy Sem. Teor. Mash. Mekh., 14, no. 54 (1954), pp. 5–24; translated into English by F. Freudenstein, *International Journal of Mechanical Science,* 2 (1960), pp. 76–92.

169. Lindholm, J.C., "A Survey of the Graphical Techniques in Designing for Specific Input–Output Relationships of a Four-Bar Mechanism," *Proceedings of the First Applied Mechanisms Conference (Stillwater, Oklahoma, July 1969),* pp. 35–1 to 35–6. Stillwater, Oklahoma: Oklahoma State University, 1969.

170. ———, "Design for Path Generation—Point Position Reduction," in *Monograph on Mechanism Design,* Paper No. 38, 7 pgs. New York: McGraw-Hill Book Company; NSF Report No. GK36624, 1977.

171. Loerch, R.J., A.G. Erdman, and G.N. Sandor, "On the Existence of Circle-Point and Center-Point Circles for Three-Precision Point Dyad Synthesis," *Journal of Mechanical Design,* (October 1979), pp. 554–62.

172. ———, A.G. Erdman, G.N. Sandor, and A. Midha, "Synthesis of Four-Bar Linkages with Specified Ground Pivots," *Proceedings of the Fourth Applied Mechanisms Conference (Chicago, November 1975),* pp. 10.1–10.6. Stillwater, Oklahoma: Oklahoma State University, 1975.

173. Lowen, G.G., and R.S. Berkof, "Determination of Force Balanced Four-Bar Linkages with Optimum Shaking Moment Characteristics," *Journal of Engineering for Industry,* 93, no. 1 (1971), pp. 39–46.

174. ______, and R.S. Berkof, "Survey of Investigations into the Balancing of Linkages," *Journal of Mechanisms,* 3, no. 4 (1968), pp. 221–31.

175. ______, and W.G. Jandrasits, "Survey of Investigations into the Dynamic Behavior of Mechanisms Containing Links with Distributed Mass and Elasticity," *Journal of Mechanism and Machine Theory,* 7 (1972), pp. 3–17.

176. ______, F.R. Tepper, and R.S. Berkof, "The Qualitative Influence of Complete Force Balancing on the Forces and Moments of Certain Families of Four-Bar Linkages," *Journal of Mechanism and Machine Theory,* 9 (1974), pp. 299–323.

177. Mabie, H.H., and F.W. Ocvirk, *Kinematics and Dynamics of Machinery,* 3rd ed. New York: John Wiley & Sons, Inc., 1978.

178. Martin, H.C., and G.F. Carey, *Introduction to Finite Element Analysis.* New York: McGraw-Hill Book Company, 1973.

179. Matthew, G.K., and D. Tesar, "Synthesis of Spring Parameters to Balance General Forcing Functions in Planar Mechanisms," *Journal of Engineering for Industry,* 99 (May 1977), pp. 347–52.

180. ______, and D. Tesar, "Synthesis of Spring Parameters to Satisfy Specified Energy Levels in Planar Mechanisms," *Journal of Engineering for Industry,* 99 (May 1977), pp. 341–46.

181. McGovern, J.F., and G.N. Sandor, "Kinematic Synthesis of Adjustable Mechanisms, Part I: Function Generation; Part II: Path Generation," *Journal of Engineering for Industry,* 95, no. 2 (May 1973), pp. 417–29.

182. McLarnan, C.W., "A Linkage Synthesis with Minimum Error," *Journal of Mechanisms,* 3, no. 2 (Summer 1968), pp. 101–5.

183. Meck, J.L., *Matrix Structural Analysis.* New York: McGraw-Hill Book Company, 1971.

184. Meyer zur Capellen, W., "Bending Vibrations in the Coupler of an Oscillating Crank Mechanism," *Osterreichisches Ingenieur Archiv,* 16, no. 4 (1962), pp. 341–48.

185. ______, "Kinematics—A Survey in Retrospect and Prospect," *Journal of Mechanisms,* 1 (1966), pp. 211–28.

186. Midha, A., "Creativity in the Classroom—A Collection of Case Studies in Mechanisms Design Methods," *Proceedings of the Seventh Applied Mechanisms Conference* (*Kansas City, Missouri, December 1981*), pp. XII–1 to XII–15. Stillwater, Oklahoma: Oklahoma State University, 1981.

187. ______, A.G. Erdman, and D.A. Frohrib, "A Closed-Form Numerical Algorithm for the Periodic Response of High-Speed Elastic Linkages," *Journal of Mechanical Design,* 101, no. 1 (January 1979), pp. 154–62.

188. ______, A.G. Erdman, and D.A. Frohrib, "A Computationally Efficient Numerical Algorithm for the Transient Response of High-Speed Elastic Linkages," *Journal of Mechanical Design,* 101, no. 1 (January 1979), pp. 138–48.

189. ______, A.G. Erdman, and D.A. Frohrib, "An Approximate Method for the Dynamic Analysis of Elastic Linkages," *Journal of Engineering for Industry,* 99, no. 2 (May 1977), pp. 449–55.

190. ———, A.G. Erdman, and D.A. Frohrib, "Finite Element Approach to Mathematical Modeling of High-Speed Elastic Linkages," *Journal of Mechanism and Machine Theory,* 13 (1978), pp. 603–18.
191. ———, A.G. Erdman, and D.A. Frohrib, "Finite Element Approach to Mathematical Modeling of High-Speed Elastic Linkages," *Proceedings of the Fifth Applied Mechanisms Conference (Oklahoma City, Oklahoma, November 1977).* Stillwater, Oklahoma: Oklahoma State University, 1977.
192. ———, A.G. Erdman, G.N. Sandor, and D.A. Frohrib, "An Alternate Computationally Efficient and Conservative Method for Kineto-elastodynamic Analysis of Mechanisms," *Proceedings of the Fourth Applied Mechanisms Conference (Chicago, November 1975),* pp. 19.1–19.19. Stillwater, Oklahoma: Oklahoma State University, 1975.
193. Modrey, J., "Analysis of Complex Kinematic Chains with Influence Coefficients," *Journal of Applied Mechanics,* 81E (1959), pp. 184–88.
194. Morris, C.M., and C.E. Crede, "Numerical Methods of Analysis," *Shock and Vibration Handbook,* 1st ed., Vol. 2, pp. 28–32 to 28–36. New York: McGraw-Hill Book Company, 1961.
195. Neubauer, A.H., Jr., R. Cohen, and A.S. Hall, Jr., "An Analytical Study of the Dynamics of an Elastic Linkage," *Journal of Engineering for Industry,* 88, no. 3 (August 1966), pp. 311–17.
196. Newman, W.M., and R.F. Sproull, *Principles of Interactive Computer Graphics,* 2nd ed. New York: McGraw-Hill Book Company, 1979.
197. Newmark, N.M., "A Method of Computation for Structural Dynamics," *Journal of the Engineering Mechanics Division,* 85, no. EM3 (1959), pp. 67–94.
198. Nickell, R.E., "Direct Integration Methods in Structural Dynamics," *Journal of the Engineering Mechanics Division,* 99, no. EM2 (1973), pp. 303–17.
199. Oakberg, T.G., "The Analysis of Frames with Shear Walls by Finite Elements," Doctoral dissertation, Stanford University, 1967.
200. Paul, B., "Analytical Dynamics of Mechanisms—A Computer Oriented Overview," *Journal of Mechanism and Machine Theory,* 10, no. 6 (1975), pp. 481–507.
201. Paul, R.P., *Robot Manipulators: Mathematics, Programming, and Control.* Cambridge, Massachusetts: The MIT Press, 1981.
202. Paulson, W.C., "Finite Element Stress Analysis," *Machine Design,* 43, no. 24 (September 1971), pp. 46–94.
203. Peterson, D., "Linkage Design Using Analysis and Interactive Computer Graphics," in *Monograph on Mechanism Design.* Papers Nos. 76 and 77, 9 pgs. New York: McGraw-Hill Book Company; NSF Report No. GK36624, 1977.
204. Peterson, J.A., and E.W. Nelson, "Operator for a Casement-Type Window," U.S. Patent No. 4,253,276, March 3, 1981.
205. Pian, T.H., and P. Tong, "Finite Element Methods in Continuum Mechanics," in *Advances in Applied Mechanics,* Vol. 12, pp. 1–58. New York: Academic Press, Inc., 1972.
206. Pouliot, H.N., W.R. Delameter, and C.W. Robinson, "A Variable-Displacement Spark Engine," SAE Paper No. 770114, 1977.
207. Primrose, E.J.F., F. Freudenstein, and G.N. Sandor, "Finite Burmester Theory in Plane Kinematics," *Journal of Applied Mechanics,* 31, no. 4 (December 1964), pp. 683–93.
208. Przemieniecki, J.S., *Theory of Matrix Structural Analysis.* New York: McGraw-Hill Book Company, 1968.

209. Quinn, B.E., "Energy Method for Determining Dynamic Characteristics of Mechanisms," *Journal of Applied Mechanics,* 16E (1949), pp. 283–88.

210. Radcliffe, C.W., "Kinematics in Biomechanics Research," *Proceedings of the National Science Foundation Workshop on New Directions for Kinematics Research* (*Stanford University, August 1976*), pp. 174–98.

211. ______, "Polycentric Linkages as Prosthetic Knee Mechanisms for the Through-Knee Amputee," *Proceedings of the World Congress of ISPO, INTERBOR, and APO* (*Montreux, Switzerland, October 1974*).

212. Rao, A.V.M., and G.N. Sandor, "Closed Form Synthesis of Four-Bar Path Generators by Linear Superposition," *Proceedings of the Third World Congress on the Theory of Machines and Mechanisms* (*Dubrovnik, Yugoslavia, September 1971*), pp. 383–94.

213. ______, and G.N. Sandor, "Extension of Freudenstein's Equation to Geared Linkages," *Journal of Engineering for Industry,* 93, no. 1 (February 1971), pp. 201–10.

214. ______, and G.N. Sandor, "Synthesis of Function Generating Mechanisms with Scale Factors as Unknown Design Parameters," *Transactions of the International Symposium on Linkages and Computer Design Methods* (*Bucharest, June 1973*), Vol. A-44, pp. 602–23.

215. ______, A.G. Erdman, G.N. Sandor, et al., "Synthesis of Multi-loop, Dual Purpose Planar Mechanisms Utilizing Burmester Theory," *Proceedings of the Second Applied Mechanisms Conference* (*Stillwater, Oklahoma, October 1971*), pp. 7.1–7.23. Stillwater, Oklahoma: Oklahoma State University, 1971.

216. ______, G.N. Sandor, and Steven N. Kramer, "Geared Six-Bar Design," *Proceedings of the Second Applied Mechanisms Conference* (*Stillwater, Oklahoma, October 1971*), pp. 25.1–25.13. Stillwater, Oklahoma: Oklahoma State University, 1971.

217. ______, G.N. Sandor, D. Kohli, and A.H. Soni, "Closed-Form Synthesis of Spatial Function Generating Mechanisms for Maximum Number of Precision Points," *Journal of Engineering for Industry,* 95, no. 3 (August 1973), pp. 725–36.

218. Reklaitis, G.V., A. Ravindran, and K.M. Ragsdell, *Engineering Optimization: Methods and Application.* New York: John Wiley & Sons, Inc., 1983.

219. Reuleaux, F., *The Kinematics of Machinery.* Translated by Alex B.W. Kennedy. New York: Macmillan & Co., 1876.

220. Richardson, M.M., and A.G. Erdman, "Computer-Aided Mechanism Synthesis Using the LINCAGES Package," *Proceedings of the First International Conference on Applied Modeling and Simulation in Industry,* Vol. 1, pp. 260–66. Paris, France: AMSE, 1981.

221. Roark, R.J., *Formulas for Stress and Strain.* New York: McGraw-Hill Book Company, 1938.

222. Rogers, D.F., and J.A. Adams, *Mathematical Elements for Computer Graphics.* New York: McGraw-Hill Book Company, 1976.

223. Rooney, J., and J. Duffy, "On the Closures of Spatial Mechanisms," ASME Paper No. 72-Mech-77, 1972.

224. Rose, P.S., and G.N. Sandor, "Direct Analytical Synthesis of Four-Bar Function Generators with Optimal Structural Error," *Journal of Engineering for Industry,* 95, no. 2 (May 1973), pp. 563–71.

225. Roth, B., "On the Screw Axis and Other Special Lines Associated with Spatial Displacements of a Rigid Body," *Journal of Engineering for Industry,* 89, no. 1 (1967), pp. 102–10.

226. ———, F. Freudenstein, and G.N. Sandor, "Synthesis of Four-Bar Path-Generating Characteristics," *Transactions of the Seventh Conference on Mechanisms (Purdue University, October 1962)*, pp. 46–8. West Lafayette, Indiana: Purdue University, 1962.

227. Rubel, A.J., and R.E. Kaufman, "KINSYN III: A New Human-Engineered System for Interactive Computer-Aided Design of Planar Linkages," *Journal of Engineering for Industry,* 99, no. 2 (May 1977).

228. Sadler, J.P., "A Lumped Parameter Approach to Kineto-Elastodynamic Analysis of Mechanisms," Doctoral dissertation, Rensselaer Polytechnic Institute, 1972.

229. ———, "On the Analytical Lumped-Mass Model of an Elastic Four-Bar Mechanism," *Journal of Engineering for Industry,* 97, no. 2 (May 1975), pp. 561–65.

230. ———, and G.N. Sandor, "Kineto-elastodynamic Harmonic Analysis of Four-Bar Path Generating Mechanisms," ASME Paper No. 70-Mech-61, 1970.

231. ———, and G.N. Sandor, "A Lumped Parameter Approach to Vibration and Stress Analysis of Elastic Linkages," *Journal of Engineering for Industry,* 95, no. 2 (May 1973), pp. 549–57.

232. ———, and G.N. Sandor, "Non-linear Vibration Analysis of Elastic Four-Bar Linkages," *Journal of Engineering for Industry,* 96, no. 2 (May 1976), pp. 411–19.

233. ———, and R. W. Wayne, "Balancing of Mechanisms by Nonlinear Programming," *Proceedings of the Third Applied Mechanisms Conference (Stillwater, Oklahoma, November 1973)*, pp. 29–1 to 29–17. Stillwater, Oklahoma: Oklahoma State University, 1973.

234. Sanders, J.R., and D. Tesar, "The Analytical and Experimental Evaluation of Vibratory Oscillations in Realistically Proportioned Mechanisms," *Journal of Mechanical Design,* 100, no. 4 (October 1978), pp. 762–68.

235. Sandor, G.N., "A General Complex-Number Method for Plane Kinematic Synthesis with Applications," Doctoral dissertation, Columbia University in the City of New York, 1959; University Microfilms, Ann Arbor, Michigan, 305 pgs. Library of Congress Card No. MIC 59–2596.

236. ———, "On Computer-Aided Graphical Kinematic Synthesis," Technical Seminar Series, Princeton University, 1962.

237. ———, "On Infinitesimal Cycloidal Kinematic Theory of Planar Motion," *Journal of Applied Mechanics,* 33, no. 4 (December 1966), pp. 927–33.

238. ———, "On the Existence of a Cycloidal Burmester Theory in Planar Kinematics," *Journal of Applied Mechanics,* 31, no. 4 (December 1964), pp. 694–99.

239. ———, "On the Loop Equations in Kinematics," *Transactions of the Seventh Conference on Mechanisms (Purdue University, October 1962)*, pp. 49–56. West Lafayette, Indiana: Purdue University, 1962.

240. ———, "Principles of a General Quaternion-Operator Method of Spatial Kinematic Synthesis," *Journal of Applied Mechanics,* 35, no. 1 (March 1968), pp. 40–6.

241. ———, and K.E. Bisshopp, "On a General Method of Spatial Kinematic Synthesis by Means of a Stretch Rotation Tensor," *Journal of Engineering for Industry,* 91, no. 1 (February 1969), pp. 115–22.

242. ———, and A.G. Erdman, "Kineto-elastodynamics—A Frontier in Mechanism Design," *Mechanical Engineering News,* 7 (November 1970), pp. 27–8.

243. ———, and F. Freudenstein, "Higher-Order Plane Motion Theories in Kinematic Synthesis," *Journal of Engineering for Industry,* 89, no. 2 (May 1967), pp. 223–30.

244. ______, and Dan Perju, "Contributions to the Kinematic Synthesis of Adjustable Mechanisms," *Transactions of the International Symposium on Linkages and Computer Design Methods (Bucharest, Romania, June 1973)*, Vol. A-46, pp. 636–50.

245. ______, and Donald R. Wilt, "Synthesis of a Geared Four-Link Mechanism," *Proceedings of the Second International Congress on the Theory of Machines and Mechanisms (Zakopane, Poland, September 1969)*, Vol. 2, pp. 222–32; *Journal of Mechanisms*, no. 4 (Winter 1969), pp. 291–302.

246. ______, A.G. Erdman, L. Hunt, and E. Raghavacharyulu, "New Complex-Number Form of the Cubic of Stationary Curvature in a Computer-Oriented Treatment of Planar Path-Curvature Theory for Higher-Pair Rolling Contact," *Journal of Mechanical Design*, 104, pp. 233–38.

247. ______, A.G. Erdman, L. Hunt, and E. Raghavacharyulu, "New Complex-Number Form of the Euler–Savary Equation in a Computer-Oriented Treatment of Planar Path-Curvature Theory for Higher-Pair Rolling Contact," *Journal of Mechanical Design*, 104, pp. 227–32.

248. ______, A.G. Erdman, and E. Raghavacharyulu, "Coriolis Acceleration Analysis in Planar Mechanisms—A Complex Number Approach," *Journal of Mechanism and Machine Theory*, 17, no. 6 (1982), pp. 405–15.

249. ______, I. Imam, and A.G. Erdman, "Applied Kineto-elastodynamics," *Proceedings of the Second Applied Mechanisms Conference (Stillwater, Oklahoma, October 1971)*, pp. 21.1–21.17. Stillwater, Oklahoma: Oklahoma State University, 1971.

250. ______, R.E. Kaufman, A.G. Erdman, et al., "Kinematic Synthesis of Geared Linkages," *Journal of Mechanisms*, 5, no. 1 (Spring 1970).

251. ______, D. Kohli, X. Zhuang, and C. Reinholtz, "Synthesis of a Four-Link Spatial Motion Generator," ASME Paper No. 82-DET-130, 1982.

252. ______, J.F. McGovern, and C.Z. Smith, "The Design of Four-Bar Path Generating Linkages by Fifth-Order Path Approximation in the Vicinity of a Single Point," *Proceedings of the Institution of Mechanical Engineers (London, September 1973)*, pp. 65–77.

253. ______, A.V. Mohan Rao, and A.G. Erdman, "A General Complex-Number Method of Synthesis and Analysis of Mechanisms Containing Prismatic and Revolute Pairs," *Proceedings of the Third World Congress on the Theory of Machines and Mechanisms (Dubrovnik, Yugoslavia, September 1971)*, Vol. D, pp. 237–49. Belgrade: Yugoslavian Committee on the Theory of Machines and Mechanisms, 1972.

254. Sauer, B., B. Williams, and A.G. Erdman, "Integration of Computer Graphics and Spatial Mechanism Analysis," *Proceedings of the Seventh Applied Mechanisms Conference (Kansas City, Missouri, 1981)*, Paper No. XXIII. Stillwater, Oklahoma: Oklahoma State University, 1981.

255. Scott, J.E., *Introduction to Interactive Computer Graphics*. New York: John Wiley & Sons, Inc., 1982.

256. *Selby Furniture Hardware Co. Inc. Catalog*, 15/21 East 22nd Street, New York, New York 10010.

257. Sheth, P.N., "A Digital Computer Based Simulation Procedure for Multiple Degree of Freedom Mechanical Systems with Geometric Constraints," Doctoral dissertation, University of Wisconsin, 1972; University Microfilm No. 73–2565.

258. ———, and J.J. Uicker, "IMP (Integrated Mechanisms Program), A Computer-Aided Design Analysis System for Mechanisms and Linkages," *Journal of Engineering for Industry,* 94, no. 2 (May 1972), pp. 454–64.

259. Shigley, J.E., *Kinematic Analysis of Mechanisms.* New York: McGraw-Hill Book Company, 1969.

260. Shimojima, Hiroshi, Kiyoshi Ogawa, and Toru Kawano, "A Transmissibility for Single-Loop Spatial Mechanisms," *Bulletin of the JSME,* 22, no. 165 (March 1979).

261. Shoup, T.E., *A Practical Guide to Computer Methods for Engineers.* Englewood Cliffs, New Jersey: Prentice-Hall, Inc., 1979.

262. ———, and J.M. Herrera, "Design of Double Boom Cranes for Optimum Load Capacity," *Proceedings of the Fifth Applied Mechanisms Conference (Oklahoma City, Oklahoma, November 1977),* Paper No. 30. Stillwater, Oklahoma: Oklahoma State University, 1977.

263. ———, and C.W. McLarnan, "On the Use of the Undulating Elastica for the Analysis of Flexible Link Mechanisms," *Journal of Engineering for Industry,* 93B (1971), pp. 263–67.

264. ———, and B.J. Pehan, "Design of Four-Bar Mechanisms for Optimum Transmission Angle and Optimum Structural Error," *Proceedings of the Second Applied Mechanisms Conference (Stillwater, Oklahoma, October 1971),* pp. 4.1–4.9. Stillwater, Oklahoma: Oklahoma State University, 1971.

265. Showlater, G., R. Giese, and A.G. Erdman, "Synthesis of Skylight Mechanisms," *Proceedings of the Seventh Applied Mechanisms Conference (Kansas City, Missouri, 1981),* Paper No. XXXVII. Stillwater, Oklahoma: Oklahoma State University, 1981.

266. Skreiner, M., "Acceleration Analysis of Spatial Linkages Using Axodes and the Instantaneous Screw Axis," *Journal of Engineering for Industry,* 89B, (1967), pp. 97–101.

267. Smith, M.R., "Dynamic Analysis and Balancing of Linkages with Interactive Computer Graphics," *Computer Aided Design,* 7, no. 1 (January 1975), pp. 15–19.

268. Soni, A.H., *Mechanism Synthesis and Analysis.* New York: McGraw-Hill Book Company, 1974.

269. ———, and L. Harrisberger, "Application of (3 × 3) Screw Matrix to Kinematic and Dynamic Analysis of Mechanisms," *VDI Berichte,* (1968).

270. ———, and M. Huang, "Synthesis of Four Link Space Mechanisms via Extension of Point-Position-Reduction Technique," *Journal of Engineering for Industry,* 93, no. 1 (1971), pp. 85–9.

271. Stepanoff, B.I., "Design of Spatial Transmission Mechanisms with Lower Pairs," *Akademiya Nauk SSSR,* 45 (1951).

272. Struble, K.R., J.E. Gustafson, and A.G. Erdman, "Case Study: Synthesis of a Four-Bar Linkage to Pick and Place Filters Using the LINCAGES Computer Package," *Proceedings of the Fifth Applied Mechanisms Conference (Oklahoma City, Oklahoma, November 1977).* Stillwater, Oklahoma: Oklahoma State University, 1977.

273. Sturm, A.J., A.G. Erdman, and S.H. Wang, "Design and Analysis of an Industrial (3P3R) Robot," ASME Paper No. 82-DET-39, 1982.

274. Suh, C.H., "Design of Space Mechanisms for Function Generation," *Journal of Engineering for Industry,* 90, no. 3 (1968), pp. 507–13.

275. ———, "Design of Space Mechanisms for Rigid Body Guidance," *Journal of Engineering for Industry,* 90, no. 3 (1968), pp. 499–506.

276. ______, "Synthesis and Analysis of Space Mechanisms with the Use of the Displacement Theory," Ph.D. dissertation, University of California—Berkeley, 1966.

277. ______, and C.W. Radcliffe, *Kinematics and Mechanisms Design.* New York: John Wiley & Sons, Inc., 1978.

278. Sutherland, G.H., "Analytical and Experimental Investigation of a High-Speed Elastic-Membered Linkage," *Journal of Engineering for Industry,* 98, no. 3 (August 1976), pp. 788–99.

279. Tacheny, J., A.G. Erdman, and D.L. Hagen, "Experimental Determination of Mechanism Time Response," *Proceedings of the Fifth World Congress on the Theory of Machines and Mechanisms (Montreal, July 1979),* pp. 130–38.

280. Tanner, W.R., ed., *Industrial Robots.* Dearborn, Michigan: Society of Manufacturing Engineers, 1979.

281. Tao, D.C., *Applied Linkage Synthesis.* Reading, Massachusetts: Addison-Wesley Publishing Company, Inc., 1964.

282. ______, *Fundamentals of Applied Kinematics.* Reading, Massachusetts: Addison-Wesley Publishing Company, Inc., 1967.

283. Tepper, F.R., and G.G. Lowen, "General Theorems Concerning Full Force Balancing of Planar Linkages by Internal Mass Redistribution," *Journal of Engineering for Industry,* 94, no. 3 (August 1972), pp. 789–96.

284. ______, and G.G. Lowen, "A New Criterion for Evaluating the RMS Shaking Moment in Unbalanced Planar Mechanisms," *Proceedings of the Third Applied Mechanisms Conference (Stillwater, Oklahoma, November 1973),* p. 11–1. Stillwater, Oklahoma: Oklahoma State University, 1973.

285. ______, and G.G. Lowen, "On the Distribution of the RMS Shaking Moment of Unbalanced Planar Mechanisms: Theory of Isomomental Ellipses," ASME Paper No. 72-Mech-4, 1972.

286. Tesar, D., and J.W. Sparks, "Multiply Separated Position Synthesis," ASME Paper No. 68-MECH-66, 1968.

287. Thompson, B.S., "The Analsysis of an Elastic Four-Bar Linkage on a Vibrating Foundation Using a Variational Method," ASME Paper No. 79-DET-64, 1979.

288. Thompson, W.T., *Vibration Theory and Applications.* New York: Prentice-Hall, Inc., 1965.

289. Timoshenko, S., and D.G. Young, *Advanced Dynamics.* New York: McGraw-Hill Book Company, 1948.

290. ______, and D.G. Young, *Engineering Mechanics.* New York: McGraw-Hill Book Company, 1940.

291. Tobias, J.R., "The Design of Planar Mechanisms with Distributed Flexibility and Inertia," Doctoral dissertation, University of Minnesota, 1970.

292. Torfason, L.E., and A.K. Sharma, "Analysis of Spatial RRGRR Mechanisms by the Method of Generated Surface," *Journal of Engineering for Industry,* 95, no. 3 (August 1973), pp. 704–8.

293. Tsai, L.W., and B. Roth, "Design of Dyads with Helical, Cylindrical, Spherical, Revolute, and Prismatic Joints," *Journal of Mechanism and Machine Theory,* 7 (1972), pp. 85–102.

294. Turcic, D.A., "A General Approach to the Dynamic Analysis of Elastic Mechanism Systems," Doctoral dissertation, The Pennsylvania State University, 1982.

295. Turner, J., and A.G. Erdman, "Design of a Mechanism Clock," *Proceedings of the Fourth Applied Mechanisms Conference, (Chicago, November 1975)*, pp. 2–1 to 2–5. Stillwater, Oklahoma: Oklahoma State University, 1975.

296. Uicker, J.J., Jr., "Dynamic Behavior of Spatial Linkages—1. Exact Equations of Motion," *Journal of Engineering for Industry,* 91B (1969), pp. 251–65.

297. ———, "Dynamic Force Analysis of Spatial Linkages," *Journal of Applied Mechanics,* 98E (1967), pp. 418–24.

298. ———, J. Denavit, and R.S. Hartenberg, "An Iterative Method for the Displacement Analysis of Spatial Mechanisms," *Journal of Applied Mechanics,* 86, no. 2 (June 1964), pp. 309–14.

299. Ural, Oktay, *Finite Element Method: Basic Concepts and Applications.* New York: Intext Education Publishers, 1973.

300. Van Klompenburg, M.G., J.A. Peterson, and E.W. Nelson, "Dual Arm Operator for a Casement-Type Window," U.S. Patent No. 4,241,541, December 30, 1980.

301. Vierck, R.K., *Vibration Analysis,* 2nd ed. New York: Harper & Row Publishers, Inc., 1979.

302. Viscomi, B.V., and R.S. Ayre, "Nonlinear Dynamic Response of Elastic Slider-Crank Mechanism," *Journal of Engineering for Industry,* 93, no. 1 (February 1971), pp. 251–62.

303. Waldron, K.J., "Elimination of the Branch Problem in Graphical Burmester Mechanism Synthesis for Four Finitely Separated Positions," *Journal of Engineering for Industry,* 98 (1976), pp. 176–82.

304. ———, "Graphical Solution of the Branch and Order Problems of Linkage Synthesis for Multiply Separated Positions," ASME Paper No. 76-DET-16, 1976.

305. ———, "Improved Solutions of the Branch and Order Problems of Burmester Linkage Synthesis," *Journal of Mechanism and Machine Theory,* 13 (1978), pp. 199–207.

306. ———, "Location of Burmester Synthesis Solutions with Fully Rotatable Cranks," *Journal of Mechanism and Machine Theory,* 13 (1978), pp. 125–37.

307. ———, and S.M. Song, "Theoretical and Numerical Improvements to an Interactive Linkage Design Program, RECSYN," *Proceedings of the Seventh Applied Mechanisms Conference (Kansas City, Missouri, December 1981)*, pp. 8.1–8.8. Stillwater, Oklahoma: Oklahoma State University, 1981.

308. ———, and W.H. Sun, "Graphical Transmission Angle Control in Planar Linkage Synthesis," *Proceedings of the Sixth Applied Mechanisms Conference (Denver, Colorado, October 1979)*, pp. 34.1–34.8. Stillwater, Oklahoma: Oklahoma State University, 1979.

309. Wallace, D.M., and F. Freudenstein, "The Displacement Analysis of the Generalized Tracta Coupling," *Journal of Applied Mechanics,* 37 (September 1970), pp. 713–19.

310. Wilde, D.J., "Jacobians in Constrained Nonlinear Optimization," *Operations Research,* 13, no. 5 (September 1965), pp. 848–56.

311. Wiley, J.C., B.E. Romiz, N. Orlandea, T.A. Berenyi, and D.W. Smith, "Automated Simulation and Display of Mechanisms and Vehicle Behavior," *Proceedings of the Fifth World Congress on the Theory of Machines and Mechanisms (1979)*, pp. 680–83.

312. Williams, R.J., and S. Rupprecht, "Dynamic Force Analysis of Planar Mechanisms," *Proceedings of the Sixth Applied Mechanisms Conference (Denver, Colorado, October 1979)*, Paper No. XLIII, 9 pgs. Stillwater, Oklahoma: Oklahoma State University, 1979.

313. Wilson, E.L., and R.W. Clough, "Dynamic Response by Step-by-Step Matrix Analysis," *Symposium on the Use of Computers in Civil Engineering (Lisbon, Portugal, October 1962)*, pp. 45.1–45.14.

314. ______, I. Farhoomand, and K.J. Bathe, "Nonlinear Dynamic Analysis of Complex Structures," *International Journal of Earthquake Engineering and Structural Dynamics,* 1, no. 3 (1973), pp. 241–52.

315. Wilson, E.L., and J. Penzien, "Evaluation of Orthogonal Damping Matrices," *International Journal of Numerical Methods in Engineering,* 4, no. 1 (1972), pp. 5–10.

316. Winfrey, R.C., "Dynamic Analysis of Elastic Link Mechanisms by Reduction of Coordinates," *Journal of Engineering for Industry,* 94 (November 1972), pp. 577–82.

317. ______, "Dynamics of Mechanisms with Elastic Links," Doctoral dissertation, University of California—Los Angeles, 1969.

318. ______, "Elastic Link Mechanism Dynamics," *Journal of Engineering for Industry,* 93, no. 1 (February 1971), pp. 268–72.

319. ______, "Multidegree-of-Freedom Elastic Systems Having Multiple Clearances," *Shock Vibration Bulletin,* 43, Pt. 2, pp. 23–30. Washington, D.C.: Shock and Vibration Information Center, U.S. Naval Research Lab, 1973.

320. ______, R.V. Anderson, and C.W. Gnilka, "Analysis of Elastic Machinery with Clearances," *Journal of Engineering for Industry,* 95, no. 3 (August 1973), pp. 695–703.

321. Wolford, J.C., "The Application of Chace's Vector Equations to the Computer Generation of Burmester Curves," *Proceedings of the Sixth Applied Mechanisms Conference (Denver, Colorado, October 1979)*, pp. IV–1 to IV–17. Stillwater, Oklahoma: Oklahoma State University, 1979.

322. Yang, A., "A Brief Survey of Space Mechanisms," *Proceedings of Design Technology Conference (New York, 1974)*, pp. 315–22.

323. ______, "Displacement Analysis of Spatial Five-Link Mechanisms Using (3 × 3) Matrices with Dual-Number Elements," *Journal of Engineering for Industry,* 91, no. 1 (1969), pp. 152–57.

324. ______, and F. Freudenstein, "Application of Dual-Number Quaternion Algebra to the Analysis of Spatial Mechanisms," *Journal of Applied Mechanics,* 86, no. 2 (June 1964), pp. 300–308.

325. Young, J.F., *Robotics.* New York: John Wiley & Sons, Inc., 1973.

326. Yuan, M.S.C., and F. Freudenstein, "Kinematic Analysis of Spatial Mechanisms by Means of Screw Coordinates, Part I," *Transactions of the ASME, Series B,* 93, no. 1 (1971), pp. 61–6.

327. ______, F. Freudenstein, and L. Wood, "Kinematic Analysis of Spatial Mechanisms by Means of Screw Coordinates, Part 2," *Trans. ASME, Series B,* 93, no. 1 (1971), pp. 67–73.

328. Zienkiewicz, O.C., *The Finite Element Method in Engineering Science.* New York: McGraw-Hill Book Company, 1971.

329. ______, "The Finite Element Method: From Intuition to Generality," *Applied Mechanics Reviews,* 23, no. 3 (March 1970), pp. 249–56.

330. Zorzi, E.S., "Dynamic Response of a Layered Viscoelastic Damped Slider-Crank Mechanism," Doctoral dissertation, University of Minnesota, 1973.

# Index